One Thousand Exercises in Probability

Time and a fox turning the wheel of fortune
Woodcut by Albrecht Dürer (*c.* 1526)

One Thousand Exercises in Probability

THIRD EDITION

GEOFFREY R. GRIMMETT
Statistical Laboratory, Cambridge University

and

DAVID R. STIRZAKER
Mathematical Institute, Oxford University

Great Clarendon Street, Oxford, OX2 6DP,
United Kingdom

Oxford University Press is a department of the University of Oxford. It furthers the University's objective of excellence in research, scholarship, and education by publishing worldwide. Oxford is a registered trade mark of Oxford University Press in the UK and in certain other countries

Third Edition published in 2020

Published in the United States of America by Oxford University Press
198 Madison Avenue, New York, NY 10016, United States of America

British Library Cataloguing in Publication Data
Data available

Library of Congress Control Number: 2020938275

ISBN 978–0–19–884761–8
ISBN 978–0–19–884762–5 (set with Probability and Random Processes, 4e)

Printed in Great Britain by
Bell & Bain Ltd., Glasgow

Life is good for only two things, discovering mathematics and teaching it.

Siméon Poisson

In mathematics you don't understand things, you just get used to them.

John von Neumann

Probability is the bane of the age.

Anthony Powell
Casanova's Chinese Restaurant

The traditional professor writes a, says b, and means c; but it should be d.

George Pólya

To many persons the mention of Probability suggests little else than the notion of a set of rules, very ingenious and profound rules no doubt, with which mathematicians amuse themselves by setting and solving puzzles.

John Venn
The Logic of Chance

Preface to the Third Edition

This volume contains more than 1300 exercises in probability and random processes together with their solutions. Apart from being a volume of worked exercises in its own right, it is also a solutions manual for exercises and problems appearing in the fourth edition of our textbook *Probability and Random Processes*, published by Oxford University Press in 2020, henceforth referred to as PRP. These exercises are not merely for drill, but complement and illustrate the text of PRP, or are entertaining, or both. The current edition extends the previous edition by the inclusion of numerous new exercises, and several new sections devoted to further topics in aspects of stochastic processes. Since many exercises have multiple parts, the total number of interrogatives exceeds 3000.

Despite being intended in part as a companion to PRP, the present volume is as self-contained as reasonably possible. Where knowledge of a substantial chunk of bookwork is unavoidable, the reader is provided with a reference to the relevant passage in PRP. Expressions such as 'clearly' appear frequently in the solutions. Although we do not use such terms in their Laplacian sense to mean 'with difficulty', to call something 'clear' is not to imply that explicit verification is necessarily free of tedium.

The table of contents reproduces that of PRP. The covered range of topics is broad, beginning with the elementary theory of probability and random variables, and continuing, via chapters on Markov chains and convergence, to extensive sections devoted to stationarity and ergodic theory, renewals, queues, martingales, and diffusions, including an introduction to the pricing of options. Generally speaking, *exercises* are questions which test knowledge of particular pieces of theory, while *problems* are less specific in their requirements. There are questions of all standards, the great majority being elementary or of intermediate difficulty. We have found some of the later ones to be rather tricky, but have refrained from magnifying any difficulty by adding asterisks or equivalent devices. To those using this book for self-study, our advice is not to attempt more than a respectable fraction of these at a first read.

We offer two *caveats* to readers. While a great deal of care has been devoted to ensuring these exercises are as correct as possible, there inevitably remain a few slips which have so far escaped detection, and in this regard the readers' patience is invited. Secondly, there will always be debate on just what constitutes a proper or full solution. We have aimed at conveying the nubs of arguments, with as many details as needed to satisfy most readers.

We pay tribute to all those anonymous pedagogues whose examination papers, work assignments, and textbooks have been so influential in the shaping of this collection. To them and to their successors we wish, in turn, much happy plundering. If you detect errors or have potential improvements to propose, try to keep them secret, except from us.

Cambridge and Oxford G.R.G.
April 2020 D.R.S.

Note on the Frontispiece

Fortuna, the goddess of chance and luck, is often depicted with a wheel symbolizing the uncertainty occasioned by the passing of time; that is, the ups and downs of life. In this image, the Wheel of Fortune is being turned not only by Time but also by a fox. From Aesop's fables (*c.*600 BC) to the mediaeval story-cycles of Reynard the Fox (*c.*1100 AD), the fox has symbolized craft, cunning, and trickery, often actuated by malice. This particular Wheel of Fortune may thus suggest that the workings of chance cannot be expected to give necessarily just or fair outcomes.

A further discussion of the iconography is provided in Roberts 1998.

Contents

1

Events and their probabilities

1.2 Exercises. Events as sets

1. Let $\{A_i : i \in I\}$ be a collection of sets. Prove 'De Morgan's Laws'†:

$$\left(\bigcup_i A_i\right)^{\mathrm{c}} = \bigcap_i A_i^{\mathrm{c}}, \qquad \left(\bigcap_i A_i\right)^{\mathrm{c}} = \bigcup_i A_i^{\mathrm{c}}.$$

2. Let A and B belong to some σ-field $\mathcal{F}$. Show that $\mathcal{F}$ contains the sets $A \cap B$, $A \setminus B$, and $A \bigtriangleup B$.

3. A conventional knock-out tournament (such as that at Wimbledon) begins with 2^n competitors and has n rounds. There are no play-offs for the positions $2, 3, \dots, 2^n - 1$, and the initial table of draws is specified. Give a concise description of the sample space of all possible outcomes.

4. Let $\mathcal{F}$ be a σ-field of subsets of Ω and suppose that $B \in \mathcal{F}$. Show that $\mathcal{G} = \{A \cap B : A \in \mathcal{F}\}$ is a σ-field of subsets of B.

5. Which of the following are identically true? For those that are not, say when they are true.

(a) $A \cup (B \cap C) = (A \cup B) \cap (A \cup C)$;

(b) $A \cap (B \cap C) = (A \cap B) \cap C$;

(c) $(A \cup B) \cap C = A \cup (B \cap C)$;

(d) $A \setminus (B \cap C) = (A \setminus B) \cup (A \setminus C)$.

1.3 Exercises. Probability

1. Let A and B be events with probabilities $\mathbb{P}(A) = \frac{3}{4}$ and $\mathbb{P}(B) = \frac{1}{3}$. Show that $\frac{1}{12} \le \mathbb{P}(A \cap B) \le \frac{1}{3}$, and give examples to show that both extremes are possible. Find corresponding bounds for $\mathbb{P}(A \cup B)$.

2. A fair coin is tossed repeatedly. Show that, with probability one, a head turns up sooner or later. Show similarly that any given finite sequence of heads and tails occurs eventually with probability one. Explain the connection with Murphy's Law.

3. Six cups and saucers come in pairs: there are two cups and saucers which are red, two white, and two with stars on. If the cups are placed randomly onto the saucers (one each), find the probability that no cup is upon a saucer of the same pattern.

---

†Augustus De Morgan is well known for having given the first clear statement of the principle of mathematical induction. He applauded probability theory with the words: "The tendency of our study is to substitute the satisfaction of mental exercise for the pernicious enjoyment of an immoral stimulus".

4. Let $A_1, A_2, \ldots, A_n$ be events where $n \geq 2$, and prove that

$$\mathbb{P}\Big(\bigcup_{i=1}^{n} A_i\Big) = \sum_i \mathbb{P}(A_i) - \sum_{i<j} \mathbb{P}(A_i \cap A_j) + \sum_{i<j<k} \mathbb{P}(A_i \cap A_j \cap A_k)$$
$$- \cdots + (-1)^{n+1}\mathbb{P}(A_1 \cap A_2 \cap \cdots \cap A_n).$$

In each packet of Corn Flakes may be found a plastic bust of one of the last five Vice-Chancellors of Cambridge University, the probability that any given packet contains any specific Vice-Chancellor being $\frac{1}{5}$, independently of all other packets. Show that the probability that each of the last three Vice-Chancellors is obtained in a bulk purchase of six packets is $1 - 3(\frac{4}{5})^6 + 3(\frac{3}{5})^6 - (\frac{2}{5})^6$.

5. Let $A_r, r \geq 1$, be events such that $\mathbb{P}(A_r) = 1$ for all r. Show that $\mathbb{P}\big(\bigcap_{r=1}^{\infty} A_r\big) = 1$.

6. You are given that at least one of the events A_r, $1 \leq r \leq n$, is certain to occur, but certainly no more than two occur. If $\mathbb{P}(A_r) = p$, and $\mathbb{P}(A_r \cap A_s) = q, r \neq s$, show that $p \geq 1/n$ and $q \leq 2/n$.

7. You are given that at least one, but no more than three, of the events A_r, $1 \leq r \leq n$, occur, where $n \geq 3$. The probability of at least two occurring is $\frac{1}{2}$. If $\mathbb{P}(A_r) = p$, $\mathbb{P}(A_r \cap A_s) = q$, $r \neq s$, and $\mathbb{P}(A_r \cap A_s \cap A_t) = x$, $r < s < t$, show that $p \geq 3/(2n)$, and $q \leq 4/n$.

1.4 Exercises. Conditional probability

1. Prove that $\mathbb{P}(A \mid B) = \mathbb{P}(B \mid A)\mathbb{P}(A)/\mathbb{P}(B)$ whenever $\mathbb{P}(A)\mathbb{P}(B) \neq 0$. Show that, if $\mathbb{P}(A \mid B) > \mathbb{P}(A)$, then $\mathbb{P}(B \mid A) > \mathbb{P}(B)$.

2. For events $A_1, A_2, \ldots, A_n$ satisfying $\mathbb{P}(A_1 \cap A_2 \cap \cdots \cap A_{n-1}) > 0$, prove that

$$\mathbb{P}(A_1 \cap A_2 \cap \cdots \cap A_n) = \mathbb{P}(A_1)\mathbb{P}(A_2 \mid A_1)\mathbb{P}(A_3 \mid A_1 \cap A_2) \cdots \mathbb{P}(A_n \mid A_1 \cap A_2 \cap \cdots \cap A_{n-1}).$$

3. A man possesses five coins, two of which are double-headed, one is double-tailed, and two are normal. He shuts his eyes, picks a coin at random, and tosses it. What is the probability that the lower face of the coin is a head?

He opens his eyes and sees that the coin is showing heads; what is the probability that the lower face is a head? He shuts his eyes again, and tosses the coin again. What is the probability that the lower face is a head? He opens his eyes and sees that the coin is showing heads; what is the probability that the lower face is a head?

He discards this coin, picks another at random, and tosses it. What is the probability that it shows heads?

4. What do you think of the following 'proof' by Lewis Carroll that an urn cannot contain two balls of the same colour? Suppose that the urn contains two balls, each of which is either black or white; thus, in the obvious notation, $\mathbb{P}(\text{BB}) = \mathbb{P}(\text{BW}) = \mathbb{P}(\text{WB}) = \mathbb{P}(\text{WW}) = \frac{1}{4}$. We add a black ball, so that $\mathbb{P}(\text{BBB}) = \mathbb{P}(\text{BBW}) = \mathbb{P}(\text{BWB}) = \mathbb{P}(\text{BWW}) = \frac{1}{4}$. Next we pick a ball at random; the chance that the ball is black is (using conditional probabilities) $1 \cdot \frac{1}{4} + \frac{2}{3} \cdot \frac{1}{4} + \frac{2}{3} \cdot \frac{1}{4} + \frac{1}{3} \cdot \frac{1}{4} = \frac{2}{3}$. However, if there is probability $\frac{2}{3}$ that a ball, chosen randomly from three, is black, then there must be two black and one white, which is to say that originally there was one black and one white ball in the urn.

5. **The Monty Hall problem: goats and cars.** (a) In a game show; you have to choose one of three doors. One conceals a new car, two conceal old goats. You choose, but your chosen door is not opened immediately. Instead the presenter opens another door, which reveals a goat. He offers you the opportunity to change your choice to the third door (unopened and so far unchosen). Let p be the (conditional) probability that the third door conceals the car. The presenter's protocol is:

(i) he is determined to show you a goat; with a choice of two, he picks one at random. Show $p = \frac{2}{3}$.

(ii) he is determined to show you a goat; with a choice of two goats (Bill and Nan, say) he shows you Bill with probability b. Show that, given you see Bill, the probability is $1/(1+b)$.

(iii) he opens a door chosen at random irrespective of what lies behind. Show $p = \frac{1}{2}$.

(b) Show that, for $\alpha \in [\frac{1}{2}, \frac{2}{3}]$, there exists a protocol such that $p = \alpha$. Are you well advised to change your choice to the third door?

(c) In a variant of this question, the presenter is permitted to open the first door chosen, and to reward you with whatever lies behind. If he chooses to open another door, then this door invariably conceals a goat. Let p be the probability that the unopened door conceals the car, conditional on the presenter having chosen to open a second door. Devise protocols to yield the values $p = 0$, $p = 1$, and deduce that, for any $\alpha \in [0, 1]$, there exists a protocol with $p = \alpha$.

6. The prosecutor's fallacy†**.** Let G be the event that an accused is guilty, and T the event that some testimony is true. Some lawyers have argued on the assumption that $\mathbb{P}(G \mid T) = \mathbb{P}(T \mid G)$. Show that this holds if and only if $\mathbb{P}(G) = \mathbb{P}(T)$.

7. Urns. There are n urns of which the rth contains $r-1$ red balls and $n-r$ magenta balls. You pick an urn at random and remove two balls at random without replacement. Find the probability that:

(a) the second ball is magenta;

(b) the second ball is magenta, given that the first is magenta.

8. Boys and girls, Example (1.4.3) revisited. Consider a family of two children in a population in which each child is equally likely to be male as female; each child has red hair with probability r; these characteristics are independent of each other and occur independently between children.

What is the probability that both children are boys given that the family contains at least one red-haired boy? Show that the probability that both are boys, given that one is a boy born on a Monday, is $13/27$.

1.5 Exercises. Independence

1. Let A and B be independent events; show that A^c, B are independent, and deduce that A^c, B^c are independent.

2. We roll a die n times. Let A_{ij} be the event that the ith and jth rolls produce the same number. Show that the events $\{A_{ij} : 1 \le i < j \le n\}$ are pairwise independent but not independent.

3. A fair coin is tossed repeatedly. Show that the following two statements are equivalent:

(a) the outcomes of different tosses are independent,

(b) for any given finite sequence of heads and tails, the chance of this sequence occurring in the first m tosses is 2^{-m}, where m is the length of the sequence.

4. Let $\Omega = \{1, 2, \dots, p\}$ where p is prime, $\mathcal{F}$ be the set of all subsets of Ω, and $\mathbb{P}(A) = |A|/p$ for all $A \in \mathcal{F}$. Show that, if A and B are independent events, then at least one of A and B is either $\varnothing$ or Ω.

5. Show that the conditional independence of A and B given C neither implies, nor is implied by, the independence of A and B. For which events C is it the case that, for all A and B, the events A and B are independent if and only if they are conditionally independent given C?

6. Safe or sorry? Some form of prophylaxis is said to be 90 per cent effective at prevention during one year's treatment. If the degrees of effectiveness in different years are independent, show that the treatment is more likely than not to fail within 7 years.

†The prosecution made this error in the famous Dreyfus case of 1894.

7. Families. Jane has three children, each of which is equally likely to be a boy or a girl independently of the others. Define the events:

$$A = \{\text{all the children are of the same sex}\},$$
$$B = \{\text{there is at most one boy}\},$$
$$C = \{\text{the family includes a boy and a girl}\}.$$

(a) Show that A is independent of B, and that B is independent of C.

(b) Is A independent of C?

(c) Do these results hold if boys and girls are not equally likely?

(d) Do these results hold if Jane has four children?

8. Galton's paradox. You flip three fair coins. At least two are alike, and it is an evens chance that the third is a head or a tail. Therefore $\mathbb{P}(\text{all alike}) = \frac{1}{2}$. Do you agree?

9. Two fair dice are rolled. Show that the event that their sum is 7 is independent of the score shown by the first die.

10. Let X and Y be the scores on two fair dice taking values in the set $\{1, 2, \dots, 6\}$. Let $A_1 = \{X + Y = 9\}$, $A_2 = \{X \in \{1, 2, 3\}\}$, and $A_3 = \{X \in \{3, 4, 5\}\}$. Show that

$$\mathbb{P}(A_1 \cap A_2 \cap A_3) = \mathbb{P}(A_1)\mathbb{P}(A_2)\mathbb{P}(A_3).$$

Are these three events independent?

1.7 Exercises. Worked examples

1. There are two roads from A to B and two roads from B to C. Each of the four roads is blocked by snow with probability p, independently of the others. Find the probability that there is an open road from A to B given that there is no open route from A to C.

If, in addition, there is a direct road from A to C, this road being blocked with probability p independently of the others, find the required conditional probability.

2. Calculate the probability that a hand of 13 cards dealt from a normal shuffled pack of 52 contains exactly two kings and one ace. What is the probability that it contains exactly one ace given that it contains exactly two kings?

3. A symmetric random walk takes place on the integers $0, 1, 2, \dots, N$ with absorbing barriers at 0 and N, starting at k. Show that the probability that the walk is never absorbed is zero.

4. The so-called 'sure thing principle' asserts that if you prefer x to y given C, and also prefer x to y given C^c, then you surely prefer x to y. Agreed?

5. A pack contains m cards, labelled $1, 2, \dots, m$. The cards are dealt out in a random order, one by one. Given that the label of the kth card dealt is the largest of the first k cards dealt, what is the probability that it is also the largest in the pack?

6. A group of $2b$ friends meet for a bridge soirée. There are m men and $2b - m$ women where $2 \le m \le b$. The group divides into b teams of pairs, formed uniformly at random. What is the probability that no pair comprises 2 men?

1.8 Problems

1. A traditional fair die is thrown twice. What is the probability that:

(a) a six turns up exactly once?

(b) both numbers are odd?

(c) the sum of the scores is 4?

(d) the sum of the scores is divisible by 3?

2. A fair coin is thrown repeatedly. What is the probability that on the nth throw:

(a) a head appears for the first time?

(b) the numbers of heads and tails to date are equal?

(c) exactly two heads have appeared altogether to date?

(d) at least two heads have appeared to date?

3. Let $\mathcal{F}$ and $\mathcal{G}$ be σ-fields of subsets of Ω.

(a) Use elementary set operations to show that $\mathcal{F}$ is closed under countable intersections; that is, if $A_1, A_2, \dots$ are in $\mathcal{F}$, then so is $\bigcap_i A_i$.

(b) Let $\mathcal{H} = \mathcal{F} \cap \mathcal{G}$ be the collection of subsets of Ω lying in both $\mathcal{F}$ and $\mathcal{G}$. Show that $\mathcal{H}$ is a σ-field.

(c) Show that $\mathcal{F} \cup \mathcal{G}$, the collection of subsets of Ω lying in either $\mathcal{F}$ or $\mathcal{G}$, is not necessarily a σ-field.

4. Describe the underlying probability spaces for the following experiments:

(a) a biased coin is tossed three times;

(b) two balls are drawn without replacement from an urn which originally contained two ultramarine and two vermilion balls;

(c) a biased coin is tossed repeatedly until a head turns up.

5. Show that the probability that *exactly* one of the events A and B occurs is

$$\mathbb{P}(A) + \mathbb{P}(B) - 2\mathbb{P}(A \cap B).$$

6. Prove that $\mathbb{P}(A \cup B \cup C) = 1 - \mathbb{P}(A^c \mid B^c \cap C^c)\mathbb{P}(B^c \mid C^c)\mathbb{P}(C^c)$.

7. (a) If A is independent of itself, show that $\mathbb{P}(A)$ is 0 or 1.

(b) If $\mathbb{P}(A)$ is 0 or 1, show that A is independent of all events B.

8. Let $\mathcal{F}$ be a σ-field of subsets of Ω, and suppose $\mathbb{P} : \mathcal{F} \to [0, 1]$ satisfies: (i) $\mathbb{P}(\Omega) = 1$, and (ii) $\mathbb{P}$ is additive, in that $\mathbb{P}(A \cup B) = \mathbb{P}(A) + \mathbb{P}(B)$ whenever $A \cap B = \varnothing$. Show that $\mathbb{P}(\varnothing) = 0$.

9. Suppose $(\Omega, \mathcal{F}, \mathbb{P})$ is a probability space and $B \in \mathcal{F}$ satisfies $\mathbb{P}(B) > 0$. Let $\mathbb{Q} : \mathcal{F} \to [0, 1]$ be defined by $\mathbb{Q}(A) = \mathbb{P}(A \mid B)$. Show that $(\Omega, \mathcal{F}, \mathbb{Q})$ is a probability space. If $C \in \mathcal{F}$ and $\mathbb{Q}(C) > 0$, show that $\mathbb{Q}(A \mid C) = \mathbb{P}(A \mid B \cap C)$; discuss.

10. Let $B_1, B_2, \dots$ be a partition of the sample space Ω, each B_i having positive probability, and show that

$$\mathbb{P}(A) = \sum_{j=1}^{\infty} \mathbb{P}(A \mid B_j)\mathbb{P}(B_j).$$

11. Prove **Boole's inequalities**:

$$\mathbb{P}\left(\bigcup_{i=1}^{n} A_i\right) \le \sum_{i=1}^{n} \mathbb{P}(A_i), \qquad \mathbb{P}\left(\bigcap_{i=1}^{n} A_i\right) \ge 1 - \sum_{i=1}^{n} \mathbb{P}(A_i^c).$$

12. Prove that

$$\mathbb{P}\Big(\bigcap_1^n A_i\Big) = \sum_i \mathbb{P}(A_i) - \sum_{i<j} \mathbb{P}(A_i \cup A_j) + \sum_{i<j<k} \mathbb{P}(A_i \cup A_j \cup A_k) - \cdots - (-1)^n \mathbb{P}(A_1 \cup A_2 \cup \cdots \cup A_n).$$

13. Let $A_1, A_2, \dots, A_n$ be events, and let N_k be the event that exactly k of the A_i occur. Prove the result sometimes referred to as **Waring's theorem**:

$$\mathbb{P}(N_k) = \sum_{i=0}^{n-k} (-1)^i \binom{k+i}{k} S_{k+i}, \text{ where } S_j = \sum_{i_1<i_2<\cdots<i_j} \mathbb{P}(A_{i_1} \cap A_{i_2} \cap \cdots \cap A_{i_j}).$$

Use this result to find an expression for the probability that a purchase of six packets of Corn Flakes yields exactly three distinct busts (see Exercise (1.3.4)).

14. Prove **Bayes's formula**: if $A_1, A_2, \dots, A_n$ is a partition of Ω, each A_i having positive probability, then

$$\mathbb{P}(A_j \mid B) = \frac{\mathbb{P}(B \mid A_j)\mathbb{P}(A_j)}{\sum_1^n \mathbb{P}(B \mid A_i)\mathbb{P}(A_i)}.$$

15. A random number N of dice is thrown. Let A_i be the event that $N = i$, and assume that $\mathbb{P}(A_i) = 2^{-i}, i \ge 1$. The sum of the scores is S. Find the probability that:

(a) $N = 2$ given $S = 4$;

(b) $S = 4$ given N is even;

(c) $N = 2$, given that $S = 4$ and the first die showed 1;

(d) the largest number shown by any die is r, where S is unknown.

16. Let $A_1, A_2, \dots$ be a sequence of events. Define

$$B_n = \bigcup_{m=n}^{\infty} A_m, \quad C_n = \bigcap_{m=n}^{\infty} A_m.$$

Clearly $C_n \subseteq A_n \subseteq B_n$. The sequences $\{B_n\}$ and $\{C_n\}$ are decreasing and increasing respectively with limits

$$\lim B_n = B = \bigcap_n B_n = \bigcap_n \bigcup_{m \ge n} A_m, \qquad \lim C_n = C = \bigcup_n C_n = \bigcup_n \bigcap_{m \ge n} A_m.$$

The events B and C are denoted $\limsup_{n\to\infty} A_n$ and $\liminf_{n\to\infty} A_n$ respectively. Show that

(a) $B = \{\omega \in \Omega : \omega \in A_n$ for infinitely many values of $n\}$,

(b) $C = \{\omega \in \Omega : \omega \in A_n$ for all but finitely many values of $n\}$.

We say that the sequence $\{A_n\}$ converges to a limit $A = \lim A_n$ if B and C are the same set A. Suppose that $A_n \to A$ and show that

(c) A is an event, in that $A \in \mathcal{F}$,

(d) $\mathbb{P}(A_n) \to \mathbb{P}(A)$.

17. In Problem (1.8.16) above, show that B and C are independent whenever B_n and C_n are independent for all n. Deduce that if this holds and furthermore $A_n \to A$, then $\mathbb{P}(A)$ equals either zero or one.

18. Show that the assumption that $\mathbb{P}$ is *countably* additive is equivalent to the assumption that $\mathbb{P}$ is continuous. That is to say, show that if a function $\mathbb{P} : \mathcal{F} \to [0, 1]$ satisfies $\mathbb{P}(\varnothing) = 0$, $\mathbb{P}(\Omega) = 1$, and $\mathbb{P}(A \cup B) = \mathbb{P}(A) + \mathbb{P}(B)$ whenever $A, B \in \mathcal{F}$ and $A \cap B = \varnothing$, then $\mathbb{P}$ is countably additive (in the sense of satisfying Definition (1.3.1b)) if and only if $\mathbb{P}$ is continuous (in the sense of Lemma (1.3.5)).

19. Anne, Betty, Chloë, and Daisy were all friends at school. Subsequently each of the $\binom{4}{2} = 6$ subpairs meet up; at each of the six meetings the pair involved quarrel with some fixed probability p, or become firm friends with probability $1 - p$. Quarrels take place independently of each other. In future, if any of the four hears a rumour, then she tells it to her firm friends only. If Anne hears a rumour, what is the probability that:

(a) Daisy hears it?

(b) Daisy hears it if Anne and Betty have quarrelled?

(c) Daisy hears it if Betty and Chloë have quarrelled?

(d) Daisy hears it if she has quarrelled with Anne?

20. A biased coin is tossed repeatedly. Each time there is a probability p of a head turning up. Let p_n be the probability that an even number of heads has occurred after n tosses (zero is an even number). Show that $p_0 = 1$ and that $p_n = p(1 - p_{n-1}) + (1 - p)p_{n-1}$ if $n \geq 1$. Solve this difference equation.

21. A biased coin is tossed repeatedly. Find the probability that there is a run of r heads in a row before there is a run of s tails, where r and s are positive integers.

22. (a) A bowl contains twenty cherries, exactly fifteen of which have had their stones removed. A greedy pig eats five whole cherries, picked at random, without remarking on the presence or absence of stones. Subsequently, a cherry is picked randomly from the remaining fifteen.

(i) What is the probability that this cherry contains a stone?

(ii) Given that this cherry contains a stone, what is the probability that the pig consumed at least one stone?

(b) 100 contestants buy numbered lottery tickets for a reverse raffle, in which the last ticket drawn from an urn is the winner. Halfway through the draw, the Mistress of Ceremonies discovers that 10 tickets have inadvertently not been added to the urn, so she adds them, and continues the draw. Is the lottery fair?

23. The **'ménages'** problem poses the following question. Some consider it to be desirable that men and women alternate when seated at a circular table. If n heterosexual couples are seated randomly according to this rule, show that the probability that nobody sits next to his or her partner is

$$\frac{1}{n!}\sum_{k=0}^{n}(-1)^k \frac{2n}{2n-k}\binom{2n-k}{k}(n-k)!$$

You may find it useful to show first that the number of ways of selecting k non-overlapping pairs of adjacent seats is $\binom{2n-k}{k}2n(2n-k)^{-1}$.

24. An urn contains b blue balls and r red balls. They are removed at random and not replaced. Show that the probability that the first red ball drawn is the $(k+1)$th ball drawn equals $\binom{r+b-k-1}{r-1}\Big/\binom{r+b}{b}$. Find the probability that the last ball drawn is red.

25. An urn contains a azure balls and c carmine balls, where $ac \neq 0$. Balls are removed at random and discarded until the first time that a ball (B, say) is removed having a different colour from its predecessor. The ball B is now replaced and the procedure restarted. This process continues until the last ball is drawn from the urn. Show that this last ball is equally likely to be azure or carmine.

26. Protocols. A pack of four cards contains one spade, one club, and the two red aces. You deal two cards faces downwards at random in front of a truthful friend. She inspects them and tells you that one of them is the ace of hearts. What is the chance that the other card is the ace of diamonds? Perhaps $\frac{1}{3}$?

Suppose that your friend's protocol was:

(a) with no red ace, say "no red ace",

(b) with the ace of hearts, say "ace of hearts",

(c) with the ace of diamonds but not the ace of hearts, say "ace of diamonds".

Show that the probability in question is $\frac{1}{3}$.

Devise a possible protocol for your friend such that the probability in question is zero.

27. Eddington's controversy. Four witnesses, A, B, C, and D, at a trial each speak the truth with probability $\frac{1}{3}$ independently of each other. In their testimonies, A claimed that B denied that C declared that D lied. What is the (conditional) probability that D told the truth? [This problem seems to have appeared first as a parody in a university magazine of the 'typical' Cambridge Philosophy Tripos question.]

28. The probabilistic method. 10 per cent of the surface of a sphere is coloured blue, the rest is red. Show that, irrespective of the manner in which the colours are distributed, it is possible to inscribe a cube in S with all its vertices red.

29. Repulsion. The event A is said to be repelled by the event B if $\mathbb{P}(A \mid B) < \mathbb{P}(A)$, and to be attracted by B if $\mathbb{P}(A \mid B) > \mathbb{P}(A)$. Show that if B attracts A, then A attracts B, and B^{c} repels A.

If A attracts B, and B attracts C, does A attract C?

30. Birthdays. At a lecture, there a m students born on independent days in 2007.

(a) With $2 \le m \le 365$, show that the probability that at least two of them share a birthday is $p = 1 - (365)!/\{(365-m)!\,365^m\}$. Show that $p > \frac{1}{2}$ when $m = 23$.

(b) With $2 \le m \le 366$, find the probability p_1 that exactly one pair of individuals share a birthday, with no others sharing.

(c) Suppose m students are born on independent random days on the planet Magrathea, whose year has $M \gg m$ days. Show that the probability p_0 that no two students share a birthday is approximately $\exp\big(-\frac{1}{2}m(m-1)/M\big)$ for large M.

31. Lottery. You choose r of the first n positive integers, and a lottery chooses a random subset L of the same size. What is the probability that:

(a) L includes no consecutive integers?

(b) L includes exactly one pair of consecutive integers?

(c) the numbers in L are drawn in increasing order?

(d) your choice of numbers is the same as L?

(e) there are exactly k of your numbers matching members of L?

32. Bridge. During a game of bridge, you are dealt at random a hand of thirteen cards. With an obvious notation, show that $\mathbb{P}(4S, 3H, 3D, 3C) \simeq 0.026$ and $\mathbb{P}(4S, 4H, 3D, 2C) \simeq 0.018$. However if suits are not specified, so numbers denote the shape of your hand, show that $\mathbb{P}(4, 3, 3, 3) \simeq 0.11$ and $\mathbb{P}(4, 4, 3, 2) \simeq 0.22$.

33. Poker. During a game of poker, you are dealt a five-card hand at random. With the convention that aces may count high or low, show that:

$$\mathbb{P}(\text{1 pair}) \simeq 0.423, \qquad \mathbb{P}(\text{2 pairs}) \simeq 0.0475, \qquad \mathbb{P}(\text{3 of a kind}) \simeq 0.021,$$
$$\mathbb{P}(\text{straight}) \simeq 0.0039, \qquad \mathbb{P}(\text{flush}) \simeq 0.0020, \qquad \mathbb{P}(\text{full house}) \simeq 0.0014,$$
$$\mathbb{P}(\text{4 of a kind}) \simeq 0.00024, \qquad \mathbb{P}(\text{straight flush}) \simeq 0.000015.$$

34. Poker dice. There are five dice each displaying 9, 10, J, Q, K, A. Show that, when rolled:

$$\mathbb{P}(1 \text{ pair}) \simeq 0.46, \quad \mathbb{P}(2 \text{ pairs}) \simeq 0.23, \quad \mathbb{P}(3 \text{ of a kind}) \simeq 0.15,$$
$$\mathbb{P}(\text{no 2 alike}) \simeq 0.093, \quad \mathbb{P}(\text{full house}) \simeq 0.039, \quad \mathbb{P}(4 \text{ of a kind}) \simeq 0.019,$$
$$\mathbb{P}(5 \text{ of a kind}) \simeq 0.0008.$$

35. You are lost in the National Park of **Bandrika**†. Tourists comprise two-thirds of the visitors to the park, and give a correct answer to requests for directions with probability $\frac{3}{4}$. (Answers to repeated questions are independent, even if the question and the person are the same.) If you ask a Bandrikan for directions, the answer is always false.

(a) You ask a passer-by whether the exit from the Park is East or West. The answer is East. What is the probability this is correct?

(b) You ask the same person again, and receive the same reply. Show the probability that it is correct is $\frac{1}{2}$.

(c) You ask the same person again, and receive the same reply. What is the probability that it is correct?

(d) You ask for the fourth time, and receive the answer East. Show that the probability it is correct is $\frac{27}{70}$.

(e) Show that, had the fourth answer been West instead, the probability that East is nevertheless correct is $\frac{9}{10}$.

36. Mr Bayes goes to Bandrika. Tom is in the same position as you were in the previous problem, but he has reason to believe that, with probability ϵ, East is the correct answer. Show that:

(a) whatever answer first received, Tom continues to believe that East is correct with probability ϵ,

(b) if the first two replies are the same (that is, either WW or EE), Tom continues to believe that East is correct with probability ϵ,

(c) after three like answers, Tom will calculate as follows, in the obvious notation:

$$\mathbb{P}(\text{East correct} \mid \text{EEE}) = \frac{9\epsilon}{11 - 2\epsilon}, \quad \mathbb{P}(\text{East correct} \mid \text{WWW}) = \frac{11\epsilon}{9 + 2\epsilon}.$$

Evaluate these when $\epsilon = \frac{9}{20}$.

37. Bonferroni's inequality. Show that

$$\mathbb{P}\left(\bigcup_{r=1}^{n} A_r\right) \geq \sum_{r=1}^{n} \mathbb{P}(A_r) - \sum_{r<k} \mathbb{P}(A_r \cap A_k).$$

38. Kounias's inequality. Show that

$$\mathbb{P}\left(\bigcup_{r=1}^{n} A_r\right) \leq \min_k \left\{ \sum_{r=1}^{n} \mathbb{P}(A_r) - \sum_{r:r\neq k} \mathbb{P}(A_r \cap A_k) \right\}.$$

†A fictional country made famous in the Hitchcock film 'The Lady Vanishes'.

39. The lost boarding pass†. The n passengers for a Bell-Air flight in an airplane with n seats have been told their seat numbers. They get on the plane one by one. The first person sits in the wrong seat. Subsequent passengers sit in their assigned seats whenever they find them available, or otherwise in a randomly chosen empty seat. What is the probability that the last passenger finds his or her assigned seat to be free?

What is the answer if the first person sits in a seat chosen uniformly at random from the n available?

40. Flash's problem. A number n of spaceships land independently and uniformly at random on the surface of planet Mongo. Each ship controls the hemisphere of which it is the centre. Show that the probability that every point on Mongo is controlled by at least one ship is $1 - 2^{-n}(n^2 - n + 2)$. [Hint: n great circles almost surely partition the surface of the sphere into $n^2 - n + 2$ disjoint regions.]

41. Let X be uniformly distributed on $\{1, 2, \ldots, n-1\}$, where $n \geq 2$. Given X, a team of size X is selected at random from a pool of n players (including you), each such subset of size X being equally likely. Call the selected team A, and the remainder team B.

(a) What is the probability that your team has size k?

(b) Each team picks a captain uniformly at random from its members. What is the probability your team has size k given that you are chosen as captain?

42. Alice and Bob flip a fair coin in turn. A wins if she gets a head, provided her preceding flip was a tail; B wins if he gets a tail, provided his preceding flip was a head. Let $n \geq 3$. Show that the probability the game ends on the nth flip is $(n+1)(n-1)/2^{n+2}$ if n is odd, and $(n+2)(n-2)/2^{n+2}$ if even.

What is the probability that A wins the game?

43. A coin comes up heads with probability $p \in (0, 1)$. Let $k \geq 1$, and let ρ_m be the probability that, in m (≥ 1) coin flips, the longest run of consecutive heads has length strictly less than k. Show that

$$\rho_m - \rho_{m-1} + (1-p)p^k \rho_{m-k-1} = 0, \qquad m \geq k+1,$$

and find ρ_m when $k = 2$.

---

†The authors learned of this problem from David Bell in 2000 or earlier.

2

Random variables and their distributions

2.1 Exercises. Random variables

1. Let X be a random variable on a given probability space, and let $a \in \mathbb{R}$. Show that
(a) aX is a random variable,
(b) $X - X = 0$, the random variable taking the value 0 always, and $X + X = 2X$.

2. A random variable X has distribution function F. What is the distribution function of $Y = aX + b$, where a and b are real constants?

3. A fair coin is tossed n times. Show that, under reasonable assumptions, the probability of exactly k heads is $\binom{n}{k}(\frac{1}{2})^n$. What is the corresponding quantity when heads appears with probability p on each toss?

4. Show that if F and G are distribution functions and $0 \le \lambda \le 1$ then $\lambda F + (1-\lambda)G$ is a distribution function. Is the product FG a distribution function?

5. Let F be a distribution function and r a positive integer. Show that the following are distribution functions: (a) $F(x)^r$, (b) $1 - \{1 - F(x)\}^r$, (c) $F(x) + \{1 - F(x)\}\log\{1 - F(x)\}$, (d) $\{F(x) - 1\}e + \exp\{1 - F(x)\}$.

6. Uniform distribution. A random variable that is equally likely to take any value in a finite set S is said to have the *uniform distribution* on S. If U is such a random variable and $\varnothing \ne R \subseteq S$, show that the distribution of U conditional on $\{U \in R\}$ is uniform on R.

2.2 Exercises. The law of averages

1. You wish to ask each of a large number of people a question to which the answer "yes" is embarrassing. The following procedure is proposed in order to determine the embarrassed fraction of the population. As the question is asked, a coin is tossed out of sight of the questioner. If the answer would have been "no" and the coin shows heads, then the answer "yes" is given. Otherwise people respond truthfully. What do you think of this procedure?

2. A coin is tossed repeatedly and heads turns up on each toss with probability p. Let H_n and T_n be the numbers of heads and tails in n tosses. Show that, for $\epsilon > 0$,

$$\mathbb{P}\left(2p - 1 - \epsilon \le \frac{1}{n}(H_n - T_n) \le 2p - 1 + \epsilon\right) \to 1 \qquad \text{as } n \to \infty.$$

3. Let $\{X_r : r \ge 1\}$ be observations which are independent and identically distributed with unknown distribution function F. Describe and justify a method for estimating $F(x)$.

2.3 Exercises. Discrete and continuous variables

1. Let X be a random variable with distribution function F, and let $a = (a_m : -\infty < m < \infty)$ be a strictly increasing sequence of real numbers satisfying $a_{-m} \to -\infty$ and $a_m \to \infty$ as $m \to \infty$. Define $G(x) = \mathbb{P}(X \le a_m)$ when $a_{m-1} \le x < a_m$, so that G is the distribution function of a discrete random variable. How does the function G behave as the sequence a is chosen in such a way that $\sup_m |a_m - a_{m-1}|$ becomes smaller and smaller?

2. Let X be a random variable and let $g : \mathbb{R} \to \mathbb{R}$ be continuous and strictly increasing. Show that $Y = g(X)$ is a random variable.

3. Let X be a random variable with distribution function

$$\mathbb{P}(X \le x\} = \begin{cases} 0 & \text{if } x \le 0, \\ x & \text{if } 0 < x \le 1, \\ 1 & \text{if } x > 1. \end{cases}$$

Let F be a distribution function which is continuous and strictly increasing. Show that $Y = F^{-1}(X)$ is a random variable having distribution function F. Is it necessary that F be continuous and/or strictly increasing?

4. Show that, if f and g are density functions, and $0 \le \lambda \le 1$, then $\lambda f + (1 - \lambda)g$ is a density. Is the product fg a density function?

5. Which of the following are density functions? Find c and the corresponding distribution function F for those that are.

(a) $f(x) = \begin{cases} cx^{-d} & x > 1, \\ 0 & \text{otherwise.} \end{cases}$

(b) $f(x) = ce^x(1 + e^x)^{-2}$, $x \in \mathbb{R}$.

2.4 Exercises. Worked examples

1. Let X be a random variable with a continuous distribution function F. Find expressions for the distribution functions of the following random variables:

(a) X^2, (b) $\sqrt{X}$,
(c) $\sin X$, (d) $G^{-1}(X)$,
(e) $F(X)$, (f) $G^{-1}(F(X))$,

where G is a continuous and strictly increasing function.

2. **Truncation**. Let X be a random variable with distribution function F, and let $a < b$. Sketch the distribution functions of the 'truncated' random variables Y and Z given by

$$Y = \begin{cases} a & \text{if } X < a, \\ X & \text{if } a \le X \le b, \\ b & \text{if } X > b, \end{cases} \qquad Z = \begin{cases} X & \text{if } |X| \le b, \\ 0 & \text{if } |X| > b. \end{cases}$$

Indicate how these distribution functions behave as $a \to -\infty$, $b \to \infty$.

2.5 Exercises. Random vectors

1. A fair coin is tossed twice. Let X be the number of heads, and let W be the indicator function of the event $\{X = 2\}$. Find $\mathbb{P}(X = x, W = w)$ for all appropriate values of x and w.

2. Let X be a Bernoulli random variable, so that $\mathbb{P}(X = 0) = 1 - p$, $\mathbb{P}(X = 1) = p$. Let $Y = 1 - X$ and $Z = XY$. Find $\mathbb{P}(X = x, Y = y)$ and $\mathbb{P}(X = x, Z = z)$ for $x, y, z \in \{0, 1\}$.

3. The random variables X and Y have joint distribution function

$$F_{X,Y}(x, y) = \begin{cases} 0 & \text{if } x < 0, \\ (1 - e^{-x})\left(\dfrac{1}{2} + \dfrac{1}{\pi}\tan^{-1} y\right) & \text{if } x \geq 0. \end{cases}$$

Show that X and Y are (jointly) continuously distributed.

4. Let X and Y have joint distribution function F. Show that

$$\mathbb{P}(a < X \leq b,\ c < Y \leq d) = F(b, d) - F(a, d) - F(b, c) + F(a, c)$$

whenever $a < b$ and $c < d$.

5. Let X, Y be discrete random variables taking values in the integers, with joint mass function f. Show that, for integers x, y,

$$\begin{aligned} f(x, y) = \mathbb{P}(X \geq x,\ Y \leq y) - \mathbb{P}(X \geq x + 1,\ Y \leq y) \\ - \mathbb{P}(X \geq x,\ Y \leq y - 1) + \mathbb{P}(X \geq x + 1,\ Y \leq y - 1). \end{aligned}$$

Hence find the joint mass function of the smallest and largest numbers shown in r rolls of a fair die.

6. Is the function $F(x, y) = 1 - e^{-xy}$, $0 \leq x, y < \infty$, the joint distribution function of some pair of random variables?

2.7 Problems

1. Each toss of a coin results in a head with probability p. The coin is tossed until the first head appears. Let X be the total number of tosses. What is $\mathbb{P}(X > m)$? Find the distribution function of the random variable X.

2. (a) Show that any discrete random variable may be written as a linear combination of indicator variables.

(b) Show that any random variable may be expressed as the limit of an increasing sequence of discrete random variables.

(c) Show that the limit of any increasing convergent sequence of random variables is a random variable.

3. (a) Show that, if X and Y are random variables on a probability space $(\Omega, \mathcal{F}, \mathbb{P})$, then so are $X + Y$, XY, and $\min\{X, Y\}$.

(b) Show that the set of all random variables on a given probability space $(\Omega, \mathcal{F}, \mathbb{P})$ constitutes a vector space over the reals. If Ω is finite, write down a basis for this space.

4. Let X have distribution function

$$F(x) = \begin{cases} 0 & \text{if } x < 0, \\ \frac{1}{2}x & \text{if } 0 \leq x \leq 2, \\ 1 & \text{if } x > 2, \end{cases}$$

and let $Y = X^2$. Find
(a) $\mathbb{P}(\frac{1}{2} \le X \le \frac{3}{2})$, (b) $\mathbb{P}(1 \le X < 2)$,
(c) $\mathbb{P}(Y \le X)$, (d) $\mathbb{P}(X \le 2Y)$,
(e) $\mathbb{P}(X + Y \le \frac{3}{4})$, (f) the distribution function of $Z = \sqrt{X}$.

5. Let X have distribution function

$$F(x) = \begin{cases} 0 & \text{if } x < -1, \\ 1 - p & \text{if } -1 \le x < 0, \\ 1 - p + \frac{1}{2}xp & \text{if } 0 \le x \le 2, \\ 1 & \text{if } x > 2. \end{cases}$$

Sketch this function, and find: (a) $\mathbb{P}(X = -1)$, (b) $\mathbb{P}(X = 0)$, (c) $\mathbb{P}(X \ge 1)$.

6. Buses arrive at ten minute intervals starting at noon. A man arrives at the bus stop a random number X minutes after noon, where X has distribution function

$$\mathbb{P}(X \le x) = \begin{cases} 0 & \text{if } x < 0, \\ x/60 & \text{if } 0 \le x \le 60, \\ 1 & \text{if } x > 60. \end{cases}$$

What is the probability that he waits less than five minutes for a bus?

7. Airlines find that each passenger who reserves a seat fails to turn up with probability $\frac{1}{10}$ independently of the other passengers. EasyPeasy Airlines always sell 10 tickets for their 9 seat aeroplane while RyeLoaf Airways always sell 20 tickets for their 18 seat aeroplane. Which is more often over-booked?

8. A fairground performer claims the power of telekinesis. The crowd throws coins and he wills them to fall heads up. He succeeds five times out of six. What chance would he have of doing at least as well if he had no supernatural powers?

9. Express the distribution functions of

$$X^+ = \max\{0, X\}, \quad X^- = -\min\{0, X\}, \quad |X| = X^+ + X^-, \quad -X,$$

in terms of the distribution function F of the random variable X.

10. Show that $F_X(x)$ is continuous at $x = x_0$ if and only if $\mathbb{P}(X = x_0) = 0$.

11. The real number m is called a *median* of the distribution function F whenever $\lim_{y \uparrow m} F(y) \le \frac{1}{2} \le F(m)$.

(a) Show that every distribution function F has at least one median, and that the set of medians of F is a closed interval of $\mathbb{R}$.

(b) Show, if F is continuous, that $F(m) = \frac{1}{2}$ for any median m.

12. Loaded dice.

(a) Show that it is not possible to weight two dice in such a way that the sum of the two numbers shown by these loaded dice is equally likely to take any value between 2 and 12 (inclusive).

(b) Given a fair die and a loaded die, show that the sum of their scores, modulo 6, has the same distribution as a fair die, irrespective of the loading.

13. A function $d : S \times S \to \mathbb{R}$ is called a *metric* on S if:

(i) $d(s, t) = d(t, s) \ge 0$ for all $s, t \in S$,

(ii) $d(s, t) = 0$ if and only if $s = t$, and

(iii) $d(s, t) \le d(s, u) + d(u, t)$ for all $s, t, u \in S$.

(a) **Lévy metric.** Let F and G be distribution functions and define the *Lévy metric*

$$d_{\mathrm{L}}(F,G) = \inf\Big\{\epsilon > 0 : G(x-\epsilon) - \epsilon \le F(x) \le G(x+\epsilon) + \epsilon \text{ for all } x\Big\}.$$

Show that d_{L} is indeed a metric on the space of distribution functions.

(b) **Total variation distance.** Let X and Y be integer-valued random variables, and let

$$d_{\mathrm{TV}}(X,Y) = \sum_k \big|\mathbb{P}(X=k) - \mathbb{P}(Y=k)\big|.$$

Show that d_{TV} satisfies (i) and (iii) with S the space of integer-valued random variables, and that $d_{\mathrm{TV}}(X,Y) = 0$ if and only if X and Y have the same distribution. Thus d_{TV} is a metric on the space of equivalence classes of S with equivalence relation given by $X \sim Y$ if X and Y have the same distribution. We call d_{TV} the *total variation distance*.

Show that

$$d_{\mathrm{TV}}(X,Y) = 2\sup_{A\subseteq\mathbb{Z}} \big|\mathbb{P}(X\in A) - \mathbb{P}(Y\in A)\big|.$$

14. Ascertain in the following cases whether or not F is the joint distribution function of some pair (X, Y) of random variables. If your conclusion is affirmative, find the distribution functions of X and Y separately.

(a)
$$F(x,y) = \begin{cases} 1 - e^{-x-y} & \text{if } x, y \ge 0, \\ 0 & \text{otherwise.} \end{cases}$$

(b)
$$F(x,y) = \begin{cases} 1 - e^{-x} - xe^{-y} & \text{if } 0 \le x \le y, \\ 1 - e^{-y} - ye^{-y} & \text{if } 0 \le y \le x, \\ 0 & \text{otherwise.} \end{cases}$$

15. It is required to place in order n books $B_1, B_2, \dots, B_n$ on a library shelf in such a way that readers searching from left to right waste as little time as possible on average. Assuming that each reader requires book B_i with probability p_i, find the ordering of the books which minimizes $\mathbb{P}(T \ge k)$ for all k, where T is the (random) number of titles examined by a reader before discovery of the required book.

16. Transitive coins. Three coins each show heads with probability $\frac{3}{5}$ and tails otherwise. The first counts 10 points for a head and 2 for a tail, the second counts 4 points for both head and tail, and the third counts 3 points for a head and 20 for a tail.

You and your opponent each choose a coin; you cannot choose the same coin. Each of you tosses your coin and the person with the larger score wins £10^{10}. Would you prefer to be the first to pick a coin or the second?

17. Before the development of radar, inertial navigation, and GPS, flying to isolated islands (for example, from Los Angeles to Hawaii) was somewhat 'hit or miss'. In heavy cloud or at night it was necessary to fly by dead reckoning, and then to search the surface. With the aid of a radio, the pilot had a good idea of the correct great circle along which to search, but could not be sure which of the two directions along this great circle was correct (since a strong tailwind could have carried the plane over its target). When you are the pilot, you calculate that you can make n searches before your plane will run out of fuel. On each search you will discover the island with probability p (if it is indeed in the direction of the search) independently of the results of other searches; you estimate initially that there is probability α that the island is ahead of you. What policy should you adopt in deciding the directions of your various searches in order to maximize the probability of locating the island?

18. Eight pawns are placed randomly on a chessboard, no more than one to a square. What is the probability that:

(a) they are in a straight line (do not forget the diagonals)?

(b) no two are in the same row or column?

19. Which of the following are distribution functions? For those that are, give the corresponding density function f.

(a) $F(x) = \begin{cases} 1 - e^{-x^2} & x \geq 0, \\ 0 & \text{otherwise.} \end{cases}$

(b) $F(x) = \begin{cases} e^{-1/x} & x > 0, \\ 0 & \text{otherwise.} \end{cases}$

(c) $F(x) = e^x/(e^x + e^{-x}),\ x \in \mathbb{R}$.

(d) $F(x) = e^{-x^2} + e^x/(e^x + e^{-x}),\ x \in \mathbb{R}$.

20. (a) If U and V are jointly continuous, show that $\mathbb{P}(U = V) = 0$.

(b) Let X be uniformly distributed on $(0, 1)$, and let $Y = X$. Then X and Y are continuous, and $\mathbb{P}(X = Y) = 1$. Is there a contradiction here?

21. Continued fractions. Let X be uniformly distributed on the interval $[0, 1]$, and express X as a continued fraction thus:

$$X = \cfrac{1}{Y_1 + \cfrac{1}{Y_2 + \cfrac{1}{Y_3 + \cdots}}}.$$

Show that the joint mass function of Y_1 and Y_2 is

$$f(u, v) = \frac{1}{(uv+1)(uv+u+1)}, \qquad u, v = 1, 2, \ldots.$$

22. Let V be a vector space of dimension n over a finite field $\mathbb{F}$ with q elements. Let $X_1, X_2, \ldots, X_m$ be independent random variables, each uniformly distributed on V.

(a) Let $a_i \in \mathbb{F}$, $i = 1, 2, \ldots, m$, be not all zero. Show that the linear combination $Z = \sum_i a_i X_i$ is uniformly distributed on V.

(b) Let p_m be the probability that $X_1, X_2, \ldots, X_m$ are linearly dependent. Show that, if $m \leq n + 1$,

$$q^{-(n-m-1)} \leq p_m \leq q^{-(n-m)}, \qquad m = 1, 2, \ldots, n + 1.$$

23. Modes. A random variable X with distribution function F is said to be *unimodal*† about a mode M if F is convex on $(-\infty, M)$ and concave on (M, ∞). Show that, if F is unimodal about M, then the following hold.

(a) F is absolutely continuous, except possibly for an atom at M.

(b) If F is differentiable, then it has a density that is non-decreasing on $(-\infty, M)$ and non-increasing on (M, ∞), and furthermore, the set of modes of F is a closed bounded interval. [Cf. Problem (2.7.11).]

(c) If the distribution functions F and G are unimodal about the same mode M, then $aF + (1 - a)G$ is unimodal about M for any $0 < a < 1$.

---

†It is a source of potential confusion that the word 'mode' is used in several contexts. A function is sometimes said to be unimodal if it has a unique maximum. The word mode is also used for the value(s) of x at which a mass function (or density function) $f(x)$ is maximized, and even on occasion the locations of its local maxima.

3

Discrete random variables

3.1 Exercises. Probability mass functions

1. For what values of the constant C do the following define mass functions on the positive integers $1, 2, \dots$?

(a) Geometric: $f(x) = C2^{-x}$.

(b) Logarithmic: $f(x) = C2^{-x}/x$.

(c) Inverse square: $f(x) = Cx^{-2}$.

(d) 'Modified' Poisson: $f(x) = C2^x/x!$.

2. For a random variable X having (in turn) each of the four mass functions of Exercise (3.1.1), find:

(i) $\mathbb{P}(X > 1)$,

(ii) the most probable value of X,

(iii) the probability that X is even.

3. We toss n coins, and each one shows heads with probability p, independently of each of the others. Each coin which shows heads is tossed again. What is the mass function of the number of heads resulting from the second round of tosses?

4. Let S_k be the set of positive integers whose base-10 expansion contains exactly k elements (so that, for example, $1024 \in S_4$). A fair coin is tossed until the first head appears, and we write T for the number of tosses required. We pick a random element, N say, from S_T, each such element having equal probability. What is the mass function of N?

5. Log-convexity. (a) Show that, if X is a binomial or Poisson random variable, then the mass function $f(k) = \mathbb{P}(X = k)$ has the property that $f(k-1)f(k+1) \le f(k)^2$.

(b) Show that, if $f(k) = 90/(\pi k)^4$, $k \ge 1$, then $f(k-1)f(k+1) \ge f(k)^2$.

(c) Find a mass function f such that $f(k)^2 = f(k-1)f(k+1)$, $k \ge 1$.

3.2 Exercises. Independence

1. Let X and Y be independent random variables, each taking the values -1 or 1 with probability $\frac{1}{2}$, and let $Z = XY$. Show that X, Y, and Z are pairwise independent. Are they independent?

2. Let X and Y be independent random variables taking values in the positive integers and having the same mass function $f(x) = 2^{-x}$ for $x = 1, 2, \dots$. Find:

(a) $\mathbb{P}(\min\{X, Y\} \le x)$, (b) $\mathbb{P}(Y > X)$,

(c) $\mathbb{P}(X = Y)$, (d) $\mathbb{P}(X \ge kY)$, for a given positive integer k,

(e) $\mathbb{P}(X \text{ divides } Y)$, (f) $\mathbb{P}(X = rY)$, for a given positive rational r.

3. Let X_1, X_2, X_3 be independent random variables taking values in the positive integers and having mass functions given by $\mathbb{P}(X_i = x) = (1 - p_i)p_i^{x-1}$ for $x = 1, 2, \ldots$, and $i = 1, 2, 3$.

(a) Show that

$$\mathbb{P}(X_1 < X_2 < X_3) = \frac{(1 - p_1)(1 - p_2)p_2 p_3^2}{(1 - p_2 p_3)(1 - p_1 p_2 p_3)}.$$

(b) Find $\mathbb{P}(X_1 \le X_2 \le X_3)$.

4. Three players, A, B, and C, take turns to roll a die; they do this in the order ABCABCA....

(a) Show that the probability that, of the three players, A is the first to throw a 6, B the second, and C the third, is 216/1001.

(b) Show that the probability that the first 6 to appear is thrown by A, the second 6 to appear is thrown by B, and the third 6 to appear is thrown by C, is 46656/753571.

5. Let X_r, $1 \le r \le n$, be independent random variables which are symmetric about 0; that is, X_r and $-X_r$ have the same distributions. Show that, for all x, $\mathbb{P}(S_n \ge x) = \mathbb{P}(S_n \le -x)$ where $S_n = \sum_{r=1}^{n} X_r$.

Is the conclusion necessarily true without the assumption of independence?

3.3 Exercises. Expectation

1. Is it generally true that $\mathbb{E}(1/X) = 1/\mathbb{E}(X)$? Is it ever true that $\mathbb{E}(1/X) = 1/\mathbb{E}(X)$?

2. **Coupons.** Every package of some intrinsically dull commodity includes a small and exciting plastic object. There are c different types of object, and each package is equally likely to contain any given type. You buy one package each day.

(a) Find the mean number of days which elapse between the acquisitions of the jth new type of object and the $(j + 1)$th new type.

(b) Find the mean number of days which elapse before you have a full set of objects.

3. Each member of a group of n players rolls a die.

(a) For any pair of players who throw the same number, the group scores 1 point. Find the mean and variance of the total score of the group.

(b) Find the mean and variance of the total score if any pair of players who throw the same number scores that number.

4. **St Petersburg paradox†.** A fair coin is tossed repeatedly. Let T be the number of tosses until the first head. You are offered the following prospect, which you may accept on payment of a fee. If $T = k$, say, then you will receive £2^k. What would be a 'fair' fee to ask of you?

5. Let X have mass function

$$f(x) = \begin{cases} \{x(x+1)\}^{-1} & \text{if } x = 1, 2, \ldots, \\ 0 & \text{otherwise,} \end{cases}$$

and let $\alpha \in \mathbb{R}$. For what values of α is it the case‡ that $\mathbb{E}(X^\alpha) < \infty$?

---

†This problem was mentioned by Nicholas Bernoulli in 1713, and Daniel Bernoulli wrote about the question for the Academy of St Petersburg.

‡If α is not integral, than $\mathbb{E}(X^\alpha)$ is called the *fractional moment of order* α of X. A point concerning notation: for real α and complex $x = re^{i\theta}$, x^α should be interpreted as $r^\alpha e^{i\theta\alpha}$, so that $|x^\alpha| = r^\alpha$. In particular, $\mathbb{E}(|X^\alpha|) = \mathbb{E}(|X|^\alpha)$.

6. Show that $\mathrm{var}(a + X) = \mathrm{var}(X)$ for any random variable X and constant a.

7. Arbitrage. Suppose you find a warm-hearted bookmaker offering payoff odds of $\pi(k)$ against the kth horse in an n-horse race where $\sum_{k=1}^{n}\{\pi(k)+1\}^{-1} < 1$. Show that you can distribute your bets in such a way as to ensure you win.

8. You roll a conventional fair die repeatedly. If it shows 1, you must stop, but you may choose to stop at any prior time. Your score is the number shown by the die on the final roll. What stopping strategy yields the greatest expected score? What strategy would you use if your score were the square of the final roll?

9. Continuing with Exercise (3.3.8), suppose now that you lose c points from your score each time you roll the die. What strategy maximizes the expected final score if $c = \frac{1}{3}$? What is the best strategy if $c = 1$?

10. Random social networks. Let $G = (V, E)$ be a random graph with $m = |V|$ vertices and edge-set E. Write d_v for the degree of vertex v, that is, the number of edges meeting at v. Let Y be a uniformly chosen vertex, and Z a uniformly chosen neighbour of Y.

(a) Show that $\mathbb{E}d_Z \geq \mathbb{E}d_Y$.

(b) Interpret this inequality when the vertices represent people, and the edges represent friendship.

11. The gambler Lester Savage makes up to three successive bets that a fair coin flip will show heads. He places a stake on each bet, which, if a head is shown, pays him back twice the stake. If a tail is shown, he loses his stake.

His stakes are determined as follows. Let $x > y > z > 0$. He bets x on the first flip; if it shows heads he quits, otherwise he continues. If he continues, he bets y on the second flip; if it shows heads he quits, otherwise he continues. If he continues, he bets z on the third flip.

Let G be his accumulated gain (positive or negative). List the possible values of G and their probabilities. Show that $\mathbb{E}(G) = 0$, and find $\mathrm{var}(G)$ and $\mathbb{P}(G < 0)$.

Lester decides to stick with the three numbers x, y, z but to vary their order. How should he place his bets in order to simultaneously minimize both $\mathbb{P}(G < 0)$ and $\mathrm{var}(G)$? Explain.

12. Quicksort†. A set of n different words is equally likely to be in any of the $n!$ possible orders. It is decided to place them in lexicographic order using the following algorithm.

(i) Compare the first word w with the others, and find the set of earlier words and the set of later words.

(ii) Iterate the procedure on each of the two sets thus obtained.

(iii) Continue until the final ordering is achieved.

(a) Give an expression for the mean number c_n of comparisons required.

(b) Show that $c_n = 2n(\log n + \gamma - 2) + \mathrm{O}(1)$ as $n \to \infty$, where γ is Euler's constant.

(c) Let n be replaced by a random variable N with mass function

$$\mathbb{P}(N = n) = \frac{A}{(n-1)n(n+1)}, \qquad n \geq 2,$$

for suitable A. Show that the mean number of comparisons is 4.

---

†Invented by C. A. R. Hoare in 1959.

3.4 Exercises. Indicators and matching

1. (a) A biased coin is tossed n times, and heads shows with probability p on each toss. A *run* is a sequence of throws which result in the same outcome, so that, for example, the sequence HHTHTTH contains five runs. Show that the expected number of runs is $1+2(n-1)p(1-p)$. Find the variance of the number of runs.

(b) Let h heads and t tails be arranged randomly in a line. Find the mean and variance of the number of runs of heads

2. An urn contains n balls numbered $1, 2, \ldots, n$. We remove k balls at random (without replacement) and add up their numbers. Find the mean and variance of the total.

3. Of the $2n$ people in a given collection of n couples, exactly m die. Assuming that the m have been picked at random, find the mean number of surviving couples. This problem was formulated by Daniel Bernoulli in 1768.

4. Urn R contains n red balls and urn B contains n blue balls. At each stage, a ball is selected at random from each urn, and they are swapped. Show that the mean number of red balls in urn R after stage k is $\frac{1}{2}n\{1+(1-2/n)^k\}$. This 'diffusion model' was described by Daniel Bernoulli in 1769.

5. Consider a square with diagonals, with distinct source and sink. Each edge represents a component which is working correctly with probability p, independently of all other components. Write down an expression for the Boolean function which equals 1 if and only if there is a working path from source to sink, in terms of the indicator functions X_i of the events {edge i is working} as i runs over the set of edges. Hence calculate the reliability of the network.

6. A system is called a 'k out of n' system if it contains n components and it works whenever k or more of these components are working. Suppose that each component is working with probability p, independently of the other components, and let X_c be the indicator function of the event that component c is working. Find, in terms of the X_c, the indicator function of the event that the system works, and deduce the reliability of the system.

7. The probabilistic method. Let $G=(V,E)$ be a finite graph. For any set W of vertices and any edge $e \in E$, define the indicator function

$$I_W(e) = \begin{cases} 1 & \text{if } e \text{ connects } W \text{ and } W^c, \\ 0 & \text{otherwise.} \end{cases}$$

Set $N_W = \sum_{e\in E} I_W(e)$. Show that there exists $W \subseteq V$ such that $N_W \geq \frac{1}{2}|E|$.

8. A total of n bar magnets are placed end to end in a line with random independent orientations. Adjacent like poles repel, ends with opposite polarities join to form blocks. Let X be the number of blocks of joined magnets. Find $\mathbb{E}(X)$ and $\text{var}(X)$.

9. Matching. (a) Use the inclusion–exclusion formula (3.4.2) to derive the result of Example (3.4.3), namely: in a random permutation of the first n integers, the probability that exactly r retain their original positions is

$$\frac{1}{r!}\left(\frac{1}{2!} - \frac{1}{3!} + \cdots + \frac{(-1)^{n-r}}{(n-r)!}\right).$$

(b) Let d_n be the number of derangements of the first n integers (that is, rearrangements with no integers in their original positions). Show that $d_{n+1} = nd_n + nd_{n-1}$ for $n \geq 2$. Deduce the result of part (a).

(c) Given that exactly m integers retain their original positions, find the probability that the integer 1 remains in first place.

10. Birthdays. In a lecture audience, there are n students born in 2011, and they were born on independent, uniformly distributed days. Calculate the mean of the number B of pairs of students sharing a birthday, and show that $\mathbb{E}(B) > 1$ if and only if $n \geq 28$. Compare this with the result of Problem (1.8.30). Find the variance of B.

11. Inaba's theorem. Show that any set of 10 points in the plane $\mathbb{R}^2$ can be covered by a suitable placement of disjoint open unit disks. [Hint: Consider an infinite array of unit disks whose centres form a triangular lattice.]

12. Days are either wet or dry, and, given today's weather, tomorrow's is the same as today's with probability p, and different otherwise. Let w_n be the probability that the weather n days into the future from today will be wet. Show that $w_{n+1} = 1 - p + (2p - 1)w_{n-1}$, and find w_n. What is the mean number of wet days in the next week?

13. An urn contains b balls of which g are green. Balls are sampled from the urn at random, one by one. After a ball is sampled, its colour is noted, and it is discarded. Find the mean and variance of the number of green balls in a sample of size n $(\leq b)$.

14. Ménages (1.8.23) revisited. Let n (≥ 2) heterosexual couples be seated randomly at a circular table, subject only to the rule that the sexes alternate. There is no requirement that couples sit together. Let X be the number of couples seated adjacently. Show that $\mathbb{E}(X) = 2$ and $\text{var}(X) = 2 - 2/(n-1)$.

3.5 Exercises. Examples of discrete variables

1. De Moivre trials. Each trial may result in any of t given outcomes, the ith outcome having probability p_i. Let N_i be the number of occurrences of the ith outcome in n independent trials. Show that

$$\mathbb{P}(N_i = n_i \text{ for } 1 \leq i \leq t) = \frac{n!}{n_1!\, n_2! \cdots n_t!}\, p_1^{n_1} p_2^{n_2} \cdots p_t^{n_t}$$

for any collection $n_1, n_2, \dots, n_t$ of non-negative integers with sum n. The vector N is said to have the *multinomial distribution*.

2. In your pocket is a random number N of coins, where N has the Poisson distribution with parameter λ. You toss each coin once, with heads showing with probability p each time. Show that the total number of heads has the Poisson distribution with parameter λp.

3. Let X be Poisson distributed where $\mathbb{P}(X = n) = p_n(\lambda) = \lambda^n e^{-\lambda}/n!$ for $n \geq 0$. Show that $\mathbb{P}(X \leq n) = 1 - \int_0^\lambda p_n(x)\, dx$.

4. Capture–recapture. A population of b animals has had a number a of its members captured, marked, and released. Let X be the number of animals it is necessary to recapture (without re-release) in order to obtain m marked animals. Show that

$$\mathbb{P}(X = n) = \frac{a}{b}\binom{a-1}{m-1}\binom{b-a}{n-m}\bigg/\binom{b-1}{n-1},$$

and find $\mathbb{E}X$. This distribution has been called *negative hypergeometric*.

5. Compound Poisson distribution. Let Λ be a positive random variable with density function f and distribution function F, and let Y have the Poisson distribution with parameter Λ. Show for $n = 0, 1, 2, \dots$ that

$$\mathbb{P}(Y \leq n) = \int_0^\infty p_n(\lambda) F(\lambda)\, d\lambda, \qquad \mathbb{P}(Y > n) = \int_0^\infty p_n(\lambda)[1 - F(\lambda)]\, d\lambda,$$

where $p_n(\lambda) = e^{-\lambda}\lambda^n/n!$.

3.6 Exercises. Dependence

1. Show that the collection of random variables on a given probability space and having finite variance forms a vector space over the reals.

2. Find the marginal mass functions of the multinomial distribution of Exercise (3.5.1).

3. Let X and Y be discrete random variables with joint mass function

$$f(x, y) = \frac{C}{(x+y-1)(x+y)(x+y+1)}, \qquad x, y = 1, 2, 3, \dots.$$

Find the marginal mass functions of X and Y, calculate C, and also the covariance of X and Y.

4. Let X and Y be discrete random variables with mean 0, variance 1, and covariance ρ. Show that $\mathbb{E}(\max\{X^2, Y^2\}) \le 1 + \sqrt{1-\rho^2}$.

5. Mutual information. Let X and Y be discrete random variables with joint mass function f.

(a) Show that $\mathbb{E}(\log f_X(X)) \ge \mathbb{E}(\log f_Y(X))$.

(b) Show that the *mutual information*

$$I = \mathbb{E}\left(\log\left\{\frac{f(X, Y)}{f_X(X) f_Y(Y)}\right\}\right)$$

satisfies $I \ge 0$, with equality if and only if X and Y are independent.

6. Voter paradox. Let X, Y, Z be discrete random variables with the property that their values are distinct with probability 1. Let $a = \mathbb{P}(X > Y)$, $b = \mathbb{P}(Y > Z)$, $c = \mathbb{P}(Z > X)$.

(a) Show that $\min\{a, b, c\} \le \frac{2}{3}$, and give an example where this bound is attained.

(b) Show that, if X, Y, Z are independent and identically distributed, then $a = b = c = \frac{1}{2}$.

(c) Find $\min\{a, b, c\}$ and $\sup_p \min\{a, b, c\}$ when $\mathbb{P}(X = 0) = 1$, and Y, Z are independent with $\mathbb{P}(Z = 1) = \mathbb{P}(Y = -1) = p$, $\mathbb{P}(Z = -2) = \mathbb{P}(Y = 2) = 1 - p$. Here, $\sup_p$ denotes the supremum as p varies over $[0, 1]$.

[Part (a) is related to de Condorcet's observation that, in an election, it is possible for more than half of the voters to prefer candidate A to candidate B, more than half B to C, and more than half C to A.]

7. Benford's distribution, or the law of anomalous numbers. If one picks a numerical entry at random from an almanac, or the annual accounts of a corporation, the first two significant digits, X, Y, are found to have approximately the joint mass function

$$f(x, y) = \log_{10}\left(1 + \frac{1}{10x + y}\right), \qquad 1 \le x \le 9,\ 0 \le y \le 9.$$

Find the mass function of X and an approximation to its mean. [A heuristic explanation for this phenomenon may be found in the second of Feller's volumes published in 1971. See also Berger and Hill 2015.]

8. Let X and Y have joint mass function

$$f(j, k) = \frac{c(j+k)a^{j+k}}{j!\,k!}, \qquad j, k \ge 0,$$

where a is a constant. Find c, $\mathbb{P}(X = j)$, $\mathbb{P}(X + Y = r)$, and $\mathbb{E}(X)$.

9. Correlation. Let X, Y, Z be non-degenerate and independent random variables. By considering $U = X + Y$, $V = Y + Z$, $W = Z - X$, or otherwise, show that having positive correlation is not a transitive relation.

10. Cauchy–Schwarz inequality. Use the identity $a^2d^2 + b^2c^2 - 2abcd = (ad - bc)^2$ to prove the Cauchy–Schwarz inequality.

11. Cantelli, or one-sided Chebyshov inequality. Show that

$$\mathbb{P}\big(X - \mathbb{E}(X) > t\big) \le \frac{\operatorname{var}(X)}{t^2 + \operatorname{var}(X)}, \qquad t > 0.$$

3.7 Exercises. Conditional distributions and conditional expectation

1. Show the following:

(a) $\mathbb{E}(aY + bZ \mid X) = a\mathbb{E}(Y \mid X) + b\mathbb{E}(Z \mid X)$ for $a, b \in \mathbb{R}$,

(b) $\mathbb{E}(Y \mid X) \ge 0$ if $Y \ge 0$,

(c) $\mathbb{E}(1 \mid X) = 1$,

(d) if X and Y are independent then $\mathbb{E}(Y \mid X) = \mathbb{E}(Y)$,

(e) ('pull-through property') $\mathbb{E}(Yg(X) \mid X) = g(X)\mathbb{E}(Y \mid X)$ for any suitable function g,

(f) ('tower property') $\mathbb{E}\{\mathbb{E}(Y \mid X, Z) \mid X\} = \mathbb{E}(Y \mid X) = \mathbb{E}\{\mathbb{E}(Y \mid X) \mid X, Z\}$.

2. Uniqueness of conditional expectation. Suppose that X and Y are discrete random variables, and that $\phi(X)$ and $\psi(X)$ are two functions of X satisfying

$$\mathbb{E}\big(\phi(X)g(X)\big) = \mathbb{E}\big(\psi(X)g(X)\big) = \mathbb{E}\big(Yg(X)\big)$$

for any function g for which all the expectations exist. Show that $\phi(X)$ and $\psi(X)$ are almost surely equal, in that $\mathbb{P}(\phi(X) = \psi(X)) = 1$.

3. Suppose that the conditional expectation of Y given X is defined as the (almost surely) unique function $\psi(X)$ such that $\mathbb{E}(\psi(X)g(X)) = \mathbb{E}(Yg(X))$ for all functions g for which the expectations exist. Show (a)–(f) of Exercise (3.7.1) above (with the occasional addition of the expression 'with probability 1').

4. Conditional variance formula. How should we define $\operatorname{var}(Y \mid X)$, the conditional variance of Y given X? Show that $\operatorname{var}(Y) = \mathbb{E}(\operatorname{var}(Y \mid X)) + \operatorname{var}(\mathbb{E}(Y \mid X))$.

5. The lifetime of a machine (in days) is a random variable T with mass function f. Given that the machine is working after t days, what is the mean subsequent lifetime of the machine when:

(a) $f(x) = (N+1)^{-1}$ for $x \in \{0, 1, \dots, N\}$,

(b) $f(x) = 2^{-x}$ for $x = 1, 2, \dots$.

(The first part of Problem (3.11.13) may be useful.)

6. Let $X_1, X_2, \dots$ be identically distributed random variables with mean μ and variance σ^2, and let N be a random variable taking values in the non-negative integers and independent of the X_i. Let $S = X_1 + X_2 + \cdots + X_N$. Show that $\mathbb{E}(S \mid N) = \mu N$, and deduce that $\mathbb{E}(S) = \mu\mathbb{E}(N)$. Find $\operatorname{var} S$ in terms of the first two moments of N, using the conditional variance formula of Exercise (3.7.4).

7. A factory has produced n robots, each of which is faulty with probability ϕ. To each robot a test is applied which detects the fault (if present) with probability δ. Let X be the number of faulty robots, and Y the number detected as faulty. Assuming the usual independence, show that

$$\mathbb{E}(X \mid Y) = \big\{n\phi(1-\delta) + (1-\phi)Y\big\}/(1 - \phi\delta).$$

8. Families. Each child is equally likely to be male or female, independently of all other children.
(a) Show that, in a family of predetermined size, the expected number of boys equals the expected number of girls. Was the assumption of independence necessary?
(b) A randomly selected child is male; does the expected number of his brothers equal the expected number of his sisters? What happens if you do not require independence?

9. Let X and Y be independent with mean μ. Explain the error in the following equation:

$$\text{`}\mathbb{E}(X \mid X+Y=z) = \mathbb{E}(X \mid X = z-Y) = \mathbb{E}(z-Y) = z-\mu\text{'}.$$

10. A coin shows heads with probability p. Let X_n be the number of flips required to obtain a run of n consecutive heads. Show that $\mathbb{E}(X_n) = \sum_{k=1}^{n} p^{-k}$.

11. Conditional covariance. Give a definition of the *conditional covariance* $\operatorname{cov}(X, Y \mid Z)$. Show that

$$\operatorname{cov}(X,Y) = \mathbb{E}\big(\operatorname{cov}(X,Y \mid Z)\big) + \operatorname{cov}\big(\mathbb{E}(X \mid Z), \mathbb{E}(Y \mid Z)\big).$$

12. An urn contains initially b blue balls and r red balls, where $b, r \ge 2$. Balls are drawn one by one without replacement. Show that the mean number of draws until the first colour drawn is first repeated equals 3.

13. (a) Let X be uniformly distributed on $\{0, 1, \dots, n\}$. Show that $\operatorname{var}(X) = \frac{1}{12}n(n+2)$.
(b) A student sits two examinations, gaining X and Y marks, respectively. In the interests of economy and fairness, the examiner determines that X shall be uniformly distributed on $\{0, 1, \dots, n\}$, and that, conditional on $X = k$, Y shall have the binomial $\text{bin}(n, k/n)$ distribution.

(i) Show that $\mathbb{E}(Y) = \frac{1}{2}n$, and $\operatorname{var}(Y) = \frac{1}{12}(n^2+4n-2)$.
(ii) Find $\mathbb{E}(X+Y)$ and $\operatorname{var}(X+Y)$.

14. Let X, Y be discrete integer-valued random variables with the joint mass function

$$f(x,y) = \frac{\lambda^y e^{-2\lambda}}{x!\,(y-x)!}, \qquad 0 \le x \le y < \infty.$$

Show that X and Y are each Poisson-distributed, and that the conditional distribution of X given Y is binomial.

3.8 Exercises. Sums of random variables

1. Let X and Y be independent variables, X being equally likely to take any value in $\{0, 1, \dots, m\}$, and Y similarly in $\{0, 1, \dots, n\}$. Find the mass function of $Z = X + Y$. The random variable Z is said to have the *trapezoidal distribution*.

2. Let X and Y have the joint mass function

$$f(x,y) = \frac{C}{(x+y-1)(x+y)(x+y+1)}, \qquad x, y = 1, 2, 3, \dots.$$

Find the mass functions of $U = X + Y$ and $V = X - Y$.

3. Let X and Y be independent geometric random variables with respective parameters α and β. Show that

$$\mathbb{P}(X+Y=z) = \frac{\alpha\beta}{\alpha-\beta}\{(1-\beta)^{z-1} - (1-\alpha)^{z-1}\}.$$

4. Let $\{X_r : 1 \le r \le n\}$ be independent geometric random variables with parameter p. Show that $Z = \sum_{r=1}^{n} X_r$ has a negative binomial distribution. [Hint: No calculations are necessary.]

5. Pepys's problem†. Sam rolls $6n$ dice once; he needs at least n sixes. Isaac rolls $6(n+1)$ dice; he needs at least $n+1$ sixes. Who is more likely to obtain the number of sixes he needs?

6. Stein–Chen equation. Let N be Poisson distributed with parameter λ. Show that, for any function g such that the expectations exist, $\mathbb{E}(Ng(N)) = \lambda\mathbb{E}g(N+1)$. More generally, if $S = \sum_{r=1}^{N} X_r$, where $\{X_r : r \ge 0\}$ are independent identically distributed non-negative integer-valued random variables, show that

$$\mathbb{E}\big(Sg(S)\big) = \lambda\mathbb{E}\big(g(S + X_0)X_0\big).$$

7. Random sum. Let $S = \sum_{i=1}^{N} X_i$, where the X_i, $i \ge 1$, are independent, identically distributed random variables with mean μ and variance σ^2, and N is positive, integer-valued, and independent of the X_i. Show that $\mathbb{E}(S) = \mu\mathbb{E}(N)$, and

$$\operatorname{var}(S) = \sigma^2\mathbb{E}(N) + \mu^2 \operatorname{var}(N).$$

8. Let X and Y be independent random variables with the geometric distributions $f_X(k) = pq^{k-1}$, $f_Y(k) = \lambda\mu^{k-1}$, for $k \ge 1$, where $p + q = \lambda + \mu = 1$ and $q \ne \mu$. Write down $\mathbb{P}(X + Y = n+1,\ X = k)$, and hence find the distribution of $X + Y$, and the conditional distribution of X given that $X + Y = n + 1$. Does anything special occur when $q = \mu$?

3.9 Exercises. Simple random walk

1. Let T be the time which elapses before a simple random walk is absorbed at either of the absorbing barriers at 0 and N, having started at k where $0 \le k \le N$. Show that $\mathbb{P}(T < \infty) = 1$ and $\mathbb{E}(T^k) < \infty$ for all $k \ge 1$.

2. For simple random walk S with absorbing barriers at 0 and N, let W be the event that the particle is absorbed at 0 rather than at N, and let $p_k = \mathbb{P}(W \mid S_0 = k)$. Show that, if the particle starts at k where $0 < k < N$, the conditional probability that the first step is rightwards, given W, equals pp_{k+1}/p_k. Deduce that the mean duration J_k of the walk, conditional on W, satisfies the equation

$$pp_{k+1}J_{k+1} - p_kJ_k + (p_k - pp_{k+1})J_{k-1} = -p_k, \qquad \text{for } 0 < k < N,$$

subject to the convention that $p_N J_N = 0$. Show that we may take as boundary condition $J_0 = 0$. Find J_k in the symmetric case, when $p = \frac{1}{2}$.

3. With the notation of Exercise (3.9.2), suppose further that at any step the particle may remain where it is with probability r where $p + q + r = 1$. Show that J_k satisfies

$$pp_{k+1}J_{k+1} - (1-r)p_kJ_k + qp_{k-1}J_{k-1} = -p_k$$

and that, when $\rho = q/p \ne 1$,

$$J_k = \frac{1}{p-q} \cdot \frac{1}{\rho^k - \rho^N}\left\{k(\rho^k + \rho^N) - \frac{2N\rho^N(1-\rho^k)}{1-\rho^N}\right\}.$$

---

†Pepys put a simple version of this problem to Newton in 1693, but was reluctant to accept the correct reply he received.

4. **Problem of the points.** A coin is tossed repeatedly, heads turning up with probability p on each toss. Player A wins the game if m heads appear before n tails have appeared, and player B wins otherwise. Let p_{mn} be the probability that A wins the game. Set up a difference equation for the p_{mn}. What are the boundary conditions?

5. Consider a simple random walk on the set $\{0, 1, 2, \dots, N\}$ in which each step is to the right with probability p or to the left with probability $q = 1 - p$. Absorbing barriers are placed at 0 and N. Show that the number X of positive steps of the walk before absorption satisfies

$$\mathbb{E}(X) = \tfrac{1}{2}\{D_k - k + N(1 - p_k)\}$$

where D_k is the mean number of steps until absorption having started at k, and p_k is the probability of absorption at 0.

6. **Gambler's ruin revisited.** Let D_k be the duration of a random walk on $\{0, 1, 2, \dots, a\}$ with absorbing barriers at 0 and a, and started at k, with steps X_i satisfying

$$\mathbb{P}(X_i = 1) = \mathbb{P}(X_i = -1) = p, \quad \mathbb{P}(X_i = 0) = 1 - 2p.$$

(a) When $p = \frac{1}{2}$, show that

$$\operatorname{var}(D_k) = \tfrac{1}{3}k(a-k)\{(a-k)^2 + k^2 - 2\}.$$

(b) Deduce (without lengthy calculation) that, for $p < \frac{1}{2}$,

$$\operatorname{var}(D_k) = \frac{k(a-k)}{(2p)^2}\left[1 - 2p + \tfrac{1}{3}\{(a-k)^2 + k^2 - 2\}\right].$$

7. **Returns and visits by random walk.** Consider a simple symmetric random walk on the set $\{0, 1, 2, \dots, a\}$ with absorbing barriers at 0 and a, and starting at k where $0 < k < a$. Let r_k be the probability the walk ever returns to k, and let v_x be the mean number of visits to point x before absorption. Find r_k, and hence show that

$$v_x = \begin{cases} 2x(a-k)/a & \text{for } 0 < x < k, \\ 2k(a-x)/a & \text{for } k < x < a. \end{cases}$$

8. (a) "Millionaires should always gamble, poor men never" [J. M. Keynes].

(b) "If I wanted to gamble, I would buy a casino" [P. Getty].

(c) "That the chance of gain is naturally overvalued, we may learn from the universal success of lotteries" [Adam Smith, 1776].

Discuss.

3.10 Exercises. Random walk: counting sample paths

1. Consider a symmetric simple random walk S with $S_0 = 0$. Let $T = \min\{n \geq 1 : S_n = 0\}$ be the time of the first return of the walk to its starting point. Show that

$$\mathbb{P}(T = 2n) = \frac{1}{2n-1}\binom{2n}{n}2^{-2n},$$

and deduce that $\mathbb{E}(T^\alpha) < \infty$ if and only if $\alpha < \frac{1}{2}$. You may need Stirling's formula: $n! \sim n^{n+\frac{1}{2}}e^{-n}\sqrt{2\pi}$.

2. For a symmetric simple random walk starting at 0, show that the mass function of the maximum satisfies $\mathbb{P}(M_n = r) = \mathbb{P}(S_n = r) + \mathbb{P}(S_n = r + 1)$ for $r \geq 0$.

3. For a symmetric simple random walk starting at 0, show that the probability that the first visit to S_{2n} takes place at time $2k$ equals the product $\mathbb{P}(S_{2k} = 0)\mathbb{P}(S_{2n-2k} = 0)$, for $0 \leq k \leq n$.

4. Samuels' theorem. A simple random walk on the integers $\mathbb{Z}$ moves one step rightwards with probability p and otherwise one step leftwards, where $p \in (0, 1)$. Suppose it starts at 0 and has absorbing barriers at $\pm a$. Show that the time and place of absorption are independent.

5. Hitting time theorem. Let $\{X_m : m \geq 1\}$ be independent, identically distributed random variables taking integer values such that $\mathbb{P}(X_1 \geq -1) = 1$. Let S_n be the (generalized) random walk given by $S_n = k + X_1 + X_2 + \cdots + X_n$, where $k \geq 0$ is given, and let $T = \inf\{n \geq 0 : S_n = 0\}$ be the hitting time of 0.

Show by induction that $\mathbb{P}(T = n) = (k/n)\mathbb{P}(S_n = 0)$ when $n \geq 1, k \geq 0$.

3.11 Problems

1. (a) Let X and Y be independent discrete random variables, and let $g, h : \mathbb{R} \to \mathbb{R}$. Show that $g(X)$ and $h(Y)$ are independent.

(b) Show that two discrete random variables X and Y are independent if and only if $f_{X,Y}(x, y) = f_X(x)f_Y(y)$ for all $x, y \in \mathbb{R}$.

(c) More generally, show that X and Y are independent if and only if $f_{X,Y}(x, y)$ can be factorized as the product $g(x)h(y)$ of a function of x alone and a function of y alone.

2. Show that if $\text{var}(X) = 0$ then X is almost surely constant; that is, there exists $a \in \mathbb{R}$ such that $\mathbb{P}(X = a) = 1$. (First show that if $\mathbb{E}(X^2) = 0$ then $\mathbb{P}(X = 0) = 1$.)

3. (a) Let X be a discrete random variable and let $g : \mathbb{R} \to \mathbb{R}$. Show that, when the sum is absolutely convergent,

$$\mathbb{E}(g(X)) = \sum_x g(x)\mathbb{P}(X = x).$$

(b) If X and Y are independent and $g, h : \mathbb{R} \to \mathbb{R}$, show that $\mathbb{E}(g(X)h(Y)) = \mathbb{E}(g(X))\mathbb{E}(h(Y))$ whenever these expectations exist.

4. Let $\Omega = \{\omega_1, \omega_2, \omega_3\}$, with $\mathbb{P}(\omega_1) = \mathbb{P}(\omega_2) = \mathbb{P}(\omega_3) = \frac{1}{3}$. Define $X, Y, Z : \Omega \to \mathbb{R}$ by

$$\begin{aligned} X(\omega_1) &= 1, & X(\omega_2) &= 2, & X(\omega_3) &= 3, \\ Y(\omega_1) &= 2, & Y(\omega_2) &= 3, & Y(\omega_3) &= 1, \\ Z(\omega_1) &= 2, & Z(\omega_2) &= 2, & Z(\omega_3) &= 1. \end{aligned}$$

Show that X and Y have the same mass functions. Find the mass functions of $X + Y$, XY, and X/Y. Find the conditional mass functions $f_{Y|Z}$ and $f_{Z|Y}$.

5. For what values of k and α is f a mass function, where:

(a) $f(n) = k/\{n(n + 1)\}$, $n = 1, 2, \ldots$,

(b) $f(n) = kn^\alpha$, $n = 1, 2, \ldots$ (*zeta* or *Zipf distribution*)?

6. Let X and Y be independent Poisson variables with respective parameters λ and μ. Show that:

(a) $X + Y$ is Poisson, parameter $\lambda + \mu$,

(b) the conditional distribution of X, given $X + Y = n$, is binomial, and find its parameters.

7. If X is geometric, show that $\mathbb{P}(X = n+k \mid X > n) = \mathbb{P}(X = k)$ for $k, n \geq 1$. Why do you think that this is called the 'lack of memory' property? Does any other distribution on the positive integers have this property?

8. Show that the sum of two independent binomial variables, $\text{bin}(m, p)$ and $\text{bin}(n, p)$ respectively, is $\text{bin}(m + n, p)$.

9. Let N be the number of heads occurring in n tosses of a biased coin. Write down the mass function of N in terms of the probability p of heads turning up on each toss. Prove and utilize the identity

$$\sum_i \binom{n}{2i} x^{2i} y^{n-2i} = \tfrac{1}{2}\{(x+y)^n + (y-x)^n\}$$

in order to calculate the probability p_n that N is even. Compare with Problem (1.8.20).

10. An urn contains N balls, b of which are blue and r $(= N - b)$ of which are red. A random sample of n balls is withdrawn without replacement from the urn. Show that the number B of blue balls in this sample has the mass function

$$\mathbb{P}(B = k) = \binom{b}{k}\binom{N-b}{n-k} \Big/ \binom{N}{n}.$$

This is called the *hypergeometric distribution* with parameters N, b, and n. Show further that if N, b, and r approach ∞ in such a way that $b/N \to p$ and $r/N \to 1 - p$, then

$$\mathbb{P}(B = k) \to \binom{n}{k} p^k (1-p)^{n-k}.$$

You have shown that, for small n and large N, the distribution of B barely depends on whether or not the balls are replaced in the urn immediately after their withdrawal.

11. Let X and Y be independent $\text{bin}(n, p)$ variables, and let $Z = X + Y$. Show that the conditional distribution of X given $Z = N$ is the hypergeometric distribution of Problem (3.11.10).

12. Suppose X and Y take values in $\{0, 1\}$, with joint mass function $f(x, y)$. Write $f(0, 0) = a$, $f(0, 1) = b$, $f(1, 0) = c$, $f(1, 1) = d$, and find necessary and sufficient conditions for X and Y to be: (a) uncorrelated, (b) independent.

13. Tail sum for expectation.

(a) If X takes non-negative integer values show that

$$\mathbb{E}(X) = \sum_{n=0}^{\infty} \mathbb{P}(X > n).$$

(b) An urn contains b blue and r red balls. Balls are removed at random until the first blue ball is drawn. Show that the expected number drawn is $(b + r + 1)/(b + 1)$.

(c) The balls are replaced and then removed at random until all the remaining balls are of the same colour. Find the expected number remaining in the urn.

(d) Let X and Y be independent random variables taking values in the non-negative integers, with finite means. Let $U = \min\{X, Y\}$ and $V = \max\{X, Y\}$. Show that

$$\mathbb{E}(U) = \sum_{r=1}^{\infty} \mathbb{P}(X \geq r)\mathbb{P}(Y \geq r),$$

$$\mathbb{E}(V) = \sum_{r=1}^{\infty} \Big[\mathbb{P}(X \geq r) + \mathbb{P}(Y \geq r) - \mathbb{P}(X \geq r)\mathbb{P}(Y \geq r)\Big],$$

$$\mathbb{E}(UV) = \sum_{r,s=1}^{\infty} \mathbb{P}(X \geq r)\mathbb{P}(Y \geq s).$$

(e) Let X take values in the non-negative integers. Show that

$$\mathbb{E}(X^2) = \mathbb{E}(X) + 2\sum_{r=0}^{\infty} r\mathbb{P}(X > r) = \sum_{r=0}^{\infty}(2r+1)\mathbb{P}(X > r),$$

and find a similar formula for $\mathbb{E}(X^3)$.

14. Let $X_1, X_2, \dots, X_n$ be independent random variables, and suppose that X_k is Bernoulli with parameter p_k. Show that $Y = X_1 + X_2 + \cdots + X_n$ has mean and variance given by

$$\mathbb{E}(Y) = \sum_{1}^{n} p_k, \quad \mathrm{var}(Y) = \sum_{1}^{n} p_k(1 - p_k).$$

Show that, for $\mathbb{E}(Y)$ fixed, $\mathrm{var}(Y)$ is a maximum when $p_1 = p_2 = \cdots = p_n$. That is to say, the variation in the sum is greatest when individuals are most alike. Is this contrary to intuition?

15. Let $\mathbf{X} = (X_1, X_2, \dots, X_n)$ be a vector of random variables. The *covariance matrix* $\mathbf{V}(\mathbf{X})$ of $\mathbf{X}$ is defined to be the symmetric n by n matrix with entries $(v_{ij} : 1 \leq i, j \leq n)$ given by $v_{ij} = \mathrm{cov}(X_i, X_j)$. Show that $|\mathbf{V}(\mathbf{X})| = 0$ if and only if the X_i are linearly dependent with probability one, in that $\mathbb{P}(a_1X_1 + a_2X_2 + \cdots + a_nX_n = b) = 1$ for some $\mathbf{a}$ and b. ($|\mathbf{V}|$ denotes the determinant of $\mathbf{V}$.)

16. Let X and Y be independent Bernoulli random variables with parameter $\frac{1}{2}$. Show that $X + Y$ and $|X - Y|$ are dependent though uncorrelated.

17. A secretary drops n matching pairs of letters and envelopes down the stairs, and then places the letters into the envelopes in a random order. Use indicators to show that the number X of correctly matched pairs has mean and variance 1 for all $n \geq 2$. Show that the mass function of X converges to a Poisson mass function as $n \to \infty$.

18. Let $\mathbf{X} = (X_1, X_2, \dots, X_n)$ be a vector of independent random variables each having the Bernoulli distribution with parameter p. Let $f : \{0, 1\}^n \to \mathbb{R}$ be *increasing*, which is to say that $f(\mathbf{x}) \leq f(\mathbf{y})$ whenever $x_i \leq y_i$ for each i.

(a) Let $e(p) = \mathbb{E}(f(\mathbf{X}))$. Show that $e(p_1) \leq e(p_2)$ if $p_1 \leq p_2$.

(b) **FKG inequality**†. Let f and g be increasing functions from $\{0, 1\}^n$ into $\mathbb{R}$. Show by induction on n that $\mathrm{cov}(f(\mathbf{X}), g(\mathbf{X})) \geq 0$.

19. Let $R(p)$ be the reliability function of a network G with a given source and sink, each edge of which is working with probability p, and let A be the event that there exists a working connection from source to sink. Show that

$$R(p) = \sum_{\omega} I_A(\omega) p^{N(\omega)}(1 - p)^{m - N(\omega)}$$

†Named after C. Fortuin, P. Kasteleyn, and J. Ginibre 1971, but due in this form to T. E. Harris 1960.

where ω is a typical realization (i.e. outcome) of the network, $N(\omega)$ is the number of working edges of ω, and m is the total number of edges of G.

Deduce that $R'(p) = \operatorname{cov}(I_A, N)/\{p(1-p)\}$, and hence that

$$\frac{R(p)(1-R(p))}{p(1-p)} \le R'(p) \le \sqrt{\frac{mR(p)(1-R(p))}{p(1-p)}}.$$

20. Let $R(p)$ be the reliability function of a network G, each edge of which is working with probability p.

(a) Show that $R(p_1 p_2) \le R(p_1)R(p_2)$ if $0 \le p_1, p_2 \le 1$.

(b) Show that $R(p^\gamma) \le R(p)^\gamma$ for all $0 \le p \le 1$ and $\gamma \ge 1$.

21. DNA fingerprinting. In a certain style of detective fiction, the sleuth is required to declare "the criminal has the unusual characteristics ...; find this person and you have your man". Assume that any given individual has these unusual characteristics with probability 10^{-7} independently of all other individuals, and that the city in question contains 10^7 inhabitants. Calculate the expected number of such people in the city.

(a) Given that the police inspector finds such a person, what is the probability that there is at least one other?

(b) If the inspector finds two such people, what is the probability that there is at least one more?

(c) How many such people need be found before the inspector can be reasonably confident that he has found them all?

(d) For the given population, how improbable should the characteristics of the criminal be, in order that he (or she) be specified uniquely?

22. In 1710, J. Arbuthnot observed that male births had exceeded female births in London for 82 successive years. Arguing that this showed the two sexes cannot be equally likely, since 2^{-82} is very small, he attributed this run of masculinity to Divine Providence. Let us assume that each birth results in a girl with probability $p = 0.485$, and that the outcomes of different confinements are independent of each other. Ignoring the possibility of twins (and so on), show that the probability that girls outnumber boys in $2n$ live births is no greater than $\binom{2n}{n} p^n q^n \{q/(q-p)\}$, where $q = 1 - p$. Suppose that 20,000 children are born in each of 82 successive years. Show that the probability that boys outnumber girls every year is at least 0.99. You may need Stirling's formula.

23. Consider a symmetric random walk with an absorbing barrier at N and a reflecting barrier at 0 (so that, when the particle is at 0, it moves to 1 at the next step). Let $\alpha_k(j)$ be the probability that the particle, having started at k, visits 0 exactly j times before being absorbed at N. We make the convention that, if $k = 0$, then the starting point counts as one visit. Show that

$$\alpha_k(j) = \frac{N-k}{N^2}\left(1 - \frac{1}{N}\right)^{j-1}, \qquad j \ge 1,\ 0 \le k \le N.$$

24. Problem of the points (3.9.4). A coin is tossed repeatedly, heads turning up with probability p on each toss. Player A wins the game if heads appears at least m times before tails has appeared n times; otherwise player B wins the game. Find the probability that A wins the game.

25. A coin is tossed repeatedly, heads appearing on each toss with probability p. A gambler starts with initial fortune k (where $0 < k < N$); he wins one point for each head and loses one point for each tail. If his fortune is ever 0 he is bankrupted, whilst if it ever reaches N he stops gambling to buy a Jaguar. Suppose that $p < \frac{1}{2}$. Show that the gambler can increase his chance of winning by doubling the stakes. You may assume that k and N are even.

What is the corresponding strategy if $p \ge \frac{1}{2}$?

26. A compulsive gambler is never satisfied. At each stage he wins £1 with probability p and loses £1 otherwise. Find the probability that he is ultimately bankrupted, having started with an initial fortune of £k.

27. Range of random walk. Let $\{X_n : n \geq 1\}$ be independent, identically distributed random variables taking integer values. Let $S_0 = 0$, $S_n = \sum_{i=1}^{n} X_i$. The *range* R_n of $S_0, S_1, \ldots, S_n$ is the number of distinct values taken by the sequence. Show that $\mathbb{P}(R_n = R_{n-1}+1) = \mathbb{P}(S_1 S_2 \cdots S_n \neq 0)$, and deduce that, as $n \to \infty$,

$$\frac{1}{n}\mathbb{E}(R_n) \to \mathbb{P}(S_k \neq 0 \text{ for all } k \geq 1).$$

Hence show that, for the simple random walk, $n^{-1}\mathbb{E}(R_n) \to |p - q|$ as $n \to \infty$.

28. Arc sine law for maxima. Consider a symmetric random walk S starting from the origin, and let $M_n = \max\{S_i : 0 \leq i \leq n\}$. Show that, for $i = 2k, 2k+1$, the probability that the walk reaches M_{2n} for the first time at time i equals $\frac{1}{2}\mathbb{P}(S_{2k} = 0)\mathbb{P}(S_{2n-2k} = 0)$.

29. Let S be a symmetric random walk with $S_0 = 0$, and let N_n be the number of points that have been visited by S exactly once up to time n. Show that $\mathbb{E}(N_n) = 2$.

30. Family planning. Consider the following fragment of verse entitled 'Note for the scientist'.

> People who have three daughters try for more,
> And then its fifty–fifty they'll have four,
> Those with a son or sons will let things be,
> Hence all these surplus women, QED.

(a) What do you think of the argument?

(b) Show that the mean number of children of either sex in a family whose fertile parents have followed this policy equals 1. (You should assume that each delivery yields exactly one child whose sex is equally likely to be male or female.) Discuss.

31. Dirichlet distribution. Let $\beta > 1$, let $p_1, p_2, \ldots$ denote the prime numbers, and let $N(1), N(2), \ldots$ be independent random variables, $N(i)$ having mass function $\mathbb{P}(N(i) = k) = (1 - \gamma_i)\gamma_i^k$ for $k \geq 0$, where $\gamma_i = p_i^{-\beta}$ for all i. Show that $M = \prod_{i=1}^{\infty} p_i^{N(i)}$ is a random integer with mass function $\mathbb{P}(M = m) = Cm^{-\beta}$ for $m \geq 1$ (this may be called the *Dirichlet distribution*), where C is a constant satisfying

$$C = \prod_{i=1}^{\infty}\left(1 - \frac{1}{p_i^{\beta}}\right) = \left(\sum_{m=1}^{\infty} \frac{1}{m^{\beta}}\right)^{-1}.$$

32. $N+1$ plates are laid out around a circular dining table, and a hot cake is passed between them in the manner of a symmetric random walk: each time it arrives on a plate, it is tossed to one of the two neighbouring plates, each possibility having probability $\frac{1}{2}$. The game stops at the moment when the cake has visited every plate at least once. Show that, with the exception of the plate where the cake began, each plate has probability $1/N$ of being the last plate visited by the cake.

33. Simplex algorithm†. There are $\binom{n}{m}$ points ranked in order of merit with no matches. You seek to reach the best, B. If you are at the jth best, you step to any one of the $j-1$ better points, with equal probability of stepping to each. Let r_j be the expected number of steps to reach B from the jth best vertex. Show that $r_j = \sum_{k=1}^{j-1} k^{-1}$. Give an asymptotic expression for the expected time to reach B from the worst vertex, for large m, n.

†Due to George Dantzig (1914–2005), not to be confused with David van Dantzig.

34. Dimer problem. There are n unstable molecules in a row, $m_1, m_2, \dots, m_n$. One of the $n-1$ pairs of neighbours, chosen at random, combines to form a stable dimer; this process continues until there remain U_n isolated molecules no two of which are adjacent. Show that the probability that m_1 remains isolated is $\sum_{r=0}^{n-1}(-1)^r/r! \to e^{-1}$ as $n \to \infty$. Deduce that $\lim_{n\to\infty} n^{-1}\mathbb{E}U_n = e^{-2}$.

35. Poisson approximation. Let $\{I_r : 1 \le r \le n\}$ be independent Bernoulli random variables with respective parameters $\{p_r : 1 \le r \le n\}$ satisfying $p_r \le c < 1$ for all r and some c. Let $\lambda = \sum_{r=1}^n p_r$ and $X = \sum_{r=1}^n X_r$. Show that

$$\mathbb{P}(X = k) = \frac{\lambda^k e^{-\lambda}}{k!}\left\{1 + \mathrm{O}\left(\lambda \max_r p_r + \frac{k^2}{\lambda}\max p_r\right)\right\}.$$

36. Sampling. The length of the tail of the rth member of a troop of N chimeras is x_r. A random sample of n chimeras is taken (without replacement) and their tails measured. Let I_r be the indicator of the event that the rth chimera is in the sample. Set

$$X_r = x_r I_r, \quad \overline{Y} = \frac{1}{n}\sum_{r=1}^N X_r, \quad \mu = \frac{1}{N}\sum_{r=1}^N x_r, \quad \sigma^2 = \frac{1}{N}\sum_{r=1}^N (x_r - \overline{x})^2.$$

Show that $\mathbb{E}(\overline{Y}) = \mu$, and $\mathrm{var}(\overline{Y}) = (N-n)\sigma^2/\{n(N-1)\}$.

37. Berkson's fallacy. Any individual in a group G contracts a certain disease C with probability γ; such individuals are hospitalized with probability c. Independently of this, anyone in G may be in hospital with probability a, for some other reason. Let X be the number in hospital, and Y the number in hospital who have C (including those with C admitted for any other reason). Show that the correlation between X and Y is

$$\rho(X,Y) = \sqrt{\frac{\gamma p}{1-\gamma p}\cdot\frac{(1-a)(1-\gamma c)}{a+\gamma c - a\gamma c}},$$

where $p = a + c - ac$.

It has been stated erroneously that, when $\rho(X,Y)$ is near unity, this is evidence for a causal relation between being in G and contracting C.

38. A telephone sales company attempts repeatedly to sell new kitchens to each of the N families in a village. Family i agrees to buy a new kitchen after it has been solicited K_i times, where the K_i are independent identically distributed random variables with mass function $f(n) = \mathbb{P}(K_i = n)$. The value ∞ is allowed, so that $f(\infty) \ge 0$. Let X_n be the number of kitchens sold at the nth round of solicitations, so that $X_n = \sum_{i=1}^N I_{\{K_i = n\}}$. Suppose that N is a random variable with the Poisson distribution with parameter ν.

(a) Show that the X_n are independent random variables, X_r having the Poisson distribution with parameter $\nu f(r)$.

(b) The company loses heart after the Tth round of calls, where $T = \inf\{n : X_n = 0\}$. Let $S = X_1 + X_2 + \cdots + X_T$ be the number of solicitations made up to time T. Show further that $\mathbb{E}(S) = \nu\mathbb{E}(F(T))$ where $F(k) = f(1) + f(2) + \cdots + f(k)$.

39. A particle performs a random walk on the non-negative integers as follows. When at the point n (> 0) its next position is uniformly distributed on the set $\{0, 1, 2, \dots, n+1\}$. When it hits 0 for the first time, it is absorbed. Suppose it starts at the point a.

(a) Find the probability that its position never exceeds a, and prove that, with probability 1, it is absorbed ultimately.

(b) Find the probability that the final step of the walk is from 1 to 0 when $a = 1$.

(c) Find the expected number of steps taken before absorption when $a = 1$.

40. Let G be a finite graph with neither loops nor multiple edges, and write d_v for the degree of the vertex v. An *independent set* is a set of vertices no pair of which is joined by an edge. Let $\alpha(G)$ be the size of the largest independent set of G. Use the probabilistic method to show that $\alpha(G) \geq \sum_v 1/(d_v + 1)$. [This conclusion is sometimes referred to as *Turán's theorem*.]

41. Kelly betting, or proportional investment. A gambler (or 'investor') makes a sequence of bets of the following type: on each bet, for a given stake S, the return is either the loss of the stake with probability q $(= 1 - p)$ or a win totalling $(1 + r)S$ with probability p. (Assume the usual independence.) The entry fee is cS where $c < r$. Show that the mean gain per play for stake S is gS, where $g = pr - q - c$ (with a negative value indicating a loss).

The gambler decides to bet a fixed fraction f of her current fortune at each stage. That is, given a current fortune F, she bets fF for some fixed f. Show that her resulting fortune is

$$F' = F\{1 + f[(1 + r)I - (1 + c)]\},$$

where I is the indicator function of a win.

Suppose $p > (1 + c)/(1 + r)$. The gambler considers two policies for choosing f: for given F,

(a) maximize $\mathbb{E}(F')$,

(b) maximize $\mathbb{E}(\log F')$.

Find an expression for f in each case. Show that $f_a > f_b$, and explain why a cautious gambler may prefer (b) to (a) even though this entails a slower rate of growth of her expected fortune.

42. Random adding sequence. Let $x_1, x_2, \dots, x_r$ be given reals with $r \geq 2$, and let the sequence $\{X_n : n \geq 1\}$ of random variables be given as follows. First, $X_n = x_n$ for $n \leq r$. For $n \geq r$, we set $X_{n+1} = X_{U_n} + X_{V_n}$ where U_n, V_n are uniformly distributed on $\{1, 2, \dots, n\}$, and the family $\{U_n, V_n : n \geq r\}$ are independent. Show that

$$\mathbb{E}(X_n) = \frac{2n}{r(r+1)} \sum_{k=1}^{r} x_k.$$

Enthusiasts may care to show that, when $r = 1 = x_1$ and $n \to \infty$,

$$\frac{1}{n^2}\mathbb{E}(X_n^2) \to \frac{1}{2\pi} \sinh \pi.$$

43. Random subtracting sequence. Let $X_1 = 1$. For $n \geq 1$, let $X_{n+1} = X_{U_n} - X_{V_n}$ where U_n, V_n are uniformly distributed on $\{1, 2, \dots, n\}$, and the family $\{U_n, V_n : n \geq 1\}$ are independent. Show that

$$\frac{1}{n} \operatorname{var}(X_n) \to -\frac{\sin(\pi\sqrt{3})}{\pi\sqrt{3}} \qquad \text{as } n \to \infty.$$

You may find it useful to recall Euler's sine formula:

$$\sin(\pi x) = \pi x \prod_{n=1}^{\infty} \left(1 - \frac{x^2}{n^2}\right).$$

44. Bilbo baffled. Gollum has concealed the ring of power in a box chosen randomly from a row of $n \geq 1$ such boxes. Bilbo opens a box at random. If the ring is not there, his occult powers are sufficient for him to learn whether the ring lies to the right or the left, and he opens further boxes accordingly. Find an expression for the mean number b_n of boxes opened before finding the ring, and deduce that $b_n \sim 2 \log n$ as $n \to \infty$.

45. Bilbo reloaded. In a variant of the previous problem, we have $n = 2^r - 1$, and Bilbo invariably chooses the middle box. Find the mean number m_r of boxes inspected, and the asymptotics of m_r as $r \to \infty$.

46. Fairies. A bad fairy has cursed you. A good fairy has concealed the magic word that cancels the curse in one of n numbered boxes, and she has told you that it is in box i with probability p_i, for $i = 1, 2, \dots, n$. Each day you are permitted to look in one box.

(a) Assume that, each day, you inspect a box chosen at random, box i being chosen with probability c_i, and, furthermore, boxes chosen on different days are independent. Find the mean number of days that elapse before your release from the curse, and find the mass function c that minimizes this mean value.

(b) Suppose now that you remember the results of your previous failed searches. What now is your optimal policy, and what is the mean number of days that elapse?

(c) After each search, the bad fairy removes the magic word, which is immediately replaced by the good fairy in an independently chosen box, with the same distribution at each replacement. What now is your optimal policy? Find the mean number of elapsed days.

47. Duration of play. Gwen and John play 'best of $2n + 1$ games', and play stops as soon as either has won $n + 1$ games. Gwen wins each game with probability $\gamma \in (0, 1)$ and John otherwise (with probability $\delta = 1 - \gamma$). Different games have independent winners. Write down an expression for the probability f_r that Gwen wins r games altogether, given that John has won the match, and deduce that $rf_r = (n + r)\gamma f_{r-1}$ for $0 < r \le n$.

Hence or otherwise, prove that the mean total number T_n of games in the match is

$$T_n = (n+1)\left(\frac{\gamma(1-P_n)}{\delta} + \frac{\delta P_n}{\gamma} + 1\right) - (2n+1)\binom{2n}{n}(\gamma\delta)^n,$$

where P_n is the probability that Gwen wins the match.

When $p = \frac{1}{2}$, show that

$$T_n = 2n - 2\sqrt{n/\pi} + 2 + O(n^{-1/2}), \qquad \text{as } n \to \infty.$$

48. Stirling numbers, Bell numbers, Dobinski's formula. Let $S(n, k)$ be the number of ways to partition $N = \{1, 2, \dots, n\}$ into k non-empty parts. Suppose each element of N is coloured with one of c distinct colours. Prove that

$$c^n = \sum_{k=1}^{n} S(n,k)c(c-1)\cdots(c-k+1).$$

Deduce that the nth moment of the Poisson distribution with parameter 1 equals the number b_n of ways to partition N, that is, $b_n = \sum_{k=1}^{n} S(n, k)$.

49. Server advantage? Let $\beta, \gamma \in [0, 1]$. Bertha and Harold play a game of rackets. Bertha wins the point with probability β when she serves, and Harold wins with probability γ when he serves. The first player to win n points wins the game, and Bertha serves first. Consider the following rules for changing server.

(a) Service alternates between players.

(b) Service is retained until a point is lost, and then passes to the other player.

(c) Service is retained until a point is won, and then passes to the other player.

(d) Bertha serves the first n points, and then Harold serves any further points required to decide the outcome.

Show that $\mathbb{P}$(Bertha wins) is the same for all four rules.

50. Entropy. Let X, Y be discrete random variables with respective mass functions f_X, f_Y, and joint mass function $f_{X,Y}$. Define the:

$$\begin{aligned}\text{entropy of } X \ &: H(X) = -\mathbb{E}\big(\log f_X(X)\big),\\ \text{joint entropy of } X \text{ given } Y \ &: H(X, Y) = -\mathbb{E}\big(\log f_{X,Y}(X, Y)\big),\\ \text{conditional entropy of } X \text{ given } Y \ &: H(X \mid Y) = -\mathbb{E}\big(\log f_{X\mid Y}(X \mid Y)\big).\end{aligned}$$

[It is normal to use logarithms to base 2 in information theory, but we use natural logarithms here.]

(a) Show that $H(X + a) = H(X)$ for $a \in \mathbb{R}$.

(b) Show that $H(X) - H(X \mid Y) = I(X; Y)$, where I is the mutual information of Exercise (3.6.5).

(c) Show when X and Y are independent that $H(X, Y) = H(X) + H(Y)$.

(d) Show that the entropy of the binomial distribution bin(n, p) is non-decreasing in n.

(e) Find the entropy of the geometric distribution, parameter p, and show it is decreasing in p.

51. (a) Show that the entropy $H(\lambda)$, as defined in Problem (3.11.50), of the Poisson distribution with parameter λ (using natural logarithms) is given by

$$H(\lambda) = \lambda - \lambda \log \lambda + e^{-\lambda} \sum_{m=0}^{\infty} \frac{\lambda^m \log(m!)}{m!}.$$

(b) Recalling Exercise (3.6.5) and Example (3.7.5), show that the mutual information of the number N of hens and the number K of chicks is $I(N; K) = H(\lambda) - H(\lambda(1 - p))$. Deduce that $H(\lambda)$ is increasing in λ.

(c) Spend a short time seeking to show the last statement directly from the expression in part (a).

52. Dirichlet distribution revisited. Let $\beta > 1$, and let X, Y be independent random variables with the Dirichlet distribution with parameter β of Problem (3.11.31).

(a) Show that the events $E_p = \{X \text{ is divisible by } p\}$ are independent for p prime.

(b) Deduce Euler's formula

$$\prod_{p \text{ prime}} \left(1 - \frac{1}{p^\beta}\right) = \frac{1}{\zeta(\beta)},$$

where $\zeta(\beta)$ is the Riemann zeta function, $\zeta(\beta) = \sum_{m=1}^{\infty} m^{-\beta}$, and $\beta > 1$.

(c) Show that the probability that X is 'square-free' (that is, indivisible by any perfect square other than 1) equals $1/\zeta(2\beta)$.

(d) Let H be the highest common factor of X and Y. Prove that

$$\mathbb{P}(H = m) = \frac{m^{-2\beta}}{\zeta(2\beta)}, \qquad m = 1, 2, \ldots.$$

53. Strict Oxford secrets. A *strict Oxford secret* is a secret which you may tell to no more than one other person†. Of a group of $n + 1$ Oxonians, one learns a strict secret. In accordance with the rules, she tells it to one other person selected uniformly at random from the rest of the group. Each confidante tells the secret to exactly one person picked at random from the rest of the group, excluding

†Oliver Franks (1905–1992) defined a secret in the Oxford sense as one that you can tell to no more than one person *at a time*.

the person from whom they heard the secret. When someone who already knows the secret hears it repeated, the entire process ceases.

Let S be the total number of people who eventually know the secret. Find the distribution of S, and show that

$$\frac{1}{\sqrt{n}}\mathbb{E}(S) \to \sqrt{\pi/2}, \quad \frac{1}{n}\operatorname{var}(S) \to \tfrac{1}{2}(4-\pi), \qquad \text{as } n \to \infty.$$

With $\underline{r}$ denoting $n!/(n-r)!$, you may find it useful that

$$\sum_{r=1}^{\infty} \frac{n^{\underline{r}}}{n^r} \sim \sqrt{\pi n/2}, \qquad \sum_{r=1}^{\infty} r\frac{n^{\underline{r}}}{n^r} \sim n.$$

54. Transposition shuffle. From a pack of n cards, two different cards are selected randomly, and they are transposed. Let p_r be the probability that any given card (say, the top card) is in its original position after $r > 0$ such independent transpositions.

(a) Show that

$$p_r = \frac{1}{n} + \frac{n-1}{n}\left(\frac{n-3}{n-1}\right)^r.$$

(b) Find $\mathbb{E}(C_r)$, where C_r is the number of cards in their original place after r random transpositions.

(c) Show that, for large n, the number r of transpositions needed for $\mathbb{E}(C_r) \approx 2$ is approximately $\frac{1}{2}n \log n$.

55. Random walk on the d-cube. The d-cube C_d is the graph with vertex-set $\{0, 1\}^d$, and an edge between two vertices $\mathbf{x} = (x_1, x_2, \dots, x_d)$ and $\mathbf{y} = (y_1, y_2, \dots, y_d)$ if and only if $\sum_i |x_i - y_i| = 1$. A particle pursues a random walk on C_d. At each epoch of time, it moves from its current position to a neighbour chosen uniformly at random, with the usual independence. Two vertices are called 'antipodal' if the graph-distance between them equals d.

Show that the mean first passage time m_d of the walker between two antipodal vertices satisfies $\mu_d \sim 2^d$ as $d \to \infty$.

56. The lost boarding pass, Problem (1.8.39) revisited.

(a) A particle performs a type of random walk on the set $S = \{1, 2, \dots, n\}$. Let X_r be the particle's position at time r. Given that $X_r = x$, X_{r+1} is chosen uniformly at random from the set $\{1\} \cup \{x+1, x+2, \dots, n\}$. The particle stops moving at the first instant it arrives at either 1 or n (so that 1 and n are 'absorbing', but absorption does not occur at time 0 even if $X_0 \in \{1, n\}$). The point $m \in \{1, 2, \dots, n\}$ is said to be *hit* if $X_r = m$ for some $r \geq 1$. Show that

$$\mathbb{P}(m \text{ is hit} \mid X_0 = 1) = \begin{cases} \dfrac{1}{2} & \text{if } m = 1, \\[2ex] \dfrac{1}{n-m+2} & \text{if } m \geq 2. \end{cases}$$

(b) The n passengers on a flight in an airplane with n seats have been told their seat numbers. They board the plane one by one. The first person sits in a seat chosen uniformly at random from the n seats then available. Subsequent passengers sit in their assigned seats whenever they find them available, or otherwise in a randomly chosen empty seat. For $m \geq 2$, what is the probability that the mth passenger finds his or her assigned seat to be already occupied?

57. Paley–Zygmund 'second moment' inequalities.

(a) Let X be a random variable with $\mathbb{E}(X) > 0$ and $0 < \mathbb{E}(X^2) < \infty$. Show that

$$\mathbb{P}(X > a\mathbb{E}(X)) \geq \frac{(1-a)^2\mathbb{E}(X)^2}{\mathbb{E}(X^2)} \qquad \text{for } 0 \leq a \leq 1.$$

(b) Deduce that, if $\mathbb{P}(X \geq 0) = 1$,

$$\mathbb{P}(X = 0) \leq \frac{\operatorname{var}(X)}{\mathbb{E}(X^2)} \leq \frac{\operatorname{var}(X)}{\mathbb{E}(X)^2}.$$

(c) Let $A_1, A_2, \ldots, A_n$ be events, and let $X = \sum_{r=1}^{n} I_{A_r}$ be the sum of their indicator functions. Show that

$$\mathbb{P}(X = 0) \leq \frac{1}{\mathbb{E}(X)} + \frac{1}{\mathbb{E}(X)^2} \sum\nolimits^{*} \mathbb{P}(A_r \cap A_s),$$

where the summation $\sum^*$ is over all distinct unordered pairs r, s such that A_r and A_s are not independent.

4

Continuous random variables

4.1 Exercises. Probability density functions

1. For what values of the parameters are the following functions probability density functions?

(a) $f(x) = C\{x(1-x)\}^{-\frac{1}{2}}$, $0 < x < 1$, the density function of the 'arc sine law'.

(b) $f(x) = C\exp(-x - e^{-x})$, $x \in \mathbb{R}$, the density function of the 'extreme-value distribution'.

(c) $f(x) = C(1+x^2)^{-m}$, $x \in \mathbb{R}$.

2. Find the density function of $Y = aX$, where $a > 0$, in terms of the density function of X. Show that the continuous random variables X and $-X$ have the same distribution function if and only if $f_X(x) = f_X(-x)$ for all $x \in \mathbb{R}$.

3. If f and g are density functions of random variables X and Y, show that $\alpha f + (1-\alpha)g$ is a density function for $0 \le \alpha \le 1$, and describe a random variable of which it is the density function.

4. **Survival.** Let X be a positive random variable with density function f and distribution function F. Define the *hazard function* $H(x) = -\log[1 - F(x)]$ and the *hazard rate*

$$r(x) = \lim_{h \downarrow 0} \frac{1}{h}\mathbb{P}(X \le x + h \mid X > x), \qquad x \ge 0.$$

Show that:

(a) $r(x) = H'(x) = f(x)/\{1 - F(x)\}$,

(b) If $r(x)$ increases with x then $H(x)/x$ increases with x,

(c) $H(x)/x$ increases with x if and only if $[1 - F(x)]^\alpha \le 1 - F(\alpha x)$ for all $0 \le \alpha \le 1$,

(d) If $H(x)/x$ increases with x, then $H(x+y) \ge H(x) + H(y)$ for all $x, y \ge 0$.

4.2 Exercises. Independence

1. I am selling my house, and have decided to accept the first offer exceeding £K. Assuming that offers are independent random variables with common distribution function F, find the expected number of offers received before I sell the house.

2. Let X and Y be independent random variables with common distribution function F and density function f. Show that $V = \max\{X, Y\}$ has distribution function $\mathbb{P}(V \le x) = F(x)^2$ and density function $f_V(x) = 2f(x)F(x)$, $x \in \mathbb{R}$. Find the density function of $U = \min\{X, Y\}$.

3. The annual rainfall figures in Bandrika are independent identically distributed continuous random variables $\{X_r : r \ge 1\}$. Find the probability that:

(a) $X_1 < X_2 < X_3 < X_4$,
(b) $X_1 > X_2 < X_3 < X_4$.

4. Let $\{X_r : r \geq 1\}$ be independent and identically distributed with distribution function F satisfying $F(y) < 1$ for all y, and let $Y(y) = \min\{k : X_k > y\}$. Show that

$$\lim_{y\to\infty} \mathbb{P}\big(Y(y) \leq \mathbb{E}Y(y)\big) = 1 - e^{-1}.$$

5. Peripheral points. Let $P_i = (X_i, Y_i)$, $1 \leq i \leq n$, be independent, uniformly distributed points in the unit square $[0,1]^2$. A point P_i is called *peripheral* if, for all $r = 1, 2, \dots, n$, either $X_r \leq X_i$ or $Y_r \leq Y_i$, or both. Show that the mean number of peripheral points is $n\left(\frac{3}{4}\right)^{n-1}$.

6. Let U and V be independent and uniformly distributed on the interval $[0, 1]$.

(a) Show that

$$\mathbb{P}(x < V < U^2) = \tfrac{1}{3} - x + \tfrac{2}{3}x^{3/2}, \qquad x \in [0,1).$$

(b) Find the conditional density function of V given $U^2 > V$.

(c) Find the probability that the equation $x^2 + 2Ux + V = 0$ has two distinct real roots.

(d) Given that the two roots R_1 and R_2 are real and distinct, find the probability that both roots have absolute value less than 1.

7. Random hemispheres.

(a) We select n points independently and uniformly at random on the perimeter of a circle. What is the probability that they all lie within some semicircle?

(b) This time we place our n points uniformly on the surface of a sphere in $\mathbb{R}^3$. Show that they all lie within some hemisphere with probability $(n^2 - n + 2)2^{-n}$.

4.3 Exercises. Expectation

1. For what values of α is $\mathbb{E}(|X|^\alpha)$ finite, if the density function of X is:

(a) $f(x) = e^{-x}$ for $x \geq 0$,

(b) $f(x) = C(1 + x^2)^{-m}$ for $x \in \mathbb{R}$?

If α is not integral, then $\mathbb{E}(|X|^\alpha)$ is called the *fractional moment of order* α of X, whenever the expectation is well defined; see Exercise (3.3.5).

2. Let $X_1, X_2, \dots, X_n$ be independent identically distributed random variables, and let $S_m = X_1 + X_2 + \dots + X_m$. Assume that $\mathbb{P}(S_n = 0) = 0$. Show that, if $m \leq n$, then $\mathbb{E}(S_m/S_n) = m/n$.

3. Let X be a non-negative random variable with density function f. Show that

$$\mathbb{E}(X^r) = \int_0^\infty r x^{r-1} \mathbb{P}(X > x)\, dx$$

for any $r \geq 1$ for which the expectation is finite.

4. Mallows's inequality. Show that the mean μ, median m, and variance σ^2 of the continuous random variable X satisfy $(\mu - m)^2 \leq \sigma^2$. [It can be shown that $|\mu - m| \leq \sigma\sqrt{0.6}$. See Basu and Dasgupta 1997.]

5. Let X be a random variable with mean μ and continuous distribution function F.

(a) Show that

$$\int_{-\infty}^{a} F(x)\, dx = \int_a^\infty [1 - F(x)]\, dx,$$

if and only if $a = \mu$.

(b) Show that the set of medians of X is the set of all a for which $\mathbb{E}|X - a|$ is a minimum.

6. Let $X_1, X_2, \dots, X_n$ be non-negative random variables with finite means. Show that

$$\mathbb{E}(\max_j X_j) = \sum_j \mathbb{E}(X_j) - \sum_{i<j} \mathbb{E}(\min\{X_i, X_j\}) + \cdots + (-1)^{n+1}\mathbb{E}(\min_j X_j).$$

7. Tails and moments. If X is a continuous random variable and $\mathbb{E}(X^r)$ exists, where $r \geq 1$ is an integer, show that

$$\int_0^\infty x^{r-1}\mathbb{P}(|X| > x)\,dx < \infty, \qquad \text{and} \qquad x^r\mathbb{P}(|X| > x) \to 0 \quad \text{as } x \to \infty.$$

8. Integral inequality. Let X be a random variable taking values in $[0, a]$, with density function f satisfying $f(x) \leq b$ for all x. Show that $\mathbb{E}(X) \geq (2b)^{-1}$. More generally, show for $n \geq 1$ that $\mathbb{E}(X^n) \geq 1/[(n+1)b^n]$.

[Hint: You may find a form of Steffensen's integral inequality to be useful, namely: for integrable functions g and h satisfying $0 \leq h(x) \leq 1$ and g increasing on $[0, a]$, we have

$$\int_0^s g(x)\,dx \leq \int_0^a g(x)h(x)\,dx,$$

where $s = \int_0^a h(x)\,dx$.]

9. Let X be continuous with finite variance. Show that $g(a) = \mathbb{E}((X - a)^2)$ is a minimum when $a = \mathbb{E}(X)$.

10. Archery. A target comprising a disc with centre O and unit radius is hit by n arrows. They strike independent spots on the target, and for each the chance of hitting within distance $a \in (0, 1)$ of O is a^2. Let R be the radius of the smallest circle centred at O that covers all the arrow strikes. Show that R has density $f(r) = 2nr^{2n-1}$, $0 \leq r \leq 1$, and find the mean area of this circle.

The arrow furthest from O falls out. Show that the area of the smallest circle centred at O and covering the remaining arrows is $(n-1)\pi/(n+1)$.

11. Johnson–Rogers inequality. Let X be continuous with variance σ^2, and also unimodal about a mode M. Show that $|M - \mathbb{E}X| \leq \sigma\sqrt{3}$, with equality if and only if X is uniformly distributed. [You may use without proof the fact that X, with distribution function F, is unimodal about 0 if and only if there exist independent random variables U and Y, where U is uniformly distributed on $(0, 1)$, such that UY has distribution function F; this is called Khinchin's representation.]

What can you say about $|M - m|$ where m is a median of X?

12. A football is placed at a uniformly distributed position W in the unit interval $[0, 1]$, and kicked with a non-zero random velocity V having a distribution that is symmetric about 0. Assuming that V and W are independent, show (neglecting air resistance) that the time T at which the football hits an endpoint of the interval is unimodal.

13. Continuation. Here are two further questions concerning the football of the previous exercise.

(a) Suppose that the football's speed $|V|$ has density function $g(u) = u^{-3}e^{-1/u}$ for $u > 0$. Show that the hitting time T is exponentially distributed with parameter 1.

(b) Find the density function of T when V has the so-called Laplace density function $g(v) = \frac{1}{2}e^{-|v|}$.

4.4 Exercises. Examples of continuous variables

1. Prove that the gamma function satisfies $\Gamma(t) = (t-1)\Gamma(t-1)$ for $t > 1$, and deduce that $\Gamma(n) = (n-1)!$ for $n = 1, 2, \ldots$. Show that $\Gamma(\frac{1}{2}) = \sqrt{\pi}$ and deduce a closed form for $\Gamma(n+\frac{1}{2})$ for $n = 0, 1, 2, \ldots$.

2. Show, as in paragraph (4.4.8), that the beta function satisfies $B(a, b) = \Gamma(a)\Gamma(b)/\Gamma(a+b)$.

3. Let X have the uniform distribution on $[0, 1]$. For what function g does $Y = g(X)$ have the exponential distribution with parameter 1?

4. Find the distribution function of a random variable X with the Cauchy distribution. For what values of α does $|X|$ have a finite (possibly fractional) moment of order α?

5. **Log-normal distribution.** Let $Y = e^X$ where X has the $N(0, 1)$ distribution. Find the density function of Y.

6. **Stein's identity.** Let X be $N(\mu, \sigma^2)$. Show that $\mathbb{E}\{(X-\mu)g(X)\} = \sigma^2\mathbb{E}(g'(X))$ when both sides exist.

7. With the terminology of Exercise (4.1.4), find the hazard rate when:

(a) X has the Weibull distribution, $\mathbb{P}(X > x) = \exp(-\alpha x^{\beta-1})$, $x \ge 0$,

(b) X has the exponential distribution with parameter λ,

(c) X has density function $\alpha f + (1-\alpha)g$, where $0 < \alpha < 1$ and f and g are the densities of exponential variables with respective parameters λ and μ. What happens to this last hazard rate $r(x)$ in the limit as $x \to \infty$?

8. **Mills's ratio.** (a) For the standard normal density function $\phi(x)$, show that $\phi'(x) + x\phi(x) = 0$. Hence show that *Mills's ratio* $M(x) = (1-\Phi(x))/\phi(x)$ satisfies

$$\frac{1}{x} - \frac{1}{x^3} < M(x) < \frac{1}{x} - \frac{1}{x^3} + \frac{3}{x^5}, \qquad x > 0.$$

Here, Φ denotes the $N(0, 1)$ distribution function.

(b) Let X have the $N(\mu, \sigma^2)$ distribution where $\sigma^2 > 0$, and show that

$$\mathbb{E}(X \mid X > c) = \mu + \frac{\sigma}{M((c-\mu)/\sigma)},$$
$$\mathbb{E}(X \mid X < c) = \mu - \frac{\sigma}{M((c-\mu)/\sigma)}.$$

9. **Ordered exponentials.** Let U, V, W be independent, exponentially distributed random variables with respective parameters λ, μ, ν. Show that

$$\mathbb{P}(U \le V \le W) = \frac{\lambda\mu\nu}{\nu(\nu+\mu)(\nu+\mu+\lambda)}.$$

10. Let U and X be independent, where U is uniform on $(0, 1)$ and X is exponentially distributed with parameter λ. Show that $\mathbb{E}(\min\{U, X\}) = \lambda^{-1}e^{-\lambda} - \lambda^{-2}(1-e^{-\lambda})$.

11. **Pareto distribution.** (a) Let X be uniformly distributed on $[0, 1]$, and let $a > 0$. Find the distribution of $Y = aX/(1-X)$.

(b) Let $Y_1, Y_2, \ldots, Y_n$ be independent random variables with the same distribution as Y. Show that $S = \min\{Y_1, Y_2, \ldots, Y_n\}$ has the Pareto density function

$$f(s) = \frac{n}{a}\left(\frac{a}{a+s}\right)^{n+1}, \qquad s > 0,$$

and find its mean value.

(c) Find the distribution function of $T = \max\{Y_1, Y_2, \dots, Y_n\}$.

12. Arc sine distribution. Let X have the Cauchy distribution, and show that $Y = 1/(1+X^2)$ has the arc sine density function

$$f_Y(y) = \frac{2}{\pi\sqrt{1-y^2}}, \qquad 0 \le y \le 1.$$

13. Let $X_1, X_2, \dots$ be independent and uniformly distributed on $[-c, c]$.

(a) Find expressions for the probability that

(i) $X_i \ge b$ for $1 \le i \le n$,

(ii) $X_i \le b$ for $1 \le i \le n$,

where $b \in [-c, c]$.

(b) Show that the median Z of $X_1, X_2, \dots, X_{2n+1}$ (that is, the middle value) has density function

$$f_Z(z) = \frac{(2n+1)!}{(n!)^2} \cdot \frac{(c^2 - z^2)^n}{(2c)^{2n+1}}, \qquad z \in [-c, c].$$

(c) Deduce the value of the integral $\int_{-c}^{c} (c^2 - z^2)\, dz$.

(d) Calculate the mean and variance of Z.

14. Beta distribution of the second kind. Let X have the beta distribution $\beta(a, b)$. Show that $Y = X/(1-X)$ has density function

$$f(y) = \frac{\Gamma(a+b)}{\Gamma(a)\Gamma(b)} \cdot \frac{y^{a-1}}{(1+y)^{a+b}}, \qquad y > 0.$$

This is called the *beta distribution of the second kind*. Find $\mathbb{E}(Y^n)$ where $1 \le n < b$.

15. Let Z have the $\Gamma(1, t)$ distribution of paragraph (4.4.6). Show that

$$\mathbb{E}\big(|Z - t|/\sqrt{t}\big) = \frac{2e^{-t} t^{t-\frac{1}{2}}}{\Gamma(t)}.$$

4.5 Exercises. Dependence

1. Clarke's example. Let

$$f(x, y) = \frac{|x|}{\sqrt{8\pi}} \exp\{-|x| - \tfrac{1}{2}x^2y^2\}, \qquad x, y \in \mathbb{R}.$$

Show that f is a continuous joint density function, but that the (first) marginal density function $g(x) = \int_{-\infty}^{\infty} f(x, y)\, dy$ is not continuous. Let $Q = \{q_n : n \ge 1\}$ be a set of real numbers, and define

$$f_Q(x, y) = \sum_{n=1}^{\infty} (\tfrac{1}{2})^n f(x - q_n, y).$$

Show that f_Q is a continuous joint density function whose first marginal density function is discontinuous at the points in Q. Can you construct a continuous joint density function whose first marginal density function is continuous nowhere?

2. Buffon's needle revisited. Two grids of parallel lines are superimposed: the first grid contains lines distance a apart, and the second contains lines distance b apart which are perpendicular to those of the first set. A needle of length r ($< \min\{a, b\}$) is dropped at random. Show that the probability it intersects a line equals $r(2a + 2b - r)/(\pi ab)$.

3. Buffon's cross. The plane is ruled by the lines $y = n$, for $n = 0, \pm 1, \dots$, and onto this plane we drop a cross formed by welding together two unit needles perpendicularly at their midpoints.

(a) Let Z be the number of intersections of the cross with the grid of parallel lines. Show that $\mathbb{E}(Z/2) = 2/\pi$ and that

$$\operatorname{var}(Z/2) = \frac{3 - \sqrt{2}}{\pi} - \frac{4}{\pi^2}.$$

(b) If you had the choice of using either a needle of unit length, or the cross, in estimating $2/\pi$, which would you use?

(c) Would it be preferable to use a unit needle on the grid of Exercise (4.5.2) with $a = b = 1$?

4. Let X and Y be independent random variables each having the uniform distribution on $[0, 1]$. Let $U = \min\{X, Y\}$ and $V = \max\{X, Y\}$. Find $\mathbb{E}(U)$, and hence calculate $\operatorname{cov}(U, V)$.

5. (a) Let X and Y be independent continuous random variables. Show that

$$\mathbb{E}\big(g(X)h(Y)\big) = \mathbb{E}(g(X))\mathbb{E}(h(Y)),$$

whenever these expectations exist. If X and Y have the exponential distribution with parameter 1, find $\mathbb{E}\{\exp(\frac{1}{2}(X + Y))\}$.

(b) Let X have finite variance. Show that $2\operatorname{var}(X) = \mathbb{E}\big((X - Y)^2\big)$ where Y is independent and distributed as X.

6. Three points A, B, C are chosen independently at random on the circumference of a circle. Let $b(x)$ be the probability that at least one of the angles of the triangle ABC exceeds $x\pi$. Show that

$$b(x) = \begin{cases} 1 - (3x - 1)^2 & \text{if } \frac{1}{3} \le x \le \frac{1}{2}, \\ 3(1 - x)^2 & \text{if } \frac{1}{2} \le x \le 1. \end{cases}$$

Hence find the density and expectation of the largest angle in the triangle.

7. Let $\{X_r : 1 \le r \le n\}$ be independent and identically distributed with finite variance, and define $\overline{X} = n^{-1}\sum_{r=1}^{n} X_r$. Show that $\operatorname{cov}(\overline{X}, X_r - \overline{X}) = 0$.

8. Let X and Y be independent random variables with finite variances, and let $U = X + Y$ and $V = XY$. Under what condition are U and V uncorrelated?

9. Let X and Y be independent continuous random variables, and let U be independent of X and Y taking the values ± 1 with probability $\frac{1}{2}$. Define $S = UX$ and $T = UY$. Show that S and T are in general dependent, but S^2 and T^2 are independent.

10. Let X, Y, Z be independent and identically distributed continuous random variables. Show that $\mathbb{P}(X > Y) = \mathbb{P}(Z > Y) = \frac{1}{2}$. What is $\mathbb{P}(Z > Y \mid X > Y)$?

11. Hoeffding's identity. Let (X, Y) and (U, V) be independent random vectors with common distribution function $F(x, y)$ and marginal distribution functions $F_X(x)$ and $F_Y(y)$. Show that, when $|\operatorname{cov}(X, Y)| < \infty$, we have that $\mathbb{E}\{(X - U)(Y - V)\} = 2\operatorname{cov}(X, Y)$ and

$$\operatorname{cov}(X, Y) = \iint_{\mathbb{R}^2} \big[F(x, y) - F_X(x)F_Y(y)\big]\, dx\, dy.$$

12. Let the pair X, Y have the bivariate normal distribution with means 0, variances 1, and correlation ρ. Show that

$$\mathbb{E}(\max\{X, Y\}) = \sqrt{\frac{1-\rho}{\pi}}.$$

13. Let X be exponentially distributed with parameter λ. Let N be the greatest integer not greater than X, and set $M = X - N$. Show that M and N are independent. Find the density function of M and the distribution of N.

14. Let X and Y be independent $N(0, 1)$ random variables. Show, if $2a < 1$, $2b < 1$, and $4b^2 < (1-2a)(1-2c)$, that

$$\mathbb{E}(\exp\{aX^2 + 2bXY + cY^2\}) = \frac{1}{\sqrt{(1-2a)(1-2c) - 4b^2}}.$$

15. Contingency coefficient. Let X, Y be random variables with joint density function $f(x, y)$ and marginal densities g, h. Their *contingency coefficient* is given as

$$\phi^2 := \mathbb{E}\left(\frac{f(X, Y)}{g(X)h(Y)}\right) - 1.$$

If X, Y are bivariate normal with correlation $\rho \in (-1, 1)$, show that $1 + \phi^2 = 1/(1-\rho^2)$. Show in addition that their mutual information (see Exercise (3.6.5)) is $I = -\frac{1}{2}\log(1-\rho^2)$.

16. Let X be uniformly distributed on $[-1, 1]$. Are the random variables $Z_n = \cos(n\pi X)$, $n = 1, 2, \dots$, correlated? Are they independent? Explain your answers.

17. Mean absolute difference. The mean absolute difference, or MAD, of two independent random variables X, Y, with common distribution function F, is given by $\text{MAD} = \mathbb{E}|X - Y|$. Show that, for non-negative random variables X, Y,

$$\text{MAD} = 2\left\{\mathbb{E}(X) - \int_0^\infty (1 - F(x))^2\, dx\right\}.$$

More generally, for X and Y taking values in $\mathbb{R}$, with a common continuous distribution function with unique inverse Q, show that

$$\text{MAD} = \iint_{(0,1)^2} |Q(u) - Q(v)|\, du\, dv.$$

Find MAD for the following distributions:

(a) uniform on $[0, 1]$,
(b) exponential with parameter λ,
(c) Pareto with $F(x) = 1 - x^{-a}$, $x > 1$,
(d) normal $N(0, 1)$.

18. Absolute normals. Let X, Y be independent $N(0, 1)$ random variables, and $U = \min\{|X|, |Y|\}$, $V = \max\{|X|, |Y|\}$. Show that $\mathbb{E}(U/V) = (4/\pi)\log\sqrt{2}$.

19. Another random triangle. A point $P = (X, Y)$ is selected uniformly at random inside a triangle Δ with corners $(1, 0)$, $(1, 1)$, $(0, 1)$. Show that a triangle ABC with BC$= X$, CA$= Y$, AB$= 2-X-Y$ can always be constructed, and prove that the angle $\widehat{\text{ABC}}$ is obtuse with probability $3 - 4\log 2$.

20. Random variables X, Y have joint density function $f(x, y) = 3\{(x + y) - (x^2 + y^2)\}$ for $x, y \in [0, 1]$. Show that X and Y are uncorrelated but dependent.

4.6 Exercises. Conditional distributions and conditional expectation

1. A point is picked uniformly at random on the surface of a unit sphere. Writing Θ and Φ for its longitude and latitude, find the conditional density functions of Θ given Φ, and of Φ given Θ.

2. Show that the conditional expectation $\psi(X) = \mathbb{E}(Y \mid X)$ satisfies $\mathbb{E}(\psi(X)g(X)) = \mathbb{E}(Yg(X))$, for any function g for which both expectations exist.

3. Construct an example of two random variables X and Y for which $\mathbb{E}(Y) = \infty$ but such that $\mathbb{E}(Y \mid X) < \infty$ almost surely.

4. Find the conditional density function and expectation of Y given X when they have joint density function:

(a) $f(x, y) = \lambda^2 e^{-\lambda y}$ for $0 \le x \le y < \infty$,

(b) $f(x, y) = xe^{-x(y+1)}$ for $x, y \ge 0$.

5. Let Y be distributed as bin(n, X), where X is a random variable having a beta distribution on $[0, 1]$ with parameters a and b. Describe the distribution of Y, and find its mean and variance. What is the distribution of Y in the special case when X is uniform?

6. Let $\{X_r : r \ge 1\}$ be independent and uniformly distributed on $[0, 1]$. Let $0 < x < 1$ and define

$$N = \min\{n \ge 1 : X_1 + X_2 + \cdots + X_n > x\}.$$

Show that $\mathbb{P}(N > n) = x^n/n!$, and hence find the mean and variance of N.

7. Let X and Y be random variables with correlation ρ. Show that $\mathbb{E}(\text{var}(Y \mid X)) \le (1 - \rho^2)\,\text{var}\,Y$.

8. Let X, Y, Z be independent and exponential random variables with respective parameters λ, μ, ν. Find $\mathbb{P}(X < Y < Z)$.

9. Let X and Y have the joint density $f(x, y) = cx(y - x)e^{-y}$, $0 \le x \le y < \infty$.

(a) Find c.

(b) Show that:

$$f_{X|Y}(x \mid y) = 6x(y - x)y^{-3}, \qquad 0 \le x \le y,$$
$$f_{Y|X}(y \mid x) = (y - x)e^{x-y}, \qquad 0 \le x \le y < \infty.$$

(c) Deduce that $\mathbb{E}(X \mid Y) = \frac{1}{2}Y$ and $\mathbb{E}(Y \mid X) = X + 2$.

10. Let $\{X_r : r \ge 0\}$ be independent and identically distributed random variables with density function f and distribution function F. Let $N = \min\{n \ge 1 : X_n > X_0\}$ and $M = \min\{n \ge 1 : X_0 \ge X_1 \ge \cdots \ge X_{n-1} < X_n\}$. Show that X_N has distribution function $F + (1 - F)\log(1 - F)$, and find $\mathbb{P}(M = m)$.

11. Let the point $P = (X, Y)$ be uniformly distributed in the square $S = [0, 1]^2$, and denote by $f(x \mid D)$ the density function of X conditional on the event D that P lies on the diagonal of S through the origin.

(a) Let $U = X - Y$. Find $f(x \mid D)$ by conditioning on $U = 0$.

(b) Let $V = Y/X$. Find $f(x \mid D)$ by conditioning on $V = 1$.

(c) Explain.

12. **Threshold game.** (a) Let X and Y be independent with density functions f_X, f_Y and distribution functions F_X, F_Y. Show that

$$\mathbb{P}(X < Y) = \int_{-\infty}^{\infty} F_X(y) f_Y(y)\, dy.$$

(b) A and B play a game as follows. Each creates a random variable, denoted respectively U_A and U_B, which is uniformly distributed on $[0, 1]$; assume U_A and U_B are independent, and neither player knows the opponent's value. Each player has the option to (once) discard their number and to resample from the uniform distribution. After any such choices, the two numbers are compared and the larger wins. Show that the strategy 'replace your number if and only if it is less than $\frac{1}{2}(\sqrt{5}-1)$' ensures a win with probability at least $\frac{1}{2}$.

13. Record times. Let $X_1, X_2, \dots$ be independent, each with density function $f : \mathbb{R} \to [0, \infty)$. The index $r > 1$ is called a *record time* if $X_r > \max\{X_1, X_2, \dots, X_{r-1}\}$, and $r = 1$ is called a record time by convention. Let A_r be the event that r is a record time.

(a) Show that the A_r are independent events with $\mathbb{P}(A_r) = 1/r$.

(b) Show that the number R_n of record times up to time n has variance

$$\text{var}(R_n) = \sum_{r=1}^{n} (r^{-1} - r^{-2}).$$

(c) Show that the first record time T after $r = 1$ has mean $\mathbb{E}(T) = \infty$.

14. Stick breaking. A unit stick is broken uniformly at random, and the larger piece is broken again uniformly at random. Show that the probability the three pieces may be used to form a triangle is $2\log 2 - 1$.

15. Let $U_1, U_2, \dots, U_{n+1}$ be independent random variables with the uniform distribution on $[0, 1]$, and let

$$U_{(1)} = \min_{1 \le r \le n} U_r, \qquad U_{(n)} = \max_{1 \le r \le n} U_r.$$

Show that $\mathbb{E}(U_{(1)}) = 1/(n+1)$ and deduce that

$$\mathbb{P}\big(U_{(n)} - U_{(1)} \le U_{n+1}\big) = \frac{2}{n+1}.$$

4.7 Exercises. Functions of random variables

1. Let X, Y, and Z be independent and uniformly distributed on $[0, 1]$. Find the joint density function of XY and Z^2, and show that $\mathbb{P}(XY < Z^2) = \frac{5}{9}$.

2. Let X and Y be independent exponential random variables with parameter 1. Find the joint density function of $U = X + Y$ and $V = X/(X+Y)$, and deduce that V is uniformly distributed on $[0, 1]$.

3. Let X be uniformly distributed on $[0, \frac{1}{2}\pi]$. Find the density function of $Y = \sin X$.

4. Find the density function of $Y = \sin^{-1} X$ when:

(a) X is uniformly distributed on $[0, 1]$,

(b) X is uniformly distributed on $[-1, 1]$.

5. Normal orthant probability. Let X and Y have the bivariate normal density function

$$f(x, y) = \frac{1}{2\pi\sqrt{1-\rho^2}} \exp\left\{-\frac{1}{2(1-\rho^2)}(x^2 - 2\rho xy + y^2)\right\}.$$

Show that X and $Z = (Y - \rho X)/\sqrt{1-\rho^2}$ are independent $N(0, 1)$ variables, and deduce that

$$\mathbb{P}(X > 0,\ Y > 0) = \frac{1}{4} + \frac{1}{2\pi}\sin^{-1}\rho.$$

6. Let X and Y have the standard bivariate normal density function of Exercise (4.7.5), and define $Z = \max\{X, Y\}$. Show that $\mathbb{E}(Z) = \sqrt{(1-\rho)/\pi}$, and $\mathbb{E}(Z^2) = 1$.

7. Let X and Y be independent exponential random variables with parameters λ and μ. Show that $Z = \min\{X, Y\}$ is independent of the event $\{X < Y\}$. Find:

(a) $\mathbb{P}(X = Z)$,

(b) the distributions of $U = \max\{X - Y, 0\}$, denoted $(X - Y)^+$, and $V = \max\{X, Y\} - \min\{X, Y\}$,

(c) $\mathbb{P}(X \le t < X + Y)$ where $t > 0$.

8. A point (X, Y) is picked at random uniformly in the unit circle. Find the joint density of R and X, where $R^2 = X^2 + Y^2$.

9. A point (X, Y, Z) is picked uniformly at random inside the unit ball of $\mathbb{R}^3$. Find the joint density of Z and R, where $R^2 = X^2 + Y^2 + Z^2$.

10. Let X and Y be independent and exponentially distributed with parameters λ and μ. Find the joint distribution of $S = X + Y$ and $R = X/(X + Y)$. What is the density of R?

11. Find the density of $Y = a/(1 + X^2)$, where X has the Cauchy distribution.

12. Let (X, Y) have the bivariate normal density of Exercise (4.7.5) with $0 \le \rho < 1$. Show that

$$[1 - \Phi(a)][1 - \Phi(c)] \le \mathbb{P}(X > a,\ Y > b) \le [1 - \Phi(a)][1 - \Phi(c)] + \frac{\rho\phi(b)[1 - \Phi(d)]}{\phi(a)},$$

where $c = (b - \rho a)/\sqrt{1-\rho^2}$, $d = (a - \rho b)/\sqrt{1-\rho^2}$, and ϕ and Φ are the density and distribution function of the $N(0, 1)$ distribution.

13. Let X have the Cauchy distribution. Show that $Y = X^{-1}$ has the Cauchy distribution also. Find another non-trivial distribution with this property of invariance.

14. Let X and Y be independent and gamma distributed as $\Gamma(\lambda, \alpha)$, $\Gamma(\lambda, \beta)$ respectively.

(a) Show that $W = X + Y$ and $Z = X/(X + Y)$ are independent, and that Z has the beta distribution with parameters α, β.

(b) Show that $R = X/Y$ has a beta distribution of the second kind (see Exercise (4.4.14)).

15. Frailty. Let X, Y be independent, positive, continuous random variables, such that $1 - F_Y(y) = (1 - F_X(y))^\lambda$ for $y > 0$. The positive parameter λ may be called the *frailty* or *proportional hazard*. In more general models, λ may itself be a random variable.

Show that

$$\mathbb{P}(X > Y) = \lambda\mathbb{P}(Y > X) = \frac{\lambda}{1+\lambda}.$$

16. Rayleigh distribution. Let X and Y be independent random variables, where X has an arc sine distribution and Y a Rayleigh distribution:

$$f_X(x) = \frac{1}{\pi\sqrt{1-x^2}}, \quad |x| < 1, \qquad f_Y(y) = ye^{-\frac{1}{2}y^2}, \quad y > 0.$$

Write down the joint density function of the pair (Y, XY), and deduce that XY has the standard normal distribution.

17. Binary expansions. Let U be uniformly distributed on the interval $(0, 1)$.

(a) Let S be a (measurable) subset of $(0, 1)$ with strictly positive measure (length). Show that the conditional distribution of U, given that $U \in S$, is uniform on S.

(b) Let $V = \sqrt{U}$, and write the binary expansions of U and V as $U = \sum_{r=1}^{\infty} U_r 2^{-r}$ and $V = \sum_{r=1}^{\infty} V_r 2^{-r}$. Show that U_r and U_s are independent for $r \neq s$, while $\text{cov}(V_1, V_2) = -\frac{1}{32}$. Prove that $\lim_{n\to\infty} \mathbb{P}(V_r = 1) = \frac{1}{2}$.

18. Let (X, Y) have the standard bivariate normal distribution with correlation ρ.

(a) Let $\rho = 0$. By changing from Cartesian to polar coordinates, or otherwise, show that $Z = Y/X$ has the Cauchy distribution.

(b) Let $\rho > 0$. Show that $Z = Y/X$ has density function

$$f(z) = \frac{\sqrt{1-\rho^2}}{\pi(1 - 2\rho z + z^2)}, \qquad z \in \mathbb{R}.$$

19. Inverse Mills's ratio. Let (X, Y) have the standard bivariate normal distribution with correlation ρ. Show that

$$\mathbb{E}(Y \mid X > x) = \rho \frac{\phi(x)}{\Phi(-x)},$$

where ϕ and Φ are the $N(0, 1)$ density and distribution functions.

20. Let (X, Y) have the standard bivariate normal distribution with density function f and correlation ρ. Show that the probability that (X, Y) lies in the interior of the ellipse $f(x, y) = k$ is $1 - \exp\{-A/(2\pi\sqrt{1-\rho^2})\}$, where A is the area of the ellipse.

21. Let $R = X/Y$ where X, Y are independent and uniformly distributed on $(0, 1)$. Find the probability that the integer closest to R is odd.

22. Let (X, Y) have the standard bivariate normal distribution with correlation ρ and density function f. Show that $f(X, Y)$ is uniformly distributed on the interval $(0, \zeta)$ where $\zeta = 1/\{2\pi\sqrt{1-\rho^2}\}$.

23. A number n of friends each visit Old Slaughter's coffee house independently and uniformly at random during their lunch break from noon to 1pm. Each leaves after δ hours (or at 1pm if that is sooner), where $\delta < 1/(n-1)$. Show that the probability that none of them meet inside is $(1-(n-1)\delta)^n$.

24. Let X and Y be independent random variables with the uniform distribution on $(0, 1)$. Find the joint density function of $W = XY$ and $Z = Y/X$, and deduce their marginal density functions.

25. Stein's identity. (a) Let X have the $N(\mu, \sigma^2)$ distribution with $\sigma^2 > 0$. Show that, for suitable $g : \mathbb{R} \to \mathbb{R}$,

$$\mathbb{E}\{g(X)(X - \mathbb{E}X)\} = \text{var}(X)\mathbb{E}(g'(X)),$$

when both sides exist.

(b) More generally, if X and Y have a bivariate normal distribution, show that

$$\text{cov}(g(X), Y) = \mathbb{E}(g'(X))\text{cov}(X, Y).$$

(c) Let X be a random variable such that $\mathbb{E}(Xg(X)) = \mathbb{E}(g'(X))$ for all appropriate smooth functions g satisfying $g(x)e^{-\frac{1}{2}x^2} \to 0$ as $x \to -\infty$. Show that X has the $N(0, 1)$ distribution.

26. Chernoff–Cacoullos inequalities. Let X have the $N(0, 1)$ distribution, and let G be a function with derivative g. Show that

$$\{\mathbb{E}g(X)\}^2 \le \text{var}\, G(X) \le \mathbb{E}(g(X)^2).$$

27. The joint density function of the pair (X, Y) is

$$f(x, y) = \tfrac{2}{3}(x + y)e^{-x}, \qquad x \in (0, \infty),\ y \in (0, 1).$$

Find the joint density function of the pair $U = X$, and $V = X + Y$, and deduce the density function of V.

4.8 Exercises. Sums of random variables

1. Let X and Y be independent variables having the exponential distribution with parameters λ and μ respectively. Find the density function of $X + Y$.

2. Let X and Y be independent variables with the Cauchy distribution. Find the density function of $\alpha X + \beta Y$ where $\alpha\beta \neq 0$. (Do you know about contour integration?)

3. Find the density function of $Z = X + Y$ when X and Y have joint density function $f(x, y) = \frac{1}{2}(x + y)e^{-(x+y)}$, $x, y \geq 0$.

4. **Hypoexponential distribution.** Let $\{X_r : r \geq 1\}$ be independent exponential random variables with respective parameters $\{\lambda_r : r \geq 1\}$ no two of which are equal. Find the density function of $S_n = \sum_{r=1}^{n} X_r$. [Hint: Use induction.]

5. (a) Let X, Y, Z be independent and uniformly distributed on $[0, 1]$. Find the density function of $X + Y + Z$.

(b) If $\{X_r : r \geq 1\}$ are independent and uniformly distributed on $[0, 1]$, show that the density of $S_n = \sum_{r=1}^{n} X_r$ at any point $x \in (0, n)$ is a polynomial in x of degree $n - 1$. Show in particular that the density function f_n of S_n satisfies $f_n(x) = x^{n-1}/(n-1)!$ for $x \in [0, 1]$.

(c) Let $n \geq 3$. What is the probability that the $X_1, X_2, \dots, X_n$ of part (b) can be the lengths of the edges of an n-gon?

6. For independent identically distributed random variables X and Y, show that $U = X + Y$ and $V = X - Y$ are uncorrelated but not necessarily independent. Show that U and V are independent if X and Y are $N(0, 1)$.

7. Let X and Y have a bivariate normal density with zero means, variances σ^2, τ^2, and correlation ρ. Show that:

(a) $\mathbb{E}(X \mid Y) = \dfrac{\rho\sigma}{\tau} Y$,

(b) $\operatorname{var}(X \mid Y) = \sigma^2(1 - \rho^2)$,

(c) $\mathbb{E}(X \mid X + Y = z) = \dfrac{(\sigma^2 + \rho\sigma\tau)z}{\sigma^2 + 2\rho\sigma\tau + \tau^2}$,

(d) $\operatorname{var}(X \mid X + Y = z) = \dfrac{\sigma^2\tau^2(1 - \rho^2)}{\tau^2 + 2\rho\sigma\tau + \sigma^2}$.

8. Let X and Y be independent $N(0, 1)$ random variables, and let $Z = X + Y$. Find the distribution and density of Z given that $X > 0$ and $Y > 0$. Show that

$$\mathbb{E}(Z \mid X > 0, Y > 0) = 2\sqrt{2/\pi}.$$

9. Let X and Y be independent $N(0, 1)$ random variables. Find the joint density function of $U = aX + bY$ and $V = bX - aY$, and hence show that U is $N(0, a^2 + b^2)$. Deduce a proof of the last part of Example (4.8.3).

10. Let X and Y have joint distribution function F, with marginals F_X and F_Y. Show that, if $\mu = \mathbb{E}(X + Y)$ is well defined, then its value is determined by knowledge of F_X and F_Y.

11. Let $S = U + V + W$ be the sum of three independent random variables with the uniform distribution on $(0, 1)$.

(a) Show that $\mathbb{E}\{\operatorname{var}(U \mid S)\} = \frac{1}{18}$.

(b) Find $v(s) = \operatorname{var}(U \mid S = s)$ for $0 < s < 3$.

4.9 Exercises. Multivariate normal distribution

1. A symmetric matrix is called *non-negative* (respectively *positive*) *definite* if its eigenvalues are non-negative (respectively strictly positive). Show that a non-negative definite symmetric matrix $\mathbf{V}$ has a square root, in that there exists a symmetric matrix $\mathbf{W}$ satisfying $\mathbf{W}^2 = \mathbf{V}$. Show further that $\mathbf{W}$ is non-singular if and only if $\mathbf{V}$ is positive definite.

2. If $\mathbf{X}$ is a random vector with the $N(\boldsymbol{\mu}, \mathbf{V})$ distribution where $\mathbf{V}$ is non-singular, show that $\mathbf{Y} = (\mathbf{X} - \boldsymbol{\mu})\mathbf{W}^{-1}$ has the $N(\mathbf{0}, \mathbf{I})$ distribution, where $\mathbf{I}$ is the identity matrix and $\mathbf{W}$ is a symmetric matrix satisfying $\mathbf{W}^2 = \mathbf{V}$. The random vector $\mathbf{Y}$ is said to have the *standard* multivariate normal distribution.

3. Let $\mathbf{X} = (X_1, X_2, \dots, X_n)$ have the $N(\boldsymbol{\mu}, \mathbf{V})$ distribution, and show that $Y = a_1X_1 + a_2X_2 + \cdots + a_nX_n$ has the (univariate) $N(\mu, \sigma^2)$ distribution where

$$\mu = \sum_{i=1}^{n} a_i \mathbb{E}(X_i), \qquad \sigma^2 = \sum_{i=1}^{n} a_i^2 \operatorname{var}(X_i) + 2\sum_{i<j} a_i a_j \operatorname{cov}(X_i, X_j).$$

4. Let X and Y have the bivariate normal distribution with zero means, unit variances, and correlation ρ. Find the joint density function of $X + Y$ and $X - Y$, and their marginal density functions.

5. Let X have the $N(0, 1)$ distribution and let $a > 0$. Show that the random variable Y given by

$$Y = \begin{cases} X & \text{if } |X| < a \\ -X & \text{if } |X| \geq a \end{cases}$$

has the $N(0, 1)$ distribution, and find an expression for $\rho(a) = \operatorname{cov}(X, Y)$ in terms of the density function ϕ of X. Does the pair (X, Y) have a bivariate normal distribution?

6. Let $\{Y_r : 1 \leq r \leq n\}$ be independent $N(0, 1)$ random variables, and define $X_j = \sum_{r=1}^{n} c_{jr} Y_r$, $1 \leq r \leq n$, for constants c_{jr}. Show that

$$\mathbb{E}(X_j \mid X_k) = \left(\frac{\sum_r c_{jr} c_{kr}}{\sum_r c_{kr}^2} \right) X_k.$$

What is $\operatorname{var}(X_j \mid X_k)$?

7. Let the vector $(X_r : 1 \leq r \leq n)$ have a multivariate normal distribution with covariance matrix $\mathbf{V} = (v_{ij})$. Show that, conditional on the event $\sum_1^n X_r = x$, X_1 has the $N(a, b)$ distribution where $a = (\rho s/t)x$, $b = s^2(1 - \rho^2)$, and $s^2 = v_{11}$, $t^2 = \sum_{ij} v_{ij}$, $\rho = \sum_i v_{i1}/(st)$.

8. Let X, Y, and Z have a standard trivariate normal distribution centred at the origin, with zero means, unit variances, and correlation coefficients ρ_1, ρ_2, and ρ_3. Show that

$$\mathbb{P}(X > 0, Y > 0, Z > 0) = \frac{1}{8} + \frac{1}{4\pi}\{\sin^{-1}\rho_1 + \sin^{-1}\rho_2 + \sin^{-1}\rho_3\}.$$

9. Let X, Y, Z have the standard trivariate normal density of Exercise (4.9.8), with $\rho_1 = \rho(X, Y)$. Show that

$$\mathbb{E}(Z \mid X, Y) = \{(\rho_3 - \rho_1\rho_2)X + (\rho_2 - \rho_1\rho_3)Y\}/(1 - \rho_1^2),$$
$$\operatorname{var}(Z \mid X, Y) = \{1 - \rho_1^2 - \rho_2^2 - \rho_3^2 + 2\rho_1\rho_2\rho_3\}/(1 - \rho_1^2).$$

10. Rotated normals. Let X and Y have the standard bivariate normal distribution with correlation $\rho \neq 0$, and let $U = X\cos\theta + Y\sin\theta$ and $V = -X\sin\theta + Y\cos\theta$, where $\theta \in [0, \pi)$. Is there a value of θ such that U and V are independent?

11. Let X have the $N(0, 1)$ distribution, and let $Y = BX$ where B is independent of X with mass function $\mathbb{P}(B = 1) = \mathbb{P}(B = -1) = \frac{1}{2}$. Show that the pair (X, Y) has a singular distribution (in the sense that the support of the joint density function has zero (Lebesgue) area in the plane), but is not singular bivariate normal.

4.10 Exercises. Distributions arising from the normal distribution

1. Let X_1 and X_2 be independent variables with the $\chi^2(m)$ and $\chi^2(n)$ distributions respectively. Show that $X_1 + X_2$ has the $\chi^2(m + n)$ distribution.

2. Show that the mean of the $t(r)$ distribution is 0, and that the mean of the $F(r, s)$ distribution is $s/(s - 2)$ if $s > 2$. What happens if $s \le 2$?

3. Show that the $t(1)$ distribution and the Cauchy distribution are the same.

4. Let X and Y be independent variables having the exponential distribution with parameter 1. Show that X/Y has an F distribution. Which?

5. Show the independence of the sample mean and sample variance of an independent sample from the $N(\mu, \sigma^2)$ distribution. This may be done by either or both of: (i) the result of Exercise (4.5.7), (ii) induction on n.

6. Let $\{X_r : 1 \le r \le n\}$ be independent $N(0, 1)$ variables. Let $\Psi \in [0, \pi]$ be the angle between the vector $(X_1, X_2, \dots, X_n)$ and some fixed vector in $\mathbb{R}^n$. Show that Ψ has density $f(\psi) = (\sin\psi)^{n-2}/B(\frac{1}{2}, \frac{1}{2}n - \frac{1}{2})$, $0 \le \psi < \pi$, where B is the beta function.

7. Let $X_1, X_2, \dots, X_n$ be independent $N(0, 1)$ random variables, and let $V_n \ge 0$ be given by $V_n^2 = \sum_{i=1}^n X_i^2$. Show that the random vector $Y = (X_1, X_2, \dots, X_n)/V_n$ is uniformly distributed on the unit sphere. Deduce the result of Exercise (4.10.6).

4.11 Exercises. Sampling from a distribution

1. Uniform distribution. If U is uniformly distributed on $[0, 1]$, what is the distribution of $X = \lfloor nU \rfloor + 1$?

2. Random permutation. Given the first n integers in any sequence S_0, proceed thus:

(a) pick any position P_0 from $\{1, 2, \dots, n\}$ at random, and swap the integer in that place of S_0 with the integer in the nth place of S_0, yielding S_1.

(b) pick any position P_1 from $\{1, 2, \dots, n - 1\}$ at random, and swap the integer in that place of S_1 with the integer in the $(n - 1)$th place of S_1, yielding S_2,

(c) at the $(r - 1)$th stage the integer in position P_{r-1}, chosen randomly from $\{1, 2, \dots, n - r + 1\}$, is swapped with the integer at the $(n - r + 1)$th place of the sequence S_{r-1}.

Show that S_{n-1} is equally likely to be any of the $n!$ permutations of $\{1, 2, \dots, n\}$.

3. Gamma distribution. Use the rejection method to sample from the gamma density $\Gamma(\lambda, t)$ where t (≥ 1) may not be assumed integral. [Hint: You might want to start with an exponential random variable with parameter $1/t$.]

4. Beta distribution. Show how to sample from the beta density $\beta(\alpha, \beta)$ where $\alpha, \beta \ge 1$. [Hint: Use Exercise (4.11.3).]

5. Describe three distinct methods of sampling from the density $f(x) = 6x(1 - x)$, $0 \le x \le 1$.

6. Aliasing method. A finite real vector is called a *probability vector* if it has non-negative entries with sum 1. Show that a probability vector $\mathbf{p}$ of length n may be written in the form

$$\mathbf{p} = \frac{1}{n-1}\sum_{r=1}^{n} \mathbf{v}_r,$$

where each $\mathbf{v}_r$ is a probability vector with at most two non-zero entries. Describe a method, based on this observation, for sampling from $\mathbf{p}$ viewed as a probability mass function.

7. Box–Muller normals. Let U_1 and U_2 be independent and uniformly distributed on $[0, 1]$, and let $T_i = 2U_i - 1$. Show that, conditional on the event that $R = \sqrt{T_1^2 + T_2^2} \le 1$,

$$X = \frac{T_1}{R}\sqrt{-2\log R^2}, \quad Y = \frac{T_2}{R}\sqrt{-2\log R^2},$$

are independent standard normal random variables.

8. Let U be uniform on $[0, 1]$ and $0 < q < 1$. Show that $X = 1 + \lfloor \log U / \log q \rfloor$ has a geometric distribution.

9. A point (X, Y) is picked uniformly at random in the semicircle $x^2 + y^2 \le 1$, $x \ge 0$. What is the distribution of $Z = Y/X$?

10. Hazard-rate technique. Let X be a non-negative integer-valued random variable with $h(r) = \mathbb{P}(X = r \mid X \ge r)$. If $\{U_i : i \ge 0\}$ are independent and uniform on $[0, 1]$, show that $Z = \min\{n : U_n \le h(n)\}$ has the same distribution as X.

11. Antithetic variables†. Let $g(x_1, x_2, \dots, x_n)$ be an increasing function in all its variables, and let $\{U_r : r \ge 1\}$ be independent and identically distributed random variables having the uniform distribution on $[0, 1]$. Show that

$$\operatorname{cov}\{g(U_1, U_2, \dots, U_n), g(1 - U_1, 1 - U_2, \dots, 1 - U_n)\} \le 0.$$

[Hint: Use the FKG inequality of Problem (3.11.18).] Explain how this can help in the efficient estimation of $I = \int_0^1 g(\mathbf{x})\,d\mathbf{x}$.

12. Importance sampling. We wish to estimate $I = \int g(x) f_X(x)\,dx = \mathbb{E}(g(X))$, where either it is difficult to sample from the density f_X, or $g(X)$ has a very large variance. Let f_Y be equivalent to f_X, which is to say that, for all x, $f_X(x) = 0$ if and only if $f_Y(x) = 0$. Let $\{Y_i : 0 \le i \le n\}$ be independent random variables with density function f_Y, and define

$$J = \frac{1}{n}\sum_{r=1}^{n} \frac{g(Y_r) f_X(Y_r)}{f_Y(Y_r)}.$$

Show that:

(a) $\mathbb{E}(J) = I = \mathbb{E}\left[\dfrac{g(Y) f_X(Y)}{f_Y(Y)}\right]$,

(b) $\operatorname{var}(J) = \dfrac{1}{n}\left[\mathbb{E}\left(\dfrac{g(Y)^2 f_X(Y)^2}{f_Y(Y)^2}\right) - I^2\right]$,

(c) $J \xrightarrow{\text{a.s.}} I$ as $n \to \infty$. (See Chapter 7 for an account of convergence.)

†A technique invented by J. M. Hammersley and K. W. Morton in 1956.

The idea here is that f_Y should be easy to sample from, and chosen if possible so that var J is much smaller than $n^{-1}[\mathbb{E}(g(X)^2) - I^2]$. The function f_Y is called the *importance density*.

13. Construct two distinct methods of sampling from the arc sine density function

$$f(x) = \frac{2}{\pi\sqrt{1-x^2}}, \qquad 0 \le x \le 1.$$

14. Marsaglia's method. Let (X, Y) be a random point chosen uniformly on the unit disk. Show that

$$(U, V, W) = \left(2X\sqrt{1-X^2-Y^2}, 2Y\sqrt{1-X^2-Y^2}, 1-2(X^2+Y^2)\right)$$

is uniformly distributed on the unit sphere.

15. Acceptance–complement method. It is desired to sample from a density function f_X satisfying $f_X = f_1 + f_2$ where $f_1, f_2 : \mathbb{R} \to [0, \infty)$, and $f_1 \le f_Y$ for some random variable Y. Let U be uniformly distributed on $[0, 1]$, and independent of Y. From Y and U, we generate another random variable Z as follows. If $U > f_1(Y)/f_Y(Y)$, let Z have density $f_2(z)/\int_{-\infty}^{\infty} f_2(u)\,du$; otherwise set $Z = Y$. Show that Z has density function f_X.

16. Hardy–Littlewood transform. The function F^{-1} defined in Theorem (4.11.1) may be called the *quantile function* of the random variable in question.

(a) For a random variable U with the uniform distribution on $(0, 1)$, and a continuously increasing function $G : (0, 1) \to \mathbb{R}$, show that the quantile function of $G(U)$ is G.

(b) For a random variable X with distribution function F and density function f, define $H_X(v) = \mathbb{E}(F^{-1}(U) \mid U > v)$, where U is as in part (a). The *Hardy–Littlewood transform* of X is defined as $H_X(V)$, where V is uniformly distributed on $(0, 1)$ and independent of U. Show that the quantile function of $H_X(V)$ is $(1-v)^{-1}\int_v^{\infty} F^{-1}(w)\,dw$, for $v \in (0, 1)$.

17. Generalized inverse transform†. For an arbitrary (not necessarily continuous) distribution function $F = F_X$, define

$$\begin{aligned} G(x, u) &= \mathbb{P}(X < x) + u\mathbb{P}(X = x) \\ &= F(x-) + u(F(x) - F(x-)), \qquad u \in [0, 1]. \end{aligned}$$

Let U be uniformly distributed on $(0, 1)$ and independent of X. We call $G(X, U)$ the *generalized distributional transform* of X. Show that $W = G(X, U)$ is uniformly distributed on $(0, 1)$, and that the random variable $F^{-1}(W)$ has distribution function F.

18. Copulas. Let $(X_1, X_2, \dots, X_n)$ be a random vector with joint distribution function F and continuous marginal distribution functions $F_1, F_2, \dots, F_n$. Show that there exists a 'copula-form' distribution function $G(u_1, u_2, \dots, u_n)$ such that $F(x_1, x_2, \dots, x_n) = G(F_1(x_1), F_2(x_2), \dots, F_n(x_n))$. [A 'copula-form' distribution function is one with the property that all its marginals are uniform on [0, 1].]

4.12 Exercises. Coupling and Poisson approximation

1. Show that X is stochastically larger than Y if and only if $\mathbb{E}(u(X)) \ge \mathbb{E}(u(Y))$ for any non-decreasing function u for which the expectations exist.

2. Let X and Y be Poisson distributed with respective parameters λ and μ. Show that X is stochastically larger than Y if $\lambda \ge \mu$.

† Also known in this context as the 'quantile function'.

3. Show that the total variation distance between two discrete variables X, Y satisfies

$$d_{TV}(X, Y) = 2 \sup_{A \subseteq \mathbb{R}} \left| \mathbb{P}(X \in A) - \mathbb{P}(Y \in A) \right|.$$

4. Maximal coupling. Show for discrete random variables X, Y that $\mathbb{P}(X = Y) \le 1 - \frac{1}{2} d_{TV}(X, Y)$, where d_{TV} denotes total variation distance.

5. Maximal coupling continued. Show that equality is possible in the inequality of Exercise (4.12.4) in the following sense. For any pair X, Y of discrete random variables, there exists a pair X', Y' having the same marginal distributions as X, Y such that $\mathbb{P}(X' = Y') = 1 - \frac{1}{2} d_{TV}(X, Y)$.

6. Let X and Y be indicator variables with $\mathbb{E}X = p$, $\mathbb{E}Y = q$. What is the maximum possible value of $\mathbb{P}(X = Y)$, as a function of p, q? Explain how X, Y need to be distributed in order that $\mathbb{P}(X = Y)$ be: (a) maximized, (b) minimized.

7. Stop-loss ordering. The random variable X is said to be smaller than Y in *stop-loss order* if $\mathbb{E}((X - a)^+) \le \mathbb{E}((Y - a)^+)$ for all $a \in \mathbb{R}$, in which case we write $X \le_{sl} Y$. Show that $X \le_{sl} Y$ is equivalent to: $\mathbb{E}(X^+), \mathbb{E}(Y^+) < \infty$ and $\mathbb{E}(c(X)) \le \mathbb{E}(c(Y))$ for all increasing convex functions c.

8. Convex ordering. The random variable X is said to be smaller than Y in *convex order* if $\mathbb{E}(u(X)) \le \mathbb{E}(u(Y))$ for all convex functions u such that the expectations exist, in which case we write $X \le_{cx} Y$.

(a) Show that, if $X \le_{cx} Y$, then $\mathbb{E}X = \mathbb{E}Y$ and $\text{var}(X) \le \text{var}(Y)$.

(b) Show that $X \le_{cx} Y$ if and only if $\mathbb{E}X = \mathbb{E}Y$ and $X \le_{sl} Y$.

4.13 Exercises. Geometrical probability

With apologies to those who prefer their exercises better posed ...

1. Pick two points A and B independently at random on the circumference of a circle C with centre O and unit radius. Let Π be the length of the perpendicular from O to the line AB, and let Θ be the angle AB makes with the horizontal. Show that (Π, Θ) has joint density

$$f(p, \theta) = \frac{1}{\pi^2 \sqrt{1 - p^2}}, \quad 0 \le p \le 1, \ 0 \le \theta < 2\pi.$$

2. Let S_1 and S_2 be disjoint convex shapes with boundaries of length $b(S_1)$, $b(S_2)$, as illustrated in the figure beneath. Let $b(H)$ be the length of the boundary of the convex hull of S_1 and S_2, incorporating their exterior tangents, and $b(X)$ the length of the crossing curve using the interior tangents to loop round S_1 and S_2. Show that the probability that a random line crossing S_1 also crosses S_2 is $\{b(X) - b(H)\}/b(S_1)$. (See Example (4.13.2) for an explanation of the term 'random line'.) How is this altered if S_1 and S_2 are not disjoint?

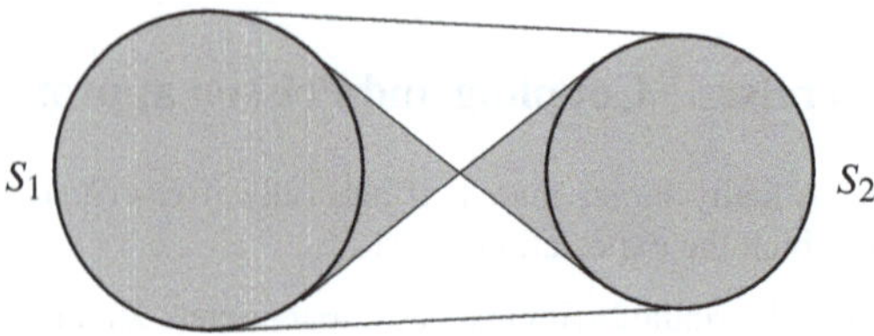

The circles are the shapes S_1 and S_2. The shaded regions are denoted A and B, and $b(X)$ is the sum of the perimeter lengths of A and B.

3. Let S_1 and S_2 be convex figures such that $S_2 \subseteq S_1$. Show that the probability that two independent random lines λ_1 and λ_2, crossing S_1, meet within S_2 is $2\pi|S_2|/b(S_1)^2$, where $|S_2|$ is the area of S_2 and $b(S_1)$ is the length of the boundary of S_1. (See Example (4.13.2) for an explanation of the term 'random line'.)

4. Let Z be the distance between two points picked independently at random in a disk of radius a. Show that $\mathbb{E}(Z) = 128a/(45\pi)$, and $\mathbb{E}(Z^2) = a^2$.

5. Pick two points A and B independently at random in a ball with centre O. Show that the probability that the angle $\widehat{\mathrm{AOB}}$ is obtuse is $\frac{5}{8}$. Compare this with the corresponding result for two points picked at random in a circle.

6. A triangle is formed by A, B, and a point P picked at random in a set S with centre of gravity G. Show that $\mathbb{E}|\mathrm{ABP}| = |\mathrm{ABG}|$.

7. A point D is fixed on the side BC of the triangle ABC. Two points P and Q are picked independently at random in ABD and ADC respectively. Show that $\mathbb{E}|\mathrm{APQ}| = |\mathrm{AG_1G_2}| = \frac{2}{9}|\mathrm{ABC}|$, where $\mathrm{G_1}$ and $\mathrm{G_2}$ are the centres of gravity of ABD and ADC.

8. From the set of all triangles that are similar to the triangle ABC, similarly oriented, and inside ABC, one is selected uniformly at random. Show that its mean area is $\frac{1}{10}|\mathrm{ABC}|$.

9. Two points X and Y are picked independently at random in the interval $(0, a)$. By varying a, show that $F(z, a) = \mathbb{P}(|X - Y| \le z)$ satisfies

$$\frac{\partial F}{\partial a} + \frac{2}{a}F = \frac{2z}{a^2}, \qquad 0 \le z \le a,$$

and hence find $F(z, a)$. Let $r \ge 1$, and show that $m_r(a) = \mathbb{E}(|X - Y|^r)$ satisfies

$$a\frac{dm_r}{da} = 2\left\{\frac{a^r}{r+1} - m_r\right\}.$$

Hence find $m_r(a)$.

10. Lines are laid down independently at random on the plane, dividing it into polygons. Show that the average number of sides of this set of polygons is 4. [Hint: Consider n random great circles of a sphere of radius R; then let R and n increase.]

11. A point P is picked at random in the triangle ABC. The lines AP, BP, CP, produced, meet BC, AC, AB respectively at L, M, N. Show that $\mathbb{E}|\mathrm{LMN}| = (10 - \pi^2)|\mathrm{ABC}|$.

12. Sylvester's problem. If four points are picked independently at random inside the triangle ABC, show that the probability that no one of them lies inside the triangle formed by the other three is $\frac{2}{3}$.

13. If three points P, Q, R are picked independently at random in a disk of radius a, show that $\mathbb{E}|\mathrm{PQR}| = 35a^2/(48\pi)$. [You may find it useful that $\int_0^\pi \int_0^\pi \sin^3 x \sin^3 y \sin|x - y|\,dx\,dy = 35\pi/128$.]

14. Two points A and B are picked independently at random inside a disk C. Show that the probability that the circle having centre A and radius |AB| lies inside C is $\frac{1}{6}$.

15. Two points A and B are picked independently at random inside a ball S. Show that the probability that the sphere having centre A and radius |AB| lies inside S is $\frac{1}{20}$.

16. Pick two points independently and uniformly at random on the surface of the sphere of $\mathbb{R}^3$ with radius 1. Show that the density of the Euclidean distance D between them is $f(d) = \frac{1}{2}d$ for $d \in (0, 2)$.

17. Two points are chosen independently and uniformly at random on the perimeter (including the diameter) of a semicircle with unit radius. What is the probability of the event D that exactly one of them lies on the diameter, and of the event N that neither lies on the diameter?

Let A be the area of the triangle formed by the two points and the midpoint of the diameter. Show that $\mathbb{E}(A \mid D) = 1/(2\pi)$, and $\mathbb{E}(A \mid N) = 1/\pi$. Hence or otherwise show that $\mathbb{E}(A) = 1/(2+\pi)$.

18. Stevens's solution of Jeffreys's bicycle wheel problem. After passing over a stretch of road strewn with tacks, a cyclist looks repeatedly at the front wheel to check whether the tyre has picked up a tack. One inspection of the wheel covers a fraction x of the tyre. After n independent such inspections, each uniformly positioned on the wheel, show that the probability of having inspected the entire tyre is $1 - \binom{n}{1}(1-x)^{n-1} + \binom{n}{2}(1-2x)^{n-1} - \cdots$, where the series terminates at the term in $1 - kx$ where $k = \lfloor 1/x \rfloor$ is the integer part of $1/x$.

4.14 Problems

1. (a) Show that $\int_{-\infty}^{\infty} e^{-x^2}\,dx = \sqrt{\pi}$, and deduce that

$$f(x) = \frac{1}{\sigma\sqrt{2\pi}} \exp\left\{-\frac{(x-\mu)^2}{2\sigma^2}\right\}, \qquad -\infty < x < \infty,$$

is a density function if $\sigma > 0$.

(b) Calculate the mean and variance of a standard normal variable.

(c) Show that the $N(0, 1)$ distribution function Φ satisfies

$$(x^{-1} - x^{-3})e^{-\frac{1}{2}x^2} < \sqrt{2\pi}[1 - \Phi(x)] < x^{-1}e^{-\frac{1}{2}x^2}, \qquad x > 0.$$

These bounds are of interest because Φ has no closed form.

(d) Let X be $N(0, 1)$, and $a > 0$. Show that $\mathbb{P}(X > x + a/x \mid X > x) \to e^{-a}$ as $x \to 0$.

2. Let X be continuous with density function $f(x) = C(x - x^2)$, where $\alpha < x < \beta$ and $C > 0$.

(a) What are the possible values of α and β?

(b) What is C?

3. Let X be a random variable which takes non-negative values only. Show that

$$\sum_{i=1}^{\infty}(i-1)I_{A_i} \le X < \sum_{i=1}^{\infty} i I_{A_i},$$

where $A_i = \{i - 1 \le X < i\}$. Deduce that

$$\sum_{i=1}^{\infty}\mathbb{P}(X \ge i) \le \mathbb{E}(X) < 1 + \sum_{i=1}^{\infty}\mathbb{P}(X \ge i).$$

4. (a) Let X have a continuous distribution function F. Show that

(i) $F(X)$ is uniformly distributed on $[0, 1]$,

(ii) $-\log F(X)$ is exponentially distributed.

(b) A straight line l touches a circle with unit diameter at the point P which is diametrically opposed on the circle to another point Q. A straight line QR joins Q to some point R on l. If the angle $\widehat{PQR}$ between the lines PQ and QR is a random variable with the uniform distribution on $[-\frac{1}{2}\pi, \frac{1}{2}\pi]$, show that the length of PR has the Cauchy distribution (this length is measured positive or negative depending upon which side of P the point R lies).

(c) Let the net scores in the two halves of a game between teams A and B be independent and identically distributed random variables X, Y that are symmetric about 0 and have a continuous distribution function F. Team A wins (respectively, loses) if $X+Y > 0$ (respectively, $X+Y \leq 0$). Find the probability that A wins conditional on the half-time score X.

5. Lack of memory property. (a) Let $g : [0, \infty) \to (0, \infty)$ be such that $g(s+t) = g(s)g(t)$ for $s, t \geq 0$. If g is monotone, show that $g(s) = e^{\mu s}$ for some $\mu \in \mathbb{R}$.

(b) Let X have an exponential distribution. Show that $\mathbb{P}(X > s + x \mid X > s) = \mathbb{P}(X > x)$, for $x, s \geq 0$. This is the 'lack of memory' property again. Show that the exponential distribution is the only continuous distribution with this property.

(c) The height M of the largest tidal surge in a certain estuary has the exponential distribution with parameter 1, and it costs h to build a barrier with height h in order to protect an upstream city. If $M \leq h$, the surge costs nothing; if $M > h$, the surge entails an extra cost of $a + b(M - h)$. Show that, for $a + b \geq 1$, the expected total cost is minimal when $h = \log(a + b)$. What if $a + b < 1$?

6. Show that X and Y are independent continuous variables if and only if their joint density function f factorizes as the product $f(x, y) = g(x)h(y)$ of functions of the single variables x and y alone.

Explain why the density function $f(x, y) = Ce^{-x-y}$ (for $x, y \geq 0$, $x + y > 1$) does not provide a counterexample to this assertion.

7. Let X and Y have joint density function $f(x, y) = 2e^{-x-y}$, $0 < x < y < \infty$. Are they independent? Find their marginal density functions and their covariance.

8. Bertrand's paradox extended. A chord of the unit circle is picked at random. What is the probability that an equilateral triangle with the chord as base can fit inside the circle if:

(a) the chord passes through a point P picked uniformly in the disk, and the angle it makes with a fixed direction is uniformly distributed on $[0, 2\pi)$,

(b) the chord passes through a point P picked uniformly at random on a randomly chosen radius, and the angle it makes with the radius is uniformly distributed on $[0, 2\pi)$.

9. Monte Carlo. It is required to estimate $J = \int_0^1 g(x)\,dx$ where $0 \leq g(x) \leq 1$ for all x, as in Example (2.6.3). Let X and Y be independent random variables with common density function $f(x) = 1$ if $0 < x < 1$, $f(x) = 0$ otherwise. Let $U = I_{\{Y \leq g(X)\}}$, the indicator function of the event that $Y \leq g(X)$, and let $V = g(X)$, $W = \frac{1}{2}\{g(X) + g(1-X)\}$. Show that $\mathbb{E}(U) = \mathbb{E}(V) = \mathbb{E}(W) = J$, and that $\text{var}(W) \leq \text{var}(V) \leq \text{var}(U)$, so that, of the three, W is the most 'efficient' estimator of J.

10. Let $X_1, X_2, \ldots, X_n$ be independent exponential variables, parameter λ. Show by induction that $S = X_1 + X_2 + \cdots + X_n$ has the $\Gamma(\lambda, n)$ distribution.

11. Let X and Y be independent variables, $\Gamma(\lambda, m)$ and $\Gamma(\lambda, n)$ respectively.

(a) Use the result of Problem (4.14.10) to show that $X + Y$ is $\Gamma(\lambda, m + n)$ when m and n are integral (the same conclusion is actually valid for non-integral m and n).

(b) Find the joint density function of $X + Y$ and $X/(X + Y)$, and deduce that they are independent.

(c) If Z is Poisson with parameter λt, and m is integral, show that $\mathbb{P}(Z < m) = \mathbb{P}(X > t)$.

(d) If $0 < m < n$ and B is independent of Y with the beta distribution with parameters m and $n - m$, show that YB has the same distribution as X.

12. Let $X_1, X_2, \ldots, X_n$ be independent $N(0, 1)$ variables.

(a) Show that X_1^2 is $\chi^2(1)$.

(b) Show that $X_1^2 + X_2^2$ is $\chi^2(2)$ by expressing its distribution function as an integral and changing to polar coordinates.

(c) More generally, show that $X_1^2 + X_2^2 + \cdots + X_n^2$ is $\chi^2(n)$.

13. Let X and Y have the bivariate normal distribution with means μ_1, μ_2, variances σ_1^2, σ_2^2, and correlation ρ. Show that

(a) $\mathbb{E}(X \mid Y) = \mu_1 + \rho\sigma_1(Y - \mu_2)/\sigma_2$,

(b) the variance of the conditional density function $f_{X|Y}$ is $\text{var}(X \mid Y) = \sigma_1^2(1 - \rho^2)$.

14. Let X and Y have joint density function f. Find the density function of Y/X.

15. Let X and Y be independent variables with common density function f. Show that $\tan^{-1}(Y/X)$ has the uniform distribution on $(-\frac{1}{2}\pi, \frac{1}{2}\pi)$ if and only if

$$\int_{-\infty}^{\infty} f(x)f(xy)|x|\,dx = \frac{1}{\pi(1+y^2)}, \qquad y \in \mathbb{R}.$$

Verify that this is valid if either f is the $N(0, 1)$ density function or $f(x) = a(1 + x^4)^{-1}$ for some constant a.

16. Rayleigh distribution. Let X and Y be independent $N(0, 1)$ variables, and think of (X, Y) as a random point in the plane. Change to polar coordinates (R, Θ) given by $R^2 = X^2+Y^2$, $\tan\Theta = Y/X$; show that R^2 is $\chi^2(2)$, $\tan\Theta$ has the Cauchy distribution, and R and Θ are independent. Find the density of R.

Find $\mathbb{E}(X^2/R^2)$ and

$$\mathbb{E}\left\{\frac{\min\{|X|, |Y|\}}{\max\{|X|, |Y|\}}\right\}.$$

17. If X and Y are independent random variables, show that $U = \min\{X, Y\}$ and $V = \max\{X, Y\}$ have distribution functions

$$F_U(u) = 1 - \{1 - F_X(u)\}\{1 - F_Y(u)\}, \qquad F_V(v) = F_X(v)F_Y(v).$$

Let X and Y be independent exponential variables, parameter 1. Show that

(a) U is exponential, parameter 2,

(b) V has the same distribution as $X + \frac{1}{2}Y$. Hence find the mean and variance of V.

18. Let X and Y be independent variables having the exponential distribution with parameters λ and μ respectively. Let $U = \min\{X, Y\}$, $V = \max\{X, Y\}$, and $W = V - U$.

(a) Find $\mathbb{P}(U = X) = \mathbb{P}(X \le Y)$.

(b) Show that U and W are independent.

19. Let X and Y be independent non-negative random variables with continuous density functions on $(0, \infty)$.

(a) If, given $X + Y = u$, X is uniformly distributed on $[0, u]$ whatever the value of u, show that X and Y have the exponential distribution.

(b) If, given that $X + Y = u$, X/u has a given beta distribution (parameters α and β, say) whatever the value of u, show that X and Y have gamma distributions.

You may need the fact that the only non-negative continuous solutions of the functional equation $g(s + t) = g(s)g(t)$ for $s, t \ge 0$, with $g(0) = 1$, are of the form $g(s) = e^{\mu s}$. Remember Problem (4.14.5).

20. Show that it cannot be the case that $U = X + Y$ where U is uniformly distributed on $[0, 1]$ and X and Y are independent and identically distributed. You should not assume that X and Y are continuous variables.

21. Order statistics. Let $X_1, X_2, \dots, X_n$ be independent identically distributed variables with a common density function f. Such a collection is called a *random sample*. For each $\omega \in \Omega$, arrange the sample values $X_1(\omega), \dots, X_n(\omega)$ in non-decreasing order $X_{(1)}(\omega) \le X_{(2)}(\omega) \le \cdots \le X_{(n)}(\omega)$, where

(1), (2), ..., (n) is a (random) permutation of $1, 2, \ldots, n$. The new variables $X_{(1)}, X_{(2)}, \ldots, X_{(n)}$ are called the *order statistics*. Show, by a symmetry argument, that the joint distribution function of the order statistics satisfies

$$\mathbb{P}(X_{(1)} \le y_1, \ldots, X_{(n)} \le y_n) = n!\,\mathbb{P}(X_1 \le y_1, \ldots, X_n \le y_n,\ X_1 < X_2 < \cdots < X_n)$$
$$= \int \cdots \int_{\substack{x_1 \le y_1 \\ x_2 \le y_2 \\ \vdots \\ x_n \le y_n}} L(x_1, \ldots, x_n) n!\, f(x_1) \cdots f(x_n)\, dx_1 \cdots dx_n$$

where L is given by

$$L(\mathbf{x}) = \begin{cases} 1 & \text{if } x_1 < x_2 < \cdots < x_n, \\ 0 & \text{otherwise}, \end{cases}$$

and $\mathbf{x} = (x_1, x_2, \ldots, x_n)$. Deduce that the joint density function of $X_{(1)}, \ldots, X_{(n)}$ is $g(\mathbf{y}) = n!\, L(\mathbf{y}) f(y_1) \cdots f(y_n)$.

22. Find the marginal density function of the kth order statistic $X_{(k)}$ of a sample with size n:
(a) by integrating the result of Problem (4.14.21),
(b) directly.

23. Find the joint density function of the order statistics of n independent uniform variables on $[0, T]$.

24. Let $X_1, X_2, \ldots, X_n$ be independent and uniformly distributed on $[0, 1]$, with order statistics $X_{(1)}, X_{(2)}, \ldots, X_{(n)}$.
(a) Show that, for fixed k, the density function of $nX_{(k)}$ converges as $n \to \infty$, and find and identify the limit function.
(b) Show that $\log X_{(k)}$ has the same distribution as $-\sum_{i=k}^{n} i^{-1} Y_i$, where the Y_i are independent random variables having the exponential distribution with parameter 1.
(c) Show that $Z_1, Z_2, \ldots, Z_n$, defined by $Z_k = (X_{(k)}/X_{(k+1)})^k$ for $k < n$ and $Z_n = (X_{(n)})^n$, are independent random variables with the uniform distribution on $[0, 1]$.

25. Let X_1, X_2, X_3 be independent variables with the uniform distribution on $[0, 1]$. What is the probability that rods of lengths X_1, X_2, and X_3 may be used to make a triangle? Generalize your answer to n rods used to form a polygon.

26. Stick breaking. Let X_1 and X_2 be independent variables with the uniform distribution on $[0, 1]$. A stick of unit length is broken at points distance X_1 and X_2 from one of the ends. What is the probability that the three pieces may be used to make a triangle? Generalize your answer to a stick broken in n places.

27. Let X, Y be a pair of jointly continuous variables.
(a) **Hölder's inequality.** Show that if $p, q > 1$ and $p^{-1} + q^{-1} = 1$ then

$$\mathbb{E}|XY| \le \{\mathbb{E}|X^p|\}^{1/p} \{\mathbb{E}|Y^q|\}^{1/q}.$$

Set $p = q = 2$ to deduce the Cauchy–Schwarz inequality $\mathbb{E}(XY)^2 \le \mathbb{E}(X^2)\mathbb{E}(Y^2)$.
(b) **Minkowski's inequality.** Show that, if $p \ge 1$, then

$$\{\mathbb{E}(|X + Y|^p)\}^{1/p} \le \{\mathbb{E}|X^p|\}^{1/p} + \{\mathbb{E}|Y^p|\}^{1/p}.$$

Note that in both cases your proof need not depend on the continuity of X and Y; deduce that the same inequalities hold for discrete variables.

28. Let Z be a random variable. Choose X and Y appropriately in the Cauchy–Schwarz (or Hölder) inequality to show that $g(p) = \log \mathbb{E}|Z^p|$ is a convex function of p on the interval of values of p such that $\mathbb{E}|Z^p| < \infty$. Deduce **Lyapunov's inequality**:

$$\{\mathbb{E}|Z^r|\}^{1/r} \geq \{\mathbb{E}|Z^s|\}^{1/s} \quad \text{whenever } r \geq s > 0.$$

You have shown in particular that, if Z has finite rth moment, then Z has finite sth moment for all positive $s \leq r$.

Show more generally that, if $r_1, r_2, \ldots, r_n$ are real numbers such that $\mathbb{E}|X^{r_k}| < \infty$ and $s = n^{-1} \sum_{k=1}^{n} r_k$, then

$$\left\{ \prod_{k=1}^{n} \mathbb{E}|X^{r_k}| \right\} - \{\mathbb{E}(|X|^s\} \geq 0.$$

29. Show that, using the obvious notation, $\mathbb{E}\{\mathbb{E}(X \mid Y, Z) \mid Y\} = \mathbb{E}(X \mid Y)$.

30. Rényi's parking problem. Motor cars of unit length park randomly in a street in such a way that the centre of each car, in turn, is positioned uniformly at random in the space available to it. Let $m(x)$ be the expected number of cars which are able to park in a street of length x. Show that

$$m(x+1) = \frac{1}{x} \int_0^x \{m(y) + m(x-y) + 1\}\, dy.$$

It is possible to deduce that $m(x)$ is about as big as $\frac{3}{4}x$ when x is large.

31. Buffon's needle revisited: Buffon's noodle.

(a) A plane is ruled by the lines $y = nd$ ($n = 0, \pm 1, \ldots$). A needle with length L ($< d$) is cast randomly onto the plane. Show that the probability that the needle intersects a line is $2L/(\pi d)$.

(b) Now fix the needle and let C be a circle diameter d centred at the midpoint of the needle. Let λ be a line whose direction and distance from the centre of C are independent and uniformly distributed on $[0, 2\pi]$ and $[0, \frac{1}{2}d]$ respectively. This is equivalent to 'casting the ruled plane at random'. Show that the probability of an intersection between the needle and λ is $2L/(\pi d)$.

(c) Let S be a curve within C having finite length $L(S)$. Use indicators to show that the expected number of intersections between S and λ is $2L(S)/(\pi d)$.

This type of result is used in stereology, which seeks knowledge of the contents of a cell by studying its cross sections.

32. Buffon's needle ingested. In the excitement of calculating π, Mr Buffon (no relation) inadvertently swallows the needle and is X-rayed. If the needle exhibits no preference for direction in the gut, what is the distribution of the length of its image on the X-ray plate? If he swallowed Buffon's cross (see Exercise (4.5.3)) also, what would be the joint distribution of the lengths of the images of the two arms of the cross?

33. Let $X_1, X_2, \ldots, X_n$ be independent exponential variables with parameter λ, and let $X_{(1)} \leq X_{(2)} \leq \cdots \leq X_{(n)}$ be their order statistics. Show that

$$Y_1 = nX_{(1)}, \quad Y_r = (n+1-r)(X_{(r)} - X_{(r-1)}), \qquad 1 < r \leq n$$

are also independent and have the same joint distribution as the X_i.

34. Let $X_{(1)}, X_{(2)}, \ldots, X_{(n)}$ be the order statistics of a family of independent variables with common continuous distribution function F. Show that

$$Y_n = \{F(X_{(n)})\}^n, \quad Y_r = \left\{ \frac{F(X_{(r)})}{F(X_{(r+1)})} \right\}^r, \quad 1 \leq r < n,$$

are independent and uniformly distributed on [0, 1]. This is equivalent to Problem (4.14.33). Why?

35. Secretary/marriage problem. You are permitted to inspect the n prizes at a fête in a given order, at each stage either rejecting or accepting the prize under consideration. There is no recall, in the sense that no rejected prize may be accepted later. It may be assumed that, given complete information, the prizes may be ranked in a strict order of preference, and that the order of presentation is independent of this ranking. Find the strategy which maximizes the probability of accepting the best prize, and describe its behaviour when n is large.

36. Fisher's spherical distribution. Let $R^2 = X^2+Y^2+Z^2$ where X, Y, Z are independent normal random variables with means λ, μ, ν, and common variance σ^2, where $(\lambda, \mu, \nu) \neq (0, 0, 0)$. Show that the conditional density of the point (X, Y, Z) given $R = r$, when expressed in spherical polar coordinates relative to an axis in the direction $\mathbf{e} = (\lambda, \mu, \nu)$, is of the form

$$f(\theta, \phi) = \frac{a}{4\pi \sinh a} e^{a\cos\theta} \sin\theta, \quad 0 \le \theta < \pi,\ 0 \le \phi < 2\pi,$$

where $a = r|\mathbf{e}|$.

37. Let ϕ be the $N(0, 1)$ density function, and define the functions H_n, $n \ge 0$, by $H_0 = 1$, and $(-1)^n H_n\phi = \phi^{(n)}$, the nth derivative of ϕ. Show that:

(a) $H_n(x)$ is a polynomial of degree n having leading term x^n, and

$$\int_{-\infty}^{\infty} H_m(x)H_n(x)\phi(x)\,dx = \begin{cases} 0 & \text{if } m \neq n, \\ n! & \text{if } m = n. \end{cases}$$

(b) $\displaystyle\sum_{n=0}^{\infty} H_n(x)\frac{t^n}{n!} = \exp(tx - \tfrac{1}{2}t^2)$.

38. Lancaster's theorem. Let X and Y have a standard bivariate normal distribution with zero means, unit variances, and correlation coefficient ρ, and suppose $U = u(X)$ and $V = v(Y)$ have finite variances. Show that $|\rho(U, V)| \le |\rho|$. [Hint: Use Problem (4.14.37) to expand the functions u and v. You may assume that u and v lie in the linear span of the H_n.]

39. Let $X_{(1)}, X_{(2)}, \dots, X_{(n)}$ be the order statistics of n independent random variables, uniform on [0, 1]. Show that:

(a) $\mathbb{E}(X_{(r)}) = \dfrac{r}{n+1}$, (b) $\operatorname{cov}(X_{(r)}, X_{(s)}) = \dfrac{r(n-s+1)}{(n+1)^2(n+2)}$ for $r \le s$.

40. (a) Let X, Y, Z be independent $N(0, 1)$ variables, and set $R = \sqrt{X^2+Y^2+Z^2}$. Show that X^2/R^2 has a beta distribution with parameters $\frac{1}{2}$ and 1, and is independent of R^2.

(b) Let X, Y, Z be independent and uniform on $[-1, 1]$ and set $R = \sqrt{X^2+Y^2+Z^2}$. Find the density of X^2/R^2 given that $R^2 \le 1$.

41. (a) **Skew normal distribution.** Let ϕ and Φ be the density and distribution functions of the random variable X with the standard normal distribution. Show, for $\lambda \in \mathbb{R}$, that $g(x) = 2\phi(x)\Phi(\lambda x)$, $x \in \mathbb{R}$, is the density function of some random variable (denoted by Y). Show that $|X|$ and $|Y|$ have the same distributions.

Let X have the $N(0, 1)$ distribution and be independent of Y, and define $U = (X+\lambda|Y|)/\sqrt{1+\lambda^2}$. Write down the joint density of U and $V = |Y|$, and deduce that U has density function g.

(b) **General skew distributions.** For $i = 1, 2$, let X_i have a density function f_i, which is symmetric about 0 with distribution function F_i.

(i) Let $\lambda \in \mathbb{R}$, and show that $g_1(x) = 2f_1(x)F_2(\lambda x)$ and $g_2(x) = 2f_2(x)F_1(\lambda x)$ are density functions.

(ii) Let X_1 and X_2 be independent. Find the conditional density function of X_2 given that $X_1 < \lambda X_2$.

42. The six coordinates (X_i, Y_i), $1 \le i \le 3$, of three points A, B, C in the plane are independent $N(0, 1)$. Show that the the probability that C lies inside the circle with diameter AB is $\frac{1}{4}$.

43. The coordinates (X_i, Y_i, Z_i), $1 \le i \le 3$, of three points A, B, C are independent $N(0, 1)$. Show that the probability that C lies inside the sphere with diameter AB is $\dfrac{1}{3} - \dfrac{\sqrt{3}}{4\pi}$.

44. Skewness. Let X have variance σ^2 and write $m_k = \mathbb{E}(X^k)$. Define the *skewness* of X by $\operatorname{skw}(X) = \mathbb{E}[(X - m_1)^3]/\sigma^3$. Show that:

(a) $\operatorname{skw}(X) = (m_3 - 3m_1m_2 + 2m_1^3)/\sigma^3$,

(b) $\operatorname{skw}(S_n) = \operatorname{skw}(X_1)/\sqrt{n}$, where $S_n = \sum_{r=1}^n X_r$ is a sum of independent identically distributed random variables,

(c) $\operatorname{skw}(X) = (1 - 2p)/\sqrt{npq}$, when X is $\operatorname{bin}(n, p)$ where $p + q = 1$,

(d) $\operatorname{skw}(X) = 1/\sqrt{\lambda}$, when X is Poisson with parameter λ,

(e) $\operatorname{skw}(X) = 2/\sqrt{t}$, when X is gamma $\Gamma(\lambda, t)$, and t is integral.

45. Kurtosis. Let X have variance σ^2 and $\mathbb{E}(X^k) = m_k$. Define the *kurtosis* of X by $\operatorname{kur}(X) = \mathbb{E}[(X - m_1)^4]/\sigma^4$. Show that:

(a) $\operatorname{kur}(X) = 3$, when X is $N(\mu, \sigma^2)$,

(b) $\operatorname{kur}(X) = 9$, when X is exponential with parameter λ,

(c) $\operatorname{kur}(X) = 3 + \lambda^{-1}$, when X is Poisson with parameter λ,

(d) $\operatorname{kur}(S_n) = 3 + \{\operatorname{kur}(X_1) - 3\}/n$, where $S_n = \sum_{r=1}^n X_r$ is a sum of independent identically distributed random variables.

46. Extreme value. Fisher–Gumbel–Tippett distribution. Let X_r, $1 \le r \le n$, be independent and exponentially distributed with parameter 1. Show that $X_{(n)} = \max\{X_r : 1 \le r \le n\}$ satisfies

$$\lim_{n\to\infty} \mathbb{P}(X_{(n)} - \log n \le x) = \exp(-e^{-x}), \qquad -\infty < x < \infty.$$

Hence show that $\int_0^\infty \{1 - \exp(-e^{-x}) - \exp(-e^{x})\}\,dx = \gamma$ where γ is Euler's constant.

47. Squeezing. Let S and X have density functions satisfying $b(x) \le f_S(x) \le a(x)$ and $f_S(x) \le f_X(x)$. Let U be uniformly distributed on $[0, 1]$ and independent of X. Given the value X, we implement the following algorithm:

if $Uf_X(X) > a(X)$, reject X;
otherwise: if $Uf_X(X) < b(X)$, accept X;
otherwise: if $Uf_X(X) \le f_S(X)$, accept X;
otherwise: reject X.

Show that, conditional on ultimate acceptance, X is distributed as S. Explain when you might use this method of sampling.

48. Let X, Y, and $\{U_r : r \ge 1\}$ be independent random variables, where:

$$\mathbb{P}(X = x) = (e - 1)e^{-x}, \ \mathbb{P}(Y = y) = \frac{1}{(e - 1)y!} \text{ for } x, y = 1, 2, \ldots,$$

and the U_r are uniform on $[0, 1]$. Let $M = \max\{U_1, U_2, \dots, U_Y\}$, and show that $Z = X - M$ is exponentially distributed.

49. Let U and V be independent and uniform on $[0, 1]$. Set $X = -\alpha^{-1}\log U$ and $Y = -\log V$ where $\alpha > 0$.

(a) Show that, conditional on the event $Y \geq \frac{1}{2}(X-\alpha)^2$, X has density function $f(x) = \sqrt{2/\pi}e^{-\frac{1}{2}x^2}$ for $x > 0$.

(b) In sampling from the density function f, it is decided to use a rejection method: for given $\alpha > 0$, we sample U and V repeatedly, and we accept X the first time that $Y \geq \frac{1}{2}(X-\alpha)^2$. What is the optimal value of α?

(c) Describe how to use these facts in sampling from the $N(0, 1)$ distribution.

50. Let S be a semicircle of unit radius on a diameter D.

(a) A point P is picked at random on D. If X is the distance from P to S along the perpendicular to D, show $\mathbb{E}(X) = \pi/4$.

(b) A point Q is picked at random on S. If Y is the perpendicular distance from Q to D, show $\mathbb{E}(Y) = 2/\pi$.

51. (Set for the Fellowship examination of St John's College, Cambridge in 1858.) 'A large quantity of pebbles lies scattered uniformly over a circular field; compare the labour of collecting them one by one:

(i) at the centre O of the field,

(ii) at a point A on the circumference.'

To be precise, if L_{O} and L_{A} are the respective labours per stone, show that $\mathbb{E}(L_{\text{O}}) = \frac{2}{3}a$ and $\mathbb{E}(L_{\text{A}}) = 32a/(9\pi)$ for some constant a.

(iii) Suppose you take each pebble to the nearer of two points A or B at the ends of a diameter. Show in this case that the labour per stone satisfies

$$\mathbb{E}(L_{\text{AB}}) = \frac{4a}{3\pi}\left\{\frac{16}{3} - \frac{17}{6}\sqrt{2} + \frac{1}{2}\log(1+\sqrt{2})\right\} \simeq 1.13 \times \frac{2}{3}a.$$

(iv) Finally suppose you take each pebble to the nearest vertex of an equilateral triangle ABC inscribed in the circle. Why is it obvious that the labour per stone now satisfies $\mathbb{E}(L_{\text{ABC}}) < \mathbb{E}(L_{\text{O}})$? Enthusiasts are invited to calculate $\mathbb{E}(L_{\text{ABC}})$.

52. The lines L, M, and N are parallel, and P lies on L. A line picked at random through P meets M at Q. A line picked at random through Q meets N at R. What is the density function of the angle Θ that RP makes with L? [Hint: Recall Exercise (4.8.2) and Problem (4.14.4).]

53. Let Δ denote the event that you can form a triangle with three given parts of a rod R.

(a) R is broken at two points chosen independently and uniformly. Show that $\mathbb{P}(\Delta) = \frac{1}{4}$.

(b) R is broken in two uniformly at random, the longer part is broken in two uniformly at random. Show that $\mathbb{P}(\Delta) = \log(4/e)$.

(c) R is broken in two uniformly at random, a randomly chosen part is broken into two equal parts. Show that $\mathbb{P}(\Delta) = \frac{1}{2}$.

(d) In case (c) show that, given Δ, the triangle is obtuse with probability $3 - 2\sqrt{2}$.

54. You break a rod at random into two pieces. Let R be the ratio of the lengths of the shorter to the longer piece. Find the density function f_R, together with the mean and variance of R.

55. Let R be the distance between two points picked at random inside a square of side a. Show that $\mathbb{E}(R^2) = \frac{1}{3}a^2$, and that R^2/a^2 has density function

$$f(r) = \begin{cases} r - 4\sqrt{r} + \pi & \text{if } 0 \le r \le 1, \\ 4\sqrt{r-1} - 2 - r + 2\sin^{-1}\sqrt{r^{-1}} - 2\sin^{-1}\sqrt{1-r^{-1}} & \text{if } 1 \le r \le 2. \end{cases}$$

56. Show that a sheet of paper of area A cm^2 can be placed on the square lattice with period 1 cm in such a way that at least $\lceil A \rceil$ points are covered.

57. Show that it is possible to position a convex rock of surface area S in sunlight in such a way that its shadow has area at least $\frac{1}{4}S$.

58. Dirichlet distribution. Let $\{X_r : 1 \le r \le k+1\}$ be independent $\Gamma(\lambda, \beta_r)$ random variables (respectively).

(a) Show that $Y_r = X_r/(X_1 + \cdots + X_r)$, $2 \le r \le k+1$, are independent random variables.

(b) Show that $Z_r = X_r/(X_1 + \cdots + X_{k+1})$, $1 \le r \le k$, have the joint *Dirichlet density*

$$\frac{\Gamma(\beta_1 + \cdots + \beta_{k+1})}{\Gamma(\beta_1)\cdots\Gamma(\beta_{k+1})} z_1^{\beta_1 - 1} z_2^{\beta_2 - 1} \cdots z_k^{\beta_k - 1} (1 - z_1 - z_2 - \cdots - z_k)^{\beta_{k+1} - 1}.$$

59. Hotelling's theorem. Let $\mathbf{X}_r = (X_{1r}, X_{2r}, \ldots, X_{mr})$, $1 \le r \le n$, be independent multivariate normal random vectors having zero means and the same covariance matrix $\mathbf{V} = (v_{ij})$. Show that the two random variables

$$S_{ij} = \sum_{r=1}^{n} X_{ir}X_{jr} - \frac{1}{n}\sum_{r=1}^{n} X_{ir} \sum_{r=1}^{n} X_{jr}, \quad T_{ij} = \sum_{r=1}^{n-1} X_{ir}X_{jr},$$

are identically distributed.

60. Choose P, Q, and R independently at random in the square $S(a)$ of side a. Show that $\mathbb{E}|\mathrm{PQR}| = 11a^2/144$. Deduce that four points picked at random in a parallelogram form a convex quadrilateral with probability $(\frac{5}{6})^2$.

61. Choose P, Q, and R uniformly at random within the convex region C illustrated beneath. By considering the event that the convex hull of four randomly chosen points is a triangle, or otherwise, show that the mean area of the shaded region is three times the mean area of the triangle PQR.

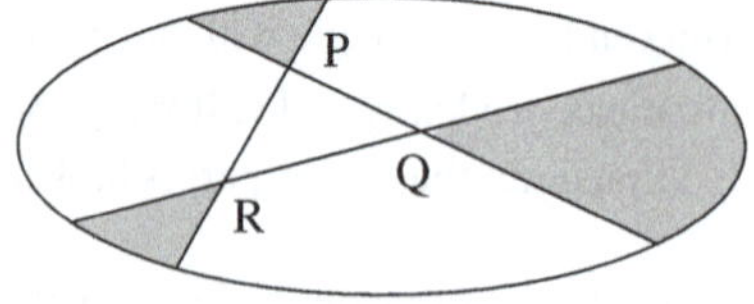

62. Multivariate normal sampling. Let $\mathbf{V}$ be a positive-definite symmetric $n \times n$ matrix, and $\mathbf{L}$ a lower-triangular matrix such that $\mathbf{V} = \mathbf{L}'\mathbf{L}$; this is called the *Cholesky decomposition* of $\mathbf{V}$. Let $\mathbf{X} = (X_1, X_2, \ldots, X_n)$ be a vector of independent random variables distributed as $N(0, 1)$. Show that the vector $\mathbf{Z} = \boldsymbol{\mu} + \mathbf{XL}$ has the multivariate normal distribution with mean vector $\boldsymbol{\mu}$ and covariance matrix $\mathbf{V}$.

63. Verifying matrix multiplications. We need to decide whether or not $\mathbf{AB} = \mathbf{C}$ where $\mathbf{A}, \mathbf{B}, \mathbf{C}$ are given $n \times n$ matrices, and we adopt the following random algorithm. Let $\mathbf{x}$ be a random $\{0, 1\}^n$-valued vector, each of the 2^n possibilities being equally likely. If $(\mathbf{AB} - \mathbf{C})\mathbf{x} = \mathbf{0}$, we decide that $\mathbf{AB} = \mathbf{C}$, and otherwise we decide that $\mathbf{AB} \neq \mathbf{C}$. Show that

$$\mathbb{P}\big(\text{the decision is correct}\big)\begin{cases} = 1 & \text{if } \mathbf{AB} = \mathbf{C}, \\ \geq \frac{1}{2} & \text{if } \mathbf{AB} \neq \mathbf{C}. \end{cases}$$

Describe a similar procedure which results in an error probability which may be made as small as desired.

64. Coupon collecting, Exercise (3.3.2), revisited. Each box of cereal contains a worthless and inedible object. The objects in different boxes are independent and equally likely to be any of the n available types. Let T_n be the number of boxes opened before collection of a full set.

(a) Use the result of Exercise (4.3.6) to show that

$$\frac{1}{n}\mathbb{E}(T_n) = \sum_{r=1}^{n} \frac{(-1)^{r+1}}{r}\binom{n}{r} = \sum_{r=1}^{n} \frac{1}{r}.$$

(b) Prove the above combinatorial identity directly.

(c) By considering $\mathbb{E}(T_n^2)$, show that

$$2n \sum_{r=1}^{n} \frac{(-1)^{r+1}}{r^2}\binom{n}{r} = \sum_{r=1}^{n} \frac{1}{r^2} + \frac{n-1}{n}\sum_{r=1}^{n} \frac{1}{r} + \left(\sum_{r=1}^{n} \frac{1}{r}\right)^2.$$

65. Points P and Q are picked independently and uniformly at random in the triangle ABC, in such a way that the straight line passing through P and Q divides ABC into a triangle T and a quadrilateral R. Show that the ratio of the mean area of T to that of R is 4 : 5.

66. Let X and Y be independent, identically distributed random variables with finite means. Show that

$$\mathbb{E}|X - Y| \leq \mathbb{E}|X + Y|,$$

with equality if and only if their common distribution is symmetric about 0.

67. Malmquist's theorem. Simulated order statistics. Let $U_1, U_2, \dots, U_n$ be independent and identically distributed on $[0, 1]$. Define $X_1, X_2, \dots, X_n$ recursively as follows:

$$X_1 = U_1^{1/n},\ X_2 = X_1 U_2^{1/(n-1)}, \dots,\ X_j = X_{j-1} U_j^{1/(n-j+1)}, \dots,\ X_n = X_{n-1} U_n.$$

Show that the random variables $X_n < X_{n-1} < \cdots < X_1$ have the same joint distribution as the order statistics $U_{(1)}, U_{(2)}, \dots, U_{(n)}$. You may do this in two different ways: first by construction, and second by using the change of variables formula.

68. Semi-moment. The value X of a financial index in one year's time has the normal distribution with mean μ and standard deviation $\sigma > 0$. You possess a derivative contract that will pay you $\max\{X - a, 0\}$, where a is a predetermined constant. Show that the expected payout V of the contract is given by $V = \sigma\{\phi(y) - y(1 - \Phi(y))\}$, where ϕ and Φ are the $N(0, 1)$ density and distribution functions, and $y = (a - \mu)/\sigma$. Deduce that, for large positive y,

$$V \approx \frac{\sigma^3}{(a-\mu)^2}\phi((a-\mu)/\sigma).$$

69. Gauss's inequality. Let g be the density function of a positive, unimodal random variable X with finite variance, with a unique mode at 0 (so that g is non-increasing on $[0, \infty)$, recall the definitions of Problem (2.7.23)). For any x such that $g(x) > 0$, let $y = x + g(x)^{-1}\int_x^\infty g(v)\,dv$.

(a) Prove that, for $0 < x < y < \infty$, we have that $(y - x)x^2 \le \frac{4}{9}\int_0^y v^2\,dv$.

(b) Deduce that

$$\mathbb{P}(|X| > x) \le \frac{4}{9x^2}\mathbb{E}(X^2), \qquad x > 0.$$

(c) Prove the inequality of (b) for a continuous random variable X with a unique mode at 0.

70. Bruss's odds rule. Let I_i be the indicator function of success in the ith of n independent trials. Let $p_i = \mathbb{E}I_i = 1 - q_i > 0$, and let $r_i = p_i/q_i$ be the ith 'odds ratio'. Let $R_k = \sum_{i=k}^n r_i$, and for the moment assume that $p_i < 1$ for all i.

(a) Show that the probability of exactly one success after the kth trial is $\sigma_k = R_{k+1}\prod_{i=k+1}^n q_i$.

(b) Prove that σ_k is a unimodal function of k, in that there is a unique m such that σ_k is greatest either for $k = m$ or for $k \in \{m, m+1\}$.

(c) It is desired to stop the process at the last success. Show that the optimal rule for achieving this is to stop at the first success at time τ or later, where $\tau = \max\{1, \max\{k : R_k \ge 1\}\}$.

(d) What can be said if $p_i = 1$ for some i?

(e) Show that the probability of stopping at the final success is $\sigma_\tau = R_\tau \prod_{i=\tau}^n q_i$.

(f) Use this result to solve the marriage problem (4.14.35).

71. Prophet inequality. Let $X_0, X_1, \dots, X_n$ be non-negative, independent random variables. Their values are revealed to you in order, and you are required to stop the process at some time T, and exit with X_T. Your target is to maximize $\mathbb{E}(X_T)$, and you are permitted to choose any 'stopping strategy' T that depends only on the past and present values $X_0, X_1, \dots, X_T$. (Such a random variable is called a 'stopping time', see Exercise (6.1.6).)

Show that $\mathbb{E}(\max_r X_r) \le 2\sup_T \mathbb{E}(X_T)$, where $\sup_T$ is the supremum over all stopping strategies.

72. Stick breaking. A stick of length s is broken at n points chosen independently and uniformly in the interval $[0, s]$. Let the order statistics of the resulting lengths of sticks be $S_1 < S_2 < \cdots < S_{n+1}$. Fix $y > 0$ and write $p_n(s, y) = \mathbb{P}(S_1 > y)$. Show that

$$p_{n+1}(s, y) = \frac{n+1}{s^{n+1}}\int_0^{s-y} x^n p_n(x, y)\,dx.$$

Deduce that

(i) $p_n(s, y) = s^{-n}\{(s - (n+1)y)^+\}^n$ for $n \ge 1$, where $x^+ = \max\{0, x\}$.

(ii) $\mathbb{E}S_1 = s/(n+1)^2$.

(iii) $\mathbb{E}S_r = s(H_{n+1} - H_{n-r+1})/(n+1)$, where $H_k = \sum_{r=1}^k r^{-1}$.

73. Let $\alpha > -1$, and let X and Y have joint density function $f(x, y) = cx^\alpha$ for $x, y \in (0, 1)$, $x + y > 1$. We set $f(x, y) = 0$ for other pairs x, y.

(a) Find the value of c and the joint distribution function of X, Y.

(b) Show that it is possible to construct a triangle with side-lengths $X, Y, 2 - X - Y$, with probability one.

(c) Show that the angle opposite the side with length Y is obtuse with probability

$$\pi = c\int_0^1 \frac{x^{\alpha+1} - x^{\alpha+2}}{2 - x}\,dx,$$

and find π when $\alpha = 0$.

74. Size-biasing on the freeway. You are driving at constant speed v along a vast multi-lane highway, and other traffic in the same direction is moving at independent random speeds having the same distribution as a random variable X with density function f and finite variance.

(a) Show that the expected value of the speed of vehicles passing you or being passed by you is

$$m(v) = \frac{\mathbb{E}\{X|X - v|\}}{\mathbb{E}|X - v|}.$$

(b) Suppose that f is symmetric about its mean μ, and that $v < \mu$. Show that the difference $m(v) - \mu$ is positive, and that it has a maximum value for some value $v = v_{\max} < \mu$.

(c) What can be said if $v > \mu$? Discuss.

(d) Suppose X is uniformly distributed on the interval $\{60, 61, \dots, 80\}$. Show for $v = 63$ that $m(v) \approx 74$, while for $v = 77$ we have $m(v) \approx 66$.

5

Generating functions and their applications

5.1 Exercises. Generating functions

1. Find the generating functions of the following mass functions, and state where they converge. Hence calculate their means and variances.

(a) $f(m) = \binom{n+m-1}{m} p^n (1-p)^m$, for $m \geq 0$.

(b) $f(m) = \{m(m+1)\}^{-1}$, for $m \geq 1$.

(c) $f(m) = (1-p)p^{|m|}/(1+p)$, for $m = \dots, -1, 0, 1, \dots$.

The constant p satisfies $0 < p < 1$.

2. Let X (≥ 0) have probability generating function G and write $t(n) = \mathbb{P}(X > n)$ for the 'tail' probabilities of X. Show that the generating function of the sequence $\{t(n) : n \geq 0\}$ is $T(s) = (1 - G(s))/(1-s)$. Show that $\mathbb{E}(X) = T(1)$ and $\text{var}(X) = 2T'(1) + T(1) - T(1)^2$.

3. Let $G_{X,Y}(s,t)$ be the joint probability generating function of X and Y. Show that $G_X(s) = G_{X,Y}(s,1)$ and $G_Y(t) = G_{X,Y}(1,t)$. Show that

$$\mathbb{E}(XY) = \left.\frac{\partial^2}{\partial s\, \partial t} G_{X,Y}(s,t)\right|_{s=t=1}.$$

4. Find the joint generating functions of the following joint mass functions, and state for what values of the variables the series converge.

(a) $f(j,k) = (1-\alpha)(\beta - \alpha)\alpha^j \beta^{k-j-1}$, for $0 \leq k \leq j$, where $0 < \alpha < 1$, $\alpha < \beta$.

(b) $f(j,k) = (e-1)e^{-(2k+1)}k^j/j!$, for $j, k \geq 0$.

(c) $f(j,k) = \binom{k}{j} p^{j+k}(1-p)^{k-j} / \left[k \log\{1/(1-p)\}\right]$, for $0 \leq j \leq k$, $k \geq 1$, where $0 < p < 1$.

Deduce the marginal probability generating functions and the covariances.

5. A coin is tossed n times, and heads turns up with probability p on each toss. Assuming the usual independence, show that the joint probability generating function of the numbers H and T of heads and tails is $G_{H,T}(x,y) = \{px + (1-p)y\}^n$. Generalize this conclusion to find the joint probability generating function of the multinomial distribution of Exercise (3.5.1).

6. Let X have the binomial distribution $\text{bin}(n, U)$, where U is uniform on $(0,1)$. Show that X is uniformly distributed on $\{0, 1, 2, \dots, n\}$.

7. Show that

$$G(x,y,z,w) = \tfrac{1}{8}(xyzw + xy + yz + zw + zx + yw + xz + 1)$$

is the joint generating function of four variables that are pairwise and triplewise independent, but are nevertheless *not* independent.

8. Let $p_r > 0$ and $a_r \in \mathbb{R}$ for $1 \le r \le n$. Which of the following is a moment generating function, and for what random variable?

$$\text{(a) } M(t) = 1 + \sum_{r=1}^{n} p_r t^r, \qquad \text{(b) } M(t) = \sum_{r=1}^{n} p_r e^{a_r t}.$$

9. Let G_1 and G_2 be probability generating functions, and suppose that $0 \le \alpha \le 1$. Show that $G_1 G_2$, and $\alpha G_1 + (1-\alpha)G_2$ are probability generating functions. Is $G(\alpha s)/G(\alpha)$ necessarily a probability generating function?

10. Let $X_1, X_2, \ldots$ be independent, continuous random variables with common distribution function F, that are independent of the positive integer-valued random variable Z. Define the maximum $M = \max\{X_1, X_2, \ldots, X_Z\}$. Show that

$$\mathbb{E}(Z \mid M = m) = 1 + \frac{F(m)G''(F(m))}{G'(F(m))},$$

where G is the probability generating function of Z.

11. Truncated geometric distribution. Let $0 < p = 1 - q < 1$, and let X have the geometric mass function $f(x) = q^{x-1}p$ for $x = 1, 2, \ldots$. Find the probability generating function of $Y = \min\{n, X\}$ for fixed $n \ge 1$, and show that $\mathbb{E}(Y) = (1 - q^n)/p$.

12. Van Dantzig's collective marks. Let X_i be the number of balls in the ith of a sequence of bins, and assume the X_i are independent with common probability generating function G_X. There are N such bins, where N is independent of the X_i with probability generating function G_N. Each ball is 'unmarked' with probability u, and marked otherwise, with marks appearing independently.

(a) Show that the probability π, that all the balls in the ith bin are unmarked, satisfies $\pi = G_X(u)$.

(b) Deduce that the probability that all the balls are unmarked equals $G_N(G_X(u))$, and deduce the random sum formula of Theorem (5.1.25).

(c) Find the mean and variance of the total number T of balls in terms of the moments of X and N, and compare the argument with the method of Exercise (3.7.4).

(d) Find the probability generating function of the total number U of unmarked balls, and deduce the mean and variance of U.

5.2 Exercises. Some applications

1. Let X be the number of events in the sequence $A_1, A_2, \ldots, A_n$ which occur. Let $S_m = \mathbb{E}\binom{X}{m}$, the mean value of the random binomial coefficient $\binom{X}{m}$, and show that

$$\mathbb{P}(X \ge i) = \sum_{j=i}^{n} (-1)^{j-i} \binom{j-1}{i-1} S_j, \qquad \text{for } 1 \le i \le n,$$

$$\text{where} \quad S_m = \sum_{j=m}^{n} \binom{j-1}{m-1} \mathbb{P}(X \ge j), \qquad \text{for } 1 \le m \le n.$$

2. Each person in a group of n people chooses another at random. Find the probability:

(a) that exactly k people are chosen by nobody,

(b) that at least k people are chosen by nobody.

3. Compounding.

(a) Let X have the Poisson distribution with parameter Y, where Y has the Poisson distribution with parameter μ. Show that $G_{X+Y}(x) = \exp\{\mu(xe^{x-1} - 1)\}$.

(b) Let $X_1, X_2, \dots$ be independent identically distributed random variables with the *logarithmic* mass function

$$f(k) = \frac{(1-p)^k}{k\log(1/p)}, \qquad k \geq 1,$$

where $0 < p < 1$. If N is independent of the X_i and has the Poisson distribution with parameter μ, show that $Y = \sum_{i=1}^{N} X_i$ has a type of negative binomial distribution.

4. Let X have the binomial distribution with parameters n and p, and show that

$$\mathbb{E}\left(\frac{1}{1+X}\right) = \frac{1-(1-p)^{n+1}}{(n+1)p}.$$

Find the limit of this expression as $n \to \infty$ and $p \to 0$, the limit being taken in such a way that $np \to \lambda$ where $0 < \lambda < \infty$. Comment.

5. A coin is tossed repeatedly, and heads turns up with probability p on each toss. Let h_n be the probability of an even number of heads in the first n tosses, with the convention that 0 is an even number. Find a difference equation for the h_n and deduce that they have generating function $\frac{1}{2}\{(1+2ps-s)^{-1} + (1-s)^{-1}\}$.

6. An unfair coin is flipped repeatedly, where $\mathbb{P}(\mathrm{H}) = p = 1-q$. Let X be the number of flips until HTH first appears, and Y the number of flips until either HTH or THT appears. Show that $\mathbb{E}(s^X) = (p^2qs^3)/(1-s+pqs^2-pq^2s^3)$ and find $\mathbb{E}(s^Y)$.

7. Matching again. The pile of (by now dog-eared) letters is dropped again and enveloped at random, yielding X_n matches. Show that $\mathbb{P}(X_n = j) = (j+1)\mathbb{P}(X_{n+1} = j+1)$. Deduce that the derivatives of the $G_n(s) = \mathbb{E}(s^{X_n})$ satisfy $G'_{n+1} = G_n$, and hence derive the conclusion of Example (3.4.3), namely:

$$\mathbb{P}(X_n = r) = \frac{1}{r!}\left(\frac{1}{2!} - \frac{1}{3!} + \cdots + \frac{(-1)^{n-r}}{(n-r)!}\right).$$

8. Let X have a Poisson distribution with parameter Λ, where Λ is exponential with parameter μ. Show that X has a geometric distribution.

9. Coupons. Recall from Exercise (3.3.2) that each packet of an overpriced commodity contains a worthless plastic object. There are four types of object, and each packet is equally likely to contain any of the four. Let T be the number of packets you open until you first have the complete set. Find $\mathbb{E}(s^T)$ and $\mathbb{P}(T = k)$.

10. Library books. A library permits a reader to hold at most m books at any one time. Your holding before a visit is B books, having the binomial $\text{bin}(m, p)$ distribution. On each visit, a held book is retained with probability r, and returned otherwise; given the number R of retained books, you borrow a further N books with the $\text{bin}(m-R, \alpha)$ distribution. Assuming the usual independence, find the distribution of the number $A = R + N$ of books held after a visit. Show that, after a large number of visits, the number of books held has approximately the $\text{bin}(m, s)$ distribution where $s = \alpha/[1 - r(1-\alpha)]$. You may assume that $0 < \alpha, p, r < 1$.

11. Matching yet again, Exercise (5.2.7) revisited. Each envelope is addressed correctly with probability t, and incorrectly otherwise (assume the usual independence). It is decided that an incorrectly addressed envelope cannot count as a match even when containing the correct letter. Show that the number Y_n of correctly addressed matches satisfies

$$\mathbb{P}(Y_n = y) = \frac{t^y}{y!}\sum_{k=0}^{n-y}\frac{(-1)^k t^k}{k!} \to \frac{e^{-t}t^y}{y!} \qquad \text{as } n \to \infty, \text{ for } y = 0, 1, 2, \dots.$$

Note the Poisson limit distribution.

12. Sampling discrete random variables. A random variable is called *simple* if it may take only finitely many values. Show that a simple random variable X may be simulated using a finite sequence of Bernoulli random variables, in the sense that there exist independent Bernoulli random variables $\{B_j : j = 1, 2, \dots, m\}$ with respective parameters b_j, and real numbers $\{r_j\}$, such that $Y = \sum_j r_j B_j$ has the same distribution as X.

13. General Bonferroni inequalities. (a) For a generating function $A(z) = \sum_{i=0}^{\infty} a_i z^i$, show that $a_0 + a_1 + \cdots + a_m$ is the coefficient of z^m in $B(z) := A(z)(1 - z^{m+1})/(1 - z)$.

(b) With the definitions and notation of Example (5.2.11), show that, for r an even integer,

$$S_i - \binom{i+1}{i} S_{i+1} + \cdots - \binom{i+r+1}{i} S_{i+r+1}$$
$$\le \mathbb{P}(X = i) \le S_i - \binom{i+1}{i} S_{i+1} + \cdots + \binom{i+r}{i} S_{i+r}.$$

5.3 Exercises. Random walk

1. For a simple random walk S with $S_0 = 0$ and $p = 1 - q < \frac{1}{2}$, show that the maximum $M = \max\{S_n : n \ge 0\}$ satisfies $\mathbb{P}(M \ge r) = (p/q)^r$ for $r \ge 0$.

2. Use generating functions to show that, for a symmetric random walk,

(a) $2k f_0(2k) = \mathbb{P}(S_{2k-2} = 0)$ for $k \ge 1$, and

(b) $\mathbb{P}(S_1 S_2 \cdots S_{2n} \ne 0) = \mathbb{P}(S_{2n} = 0)$ for $n \ge 1$.

3. A particle performs a random walk on the corners of the square ABCD. At each step, the probability of moving from corner c to corner d equals ρ_{cd}, where

$$\rho_{\mathrm{AB}} = \rho_{\mathrm{BA}} = \rho_{\mathrm{CD}} = \rho_{\mathrm{DC}} = \alpha, \qquad \rho_{\mathrm{AD}} = \rho_{\mathrm{DA}} = \rho_{\mathrm{BC}} = \rho_{\mathrm{CB}} = \beta,$$

and $\alpha, \beta > 0$, $\alpha + \beta = 1$. Let $G_{\mathrm{A}}(s)$ be the generating function of the sequence $(p_{\mathrm{AA}}(n) : n \ge 0)$, where $p_{\mathrm{AA}}(n)$ is the probability that the particle is at A after n steps, having started at A. Show that

$$G_{\mathrm{A}}(s) = \frac{1}{2}\left\{\frac{1}{1 - s^2} + \frac{1}{1 - |\beta - \alpha|^2 s^2}\right\}.$$

Hence find the probability generating function of the time of the first return to A.

4. A particle performs a symmetric random walk in two dimensions starting at the origin: each step is of unit length and has equal probability $\frac{1}{4}$ of being northwards, southwards, eastwards, or westwards. The particle first reaches the line $x + y = m$ at the point (X, Y) and at the time T. Find the probability generating functions of T and $X - Y$, and state where they converge.

5. Derive the arc sine law for sojourn times, Theorem (3.10.21), using generating functions. That is to say, let L_{2n} be the length of time spent (up to time $2n$) by a simple symmetric random walk to the right of its starting point. Show that

$$\mathbb{P}(L_{2n} = 2k) = \mathbb{P}(S_{2k} = 0)\mathbb{P}(S_{2n-2k} = 0) \qquad \text{for } 0 \le k \le n.$$

6. Let $\{S_n : n \ge 0\}$ be a simple symmetric random walk with $S_0 = 0$, and let $T = \min\{n > 0 : S_n = 0\}$. Show that

$$\mathbb{E}\big(\min\{T, 2m\}\big) = 2\mathbb{E}|S_{2m}| = 4m\mathbb{P}(S_{2m} = 0) \qquad \text{for } m \ge 0.$$

7. Let $S_n = \sum_{r=0}^{n} X_r$ be a left-continuous random walk on the integers with a retaining barrier at zero. More specifically, we assume that the X_r are identically distributed integer-valued random variables with $X_1 \geq -1$, $\mathbb{P}(X_1 = 0) \neq 0$, and

$$S_{n+1} = \begin{cases} S_n + X_{n+1} & \text{if } S_n > 0, \\ S_n + X_{n+1} + 1 & \text{if } S_n = 0. \end{cases}$$

Show that the distribution of S_0 may be chosen in such a way that $\mathbb{E}(z^{S_n}) = \mathbb{E}(z^{S_0})$ for all n, if and only if $\mathbb{E}(X_1) < 0$, and in this case

$$\mathbb{E}(z^{S_n}) = \frac{(1-z)\mathbb{E}(X_1)\mathbb{E}(z^{X_1})}{1-\mathbb{E}(z^{X_1})}.$$

8. Consider a simple random walk starting at 0 in which each step is to the right with probability p $(= 1 - q)$. Let T_b be the number of steps until the walk first reaches b where $b > 0$. Show that $\mathbb{E}(T_b \mid T_b < \infty) = b/|p - q|$.

9. Conditioned random walk. Let $S = \{S_k : k = 0, 1, 2, \dots\}$ be a simple random walk on the non-negative integers, with $S_0 = i$ and an absorbing barrier at 0. A typical jump X has mass function $\mathbb{P}(X = 1) = p$ and $\mathbb{P}(X = -1) = q = 1 - p$, where $p \in (\frac{1}{2}, 1)$. Let H be the event that the walk is ultimately absorbed at 0. Show that, conditional on H, the walk has the same distribution as a simple random walk $W = \{W_k : k = 0, 1, 2, \dots\}$ for which a typical jump Y satisfies $\mathbb{P}(Y = 1) = q$, $\mathbb{P}(Y = -1) = p$.

5.4 Exercises. Branching processes

1. Let Z_n be the size of the nth generation in an ordinary branching process with $Z_0 = 1$, $\mathbb{E}(Z_1) = \mu$, and $\text{var}(Z_1) > 0$. Show that $\mathbb{E}(Z_n Z_m) = \mu^{n-m}\mathbb{E}(Z_m^2)$ for $m \leq n$. Hence find the correlation coefficient $\rho(Z_m, Z_n)$ in terms of μ.

2. Consider a branching process with generation sizes Z_n satisfying $Z_0 = 1$ and $\mathbb{P}(Z_1 = 0) = 0$. Pick two individuals at random (with replacement) from the nth generation and let L be the index of the generation which contains their most recent common ancestor. Show that $\mathbb{P}(L \geq r) \geq \mathbb{E}(Z_r^{-1})$ for $0 \leq r < n$. Show when $r \neq 0$ that equality holds if and only if Z_1 is a.s. constant. What can be said if $\mathbb{P}(Z_1 = 0) > 0$?

3. Consider a branching process whose family sizes have the geometric mass function $f(k) = qp^k$, $k \geq 0$, where $p + q = 1$, and let Z_n be the size of the nth generation. Let $T = \min\{n : Z_n = 0\}$ be the extinction time, and suppose that $Z_0 = 1$. Find $\mathbb{P}(T = n)$. For what values of p is it the case that $\mathbb{E}(T) < \infty$?

4. Let Z_n be the size of the nth generation of a branching process, and assume $Z_0 = 1$. Find an expression for the generating function G_n of Z_n, in the cases when Z_1 has generating function:

(a) $G(s) = 1 - \alpha(1 - s)^\beta$, $0 < \alpha, \beta < 1$.

(b) $G(s) = f^{-1}\{P(f(s))\}$, where P is a probability generating function, and f is a suitable function satisfying $f(1) = 1$.

(c) Suppose in the latter case that $f(x) = x^m$ and $P(s) = s\{\gamma - (\gamma - 1)s\}^{-1}$ where $\gamma > 1$. Calculate the answer explicitly.

5. Branching with immigration. Each generation of a branching process (with a single progenitor) is augmented by a random number of immigrants who are indistinguishable from the other members of the population. Suppose that the numbers of immigrants in different generations are independent of each other and of the past history of the branching process, each such number having probability generating function $H(s)$. Show that the probability generating function G_n of the size of the nth generation satisfies $G_{n+1}(s) = G_n(G(s))H(s)$, where G is the probability generating function of a typical family of offspring.

6. Let Z_n be the size of the nth generation in a branching process with $\mathbb{E}(s^{Z_1}) = (2-s)^{-1}$ and $Z_0 = 1$. Let V_r be the total number of generations of size r. Show that $\mathbb{E}(V_1) = \frac{1}{6}\pi^2$, and $\mathbb{E}(2V_2 - V_3) = \frac{1}{6}\pi^2 - \frac{1}{90}\pi^4$.

7. Let T be the total number of individuals in a branching process with family-size distribution $\text{bin}(2, p)$, where $p \neq \frac{1}{2}$. Show that

$$\mathbb{E}(T \mid T < \infty) = \frac{1}{|2p-1|}.$$

What is the family-size distribution conditional on $T < \infty$?

8. Let Z be a branching process with $Z_0 = 1$, $\mathbb{E}(Z_1) = \mu > 1$, and $\text{var}(Z_1) = \sigma^2$. Use the Paley–Zygmund inequality to show that the extinction probability $\eta_n = \mathbb{P}(Z_n = 0)$ satisfies

$$\eta_n \le \frac{\sigma^2}{\mu(\mu-1)}(1 - \mu^{-n}).$$

5.5 Exercises. Age-dependent branching processes

1. Let $Z(t)$ be the population-size at time t in an age-dependent branching process, the lifetime distribution of which is exponential with parameter λ. If $Z(0) = 1$, show that the probability generating function $G_t(s)$ of $Z(t)$ satisfies

$$\frac{\partial}{\partial t} G_t(s) = \lambda\big\{G(G_t(s)) - G_t(s)\big\},$$

where G is the probability generating function of a typical family-size. Show in the case of 'exponential binary fission', when $G(s) = s^2$, that

$$G_t(s) = \frac{se^{-\lambda t}}{1 - s(1 - e^{-\lambda t})}$$

and hence derive the probability mass function of the population size $Z(t)$ at time t.

2. Solve the differential equation of Exercise (5.5.1) when $\lambda = 1$ and $G(s) = \frac{1}{2}(1 + s^2)$, to obtain

$$G_t(s) = \frac{2s + t(1-s)}{2 + t(1-s)}.$$

Hence find $\mathbb{P}(Z(t) \ge k)$, and deduce that

$$\mathbb{P}\big(Z(t)/t \ge x \,\big|\, Z(t) > 0\big) \to e^{-2x} \quad \text{as } t \to \infty.$$

5.6 Exercises. Expectation revisited

1. (a) **Jensen's inequality.** A function $u : \mathbb{R} \to \mathbb{R}$ is called *convex* if, for $a \in \mathbb{R}$, there exists $\lambda = \lambda(a)$ such that $u(x) \geq u(a) + \lambda(x - a)$ for all x. Draw a diagram to illustrate this definition†. The convex function u is called *strictly convex* if $\lambda(a)$ is strictly increasing in a.

(i) Show that, if u is convex and X is a random variable with finite mean, then $\mathbb{E}(u(X)) \geq u(\mathbb{E}X)$.

(ii) Show further that, if u is strictly convex and $\mathbb{E}(u(X)) = u(\mathbb{E}X)$, then X is a.s. constant.

(b) The *entropy* of a probability density function f is defined by $H(f) = -\int_{\mathbb{R}} f(x)\log f(x)\,dx$, and the *support* of f is $S(f) = \{x \in \mathbb{R} : f(x) > 0\}$. Show that, among density functions with support $\mathbb{R}$, and with finite mean μ and variance $\sigma^2 > 0$, the normal $N(\mu, \sigma^2)$ density function, and no other, has maximal entropy.

2. Let $X_1, X_2, \dots$ be random variables satisfying $\mathbb{E}\big(\sum_{i=1}^{\infty} |X_i|\big) < \infty$. Show that

$$\mathbb{E}\left(\sum_{i=1}^{\infty} X_i\right) = \sum_{i=1}^{\infty} \mathbb{E}(X_i).$$

3. Let $\{X_n\}$ be a sequence of random variables satisfying $X_n \leq Y$ a.s. for some Y with $\mathbb{E}|Y| < \infty$. Show that

$$\mathbb{E}\left(\limsup_{n\to\infty} X_n\right) \geq \limsup_{n\to\infty} \mathbb{E}(X_n).$$

4. Suppose that $\mathbb{E}|X^r| < \infty$ where $r > 0$. Deduce that $x^r\mathbb{P}(|X| \geq x) \to 0$ as $x \to \infty$. Conversely, suppose that $x^r\mathbb{P}(|X| \geq x) \to 0$ as $x \to \infty$ where $r \geq 0$, and show that $\mathbb{E}|X^s| < \infty$ for $0 \leq s < r$.

5. Show that $\mathbb{E}|X| < \infty$ if and only if the following holds: for all $\epsilon > 0$, there exists $\delta > 0$, such that $\mathbb{E}(|X|I_A) < \epsilon$ for all A such that $\mathbb{P}(A) < \delta$.

6. Let $M = \max\{X, Y\}$ where X, Y have some joint distribution. Show that $\mathrm{var}(M) \leq \mathrm{var}(X) + \mathrm{var}(Y)$.

7. Let $A_1, A_2, \dots, A_n$ be events, and let S be the number of them which occur. Show that

$$\mathbb{P}(S > 0) \geq \sum_{r=1}^{n} \frac{\mathbb{P}(I_r = 1)}{\mathbb{E}(S \mid I_r = 1)},$$

where I_r is the indicator function of A_r.

5.7 Exercises. Characteristic functions

1. Find two dependent random variables X and Y such that $\phi_{X+Y}(t) = \phi_X(t)\phi_Y(t)$ for all t.

2. If ϕ is a characteristic function, show that $\mathrm{Re}\{1 - \phi(t)\} \geq \frac{1}{4}\mathrm{Re}\{1 - \phi(2t)\}$, and deduce that $1 - |\phi(2t)| \leq 8\{1 - |\phi(t)|\}$.

3. The **cumulant generating function** $K_X(\theta)$ of the random variable X is defined by $K_X(\theta) = \log \mathbb{E}(e^{\theta X})$, the logarithm of the moment generating function of X. If the latter is finite in a neighbourhood of the origin, then K_X has a convergent Taylor expansion:

$$K_X(\theta) = \sum_{n=1}^{\infty} \frac{1}{n!} k_n(X)\theta^n$$

†There is room for debate about the 'right' definition of a convex function. We adopt the above definition since it is convenient for our uses, and is equivalent to the more usual one.

and $k_n(X)$ is called the nth *cumulant* (or *semi-invariant*) of X.

(a) Express $k_1(X)$, $k_2(X)$, and $k_3(X)$ in terms of the moments of X.

(b) If X and Y are independent random variables, show that $k_n(X+Y) = k_n(X) + k_n(Y)$.

4. Let X be $N(0,1)$, and show that the cumulants of X are $k_2(X) = 1$, $k_m(X) = 0$ for $m \neq 2$.

5. The random variable X is said to have a *lattice distribution* if there exist a and b such that X takes values in the set $L(a,b) = \{a + bm : m = 0, \pm 1, \dots\}$. The *span* of such a variable X is the maximal value of b for which there exists a such that X takes values in $L(a,b)$.

(a) Suppose that X has a lattice distribution with span b. Show that $|\phi_X(2\pi/b)| = 1$, and that $|\phi_X(t)| < 1$ for $0 < t < 2\pi/b$.

(b) Suppose that $|\phi_X(\theta)| = 1$ for some $\theta \neq 0$. Show that X has a lattice distribution with span $2\pi k/\theta$ for some integer k.

6. Let X be a random variable with density function f. Show that $|\phi_X(t)| \to 0$ as $t \to \pm\infty$.

7. Let $X_1, X_2, \dots, X_n$ be independent variables, X_i being $N(\mu_i, 1)$, and let $Y = X_1^2 + X_2^2 + \cdots + X_n^2$. Show that the characteristic function of Y is

$$\phi_Y(t) = \frac{1}{(1-2it)^{n/2}} \exp\left(\frac{it\theta}{1-2it}\right)$$

where $\theta = \mu_1^2 + \mu_2^2 + \cdots + \mu_n^2$. The random variables Y is said to have the *non-central chi-squared distribution* with n degrees of freedom and non-centrality parameter θ, written $\chi^2(n;\theta)$.

8. Let X be $N(\mu, 1)$ and let Y be $\chi^2(n)$, and suppose that X and Y are independent. The random variable $T = X/\sqrt{Y/n}$ is said to have the *non-central t-distribution* with n degrees of freedom and non-centrality parameter μ. If U and V are independent, U being $\chi^2(m;\theta)$ and V being $\chi^2(n)$, then $F = (U/m)/(V/n)$ is said to have the *non-central F-distribution* with m and n degrees of freedom and non-centrality parameter θ, written $F(m,n;\theta)$.

(a) Show that T^2 is $F(1,n;\mu^2)$.

(b) Show that

$$\mathbb{E}(F) = \frac{n(m+\theta)}{m(n-2)} \quad \text{if } n > 2.$$

9. Let X be a random variable with density function f and characteristic function ϕ. Show, subject to an appropriate condition on f, that

$$\int_{-\infty}^{\infty} f(x)^2\,dx = \frac{1}{2\pi}\int_{-\infty}^{\infty} |\phi(t)|^2\,dt.$$

10. If X and Y are continuous random variables, show that

$$\int_{-\infty}^{\infty} \phi_X(y) f_Y(y) e^{-ity}\,dy = \int_{-\infty}^{\infty} \phi_Y(x-t) f_X(x)\,dx.$$

11. Tilted distributions. (a) Let X have distribution function F and let τ be such that $M(\tau) = \mathbb{E}(e^{\tau X}) < \infty$. Show that $F_\tau(x) = M(\tau)^{-1}\int_{-\infty}^{x} e^{\tau y}\,dF(y)$ is a distribution function, called a 'tilted distribution' of X, and find its moment generating function.

(b) Suppose X and Y are independent and $\mathbb{E}(e^{\tau X})$, $\mathbb{E}(e^{\tau Y}) < \infty$. Find the moment generating function of the tilted distribution of $X+Y$ in terms of those of X and Y.

12. Let X and Y be independent with the distributions $N(\mu, \sigma^2)$ and $N(0, \sigma^2)$, where $\sigma^2 > 0$. Show that $R = \sqrt{X^2 + Y^2}$ has density function

$$f(r) = \frac{r}{\pi\sigma^2} \exp\left\{-\frac{\mu^2 + r^2}{2\sigma^2}\right\} \int_0^\pi \exp\left\{\frac{r\mu\cos\theta}{\sigma^2}\right\} d\theta, \qquad r > 0.$$

The integral may be expressed in terms of a modified Bessel function.

13. Joint moment generating function. For each of the following joint density functions of the pair (X, Y), find the joint moment generating function $M(s, t) = \mathbb{E}(e^{sX+tY})$, and hence find $\operatorname{cov}(X, Y)$.

(a) We have that $f(x, y) = 2e^{-x-y}$ for $0 < x < y < \infty$.

(b) We have that

$$f(x, y) = \frac{x^2 + y^2 + c^2}{2\pi(2 + c^2)} \exp\{-\tfrac{1}{2}(x^2 + y^2)\}, \qquad x, y \in \mathbb{R}.$$

5.8 Exercises. Examples of characteristic functions

1. If ϕ is a characteristic function, show that $\overline{\phi}$, ϕ^2, $|\phi|^2$, $\operatorname{Re}(\phi)$ are characteristic functions. Show that $|\phi|$ is not necessarily a characteristic function.

2. Show that

$$\mathbb{P}(X \geq x) \leq \inf_{t \geq 0}\{e^{-tx} M_X(t)\},$$

where M_X is the moment generating function of X. Deduce that, if X has the $N(0, 1)$ distribution,

$$\mathbb{P}(X \geq x) \leq e^{-\frac{1}{2}x^2}, \qquad x > 0.$$

3. Let X have the $\Gamma(\lambda, m)$ distribution and let Y be independent of X with the beta distribution with parameters n and $m - n$, where m and n are non-negative integers satisfying $n \leq m$. Show that $Z = XY$ has the $\Gamma(\lambda, n)$ distribution.

4. Find the characteristic function of X^2 when X has the $N(\mu, \sigma^2)$ distribution.

5. Let $X_1, X_2, \dots$ be independent $N(0, 1)$ variables. Use characteristic functions to find the distribution of: (a) X_1^2, (b) $\sum_{i=1}^n X_i^2$, (c) X_1/X_2, (d) X_1X_2, (e) $X_1X_2 + X_3X_4$.

6. Let $X_1, X_2, \dots, X_n$ be such that, for all $a_1, a_2, \dots, a_n \in \mathbb{R}$, the linear combination $a_1X_1 + a_2X_2 + \cdots + a_nX_n$ has a normal distribution. Show that the joint characteristic function of the X_m is $\exp(i\mathbf{t}\boldsymbol{\mu}' - \frac{1}{2}\mathbf{t}\mathbf{V}\mathbf{t}')$, for an appropriate vector $\boldsymbol{\mu}$ and matrix $\mathbf{V}$. Deduce that the vector $(X_1, X_2, \dots, X_n)$ has a multivariate normal *density function* so long as $\mathbf{V}$ is invertible.

7. Let X and Y be independent $N(0, 1)$ variables, and let U and V be independent of X and Y. Show that $Z = (UX + VY)/\sqrt{U^2 + V^2}$ has the $N(0, 1)$ distribution. Formulate an extension of this result to cover the case when X and Y have a bivariate normal distribution with zero means, unit variances, and correlation ρ.

8. Let X be exponentially distributed with parameter λ. Show by elementary integration that $\mathbb{E}(e^{itX}) = \lambda/(\lambda - it)$.

9. Find the characteristic functions of the following density functions:

(a) $f(x) = \frac{1}{2}e^{-|x|}$ for $x \in \mathbb{R}$,

(b) $f(x) = \frac{1}{2}|x|e^{-|x|}$ for $x \in \mathbb{R}$.

10. Ulam's redistribution of energy. Is it possible for X, Y, and Z to have the same distribution and satisfy $X = U(Y + Z)$, where U is uniform on $[0, 1]$, and Y, Z are independent of U and of one another? (This question arises in modelling energy redistribution among physical particles.)

11. Find the joint characteristic function of two random variables having a bivariate normal distribution with zero means. (No integration is needed.)

12. Sampling from the normal distribution. Let $X_1, X_2, \dots, X_n$ be independent random variables with the $N(\mu, \sigma^2)$ distribution, where $\sigma > 0$. Let

$$\overline{X} = \frac{1}{n}\sum_{i=1}^{n} X_i, \qquad S^2 = \frac{1}{n-1}\sum_{i=1}^{n}\left(\frac{X_i - \overline{X}}{\sigma}\right)^2,$$

be the sample mean and variance. Show that $\operatorname{cov}(\overline{X}, X_i - \overline{X}) = 0$, and deduce that $\overline{X}$ and S^2 are independent.

Show that

$$\frac{(n-1)S^2}{\sigma^2} + \frac{n}{\sigma^2}(\overline{X} - \mu)^2 = \sum_{i=1}^{n}\left(\frac{X_i - \mu}{\sigma}\right)^2,$$

and use characteristic functions to prove that $(n-1)S^2/\sigma^2$ has the $\chi^2(n-1)$ distribution.

13. Isserlis's theorem. Let the vector $(X_1, X_2, \dots, X_n)$ have a multivariate normal distribution with zero means. Show that, for n odd, $\mathbb{E}(X_1X_2\cdots X_n) = 0$, while for $n = 2m$,

$$\mathbb{E}(X_1X_2\cdots X_n) = \sum_r \prod_{i_r<j_r} \mathbb{E}(X_{i_r}X_{j_r}),$$

that is, the sum over all products of expectations of m distinct pairs of variables.

14. Hypoexponential distribution, Exercise (4.8.4) revisited. Let $X_1, X_2, \dots, X_n$ be independent random variables with, respectively, the exponential distribution with parameter λ_r for $r = 1, 2, \dots, n$. Use moment generating functions to find the density function of the sum $S = X_1 + X_2 + \cdots + X_n$. Deduce that, for any set $\{\lambda_r : 1 \le r \le n\}$ of distinct positive numbers,

$$\sum_{r=1}^{n}\prod_{\substack{s=1\\ s\ne r}}^{n}\frac{\lambda_s}{\lambda_s - \lambda_r} = 1, \qquad \sum_{r=1}^{n}\frac{1}{\lambda_r}\prod_{\substack{s=1\\ s\ne r}}^{n}\frac{\lambda_s}{\lambda_s - \lambda_r} = \sum_{r=1}^{n}\frac{1}{\lambda_r}.$$

15. Let X have the $N(0, 1)$ distribution, and let $f : \mathbb{R} \to \mathbb{R}$ be sufficiently smooth. Show that

$$\mathbb{E}(e^{\theta X}f(X)) = e^{\frac{1}{2}\theta^2}\mathbb{E}(f(X + \theta)),$$

and deduce that $\mathbb{E}(Xf(X)) = \mathbb{E}(f'(X))$.

16. Normal characteristic function. Find the characteristic function of the $N(0, 1)$ distribution without using the methods of complex analysis. [Hint: Consider the derivative of the characteristic function, and use an appropriate theorem to differentiate through the integral.]

17. Size-biased distribution. Let X be a non-negative random variable with $0 < \mu = \mathbb{E}(X) < \infty$. The random variable Y is said to have the *size-biased* X distribution if $dF_Y(x) = (x/\mu)dF_X(x)$ for all x. (You may think of this as saying that $f_Y(x) \propto xf_X(x)$ if X is either discrete with mass function f_X, or continuous with density function f_X.)

Show that:

(a) the characteristic functions satisfy $\phi_Y(t) = \phi'_X(t)/(i\mu)$,

(b) when X has the $\Gamma(\lambda, r)$ distribution, Y has the $\Gamma(\lambda, r+1)$ distribution,

(c) the non-negative integer-valued random variable X has the Poisson distribution with parameter λ if and only if Y is distributed as $X+1$.

What is the size-biased X distribution when X has the binomial bin(n, p) distribution?

5.9 Exercises. Inversion and continuity theorems

1. Let X_n be a discrete random variable taking values in $\{1, 2, \dots, n\}$, each possible value having probability n^{-1}. Show that, as $n \to \infty$, $\mathbb{P}(n^{-1}X_n \le y) \to y$, for $0 \le y \le 1$.

2. Let X_n have distribution function

$$F_n(x) = x - \frac{\sin(2n\pi x)}{2n\pi}, \qquad 0 \le x \le 1.$$

(a) Show that F_n is indeed a distribution function, and that X_n has a density function.

(b) Show that, as $n \to \infty$, F_n converges to the uniform distribution function, but that the density function of F_n does not converge to the uniform density function.

3. A coin is tossed repeatedly, with heads turning up with probability p on each toss. Let N be the minimum number of tosses required to obtain k heads. Show that, as $p \downarrow 0$, the distribution function of $2Np$ converges to that of a gamma distribution.

4. If X is an integer-valued random variable with characteristic function ϕ, show that

$$\mathbb{P}(X = k) = \frac{1}{2\pi}\int_{-\pi}^{\pi} e^{-itk}\phi(t)\,dt.$$

What is the corresponding result for a random variable whose distribution is arithmetic with span λ (that is, there is probability one that X is a multiple of λ, and λ is the largest positive number with this property)?

5. Use the inversion theorem to show that

$$\int_{-\infty}^{\infty} \frac{\sin(at)\sin(bt)}{t^2}\,dt = \pi\min\{a, b\}.$$

6. Stirling's formula. Let $f_n(x)$ be a differentiable function on $\mathbb{R}$ with a a global maximum at $a > 0$, and such that $\int_0^\infty \exp\{f_n(x)\}\,dx < \infty$. Laplace's method of steepest descent (related to Watson's lemma and saddlepoint methods) asserts under mild conditions that

$$\int_0^\infty \exp\{f_n(x)\}\,dx \sim \int_0^\infty \exp\{f_n(a) + \tfrac{1}{2}(x-a)^2 f_n''(a)\}\,dx \quad \text{as } n \to \infty.$$

By setting $f_n(x) = n\log x - x$, prove Stirling's formula: $n! \sim n^n e^{-n}\sqrt{2\pi n}$.

7. Let $\mathbf{X} = (X_1, X_2, \dots, X_n)$ have the multivariate normal distribution with zero means, and covariance matrix $\mathbf{V} = (v_{ij})$ satisfying $|\mathbf{V}| > 0$ and $v_{ij} > 0$ for all i, j. Show that

$$\frac{\partial f}{\partial v_{ij}} = \begin{cases} \dfrac{\partial^2 f}{\partial x_i \partial x_j} & \text{if } i \ne j, \\[2ex] \dfrac{1}{2}\dfrac{\partial^2 f}{\partial x_i^2} & \text{if } i = j, \end{cases}$$

and deduce that $\mathbb{P}(\max_{k\le n} X_k \le u) \ge \prod_{k=1}^n \mathbb{P}(X_k \le u)$.

8. Let X_1, X_2 have a bivariate normal distribution with zero means, unit variances, and correlation ρ. Use the inversion theorem to show that

$$\frac{\partial}{\partial \rho}\mathbb{P}(X_1 > 0,\ X_2 > 0) = \frac{1}{2\pi\sqrt{1-\rho^2}}.$$

Hence find $\mathbb{P}(X_1 > 0,\ X_2 > 0)$.

9. (a) Let $X_1, X_2, \dots$ be independent, identically distributed random variables with characteristic function satisfying $\phi(t) = 1 - c|t| + \mathrm{o}(t)$ as $t \to 0$, where $c > 0$. Show that, as $n \to \infty$, the distribution of $Y_n = (X_1 + X_2 + \cdots + X_n)/(cn)$ converges to the Cauchy distribution.

(b) Let U have the uniform distribution on $[-1, 1]$, and show that

$$\phi_{1/U}(t) = 1 - |t| \int_{|t|}^{\infty} \frac{1-\cos x}{x^2}\,dx.$$

When $U_1, U_2, \dots, U_n$ are independent and distributed as U, write down the limiting distribution as $n \to \infty$ of $Y_n = (2/(n\pi))\sum_{r=1}^n U_r^{-1}$.

5.10 Exercises. Two limit theorems

1. Prove that, for $x \ge 0$, as $n \to \infty$,

(a)
$$\sum_{\substack{k:\\ |k-\frac{1}{2}n|\le\frac{1}{2}x\sqrt{n}}} \binom{n}{k} \sim 2^n \int_{-x}^{x} \frac{1}{\sqrt{2\pi}} e^{-\frac{1}{2}u^2}\,du,$$

(b)
$$\sum_{\substack{k:\\ |k-n|\le x\sqrt{n}}} \frac{n^k}{k!} \sim e^n \int_{-x}^{x} \frac{1}{\sqrt{2\pi}} e^{-\frac{1}{2}u^2}\,du.$$

2. It is well known that infants born to mothers who smoke tend to be small and prone to a range of ailments. It is conjectured that also they look abnormal. Nurses were shown selections of photographs of babies, one half of whom had smokers as mothers; the nurses were asked to judge from a baby's appearance whether or not the mother smoked. In 1500 trials the correct answer was given 910 times. Is the conjecture plausible? If so, why?

3. Let X have the $\Gamma(1, s)$ distribution; given that $X = x$, let Y have the Poisson distribution with parameter x. Find the characteristic function of Y, and show that

$$\frac{Y - \mathbb{E}(Y)}{\sqrt{\mathrm{var}(Y)}} \xrightarrow{\mathrm{D}} N(0, 1) \qquad \text{as } s \to \infty.$$

Explain the connection with the central limit theorem.

4. Let $X_1, X_2, \dots$ be independent random variables taking values in the positive integers, whose common distribution is non-arithmetic, in that $\gcd\{n : \mathbb{P}(X_1 = n) > 0\} = 1$. Prove that, for all integers x, there exist non-negative integers $r = r(x)$, $s = s(x)$, such that

$$\mathbb{P}\big(X_1 + \cdots + X_r - X_{r+1} - \cdots - X_{r+s} = x\big) > 0.$$

5. Prove the local central limit theorem for sums of random variables taking integer values. You may assume for simplicity that the summands have span 1, in that $\gcd\{|x| : \mathbb{P}(X = x) > 0\} = 1$.

6. Let $X_1, X_2, \dots$ be independent random variables having common density function $f(x) = 1/\{2|x|(\log|x|)^2\}$ for $|x| < e^{-1}$. Show that the X_i have zero mean and finite variance, and that the density function f_n of $X_1 + X_2 + \cdots + X_n$ satisfies $f_n(x) \to \infty$ as $x \to 0$. Deduce that the X_i do not satisfy the local limit theorem.

7. First-passage density. Let X have the density function $f(x) = \sqrt{2\pi x^{-3}}\exp(-\{2x\}^{-1})$, $x > 0$. Show that $\phi(is) = \mathbb{E}(e^{-sX}) = e^{-\sqrt{2s}}$, $s > 0$, and deduce that X has characteristic function

$$\phi(t) = \begin{cases} \exp\{-(1-i)\sqrt{t}\} & \text{if } t \geq 0, \\ \exp\{-(1+i)\sqrt{|t|}\} & \text{if } t \leq 0. \end{cases}$$

[Hint: Use the result of Problem (5.12.18). This distribution is called the *Lévy distribution*.]

8. Let $\{X_r : r \geq 1\}$ be independent with the distribution of the preceding Exercise (5.10.7). Let $U_n = n^{-1}\sum_{r=1}^{n} X_r$, and $T_n = n^{-1}U_n$. Show that:

(a) $\mathbb{P}(U_n < c) \to 0$ for any $c < \infty$,

(b) T_n has the same distribution as X_1.

9. A sequence of biased coins is flipped; the chance that the rth coin shows a head is Θ_r, where Θ_r is a random variable taking values in $(0, 1)$. Let X_n be the number of heads after n flips. Does X_n obey the central limit theorem when:

(a) the Θ_r are independent and identically distributed?

(b) $\Theta_r = \Theta$ for all r, where Θ is a random variable taking values in $(0, 1)$?

10. Elections. In an election with two candidates, each of the v voters is equally likely to vote for either candidate, and they vote independently of one another. Show that, when v is large, the probability that the winner was ahead when λv votes had been counted (where $\lambda \in (0, 1)$) is approximately $\frac{1}{2} + (1/\pi)\sin^{-1}\sqrt{\lambda}$.

11. Stirling's formula again. By considering the central limit theorem for the sum of independent Poisson-distributed random variables, show that

$$\frac{\sqrt{n}e^{-n}n^n}{n!} \to \frac{1}{\sqrt{2\pi}} \qquad \text{as } n \to \infty.$$

12. Size-biased distribution. Let f be a mass function on $\{1, 2, \dots\}$ with finite mean μ. Consider a large population of households with independent sizes distributed as f. An individual A is selected uniformly at random from the entire population. Show that the probability A belongs to a household of size x is approximately $xf(x)/\mu$. Find the probability generating function of this distribution in terms of that of f.

13. More size-biasing. Let $X_1, X_2, \dots$ be independent, strictly positive continuous random variables with common density function f_X and finite mean, and let $Z_r = X_r/S_n$ where $S_k = \sum_{r=1}^{k} X_r$. The Z_r give rise to a partition of $(0, 1]$ into the intervals $I_r = (S_{r-1}, S_r]/S_n$ for $r = 1, 2, \dots, n$. Let U be uniform on $(0, 1]$ and independent of the X_r, and let L be the length of the interval into which U falls. Show that L has density function $f_L(z) = (z/\mathbb{E}Z)f_Z(z)$, where $Z = Z_1$. Find the characteristic function of L in terms of that of Z.

14. Area process for random walk. Let $S_n = \sum_{r=1}^{n} X_r$ be a continuous, symmetric random walk on $\mathbb{R}$ started at 0, whose jumps are independent $N(0, 1)$ random variables X_i. Define the *area process* $A_n = \sum_{m=1}^{n} S_m$. Let $I_m = \{-1 < A_m < 1\}$, and show that, with probability 1, only finitely many of the I_m occur.

15. Outguessing machines. At Bell Labs in the early 1950s, David Hagelbarger and Claude Shannon built machines to predict whether a human coin-flipper would call heads or tails. Hagelbarger's machine (a "sequence extrapolating robot", or "SEER") was correct on 5218 trials of 9795, and Shannon's machine (a "mind-reading (?) machine") was correct on 5010 trials of 8517. In each case, what is the probability of doing at least as well by chance?

[When playing against his own machine, Shannon could beat it in around 60% of trials, in the long run.]

5.11 Exercises. Large deviations

1. A fair coin is tossed n times, showing heads H_n times and tails T_n times. Let $S_n = H_n - T_n$. Show that

$$\mathbb{P}(S_n > an)^{1/n} \to \frac{1}{\sqrt{(1+a)^{1+a}(1-a)^{1-a}}} \qquad \text{if } 0 < a < 1.$$

What happens if $a \geq 1$?

2. Show that

$$T_n^{1/n} \to \frac{4}{\sqrt{(1+a)^{1+a}(1-a)^{1-a}}}$$

as $n \to \infty$, where $0 < a < 1$ and

$$T_n = \sum_{\substack{k: \\ |k-\frac{1}{2}n|>\frac{1}{2}an}} \binom{n}{k}.$$

Find the asymptotic behaviour of $T_n^{1/n}$ where

$$T_n = \sum_{\substack{k: \\ k>n(1+a)}} \frac{n^k}{k!}, \qquad \text{where } a > 0.$$

3. Show that the moment generating function of X is finite in a neighbourhood of the origin if and only if X has exponentially decaying tails, in the sense that there exist positive constants λ and μ such that $\mathbb{P}(|X| \geq a) \leq \mu e^{-\lambda a}$ for $a > 0$. [Seen in the light of this observation, the condition of the large deviation theorem (5.11.4) is very natural].

4. Let $X_1, X_2, \dots$ be independent random variables having the Cauchy distribution, and let $S_n = X_1 + X_2 + \cdots + X_n$. Find $\mathbb{P}(S_n > an)$.

5. Chernoff inequality for Bernoulli trials. Let $S = \sum_{r=1}^{n} X_r$ be a sum of independent Bernoulli random variables X_r taking values in $\{0, 1\}$, where $\mathbb{E}(X_r) = p_r$ and $\mathbb{E}(S) = \mu > 0$. Show that

$$\mathbb{P}\big(S > (1+\epsilon)\mu\big) \leq \exp\big\{-\mu\big[(1+\epsilon)\log(1+\epsilon) - \epsilon\big]\big\}, \qquad \epsilon > 0.$$

5.12 Problems

1. A die is thrown ten times. What is the probability that the sum of the scores is 27?

2. A coin is tossed repeatedly, heads appearing with probability p on each toss.

(a) Let X be the number of tosses until the first occasion by which three heads have appeared successively. Write down a difference equation for $f(k) = \mathbb{P}(X = k)$ and solve it. Now write down an equation for $\mathbb{E}(X)$ using conditional expectation. (Try the same thing for the first occurrence of HTH).

(b) Let N be the number of heads in n tosses of the coin. Write down $G_N(s)$. Hence find the probability that: (i) N is divisible by 2, (ii) N is divisible by 3.

3. A coin is tossed repeatedly, heads occurring on each toss with probability p. Find the probability generating function of the number T of tosses before a run of n heads has appeared for the first time.

4. Find the generating function of the negative binomial mass function

$$f(k) = \binom{k-1}{r-1} p^r (1-p)^{k-r}, \qquad k = r, r+1, \dots,$$

where $0 < p < 1$ and r is a positive integer. Deduce the mean and variance.

5. For the simple random walk, show that the probability $p_0(2n)$ that the particle returns to the origin at the $(2n)$th step satisfies $p_0(2n) \sim (4pq)^n/\sqrt{\pi n}$, and use this to prove that the walk is recurrent if and only if $p = \frac{1}{2}$. You will need Stirling's formula: $n! \sim n^{n+\frac{1}{2}} e^{-n}\sqrt{2\pi}$.

6. A symmetric random walk in two dimensions is defined to be a sequence of points $\{(X_n, Y_n) : n \ge 0\}$ which evolves in the following way: if $(X_n, Y_n) = (x, y)$ then (X_{n+1}, Y_{n+1}) is one of the four points $(x \pm 1, y)$, $(x, y \pm 1)$, each being picked with equal probability $\frac{1}{4}$. If $(X_0, Y_0) = (0, 0)$:

(a) show that $\mathbb{E}(X_n^2 + Y_n^2) = n$,

(b) find the probability $p_0(2n)$ that the particle is at the origin after the $(2n)$th step, and deduce that the probability of ever returning to the origin is 1.

7. Consider the one-dimensional random walk $\{S_n\}$ given by

$$S_{n+1} = \begin{cases} S_n + 2 & \text{with probability } p, \\ S_n - 1 & \text{with probability } q = 1 - p, \end{cases}$$

where $0 < p < 1$.

(a) What is the probability of ever reaching the origin starting from $S_0 = a$ where $a > 0$?

(b) Let A_n be the mean number of points in $\{0, 1, \dots, n\}$ that the walk never visits. Find the limit $a = \lim_{n\to\infty} A_n/n$ when $p > \frac{1}{3}$, and verify for $p = \frac{1}{2}$ that $a = \frac{1}{2}(7 - 3\sqrt{5})$.

8. Let X and Y be independent variables taking values in the positive integers such that

$$\mathbb{P}(X = k \mid X + Y = n) = \binom{n}{k} p^k (1-p)^{n-k}$$

for some p and all $0 \le k \le n$. Show that X and Y have Poisson distributions.

9. In a branching process whose family sizes have mean μ and variance σ^2, find the variance of Z_n, the size of the nth generation, given that $Z_0 = 1$.

10. Waldegrave's problem. A group $\{A_1, A_2, \dots, A_r\}$ of r (> 2) people play the following game. A_1 and A_2 wager on the toss of a fair coin. The loser puts £1 in the pool, the winner goes on to play

A_3. In the next wager, the loser puts £1 in the pool, the winner goes on to play A_4, and so on. The winner of the $(r-1)$th wager goes on to play A_1, and the cycle recommences. The first person to beat all the others in sequence takes the pool.

(a) Find the probability generating function of the duration of the game.

(b) Find an expression for the probability that A_k wins.

(c) Find an expression for the expected size of the pool at the end of the game, given that A_k wins.

(d) Find an expression for the probability that the pool is intact after the nth spin of the coin.

This problem was discussed by Montmort, Bernoulli, de Moivre, Laplace, and others.

11. A branching process has a single progenitor.

(a) Show that the generating function H_n of the *total* number of individuals in the first n generations satisfies $H_n(s) = sG(H_{n-1}(s))$.

(b) Let T be the total number of individuals who ever exist, with $Q(s) = \mathbb{E}(s^T)$. Show, for $s \in [0, 1)$, that $Q(s) = sG(Q(s))$. Writing $\mu < 1$ for the mean of the family-size distribution, and $\sigma^2 > 0$ for its variance, show that:

(i) $Q(1) := \lim_{s\uparrow 1} Q(s) = 1$,

(ii) $\mathbb{E}(T) = 1/(1-\mu)$,

(iii) $\text{var}(T) = \sigma^2/(1-\mu)^3$.

(c) Find $Q(s)$ when $G(s) = p/(1-qs)$ where $0 < p = 1-q < 1$. Discuss the properties of Q in the two cases $p < q$ and $p \geq q$.

(d) Let G be as in part (c), and write $H_n(s) = y_n(s)/x_n(s)$ for appropriate polynomials x_n, y_n. Show that x_n satisfies $x_n(s) = x_{n-1}(s) - spqx_{n-2}(s)$, with $x_0 = 1$ and $x_1(s) = 1 - qs$. Deduce the form of $Q(s)$.

12. Show that the number Z_n of individuals in the nth generation of a branching process satisfies $\mathbb{P}(Z_n > N \mid Z_m = 0) \leq G_m(0)^N$ for $n < m$.

13. (a) A hen lays N eggs where N is Poisson with parameter λ. The weight of the nth egg is W_n, where $W_1, W_2, \ldots$ are independent identically distributed variables with common probability generating function $G(s)$. Show that the generating function G_W of the total weight $W = \sum_{i=1}^{N} W_i$ is given by $G_W(s) = \exp\{-\lambda + \lambda G(s)\}$. The quantity W is said to have a *compound Poisson distribution*. Show further that, for any positive integral value of n, $G_W(s)^{1/n}$ is the probability generating function of some random variable; W (or its distribution) is said to be *infinitely divisible* in this regard.

(b) Show that if $H(s)$ is the probability generating function of some infinitely divisible distribution on the non-negative integers then $H(s) = \exp\{-\lambda + \lambda G(s)\}$ for some λ (> 0) and some probability generating function $G(s)$.

(c) Can the compound Poisson distribution of W in part (a) be a Poisson distribution for any choice of G?

14. The distribution of a random variable X is called *infinitely divisible* if, for all positive integers n, there exists a sequence $Y_1^{(n)}, Y_2^{(n)}, \ldots, Y_n^{(n)}$ of independent identically distributed random variables such that X and $Y_1^{(n)} + Y_2^{(n)} + \cdots + Y_n^{(n)}$ have the same distribution.

(a) Show that the normal, Poisson, and gamma distributions are infinitely divisible.

(b) Show that the characteristic function ϕ of an infinitely divisible distribution has no real zeros, in that $\phi(t) \neq 0$ for all real t.

15. Let $X_1, X_2, \ldots$ be independent variables each taking the values 0 or 1 with probabilities $1-p$ and p, where $0 < p < 1$. Let N be a random variable taking values in the positive integers, independent of the X_i, and write $S = X_1 + X_2 + \cdots + X_N$. Write down the conditional generating function of N given that $S = N$, in terms of the probability generating function G of N. Show that N has a Poisson distribution if and only if $\mathbb{E}(x^N)^p = \mathbb{E}(x^N \mid S = N)$ for all p and x.

16. If X and Y have joint probability generating function

$$G_{X,Y}(s,t) = \mathbb{E}(s^X t^Y) = \frac{\{1-(p_1+p_2)\}^n}{\{1-(p_1 s + p_2 t)\}^n} \qquad \text{where } p_1 + p_2 \le 1,$$

find the marginal mass functions of X and Y, and the mass function of $X+Y$. Find also the conditional probability generating function $G_{X|Y}(s \mid y) = \mathbb{E}(s^X \mid Y = y)$ of X given that $Y = y$. The pair X, Y is said to have the *bivariate negative binomial distribution*.

17. If X and Y have joint probability generating function

$$G_{X,Y}(s,t) = \exp\{\alpha(s-1) + \beta(t-1) + \gamma(st-1)\}$$

find the marginal distributions of X, Y, and the distribution of $X+Y$, showing that X and Y have the Poisson distribution, but that $X+Y$ does not unless $\gamma = 0$.

18. Define

$$I(a,b) = \int_0^\infty \exp(-a^2u^2 - b^2u^{-2})\,du$$

for $a, b > 0$. Show that

(a) $I(a,b) = a^{-1}I(1,ab)$, (b) $\partial I/\partial b = -2I(1,ab)$,

(c) $I(a,b) = \sqrt{\pi}e^{-2ab}/(2a)$.

(d) If X has density function $(d/\sqrt{x})e^{-c/x-gx}$ for $x > 0$, then

$$\mathbb{E}(e^{-tX}) = d\sqrt{\frac{\pi}{g+t}}\exp\left(-2\sqrt{c(g+t)}\right), \qquad t > -g.$$

(e) If X has density function $(2\pi x^3)^{-\frac{1}{2}}e^{-1/(2x)}$ for $x > 0$, then X has moment generating function given by $\mathbb{E}(e^{-tX}) = \exp\{-\sqrt{2t}\}$, $t \ge 0$. [Note that $\mathbb{E}(X^n) = \infty$ for $n \ge 1$.]

19. Let X, Y, Z be independent $N(0,1)$ variables. Use characteristic functions and moment generating functions (Laplace transforms) to find the distributions of

(a) $U = X/Y$,

(b) $V = X^{-2}$,

(c) $W = XYZ/\sqrt{X^2Y^2 + Y^2Z^2 + Z^2X^2}$.

20. Let X have density function f and characteristic function ϕ, and suppose that $\int_{-\infty}^{\infty} |\phi(t)|\,dt < \infty$. Deduce that

$$f(x) = \frac{1}{2\pi}\int_{-\infty}^{\infty} e^{-itx}\phi(t)\,dt.$$

21. Conditioned branching process. Consider a branching process whose family sizes have the geometric mass function $f(k) = qp^k$, $k \ge 0$, where $\mu = p/q > 1$. Let Z_n be the size of the nth generation, and assume $Z_0 = 1$. Show that the conditional distribution of Z_n/μ^n, given that $Z_n > 0$, converges as $n \to \infty$ to the exponential distribution with parameter $1 - \mu^{-1}$.

22. A random variable X is called *symmetric* if X and $-X$ are identically distributed. Show that X is symmetric if and only if the imaginary part of its characteristic function is identically zero.

23. Let X and Y be independent identically distributed variables with means 0 and variances 1. Let $\phi(t)$ be their common characteristic function, and suppose that $X+Y$ and $X-Y$ are independent. Show that $\phi(2t) = \phi(t)^3\phi(-t)$, and deduce that X and Y are $N(0,1)$ variables.

More generally, suppose that X and Y are independent and identically distributed with means 0 and variances 1, and furthermore that $\mathbb{E}(X-Y \mid X+Y) = 0$ and $\text{var}(X-Y \mid X+Y) = 2$. Deduce that $\phi(s)^2 = \phi'(s)^2 - \phi(s)\phi''(s)$, and hence that X and Y are independent $N(0,1)$ variables.

24. Show that the average $Z = n^{-1}\sum_{i=1}^{n} X_i$ of n independent Cauchy variables has the Cauchy distribution too. Why does this not violate the law of large numbers?

25. Let X and Y be independent random variables each having the Cauchy density function $f(x) = \{\pi(1+x^2)\}^{-1}$, and let $Z = \frac{1}{2}(X+Y)$.

(a) Show by using characteristic functions that Z has the Cauchy distribution also.

(b) Show by the convolution formula that Z has the Cauchy density function. You may find it helpful to check first that

$$f(x)f(y-x) = \frac{f(x)+f(y-x)}{\pi(4+y^2)} + g(y)\{xf(x)+(y-x)f(y-x)\}$$

where $g(y) = 2/\{\pi y(4+y^2)\}$.

26. Let $X_1, X_2, \dots, X_n$ be independent variables with characteristic functions $\phi_1, \phi_2, \dots, \phi_n$. Describe random variables which have the following characteristic functions:

(a) $\phi_1(t)\phi_2(t)\cdots\phi_n(t)$, (b) $|\phi_1(t)|^2$,
(c) $\sum_1^n p_j\phi_j(t)$ where $p_j \ge 0$ and $\sum_1^n p_j = 1$, (d) $(2-\phi_1(t))^{-1}$,
(e) $\int_0^\infty \phi_1(ut)e^{-u}\,du$.

27. (a) Find the characteristic functions corresponding to the following density functions on $\mathbb{R}$:

(i) $1/\cosh(\pi x)$, (ii) $(1-\cos x)/(\pi x^2)$,
(iii) $\exp(-x-e^{-x})$, (iv) $\frac{1}{2}e^{-|x|}$.

Show that the mean of the 'extreme-value distribution' in part (iii) is Euler's constant γ.

(b) **Characteristic functions coincident on an interval.** Write down the density function with characteristic function $\phi(t) = \max\{0, 1-|t|/\pi\}$ for $t \in \mathbb{R}$.

Show that the periodic function $\psi(t)$ with period 2π given by $\psi(t) = \max\{0, 1-|t|/\pi\}$ for $|t| \le \pi$, is the characteristic function of the discrete random variable with mass function

$$f(0) = \frac{1}{2}, \quad f(2k+1) = f(-2k-1) = \frac{2}{\pi^2(2k+1)^2}, \qquad k = 0, 1, 2, \dots.$$

28. Which of the following are characteristic functions:

(a) $\phi(t) = 1-|t|$ if $|t| \le 1$, $\phi(t) = 0$ otherwise,
(b) $\phi(t) = (1+t^4)^{-1}$, (c) $\phi(t) = \exp(-t^4)$,
(d) $\phi(t) = \cos t$, (e) $\phi(t) = 2(1-\cos t)/t^2$.

29. Show that the characteristic function ϕ of a random variable X satisfies $|1-\phi(t)| \le \mathbb{E}|tX|$.

30. Suppose X and Y have joint characteristic function $\phi(s,t)$. Show that, subject to the appropriate conditions of differentiability,

$$i^{m+n}\mathbb{E}(X^mY^n) = \left.\frac{\partial^{m+n}\phi}{\partial s^m\,\partial t^n}\right|_{s=t=0}$$

for any positive integers m and n.

31. If X has distribution function F and characteristic function ϕ, show that for $t > 0$

(a) $$\int_{[-t^{-1},t^{-1}]} x^2\,dF \le \frac{3}{t^2}[1-\operatorname{Re}\phi(t)],$$

(b) $$\mathbb{P}\left(|X| \ge \frac{1}{t}\right) \le \frac{7}{t}\int_0^t [1-\operatorname{Re}\phi(v)]\,dv.$$

32. Let $X_1, X_2, \dots$ be independent variables which are uniformly distributed on $[0, 1]$. Let $M_n = \max\{X_1, X_2, \dots, X_n\}$ and show that $n(1 - M_n) \xrightarrow{\text{D}} X$ where X is exponentially distributed with parameter 1. You need not use characteristic functions.

33. If X is either (a) Poisson with parameter λ, or (b) $\Gamma(1, \lambda)$, show that the distribution of $Y_\lambda = (X - \mathbb{E}X)/\sqrt{\operatorname{var} X}$ approaches the $N(0, 1)$ distribution as $\lambda \to \infty$.

(c) Show that

$$e^{-n}\left(1 + n + \frac{n^2}{2!} + \cdots + \frac{n^n}{n!}\right) \to \frac{1}{2} \qquad \text{as } n \to \infty.$$

34. Coupon collecting. Recall that you regularly buy quantities of some ineffably dull commodity. To attract your attention, the manufacturers add to each packet a small object which is also dull, and in addition useless, but there are n different types. Assume that each packet is equally likely to contain any one of the different types, as usual. Let T_n be the number of packets bought before you acquire a complete set of n objects. Show that $n^{-1}(T_n - n \log n) \xrightarrow{\text{D}} T$, where T is a random variable with distribution function $\mathbb{P}(T \le x) = \exp(-e^{-x})$, $-\infty < x < \infty$.

35. Find a sequence (ϕ_n) of characteristic functions with the property that the limit given by $\phi(t) = \lim_{n\to\infty} \phi_n(t)$ exists for all t, but such that ϕ is not itself a characteristic function.

36. (a) Use generating functions to show that it is not possible to load two dice in such a way that the sum of the values which they show is equally likely to take any value between 2 and 12. Compare with your method for Problem (2.7.12).

(b) **Sicherman dice.** Show that it is possible to number the faces of two fair dice, in a different manner from two standard dice, such that the sum of the scores when rolled has the same distribution as the sum of the scores of two standard dice.

37. A biased coin is tossed N times, where N is a random variable which is Poisson distributed with parameter λ. Prove that the total number of heads shown is independent of the total number of tails. Show conversely that if the numbers of heads and tails are independent, then N has the Poisson distribution.

38. A *binary tree* is a tree (as in the section on branching processes) in which each node has exactly two descendants. Suppose that each node of the tree is coloured black with probability p, and white otherwise, independently of all other nodes. For any path π containing n nodes beginning at the root of the tree, let $B(\pi)$ be the number of black nodes in π, and let $X_n(k)$ be the number of such paths π for which $B(\pi) \ge k$. Show that there exists β_c such that

$$\mathbb{E}\{X_n(\beta n)\} \to \begin{cases} 0 & \text{if } \beta > \beta_c, \\ \infty & \text{if } \beta < \beta_c, \end{cases}$$

and show how to determine the value β_c.

Prove that

$$\mathbb{P}\big(X_n(\beta n) \ge 1\big) \to \begin{cases} 0 & \text{if } \beta > \beta_c, \\ 1 & \text{if } \beta < \beta_c. \end{cases}$$

39. Use the continuity theorem (5.9.5) to show that, as $n \to \infty$,

(a) if X_n is $\text{bin}(n, \lambda/n)$ then the distribution of X_n converges to a Poisson distribution,

(b) if Y_n is geometric with parameter $p = \lambda/n$ then the distribution of Y_n/n converges to an exponential distribution.

40. Let $X_1, X_2, \dots$ be independent random variables with zero means and such that $\mathbb{E}|X_j^3| < \infty$ for all j. Show that $S_n = X_1 + X_2 + \cdots + X_n$ satisfies $S_n/\sqrt{\text{var}(S_n)} \xrightarrow{D} N(0, 1)$ as $n \to \infty$ if

$$\sum_{j=1}^{n} \mathbb{E}|X_j^3| = \text{o}\Big(\{\text{var}(S_n)\}^{-\frac{3}{2}}\Big).$$

The following steps may be useful. Let $\sigma_j^2 = \text{var}(X_j)$, $\sigma(n)^2 = \text{var}(S_n)$, $\rho_j = \mathbb{E}|X_j^3|$, and ϕ_j and ψ_n be the characteristic functions of X_j and $S_n/\sigma(n)$ respectively.

(i) Use Taylor's theorem to show that $|\phi_j(t) - 1| \le 2t^2\sigma_j^2$ and $|\phi_j(t) - 1 + \frac{1}{2}\sigma_j^2 t^2| \le |t|^3 \rho_j$ for $j \ge 1$.

(ii) Show that $|\log(1 + z) - z| \le |z|^2$ if $|z| \le \frac{1}{2}$, where the logarithm has its principal value.

(iii) Show that $\sigma_j^3 \le \rho_j$, and deduce from the hypothesis that $\max_{1 \le j \le n} \sigma_j/\sigma(n) \to 0$ as $n \to \infty$, implying that $\max_{1 \le j \le n} |\phi_j(t/\sigma(n)) - 1| \to 0$.

(iv) Deduce an upper bound for $\left|\log \phi_j(t/\sigma(n)) - \frac{1}{2}t^2\sigma_j^2/\sigma(n)^2\right|$, and sum to obtain that $\log \psi_n(t) \to -\frac{1}{2}t^2$.

41. Let $X_1, X_2, \dots$ be independent variables each taking values $+1$ or -1 with probabilities $\frac{1}{2}$ and $\frac{1}{2}$. Show that

$$\sqrt{\frac{3}{n^3}} \sum_{k=1}^{n} kX_k \xrightarrow{D} N(0, 1) \qquad \text{as } n \to \infty.$$

42. Normal sample. Let $X_1, X_2, \dots, X_n$ be independent $N(\mu, \sigma^2)$ random variables. Define $\overline{X} = n^{-1}\sum_1^n X_i$ and $Z_i = X_i - \overline{X}$. Find the joint characteristic function of $\overline{X}, Z_1, Z_2, \dots, Z_n$, and hence prove that $\overline{X}$ and $S^2 = (n-1)^{-1}\sum_1^n (X_i - \overline{X})^2$ are independent.

43. Log-normal distribution. Let X be $N(0, 1)$, and let $Y = e^X$; Y is said to have the *log-normal* distribution. Show that the density function of Y is

$$f(x) = \frac{1}{x\sqrt{2\pi}} \exp\{-\tfrac{1}{2}(\log x)^2\}, \qquad x > 0.$$

For $|a| \le 1$, define $f_a(x) = \{1 + a\sin(2\pi \log x)\} f(x)$. Show that f_a is a density function with finite moments of all (positive) orders, none of which depends on the value of a. The family $\{f_a : |a| \le 1\}$ contains density functions which are not specified by their moments.

44. Consider a random walk whose steps are independent and identically distributed integer-valued random variables with non-zero mean. Prove that the walk is transient.

45. Recurrent events. Let $\{X_r : r \ge 1\}$ be the integer-valued identically distributed intervals between the times of a recurrent event process. Let L be the earliest time by which there has been an interval of length a containing no occurrence time. Show that, for integral a,

$$\mathbb{E}(s^L) = \frac{s^a \mathbb{P}(X_1 > a)}{1 - \sum_{r=1}^{a} s^r \mathbb{P}(X_1 = r)}.$$

46. A biased coin shows heads with probability $p\ (= 1 - q)$. It is flipped repeatedly until the first time W_n by which it has shown n consecutive heads. Let $\mathbb{E}(s^{W_n}) = G_n(s)$. Show that $G_n = psG_{n-1}/(1 - qsG_{n-1})$, and deduce that

$$G_n(s) = \frac{(1 - ps)p^n s^n}{1 - s + qp^n s^{n+1}}.$$

47. In n flips of a biased coin which shows heads with probability p $(= 1 - q)$, let L_n be the length of the longest run of heads. Show that, for $r \geq 1$,

$$1 + \sum_{n=1}^{\infty} s^n \mathbb{P}(L_n < r) = \frac{1 - p^r s^r}{1 - s + qp^r s^{r+1}}.$$

48. The random process $\{X_n : n \geq 1\}$ decays geometrically fast in that, in the absence of external input, $X_{n+1} = \frac{1}{2} X_n$. However, at any time n the process is also increased by Y_n with probability $\frac{1}{2}$, where $\{Y_n : n \geq 1\}$ is a sequence of independent exponential random variables with parameter λ. Find the limiting distribution of X_n as $n \to \infty$.

49. Let $G(s) = \mathbb{E}(s^X)$ where $X \geq 0$. Show that $\mathbb{E}\{(X+1)^{-1}\} = \int_0^1 G(s)\,ds$, and evaluate this when X is (a) Poisson with parameter λ, (b) geometric with parameter p, (c) binomial bin(n, p), (d) logarithmic with parameter p (see Exercise (5.2.3)). Is there a non-trivial choice for the distribution of X such that $\mathbb{E}\{(X+1)^{-1}\} = \{\mathbb{E}(X+1)\}^{-1}$?

50. Find the density function of $\sum_{r=1}^{N} X_r$, where $\{X_r : r \geq 1\}$ are independent and exponentially distributed with parameter λ, and N is geometric with parameter p and independent of the X_r.

51. Let X have finite non-zero variance and characteristic function $\phi(t)$. Show that

$$\psi(t) = -\frac{1}{\mathbb{E}(X^2)} \frac{d^2\phi}{dt^2}$$

is a characteristic function, and find the corresponding distribution.

52. Triangular distribution. Let X and Y have joint density function

$$f(x, y) = \tfrac{1}{4}\{1 + xy(x^2 - y^2)\}, \qquad |x| < 1,\ |y| < 1.$$

Show that $\phi_X(t)\phi_Y(t) = \phi_{X+Y}(t)$, and that X and Y are dependent. Find the probability density function of $X + Y$.

53. Exercise (4.6.6) revisited. Let $X_1, X_2, \dots$ be independent and uniformly distributed on $(0, 1)$, and let $m(x) = \mathbb{E}(N)$ where $N = \min\{n : \sum_{r=1}^{n} X_r > x\}$ for $x > 0$. Show that $m'(x) = m(x) - m(x-1)$, and deduce that the Laplace transform $m^*(s) = \int_0^\infty m(x)e^{-sx}\,dx$ is given by $m^*(s) = 1/(e^{-s} + s - 1)$ for $s \neq 0$. Hence prove that

$$m(x) = \sum_{r=0}^{\lfloor x \rfloor} \frac{(-1)^r}{r!}(x - r)^r e^{x-r}, \qquad x > 0.$$

54. Rounding error. Let $S_n = \sum_{r=1}^{n} X_r$ be a partial sum of a sequence of independent random variables with the uniform distribution on $(0, 1)$. For $x \in \mathbb{R}$, let $\{x\}$ denote the nearest integer to x, and write $R_n = \sum_{r=1}^{n} \{X_r\}$.

(a) Show that $X_r - \{X_r\}$ is uniformly distributed on $(-\frac{1}{2}, \frac{1}{2})$.

(b) Show that

$$\mathbb{P}(\{S_n\} = R_n) = 2f_{n+1}(0) = \int_{-\infty}^{\infty} \frac{1}{\pi} \left(\frac{\sin t}{t}\right)^{n+1} dt,$$

where $f_n(y)$ is the density function of the sum of n independent random variables with the uniform distribution on $(-1, 1)$.

(c) Find a similar expression for $\mathbb{P}\big(\{S_n\} - R_n = k\big)$.

55. Maxwell's molecules.

(a) Let $V = (V_1, V_2, V_3)$ be the velocity in $\mathbb{R}^3$ of a molecule M of a perfect gas, and assume that in any Cartesian coordinate system the coordinates of V are independent random variables with mean 0 and finite variance σ^2. Show that the V_i are independent with the $N(0, \sigma^2)$ distribution. When $\sigma^2 = 1$, show that $|V|$ has the *Maxwell density* $f(v) = \sqrt{2/\pi}\, v^2 e^{-\frac{1}{2}v^2}$ for $v > 0$.

(b) A physicist assumes that, initially, M is equally likely to be anywhere in the region R between two parallel planes with distance 1 apart, independently of its velocity. Assuming $\sigma^2 = 1$, show that the probability $p(t)$ that M is in R at time t is

$$p(t) = \frac{1}{\sqrt{2\pi}} \left[\int_{-1}^{1} \frac{1}{t} \exp\left\{ -\frac{x^2}{2t^2} \right\} dx - 2t \left(1 - \exp\left\{ -\frac{1}{2t^2} \right\} \right) \right].$$

(c) Deduce that the density function of the time T at which M exits R is

$$f_T(t) = \sqrt{2/\pi} \left[1 - \exp\left\{ -\frac{1}{2t^2} \right\} \right], \qquad t > 0.$$

56. Let X, Y be independent random variables with a joint distribution with circular symmetry about 0 in the x/y-plane, and with finite variances. Show that the distribution of $R = X \cos\theta + Y \sin\theta$ does not depend on the value of θ.

With the usual notation for characteristic functions, show that:

(a) $\phi_X(t\cos\theta)\phi_Y(t\sin\theta) = \phi_R(t)$,

(b) $\phi_X(t) = \phi_X(-t) = \phi_Y(t) = \phi_Y(-t)$,

(c) $\phi_X(t) = \psi(t^2)$ for some continuous real-valued function ψ,

(d) X and Y are $N(0, \sigma^2)$ for some $\sigma^2 \geq 0$.

6

Markov chains

6.1 Exercises. Markov processes

1. Show that any sequence of independent random variables taking values in the countable set S is a Markov chain. Under what condition is this chain homogeneous?

2. A die is rolled repeatedly. Which of the following are Markov chains? For those that are, supply the transition matrix.

(a) The largest number X_n shown up to the nth roll.

(b) The number N_n of sixes in n rolls.

(c) At time r, the time C_r since the most recent six.

(d) At time r, the time B_r until the next six.

3. Let $\{S_n : n \geq 0\}$ be a simple random walk with $S_0 = 0$, and show that $X_n = |S_n|$ defines a Markov chain; find the transition probabilities of this chain. Let $M_n = \max\{S_k : 0 \leq k \leq n\}$, and show that $Y_n = M_n - S_n$ defines a Markov chain. What happens if $S_0 \neq 0$?

4. Let X be a Markov chain and let $\{n_r : r \geq 0\}$ be an unbounded increasing sequence of positive integers. Show that $Y_r = X_{n_r}$ constitutes a (possibly non-homogeneous) Markov chain. Find the transition matrix of Y when $n_r = 2r$ and X is: (a) simple random walk, and (b) a branching process.

5. Let X be a Markov chain on S, and let $I : S^n \to \{0, 1\}$. Show that the distribution of $X_n, X_{n+1}, \dots$, conditional on $\{I(X_1, \dots, X_n) = 1\} \cap \{X_n = i\}$, is identical to the distribution of $X_n, X_{n+1}, \dots$ conditional on $\{X_n = i\}$.

6. **Strong Markov property.** Let X be a Markov chain on S, and let T be a random variable taking values in $\{0, 1, 2, \dots\}$ with the property that the indicator function $I_{\{T=n\}}$, of the event that $T = n$, is a function of the variables $X_1, X_2, \dots, X_n$. Such a random variable T is called a *stopping time*, and the above definition requires that it is decidable whether or not $T = n$ with a knowledge only of the past and present, $X_0, X_1, \dots, X_n$, and with no further information about the future.

Show that

$$\mathbb{P}\big(X_{T+m} = j \,\big|\, X_k = x_k \text{ for } 0 \leq k < T,\ X_T = i\big) = \mathbb{P}(X_{T+m} = j \mid X_T = i)$$

for $m \geq 0$, $i, j \in S$, and all sequences (x_k) of states.

7. Let X be a Markov chain with state space S, and suppose that $h : S \to T$ is one–one. Show that $Y_n = h(X_n)$ defines a Markov chain on T. Must this be so if h is not one–one?

8. Let X and Y be Markov chains on the set $\mathbb{Z}$ of integers.

(a) Is the sequence $Z_n = X_n + Y_n$ necessarily a Markov chain?

(b) Is Z a Markov chain if X and Y are independent chains? Give a proof or a counterexample.

(c) Show that Z is a Markov chain if X and Y are independent of one another and have independent increments.

9. Let X be a Markov chain. Which of the following are Markov chains?
(a) X_{m+r} for $r \geq 0$.
(b) X_{2m} for $m \geq 0$.
(c) The sequence of pairs (X_n, X_{n+1}) for $n \geq 0$.

10. Two-sided Markov property. Let X be a Markov chain. Show that, for $1 < r < n$,

$$\begin{aligned}\mathbb{P}(X_r = k \mid X_i = x_i \text{ for } i = 1, 2, \dots, r-1, r+1 \dots, n) \\ = \mathbb{P}(X_r = k \mid X_{r-1} = x_{r-1}, X_{r+1} = x_{r+1}).\end{aligned}$$

11. Let $\{X_n : n \geq 1\}$ be independent identically distributed integer-valued random variables. Let $S_n = \sum_{r=1}^{n} X_r$, with $S_0 = 0$, $Y_n = X_n + X_{n-1}$ with $X_0 = 0$, and $Z_n = \sum_{r=0}^{n} S_r$. Which of the following constitute Markov chains: (a) S_n, (b) Y_n, (c) Z_n, (d) the sequence of pairs (S_n, Z_n)?

12. A stochastic matrix $\mathbf{P}$ is called *doubly stochastic* if $\sum_i p_{ij} = 1$ for all j. It is called *sub-stochastic* if $\sum_i p_{ij} \leq 1$ for all j. Show that, if $\mathbf{P}$ is stochastic (respectively, doubly stochastic, sub-stochastic), then $\mathbf{P}^n$ is stochastic (respectively, doubly stochastic, sub-stochastic) for all n.

13. Lumping. Let X be a Markov chain on the finite state space S with transition matrix $\mathbf{P}$, and let $\mathcal{C} = \{C_j : j \in J\}$ be a partition of S. Let $Y_n = j$ if $X_n \in C_j$. The chain X is called *$\mathcal{C}$-lumpable* if Y is a Markov chain.

Show that X is $\mathcal{C}$-lumpable if and only if, for $a, b \in J$, $\mathbb{P}(X_{n+1} \in C_b \mid X_n = i)$ is constant for $i \in C_a$.

6.2 Exercises. Classification of states

1. Last exits. Let $l_{ij}(n) = \mathbb{P}(X_n = j, X_k \neq i \text{ for } 1 \leq k < n \mid X_0 = i)$, the probability that the chain passes from i to j in n steps without revisiting i. Writing

$$L_{ij}(s) = \sum_{n=1}^{\infty} s^n l_{ij}(n),$$

show that $P_{ij}(s) = P_{ii}(s) L_{ij}(s)$ if $i \neq j$. Deduce that the first passage times and last exit times have the same distribution for any Markov chain for which $P_{ii}(s) = P_{jj}(s)$ for all i and j. Give an example of such a chain.

2. Let X be a Markov chain containing an absorbing state s with which all other states i communicate, in the sense that $p_{is}(n) > 0$ for some $n = n(i)$. Show that all states other than s are transient.

3. Show that a state i is recurrent if and only if the mean number of visits of the chain to i, having started at i, is infinite. That is to say, i is recurrent if and only if $\sum_n p_{ii}(n) = \infty$.

4. Visits. Let $V_j = |\{n \geq 1 : X_n = j\}|$ be the number of visits of the Markov chain X to j, and define $\eta_{ij} = \mathbb{P}_i(V_j = \infty)$. Show that:

(a) $\eta_{ii} = \begin{cases} 1 & \text{if } i \text{ is recurrent,} \\ 0 & \text{if } i \text{ is transient,} \end{cases}$

(b) $\eta_{ij} = \begin{cases} \mathbb{P}_i(T_j < \infty) & \text{if } j \text{ is recurrent,} \\ 0 & \text{if } j \text{ is transient,} \end{cases}$ where $T_j = \min\{n \geq 1 : X_n = j\}$.

5. Symmetry. The distinct pair i, j of states of a Markov chain is called *symmetric* if

$$\mathbb{P}_i(T_j < T_i) = \mathbb{P}_j(T_i < T_j),$$

where $T_i = \min\{n \ge 1 : X_n = i\}$. Show that, if $X_0 = i$ and i, j is symmetric, the expected number of visits to j before the chain revisits i is 1. [Cf. the quotation following Theorem (3.10.18).]

6. Van Dantzig's collective marks. Let X be a Markov chain, and let T be a geometric random variable with $\mathbb{P}(T > n) = s^n$ for $n \ge 0$, which is independent of X. By considering the expected number of visits by X to a given state before time T, prove Theorem (6.2.3).

7. Constrained first-passage time. Let X be an ergodic Markov chain started from a, and assume X is irreducible (in that, for states i, j, there exists $m \ge 0$ such that $p_{ij}(m) > 0$). Let a, b, c be distinct states, and let $T(a, b, \neg c)$ be the time until X first visits b, with no intermediate visit to c. That is, if X visits b before c, then $T(a, b, \neg c)$ is that time, and if X visits c before b we set $T(a, b, \neg c) = \infty$. Let

$$G(a, b, \neg c; s) = \sum_{n=1}^{\infty} s^n \mathbb{P}_a\big(T(a, b, \neg c) = n\big).$$

Show that

$$G(a, b, \neg c; s) = \frac{F_{ab} - F_{ac}F_{cb}}{1 - F_{bc}F_{cb}},$$

where, for example, $F_{ab}(s)$ is the probability generating function of the first passage time T_{ab} from a to b irrespective of intermediate visits to c. Show further that

$$\mathbb{P}_a\big(T(a, b, \neg c) < \infty\big) = \frac{\mu_{ac} + \mu_{cb} - \mu_{ab}}{\mu_{bc} + \mu_{cb}},$$

where, for example, $\mu_{ab} = \mathbb{E}_a(T_{ab})$.

6.3 Exercises. Classification of chains

1. Let X be a Markov chain on $\{0, 1, 2, \dots\}$ with transition matrix given by $p_{0j} = a_j$ for $j \ge 0$, $p_{ii} = r$ and $p_{i,i-1} = 1 - r$ for $i \ge 1$. Classify the states of the chain, and find their mean recurrence times.

2. Determine whether or not the random walk on the integers having transition probabilities $p_{i,i+2} = p$, $p_{i,i-1} = 1 - p$, for all i, is recurrent.

3. Classify the states of the Markov chains with transition matrices

(a) $$\begin{pmatrix} 1-2p & 2p & 0 \\ p & 1-2p & p \\ 0 & 2p & 1-2p \end{pmatrix},$$

(b) $$\begin{pmatrix} 0 & p & 0 & 1-p \\ 1-p & 0 & p & 0 \\ 0 & 1-p & 0 & p \\ p & 0 & 1-p & 0 \end{pmatrix}.$$

In each case, calculate $p_{ij}(n)$ and the mean recurrence times of the states.

4. A particle performs a random walk on the vertices of a cube. At each step it remains where it is with probability $\frac{1}{4}$, or moves to one of its neighbouring vertices each having probability $\frac{1}{4}$. Let v and w be two diametrically opposite vertices. If the walk starts at v, find:

(a) the mean number of steps until its first return to v,

(b) the mean number of steps until its first visit to w,

(c) the mean number of visits to w before its first return to v.

5. **Visits.** With the notation of Exercise (6.2.4), show that

(a) if $i \to j$ and i is recurrent, then $\eta_{ij} = \eta_{ji} = 1$,

(b) $\eta_{ij} = 1$ if and only if $\mathbb{P}_i(T_j < \infty) = \mathbb{P}_j(T_j < \infty) = 1$.

6. **Hitting probabilities.** Let $T_A = \min\{n \geq 0 : X_n \in A\}$, where X is a Markov chain and A is a subset of the state space S, and let $\eta_j = \mathbb{P}_j(T_A < \infty)$. Show that

$$\eta_j = \begin{cases} 1 & \text{if } j \in A, \\ \sum_{k \in S} p_{jk}\eta_k & \text{if } j \notin A. \end{cases}$$

Show further that if $\mathbf{x} = (x_j : j \in S)$ is any non-negative solution of these equations then $x_j \geq \eta_j$ for all j.

7. **Mean hitting times.** In the notation of Exercise (6.3.6), let $\rho_j = \mathbb{E}_j(T_A)$. Show that

$$\rho_j = \begin{cases} 0 & \text{if } j \in A, \\ 1 + \sum_{k \in S} p_{jk}\rho_k & \text{if } j \notin A, \end{cases}$$

and that if $\mathbf{x} = (x_j : j \in S)$ is any non-negative solution of these equations then $x_j \geq \rho_j$ for all j.

8. Let X be an irreducible Markov chain and let A be a subset of the state space. Let S_r and T_r be the successive times at which the chain enters A and visits A respectively. Are the sequences $\{X_{S_r} : r \geq 1\}$, $\{X_{T_r} : r \geq 1\}$ Markov chains? What can be said about the times at which the chain exits A?

9. (a) Show that for each pair i, j of states of an irreducible aperiodic chain, there exists $N = N(i, j)$ such that $p_{ij}(r) > 0$ for all $r \geq N$.

(b) Show that there exists a function f such that, if $\mathbf{P}$ is the transition matrix of an irreducible aperiodic Markov chain with n states, then $p_{ij}(r) > 0$ for all states i, j, and all $r \geq f(n)$.

(c) Show further that $f(4) \geq 6$ and $f(n) \geq (n-1)(n-2)$.

[Hint: The postage stamp lemma asserts that, for a, b coprime, the smallest n such that all integers strictly exceeding n have the form $\alpha a + \beta b$ for some integers $\alpha, \beta \geq 0$ is $(a-1)(b-1)$.]

10. An urn initially contains n green balls and $n+2$ red balls. A ball is picked at random: if it is green then a red ball is also removed and both are discarded; if it is red then it is replaced together with an extra red and an extra green ball. This is repeated until there are no green balls in the urn. Show that the probability the process terminates is $1/(n+1)$.

Now reverse the rules: if the ball is green, it is replaced together with an extra green and an extra red ball; if it is red it is discarded along with a green ball. Show that the expected number of iterations until no green balls remain is $\sum_{j=1}^{n}(2j+1) = n(n+2)$. [Thus, a minor perturbation of a simple symmetric random walk can be positive recurrent, whereas the original is null recurrent.]

6.4 Exercises. Stationary distributions and the limit theorem

1. The proof copy of a book is read by an infinite sequence of editors checking for mistakes. Each mistake is detected with probability p at each reading; between readings the printer corrects the detected mistakes but introduces a random number of new errors (errors may be introduced even if no mistakes were detected). Assuming as much independence as usual, and that the numbers of new errors after different readings are identically distributed, find an expression for the probability generating function of the stationary distribution of the number X_n of errors after the nth editor–printer cycle,

whenever this exists. Find it explicitly when the printer introduces a Poisson-distributed number of errors at each stage.

2. Do the appropriate parts of Exercises (6.3.1)–(6.3.4) again, making use of the new techniques at your disposal.

3. **Dams.** Let X_n be the amount of water in a reservoir at noon on day n. During the 24 hour period beginning at this time, a quantity Y_n of water flows into the reservoir, and just before noon on each day exactly one unit of water is removed (if this amount can be found). The maximum capacity of the reservoir is K, and excessive inflows are spilled and lost. Assume that the Y_n are independent and identically distributed random variables and that, by rounding off to some laughably small unit of volume, all numbers in this exercise are non-negative integers. Show that (X_n) is a Markov chain, and find its transition matrix and an expression for its stationary distribution in terms of the probability generating function G of the Y_n.

Find the stationary distribution when Y has probability generating function $G(s) = p(1-qs)^{-1}$.

4. Show by example that chains which are not irreducible may have many different stationary distributions.

5. **Diagonal selection.** Let $(x_i(n) : i, n \geq 1)$ be a bounded collection of real numbers. Show that there exists an increasing sequence $n_1, n_2, \dots$ of positive integers such that $\lim_{r\to\infty} x_i(n_r)$ exists for all i. Use this result to prove that, for an irreducible Markov chain, if it is not the case that $p_{ij}(n) \to 0$ as $n \to \infty$ for all i and j, then there exists a sequence $(n_r : r \geq 1)$ and a vector $\boldsymbol{\alpha}$ ($\neq \mathbf{0}$) such that $p_{ij}(n_r) \to \alpha_j$ as $r \to \infty$ for all i and j.

6. **Random walk on a graph.** A particle performs a random walk on the vertex set of a connected graph G, which for simplicity we assume to have neither loops nor multiple edges. At each stage it moves to a neighbour of its current position, each such neighbour being chosen with equal probability. If G has η ($< \infty$) edges, show that the stationary distribution is given by $\pi_v = d_v/(2\eta)$, where d_v is the degree of vertex v.

7. Show that a random walk on the infinite binary tree is transient.

8. At each time $n = 0, 1, 2, \dots$ a number Y_n of particles enters a chamber, where $\{Y_n : n \geq 0\}$ are independent and Poisson distributed with parameter λ. Lifetimes of particles are independent and geometrically distributed with parameter p. Let X_n be the number of particles in the chamber at time n. Show that X is a Markov chain, and find its stationary distribution.

9. A random sequence of convex polygons is generated by picking two edges of the current polygon at random, joining their midpoints, and picking one of the two resulting smaller polygons at random to be the next in the sequence. Let $X_n + 3$ be the number of edges of the nth polygon thus constructed. Find $\mathbb{E}(X_n)$ in terms of X_0, and find the stationary distribution of the Markov chain X.

10. Let s be a state of an irreducible Markov chain on the non-negative integers. Show that the chain is recurrent if there exists a solution $\mathbf{y}$ to the equations $y_i \geq \sum_{j:j\neq s} p_{ij} y_j$, $i \neq s$, satisfying $y_i \to \infty$.

11. **Bow ties.** A particle performs a random walk on a bow tie ABCDE drawn beneath on the left, where C is the knot. From any vertex its next step is equally likely to be to any neighbouring vertex. Initially it is at A. Find the expected value of:

(a) the time of first return to A,

(b) the number of visits to D before returning to A,

(c) the number of visits to C before returning to A,

(d) the time of first return to A, given no prior visit by the particle to E,

(e) the number of visits to D before returning to A, given no prior visit by the particle to E.

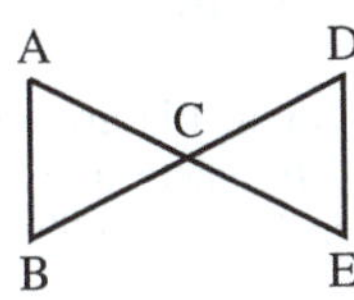

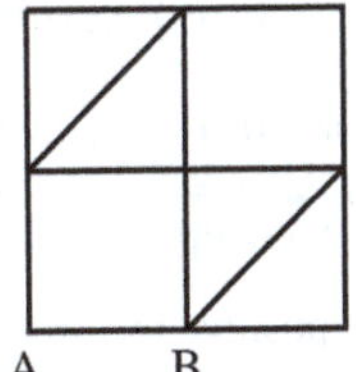

12. A particle starts at A and executes a symmetric random walk on the graph drawn above on the right. Find the expected number of visits to B before it returns to A.

13. Top-to-random shuffling. A pack contains 52 cards labelled $1, 2, \dots, 52$, and initially they are in increasing order from top to bottom. At each stage of the shuffling process, the top card is moved to one of the 52 available places determined by the other 51 cards, this place being chosen uniformly at random, independently of all previous stages. Find the mean number of stages until card 52 is first on top.

Show that, after the moment at which the card labelled 52 is inserted at random from the top, the order of the pack is uniformly distributed over the $52!$ possibilities.

14. Random-to-top shuffling. A pack contains 52 cards labelled $1, 2, \dots, 52$, and initially they are in increasing order from top to bottom. At each stage, a card is picked uniformly at random from the pack and placed on top, independently of all previous stages. Find the mean number of stages until every card has been selected at least once.

Show that, after the moment at which the final card to be selected at random is placed on top, the order of the pack is uniformly distributed over the $52!$ possibilities.

15. Quality control. Dick and Jim are writing exercises in sequence for inclusion in a textbook. Dick writes them and Jim checks them. Each exercise is faulty with probability p, independently of other exercises. Jim has two modes of operation. In Mode A, he inspects every exercise as it is produced. In Mode B, he inspects each exercise with probability $1/r$ where $r > 1$, independently of all other events.

Let $N \geq 1$. Jim operates in Mode A until he has found N consecutive non-defective exercises, at which point he changes to Mode B. He operates in Mode B until the first defective exercise is found, and then he reverts to Mode A.

Let X be the Markov chain which is in state i if Jim is operating in Mode A and the last i consecutive exercises since entering Mode A have been found to be non-defective, and is in state N if Jim is in Mode B.

(a) Write down the transition probabilities of X and find its stationary distribution.

(b) Show that the long run proportion of exercises that are inspected is $1/[1 + (r - 1)(1 - p^N)]$.

(c) Find an expression for the long run proportion of defective exercises which are not detected by Jim.

16. Exercise (3.11.39) revisited. A particle performs a random walk on the non-negative integers as follows. When at position $k \geq 0$, its next position is uniformly distributed on the set $\{0, 1, \dots, k, k+1\}$. Show that the sequence of positions forms an aperiodic, positive recurrent Markov chain, and find its stationary distribution.

Find the mean number μ of steps required to reach position 0 for the first time from position 1.

17. Let $\{X_n : n \geq 0\}$ be an irreducible Markov chain with state space S and transition matrix $\mathbf{P}$ (the chain may be either transient or recurrent). Let $k \in S$ and let $\mathbf{x}$ be a stationary measure such that $x_k = 1$. Prove that $\mathbf{x} \geq \boldsymbol{\rho}(k)$ where $\boldsymbol{\rho}(k)$ is given in equation (6.4.5). If the chain is recurrent, show that $\mathbf{x} = \boldsymbol{\rho}(k)$.

6.5 Exercises. Reversibility

1. A random walk on the set $\{0, 1, 2, \ldots, b\}$ has transition matrix given by $p_{00} = 1 - \lambda_0$, $p_{bb} = 1 - \mu_b$, $p_{i,i+1} = \lambda_i$ and $p_{i+1,i} = \mu_{i+1}$ for $0 \le i < b$, where $0 < \lambda_i, \mu_i < 1$ for all i, and $\lambda_i + \mu_i = 1$ for $1 \le i < b$. Show that this process is reversible in equilibrium.

2. Let X be an irreducible, positive recurrent, aperiodic Markov chain on the state space S.

(a) **Kolmogorov's reversibility criterion.** Show that X is reversible in equilibrium if and only if

$$p_{j_1,j_2} p_{j_2,j_3} \cdots p_{j_{n-1},j_n} p_{j_n,j_1} = p_{j_1,j_n} p_{j_n,j_{n-1}} \cdots p_{j_2,j_1}$$

for all n and all finite sequences $j_1, j_2, \ldots, j_n$ of states.

(b) **Kelly's reversibility condition.** Show that X is reversible in equilibrium if, for all distinct triples $i, j, k \in S$,

$$p_{ij} p_{jk} p_{ki} = p_{ik} p_{kj} p_{ji},$$

and in addition there exists $c \in S$ such that $p_{ic} > 0$ for all $i \ne c$.

(c) Consider a chain with $n \ge 3$ states. Show that Kolmogorov's criterion, as expressed above, may require the verification of up to $\frac{1}{2}\sum_{r=3}^{n} \binom{n}{r}(r-1)!$ equations, whereas Kelly's condition, if appropriate, requires no more than $\binom{n-1}{3}$.

(d) Show that a random walk on a finite tree is reversible in equilibrium.

3. Let X be a reversible Markov chain, and let C be a non-empty subset of the state space S. Define the Markov chain Y on S by the transition matrix $\mathbf{Q} = (q_{ij})$ where

$$q_{ij} = \begin{cases} \beta p_{ij} & \text{if } i \in C \text{ and } j \notin C, \\ p_{ij} & \text{otherwise,} \end{cases}$$

for $i \ne j$, and where β is a constant satisfying $0 < \beta < 1$. The diagonal terms q_{ii} are arranged so that $\mathbf{Q}$ is a stochastic matrix. Show that Y is reversible in equilibrium, and find its stationary distribution. Describe the situation in the limit as $\beta \downarrow 0$.

4. Can a reversible chain be periodic?

5. **Ehrenfest dog–flea model.** The dog–flea model of Example (6.5.5) is a Markov chain X on the state space $\{0, 1, \ldots, m\}$ with transition probabilities

$$p_{i,i+1} = 1 - \frac{i}{m}, \quad p_{i,i-1} = \frac{i}{m}, \quad \text{for} \quad 0 \le i \le m.$$

Show that, if $X_0 = i$,

$$\mathbb{E}\left(X_n - \frac{m}{2}\right) = \left(i - \frac{m}{2}\right)\left(1 - \frac{2}{m}\right)^n \to 0 \quad \text{as } n \to \infty.$$

6. Which of the following (when stationary) are reversible Markov chains?

(a) The chain $X = \{X_n\}$ having transition matrix $\mathbf{P} = \begin{pmatrix} 1-\alpha & \alpha \\ \beta & 1-\beta \end{pmatrix}$ where $\alpha + \beta > 0$.

(b) The chain $Y = \{Y_n\}$ having transition matrix $\mathbf{P} = \begin{pmatrix} 0 & p & 1-p \\ 1-p & 0 & p \\ p & 1-p & 0 \end{pmatrix}$ where $0 < p < 1$.

(c) $Z_n = (X_n, Y_n)$, where X_n and Y_n are independent and satisfy (a) and (b).

7. Let X_n, Y_n be independent simple random walks. Let Z_n be (X_n, Y_n) truncated to lie in the region $X_n \ge 0$, $Y_n \ge 0$, $X_n + Y_n \le a$ where a is integral. Find the stationary distribution of Z_n.

8. Show that an irreducible Markov chain with a finite state space and transition matrix $\mathbf{P}$ is reversible in equilibrium if and only if $\mathbf{P} = \mathbf{DS}$ for some symmetric matrix $\mathbf{S}$ and diagonal matrix $\mathbf{D}$ with strictly positive diagonal entries. Show further that for reversibility in equilibrium to hold, it is necessary but not sufficient that $\mathbf{P}$ has real eigenvalues.

9. **Random walk on a graph.** Let G be a finite connected graph with neither loops nor multiple edges, and let X be a random walk on G as in Exercise (6.4.6). Show that X is reversible in equilibrium.

10. Consider a random walk on the strictly positive integers with transition probabilities

$$p_{i,i-1} = \frac{1}{2} \cdot \frac{i+2}{i+1}, \qquad p_{i,i+1} = \frac{1}{2} \cdot \frac{i}{i+1}, \qquad i \geq 2,$$

and $p_{11} = \frac{3}{4}$, $p_{12} = \frac{1}{4}$. Show that the walk is positive recurrent and find the mean recurrence time of the state i.

11. An aleatory beetle performs a random walk on five vertices comprising the principal points (labelled n, e, s, w) and centre (labelled c) of a compass, with transition probabilities

$$\begin{aligned} p_{en} = p_{ws} &= \tfrac{1}{4}, \\ p_{cn} = p_{ne} = p_{ec} = p_{ce} = p_{sw} = p_{wn} &= \tfrac{1}{8}, \\ p_{nc} = p_{es} = p_{sc} = p_{cs} = p_{se} = p_{wc} &= \tfrac{1}{16}, \\ p_{cw} = p_{nw} &= \tfrac{1}{32}. \end{aligned}$$

Other moves have probability 0. Show that the mean recurrence time of the centre is $\mu_c = \frac{11}{2}$.

12. **Lazy Markov chain.** Let X be an irreducible (but not necessarily aperiodic) Markov chain on the countable state space S with transition matrix $\mathbf{P}$, with invariant distribution $\boldsymbol{\pi}$. Let $a \in (0, 1)$ and let $\mathbf{L} = a\mathbf{P} + (1 - a)\mathbf{I}$ where $\mathbf{I}$ is the identity matrix.

(a) Show that $\mathbf{L}$ is the transition matrix of an irreducible, aperiodic Markov chain Y with invariant distribution $\boldsymbol{\pi}$.

(b) Show that, if X is reversible in equilibrium, then so is Y.

6.6 Exercises. Chains with finitely many states

The first two exercises provide proofs that a Markov chain with finitely many states has a stationary distribution.

1. The Markov–Kakutani theorem asserts that, for any convex compact subset C of $\mathbb{R}^n$ and any linear continuous mapping T of C into C, T has a fixed point (in the sense that $T(x) = x$ for some $x \in C$). Use this to prove that a finite stochastic matrix has a non-negative non-zero left eigenvector corresponding to the eigenvalue 1.

2. Let $\mathbf{T}$ be a $m \times n$ matrix and let $\mathbf{v} \in \mathbb{R}^n$. Farkas's theorem asserts that exactly one of the following holds:

(i) there exists $\mathbf{x} \in \mathbb{R}^m$ such that $\mathbf{x} \geq \mathbf{0}$ and $\mathbf{xT} = \mathbf{v}$,

(ii) there exists $\mathbf{y} \in \mathbb{R}^n$ such that $\mathbf{yv}' < 0$ and $\mathbf{Ty}' \geq \mathbf{0}$.

Use this to prove that a finite stochastic matrix has a non-negative non-zero left eigenvector corresponding to the eigenvalue 1.

3. **Arbitrage.** Suppose you are betting on a race with m possible outcomes. There are n bookmakers, and a unit stake with the ith bookmaker yields t_{ij} if the jth outcome of the race occurs. A vector

$\mathbf{x} = (x_1, x_2, \dots, x_n)$, where $x_r \in (-\infty, \infty)$ is your stake with the rth bookmaker, is called a *betting scheme*. Show that exactly one of (a) and (b) holds:

(a) there exists a probability mass function $\mathbf{p} = (p_1, p_2, \dots, p_m)$ such that $\sum_{j=1}^{m} t_{ij} p_j = 0$ for all values of i,

(b) there exists a betting scheme $\mathbf{x}$ for which you surely win, that is, $\sum_{i=1}^{n} x_i t_{ij} > 0$ for all j.

4. Let X be a Markov chain with state space $S = \{1, 2, 3\}$ and transition matrix

$$\mathbf{P} = \begin{pmatrix} 1-p & p & 0 \\ 0 & 1-p & p \\ p & 0 & 1-p \end{pmatrix}$$

where $0 < p < 1$. Prove that

$$\mathbf{P}^n = \begin{pmatrix} a_{1n} & a_{2n} & a_{3n} \\ a_{3n} & a_{1n} & a_{2n} \\ a_{2n} & a_{3n} & a_{1n} \end{pmatrix}$$

where $a_{1n} + \omega a_{2n} + \omega^2 a_{3n} = (1 - p + p\omega)^n$, ω being a complex cube root of 1.

5. Let $\mathbf{P}$ be the transition matrix of a Markov chain with finite state space. Let $\mathbf{I}$ be the identity matrix, $\mathbf{U}$ the $|S| \times |S|$ matrix with all entries unity, and $\mathbf{1}$ the row $|S|$-vector with all entries unity. Let $\boldsymbol{\pi}$ be a non-negative vector with $\sum_i \pi_i = 1$. Show that $\boldsymbol{\pi}\mathbf{P} = \boldsymbol{\pi}$ if and only if $\boldsymbol{\pi}(\mathbf{I} - \mathbf{P} + \mathbf{U}) = \mathbf{1}$. Deduce that if $\mathbf{P}$ is irreducible then $\boldsymbol{\pi} = \mathbf{1}(\mathbf{I} - \mathbf{P} + \mathbf{U})^{-1}$.

6. Chess. A chess piece performs a random walk on a chessboard; at each step it is equally likely to make any one of the available moves. What is the mean recurrence time of a corner square if the piece is a: (a) king? (b) queen? (c) bishop? (d) knight? (e) rook?

7. Chess continued. A rook and a bishop perform independent symmetric random walks with synchronous steps on a 4×4 chessboard (16 squares). If they start together at a corner, show that the expected number of steps until they meet again at the same corner is 448/3.

8. Find the n-step transition probabilities $p_{ij}(n)$ for the chain X having transition matrix

$$\mathbf{P} = \begin{pmatrix} 0 & \frac{1}{2} & \frac{1}{2} \\ \frac{1}{3} & \frac{1}{4} & \frac{5}{12} \\ \frac{2}{3} & \frac{1}{4} & \frac{1}{12} \end{pmatrix}.$$

9. The 'PageRank' Markov chain. The pages on the worldwide web form a directed graph W with n vertices (representing pages) joined by directed edges (representing links). The existence of a link from i to j is denoted by $i \to j$, and the graph is specified by its adjacency matrix $L = (l_{ij})$ where $l_{ij} = 1$ if $i \to j$ and $l_{ij} = 0$ otherwise. The *out-degree* d_i (respectively, *in-degree* c_i) of vertex i is the number of links pointing away from i (respectively, towards i). Vertex i is said to *dangle* if $d_i = 0$.

The behaviour of a swiftly bored web surfer is modelled by a random walk on W. Let $b \in (0, 1)$. From any dangling vertex, the random walk moves to a randomly chosen vertex of W, each vertex having probability $1/n$. When at a non-dangling vertex i, with probability $b < 1$ the walk moves to a random linked vertex (each having probability $1/d_i$), while with probability $1 - b$ it moves to a random vertex of W (each having probability $1/n$).

(a) Show that the transition matrix $\mathbf{P}$ may be written in the form $\mathbf{P} = b\mathbf{Q} + \mathbf{v}'\mathbf{e}$, where $\mathbf{Q} = (q_{ij})$ with

$$q_{ij} = \begin{cases} 1/d_i & \text{if } i \to j, \\ 1/n & \text{if } i \text{ dangles}, \\ 0 & \text{otherwise}, \end{cases}$$

and $\mathbf{v} = (v_i)$ is a row vector with $v_i = (1-b)/n$, and $\mathbf{e}$ is a row vector with entries 1.

(b) Deduce that the stationary distribution $\boldsymbol{\pi}$ is given by $\boldsymbol{\pi} = \{(1-b)/n\}\mathbf{e}(\mathbf{I} - b\mathbf{Q})^{-1}$ where $\mathbf{I}$ is the identity matrix.

(c) Explain why the elements of $\boldsymbol{\pi}$, when rearranged in decreasing order, supply a description of the relative popularities of web pages (called 'PageRank' by Google, that being their trademark for the patented algorithm).

10. Let $\mathbf{P}$ be the transition matrix of an irreducible Markov chain on a finite state space, and let $\boldsymbol{\pi}$ be a left eigenvector of $\mathbf{P}$ corresponding to the eigenvalue 1. Show from the equation $\boldsymbol{\pi} = \boldsymbol{\pi}\mathbf{P}$ directly that the entries of $\boldsymbol{\pi}$ are either all positive or all negative, and hence prove Theorem (6.6.1d): there exists a unique distribution $\boldsymbol{\pi}$ satisfying $\boldsymbol{\pi}\mathbf{P} = \boldsymbol{\pi}$, and, furthermore, all components of $\boldsymbol{\pi}$ are strictly positive.

6.7 Exercises. Branching processes revisited

1. Let Z_n be the size of the nth generation of a branching process with $Z_0 = 1$ and $\mathbb{P}(Z_1 = k) = 2^{-k}$ for $k \geq 0$. Show directly that, as $n \to \infty$, $\mathbb{P}(Z_n \leq 2yn \mid Z_n > 0) \to 1 - e^{-2y}$, $y > 0$, in agreement with Theorem (6.7.8).

2. Let Z be a supercritical branching process with $Z_0 = 1$ and family-size generating function G. Assume that the probability η of extinction satisfies $0 < \eta < 1$. Find a way of describing the process Z, *conditioned on its ultimate extinction.*

3. Let Z_n be the size of the nth generation of a branching process with $Z_0 = 1$ and $\mathbb{P}(Z_1 = k) = qp^k$ for $k \geq 0$, where $p + q = 1$ and $p > \frac{1}{2}$. Use your answer to Exercise (6.7.2) to show that, if we condition on the ultimate extinction of Z, then the process grows in the manner of a branching process with generation sizes $\widetilde{Z}_n$ satisfying $\widetilde{Z}_0 = 1$ and $\mathbb{P}(\widetilde{Z}_1 = k) = pq^k$ for $k \geq 0$.

4. (a) Show that $\mathbb{E}(X \mid X > 0) \leq \mathbb{E}(X^2)/\mathbb{E}(X)$ for any random variable X taking non-negative values.

(b) Let Z_n be the size of the nth generation of a branching process with $Z_0 = 1$ and $\mathbb{P}(Z_1 = k) = qp^k$ for $k \geq 0$, where $p > \frac{1}{2}$. Use part (a) to show that $\mathbb{E}(Z_n/\mu^n \mid Z_n > 0) \leq 2p/(p-q)$, where $\mu = p/q$.

(c) Show that, in the notation of part (b), $\mathbb{E}(Z_n/\mu^n \mid Z_n > 0) \to p/(p-q)$ as $n \to \infty$.

6.8 Exercises. Birth processes and the Poisson process

1. Superposition. Flies and wasps land on your dinner plate in the manner of independent Poisson processes with respective intensities λ and μ. Show that the arrivals of flying objects form a Poisson process with intensity $\lambda + \mu$.

2. Thinning. Insects land in the soup in the manner of a Poisson process with intensity λ, and each such insect is green with probability p, independently of the colours of all other insects. Show that the arrivals of green insects form a Poisson process with intensity λp.

3. Let T_n be the time of the nth arrival in a Poisson process N with intensity λ, and define the excess lifetime process $E(t) = T_{N(t)+1} - t$, being the time one must wait subsequent to t before the next arrival. Show by conditioning on T_1 that

$$\mathbb{P}\big(E(t) > x\big) = e^{-\lambda(t+x)} + \int_0^t \mathbb{P}\big(E(t-u) > x\big)\lambda e^{-\lambda u}\,du.$$

Solve this integral equation in order to find the distribution function of $E(t)$. Explain your conclusion.

4. Let B be a simple birth process of paragraph (6.8.15b) with $B(0) = I$; the birth rates are $\lambda_n = n\lambda$. Write down the forward system of equations for the process and deduce that

$$\mathbb{P}\big(B(t) = k\big) = \binom{k-1}{I-1} e^{-I\lambda t}\big(1 - e^{-\lambda t}\big)^{k-I}, \qquad k \geq I.$$

Show also that $\mathbb{E}(B(t)) = Ie^{\lambda t}$ and $\operatorname{var}(B(t)) = Ie^{2\lambda t}(1 - e^{-\lambda t})$.

5. Let B be a process of simple birth with immigration (6.8.11c) with parameters λ and ν, and with $B(0) = 0$; the birth rates are $\lambda_n = n\lambda + \nu$. Write down the sequence of differential–difference equations for $p_n(t) = \mathbb{P}(B(t) = n)$. Without solving these equations, use them to show that $m(t) = \mathbb{E}(B(t))$ satisfies $m'(t) = \lambda m(t) + \nu$, and solve for $m(t)$.

6. Let N be a birth process with intensities $\lambda_0, \lambda_1, \dots$, and let $N(0) = 0$. Show that $p_n(t) = \mathbb{P}(N(t) = n)$ is given by

$$p_n(t) = \frac{1}{\lambda_n} \sum_{i=0}^{n} \lambda_i e^{-\lambda_i t} \prod_{\substack{j=0 \\ j \neq i}}^{n} \frac{\lambda_j}{\lambda_j - \lambda_i}$$

provided that $\lambda_i \neq \lambda_j$ whenever $i \neq j$.

7. Suppose that the general birth process of the previous exercise is such that $\sum_n \lambda_n^{-1} < \infty$. Show that $\lambda_n p_n(t) \to f(t)$ as $n \to \infty$ where f is the density function of the random variable $T = \sup\{t : N(t) < \infty\}$. Deduce that $\mathbb{E}(N(t) \mid N(t) < \infty)$ is finite or infinite depending on the convergence or divergence of $\sum_n n\lambda_n^{-1}$.

Find the Laplace transform of f in closed form for the case when $\lambda_n = (n + \frac{1}{2})^2$, and deduce an expression for f.

8. Traffic lights. A traffic light is green at time 0, and subsequently alternates between green and red at the instants of a Poisson process with intensity λ. Starting from time $x > 0$, let $W(x)$ be the waiting time until the light is green for the first time. Find the distribution of $W(x)$.

9. Conditional property of simple birth. Let $X = \{X(t) : t \geq 0\}$ be a simple birth process with rate λ, and let $X(0) = 1$. Let $b \geq 1$. Show that, conditional on the event $\{X(t) = b + 1\}$, the times of the b births have the same distribution as the order statistics of a random sample of size b from the density function

$$f(x) = \frac{\lambda e^{-\lambda(t-x)}}{1 - e^{-\lambda t}}, \qquad 0 \leq x \leq t.$$

10. Coincidences. Boulders fall down a chute (or couloir) at the instants of a Poisson process with intensity λ, and mountaineers ascend the chute at the instants of a Poisson process with intensity μ (the two processes are independent of one another). If a fall and an ascent occur during any interval of length c or less, it is said that a *coincidence* has occurred. Show that the time T until the first coincidence has mean

$$\mathbb{E}T = \frac{1}{\lambda + \mu}\left\{1 + \frac{\lambda^2 + \mu^2}{\lambda\mu} + 2e^{-(\lambda+\mu)c} - e^{-2(\lambda+\mu)c}\right\} \Big/ \left\{1 - e^{-2(\lambda+\mu)c}\right\}.$$

Show that $c\mathbb{E}(T) \to (2\lambda\mu)^{-1}$ as $c \downarrow 0$. Can you prove the last result directly?

11. Let S_1, S_2 be the times of the first two arrivals in a Poisson process with intensity λ, in order of arrival. Show that

$$\mathbb{P}(s < S_1 \leq t < S_2) = \lambda(t - s)e^{-\lambda t}, \qquad 0 < s < t < \infty,$$

and deduce the joint density function of S_1 and S_2.

12. Gig economy. A freelance salesman is paid R units for each sale, and commissions arrive at the instants of a Poisson process of rate λ. Living costs consume his resources at unit rate. If his initial wealth is $S > 0$, show that the probability that he ever becomes bankrupt is $a^{S/R}$, where a is the smallest $x > 0$ such that $x = e^{\lambda R(x-1)}$.

Prove that $a \in (0, 1)$ if $\lambda R > 1$, while $a = 1$ if $\lambda R < 1$.

6.9 Exercises. Continuous-time Markov chains

1. Let $\lambda\mu > 0$ and let X be a Markov chain on $\{1, 2\}$ with generator

$$\mathbf{G} = \begin{pmatrix} -\mu & \mu \\ \lambda & -\lambda \end{pmatrix}.$$

(a) Write down the forward equations and solve them for the transition probabilities $p_{ij}(t)$, $i, j = 1, 2$.

(b) Calculate $\mathbf{G}^n$ and hence find $\sum_{n=0}^{\infty}(t^n/n!)\mathbf{G}^n$. Compare your answer with that to part (a).

(c) Solve the equation $\boldsymbol{\pi}\mathbf{G} = \mathbf{0}$ in order to find the stationary distribution. Verify that $p_{ij}(t) \to \pi_j$ as $t \to \infty$.

2. As a continuation of the previous exercise, find:

(a) $\mathbb{P}(X(t) = 2 \mid X(0) = 1, X(3t) = 1)$,

(b) $\mathbb{P}(X(t) = 2 \mid X(0) = 1, X(3t) = 1, X(4t) = 1)$.

3. Jobs arrive in a computer queue in the manner of a Poisson process with intensity λ. The central processor handles them one by one in the order of their arrival, and each has an exponentially distributed runtime with parameter μ, the runtimes of different jobs being independent of each other and of the arrival process. Let $X(t)$ be the number of jobs in the system (either running or waiting) at time t, where $X(0) = 0$. Explain why X is a Markov chain, and write down its generator. Show that a stationary distribution exists if and only if $\lambda < \mu$, and find it in this case.

4. Pasta property. Let $X = \{X(t) : t \geq 0\}$ be a Markov chain having stationary distribution $\boldsymbol{\pi}$. We may sample X at the times of a Poisson process: let N be a Poisson process with intensity λ, independent of X, and define $Y_n = X(T_n)$. Show that $Y = \{Y_n : n \geq 0\}$ is a discrete-time Markov chain with the same stationary distribution as X. (This exemplifies the 'Pasta' property: Poisson arrivals see time averages.)

[The full assumption of the independence of N and X is not necessary for the conclusion. It suffices that $\{N(s) : s \geq t\}$ be independent of $\{X(s) : s \leq t\}$, a property known as 'lack of anticipation'. It is not even necessary that X be Markov; the Pasta property holds for many suitable ergodic processes.]

5. Hitting probabilities. Let X be a continuous-time Markov chain with generator $\mathbf{G}$ satisfying $g_i = -g_{ii} > 0$ for all i. Let $H_A = \inf\{t \geq 0 : X(t) \in A\}$ be the hitting time of the set A of states, and let $\eta_j = \mathbb{P}_j(H_A < \infty)$ be the chance of ever reaching A from j. By using properties of the jump chain, which you may assume to be well behaved, show that $\sum_k g_{jk}\eta_k = 0$ for $j \notin A$.

6. Mean hitting times. In continuation of the preceding exercise, let $\mu_j = \mathbb{E}_j(H_A)$. Show that the vector $\boldsymbol{\mu}$ is the minimal non-negative solution of the equations

$$\mu_j = 0 \quad \text{if } j \in A, \qquad 1 + \sum_{k \in S} g_{jk}\mu_k = 0 \quad \text{if } j \notin A.$$

7. Let X be a continuous-time Markov chain with transition probabilities $p_{ij}(t)$ and define $F_i = \inf\{t > T_1 : X(t) = i\}$ where T_1 is the time of the first jump of X. Show that, if $g_{ii} \neq 0$, then $\mathbb{P}_i(F_i < \infty) = 1$ if and only if i is recurrent.

8. Let X be the simple symmetric random walk on the integers in continuous time, so that

$$p_{i,i+1}(h) = p_{i,i-1}(h) = \tfrac{1}{2}\lambda h + o(h).$$

Show that the walk is recurrent. Let T be the time spent visiting m during an excursion from 0. Find the distribution of T.

9. Let i be a transient state of a continuous-time Markov chain X with $X(0) = i$. Show that the total time spent in state i has an exponential distribution.

10. Let X be an asymmetric simple random walk in continuous time on the non-negative integers with retention at 0, so that

$$p_{ij}(h) = \begin{cases} \lambda h + o(h) & \text{if } j = i+1,\ i \geq 0, \\ \mu h + o(h) & \text{if } j = i-1,\ i \geq 1. \end{cases}$$

Suppose that $X(0) = 0$ and $\lambda > \mu$. Show that the total time V_r spent in state r is exponentially distributed with parameter $\lambda - \mu$.

Assume now that $X(0)$ has some general distribution with probability generating function G. Find the expected amount of time spent at 0 in terms of G.

11. Let X be the continuous-time Markov chain on the finite state space S with generator $\mathbf{G} = (g_{ij})$. Show from first principles that the transition probabilities satisfy

$$p_{ij}(h) = \begin{cases} 1 + g_{ii}h + o(h) & \text{if } i = j, \\ g_{ij}h + o(h) & \text{if } i \neq j. \end{cases}$$

12. Let X be a Markov chain on the integers $\mathbb{Z}$ with generator satisfying $g_{i,i-1} = g_{i,i+1} = 2^i$ for $i \in \mathbb{Z}$, and $g_{i,j} = 0$ for other pairs (i, j) with $i \neq j$. Does X explode?

13. Growth with annihilation. A population grows subject to the threat of total annihilation. It is modelled as a Markov chain with generator $\mathbf{G} = (g_{ij})$ satisfying

$$g_{i,i+1} = \frac{1}{i+2}, \qquad i \geq 0,$$
$$g_{i,0} = \frac{1}{(i+1)(i+2)}, \qquad i \geq 1,$$

the other off-diagonal elements of $\mathbf{G}$ being 0. Show that the chain is null recurrent.

14. Skeletons. Let $Z_n = X(nh)$ where X is a Markov chain and $h > 0$. Show that i is recurrent for Z if and only if it is recurrent for X. Show that Z is irreducible if and only if X is irreducible.

6.10 Exercises. Kolmogorov equations and the limit theorem

1. Let $N < \infty$, and let $\mathcal{Q}$ be the space of $N \times N$ matrices with real entries, with norm

$$|Q| = \sup_{\mathbf{x} \neq \mathbf{0}} \frac{|Q\mathbf{x}|}{|\mathbf{x}|}, \qquad Q \in \mathcal{Q},$$

where $|\mathbf{y}|$ is the Euclidean norm of the vector $\mathbf{y}$, and the supremum is over all non-zero column vectors.
(a) Show for $Q_1, Q_2 \in \mathcal{Q}$ that

$$|Q_1 + Q_2| \leq |Q_1| + |Q_2|, \qquad |Q_1 Q_2| \leq |Q_1| \cdot |Q_2|.$$

(b) Show for $Q \in \mathcal{Q}$ that $E_n := \sum_{k=0}^{n} Q^k/k!$ converges with respect to $|\cdot|$ to a limit which we denote $E = E(Q)$.

(c) Show that $E(Q_1 + Q_2) = E(Q_1)E(Q_2)$ if Q_1 and Q_2 are commuting elements of $\mathcal{Q}$.

2. Let Y be an irreducible discrete-time Markov chain on a countably infinite state space S, having transition matrix $\mathbf{Y} = (y_{ij})$ satisfying $y_{ii} = 0$ for all states i, and with stationary distribution $\boldsymbol{\nu}$. Construct a continuous-time process X on S for which Y is the jump chain, such that X has no stationary distribution.

3. Let $X = (X(t) : t \geq 0)$ be a Markov chain on $\mathbb{Z}$ with generator $\mathbf{G} = (g_{ij})$ given by

$$g_{i,i-1} = i^2 + 1, \quad g_{i,i} = -2(i^2 + 1), \quad g_{i,i+1} = i^2 + 1, \qquad i \in \mathbb{Z}.$$

Show that X is recurrent. Is X positive recurrent?

4. Let X be a Markov chain on $\mathbb{Z}$ with generator $G = (g_{ij})$ satisfying

$$g_{i,i-1} = 3^{|i|}, \quad g_{i,i} = -3^{|i|+1}, \quad g_{i,i+1} = 2 \cdot 3^{|i|}, \qquad i \in \mathbb{Z}.$$

Show that X is transient, but has an invariant distribution. Explain.

5. Let $X = (X_t : t \geq 0)$ be a Markov chain with generator $\mathbf{G} = (g_{ij})$ on the finite state space S, and let $f : S \to \mathbb{R}$ be a function, which we identify with the vector $f = (f(i) : i \in S)$. Show that

$$\mathbf{G}f(i) = \sum_{j \in S} g_{ij}\big(f(j) - f(i)\big),$$

where $\mathbf{G}f$ denotes the standard matrix multiplication.

Show that

$$\mathbf{G}f(i) = \lim_{t \to 0} \frac{1}{t}\big[\mathbb{E}_i f(X_t) - f(i)\big], \qquad i \in S,$$

and deduce that

$$\mathbb{E}_i f(X_t) = f(i) + \int_0^t \mathbb{E}_i(\mathbf{G}f(X_s))\,ds.$$

6.11 Exercises. Birth–death processes and imbedding

1. Describe the jump chain for a birth–death process with rates λ_n and μ_n.

2. Consider an immigration–death process X, being a birth–death process with birth rates $\lambda_n = \lambda$ and death rates $\mu_n = n\mu$. Find the transition matrix of the jump chain Y, and show that it has as stationary distribution

$$\pi_n = \frac{1}{2(n!)}\left(1 + \frac{n}{\rho}\right)\rho^n e^{-\rho}$$

where $\rho = \lambda/\mu$. Explain why this differs from the stationary distribution of X.

3. Consider the birth–death process X with $\lambda_n = n\lambda$ and $\mu_n = n\mu$ for all $n \geq 0$. Suppose $X(0) = 1$ and let $\eta(t) = \mathbb{P}_1(X(t) = 0)$. Show that η satisfies the differential equation

$$\eta'(t) + (\lambda + \mu)\eta(t) = \mu + \lambda\eta(t)^2.$$

Hence find $\eta(t)$, and calculate $\mathbb{P}_1(X(t) = 0 \mid X(u) = 0)$ for $0 < t < u$.

4. For the birth–death process of the previous exercise with $\lambda < \mu$, show that the distribution of $X(t)$, conditional on the event $\{X(t) > 0\}$, converges as $t \to \infty$ to a geometric distribution.

5. Let X be a birth–death process with $\lambda_n = n\lambda$ and $\mu_n = n\mu$, and suppose $X(0) = 1$. Show that the time T at which $X(t)$ first takes the value 0 satisfies

$$\mathbb{E}(T \mid T < \infty) = \begin{cases} \dfrac{1}{\lambda}\log\left(\dfrac{\mu}{\mu-\lambda}\right) & \text{if } \lambda < \mu, \\[2ex] \dfrac{1}{\mu}\log\left(\dfrac{\lambda}{\lambda-\mu}\right) & \text{if } \lambda > \mu. \end{cases}$$

What happens when $\lambda = \mu$?

6. Let X be the birth–death process of Exercise (6.11.5) with $\lambda \neq \mu$, and let $V_r(t)$ be the total amount of time the process has spent in state $r \geq 0$, up to time t. Find the distribution of $V_1(\infty)$ and the generating function $\sum_r s^r \mathbb{E}(V_r(t))$. Hence show in two ways that $\mathbb{E}(V_1(\infty)) = [\max\{\lambda, \mu\}]^{-1}$. Show further that $\mathbb{E}(V_r(\infty)) = \lambda^{r-1} r^{-1}[\max\{\lambda, \mu\}]^{-r}$.

7. Repeat the calculations of Exercise (6.11.6) in the case $\lambda = \mu$.

8. Consider a birth–death process X with birth rates $\lambda_n > 0$ for $n \geq 0$, death rates $\mu_n > 0$ for $n > 0$, and $\mu_0 = 0$. Let $X(0) = n > 0$, and let D_n be the time until the process first takes the value $n - 1$.

(a) Show that $d_n = \mathbb{E}(D_n)$ satisfies

$$\lambda_n d_{n+1} = \mu_n d_n - 1, \qquad n \geq 1.$$

(b) Show that the moment generating function $M_n(\theta) = \mathbb{E}(e^{\theta D_n})$ satisfies

$$(\lambda_n + \mu_n - \theta)M_n(\theta) = \mu_n + \lambda_n M_n(\theta)M_{n+1}(\theta), \qquad n \geq 1.$$

9. Biofilms. In a model for a biofilm population, we assume there are n colonizable 'niches' (or 'food sources'). Let $X(t)$ be the number of occupied niches at time t, and assume X is a Markov chain that evolves as follows. The lifetime of any colony is exponentially distributed with parameter μ; if $X(t) = i$, the rate of establishment of a new colony in an empty niche is $\lambda i(n - i)$. The usual independence may be assumed.

For $X(0) = 1, 2, \ldots$, find the mean time until the population is extinct, which is to say that no niches are occupied. Discuss the implications when n is large.

6.12 Exercises. Special processes

1. Customers entering a shop are served in the order of their arrival by the single server. They arrive in the manner of a Poisson process with intensity λ, and their service times are independent exponentially distributed random variables with parameter μ. By considering the jump chain, show that the expected duration of a busy period B of the server is $(\mu - \lambda)^{-1}$ when $\lambda < \mu$. (The busy period runs from the moment a customer arrives to find the server free until the earliest subsequent time when the server is again free.)

2. Disasters. Immigrants arrive at the instants of a Poisson process of rate ν, and each independently founds a simple birth process of rate λ. At the instants of an independent Poisson process of rate δ, the population is annihilated. Find the probability generating function of the population $X(t)$, given that $X(0) = 0$.

3. More disasters. In the framework of Exercise (6.12.2), suppose that each immigrant gives rise to a simple birth–death process of rates λ and μ. Show that the mean population size stays bounded if and only if $\delta > \lambda - \mu$.

4. The queue M/G/∞. An ftp server receives clients at the times of a Poisson process with parameter λ, beginning at time 0. The ith client remains connected for a length S_i of time, where the S_i are independent identically distributed random variables, independent of the process of arrivals. Assuming that the server has an infinite capacity, show that the number of clients being serviced at time t has the Poisson distribution with parameter $\lambda \int_0^t [1 - G(x)]\,dx$, where G is the common distribution function of the S_i. Show that the mean of this distribution converges to $\lambda\mathbb{E}(S)$ as $t \to \infty$.

6.13 Exercises. Spatial Poisson processes

1. In a certain town at time $t = 0$ there are no bears. Brown bears and grizzly bears arrive as independent Poisson processes B and G with respective intensities β and γ.

(a) Show that the first bear is brown with probability $\beta/(\beta + \gamma)$.

(b) Find the probability that between two consecutive brown bears, there arrive exactly r grizzly bears.

(c) Given that $B(1) = 1$, find the expected value of the time at which the first bear arrived.

2. Campbell–Hardy theorem. Let Π be the points of a non-homogeneous Poisson process on $\mathbb{R}^d$ with intensity function λ. Let $S = \sum_{\mathbf{x}\in\Pi} g(\mathbf{x})$ where g is a (measurable) function which we assume for convenience to be non-negative.

(a) Show directly that $\mathbb{E}(S) = \int_{\mathbb{R}^d} g(\mathbf{x})\lambda(\mathbf{x})\,d\mathbf{x}$ and $\mathrm{var}(S) = \int_{\mathbb{R}^d} g(\mathbf{x})^2\lambda(\mathbf{x})\,d\mathbf{x}$, provided these integrals converge.

(b) Show that

$$\mathbb{E}(e^{-tS}) = \exp\left\{-\int_{\mathbb{R}^d}(1 - e^{-tg(\mathbf{x})})\lambda(\mathbf{x})\,d\mathbf{x}\right\}, \qquad t > 0,$$

and deduce that $\mathbb{P}(S < \infty) = 1$ if $\int_{\mathbb{R}^d} \min\{1, g(\mathbf{x})\}\lambda(\mathbf{x})\,d\mathbf{x} < \infty$.

(c) If the integral condition of part (b) holds, show that the characteristic function ϕ of S satisfies

$$\phi(t) = \exp\left\{-\int_{\mathbb{R}^d}(1 - e^{itg(\mathbf{x})})\lambda(\mathbf{x})\,d\mathbf{x}\right\}, \qquad t \in \mathbb{R},$$

3. Let Π be a Poisson process with constant intensity λ on the surface of the sphere of $\mathbb{R}^3$ with radius 1. Let P be the process given by the (X, Y) coordinates of the points projected on a plane passing through the centre of the sphere. Show that P is a Poisson process, and find its intensity function.

4. Repeat Exercise (6.13.3), when Π is a homogeneous Poisson process on the ball $\{(x_1, x_2, x_3) : x_1^2 + x_2^2 + x_3^2 \le 1\}$.

5. You stick pins in a Mercator projection of the Earth in the manner of a Poisson process with constant intensity λ. What is the intensity function of the corresponding process on the globe? What would be the intensity function on the map if you formed a Poisson process of constant intensity λ of meteorite strikes on the surface of the Earth?

6. Shocks. The rth point T_r of a Poisson process N of constant intensity λ on $\mathbb{R}_+$ gives rise to an effect $X_r e^{-\alpha(t-T_r)}$ at time $t \ge T_r$, where the X_r are independent and identically distributed with finite variance. Find the mean and variance of the total effect $S(t) = \sum_{r=1}^{N(t)} X_r e^{-\alpha(t-T_r)}$ in terms of the first two moments of the X_r, and calculate $\mathrm{cov}(S(s), S(t))$.

What is the behaviour of the correlation $\rho(S(s), S(t))$ as $s \to \infty$ with $t - s$ fixed?

7. Let N be a non-homogeneous Poisson process on $\mathbb{R}_+$ with intensity function λ. Find the joint density of the first two inter-event times, and deduce that they are not in general independent.

8. Competition lemma. Let $\{N_r(t) : r \geq 1\}$ be a collection of independent Poisson processes on $\mathbb{R}_+$ with respective constant intensities $\{\lambda_r : r \geq 1\}$, such that $\sum_r \lambda_r = \lambda < \infty$. Set $N(t) = \sum_r N_r(t)$, and let I denote the index of the process supplying the first point in N, occurring at time T. Show that

$$\mathbb{P}(I = i,\ T \geq t) = \mathbb{P}(I = i)\mathbb{P}(T \geq t) = \frac{\lambda_i}{\lambda} e^{-\lambda t}, \qquad i \geq 1.$$

9. Poisson line process. Let Π be a Poisson process on $\mathbb{R}^2 \setminus \mathbf{0}$ with intensity function $\lambda(u, v) = (u^2 + v^2)^{-3/2}$. Each point $(U, V) \in \Pi$ gives rise to a line $Ux + Vy = 1$ in the x/y-plane $\mathbb{L}$.

(a) Let (p, θ) be the polar coordinates of the foot of the perpendicular from the origin onto the line $ux + vy = 1$ in the x/y-plane $\mathbb{L}$. Express (p, θ) in terms of (u, v).

(b) Show that the Poisson line process is mapped by the map of part (a) to a uniform Poisson process on the strip $S^{\perp} = \mathbb{R} \times [0, \pi)$.

(c) Show that the line process in $\mathbb{L}$ is invariant under translations and rotations of $\mathbb{L}$.

10. Attracted by traffic. A large continent is traversed by a doubly-infinite straight freeway on which lorries are parked at the points of a Poisson process with constant intensity 1. The masses of the lorries are independent, identically distributed random variables that are independent of the parking places. Let G be the gravitational attraction due to the lorries on a pedestrian of unit mass standing beside the freeway. You may take the gravitational constant to be 1.

Show that G has characteristic function of the form $\phi(t) = \exp(-c|t|^{1/2})$ where $c > 0$. Express c in terms of the mean of a typical mass M.

6.14 Exercises. Markov chain Monte Carlo

1. Let $\mathbf{P}$ be a stochastic matrix on the finite set Θ with stationary distribution $\boldsymbol{\pi}$. Define the inner product $\langle \mathbf{x}, \mathbf{y} \rangle = \sum_{k \in \Theta} x_k y_k \pi_k$, and let $l^2(\pi) = \{\mathbf{x} \in \mathbb{R}^{\Theta} : \langle \mathbf{x}, \mathbf{x} \rangle < \infty\}$. Show, in the obvious notation, that $\mathbf{P}$ is reversible with respect to π if and only if $\langle \mathbf{x}, \mathbf{P}\mathbf{y} \rangle = \langle \mathbf{P}\mathbf{x}, \mathbf{y} \rangle$ for all $\mathbf{x}, \mathbf{y} \in l^2(\pi)$.

2. Barker's algorithm. Show that a possible choice for the acceptance probabilities in Hastings's general algorithm is

$$b_{ij} = \frac{\pi_j g_{ji}}{\pi_i g_{ij} + \pi_j g_{ji}},$$

where $\mathbf{G} = (g_{ij})$ is the proposal matrix.

3. Let S be a countable set. For each $j \in S$, the sets A_{jk}, $k \in S$, form a partition of the interval $[0, 1]$. Let $g : S \times [0, 1] \to S$ be given by $g(j, u) = k$ if $u \in A_{jk}$. The sequence $\{X_n : n \geq 0\}$ of random variables is generated recursively by $X_{n+1} = g(X_n, U_{n+1})$, $n \geq 0$, where $\{U_n : n \geq 1\}$ are independent random variables with the uniform distribution on $[0, 1]$. Show that X is a Markov chain, and find its transition matrix.

4. Dobrushin's bound. Let $\mathbf{U} = (u_{st})$ be a finite $|S| \times |T|$ stochastic matrix. *Dobrushin's ergodic coefficient* is defined to be

$$d(\mathbf{U}) = \tfrac{1}{2} \sup_{i,j \in S} \sum_{t \in T} |u_{it} - u_{jt}|.$$

(a) Show that, if $\mathbf{V}$ is a finite $|T| \times |U|$ stochastic matrix, then $d(\mathbf{UV}) \leq d(\mathbf{U}) d(\mathbf{V})$.

(b) Let X and Y be discrete-time Markov chains with the same transition matrix $\mathbf{P}$, and show that

$$\sum_k \big|\mathbb{P}(X_n = k) - \mathbb{P}(Y_n = k)\big| \leq d(\mathbf{P})^n \sum_k \big|\mathbb{P}(X_0 = k) - \mathbb{P}(Y_0 = k)\big|.$$

5. Let π be a positive mass function on the finite set Θ, and let $\mathbf{P}$ be the transition matrix of an irreducible, aperiodic Markov chain with stationary distribution π. Let $W = (W(i) : i \in \Theta)$ be a vector of random variables such that $\mathbb{P}(W(i) = j) = p_{ij}$ for $i, j \in \Theta$, and use W as an update rule in the coupling-from-the-past algorithm for sampling from π.

(a) If the $W(i)$, $i \in \Theta$, are independent, show that the coalescence time is a.s. finite.

(b) Give two examples of situations in which the coalescence time is a.s. infinite.

6. Ising model. Show that the Ising distribution of Example (6.14.2) satisfies the FKG lattice condition (6.14.20).

6.15 Problems

1. Classify the states of the discrete-time Markov chains with state space $S = \{1, 2, 3, 4\}$ and transition matrices

$$\text{(a)} \quad \begin{pmatrix} \frac{1}{3} & \frac{2}{3} & 0 & 0 \\ \frac{1}{2} & \frac{1}{2} & 0 & 0 \\ \frac{1}{4} & 0 & \frac{1}{4} & \frac{1}{2} \\ 0 & 0 & 0 & 1 \end{pmatrix} \qquad \text{(b)} \quad \begin{pmatrix} 0 & \frac{1}{2} & \frac{1}{2} & 0 \\ \frac{1}{3} & 0 & 0 & \frac{2}{3} \\ 1 & 0 & 0 & 0 \\ 0 & 0 & 1 & 0 \end{pmatrix}.$$

In case (a), calculate $f_{34}(n)$, and deduce that the probability of ultimate absorption in state 4, starting from 3, equals $\frac{2}{3}$. Find the mean recurrence times of the states in case (b).

2. A transition matrix is called *doubly stochastic* if all its column sums equal 1, that is, if $\sum_i p_{ij} = 1$ for all $j \in S$.

(a) Show that if a finite chain has a doubly stochastic transition matrix, then all its states are positive recurrent, and that if it is, in addition, irreducible and aperiodic then $p_{ij}(n) \to N^{-1}$ as $n \to \infty$, where N is the number of states.

(b) Show that, if an infinite irreducible chain has a doubly stochastic transition matrix, then its states are either all null recurrent or all transient.

3. Prove that intercommunicating states of a Markov chain have the same period.

4. (a) Show that for each pair i, j of states of an irreducible aperiodic chain, there exists $N = N(i, j)$ such that $p_{ij}(n) > 0$ for all $n \geq N$.

(b) Let X and Y be independent irreducible aperiodic chains with the same state space S and transition matrix $\mathbf{P}$. Show that the bivariate chain $Z_n = (X_n, Y_n)$, $n \geq 0$, is irreducible and aperiodic.

(c) Show that the bivariate chain Z may be reducible if X and Y are periodic.

5. Suppose $\{X_n : n \geq 0\}$ is a discrete-time Markov chain with $X_0 = i$. Let N be the total number of visits made subsequently by the chain to the state j. Show that

$$\mathbb{P}(N = n) = \begin{cases} 1 - f_{ij} & \text{if } n = 0, \\ f_{ij}(f_{jj})^{n-1}(1 - f_{jj}) & \text{if } n \geq 1, \end{cases}$$

and deduce that $\mathbb{P}(N = \infty) = 1$ if and only if $f_{ij} = f_{jj} = 1$.

6. Let i and j be two states of a discrete-time Markov chain. Show that if i communicates with j, then there is positive probability of reaching j from i without revisiting i in the meantime. Deduce that, if the chain is irreducible and recurrent, then the probability f_{ij} of ever reaching j from i equals 1 for all i and j.

7. Let $\{X_n : n \geq 0\}$ be a recurrent irreducible Markov chain on the state space S with transition matrix $\mathbf{P}$, and let $\mathbf{x}$ be a positive solution of the equation $\mathbf{x} = \mathbf{x}\mathbf{P}$.

(a) Show that

$$q_{ij}(n) = \frac{x_j}{x_i} p_{ji}(n), \qquad i, j \in S, \ n \geq 1,$$

defines the n-step transition probabilities of a recurrent irreducible Markov chain on S whose first-passage probabilities are given by

$$g_{ij}(n) = \frac{x_j}{x_i} l_{ji}(n), \qquad i \neq j, \ n \geq 1,$$

where $l_{ji}(n) = \mathbb{P}_j(X_n = i, T > n)$ and $T = \min\{m > 0 : X_m = j\}$.

(b) Show that $\mathbf{x}$ is unique up to a multiplicative constant.

(c) Let $T_j = \min\{n \geq 1 : X_n = j\}$ and define $h_{ij} = \mathbb{P}_i(T_j \leq T_i)$. Show that $x_i h_{ij} = x_j h_{ji}$ for all $i, j \in S$.

8. Renewal sequences. The sequence $u = \{u_n : n \geq 0\}$ is called a 'renewal sequence' if

$$u_0 = 1, \quad u_n = \sum_{i=1}^{n} f_i u_{n-i} \qquad \text{for } n \geq 1,$$

for some collection $f = \{f_n : n \geq 1\}$ of non-negative numbers summing to 1.

(a) Show that u is a renewal sequence if and only if there exists a Markov chain X on a countable state space S such that $u_n = \mathbb{P}(X_n = s \mid X_0 = s)$, for some recurrent $s \in S$ and all $n \geq 1$.

(b) Show that if u and v are renewal sequences then so is $\{u_n v_n : n \geq 0\}$.

9. Consider the symmetric random walk in three dimensions on the set of points $\{(x, y, z) : x, y, z = 0, \pm 1, \pm 2, \dots\}$; this process is a sequence $\{\mathbf{X}_n : n \geq 0\}$ of points such that $\mathbb{P}(\mathbf{X}_{n+1} = \mathbf{X}_n + \boldsymbol{\epsilon}) = \frac{1}{6}$ for $\boldsymbol{\epsilon} = (\pm 1, 0, 0), (0, \pm 1, 0), (0, 0, \pm 1)$. Suppose that $\mathbf{X}_0 = (0, 0, 0)$. Show that

$$\mathbb{P}(\mathbf{X}_{2n} = (0,0,0)) = \left(\frac{1}{6}\right)^{2n} \sum_{i+j+k=n} \frac{(2n)!}{(i!\,j!\,k!)^2} = \left(\frac{1}{2}\right)^{2n} \binom{2n}{n} \sum_{i+j+k=n} \left(\frac{n!}{3^n i!\,j!\,k!}\right)^2$$

and deduce by Stirling's formula that the origin is a transient state.

10. Consider the three-dimensional version of the cancer model (6.12.12). If $\kappa = 1$, are the empires of Theorem (6.12.14) inevitable in this case?

11. Let X be a discrete-time Markov chain with state space $S = \{1, 2\}$, and transition matrix

$$\mathbf{P} = \begin{pmatrix} 1-\alpha & \alpha \\ \beta & 1-\beta \end{pmatrix}.$$

Classify the states of the chain. Suppose that $\alpha\beta > 0$ and $\alpha\beta \neq 1$. Find the n-step transition probabilities and show directly that they converge to the unique stationary distribution as $n \to \infty$. For what values of α and β is the chain reversible in equilibrium?

12. Another diffusion model. N black balls and N white balls are placed in two urns so that each contains N balls. After each unit of time one ball is selected at random from each urn, and the two balls thus selected are interchanged. Let the number of black balls in the first urn denote the state of the system. Write down the transition matrix of this Markov chain and find the unique stationary distribution. Is the chain reversible in equilibrium?

13. Consider a Markov chain on the set $S = \{0, 1, 2, \dots\}$ with transition probabilities $p_{i,i+1} = a_i$, $p_{i,0} = 1 - a_i$, $i \geq 0$, where $(a_i : i \geq 0)$ is a sequence of constants which satisfy $0 < a_i < 1$ for all i. Let $b_0 = 1$, $b_i = a_0 a_1 \cdots a_{i-1}$ for $i \geq 1$. Show that the chain is

(a) recurrent if and only if $b_i \to 0$ as $i \to \infty$,

(b) positive recurrent if and only if $\sum_i b_i < \infty$,

and write down the stationary distribution if the latter condition holds.

Let A and β be positive constants and suppose that $a_i = 1 - Ai^{-\beta}$ for all large i. Show that the chain is

(c) transient if $\beta > 1$,

(d) positive recurrent if $\beta < 1$.

Finally, if $\beta = 1$ show that the chain is

(e) positive recurrent if $A > 1$,

(f) null recurrent if $A \le 1$.

14. Let X be a continuous-time Markov chain with countable state space S and semigroup $\{\mathbf{P}_t\}$. Show that $p_{ij}(t)$ is a continuous function of t. Let $g(t) = -\log p_{ii}(t)$; show that g is a continuous function, $g(0) = 0$, and $g(s+t) \le g(s) + g(t)$. We say that g is 'subadditive', and a well known theorem gives the result that

$$\lim_{t \downarrow 0} \frac{g(t)}{t} = \lambda \quad \text{exists and} \quad \lambda = \sup_{t>0} \frac{g(t)}{t} \le \infty.$$

Deduce that the limit $g_{ii} = \lim_{t\downarrow 0} t^{-1}\{p_{ii}(t) - 1\}$ exists.

15. Let X be a continuous-time Markov chain with generator $\mathbf{G} = (g_{ij})$. Show that X is irreducible if and only if for any pair i, j of distinct states there exists a sequence $i, k_1, k_2, \dots, k_n, j$ of distinct states such that $g_{i,k_1} g_{k_1,k_2} \cdots g_{k_n,j} > 0$.

16. Reversibility.

(a) Let $T > 0$ and let $X = \{X(t) : 0 \le t \le T\}$ be an irreducible, non-explosive Markov chain with stationary distribution $\boldsymbol{\pi}$, and suppose that $X(0)$ has distribution $\boldsymbol{\pi}$. Let $Y(t) = X(T-t)$ for $0 \le t \le T$. We call X *reversible (in equilibrium)* if X and Y have the same joint distributions.

(i) Show that Y is a (left-continuous) Markov chain with transition probabilities $\widehat{p}_{ij}(t) = (\pi_j/\pi_i)p_{ji}(t)$ and generator $\widehat{\mathbf{G}}$ satisfying $\pi_j \widehat{g}_{ji} = \pi_i g_{ij}$, where the $p_{ji}(t)$ and $\mathbf{G} = (g_{ij})$ are those of X. Show that Y is irreducible and non-explosive with stationary distribution $\boldsymbol{\pi}$.

(ii) Show that X is reversible in equilibrium if and only if the *detailed balance equations* $\pi_i g_{ij} = \pi_j g_{ji}$ (for all i and j) hold.

(iii) Show that a measure $\boldsymbol{\nu}$ satisfies $\boldsymbol{\nu}\mathbf{G} = \mathbf{0}$ if it satisfies the detailed balance equations.

(b) Let X be irreducible and non-explosive with stationary distribution $\boldsymbol{\pi}$, and assume $X(0)$ has distribution $\boldsymbol{\pi}$.

(i) **Kolmogorov's criterion.** Show that X is reversible if and only if, for all n and all finite sequences $k_1, k_2, \dots, k_n$ of states,

$$g_{k_1,k_2} g_{k_2,k_3} \cdots g_{k_{n-1},k_n} g_{k_n,k_1} = g_{k_1,k_n} g_{k_n,k_{n-1}} \cdots g_{k_2,k_1}.$$

(ii) **Kelly's criterion.** Show that X is reversible if, for all distinct triples $i, j, k \in S$, we have $g_{ij} g_{jk} g_{ki} = g_{ik} g_{kj} g_{ji}$, and in addition there exists $c \in S$ such that $g_{ic} > 0$ for all $i \ne c$.

(c) Show that every irreducible chain X with exactly two states is reversible in equilibrium.

(d) Show that every non-explosive birth–death process X having a stationary distribution is reversible in equilibrium.

17. Elfving's imbedding problem. Show that not every discrete-time Markov chain can be imbedded in a continuous-time chain. More precisely, let

$$\mathbf{P} = \begin{pmatrix} \alpha & 1-\alpha \\ 1-\alpha & \alpha \end{pmatrix} \quad \text{for some } 0 < \alpha < 1$$

be a transition matrix. Show that there exists a semigroup $\{\mathbf{P}_t\}$ of transition probabilities in continuous time such that $\mathbf{P}_1 = \mathbf{P}$, if and only if $\frac{1}{2} < \alpha < 1$. In this case show that $\{\mathbf{P}_t\}$ is unique and calculate it in terms of α.

18. Consider an immigration–death process $X(t)$, being a birth–death process with rates $\lambda_n = \lambda$, $\mu_n = n\mu$. Show that its generating function $G(s,t) = \mathbb{E}(s^{X(t)})$ is given by

$$G(s,t) = \{1 + (s-1)e^{-\mu t}\}^I \exp\{\rho(s-1)(1-e^{-\mu t})\}$$

where $\rho = \lambda/\mu$ and $X(0) = I$. Deduce the limiting distribution of $X(t)$ as $t \to \infty$.

19. Let N be a non-homogeneous Poisson process on $\mathbb{R}_+ = [0,\infty)$ with intensity function λ. Write down the forward and backward equations for N, and solve them.

Let $N(0) = 0$, and find the density function of the time T until the first arrival in the process. If $\lambda(t) = c/(1+t)$, show that $\mathbb{E}(T) < \infty$ if and only if $c > 1$.

20. Successive offers for my house are independent identically distributed random variables X_1, $X_2, \dots$, having density function f and distribution function F. Let $Y_1 = X_1$, let Y_2 be the first offer exceeding Y_1, and generally let Y_{n+1} be the first offer exceeding Y_n. Show that $Y_1, Y_2, \dots$ are the times of arrivals in a non-homogeneous Poisson process with intensity function $\lambda(t) = f(t)/(1-F(t))$. The Y_i are called 'record values'.

Now let Z_1 be the first offer received which is the second largest to date, and let Z_2 be the second such offer, and so on. Show that the Z_i are the arrival times of a non-homogeneous Poisson process with intensity function λ.

21. Let N be a Poisson process with constant intensity λ, and let $Y_1, Y_2, \dots$ be independent random variables with common characteristic function ϕ and density function f. The process $N^*(t) = Y_1 + Y_2 + \cdots + Y_{N(t)}$ is called a *compound* Poisson process. Y_n is the change in the value of N^* at the nth arrival of the Poisson process N. Think of it like this. A 'random alarm clock' rings at the arrival times of a Poisson process. At the nth ring the process N^* accumulates an extra quantity Y_n. Write down a forward equation for N^* and hence find the characteristic function of $N^*(t)$. Can you see directly why it has the form which you have found?

22. If the intensity function λ of a non-homogeneous Poisson process N is itself a random process, then N is called a *doubly stochastic* Poisson process (or *Cox process*).

(a) Consider the case when $\lambda(t) = \Lambda$ for all t, and Λ is a random variable taking either of two values λ_1 or λ_2, each being picked with equal probability $\frac{1}{2}$. Find the probability generating function of $N(t)$, and deduce its mean and variance.

(b) For a doubly stochastic Poisson process N, show that $\text{var}(N(t)) \geq \mathbb{E}(N(t))$.

(c) Let M be an ordinary Poisson process on the time-interval $[0,\infty)$ with constant rate 1. Let M^* be obtained from M by deleting the kth arrival for every odd value of k. Is M^* either: (i) a Poisson process, or (ii) a doubly stochastic Poisson process?

23. Show that a simple birth process X with parameter λ is a doubly stochastic Poisson process with intensity function $\lambda(t) = \lambda X(t)$.

24. Pólya's process. The Markov chain $X = \{X(t) : t \geq 0\}$ is a birth process whose intensities $\lambda_k(t)$ depend also on the time t and are given by

$$\mathbb{P}\big(X(t+h) = k+1 \,\big|\, X(t) = k\big) = \frac{1+\mu k}{1+\mu t} h + \mathrm{o}(h)$$

as $h \downarrow 0$. Show that the probability generating function $G(s,t) = \mathbb{E}(s^{X(t)})$ satisfies

$$\frac{\partial G}{\partial t} = \frac{s-1}{1+\mu t}\left\{G + \mu s \frac{\partial G}{\partial s}\right\}, \qquad 0 < s < 1.$$

Hence find the mean and variance of $X(t)$ when $X(0) = I$.

25. (a) Let X be a birth–death process with strictly positive birth rates $\lambda_0, \lambda_1, \dots$ and death rates $\mu_1, \mu_2, \dots$. Let η_i be the probability that $X(t)$ ever takes the value 0 starting from $X(0) = i$. Show that

$$\lambda_j \eta_{j+1} - (\lambda_j + \mu_j)\eta_j + \mu_j \eta_{j-1} = 0, \qquad j \geq 1,$$

and deduce that $\eta_i = 1$ for all i so long as $\sum_1^\infty e_j = \infty$ where $e_j = \mu_1\mu_2 \cdots \mu_j/(\lambda_1\lambda_2 \cdots \lambda_j)$.
(b) For the discrete-time chain on the non-negative integers with

$$p_{j,j+1} = \frac{(j+1)^2}{j^2 + (j+1)^2} \quad \text{and} \quad p_{j,j-1} = \frac{j^2}{j^2 + (j+1)^2},$$

find the probability that the chain ever visits 0, starting from 1.

26. Find a good necessary condition and a good sufficient condition for the birth–death process X of Problem (6.15.25a) to be honest.

27. Let X be a simple symmetric birth–death process with $\lambda_n = \mu_n = n\lambda$, and let T be the time until extinction. Show that

$$\mathbb{P}(T \leq x \mid X(0) = I) = \left(\frac{\lambda x}{1 + \lambda x}\right)^I,$$

and deduce that extinction is certain if $\mathbb{P}(X(0) < \infty) = 1$.

Show that $\mathbb{P}(\lambda T/I \leq x \mid X(0) = I) \to e^{-1/x}$ as $I \to \infty$.

28. Immigration–death with disasters. Let X be an immigration–death–disaster process, that is, a birth–death process with parameters $\lambda_i = \lambda$, $\mu_i = i\mu$, and with the additional possibility of 'disasters' which reduce the population to 0. Disasters occur at the times of a Poisson process with intensity δ, independently of all previous births and deaths.

(a) Show that X has a stationary distribution, and find an expression for the generating function of this distribution.
(b) Show that, in equilibrium, the mean of $X(t)$ is $\lambda/(\delta + \mu)$.

29. With any sufficiently nice (Lebesgue measurable, say) subset B of the real line $\mathbb{R}$ is associated a random variable $X(B)$ such that

(a) $X(B)$ takes values in $\{0, 1, 2, \dots\}$,
(b) if $B_1, B_2, \dots, B_n$ are disjoint then $X(B_1), X(B_2), \dots, X(B_n)$ are independent, and furthermore $X(B_1 \cup B_2) = X(B_1) + X(B_2)$,
(c) the distribution of $X(B)$ depends only on B through its Lebesgue measure ('length') $|B|$, and

$$\frac{\mathbb{P}(X(B) \geq 1)}{\mathbb{P}(X(B) = 1)} \to 1 \quad \text{as } |B| \to 0.$$

Show that X is a Poisson process.

30. Poisson forest. Let N be a Poisson process in $\mathbb{R}^2$ with constant intensity λ, and let $R_{(1)} < R_{(2)} < \cdots$ be the ordered distances from the origin of the points of the process.

(a) Show that $R_{(1)}^2, R_{(2)}^2, \dots$ are the points of a Poisson process on $\mathbb{R}_+ = [0, \infty)$ with intensity $\lambda\pi$.
(b) Show that $R_{(k)}$ has density function

$$f(r) = \frac{2\pi\lambda r(\lambda\pi r^2)^{k-1} e^{-\lambda\pi r^2}}{(k-1)!}, \qquad r > 0.$$

31. Let X be a n-dimensional Poisson process with constant intensity λ. Show that the volume of the largest (n-dimensional) sphere centred at the origin which contains no point of X is exponentially distributed. Deduce the density function of the distance R from the origin to the nearest point of X. Show that $\mathbb{E}(R) = \Gamma(1/n)/\{n(\lambda c)^{1/n}\}$ where c is the volume of the unit ball of $\mathbb{R}^n$ and Γ is the gamma function.

32. A village of $N+1$ people suffers an epidemic. Let $X(t)$ be the number of ill people at time t, and suppose that $X(0) = 1$ and X is a birth process with rates $\lambda_i = \lambda i(N+1-i)$. Let T be the length of time required until every member of the population has succumbed to the illness. Show that

$$\mathbb{E}(T) = \frac{1}{\lambda}\sum_{k=1}^{N} \frac{1}{k(N+1-k)}$$

and deduce that

$$\mathbb{E}(T) = \frac{2(\log N + \gamma)}{\lambda(N+1)} + \mathrm{O}(N^{-2})$$

where γ is Euler's constant. It is striking that $\mathbb{E}(T)$ decreases with N, for large N.

33. A particle has velocity $V(t)$ at time t, where $V(t)$ is assumed to take values in $\{n + \frac{1}{2} : n \geq 0\}$. Transitions during $(t, t+h)$ are possible as follows:

$$\mathbb{P}\big(V(t+h) = w \,\big|\, V(t) = v\big) = \begin{cases} (v+\frac{1}{2})h + \mathrm{o}(h) & \text{if } w = v+1, \\ 1 - 2vh + \mathrm{o}(h) & \text{if } w = v, \\ (v-\frac{1}{2})h + \mathrm{o}(h) & \text{if } w = v-1. \end{cases}$$

Initially $V(0) = \frac{1}{2}$. Let

$$G(s,t) = \sum_{n=0}^{\infty} s^n \mathbb{P}\big(V(t) = n + \tfrac{1}{2}\big).$$

(a) Show that

$$\frac{\partial G}{\partial t} = (1-s)^2 \frac{\partial G}{\partial s} - (1-s)G$$

and deduce that $G(s,t) = \{1 + (1-s)t\}^{-1}$.

(b) Show that the expected length $m_n(T)$ of time for which $V = n + \frac{1}{2}$ during the time interval $[0, T]$ is given by

$$m_n(T) = \int_0^T \mathbb{P}\big(V(t) = n + \tfrac{1}{2}\big)\,dt$$

and that, for fixed k, $m_k(T) - \log T \to -\sum_{i=1}^{k} i^{-1}$ as $T \to \infty$.

(c) What is the expected velocity of the particle at time t?

34. A sequence $X_0, X_1, \ldots$ of random integers is generated as follows. First, $X_0 = 0$, $X_1 = 1$. For $n \geq 1$, conditional on $X_0, X_1, \ldots, X_n$, the next value X_{n+1} is equally likely to be either $X_n + X_{n-1}$ or $|X_n - X_{n-1}|$.

(a) Is X a Markov chain?

(b) Use the Markov chain $Y_n = (X_{n-1}, X_n)$ to find the probability that X hits the value 3 before it revisits 0.

(c) Show that the probability that Y ever reaches the state (1, 1), having started at (1, 2), is $\frac{1}{2}(3 - \sqrt{5})$.

35. Take a regular hexagon and join opposite corners by straight lines meeting at the point C. A particle performs a symmetric random walk on these 7 vertices, starting at A ($\neq$ C). Find:

(a) the probability of return to A without hitting C,

(b) the expected time to return to A,

(c) the expected number of visits to C before returning to A,

(d) the expected time to return to A, given that there is no prior visit to C.

36. Diffusion, osmosis. Markov chains are defined by the following procedures at any time n:

(a) **Bernoulli model.** Two adjacent containers A and B each contain m particles; m are of type I and m are of type II. A particle is selected at random in each container. If they are of opposite types they are exchanged with probability α if the type I is in A, or with probability β if the type I is in B. Let X_n be the number of type I particles in A at time n.

(b) **Ehrenfest dog–flea model.** Two adjacent containers contain m particles in all. A particle is selected at random. If it is in A it is moved to B with probability α, if it is in B it is moved to A with probability β. Let Y_n be the number of particles in A at time n.

In each case find the transition matrix and stationary distribution of the chain.

37. Let X be an irreducible continuous-time Markov chain on the state space S with transition probabilities $p_{jk}(t)$ and unique stationary distribution $\boldsymbol{\pi}$, and write $\mathbb{P}(X(t) = j) = a_j(t)$. If $c(x)$ is a concave function, show that the function $d(t) = \sum_{j \in S} \pi_j c(a_j(t)/\pi_j)$ increases to $c(1)$ as $t \to \infty$.

The *relative entropy* (or *Kullback–Leibler divergence*) of two strictly positive probability mass functions f, g on a subset S of integers is defined as

$$D(f; g) = \sum_{i \in S} f(i) \log\big(f(i)/g(i)\big).$$

Prove that, if X has a finite state space and stationary distribution $\boldsymbol{\pi}$, the relative entropy $D(a(t); \boldsymbol{\pi})$ decreases monotonely to 0 as $t \to \infty$.

38. With the notation of the preceding problem, let $u_k(t) = \mathbb{P}(X(t) = k \mid X(0) = 0)$, and suppose the chain is reversible in equilibrium (see Problem (6.15.16)). Show that $u_0(2t) = \sum_j (\pi_0/\pi_j) u_j(t)^2$, and deduce that $u_0(t)$ decreases to π_0 as $t \to \infty$.

39. Perturbing a Poisson process. Let Π be the set of points in a Poisson process on $\mathbb{R}^d$ with constant intensity λ. Each point is displaced, where the displacements are independent and identically distributed. Show that the resulting point process is a Poisson process with intensity λ.

40. Perturbations continued. Suppose for convenience in Problem (6.15.39) that the displacements have a continuous distribution function and finite mean, and that $d = 1$. Suppose also that you are at the origin originally, and you move to a in the perturbed process. Let $\mathrm{L_R}$ be the number of points formerly on your left that are now on your right, and $\mathrm{R_L}$ the number of points formerly on your right that are now on your left. Show that $\mathbb{E}(\mathrm{L_R}) = \mathbb{E}(\mathrm{R_L})$ if and only if $a = \mu$ where μ is the mean displacement of a particle.

Deduce that if cars enter the start of a long road at the instants of a Poisson process, having independent identically distributed velocities, then, if you travel at the average speed, in the long run the rate at which you are overtaken by other cars equals the rate at which you overtake other cars.

41. Ants enter a kitchen at the instants of a Poisson process N of rate λ; they each visit the pantry and then the sink, and leave. The rth ant spends time X_r in the pantry and Y_r in the sink (and $X_r + Y_r$ in the kitchen altogether), where the vectors $V_r = (X_r, Y_r)$ and V_s are independent for $r \neq s$. At time $t = 0$ the kitchen is free of ants. Find the joint distribution of the numbers $A(t)$ of ants in the pantry and $B(t)$ of ants in the sink at time t.

Show that, as $t \to \infty$, the number of ants in the kitchen converges in distribution, provided $\mathbb{E}(X_r + Y_r) < \infty$.

Now suppose the ants arrive in pairs at the times of the Poisson process, but then separate to behave independently as above. Find the joint distribution of the numbers of ants in the two locations.

42. Let $\{X_r : r \geq 1\}$ be independent exponential random variables with parameter λ, and set $S_n = \sum_{r=1}^{n} X_r$. Show that:

(a) $Y_k = S_k/S_n$, $1 \leq k \leq n-1$, have the same distribution as the order statistics of independent variables $\{U_k : 1 \leq k \leq n-1\}$ which are uniformly distributed on $(0, 1)$,

(b) $Z_k = X_k/S_n$, $1 \le k \le n$, have the same joint distribution as the coordinates of a point $(U_1, \dots, U_n)$ chosen uniformly at random on the simplex $\sum_{r=1}^{n} u_r = 1$, $u_r \ge 0$ for all r.

43. Let X be a discrete-time Markov chain with a finite number of states and transition matrix $\mathbf{P} = (p_{ij})$ where $p_{ij} > 0$ for all i, j. Show that there exists $\lambda \in (0, 1)$ such that $|p_{ij}(n) - \pi_j| < \lambda^n$, where $\boldsymbol{\pi}$ is the stationary distribution.

44. Under the conditions of Problem (6.15.43), let $V_i(n) = \sum_{r=0}^{n-1} I_{\{X_r = i\}}$ be the number of visits of the chain to i before time n. Show that

$$\mathbb{E}\left(\left|\frac{1}{n}V_i(n) - \pi_i\right|^2\right) \to 0 \quad \text{as } n \to \infty.$$

Show further that, if f is any bounded function on the state space, then

$$\mathbb{E}\left(\left|\frac{1}{n}\sum_{r=0}^{n-1} f(X_r) - \sum_{i \in S} f(i)\pi_i\right|^2\right) \to 0.$$

45. Conditional entropy. Let A and $\mathbf{B} = (B_0, B_1, \dots, B_n)$ be a discrete random variable and vector, respectively. The *conditional entropy* of A with respect to $\mathbf{B}$ is defined as $H(A \mid \mathbf{B}) = \mathbb{E}(\mathbb{E}\{-\log f(A \mid \mathbf{B}) \mid \mathbf{B}\})$ where $f(a \mid \mathbf{b}) = \mathbb{P}(A = a \mid \mathbf{B} = \mathbf{b})$. Let X be an aperiodic Markov chain on a finite state space. Show that

$$H(X_{n+1} \mid X_0, X_1 \dots, X_n) = H(X_{n+1} \mid X_n),$$

and that

$$H(X_{n+1} \mid X_n) \to -\sum_i \pi_i \sum_j p_{ij} \log p_{ij} \qquad \text{as } n \to \infty,$$

if X is aperiodic with a unique stationary distribution $\boldsymbol{\pi}$.

46. Coupling. Let X and Y be independent recurrent birth–death processes with the same parameters (and no explosions). It is not assumed that $X_0 = Y_0$. Show that:

(a) for any $A \subseteq \mathbb{R}$, $|\mathbb{P}(X_t \in A) - \mathbb{P}(Y_t \in A)| \to 0$ as $t \to \infty$,

(b) if $\mathbb{P}(X_0 \le Y_0) = 1$, then $\mathbb{E}[g(X_t)] \le \mathbb{E}[g(Y_t)]$ for any increasing function g.

47. Resources. The number of birds in a wood at time t is a continuous-time Markov process X. Food resources impose the constraint $0 \le X(t) \le n$. Competition entails that the transition probabilities obey

$$p_{k,k+1}(h) = \lambda(n-k)h + \mathrm{o}(h), \qquad p_{k,k-1}(h) = \mu k h + \mathrm{o}(h).$$

Find $\mathbb{E}(s^{X(t)})$, together with the mean and variance of $X(t)$, when $X(0) = r$. What happens as $t \to \infty$?

48. Parrondo's paradox. A counter performs an irreducible random walk on the vertices 0, 1, 2 of the triangle in the figure beneath, with transition matrix

$$\mathbf{P} = \begin{pmatrix} 0 & p_0 & q_0 \\ q_1 & 0 & p_1 \\ p_2 & q_2 & 0 \end{pmatrix}$$

where $p_i + q_i = 1$ for all i. Show that the stationary distribution $\boldsymbol{\pi}$ has

$$\pi_0 = \frac{1 - q_2 p_1}{3 - q_1 p_0 - q_2 p_1 - q_0 p_2},$$

with corresponding formulae for π_1, π_2.

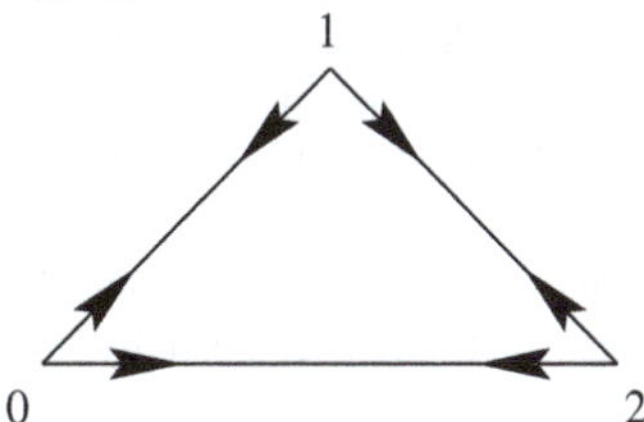

Suppose that you gain one peseta for each clockwise step of the walk, and you lose one peseta for each anticlockwise step. Show that, in equilibrium, the mean yield per step is

$$\gamma = \sum_i (2p_i - 1)\pi_i = \frac{3(2p_0p_1p_2 - p_0p_1 - p_1p_2 - p_2p_0 + p_0 + p_1 + p_2 - 1)}{3 - q_1p_0 - q_2p_1 - q_0p_2}.$$

Consider now three cases of this process:

A. We have $p_i = \frac{1}{2} - a$ for each i, where $a > 0$. Show that the mean yield per step satisfies $\gamma_A < 0$.

B. We have that $p_0 = \frac{1}{10} - a$, $p_1 = p_2 = \frac{3}{4} - a$, where $a > 0$. Show that $\gamma_B < 0$ for sufficiently small a.

C. At each step the counter is equally likely to move according to the transition probabilities of case A or case B, the choice being made independently at every step. Show that, in this case, $p_0 = \frac{3}{10} - a$, $p_1 = p_2 = \frac{5}{8} - a$. Show that $\gamma_C > 0$ for sufficiently small a.

The fact that two systematically unfavourable games may be combined to make a favourable game is called Parrondo's paradox. Such bets are not available in casinos.

49. Cars arrive at the beginning of a long road in a Poisson stream of rate λ from time $t = 0$ onwards. A car has a fixed velocity $V > 0$ which is a random variable. The velocities of cars are independent and identically distributed, and independent of the arrival process. Cars can overtake each other freely. Show that the number of cars on the first x miles of the road at time t has the Poisson distribution with parameter $\lambda\mathbb{E}[V^{-1}\min\{x, Vt\}]$.

50. Events occur at the times of a Poisson process with intensity λ, and you are offered a bet based on the process. Let $t > 0$. You are required to say the word 'now' immediately after the event which you think will be the last to occur prior to time t. You win if you succeed, otherwise you lose. If no events occur before t you lose. If you have not selected an event before time t you lose.

Consider the strategy in which you choose the first event to occur after a specified time s, where $0 < s < t$.

(a) Calculate an expression for the probability that you win using this strategy.

(b) Which value of s maximizes this probability?

(c) If $\lambda t \geq 1$, show that the probability that you win using this value of s is e^{-1}.

51. A new Oxbridge professor wishes to buy a house, and can afford to spend up to one million pounds. Declining the services of conventional estate agents, she consults her favourite internet property page on which houses are announced at the times of a Poisson process with intensity λ per day. House prices may be assumed to be independent random variables which are uniformly distributed over the interval (800,000, 2,000,000). She decides to view every affordable property announced during the next 30 days. The time spent viewing any given property is uniformly distributed over the range $(1, 2)$ hours. What is the moment generating function of the total time spent viewing houses?

52. Kemeny's constant. Let $X = \{X_n : n \geq 0\}$ be an irreducible, aperiodic Markov chain on the finite state space S, and let $h_{ij} = \mathbb{E}_i(\min\{n \geq 0 : X_n = j\})$ denote the mean hitting time of j (noting that $h_{ii} = 0$). Let $K_i = \sum_j h_{ij}\pi_j$ be the mean time to hit a state Z chosen at random according to the stationary distribution $\boldsymbol{\pi}$. Show that K_i is independent of the choice of i.

53. A professor travels between home and work on foot. She possesses a total of r umbrellas, being those at home and at work. If it is raining when she leaves either home or work, she takes an umbrella with her (if there is one available). Assume that it is raining at the start of any given walk with probability p (subject to the usual independence). Let X_n be the number of umbrellas available to her at the start of her nth walk.

(a) Explain why X is a Markov chain, and write down its transition matrix.

(b) Show that the chain has stationary distribution $\boldsymbol{\pi}$ given by

$$\pi_i = \begin{cases} \dfrac{1-p}{r+1-p} & \text{if } i = 0, \\ \dfrac{1}{r+1-p} & \text{if } i = 1, 2, \dots, r. \end{cases}$$

What proportion of walks result in her getting wet, in the long run?

(c) Let $r = 1$ and $X_1 = 1$. Calculate the mean number of walks made before she gets wet.

54. Spiders. A spider climbs a vertical spout of height h at speed 1. At the instants of a Poisson process of constant rate λ, the spider is flushed back to the bottom of the spout. It then recommences its climb. Let T be the time to reach the top, and N the number of intermediate flushes. Show that

$$\mathbb{E}(e^{-\theta T} s^N) = \frac{(\lambda+\theta)e^{-(\lambda+\theta)h}}{\lambda+\theta-\lambda s(1-e^{-(\lambda+\theta)h})}, \qquad \theta, s \in \mathbb{R}.$$

By calculating $\mathbb{E}(e^{-\theta T} \mid N = n)$ or otherwise, determine $\mathbb{E}(T \mid N = n)$ for $n \geq 0$.

55. Probability flows. Let X be an ergodic Markov chain with transition matrix $\mathbf{P}$ and stationary distribution $\boldsymbol{\pi}$. Show for any set A of states that

$$\sum_{\substack{i \in A \\ j \notin A}} \pi_i p_{ij} = \sum_{\substack{i \notin A \\ j \in A}} \pi_i p_{ij}.$$

56. Attractive hiking. A hiker of unit mass stands at the origin of the plain $\mathbb{R}^2$. Boulders with independent, identically distributed masses $M_1, M_2, \dots$ lie scattered on the plain at the points of a Poisson process with intensity 1. Let G_R be the x-component of the gravitational attraction on the hiker of the boulders within distance R of the hiker. You may take the gravitational constant to be 1.

Show that, as $R \to \infty$, G_R converges in distribution to a Cauchy distribution with characteristic function of the form $\phi(t) = e^{-c|t|}$, and express c in terms of a typical mass M.

57. Holtsmark distribution for stellar gravity. Let stars of common mass m be positioned at the points of a Poisson process with intensity 1 in $\mathbb{R}^3$. Let G_R be the x-component of the gravitational attraction due to the stars within distance R of the origin, upon a hitchhiker of unit mass at the origin. You may take the gravitational constant to be 1.

(a) Show that, as $R \to \infty$, G_R converges in distribution to a symmetric distribution with characteristic function $\phi(t) = \exp\{-c|t|^{3/2}\}$ where $c > 0$.

(b) What is the answer if the stars have independent, identically distributed random masses M_i?

7

Convergence of random variables

7.1 Exercises. Introduction

1. Let $r \geq 1$, and define $\|X\|_r = \{\mathbb{E}|X^r|\}^{1/r}$. Show that:

(a) $\|cX\|_r = |c| \cdot \|X\|_r$ for $c \in \mathbb{R}$,

(b) $\|X + Y\|_r \leq \|X\|_r + \|Y\|_r$,

(c) $\|X\|_r = 0$ if and only if $\mathbb{P}(X = 0) = 1$.

This amounts to saying that $\|\cdot\|_r$ is a norm on the set of equivalence classes of random variables on a given probability space with finite rth moment, the equivalence relation being given by $X \sim Y$ if and only if $\mathbb{P}(X = Y) = 1$.

2. Define $\langle X, Y\rangle = \mathbb{E}(XY)$ for random variables X and Y having finite variance, and define $\|X\| = \sqrt{\langle X, X\rangle}$. Show that:

(a) $\langle aX + bY, Z\rangle = a\langle X, Z\rangle + b\langle Y, Z\rangle$,

(b) $\|X + Y\|^2 + \|X - Y\|^2 = 2(\|X\|^2 + \|Y\|^2)$, the *parallelogram property*,

(c) if $\langle X_i, X_j\rangle = 0$ for all $i \neq j$ then

$$\left\|\sum_{i=1}^{n} X_i\right\|^2 = \sum_{i=1}^{n} \|X_i\|^2.$$

3. Let $\epsilon > 0$. Let $g, h : [0, 1] \to \mathbb{R}$, and define $d_\epsilon(g, h) = \int_E dx$ where $E = \{u \in [0, 1] : |g(u) - h(u)| > \epsilon\}$. Show that d_ϵ does not satisfy the triangle inequality.

4. **Lévy metric.** For two distribution functions F and G, let

$$d(F, G) = \inf\big\{\delta > 0 : F(x - \delta) - \delta \leq G(x) \leq F(x + \delta) + \delta \text{ for all } x \in \mathbb{R}\big\}.$$

Show that d is a metric on the space of distribution functions.

5. Find random variables $X, X_1, X_2, \ldots$ such that $\mathbb{E}(|X_n - X|^2) \to 0$ as $n \to \infty$, but $\mathbb{E}|X_n| = \infty$ for all n.

7.2 Exercises. Modes of convergence

1. (a) Suppose $X_n \xrightarrow{r} X$ where $r \geq 1$. Show that $\mathbb{E}|X_n^r| \to \mathbb{E}|X^r|$.

(b) Suppose $X_n \xrightarrow{1} X$. Show that $\mathbb{E}(X_n) \to \mathbb{E}(X)$. Is the converse true?

(c) Suppose $X_n \xrightarrow{2} X$. Show that $\text{var}(X_n) \to \text{var}(X)$.

2. **Dominated convergence.** Suppose $|X_n| \le Z$ for all n, where $\mathbb{E}(Z) < \infty$. Prove that if $X_n \xrightarrow{\text{P}} X$ then $X_n \xrightarrow{1} X$.

3. (a) Give a rigorous proof that $\mathbb{E}(XY) = \mathbb{E}(X)\mathbb{E}(Y)$ for any pair X, Y of independent non-negative random variables on $(\Omega, \mathcal{F}, \mathbb{P})$ with finite means. [Hint: For $k \ge 0$, $n \ge 1$, define $X_n = k/n$ if $k/n \le X < (k+1)/n$, and similarly for Y_n. Show that X_n and Y_n are independent, and $X_n \le X$, and $Y_n \le Y$. Deduce that $\mathbb{E}X_n \to \mathbb{E}X$ and $\mathbb{E}Y_n \to \mathbb{E}Y$, and also $\mathbb{E}(X_n Y_n) \to \mathbb{E}(XY)$.]

(b) Give an example to show that the product XY of dependent random variables X, Y with finite means may have $\mathbb{E}(XY) = \infty$.

4. Show that convergence in distribution is equivalent to convergence with respect to the Lévy metric of Exercise (7.1.4).

5. (a) Suppose that $X_n \xrightarrow{\text{D}} X$ and $Y_n \xrightarrow{\text{P}} c$, where c is a constant. Show that $X_n Y_n \xrightarrow{\text{D}} cX$, and that $X_n/Y_n \xrightarrow{\text{D}} X/c$ if $c \ne 0$.

(b) Suppose that $X_n \xrightarrow{\text{D}} 0$ and $Y_n \xrightarrow{\text{P}} Y$, and let $g : \mathbb{R}^2 \to \mathbb{R}$ be continuous. Show that $g(X_n, Y_n) \xrightarrow{\text{P}} g(0, Y)$.

[These results are sometimes referred to as 'Slutsky's theorem(s)'.]

6. Let $X_1, X_2, \dots$ be random variables on the probability space $(\Omega, \mathcal{F}, \mathbb{P})$. Show that the set $A = \{\omega \in \Omega : \text{the sequence } X_n(\omega) \text{ converges}\}$ is an event (that is, lies in $\mathcal{F}$), and that there exists a random variable X (that is, an $\mathcal{F}$-measurable function $X : \Omega \to \mathbb{R}$) such that $X_n(\omega) \to X(\omega)$ for $\omega \in A$.

7. Let $\{X_n\}$ be a sequence of random variables, and let $\{c_n\}$ be a sequence of reals converging to the limit c. For convergence almost surely, in rth mean, in probability, and in distribution, show that the convergence of X_n to X entails the convergence of $c_n X_n$ to cX.

8. Let $\{X_n\}$ be a sequence of independent random variables which converges in probability to the limit X. Show that X is almost surely constant.

9. **Convergence in total variation.** The sequence of discrete random variables X_n, with mass functions f_n, is said to *converge in total variation* to X with mass function f if

$$\sum_x |f_n(x) - f(x)| \to 0 \quad \text{as} \quad n \to \infty.$$

Suppose $X_n \to X$ in total variation, and $u : \mathbb{R} \to \mathbb{R}$ is bounded. Show that $\mathbb{E}(u(X_n)) \to \mathbb{E}(u(X))$.

10. Let $\{X_r : r \ge 1\}$ be independent Poisson variables with respective parameters $\{\lambda_r : r \ge 1\}$. Show that $\sum_{r=1}^{\infty} X_r$ converges or diverges almost surely according as $\sum_{r=1}^{\infty} \lambda_r$ converges or diverges.

11. **Waiting for a coincidence.** Let $X_1, X_2, \dots$ be independent random variables, uniformly distributed on $\{1, 2, \dots, m\}$, and let $I_m = \min\{r \ge 2 : X_r = X_s \text{ for some } s < r\}$ be the earliest index of a coincidence of values. Show that $I_m/\sqrt{m}$ converges in distribution as $m \to \infty$, with as limit the Rayleigh distribution with density function $f(x) = xe^{-\frac{1}{2}x^2}$, $x > 0$.

12. **Moments.** The random variable X has finite moments of all orders, and satisfies $\mathbb{P}(X > 0) = 1$, $\mathbb{P}(X > x) > 0$ for $x > 0$.

(a) Show that $c = \sum_{n=0}^{\infty} 1/\mathbb{E}(X^n)$ satisfies $c < \infty$.

(b) Let M take values in $\{1, 2, \dots\}$ with mass function $f(m) = 1/(c\mathbb{E}(X^m))$ for $m \ge 1$. Show that $\mathbb{E}(x^M) < \infty$ for all $x \ge 0$, while $\mathbb{E}(X^M) = \infty$.

7.3 Exercises. Some ancillary results

1. (a) Suppose that $X_n \xrightarrow{\text{P}} X$. Show that $\{X_n\}$ is *Cauchy convergent in probability* in that, for all $\epsilon > 0$, $\mathbb{P}(|X_n - X_m| > \epsilon) \to 0$ as $n, m \to \infty$. In what sense is the converse true?

(b) Let $\{X_n\}$ and $\{Y_n\}$ be sequences of random variables such that the pairs (X_i, X_j) and (Y_i, Y_j) have the same distributions for all i, j. If $X_n \xrightarrow{\text{P}} X$, show that Y_n converges in probability to some limit Y having the same distribution as X.

2. Show that the probability that infinitely many of the events $\{A_n : n \geq 1\}$ occur satisfies $\mathbb{P}(A_n \text{ i.o.}) \geq \limsup_{n\to\infty} \mathbb{P}(A_n)$.

3. Let $\{S_n : n \geq 0\}$ be a simple random walk which moves to the right with probability p at each step, and suppose that $S_0 = 0$. Write $X_n = S_n - S_{n-1}$.

(a) Show that $\{S_n = 0 \text{ i.o.}\}$ is not a tail event of the sequence $\{X_n\}$.

(b) Show that $\mathbb{P}(S_n = 0 \text{ i.o.}) = 0$ if $p \neq \frac{1}{2}$.

(c) Let $T_n = S_n/\sqrt{n}$, and show that

$$\Bigl\{\liminf_{n\to\infty} T_n \leq -x\Bigr\} \cap \Bigl\{\limsup_{n\to\infty} T_n \geq x\Bigr\}$$

is a tail event of the sequence $\{X_n\}$, for all $x > 0$, and deduce directly that $\mathbb{P}(S_n = 0 \text{ i.o.}) = 1$ if $p = \frac{1}{2}$.

4. Hewitt–Savage zero–one law. Let $X_1, X_2, \dots$ be independent identically distributed random variables. The event A, defined in terms of the X_n, is called *exchangeable* if A is invariant under finite permutations of the coordinates, which is to say that its indicator function I_A satisfies $I_A(X_1, X_2, \dots, X_n, \dots) = I_A(X_{i_1}, X_{i_2}, \dots, X_{i_n}, X_{n+1}, \dots)$ for all $n \geq 1$ and all permutations $(i_1, i_2, \dots, i_n)$ of $(1, 2, \dots, n)$. Show that all exchangeable events A are such that either $\mathbb{P}(A) = 0$ or $\mathbb{P}(A) = 1$.

5. Returning to the simple random walk S of Exercise (7.3.3), show that $\{S_n = 0 \text{ i.o.}\}$ is an exchangeable event with respect to the steps of the walk, and deduce from the Hewitt–Savage zero–one law that it has probability either 0 or 1.

6. Weierstrass's approximation theorem. Let $f : [0, 1] \to \mathbb{R}$ be a continuous function, and let S_n be a random variable having the binomial distribution with parameters n and x. Using the formula $\mathbb{E}(Z) = \mathbb{E}(ZI_A) + \mathbb{E}(ZI_{A^c})$ with $Z = f(x) - f(n^{-1}S_n)$ and $A = \{|n^{-1}S_n - x| > \delta\}$, show that

$$\lim_{n\to\infty} \sup_{0\leq x\leq 1} \left| f(x) - \sum_{k=0}^{n} f(k/n)\binom{n}{k} x^k (1-x)^{n-k} \right| = 0.$$

You have proved Weierstrass's approximation theorem, which states that every continuous function on [0, 1] may be approximated by a polynomial uniformly over the interval.

7. Complete convergence. A sequence $X_1, X_2, \dots$ of random variables is said to be *completely convergent* to X if

$$\sum_n \mathbb{P}(|X_n - X| > \epsilon) < \infty \qquad \text{for all } \epsilon > 0.$$

Show that, for sequences of independent variables, complete convergence is equivalent to a.s. convergence. Find a sequence of (dependent) random variables which converges a.s. but not completely.

8. Let $X_1, X_2, \dots$ be independent identically distributed random variables with common mean μ and finite variance. Show that

$$\binom{n}{2}^{-1} \sum_{1\leq i<j\leq n} X_i X_j \xrightarrow{\text{P}} \mu^2 \qquad \text{as } n \to \infty.$$

9. Let $\{X_n : n \geq 1\}$ be independent and exponentially distributed with parameter 1. Show that

$$\mathbb{P}\left(\limsup_{n\to\infty} \frac{X_n}{\log n} = 1\right) = 1.$$

10. Let $\{X_n : n \geq 1\}$ be independent $N(0, 1)$ random variables. Show that:

(a) $\mathbb{P}\left(\limsup_{n\to\infty} \frac{|X_n|}{\sqrt{\log n}} = \sqrt{2}\right) = 1$,

(b) $\mathbb{P}(X_n > a_n \text{ i.o.}) = \begin{cases} 0 & \text{if } \sum_n \mathbb{P}(X_1 > a_n) < \infty, \\ 1 & \text{if } \sum_n \mathbb{P}(X_1 > a_n) = \infty. \end{cases}$

11. Construct an example to show that the convergence in distribution of X_n to X does not imply the convergence of the unique medians of the sequence X_n.

12. (i) Let $\{X_r : r \geq 1\}$ be independent, non-negative and identically distributed with infinite mean. Show that $\limsup_{r\to\infty} X_r/r = \infty$ almost surely.

(ii) Let $\{X_r\}$ be a stationary Markov chain on the positive integers with transition probabilities

$$p_{jk} = \begin{cases} \dfrac{j}{j+2} & \text{if } k = j+1, \\ \dfrac{2}{j+2} & \text{if } k = 1. \end{cases}$$

(a) Find the stationary distribution of the chain, and show that it has infinite mean.

(b) Show that $\limsup_{r\to\infty} X_r/r \leq 1$ almost surely.

13. Let $\{X_r : 1 \leq r \leq n\}$ be independent and identically distributed with mean μ and finite variance σ^2. Let $\overline{X} = n^{-1}\sum_{r=1}^n X_r$. Show that

$$\sum_{r=1}^n (X_r - \mu) \Bigg/ \sqrt{\sum_{r=1}^n (X_r - \overline{X})^2}$$

converges in distribution to the $N(0, 1)$ distribution as $n \to \infty$.

14. For a random variable X with mean 0, variance σ^2, and $\mathbb{E}(X^4) < \infty$, show that

$$\mathbb{P}(|X| > t) \leq \frac{\mathbb{E}(X^4) - \sigma^4}{\mathbb{E}(X^4) - 2\sigma^2 t^2 + t^4}, \qquad t > 0.$$

7.4 Exercise. Laws of large numbers

1. Let $X_2, X_3, \ldots$ be independent random variables such that

$$\mathbb{P}(X_n = n) = \mathbb{P}(X_n = -n) = \frac{1}{2n\log n}, \qquad \mathbb{P}(X_n = 0) = 1 - \frac{1}{n \log n}.$$

Show that this sequence obeys the weak law but not the strong law, in the sense that $n^{-1}\sum_1^n X_i$ converges to 0 in probability but not almost surely.

2. Construct a sequence $\{X_r : r \geq 1\}$ of independent random variables with zero mean such that $n^{-1}\sum_{r=1}^n X_r \to -\infty$ almost surely, as $n \to \infty$.

3. Let N be a spatial Poisson process with constant intensity λ in $\mathbb{R}^d$, where $d \geq 2$. Let S be the ball of radius r centred at zero. Show that $N(S)/|S| \to \lambda$ almost surely as $r \to \infty$, where $|S|$ is the volume of the ball.

4. Proportional betting. In each of a sequence of independent bets, a gambler either wins 30%, or loses 25% of her current fortune, each with probability $\frac{1}{2}$. Denoting her fortune after n bets by F_n, show that $\mathbb{E}(F_n) \to \infty$ as $n \to \infty$, while $F_n \to 0$ almost surely.

5. General weak law. Let $S_n = X_1 + X_2 + \cdots + X_n$ be the sum of independent, identically distributed random variables. Let $\delta, \epsilon > 0$, and define the truncated variables $Y_j = X_j I_{\{|X_j| \leq \delta n\}}$. Let $A = \{X_j = Y_j \text{ for } j = 1, 2, \dots, n\}$, and $B = \{|S_n - n\mathbb{E}(Y_1)| \geq \epsilon n\}$. Prove that:

(a) $\mathbb{P}(B) \leq \mathbb{P}(A^c) + \mathbb{P}(B \mid A)$,

(b) $\mathbb{P}(A^c) \leq n\mathbb{P}(|X_1| > \delta n)$,

(c) $\mathbb{P}(B \mid A) \leq \mathbb{E}(Y_1^2)/(n\epsilon^2)$.

Deduce the weak law of large numbers, namely that, if $\mathbb{E}|X_1| < \infty$,

$$\mathbb{P}\big(|S_n/n - \mu| > \epsilon\big) \to 0 \qquad \text{as } n \to \infty,$$

where $\mu = \mathbb{E}(X_1)$.

7.5 Exercises. The strong law

1. Entropy. The interval [0, 1] is partitioned into n disjoint sub-intervals with lengths $p_1, p_2, \dots, p_n$, and the *entropy* of this partition is defined to be

$$h = -\sum_{i=1}^{n} p_i \log p_i .$$

Let $X_1, X_2, \dots$ be independent random variables having the uniform distribution on [0, 1], and let $Z_m(i)$ be the number of the $X_1, X_2, \dots, X_m$ which lie in the ith interval of the partition above. Show that

$$R_m = \prod_{i=1}^{n} p_i^{Z_m(i)}$$

satisfies $m^{-1} \log R_m \to -h$ almost surely as $m \to \infty$.

2. Recurrent events. Catastrophes occur at the times $T_1, T_2, \dots$ where $T_i = X_1 + X_2 + \cdots + X_i$ and the X_i are independent identically distributed positive random variables. Let $N(t) = \max\{n : T_n \leq t\}$ be the number of catastrophes which have occurred by time t. Prove that if $\mathbb{E}(X_1) < \infty$ then $N(t) \to \infty$ and $N(t)/t \to 1/\mathbb{E}(X_1)$ as $t \to \infty$, almost surely.

3. Random walk. Let $X_1, X_2, \dots$ be independent identically distributed random variables taking values in the integers $\mathbb{Z}$ and having a finite mean. Show that the Markov chain $S = \{S_n\}$ given by $S_n = \sum_1^n X_i$ is transient if $\mathbb{E}(X_1) \neq 0$.

7.6 Exercise. The law of the iterated logarithm

1. A function $\phi(x)$ is said to belong to the 'upper class' if, in the notation of this section, $\mathbb{P}(S_n > \phi(n)\sqrt{n} \text{ i.o.}) = 0$. A consequence of the law of the iterated logarithm is that $\sqrt{\alpha \log \log x}$ is in the upper class for all $\alpha > 2$. Use the first Borel–Cantelli lemma to prove the much weaker fact that $\phi(x) = \sqrt{\alpha \log x}$ is in the upper class for all $\alpha > 2$, in the special case when the X_i are independent $N(0, 1)$ variables.

7.7 Exercises. Martingales

1. Let $X_1, X_2, \dots$ be random variables such that the partial sums $S_n = X_1 + X_2 + \cdots + X_n$ determine a martingale. Show that $\mathbb{E}(X_i X_j) = 0$ if $i \neq j$.

2. Let Z_n be the size of the nth generation of a branching process with immigration, in which the family sizes have mean μ ($\neq 1$) and the mean number of immigrants in each generation is m. Suppose that $\mathbb{E}(Z_0) < \infty$, and show that

$$S_n = \mu^{-n}\left\{Z_n - m\left(\frac{1-\mu^n}{1-\mu}\right)\right\}$$

is a martingale with respect to a suitable sequence of random variables.

3. Let $X_0, X_1, X_2, \dots$ be a sequence of random variables with finite means and satisfying $\mathbb{E}(X_{n+1} \mid X_0, X_1, \dots, X_n) = aX_n + bX_{n-1}$ for $n \geq 1$, where $0 < a, b < 1$ and $a + b = 1$. Find a value of α for which $S_n = \alpha X_n + X_{n-1}$, $n \geq 1$, defines a martingale with respect to the sequence X.

4. Let X_n be the net profit to the gambler of betting a unit stake on the nth play in a casino; the X_n may be dependent, but the game is fair in the sense that $\mathbb{E}(X_{n+1} \mid X_1, X_2, \dots, X_n) = 0$ for all n. The gambler stakes Y on the first play, and thereafter stakes $f_n(X_1, X_2, \dots, X_n)$ on the $(n+1)$th play, where $f_1, f_2, \dots$ are given functions. Show that her profit after n plays is

$$S_n = \sum_{i=1}^{n} X_i f_{i-1}(X_1, X_2, \dots, X_{i-1}),$$

where $f_0 = Y$. Show further that the sequence $S = \{S_n\}$ satisfies the martingale condition $\mathbb{E}(S_{n+1} \mid X_1, X_2, \dots, X_n) = S_n$, $n \geq 1$, if Y is assumed to be known throughout.

5. A *run* in a random permutation $(\pi_1, \pi_2, \dots, \pi_n)$ of $(1, 2, \dots, n)$ is a subsequence satisfying $\pi_{r-1} > \pi_r < \pi_{r+1} < \cdots < \pi_s > \pi_{s+1}$. We set $\pi_0 = n + 1$ and $\pi_{n+1} = 0$ by convention. Let R_n be the number of runs. Show that $M_n = nR_n - \frac{1}{2}n(n+1)$ is a martingale. Find $\mathbb{E}(R_n)$ and $\mathbb{E}(R_n^2)$.

7.8 Exercises. Martingale convergence theorem

1. Kolmogorov's inequality. Let $X_1, X_2, \dots$ be independent random variables with zero means and finite variances, and let $S_n = X_1 + X_2 + \cdots + X_n$. Use the Doob–Kolmogorov inequality to show that

$$\mathbb{P}\left(\max_{1 \leq j \leq n} |S_j| > \epsilon\right) \leq \frac{1}{\epsilon^2}\sum_{j=1}^{n} \operatorname{var}(X_j) \qquad \text{for } \epsilon > 0.$$

2. Let $X_1, X_2, \dots$ be independent random variables such that $\sum_n n^{-2}\operatorname{var}(X_n) < \infty$. Use Kolmogorov's inequality to prove that

$$\sum_{i=1}^{n} \frac{X_i - \mathbb{E}(X_i)}{i} \xrightarrow{\text{a.s.}} Y \qquad \text{as } n \to \infty,$$

for some finite random variable Y, and deduce that

$$\frac{1}{n}\sum_{i=1}^{n}(X_i - \mathbb{E}X_i) \xrightarrow{\text{a.s.}} 0 \qquad \text{as } n \to \infty.$$

(You may find Kronecker's lemma to be useful: if (a_n) and (b_n) are real sequences with $b_n \uparrow \infty$ and $\sum_i a_i/b_i < \infty$, then $b_n^{-1} \sum_{i=1}^n a_i \to 0$ as $n \to \infty$.)

3. Let S be a martingale with respect to X, such that $\mathbb{E}(S_n^2) < K < \infty$ for some $K \in \mathbb{R}$. Suppose that $\mathrm{var}(S_n) \to 0$ as $n \to \infty$, and prove that $S = \lim_{n\to\infty} S_n$ exists and is constant almost surely.

7.9 Exercises. Prediction and conditional expectation

1. Let Y be uniformly distributed on $[-1, 1]$ and let $X = Y^2$.
(a) Find the best predictor of X given Y, and of Y given X.
(b) Find the best linear predictor of X given Y, and of Y given X.

2. (a) Let the pair (X, Y) have a general bivariate normal distribution. Find $\mathbb{E}(Y \mid X)$.
(b) Let $U_1, U_2, \dots, U_n$ be independent $N(0, 1)$ random variables, and let (a_i), (b_i) be real non-zero vectors. Show that $X = \sum_i a_i U_i$ and $Y = \sum_i b_i U_i$ satisfy

$$\mathbb{E}(Y \mid X) = X \frac{\sum_i a_i b_i}{\sum_i a_i^2}.$$

3. Let $X_1, X_2, \dots, X_n$ be random variables with zero means and covariance matrix $\mathbf{V} = (v_{ij})$, and let Y have finite second moment. Find the linear function h of the X_i which minimizes the mean squared error $\mathbb{E}\{(Y - h(X_1, \dots, X_n))^2\}$.

4. Verify the following properties of conditional expectation. You may assume that the relevant expectations exist.
(a) $\mathbb{E}\{\mathbb{E}(Y \mid \mathcal{G})\} = \mathbb{E}(Y)$.
(b) $\mathbb{E}(\alpha Y + \beta Z \mid \mathcal{G}) = \alpha\mathbb{E}(Y \mid \mathcal{G}) + \beta\mathbb{E}(Z \mid \mathcal{G})$ for $\alpha, \beta \in \mathbb{R}$.
(c) $\mathbb{E}(Y \mid \mathcal{G}) \ge 0$ if $Y \ge 0$.
(d) $\mathbb{E}(Y \mid \mathcal{G}) = \mathbb{E}\{\mathbb{E}(Y \mid \mathcal{H}) \mid \mathcal{G}\}$ if $\mathcal{G} \subseteq \mathcal{H}$.
(e) $\mathbb{E}(Y \mid \mathcal{G}) = \mathbb{E}(Y)$ if Y is independent of I_G for every $G \in \mathcal{G}$.
(f) **Jensen's inequality.** $g\{\mathbb{E}(Y \mid \mathcal{G})\} \le \mathbb{E}\{g(Y) \mid \mathcal{G}\}$ for all convex functions g.
(g) If $Y_n \xrightarrow{\text{a.s.}} Y$ and $|Y_n| \le Z$ a.s. where $\mathbb{E}(Z) < \infty$, then $\mathbb{E}(Y_n \mid \mathcal{G}) \xrightarrow{\text{a.s.}} \mathbb{E}(Y \mid \mathcal{G})$.
Statements (b)–(f) are of course to be interpreted 'almost surely'.

5. Let X and Y have joint mass function $f(x, y) = \{x(x + 1)\}^{-1}$ for $x = y = 1, 2, \dots$. Show that $\mathbb{E}(Y \mid X) < \infty$ while $\mathbb{E}(Y) = \infty$.

6. Let $(\Omega, \mathcal{F}, \mathbb{P})$ be a probability space and let $\mathcal{G}$ be a sub-σ-field of $\mathcal{F}$. Let H be the space of $\mathcal{G}$-measurable random variables with finite second moment.
(a) Show that H is closed with respect to the norm $\|\cdot\|_2$.
(b) Let Y be a random variable satisfying $\mathbb{E}(Y^2) < \infty$, and show the equivalence of the following two statements for any $M \in H$:
(i) $\mathbb{E}\{(Y - M)Z\} = 0$ for all $Z \in H$,
(ii) $\mathbb{E}\{(Y - M)I_G\} = 0$ for all $G \in \mathcal{G}$.

7. Maximal correlation coefficient. For possibly dependent random variables X and Y, define the *maximal correlation coefficient* $m(X, Y) = \sup \rho(f(X), g(Y))$, where ρ denotes ordinary correlation, and the supremum is over all functions f and g such that $f(X)$ and $g(Y)$ have finite non-zero variances. Show that:
(a) $m(X, Y) = 0$ if and only if X and Y are independent,

(b) $m(X, Y)^2 = \sup_g \operatorname{var}\big(\mathbb{E}(g(Y) \mid X)\big)$, where the supremum is over all functions g such that $\operatorname{var}(g(Y)) = 1$,

(c) $\widehat{f}(X)m(X, Y) = \mathbb{E}(\widehat{g}(Y) \mid X)$ a.s., where $\widehat{f}$ and $\widehat{g}$ are functions such that $m = \rho(\widehat{f}(X), \widehat{g}(Y))$.

(d) We have, a.s., that

$$\mathbb{E}\big(\mathbb{E}(\widehat{f}(X) \mid Y) \,\big|\, X\big) = m(X, Y)^2 \widehat{f}(X), \quad \mathbb{E}\big(\mathbb{E}(\widehat{g}(Y) \mid X) \,\big|\, Y\big) = m(X, Y)^2 \widehat{g}(Y).$$

(e) If the ordered triple X, Y, Z is a Markov chain, show that $m(X, Z) \le m(X, Y)m(Y, Z)$, with equality if (X, Y) and (Z, Y) are identically distributed.

(f) Deduce that, for a pair (U, V) with the standard bivariate normal distribution, $m(U, V)$ is an increasing function of the modulus $|\rho|$ of correlation. [It can be shown that $m(U, V) = |\rho|$ in this case.]

8. **Monotone correlation.** Let $\rho_{\text{mon}}(X, Y) = \sup \rho(f(X), g(Y))$, where the supremum is over all monotonic functions f and g such that $f(X)$ and $g(Y)$ have finite, non-zero variances. Show that $\rho_{\text{mon}}(X, Y) = 0$ if and only if X and Y are independent.

For random variables X, Y with finite, non-zero variances, show that $\rho(X, Y) \le \rho_{\text{mon}}(X, Y) \le m(X, Y)$, where m is the maximal correlation coefficient.

9. **Prediction.** Let (X, Y) have joint density function $f(x, y) = 2e^{-x-y}$ for $0 < x \le y < \infty$.

(a) Find the minimum mean-squared-error predictor of X given that $Y = y$.

(b) Find the minimum mean-squared-error *linear* predictor of X given that $Y = y$.

(c) Compare these.

7.10 Exercises. Uniform integrability

1. Show that the sum $\{X_n + Y_n\}$ of two uniformly integrable sequences $\{X_n\}$ and $\{Y_n\}$ gives a uniformly integrable sequence.

2. (a) Suppose that $X_n \xrightarrow{r} X$ where $r \ge 1$. Show that $\{|X_n|^r : n \ge 1\}$ is uniformly integrable, and deduce that $\mathbb{E}(X_n^r) \to \mathbb{E}(X^r)$ if r is an integer.

(b) Conversely, suppose that $\{|X_n|^r : n \ge 1\}$ is uniformly integrable where $r \ge 1$, and show that $X_n \xrightarrow{r} X$ if $X_n \xrightarrow{\text{P}} X$.

3. Let $g : [0, \infty) \to [0, \infty)$ be an increasing function satisfying $g(x)/x \to \infty$ as $x \to \infty$. Show that the sequence $\{X_n : n \ge 1\}$ is uniformly integrable if $\sup_n \mathbb{E}\{g(|X_n|)\} < \infty$.

4. Let $\{Z_n : n \ge 0\}$ be the generation sizes of a branching process with $Z_0 = 1$, $\mathbb{E}(Z_1) = 1$, $\operatorname{var}(Z_1) \ne 0$. Show that $\{Z_n : n \ge 0\}$ is not uniformly integrable.

5. **Pratt's lemma.** Suppose that $X_n \le Y_n \le Z_n$ where $X_n \xrightarrow{\text{P}} X$, $Y_n \xrightarrow{\text{P}} Y$, and $Z_n \xrightarrow{\text{P}} Z$. If $\mathbb{E}(X_n) \to \mathbb{E}(X)$ and $\mathbb{E}(Z_n) \to \mathbb{E}(Z)$, show that $\mathbb{E}(Y_n) \to \mathbb{E}(Y)$.

6. Let $\{X_n : n \ge 1\}$ be a sequence of variables satisfying $\mathbb{E}(\sup_n |X_n|) < \infty$. Show that $\{X_n\}$ is uniformly integrable.

7. Give an example of a uniformly integrable sequence $\{X_n\}$ of random variables and a σ-field $\mathcal{G}$ such that $X_n \xrightarrow{\text{a.s.}} X$ as $n \to \infty$, but $\mathbb{E}(X_n \mid \mathcal{G})$ does not converge a.s. to $\mathbb{E}(X \mid \mathcal{G})$.

7.11 Problems

1. Let X_n have density function

$$f_n(x) = \frac{n}{\pi(1+n^2x^2)}, \qquad n \geq 1.$$

With respect to which modes of convergence does X_n converge as $n \to \infty$?

2. (i) Suppose that $X_n \xrightarrow{\text{a.s.}} X$ and $Y_n \xrightarrow{\text{a.s.}} Y$, and show that $X_n + Y_n \xrightarrow{\text{a.s.}} X + Y$. Show that the corresponding result holds for convergence in rth mean and in probability, but not in distribution.

(ii) Show that if $X_n \xrightarrow{\text{a.s.}} X$ and $Y_n \xrightarrow{\text{a.s.}} Y$ then $X_n Y_n \xrightarrow{\text{a.s.}} XY$. Does the corresponding result hold for the other modes of convergence?

3. Let $g : \mathbb{R} \to \mathbb{R}$ be continuous. Show that $g(X_n) \xrightarrow{\text{P}} g(X)$ if $X_n \xrightarrow{\text{P}} X$.

4. Let $Y_1, Y_2, \dots$ be independent identically distributed variables, each of which can take any value in $\{0, 1, \dots, 9\}$ with equal probability $\frac{1}{10}$. Let $X_n = \sum_{i=1}^{n} Y_i \, 10^{-i}$. Show by the use of characteristic functions that X_n converges in distribution to the uniform distribution on $[0, 1]$. Deduce that $X_n \xrightarrow{\text{a.s.}} Y$ for some Y which is uniformly distributed on $[0, 1]$.

5. Let $N(t)$ be a Poisson process with constant intensity on $\mathbb{R}$.

(a) Find the covariance of $N(s)$ and $N(t)$.

(b) Show that N is continuous in mean square, which is to say that $\mathbb{E}\big(\{N(t+h) - N(t)\}^2\big) \to 0$ as $h \to 0$.

(c) Prove that N is continuous in probability, which is to say that $\mathbb{P}\big(|N(t+h) - N(t)| > \epsilon\big) \to 0$ as $h \to 0$, for all $\epsilon > 0$.

(d) Show that N is differentiable in probability but not in mean square.

6. Prove that $n^{-1} \sum_{i=1}^{n} X_i \xrightarrow{\text{a.s.}} 0$ whenever the X_i are independent identically distributed variables with zero means and such that $\mathbb{E}(X_1^4) < \infty$.

7. Show that $X_n \xrightarrow{\text{a.s.}} X$ whenever $\sum_n \mathbb{E}(|X_n - X|^r) < \infty$ for some $r > 0$.

8. Show that if $X_n \xrightarrow{\text{D}} X$ then $aX_n + b \xrightarrow{\text{D}} aX + b$ for any real a and b.

9. (a) **Cantelli, or one-sided Chebyshov inequality.** If X has zero mean and variance $\sigma^2 > 0$, show that

$$\mathbb{P}(X \geq t) \leq \frac{\sigma^2}{\sigma^2 + t^2} \quad \text{for } t > 0.$$

(b) Deduce that $|\mu - m| \leq \sigma$, where μ, m, and σ (> 0) are the mean, median, and standard deviation of a given distribution.

(c) Use Jensen's inequality to prove part (b) directly.

10. Show that $X_n \xrightarrow{\text{P}} 0$ if and only if

$$\mathbb{E}\left(\frac{|X_n|}{1 + |X_n|}\right) \to 0 \quad \text{as } n \to \infty.$$

11. The sequence $\{X_n\}$ is said to be *mean-square Cauchy convergent* if $\mathbb{E}\{(X_n - X_m)^2\} \to 0$ as $m, n \to \infty$. Show that $\{X_n\}$ converges in mean square to some limit X if and only if it is mean-square Cauchy convergent. Does the corresponding result hold for the other modes of convergence?

12. Suppose that $\{X_n\}$ is a sequence of uncorrelated variables with zero means and uniformly bounded variances. Show that $n^{-1}\sum_{i=1}^{n} X_i \xrightarrow{\text{m.s.}} 0$.

13. Let $X_1, X_2, \dots$ be independent identically distributed random variables with the common distribution function F, and suppose that $F(x) < 1$ for all x. Let $M_n = \max\{X_1, X_2, \dots, X_n\}$ and suppose that there exists a strictly increasing unbounded positive sequence $a_1, a_2, \dots$ such that $\mathbb{P}(M_n/a_n \le x) \to H(x)$ for some distribution function H. Let us assume that H is continuous with $0 < H(1) < 1$; substantially weaker conditions suffice but introduce extra difficulties.

(a) Show that $n[1 - F(a_n x)] \to -\log H(x)$ as $n \to \infty$ and deduce that

$$\frac{1 - F(a_n x)}{1 - F(a_n)} \to \frac{\log H(x)}{\log H(1)} \quad \text{if } x > 0.$$

(b) Deduce that if $x > 0$

$$\frac{1 - F(tx)}{1 - F(t)} \to \frac{\log H(x)}{\log H(1)} \quad \text{as } t \to \infty.$$

(c) Set $x = x_1 x_2$ and make the substitution

$$g(x) = \frac{\log H(e^x)}{\log H(1)}$$

to find that $g(x + y) = g(x)g(y)$, and deduce that

$$H(x) = \begin{cases} \exp(-\alpha x^{-\beta}) & \text{if } x \ge 0, \\ 0 & \text{if } x < 0, \end{cases}$$

for some non-negative constants α and β.

You have shown that H is the distribution function of Y^{-1}, where Y has a Weibull distribution.

14. Let $X_1, X_2, \dots, X_n$ be independent and identically distributed random variables with the Cauchy distribution. Show that $M_n = \max\{X_1, X_2, \dots, X_n\}$ is such that $\pi M_n/n$ converges in distribution, the limiting distribution function being given by $H(x) = e^{-1/x}$ if $x \ge 0$.

15. Let $X_1, X_2, \dots$ be independent and identically distributed random variables whose common characteristic function ϕ satisfies $\phi'(0) = i\mu$. Show that $n^{-1}\sum_{j=1}^{n} X_j \xrightarrow{\text{P}} \mu$.

16. Total variation distance. The *total variation distance* $d_{\text{TV}}(X, Y)$ between two random variables X and Y is defined by

$$d_{\text{TV}}(X, Y) = \sup_{u:\|u\|_\infty = 1} \big|\mathbb{E}(u(X)) - \mathbb{E}(u(Y))\big|$$

where the supremum is over all (measurable) functions $u : \mathbb{R} \to \mathbb{R}$ such that $\|u\|_\infty = \sup_x |u(x)|$ satisfies $\|u\|_\infty = 1$.

(a) If X and Y are discrete with respective masses f_n and g_n at the points x_n, show that

$$d_{\text{TV}}(X, Y) = \sum_n |f_n - g_n| = 2 \sup_{A \subseteq \mathbb{R}} \big|\mathbb{P}(X \in A) - \mathbb{P}(Y \in A)\big|.$$

(b) If X and Y are continuous with respective density functions f and g, show that

$$d_{\text{TV}}(X, Y) = \int_{-\infty}^{\infty} |f(x) - g(x)|\, dx = 2 \sup_{A \subseteq \mathbb{R}} \big|\mathbb{P}(X \in A) - \mathbb{P}(Y \in A)\big|.$$

(c) Show that $d_{\text{TV}}(X_n, X) \to 0$ implies that $X_n \to X$ in distribution, but that the converse is false.

(d) **Maximal coupling.** Show that $\mathbb{P}(X \neq Y) \geq \frac{1}{2}d_{\mathrm{TV}}(X, Y)$, and that there exists a pair X', Y' having the same marginals for which equality holds.

(e) If X_i, Y_j are independent random variables, show that

$$d_{\mathrm{TV}}\left(\sum_{i=1}^{n} X_i, \sum_{i=1}^{n} Y_i\right) \leq \sum_{i=1}^{n} d_{\mathrm{TV}}(X_i, Y_i).$$

17. Let $g : \mathbb{R} \to \mathbb{R}$ be bounded and continuous. Show that

$$\sum_{k=0}^{\infty} g(k/n) \frac{(n\lambda)^k}{k!} e^{-n\lambda} \to g(\lambda) \quad \text{as } n \to \infty.$$

18. Let X_n and Y_m be independent random variables having the Poisson distribution with parameters n and m, respectively. Show that

$$\frac{(X_n - n) - (Y_m - m)}{\sqrt{X_n + Y_m}} \xrightarrow{\mathrm{D}} N(0, 1) \quad \text{as } m, n \to \infty.$$

19. (a) Suppose that $X_1, X_2, \dots$ is a sequence of random variables, each having a normal distribution, and such that $X_n \xrightarrow{\mathrm{D}} X$. Show that X has a normal distribution, possibly degenerate.

(b) For each $n \geq 1$, let (X_n, Y_n) be a pair of random variables having a bivariate normal distribution. Suppose that $X_n \xrightarrow{\mathrm{P}} X$ and $Y_n \xrightarrow{\mathrm{P}} Y$, and show that the pair (X, Y) has a bivariate normal distribution.

20. Let $X_1, X_2, \dots$ be random variables satisfying $\mathrm{var}(X_n) < c$ for all n and some constant c. Show that the sequence obeys the weak law, in the sense that $n^{-1}\sum_1^n(X_i - \mathbb{E}X_i)$ converges in probability to 0, if the correlation coefficients satisfy either of the following:

(i) $\rho(X_i, X_j) \leq 0$ for all $i \neq j$,

(ii) $\rho(X_i, X_j) \to 0$ as $|i - j| \to \infty$.

21. Let $X_1, X_2, \dots$ be independent random variables with common density function

$$f(x) = \begin{cases} 0 & \text{if } |x| \leq 2, \\ \dfrac{c}{x^2 \log |x|} & \text{if } |x| > 2, \end{cases}$$

where c is a constant. Show that the X_i have no mean, but $n^{-1}\sum_{i=1}^n X_i \xrightarrow{\mathrm{P}} 0$ as $n \to \infty$. Show that convergence does not take place almost surely.

22. Let X_n be the Euclidean distance between two points chosen independently and uniformly from the n-dimensional unit cube. Show that $\mathbb{E}(X_n)/\sqrt{n} \to 1/\sqrt{6}$ as $n \to \infty$.

23. Let $X_1, X_2, \dots$ be independent random variables having the uniform distribution on $[-1, 1]$. Show that

$$\mathbb{P}\left(\left|\sum_{i=1}^{n} X_i^{-1}\right| > \tfrac{1}{2}n\pi\right) \to \tfrac{1}{2} \quad \text{as } n \to \infty.$$

24. Let $X_1, X_2, \dots$ be independent random variables, each X_k having mass function given by

$$\mathbb{P}(X_k = k) = \mathbb{P}(X_k = -k) = \frac{1}{2k^2},$$

$$\mathbb{P}(X_k = 1) = \mathbb{P}(X_k = -1) = \frac{1}{2}\left(1 - \frac{1}{k^2}\right) \qquad \text{if } k > 1.$$

Show that $U_n = \sum_1^n X_i$ satisfies $U_n/\sqrt{n} \xrightarrow{\text{D}} N(0,1)$ but $\text{var}(U_n/\sqrt{n}) \to 2$ as $n \to \infty$.

25. Let $X_1, X_2, \dots$ be random variables, and let $N_1, N_2, \dots$ be random variables taking values in the positive integers such that $N_k \xrightarrow{\text{P}} \infty$ as $k \to \infty$. Show that:

(i) if $X_n \xrightarrow{\text{D}} X$ and the X_n are independent of the N_k, then $X_{N_k} \xrightarrow{\text{D}} X$ as $k \to \infty$,

(ii) if $X_n \xrightarrow{\text{a.s.}} X$ then $X_{N_k} \xrightarrow{\text{P}} X$ as $k \to \infty$.

26. Stirling's formula.

(a) Let $a(k, n) = n^k/(k-1)!$ for $1 \le k \le n+1$. Use the fact that $1 - x \le e^{-x}$ if $x \ge 0$ to show that

$$\frac{a(n-k, n)}{a(n+1, n)} \le e^{-k^2/(2n)} \qquad \text{if } k \ge 0.$$

(b) Let $X_1, X_2, \dots$ be independent Poisson variables with parameter 1, and let $S_n = X_1 + \cdots + X_n$. Define the function $g : \mathbb{R} \to \mathbb{R}$ by

$$g(x) = \begin{cases} -x & \text{if } 0 \ge x \ge -M, \\ 0 & \text{otherwise,} \end{cases}$$

where M is large and positive. Show that, for large n,

$$\mathbb{E}\left(g\left\{\frac{S_n - n}{\sqrt{n}}\right\}\right) = \frac{e^{-n}}{\sqrt{n}}\{a(n+1, n) - a(n-k, n)\}$$

where $k = \lfloor Mn^{1/2} \rfloor$. Now use the central limit theorem and (a) above, to deduce Stirling's formula:

$$\frac{n!\, e^n}{n^{n+\frac{1}{2}}\sqrt{2\pi}} \to 1 \qquad \text{as } n \to \infty.$$

27. Pólya's urn. A bag contains red and green balls. A ball is drawn from the bag, its colour noted, and then it is returned to the bag together with a new ball of the same colour. Initially the bag contained one ball of each colour. If R_n denotes the number of red balls in the bag after n additions, show that $S_n = R_n/(n+2)$ is a martingale. Deduce that the ratio of red to green balls converges almost surely to some limit as $n \to \infty$.

28. Anscombe's theorem. Let $\{X_i : i \ge 1\}$ be independent identically distributed random variables with zero mean and finite positive variance σ^2, and let $S_n = \sum_1^n X_i$. Suppose that the integer-valued random process $M(t)$ satisfies $t^{-1}M(t) \xrightarrow{\text{P}} \theta$ as $t \to \infty$, where θ is a positive constant. Show that

$$\frac{S_{M(t)}}{\sigma\sqrt{\theta t}} \xrightarrow{\text{D}} N(0,1) \quad \text{and} \quad \frac{S_{M(t)}}{\sigma\sqrt{M(t)}} \xrightarrow{\text{D}} N(0,1) \quad \text{as } t \to \infty.$$

You should not assume that the process M is independent of the X_i.

29. Kolmogorov's inequality. Let $X_1, X_2, \dots$ be independent random variables with zero means, and $S_n = X_1 + X_2 + \cdots + X_n$. Let $M_n = \max_{1 \le k \le n} |S_k|$ and show that $\mathbb{E}(S_n^2 I_{A_k}) > c^2 \mathbb{P}(A_k)$ where $A_k = \{M_{k-1} \le c < M_k\}$ and $c > 0$. Deduce Kolmogorov's inequality:

$$\mathbb{P}\left(\max_{1 \le k \le n} |S_k| > c\right) \le \frac{\mathbb{E}(S_n^2)}{c^2}, \qquad c > 0.$$

30. Let $X_1, X_2, \dots$ be independent random variables with zero means, and let $S_n = X_1 + X_2 + \dots + X_n$. Using Kolmogorov's inequality or the martingale convergence theorem, show that:

(i) $\sum_{i=1}^{\infty} X_i$ converges almost surely if $\sum_{k=1}^{\infty} \mathbb{E}(X_k^2) < \infty$,

(ii) if there exists an increasing real sequence (b_n) such that $b_n \to \infty$, and satisfying the inequality $\sum_{k=1}^{\infty} \mathbb{E}(X_k^2)/b_k^2 < \infty$, then $b_n^{-1} \sum_{k=1}^{\infty} X_k \xrightarrow{\text{a.s.}} 0$ as $n \to \infty$.

31. Estimating the transition matrix. The Markov chain $X_0, X_1, \dots, X_n$ has initial distribution $f_i = \mathbb{P}(X_0 = i)$ and transition matrix $\mathbf{P}$. The *log-likelihood* function $\lambda(\mathbf{P})$ is defined as $\lambda(\mathbf{P}) = \log(f_{X_0} p_{X_0,X_1} p_{X_1,X_2} \cdots p_{X_{n-1},X_n})$. Show that:

(a) $\lambda(\mathbf{P}) = \log f_{X_0} + \sum_{i,j} N_{ij} \log p_{ij}$ where N_{ij} is the number of transitions from i to j,

(b) viewed as a function of the p_{ij}, $\lambda(\mathbf{P})$ is maximal when $p_{ij} = \widehat{p}_{ij}$ where $\widehat{p}_{ij} = N_{ij}/\sum_k N_{ik}$,

(c) if X is irreducible and ergodic then $\widehat{p}_{ij} \xrightarrow{\text{a.s.}} p_{ij}$ as $n \to \infty$.

32. Ergodic theorem in discrete time. Let X be an irreducible discrete-time Markov chain, and let μ_i be the mean recurrence time of state i. Let $V_i(n) = \sum_{r=0}^{n-1} I_{\{X_r = i\}}$ be the number of visits to i up to $n-1$, and let f be any bounded function on S. Show that:

(a) $n^{-1} V_i(n) \xrightarrow{\text{a.s.}} \mu_i^{-1}$ as $n \to \infty$,

(b) if $\mu_i < \infty$ for all i, then

$$\frac{1}{n} \sum_{r=0}^{n-1} f(X_r) \to \sum_{i \in S} f(i)/\mu_i \quad \text{as } n \to \infty.$$

33. Ergodic theorem in continuous time. Let X be an irreducible recurrent continuous-time Markov chain with generator $\mathbf{G}$ and finite mean return times m_j.

(a) Show that $\displaystyle\frac{1}{t} \int_0^t I_{\{X(s)=j\}}\, ds \xrightarrow{\text{a.s.}} \frac{1}{m_j g_j}$ as $t \to \infty$;

(b) deduce that the stationary distribution $\boldsymbol{\pi}$ satisfies $\pi_j = 1/(m_j g_j)$;

(c) show that, if f is a bounded function on S,

$$\frac{1}{t} \int_0^t f(X(s))\, ds \xrightarrow{\text{a.s.}} \sum_i \pi_i f(i) \quad \text{as } t \to \infty.$$

34. Tail equivalence. Suppose that the sequences $\{X_n : n \ge 1\}$ and $\{Y_n : n \ge 1\}$ are *tail equivalent*, which is to say that $\sum_{n=1}^{\infty} \mathbb{P}(X_n \ne Y_n) < \infty$. Show that:

(a) $\sum_{n=1}^{\infty} X_n$ and $\sum_{n=1}^{\infty} Y_n$ converge or diverge together,

(b) $\sum_{n=1}^{\infty} (X_n - Y_n)$ converges almost surely,

(c) if there exist a random variable X and a sequence a_n such that $a_n \uparrow \infty$ and $a_n^{-1} \sum_{r=1}^{n} X_r \xrightarrow{\text{a.s.}} X$, then

$$\frac{1}{a_n} \sum_{r=1}^{n} Y_r \xrightarrow{\text{a.s.}} X.$$

35. Three series theorem. Let $\{X_n : n \ge 1\}$ be independent random variables. Show that $\sum_{n=1}^{\infty} X_n$ converges a.s. if, for some $a > 0$, the following three series all converge:

(a) $\sum_n \mathbb{P}(|X_n| > a)$,

(b) $\sum_n \operatorname{var}(X_n I_{\{|X_n| \le a\}})$,

(c) $\sum_n \mathbb{E}(X_n I_{\{|X_n| \le a\}})$.

[The converse holds also, but is harder to prove.]

36. Let $\{X_n : n \geq 1\}$ be independent random variables with continuous common distribution function F. We call X_k a *record value* for the sequence if $X_k > X_r$ for $1 \leq r < k$, and we write I_k for the indicator function of the event that X_k is a record value.

(a) Show that the random variables I_k are independent.

(b) Show that $R_m = \sum_{k=1}^m I_r$ satisfies $R_m / \log m \xrightarrow{\text{a.s.}} 1$ as $m \to \infty$.

37. Random harmonic series. Let $\{X_n : n \geq 1\}$ be a sequence of independent random variables with $\mathbb{P}(X_n = 1) = \mathbb{P}(X_n = -1) = \frac{1}{2}$. Does the series $\sum_{r=1}^n X_r/r$ converge a.s. as $n \to \infty$?

38. Stirling's formula for the gamma function. Let X have the gamma distribution $\Gamma(1, s)$. By considering the integral of the density function of $Y = (X - s)/\sqrt{s}$, show that $\Gamma(s) \sim \sqrt{2\pi} s^{s-\frac{1}{2}} e^{-s}$ as $s \to \infty$. [Hint: You may find it useful that

$$\int_a^b e^{-u(x)}\, dx \leq \frac{1}{u'(a)} \int_a^b u'(x) e^{-u(x)}\, dx,$$

if $u'(x)$ is strictly positive and increasing.]

39. Random series. Let $c_1, c_2, \dots$ be reals, let $X_1, X_2, \dots$ be independent random variables with the mass function $f(1) = f(-1) = \frac{1}{2}$, and let $S_n = \sum_{r=1}^n c_r X_r$. Write

$$B_n = \sum_{r=1}^n c_r^4, \qquad D_n = \sqrt{\sum_{r=1}^n c_r^2}.$$

(a) Use characteristic functions to show that S_n/D_n converges in distribution to the $N(0, 1)$ distribution (as $n \to \infty$) if and only if $B_n/D_n^4 \to 0$. [Hint: You may use the fact that $-\frac{2}{3}\theta^4 \leq \frac{1}{2}\theta^2 + \log\cos\theta \leq -\frac{1}{12}\theta^4$ for $-\frac{1}{4}\pi \leq \theta \leq \frac{1}{4}\pi$.]

(b) Find the limit of S_n/D_n in the special case $c_r = 2^{-r}$.

40. Berge's inequality. Let X and Y be random variables with mean 0, variance 1, and correlation ρ. Show that, for $\epsilon > 0$,

$$\mathbb{P}\big(|X| \vee |Y| > \epsilon\big) \leq \frac{1}{\epsilon^2}\left(1 + \sqrt{1 - \rho^2}\right),$$

where $x \vee y = \max\{x, y\}$. [Hint: If $|t| \leq 1$, the function $g(x, y) = (x^2 - 2txy + y^2)/(\epsilon^2(1 - t^2))$ is non-negative, and moreover satisfies $g(x, y) \geq 1$ when $|x| \vee |y| \geq \epsilon$.]

41. Poisson tail, balls in bins. Let X have the Poisson distribution with parameter 1.

(a) Show that $\mathbb{P}(X \geq t) \leq e^{t-1}/t^t$ for $t \geq 1$.

(b) Deduce that the maximum M_n of n independent random variables, distributed as X, satisfies

$$\lim_{n\to\infty} \mathbb{P}\left(M_n \geq \frac{(1+a)\log n}{\log\log n}\right) = \begin{cases} 1 & \text{if } a < 0, \\ 0 & \text{if } a > 0. \end{cases}$$

8

Random processes

8.2 Exercises. Stationary processes

1. Flip–flop. Let $\{X_n\}$ be a Markov chain on the state space $S = \{0, 1\}$ with transition matrix

$$\mathbf{P} = \begin{pmatrix} 1-\alpha & \alpha \\ \beta & 1-\beta \end{pmatrix},$$

where $\alpha + \beta > 0$. Find:

(a) the correlation $\rho(X_m, X_{m+n})$, and its limit as $m \to \infty$ with n remaining fixed,

(b) $\lim_{n\to\infty} n^{-1} \sum_{r=1}^{n} \mathbb{P}(X_r = 1)$.

Under what condition is the process strongly stationary?

2. Random telegraph. Let $\{N(t) : t \geq 0\}$ be a Poisson process of intensity λ, and let T_0 be an independent random variable such that $\mathbb{P}(T_0 = \pm 1) = \frac{1}{2}$. Define $T(t) = T_0(-1)^{N(t)}$. Show that $\{T(t) : t \geq 0\}$ is stationary and find: (a) $\rho(T(s), T(s+t))$, (b) the mean and variance of $X(t) = \int_0^t T(s)\, ds$.

[The so-called Goldstein–Kac process $X(t)$ denotes the position of a particle moving with unit speed, starting from the origin along the positive x-axis, whose direction is reversed at the instants of a Poisson process.]

3. Korolyuk–Khinchin theorem. An integer-valued counting process $\{N(t) : t \geq 0\}$ with $N(0) = 0$ is called *crudely stationary* if $p_k(s,t) = \mathbb{P}(N(s+t) - N(s) = k)$ depends only on the length $t - s$ and not on the location s. It is called *simple* if, almost surely, it has jump discontinuities of size 1 only. Show that, for a simple crudely stationary process N, $\lim_{t\downarrow 0} t^{-1}\mathbb{P}(N(t) > 0) = \mathbb{E}(N(1))$.

4. Kac's ergodic formula. Let $X = \{X_n : n \geq 0\}$ be an ergodic Markov chain started in its stationary distribution $\boldsymbol{\pi}$. Let A be a subset of states, with stationary probability $\pi(A)$, and let $T_A = \min\{n \geq 1 : X_n \in A\}$. Show that $\mathbb{E}(T_A \mid X_0 \in A) = 1/\pi(A)$.

8.3 Exercises. Renewal processes

1. Let $(f_n : n \geq 1)$ be a probability distribution on the positive integers, and define a sequence $(u_n : n \geq 0)$ by $u_0 = 1$ and $u_n = \sum_{r=1}^{n} f_r u_{n-r}$, $n \geq 1$. Explain why such a sequence is called a *renewal sequence*, and show that u is a renewal sequence if and only if there exists a Markov chain U and a state s such that $u_n = \mathbb{P}(U_n = s \mid U_0 = s)$.

2. Let $\{X_i : i \geq 1\}$ be the inter-event times of a discrete renewal process on the integers. Show that the excess lifetime B_n constitutes a Markov chain. Write down the transition probabilities of the sequence $\{B_n\}$ when reversed in equilibrium. Compare these with the transition probabilities of the chain U of your solution to Exercise (8.3.1).

3. Let $(u_n : n \geq 1)$ satisfy $u_0 = 1$ and $u_n = \sum_{r=1}^{n} f_r u_{n-r}$ for $n \geq 1$, where $(f_r : r \geq 1)$ is a non-negative sequence. Show that:

(a) $v_n = \rho^n u_n$ is a renewal sequence if $\rho > 0$ and $\sum_{n=1}^{\infty} \rho^n f_n = 1$,

(b) as $n \to \infty$, $\rho^n u_n$ converges to some constant c.

4. Events occur at the times of a discrete-time renewal process N (see Example (5.2.15)). Let u_n be the probability of an event at time n, with generating function $U(s)$, and let $F(s)$ be the probability generating function of a typical inter-event time. Show that, if $|s| < 1$:

$$\sum_{r=0}^{\infty} \mathbb{E}(N(r))s^r = \frac{F(s)U(s)}{1-s} \quad \text{and} \quad \sum_{t=0}^{\infty} \mathbb{E}\left[\binom{N(t)+k}{k}\right] s^t = \frac{U(s)^k}{1-s} \quad \text{for } k \geq 0.$$

5. Prove Theorem (8.3.5): Poisson processes are the only renewal processes that are Markov chains.

8.4 Exercises. Queues

1. The two tellers in a bank each take an exponentially distributed time to deal with any customer; their parameters are λ and μ respectively. You arrive to find exactly two customers present, each occupying a teller.

(a) You take a fancy to a randomly chosen teller, and queue for that teller to be free; no later switching is permitted. Assuming any necessary independence, what is the probability p that you are the last of the three customers to leave the bank?

(b) If you choose to be served by the quicker teller, find p.

(c) Suppose you go to the teller who becomes free first. Find p.

2. Customers arrive at a desk according to a Poisson process of intensity λ. There is one clerk, and the service times are independent and exponentially distributed with parameter μ. At time 0 there is exactly one customer, currently in service. Show that the probability that the next customer arrives before time t and finds the clerk busy is

$$\frac{\lambda}{\lambda+\mu}(1 - e^{-(\lambda+\mu)t}).$$

3. Vehicles pass a crossing at the instants of a Poisson process of intensity λ; you need a gap of length at least a in order to cross. Let T be the first time at which you could succeed in crossing to the other side. Show that $\mathbb{E}(T) = (e^{a\lambda} - 1)/\lambda$, and find $\mathbb{E}(e^{\theta T})$.

Suppose there are two lanes to cross, carrying independent Poissonian traffic with respective rates λ and μ. Find the expected time to cross in the two cases when: (a) there is an island or refuge between the two lanes, (b) you must cross both in one go. Which is the greater?

4. Customers arrive at the instants of a Poisson process of intensity λ, and the single server has exponential service times with parameter μ. An arriving customer who sees n customers present (including anyone in service) will join the queue with probability $(n+1)/(n+2)$, otherwise leaving for ever. Under what condition is there a stationary distribution? Find the mean of the time spent in the queue (not including service time) by a customer who joins it when the queue is in equilibrium. What is the probability that an arrival joins the queue when in equilibrium?

5. Customers enter a shop at the instants of a Poisson process of rate 2. At the door, two representatives separately demonstrate a new corkscrew. This typically occupies the time of a customer and the representative for a period which is exponentially distributed with parameter 1, independently of arrivals and other demonstrators. If both representatives are busy, customers pass directly into the shop. No customer passes a free representative without being stopped, and all customers leave by another door. If both representatives are free at time 0, show the probability that both are busy at time t is $\frac{2}{5} - \frac{2}{3}e^{-2t} + \frac{4}{15}e^{-5t}$.

8.5 Exercises. The Wiener process

1. For a Wiener process W with $W(0) = 0$, show that

$$\mathbb{P}\big(W(s) > 0,\ W(t) > 0\big) = \frac{1}{4} + \frac{1}{2\pi}\sin^{-1}\sqrt{\frac{s}{t}} \quad \text{for } s < t.$$

Calculate $\mathbb{P}(W(s) > 0,\ W(t) > 0,\ W(u) > 0)$ when $s < t < u$.

2. Let W be a Wiener process. Show that, for $s < t < u$, the conditional distribution of $W(t)$ given $W(s)$ and $W(u)$ is normal

$$N\left(\frac{(u-t)W(s) + (t-s)W(u)}{u-s},\ \frac{(u-t)(t-s)}{u-s}\right).$$

Deduce that the conditional correlation between $W(t)$ and $W(u)$, given $W(s)$ and $W(v)$, where $s < t < u < v$, is

$$\sqrt{\frac{(v-u)(t-s)}{(v-t)(u-s)}}.$$

3. For what values of a and b is $aW_1 + bW_2$ a standard Wiener process, where W_1 and W_2 are independent standard Wiener processes?

4. Show that a Wiener process W with variance parameter σ^2 satisfies

$$\sum_{j=0}^{n-1}\big\{W((j+1)t/n) - W(jt/n)\big\}^2 \xrightarrow{\text{m.s.}} \sigma^2 t \quad \text{as } n \to \infty.$$

The process W is said to have *finite quadratic variation.*

5. Let W be a Wiener process. Which of the following define Wiener processes?
(a) $-W(t)$, (b) $\sqrt{t}W(1)$, (c) $W(2t) - W(t)$.

6. Find the distribution of $\int_0^t [W(u)/u]\,du$ where W is the Wiener process. [The integral is interpreted as the limit as $\epsilon \downarrow 0$ of the integral from ϵ to t.]

7. Wiener process in n dimensions. An n-dimensional Wiener process is a process $W(t) = (W_1(t), W_2(t), \dots, W_n(t))$ taking values in $\mathbb{R}^n$ with independent increments, started at the origin $W(0) = 0$, and such that $W(t) - W(s)$ is multivariate normal with means 0 and covariance matrix $(t-s)\mathbf{I}$ where $\mathbf{I}$ is the $n \times n$ identity matrix. Let $W(t)$ be an n-dimensional Wiener process, and let A be an orthonormal $n \times n$ matrix. Show that $AW(t)$ is an n-dimensional Wiener process.

8. Let W be a standard Wiener process. Show that, for $p \ge 0$,

$$\mathbb{E}\big(|W(t) - W(s)|^p\big) = c_p|t-s|^{p/2}, \qquad s, t \in \mathbb{R},$$

where

$$c_p = \frac{1}{\sqrt{2\pi}}\int_{-\infty}^{\infty} |y|^p e^{-\frac{1}{2}y^2}\,dy.$$

8.6 Exercises. Lévy processes and subordinators

1. Prove that the characteristic function $\phi(t,\theta) = \mathbb{E}(e^{i\theta X(t)})$ of a Lévy process X is a continuous function of t.

2. Show that a compound Poisson process is a Lévy process, and find its Lévy symbol.

3. Verify the formula of Example (8.6.6) for the Laplace exponent of the Moran gamma process.

4. Prove Theorem (8.6.7): the property of being a Lévy process is preserved under time change by an independent subordinator.

5. Let N be a Poisson process of rate 1, and T an independent random variable with the $\Gamma(1,t)$ distribution. Show that $Y = N(T)$ has probability generating function $\mathbb{E}(s^Y) = (2-s)^{-t}$ for $s < 2$.

6. Let X be a Lévy process with finite variances. Show that the following are martingales:
(a) $X(t) - \mathbb{E}(X(t))$,
(b) $Z(t)^2 - \mathbb{E}(Z(t)^2)$, where $Z(t) = X(t) - \mathbb{E}(X(t))$,
(c) $e^{i\theta X(t)}/\phi(t,\theta)$, where $\phi(t,\theta) = \mathbb{E}(e^{i\theta X(t)})$.

7. Let X be a continuous-time martingale and T an independent subordinator. Show that, subject to the moment condition $\mathbb{E}|Y(t)| < \infty$, $Y(t) = X(T(t))$ defines a martingale with respect to a suitable filtration. Show that the moment condition is valid whenever X is a positive martingale.

8.7 Exercises. Self-similarity and stability

1. Let X be stable, and let $X_1, X_2, \dots, X_n$ be a random sample distributed as X. Show that, for $n \geq 1$, there exist A_n, B_n such that $X_1 + X_2 + \cdots + X_n \stackrel{\text{D}}{=} B_n X + A_n$.

2. Prove that, for a non-degenerate, self-similar process X with stationary increments, finite variances, and Hurst exponent $H > 0$,

$$\operatorname{var}(X(t)) = t^{2H}\operatorname{var}(X(1)), \qquad t \geq 0,$$
$$\mathbb{E}(X(s)X(t)) = \tfrac{1}{2}\big(s^{2H} + t^{2H} - |t-s|^{2H}\big)\mathbb{E}(X(1)^2), \qquad s,t \geq 0.$$

3. Show that a stable distribution is infinitely divisible.

4. Show that a self-similar Lévy process either is a Wiener process or has infinite variances.

5. Prove that a distribution with characteristic function $\phi(\theta) = \exp(-|\theta|^\alpha)$, where $\alpha \in (0,2]$, is strictly α-stable and symmetric.

6. Let X and Y be independent stable random variables, where X is symmetric with exponent $\alpha \in (0,2]$, and Y is non-negative with Laplace transform $M(\theta) = \exp(-k\theta^\beta)$ for $\theta \in [0,\infty)$, where $1 \neq \beta \in (0,2]$. Show that $Z = XY^{1/\alpha}$ is symmetric and $\alpha\beta$-stable. Deduce that, for independent $N(0,1)$ random variables U, V, the ratio $Z = U/V$ has the Cauchy distribution.

8.8 Exercises. Time changes

1. Let Z be a continuous-time Markov chain, and let T be an independent subordinator. Show that $X(t) = Z(T(t))$ defines a Markov chain.

2. Let $Z = \{Z_n : n \geq 0\}$ be a discrete-time Markov chain with n-step transition probabilities $z_{ij}(n)$. Let N be a Poisson process of rate λ that is independent of Z, and set $X(t) = Z_{N(t)}$. Show that X is a continuous-time Markov chain with transition probabilities

$$p_{ij}(t) = \sum_{n=0}^{\infty} e^{-\lambda t} \frac{(\lambda t)^n}{n!} z_{ij}(n).$$

3. Let W be the standard Wiener process, and let T be an independent stable subordinator with $\mathbb{E}(e^{-\theta T(t)}) = \exp(-t\theta^{a/2})$ where $a \in (0, 2)$ and $\theta \geq 0$. Show that the time-changed process $Y(t) = W(T(t))$ is a symmetric Lévy process.

4. Markov, but not strong Markov. Let V be exponentially distributed with parameter 1, and define $X = \{X(t) : t \geq 0\}$ by: $X(t) = 0$ for $t \leq V$, and $X(t) = t - V$ for $t \geq V$.

(a) Show that X is a Markov process.

(b) Show that V is a stopping time for X.

(c) Show that the two processes $X_1(t) = X(t)$ and $X_2(t) = X(t + V)$ satisfy $X_1(0) = X_2(0) = 0$, but have different fdds.

8.10 Problems

1. Let $\{Z_n\}$ be a sequence of uncorrelated real-valued variables with zero means and unit variances, and define the 'moving average'

$$Y_n = \sum_{i=0}^{r} \alpha_i Z_{n-i},$$

for constants $\alpha_0, \alpha_1, \dots, \alpha_r$. Show that Y is stationary and find its autocovariance function.

2. Let $\{Z_n\}$ be a sequence of uncorrelated real-valued variables with zero means and unit variances. Suppose that $\{Y_n\}$ is an 'autoregressive' stationary sequence in that it satisfies $Y_n = \alpha Y_{n-1} + Z_n$, $-\infty < n < \infty$, for some real α satisfying $|\alpha| < 1$. Show that Y has autocovariance function $c(m) = \alpha^{|m|}/(1 - \alpha^2)$.

3. Let $\{X_n\}$ be a sequence of independent identically distributed Bernoulli variables, each taking values 0 and 1 with probabilities $1 - p$ and p respectively. Find the mass function of the renewal process $N(t)$ with interarrival times $\{X_n\}$.

4. Customers arrive in a shop in the manner of a Poisson process with parameter λ. There are infinitely many servers, and each service time is exponentially distributed with parameter μ. Show that the number $Q(t)$ of waiting customers at time t constitutes a birth–death process. Find its stationary distribution.

5. Let $X(t) = Y\cos(\theta t) + Z\sin(\theta t)$ where Y and Z are independent $N(0, 1)$ random variables, and let $\widetilde{X}(t) = R\cos(\theta t + \Psi)$ where R and Ψ are independent. Find distributions for R and Ψ such that the processes X and $\widetilde{X}$ have the same fdds.

6. Bartlett's theorem. Customers arrive at the entrance to a queueing system at the instants of an non-homogeneous Poisson process with rate function $\lambda(t)$. Their subsequent service histories are independent of each other, and a customer arriving at time s is in state A at time $s + t$ with probability $p(s, t)$. Show that the number of customers in state A at time t is Poisson with parameter $\int_{-\infty}^{t} \lambda(u) p(u, t - u)\, du$.

7. An insurance company receives premiums (net of costs) at unit rate per unit time, and the claims of size $X_1, X_2, \dots$ are independent random variables with common distribution function F, arriving at the instants of a Poisson process of rate 1. Show that the probability $r(y)$ of ruin (that is, assets becoming negative) for an initial capital y satisfies

$$\frac{dr(y)}{dy} = \lambda r(y) - \lambda \mathbb{P}(X_1 > y) - \lambda \int_0^y r(y-x)\, dF(x).$$

8. Fractional Brownian motion. Let X be fBM$_H$ with $\text{var}(X(1)) = 1$, where $H \in (0, 1)$. Show that:

(a) $X(t) - X(s)$ has the $N(0, |t-s|^{2H})$ distribution,

(b) for $r > 0$,

$$\mathbb{E}\big(|X(t) - X(s)|^r\big) = C|t-s|^{rH},$$

where C depends on r.

(c) X is a Wiener process if and only if $H = \frac{1}{2}$.

9. Holtsmark distribution for stellar gravity, Problem (6.15.57) revisited. Suppose stars are distributed in $\mathbb{R}^3$ in the manner of a Poisson process with intensity λ, and let G_λ be the x-coordinate of their aggregated gravitational force exerted on a unit mass at the origin.

(a) Show that $G_{\lambda+\mu} \stackrel{\text{D}}{=} G_\lambda + G_\mu$, the sum of two independent variables.

(b) Show by the inverse square law that $G_\lambda \stackrel{\text{D}}{=} \lambda^{2/3} G_1$.

(c) Comment on the relevance of the above in studying gravitational attraction in three dimensions.

10. In a Prague teashop (U Myšáka) before the Velvet Revolution of 1989, customers queue at the entrance for a blank bill. In the shop there are separate counters for coffee, sweetcakes, pretzels, milk, drinks, and ice cream, and queues form at each of these. At each service point the customers' bills are marked appropriately. There is a restricted number N of seats, and departing customers have to queue in order to pay their bills. If interarrival times and service times are exponentially distributed and the process is in equilibrium, find how much longer a greedy customer must wait if he insists on sitting down. Answers on a postcard to the authors, please.

9

Stationary processes

9.1 Exercises. Introduction

1. Let $\dots, Z_{-1}, Z_0, Z_1, Z_2, \dots$ be independent real random variables with means 0 and variances 1, and let $\alpha, \beta \in \mathbb{R}$. Show that there exists a (weakly) stationary sequence $\{W_n\}$ satisfying $W_n = \alpha W_{n-1} + \beta W_{n-2} + Z_n$, $n = \dots, -1, 0, 1, \dots$, if the (possibly complex) zeros of the quadratic equation $z^2 - \alpha z - \beta = 0$ are smaller than 1 in absolute value.

2. Let U be uniformly distributed on $[0, 1]$ with binary expansion $U = \sum_{i=1}^{\infty} X_i 2^{-i}$. Show that the sequence

$$V_n = \sum_{i=1}^{\infty} X_{i+n} 2^{-i}, \qquad n \geq 0,$$

is strongly stationary, and calculate its autocovariance function.

3. Let $\{X_n : n = \dots, -1, 0, 1, \dots\}$ be a stationary real sequence with mean 0 and autocovariance function $c(m)$.

(i) Show that the infinite series $\sum_{n=0}^{\infty} a_n X_n$ converges almost surely, and in mean square, whenever $\sum_{n=0}^{\infty} |a_n| < \infty$.

(ii) Let

$$Y_n = \sum_{k=0}^{\infty} a_k X_{n-k}, \qquad n = \dots, -1, 0, 1, \dots$$

where $\sum_{k=0}^{\infty} |a_k| < \infty$. Find an expression for the autocovariance function c_Y of Y, and show that

$$\sum_{m=-\infty}^{\infty} |c_Y(m)| < \infty.$$

4. Let $X = \{X_n : n \geq 0\}$ be a discrete-time Markov chain with countable state space S and stationary distribution $\boldsymbol{\pi}$, and suppose that X_0 has distribution $\boldsymbol{\pi}$. Show that the sequence $\{f(X_n) : n \geq 0\}$ is strongly stationary for any function $f : S \to \mathbb{R}$.

5. Let W, X, Y, Z have a multivariate normal distribution with zero means and covariance matrix

$$\begin{pmatrix} 1 & 0 & 0 & -1 \\ 0 & 1 & -1 & 0 \\ 0 & -1 & 1 & 0 \\ -1 & 0 & 0 & 1 \end{pmatrix}.$$

Let $U = W + iX$ and $V = Y + iZ$. Show that U and V are uncorrelated but not independent. Why does this not violate the conclusion of Example (4.5.9) that multivariate normal random variables are independent if and only if they are uncorrelated?

6. Let $X_n = \cos(nS + U)$ for $n \in \mathbb{Z}$, where U is uniformly distributed on $(-\pi, \pi)$, and S is independent of U with a density function g that is symmetric on its support $(-\pi, \pi)$. Show that $X = \{X_n : n \in \mathbb{Z}\}$ is weakly stationary, and find its autocorrelation function ρ_X.

Let $Y_n = Z_n + aZ_{n-1}$ for $n \in \mathbb{Z}$, where $|a| < 1$ and the Z_n are independent, identically distributed random variables with means 0 and variances 1. Is it possible to choose g in such a way that $Y = \{Y_n : n \in \mathbb{Z}\}$ has the same autocorrelation function as X? If so, how?

9.2 Exercises. Linear prediction

1. Let X be a (weakly) stationary sequence with zero mean and autocovariance function $c(m)$.

(i) Find the best linear predictor $\widehat{X}_{n+1}$ of X_{n+1} given X_n.

(ii) Find the best linear predictor $\widetilde{X}_{n+1}$ of X_{n+1} given X_n and X_{n-1}.

(iii) Find an expression for $D = \mathbb{E}\{(X_{n+1} - \widehat{X}_{n+1})^2\} - \mathbb{E}\{(X_{n+1} - \widetilde{X}_{n+1})^2\}$, and evaluate this expression when:

(a) $X_n = \cos(nU)$ where U is uniform on $[-\pi, \pi]$,

(b) X is an autoregressive scheme with $c(k) = \alpha^{|k|}$ where $|\alpha| < 1$.

2. Suppose $|a| < 1$. Does there exist a (weakly) stationary sequence $\{X_n : -\infty < n < \infty\}$ with zero means and autocovariance function

$$c(k) = \begin{cases} 1 & \text{if } k = 0, \\ \dfrac{a}{1+a^2} & \text{if } |k| = 1, \\ 0 & \text{if } |k| > 1. \end{cases}$$

Assuming that such a sequence exists, find the best linear predictor $\widehat{X}_n$ of X_n given X_{n-1}, $X_{n-2}, \ldots$, and show that the mean squared error of prediction is $(1 + a^2)^{-1}$. Verify that $\{\widehat{X}_n\}$ is (weakly) stationary.

9.3 Exercises. Autocovariances and spectra

1. Let $X_n = A\cos(n\lambda) + B\sin(n\lambda)$ where A and B are uncorrelated random variables with zero means and unit variances. Show that X is stationary with a spectrum containing exactly one point.

2. Let U be uniformly distributed on $(-\pi, \pi)$, and let V be independent of U with distribution function F. Show that $X_n = e^{i(U - Vn)}$ defines a stationary (complex) sequence with spectral distribution function F.

3. Find the autocorrelation function of the stationary process $\{X(t) : -\infty < t < \infty\}$ whose spectral density function is:

(i) $N(0, 1)$, (ii) $f(x) = \frac{1}{2}e^{-|x|}$, $-\infty < x < \infty$.

4. Let $X_1, X_2, \ldots$ be a real-valued stationary sequence with zero means and autocovariance function $c(m)$. Show that

$$\operatorname{var}\left(\frac{1}{n}\sum_{j=1}^{n} X_j\right) = c(0)\int_{(-\pi,\pi]} \left(\frac{\sin(n\lambda/2)}{n\sin(\lambda/2)}\right)^2 dF(\lambda)$$

where F is the spectral distribution function. Deduce that $n^{-1}\sum_{j=1}^{n} X_j \xrightarrow{\text{m.s.}} 0$ if and only if

$F(0) - F(0-) = 0$, and show that

$$c(0)\{F(0) - F(0-)\} = \lim_{n\to\infty} \frac{1}{n} \sum_{j=0}^{n-1} c(j).$$

5. Let $Z = \{Z_n : n \in \mathbb{Z}\}$ be a sequence of uncorrelated random variables with means 0 and variances 1, and suppose the sequence $X = \{X_n : n \in \mathbb{Z}\}$ satisfies $X_n = \phi X_{n-1} + Z_n + \theta Z_{n-1}$.

(a) For what values of ϕ, θ is X stationary?

(b) For what values of ϕ, θ does Z_n have a series representation in terms of the X_m?

(c) Show that the autocorrelation function ρ of X satisfies

$$\rho(1) = \frac{(\phi + \theta)(1 + \phi\theta)}{1 + 2\phi\theta + \theta^2},$$

and find $\rho(k)$ for $k > 1$. Hence evaluate the spectral density in a finite form.

6. Let $Z = \{Z_n : n \in \mathbb{Z}\}$ be a sequence of uncorrelated random variables with means 0 and variances 1, and suppose $X_n = \phi X_{n-1} + \theta X_{n-2} + Z_n$ for $n \in \mathbb{Z}$, where $\phi^2 + 4\theta > 0$. Suppose d_k is a real sequence such that

$$X_n = \sum_{k=0}^{\infty} d_k Z_{n-k}.$$

Show that $d_k = a_1 r_1^k + a_2 r_2^k$, where $a_1 + a_2 = 1$ and $a_1 r_1 + a_2 r_2 = \phi$. Show further that

$$a_1 = \frac{r_1}{r_1 - r_2}, \quad a_2 = \frac{-r_2}{r_1 - r_2}, \quad r_1 r_2 = -\theta, \quad r_1^2 + r_2^2 = \phi^2 + 2\theta.$$

Hence find $\mathbb{E}(X_n^2)$ in terms of ϕ and θ when $|r_1|, |r_2| < 1$.

7. Kolmogorov–Szegő formula. Let $X = \{X_n : n \in \mathbb{Z}\}$ be a stationary process with zero means and variances σ^2, and let $\widehat{X}_{n+1}$ be the best linear predictor of X_{n+1} given $\{X_r : r \le n\}$. (See Section 9.2.) The Kolmogorov–Szegő formula states that

$$\operatorname{var}(X_{n+1} - \widehat{X}_{n+1}) = \exp\left\{ \frac{1}{2\pi} \int_{-\pi}^{\pi} \log\left(2\pi\sigma^2 f(\lambda)\right) d\lambda \right\},$$

where f is the spectral density of X. Verify this for the autoregressive scheme of Example (9.3.23).

9.4 Exercises. Stochastic integration and the spectral representation

1. Let S be the spectral process of a stationary process X with zero mean and unit variance. Show that the increments of S have zero means.

2. Moving average representation. Let X be a discrete-time stationary process having zero means, continuous strictly positive spectral density function f, and with spectral process S. Let

$$Y_n = \int_{(-\pi,\pi]} \frac{e^{in\lambda}}{\sqrt{2\pi f(\lambda)}} \, dS(\lambda).$$

Show that $\dots, Y_{-1}, Y_0, Y_1, \dots$ is a sequence of uncorrelated random variables with zero means and unit variances.

Show that X_n may be represented as a moving average $X_n = \sum_{j=-\infty}^{\infty} a_j Y_{n-j}$ where the a_j are constants satisfying

$$\sqrt{2\pi f(\lambda)} = \sum_{j=-\infty}^{\infty} a_j e^{-ij\lambda} \qquad \text{for } \lambda \in (-\pi, \pi].$$

3. Gaussian process. Let X be a discrete-time stationary sequence with zero mean and unit variance, and whose fdds are of the multivariate-normal type. Show that the spectral process of X has independent increments having normal distributions.

9.5 Exercises. The ergodic theorem

1. Let $T = \{1, 2, \dots\}$ and let $\mathcal{I}$ be the set of invariant events of $(\mathbb{R}^T, \mathcal{B}^T)$. Show that $\mathcal{I}$ is a σ-field.

2. Assume that $X_1, X_2, \dots$ is a stationary sequence with autocovariance function $c(m)$. Show that

$$\operatorname{var}\left(\frac{1}{n}\sum_{i=1}^{n} X_i\right) = \frac{2}{n^2}\sum_{j=1}^{n}\sum_{i=0}^{j-1} c(i) - \frac{c(0)}{n}.$$

Assuming that $j^{-1}\sum_{i=0}^{j-1} c(i) \to \sigma^2$ as $j \to \infty$, show that

$$\operatorname{var}\left(\frac{1}{n}\sum_{i=1}^{n} X_i\right) \to \sigma^2 \qquad \text{as } n \to \infty.$$

3. Let $X_1, X_2, \dots$ be independent identically distributed random variables with zero mean and unit variance. Let

$$Y_n = \sum_{i=0}^{\infty} \alpha_i X_{n+i} \qquad \text{for } n \geq 1$$

where the α_i are constants satisfying $\sum_i \alpha_i^2 < \infty$. Use the martingale convergence theorem to show that the above summation converges almost surely and in mean square. Prove that $n^{-1}\sum_{i=1}^{n} Y_i \to 0$ a.s. and in mean, as $n \to \infty$.

9.6 Exercises. Gaussian processes

1. Show that the function $c(s, t) = \min\{s, t\}$ is positive definite. That is, show that

$$\sum_{j,k=1}^{n} c(t_k, t_j) z_j \bar{z}_k > 0$$

for all $0 \leq t_1 < t_2 < \cdots < t_n$ and all complex numbers $z_1, z_2, \dots, z_n$ at least one of which is non-zero.

2. Let $X_1, X_2, \dots$ be a stationary Gaussian sequence with zero means and unit variances which satisfies the Markov property. Find the spectral density function of the sequence in terms of the constant $\rho = \operatorname{cov}(X_1, X_2)$.

3. Show that a Gaussian process is strongly stationary if and only if it is weakly stationary.

4. Let X be a stationary Gaussian process with zero mean, unit variance, and autocovariance function $c(t)$. Find the autocovariance functions of the processes $X^2 = \{X(t)^2 : -\infty < t < \infty\}$ and $X^3 = \{X(t)^3 : -\infty < t < \infty\}$.

5. (a) Let W be a standard Wiener process, and $T : [0, \infty) \to [0, \infty)$ a non-random, right-continuous, non-decreasing function with $T(0) = 0$. Show that $X(t) := W(T(t))$ defines a Gaussian process. Find the characteristic function of $X(t)$ and the covariance function $c(s, t) = \operatorname{cov}(X(s), X(t))$.

(b) Let W be a Wiener process with $\operatorname{var}(W(t)) = 2t$, and let T be an independent α-stable subordinator, where $0 < \alpha < 1$. Show that $X(t) = W(T(t))$ is a symmetric (2α)-stable Lévy process.

6. Brownian sheet. Let $X(s, t)$ be a two-parameter, zero-mean, Gaussian process on the positive quadrant $[0, \infty)^2$ of $\mathbb{R}^2$, with covariance function $\operatorname{cov}(X(s, t), X(u, v)) = (s \wedge u)(t \wedge v)$, where $x \wedge y = \min\{x, y\}$. For any rectangle R with corners $(s, t), (u, t), (s, v), (u, v)$ where $0 \le s < u < \infty$ and $0 \le t < v < \infty$, define

$$Y(R) = X(s, t) + X(u, v) - X(s, v) - X(u, t).$$

Find $\operatorname{var}(Y(R))$, and show that Y has independent increments.

9.7 Problems

1. Let $\dots, X_{-1}, X_0, X_1, \dots$ be uncorrelated random variables with zero means and unit variances, and define

$$Y_n = X_n + \alpha \sum_{i=1}^{\infty} \beta^{i-1} X_{n-i} \qquad \text{for } -\infty < n < \infty,$$

where α and β are constants satisfying $|\beta| < 1$, $|\beta - \alpha| < 1$. Find the best linear predictor of Y_{n+1} given the entire past $Y_n, Y_{n-1}, \dots$.

2. Let $\{Y_k : -\infty < k < \infty\}$ be a stationary sequence with variance σ_Y^2, and let

$$X_n = \sum_{k=0}^{r} a_k Y_{n-k}, \qquad -\infty < n < \infty,$$

where $a_0, a_1, \dots, a_r$ are constants. Show that X has spectral density function

$$f_X(\lambda) = \frac{\sigma_Y^2}{\sigma_X^2} f_Y(\lambda) |G_a(e^{i\lambda})|^2$$

where f_Y is the spectral density function of Y, $\sigma_X^2 = \operatorname{var}(X_1)$, and $G_a(z) = \sum_{k=0}^{r} a_k z^k$.

Calculate this spectral density explicitly in the case of 'exponential smoothing', when $r = \infty$, $a_k = \mu^k(1 - \mu)$, and $0 < \mu < 1$.

3. Suppose that $\widehat{Y}_{n+1} = \alpha Y_n + \beta Y_{n-1}$ is the best linear predictor of Y_{n+1} given the entire past $Y_n, Y_{n-1}, \dots$ of the stationary sequence $\{Y_k : -\infty < k < \infty\}$. Find the spectral density function of the sequence.

4. Recurrent events, Example (5.2.15). Meteorites fall at integer times $T_1, T_2, \dots$ where $T_n = X_1 + X_2 + \dots + X_n$. We assume that the X_i are independent, $X_2, X_3, \dots$ are identically distributed, and the distribution of X_1 is such that the probability that a meteorite falls at time n is constant for all n. Let Y_n be the indicator function of the event that a meteorite falls at time n. Show that $\{Y_n\}$ is stationary and find its spectral density function in terms of the characteristic function of X_2.

5. Let $X = \{X_n : n \ge 1\}$ be given by $X_n = \cos(nU)$ where U is uniformly distributed on $[-\pi, \pi]$. Show that X is stationary but not strongly stationary. Find the autocorrelation function of X and its spectral density function.

6. (a) Let N be a Poisson process with intensity λ, and let $\alpha > 0$. Define $X(t) = N(t+\alpha) - N(t)$ for $t \ge 0$. Show that X is strongly stationary, and find its spectral density function.

(b) Let W be a Wiener process and define $X = \{X(t) : t \ge 1\}$ by $X(t) = W(t) - W(t-1)$. Show that X is strongly stationary and find its autocovariance function. Find the spectral density function of X.

7. Let $Z_1, Z_2, \dots$ be uncorrelated variables, each with zero mean and unit variance.

(a) Define the moving average process X by $X_n = Z_n + \alpha Z_{n-1}$ where α is a constant. Find the spectral density function of X.

(b) More generally, let $Y_n = \sum_{i=0}^{r} \alpha_i Z_{n-i}$, where $\alpha_0 = 1$ and $\alpha_1, \dots, \alpha_r$ are constants. Find the spectral density function of Y.

8. Show that the complex-valued stationary process $X = \{X(t) : -\infty < t < \infty\}$ has a spectral density function which is bounded and uniformly continuous whenever its autocorrelation function ρ is continuous and satisfies $\int_0^\infty |\rho(t)|\, dt < \infty$.

9. Let $X = \{X_n : n \ge 1\}$ be stationary with constant mean $\mu = \mathbb{E}(X_n)$ for all n, and such that $\text{cov}(X_1, X_n) \to 0$ as $n \to \infty$. Show that $n^{-1}\sum_{j=1}^{n} X_j \xrightarrow{\text{m.s.}} \mu$.

10. Deduce the strong law of large numbers from an appropriate ergodic theorem.

11. Let $\mathbb{Q}$ be a stationary measure on $(\mathbb{R}^T, \mathcal{B}^T)$ where $T = \{1, 2, \dots\}$. Show that $\mathbb{Q}$ is ergodic if and only if

$$\frac{1}{n}\sum_{i=1}^{n} Y_i \to \mathbb{E}(Y) \qquad \text{a.s. and in mean}$$

for all $Y : \mathbb{R}^T \to \mathbb{R}$ for which $\mathbb{E}(Y)$ exists, where $Y_i : \mathbb{R}^T \to \mathbb{R}$ is given by $Y_i(\mathbf{x}) = Y(\tau^{i-1}(\mathbf{x}))$. As usual, τ is the natural shift operator on $\mathbb{R}^T$.

12. The stationary measure $\mathbb{Q}$ on $(\mathbb{R}^T, \mathcal{B}^T)$ is called *strongly mixing* if $\mathbb{Q}(A \cap \tau^{-n}B) \to \mathbb{Q}(A)\mathbb{Q}(B)$ as $n \to \infty$, for all $A, B \in \mathcal{B}^T$; as usual, $T = \{1, 2, \dots\}$ and τ is the shift operator on $\mathbb{R}^T$. Show that every strongly mixing measure is ergodic.

13. Ergodic theorem. Let $(\Omega, \mathcal{F}, \mathbb{P})$ be a probability space, and let $T : \Omega \to \Omega$ be measurable and measure preserving (i.e. $\mathbb{P}(T^{-1}A) = \mathbb{P}(A)$ for all $A \in \mathcal{F}$). Let $X : \Omega \to \mathbb{R}$ be a random variable, and let X_i be given by $X_i(\omega) = X(T^{i-1}(\omega))$. Show that

$$\frac{1}{n}\sum_{i=1}^{n} X_i \to \mathbb{E}(X \mid \mathcal{I}) \qquad \text{a.s. and in mean}$$

where $\mathcal{I}$ is the σ-field of invariant events of T.

If T is ergodic (in that $\mathbb{P}(A)$ equals 0 or 1 whenever A is invariant), prove that $\mathbb{E}(X \mid \mathcal{I}) = \mathbb{E}(X)$ almost surely.

14. Borel's normal number theorem. Consider the probability space $(\Omega, \mathcal{F}, \mathbb{P})$ where $\Omega = [0, 1)$, $\mathcal{F}$ is the set of Borel subsets, and $\mathbb{P}$ is Lebesgue measure.

(a) Show that the shift $T : \Omega \to \Omega$ defined by $T(x) = 2x \pmod 1$ is measurable, measure preserving, and ergodic (in that $\mathbb{P}(A)$ equals 0 or 1 if $A = T^{-1}A$).

(b) Let $X : \Omega \to \mathbb{R}$ be the random variable given by the identity mapping $X(\omega) = \omega$. Show that the proportion of 1's, in the expansion of X to base 2, equals $\frac{1}{2}$ almost surely. This is sometimes called 'Borel's normal number theorem'.

(c) Deduce that, for any continuous random variable Y taking values in $[0, 1)$, the proportion of 1's in its binary expansion to n places converges a.s. to $\frac{1}{2}$ as $n \to \infty$.

(d) Let $\{X_i : i \geq 1\}$ be a sequence of independent Bernoulli random variables with parameter $p \neq \frac{1}{2}$. and set $Z = \sum_{i=1}^{\infty} X_i 2^{-i}$. Is Z a random variable? Is it continuous, or discrete, or what?

15. Let $g : \mathbb{R} \to \mathbb{R}$ be periodic with period 1, and uniformly continuous and integrable over $[0, 1]$. Define $Z_n = g\big(X + (n-1)\alpha\big)$, $n \geq 1$, where X is uniform on $[0, 1]$ and α is irrational. Show that, as $n \to \infty$,

$$\frac{1}{n}\sum_{j=1}^{n} Z_j \to \int_0^1 g(u)\,du \qquad \text{a.s.}$$

16. Let $X = \{X(t) : t \geq 0\}$ be a non-decreasing random process such that:

(a) $X(0) = 0$, X takes values in the non-negative integers,

(b) X has stationary independent increments,

(c) the sample paths $\{X(t, \omega) : t \geq 0\}$ have only jump discontinuities of unit magnitude.

Show that X is a Poisson process.

17. Let X be a continuous-time process. Show that:

(a) if X has stationary increments and $m(t) = \mathbb{E}(X(t))$ is a continuous function of t, then there exist α and β such that $m(t) = \alpha + \beta t$,

(b) if X has stationary independent increments and $v(t) = \operatorname{var}(X(t) - X(0))$ is a continuous function of t then there exists σ^2 such that $\operatorname{var}(X(s+t) - X(s)) = \sigma^2 t$ for all s.

18. A Wiener process W is called *standard* if $W(0) = 0$ and $W(1)$ has unit variance. Let W be a standard Wiener process, and let α be a positive constant. Show that:

(a) $\alpha W(t/\alpha^2)$ is a standard Wiener process,

(b) $W(t+\alpha) - W(\alpha)$ is a standard Wiener process,

(c) the process V, given by $V(t) = tW(1/t)$ for $t > 0$, $V(0) = 0$, is a standard Wiener process,

(d) the process $W(1) - W(1-t)$ is a standard Wiener process on $[0, 1]$.

19. Let W be a standard Wiener process. Show that the stochastic integrals

$$X(t) = \int_0^t dW(u), \qquad Y(t) = \int_0^t e^{-(t-u)}\,dW(u), \qquad t \geq 0,$$

are well defined, and prove that $X(t) = W(t)$, and that Y has autocovariance function $\operatorname{cov}(Y(s), Y(t)) = \frac{1}{2}(e^{-|s-t|} - e^{-s-t})$, $s < t$.

20. Let W be a standard Wiener process. Find the means of the following three processes, and the autocovariance functions in cases (b) and (c):

(a) $X(t) = |W(t)|$,

(b) $Y(t) = e^{W(t)}$,

(c) $Z(t) = \int_0^t W(u)\,du$.

(d) Which of X, Y, Z are Gaussian processes? Which of these are Markov processes?

(e) Find $\mathbb{E}(Z(t)^n)$ for $n \geq 0$.

21. Let W be a standard Wiener process. Find the conditional joint density function of $W(t_2)$ and $W(t_3)$ given that $W(t_1) = W(t_4) = 0$, where $t_1 < t_2 < t_3 < t_4$.

Show that the conditional correlation of $W(t_2)$ and $W(t_3)$ is

$$\rho = \sqrt{\frac{(t_4 - t_3)(t_2 - t_1)}{(t_4 - t_2)(t_3 - t_1)}}.$$

22. Empirical distribution function. Let $U_1, U_2, \ldots$ be independent random variables with the uniform distribution on $[0, 1]$. Let $I_j(x)$ be the indicator function of the event $\{U_j \le x\}$, and define

$$F_n(x) = \frac{1}{n}\sum_{j=1}^{n} I_j(x), \qquad 0 \le x \le 1.$$

The function F_n is called the 'empirical distribution function' of the U_j.

(a) Find the mean and variance of $F_n(x)$, and prove that $\sqrt{n}(F_n(x) - x) \xrightarrow{\text{D}} Y(x)$ as $n \to \infty$, where $Y(x)$ is normally distributed.

(b) What is the (multivariate) limit distribution of a collection of random variables of the form $\{\sqrt{n}(F_n(x_i) - x_i) : 1 \le i \le k\}$, where $0 \le x_1 < x_2 < \cdots < x_k \le 1$?

(c) Show that the autocovariance function of the asymptotic finite-dimensional distributions of $\sqrt{n}(F_n(x) - x)$, in the limit as $n \to \infty$, is the same as that of the process $Z(t) = W(t) - tW(1)$, $0 \le t \le 1$, where W is a standard Wiener process. The process Z is called a 'Brownian bridge' or 'tied-down Brownian motion'.

23. Pólya's urn revisited. An urn contains initially one red ball and one green ball. At later stages, a ball is picked from the urn uniformly at random, and is returned to the urn together with a fresh ball of the same colour. Assume the usual independence. Let X_k be the indicator function of the event that the kth ball picked is red.

(a) Show that, for $x_i \in \{0, 1\}$,

$$\mathbb{P}(X_1 = x_1, X_2 = x_2, \ldots, X_n = x_n) = \frac{r!\,(n-r)!}{(n+1)!}, \qquad x_1, x_2, \ldots, x_n \in \{0, 1\},\ r = \sum_{k=1}^{n} x_k.$$

(b) Show that $X_1, X_2, \ldots$ is a stationary sequence, and that $n^{-1}\sum_{k=1}^{n} X_k$ converges a.s. as $n \to \infty$ to some random variable R.

(c) Find the distribution of R.

10

Renewals

In the absence of indications to the contrary, $\{X_n : n \geq 1\}$ denotes the sequence of interarrival times of either a renewal process N or a delayed renewal process N^{d}. In either case, F^{d} and F are the distribution functions of X_1 and X_2 respectively, though $F^{\mathrm{d}} \neq F$ only if the renewal process is delayed. We write $\mu = \mathbb{E}(X_2)$, and shall usually assume that $0 < \mu < \infty$. The functions m and m^{d} denote the renewal functions of N and N^{d}. We write $T_n = \sum_{i=1}^{n} X_i$, the time of the nth arrival.

10.1 Exercises. The renewal equation

1. Prove that $\mathbb{E}(e^{\theta N(t)}) < \infty$ for some strictly positive θ whenever $\mathbb{E}(X_1) > 0$. [Hint: Consider the renewal process with interarrival times $X_k' = \epsilon I_{\{X_k \geq \epsilon\}}$ for some suitable ϵ.]

2. Let N be a renewal process and let W be the waiting time until the length of some interarrival time has exceeded s. That is, $W = \inf\{t : C(t) > s\}$, where $C(t)$ is the time which has elapsed (at time t) since the last arrival. Show that

$$F_W(x) = \begin{cases} 0 & \text{if } x < s, \\ 1 - F(s) + \int_0^s F_W(x-u)\,dF(u) & \text{if } x \geq s, \end{cases}$$

where F is the distribution function of an interarrival time. If N is a Poisson process with intensity λ, show that

$$\mathbb{E}(e^{\theta W}) = \frac{\lambda - \theta}{\lambda - \theta e^{(\lambda-\theta)s}} \qquad \text{for } \theta < \lambda,$$

and $\mathbb{E}(W) = (e^{\lambda s} - 1)/\lambda$. You may find it useful to rewrite the above integral equation in the form of a renewal-type equation.

3. Find an expression for the mass function of $N(t)$ in a renewal process whose interarrival times are: (a) Poisson distributed with parameter λ, (b) gamma distributed, $\Gamma(\lambda, b)$.

4. Let the times between the events of a renewal process N be uniformly distributed on $(0, 1)$. Find the mean and variance of $N(t)$ for $0 \leq t \leq 1$.

5. Let N be the renewal process with interarrival times $X_1, X_2, \dots$. Show that, for $t > 0$, the interarrival time $X_{N(t)+1}$ is stochastically larger than X_1.

6. Suppose the interarrival times X_i of a renewal process have a density function f with ordinary Laplace transform $\widehat{f}(\theta) = \int_0^\infty e^{-\theta x} f(x)\,dx$. Show that the renewal function m has Laplace transform

$$\widehat{m}(\theta) = \frac{\widehat{f}(\theta)}{\theta - \theta\widehat{f}(\theta)}, \qquad \theta > 0.$$

7. Let $r(y)$ be the ruin function of the insurance problem (8.10.7). In the notation of that problem, show that the Laplace–Stieltjes transforms $r^*(\theta)$ and $F^*(\theta)$ are related by $\lambda F^* = r^*(\theta - \lambda + \lambda F^*)$.

10.2 Exercises. Limit theorems

1. Planes land at Heathrow airport at the times of a renewal process with interarrival time distribution function F. Each plane contains a random number of people with a given common distribution and finite mean. Assuming as much independence as usual, find an expression for the rate of arrival of passengers over a long time period.

2. Let $Z_1, Z_2, \dots$ be independent identically distributed random variables with mean 0 and finite variance σ^2, and let $T_n = \sum_{i=1}^n Z_i$. Let M be a finite stopping time with respect to the Z_i such that $\mathbb{E}(M) < \infty$. Show that $\operatorname{var}(T_M) = \mathbb{E}(M)\sigma^2$.

3. Show that $\mathbb{E}(T_{N(t)+k}) = \mu(m(t)+k)$ for all $k \geq 1$, but that it is not generally true that $\mathbb{E}(T_{N(t)}) = \mu m(t)$.

4. Show that, using the usual notation, the family $\{N(t)/t : 0 \leq t < \infty\}$ is uniformly integrable. How might one make use of this observation?

5. Consider a renewal process N having interarrival times with moment generating function M, and let T be a positive random variable which is independent of N. Find $\mathbb{E}(s^{N(T)})$ when:

(a) T is exponentially distributed with parameter ν,

(b) N is a Poisson process with intensity λ, in terms of the moment generating function of T. What is the distribution of $N(T)$ in this case, if T has the gamma distribution $\Gamma(\nu, b)$?

10.3 Exercises. Excess life

1. Suppose that the distribution of the excess lifetime $E(t)$ does not depend on t. Show that the renewal process is a Poisson process.

2. Show that the current and excess lifetime processes, $C(t)$ and $E(t)$, are Markov processes.

3. Suppose that X_1 is non-arithmetic with finite mean μ.

(a) Show that $E(t)$ converges in distribution as $t \to \infty$, the limit distribution function being

$$H(x) = \int_0^x \frac{1}{\mu}[1 - F(y)]\,dy.$$

(b) Show that the rth moment of this limit distribution is given by

$$\int_0^\infty x^r\,dH(x) = \frac{\mathbb{E}(X_1^{r+1})}{\mu(r+1)},$$

assuming that this is finite.

(c) Show that

$$\mathbb{E}(E(t)^r) = \mathbb{E}\big(\{(X_1 - t)^+\}^r\big) + \int_0^t h(t-x)\,dm(x)$$

for some suitable function h to be found, and deduce by the key renewal theorem that $\mathbb{E}(E(t)^r) \to \mathbb{E}(X_1^{r+1})/\{\mu(r+1)\}$ as $t \to \infty$, assuming this limit is finite.

4. Find an expression for the mean value of the excess lifetime $E(t)$ conditional on the event that the current lifetime $C(t)$ equals x.

5. Let $M(t) = N(t) + 1$, and suppose that X_1 has finite non-zero variance σ^2.

(a) Show that $\operatorname{var}(T_{M(t)} - \mu M(t)) = \sigma^2(m(t)+1)$.

(b) In the non-arithmetic case, show that $\operatorname{var}(M(t))/t \to \sigma^2/\mu^3$ as $t \to \infty$.

10.4 Exercises. Applications

1. Find the distribution of the excess lifetime for a renewal process each of whose interarrival times is the sum of two independent exponentially distributed random variables having respective parameters λ and μ. Show that the excess lifetime has mean

$$\frac{1}{\mu} + \frac{\lambda e^{-(\lambda+\mu)t} + \mu}{\lambda(\lambda+\mu)}.$$

2. Stationary renewal and size-biasing. Let f be the density function and F the distribution function of the interarrival times $\{X_i : i \geq 2\}$ of a stationary renewal process, and μ their common mean. Let C, D, E be the current, total, and excess lifetimes in equilibrium.

(a) Show that C has density function $h(x) = (1 - F(x))/\mu$ for $x > 0$.

(b) Show that D has the *size-biased* (or *length-biased*) density function $g(y) = (y/\mu)f(y)$.

(c) Let U be uniformly distributed on $(0, 1)$ and independent of D. Show that UD has the same distribution as both C and E. Explain why this should be so.

3. Let m be the renewal function of an ordinary renewal process N whose interarrival times have finite mean.

(a) Show that the mean number of renewals in the interval $(a, b]$ is no larger than $1 + m(b - a)$.

(b) Show that there exists $A > 0$ such that $m(t) \leq A(1 + t)$ for $t \geq 0$.

(c) Suppose the interarrival times X_r are non-arithmetic with finite variance and mean μ. Show that

$$m(t) - \frac{t}{\mu} \to \frac{\sigma^2 - \mu^2}{2\mu^2} \qquad \text{as } t \to \infty.$$

[Hint: Use coupling and Blackwell's renewal theorem (10.2.5).]

10.5 Exercises. Renewal–reward processes

1. If $X(t)$ is an irreducible positive recurrent Markov chain, and $u(\cdot)$ is a bounded function on the integers, show that

$$\frac{1}{t}\int_0^t u(X(s))\,ds \xrightarrow{\text{a.s.}} \sum_{i\in S} \pi_i u(i),$$

where $\boldsymbol{\pi}$ is the stationary distribution of $X(t)$.

2. Let $M(t)$ be an alternating renewal process, with interarrival pairs $\{X_r, Y_r : r \geq 1\}$. Show that

$$\frac{1}{t}\int_0^t I_{\{M(s) \text{ is even}\}}\,ds \xrightarrow{\text{a.s.}} \frac{\mathbb{E}X_1}{\mathbb{E}X_1 + \mathbb{E}Y_1} \qquad \text{as } t \to \infty.$$

Is this limit valid for an arbitrary joint distribution of independent pairs (X_i, Y_i)?

3. Let $C(s)$ be the current lifetime (or age) of a renewal process $N(t)$ with a typical interarrival time X. Show that

$$\frac{1}{t}\int_0^t C(s)\,ds \xrightarrow{\text{a.s.}} \frac{\mathbb{E}(X^2)}{2\mathbb{E}(X)} \qquad \text{as } t \to \infty.$$

Find the corresponding limit for the excess lifetime.

4. Let j and k be distinct states of an irreducible discrete-time Markov chain X with stationary distribution $\boldsymbol{\pi}$. Show that

$$\mathbb{P}(T_j < T_k \mid X_0 = k) = \frac{1/\pi_k}{\mathbb{E}(T_j \mid X_0 = k) + \mathbb{E}(T_k \mid X_0 = j)}$$

where $T_i = \min\{n \geq 1 : X_n = i\}$ is the first passage time to the state i. [Hint: Consider the times of return to j having made an intermediate visit to k.]

5. Total life. Use the result of Exercise (10.5.2) to show that the limiting distribution function as $t \to \infty$ of the total life $D(t)$ of a renewal process is

$$F_D(y) = \int_0^y \frac{1}{\mu} x f(x)\,dx,$$

where f is the density function of a typical interarrival time X, and $\mu = \mathbb{E}(X)$. The integrand is called the *size-biased* (or *length-biased*) density for X.

6. (a) Let X be exponentially distributed with parameter λ. Show that

$$\mathbb{E}(\min\{X, d\}) = \frac{1}{\lambda}(1 - e^{-\lambda d}), \qquad d \geq 0.$$

(b) John's garage offers him the choice of two tyre replacement plans.

1. The garage undertakes to replace all the tyres of his car at the normal price whenever one of the tyres requires replacing.
2. The garage undertakes to replace all the tyres of his car at 5% of the normal price two years after they have last been replaced. However, if any tyre needs replacing earlier, the garage will then replace all the tyres at a price that is 5% higher than the normal price.

Assuming that a new tyre has an exponentially distributed lifetime with mean 8 years, determine the long run average cost per year under the two options. Which option should John choose if his car is new?

7. Uptime. A machine M is repaired at time $t = 0$, and its uptime after any repair is exponentially distributed with parameter λ, at which point it breaks down (assume the usual independence). Following any repair at time T, say, it is inspected at times $T, T+m, T+2m, \dots$, and instantly repaired if found to be broken (the inspection schedule is then restarted). Show that the long run proportion of time that M is working (the 'uptime ratio') is $m^{-1}\int_0^m e^{-\lambda x}\,dx$.

10.6 Problems

1. (a) Show that $\mathbb{P}(N(t) \to \infty \text{ as } t \to \infty) = 1$.

(b) Show that $m(t) < \infty$ if $\mu \neq 0$.

(c) More generally show that, for all $k > 0$, $\mathbb{E}(N(t)^k) < \infty$ if $\mu \neq 0$.

2. Let $v(t) = \mathbb{E}(N(t)^2)$. Show that

$$v(t) = m(t) + 2\int_0^t m(t-s)\,dm(s).$$

Find $v(t)$ when N is a Poisson process.

3. Suppose that $\sigma^2 = \text{var}(X_1) > 0$. Show that the renewal process N satisfies

$$\frac{N(t) - (t/\mu)}{\sqrt{t\sigma^2/\mu^3}} \xrightarrow{\text{D}} N(0,1), \qquad \text{as } t \to \infty.$$

4. Find the asymptotic distribution of the current life $C(t)$ of N as $t \to \infty$ when X_1 is not arithmetic.

5. Let N be a Poisson process with intensity λ. Show that the total life $D(t)$ at time t has distribution function $\mathbb{P}(D(t) \le x) = 1 - (1 + \lambda \min\{t, x\})e^{-\lambda x}$ for $x \ge 0$. Deduce that $\mathbb{E}(D(t)) = (2 - e^{-\lambda t})/\lambda$.

6. A Type 1 counter records the arrivals of radioactive particles. Suppose that the arrival process is Poisson with intensity λ, and that the counter is locked for a dead period of fixed length T after each detected arrival. Show that the detection process $\widetilde{N}$ is a renewal process with interarrival time distribution $\widetilde{F}(x) = 1 - e^{-\lambda(x-T)}$ if $x \ge T$. Find an expression for $\mathbb{P}(\widetilde{N}(t) \ge k)$.

7. Particles arrive at a Type 1 counter in the manner of a renewal process N; each detected arrival locks the counter for a dead period of random positive length. Show that

$$\mathbb{P}(\widetilde{X}_1 \le x) = \int_0^x [1 - F(x-y)]F_L(y)\,dm(y)$$

where F_L is the distribution function of a typical dead period.

8. (a) Show that $m(t) = \frac{1}{2}\lambda t - \frac{1}{4}(1 - e^{-2\lambda t})$ if the interarrival times have the gamma distribution $\Gamma(\lambda, 2)$.

(b) Radioactive particles arrive like a Poisson process, intensity λ, at a counter. The counter fails to register the nth arrival whenever n is odd but suffers no dead periods. Find the renewal function $\widetilde{m}$ of the detection process $\widetilde{N}$.

9. Show that Poisson processes are the only renewal processes with non-arithmetic interarrival times having the property that the excess lifetime $E(t)$ and the current lifetime $C(t)$ are independent for each choice of t.

10. Let N_1 be a Poisson process, and let N_2 be a renewal process which is independent of N_1 with non-arithmetic interarrival times having finite mean. Show that $N(t) = N_1(t) + N_2(t)$ is a renewal process if and only if N_2 is a Poisson process.

11. Let N be a renewal process, and suppose that F is non-arithmetic and that $\sigma^2 = \text{var}(X_1) < \infty$. Use the properties of the moment generating function $F^*(-\theta)$ of X_1 to deduce the formal expansion

$$m^*(\theta) = \frac{1}{\theta\mu} + \frac{\sigma^2 - \mu^2}{2\mu^2} + o(1) \quad \text{as } \theta \to 0.$$

Invert this Laplace–Stieltjes transform formally to obtain

$$m(t) = \frac{t}{\mu} + \frac{\sigma^2 - \mu^2}{2\mu^2} + o(1) \quad \text{as } t \to \infty.$$

Prove this rigorously by showing that

$$m(t) = \frac{t}{\mu} - F_E(t) + \int_0^t [1 - F_E(t-x)]\,dm(x),$$

where F_E is the asymptotic distribution function of the excess lifetime (see Exercise (10.3.3)), and applying the key renewal theorem. Compare the result with the renewal theorems.

12. Show that the renewal function m^{d} of a delayed renewal process satisfies

$$m^{\mathrm{d}}(t) = F^{\mathrm{d}}(t) + \int_0^t m^{\mathrm{d}}(t-x)\,dF(x).$$

Show that $v^{\mathrm{d}}(t) = \mathbb{E}(N^{\mathrm{d}}(t)^2)$ satisfies

$$v^{\mathrm{d}}(t) = m^{\mathrm{d}}(t) + 2\int_0^t m^{\mathrm{d}}(t-x)\,dm(x)$$

where m is the renewal function of the renewal process with interarrival times $X_2, X_3, \dots$.

13. Let $m(t)$ be the mean number of living individuals at time t in an age-dependent branching process with exponential lifetimes, parameter λ, and mean family size ν (> 1). Prove that $m(t) = Ie^{(\nu-1)\lambda t}$ where I is the number of initial members.

14. Alternating renewal process. The interarrival times of this process are $Z_0, Y_1, Z_1, Y_2, \ldots$, where the Y_i and Z_j are independent with respective common moment generating functions M_Y and M_Z. Let $p(t)$ be the probability that the epoch t of time lies in an interval of type Z. Show that the Laplace–Stieltjes transform p^* of p satisfies

$$p^*(\theta) = \frac{1 - M_Z(-\theta)}{1 - M_Y(-\theta)M_Z(-\theta)}.$$

15. Type 2 counters. Particles are detected by a Type 2 counter of the following sort. The incoming particles constitute a Poisson process with intensity λ. The jth particle locks the counter for a length Y_j of time, and annuls any after-effect of its predecessors. Suppose that $Y_1, Y_2, \ldots$ are independent of each other and of the Poisson process, each having distribution function G. The counter is unlocked at time 0.

Let L be the (maximal) length of the first interval of time during which the counter is locked. Show that $H(t) = \mathbb{P}(L > t)$ satisfies

$$H(t) = e^{-\lambda t}[1 - G(t)] + \int_0^t H(t-x)[1 - G(x)]\lambda e^{-\lambda x}\,dx.$$

Solve for H in terms of G, and evaluate the ensuing expression in the case $G(x) = 1 - e^{-\mu x}$ where $\mu > 0$.

16. Thinning. Consider a renewal process N, and suppose that each arrival is 'overlooked' with probability q, independently of all other arrivals. Let $M(t)$ be the number of arrivals which are detected up to time t/p where $p = 1 - q$.

(a) Show that M is a renewal process whose interarrival time distribution function F_p is given by $F_p(x) = \sum_{r=1}^{\infty} pq^{r-1}F_r(x/p)$, where F_n is the distribution function of the time of the nth arrival in the original process N.

(b) Find the characteristic function of F_p in terms of that of F, and use the continuity theorem to show that, as $p \downarrow 0$, $F_p(s) \to 1 - e^{-s/\mu}$ for $s > 0$, so long as the interarrival times in the original process have finite mean μ. Interpret!

(c) Suppose that $p < 1$, and M and N are processes with the same fdds. Show that N is a Poisson process.

17. (a) A PC keyboard has 100 different keys and a monkey is tapping them (uniformly) at random. Assuming no power failure, use the elementary renewal theorem to find the expected number of keys tapped until the first appearance of the sequence of fourteen characters 'W. Shakespeare'. Answer the same question for the sequence 'omo'.

(b) A coin comes up heads with probability p on each toss. Find the mean number of tosses until the first appearances of the sequences (i) HHH, and (ii) HTH.

18. Let N be a stationary renewal process. Let s be a fixed positive real number, and define $X(t) = N(s+t) - N(t)$ for $t \geq 0$. Show that X is a strongly stationary process.

19. Bears arrive in a village at the instants of a renewal process; they are captured and confined at a cost of \$$c$ per unit time per bear. When a given number B bears have been captured, an expedition (costing \$$d$) is organized to remove and release them a long way away. What is the long-run average cost of this policy?

20. Let $X = \{X_n : n \geq 0\}$ be an ergodic Markov chain on a finite state space S, with stationary distribution $\boldsymbol{\pi}$. Let $k \in S$, and let T be a strictly positive stopping time with finite mean such that $X_T = k$. For $j \in S$, let $V_j(k)$ be the number of visits to j by the chain started in k and stopped at time T (set $V_k(k) = 1$). Use the renewal–reward theorem to show that $\mathbb{E}(V_j(k)) = \pi_j \mathbb{E}_k(T)$.

21. Car trading. For a given type of car, costing c new, the number of years between its manufacture and being broken up for scrap is X (≥ 1), where X is a random variable with distribution function F. In its kth year after manufacture (with $k \geq 1$), depreciation has reduced its resale value to $c\lambda^k$, and in that year it costs $r\mu^{k-1}$ in repairs and maintenance, where $\mu \neq 1$. (Usually, $\lambda < 1$ and $\mu > 1$.)

You buy a new car immediately your old car either is scrapped or reaches the age of m years, whichever is the sooner. Show that, if you continue the policy indefinitely, in the long run you minimize your expected average cost by choosing m such that

$$\frac{1}{\mathbb{E}(Y)}\left\{c + \frac{r}{1-\mu} - cG(\lambda) - \frac{r}{1-\mu}G(\mu)\right\}$$

is as small as possible, where $G(s) = \mathbb{E}(s^Y)$ and $Y = \min\{m, X\}$.

11

Queues

11.2 Exercises. M/M/1

1. Consider a random walk on the non-negative integers with a reflecting barrier at 0, and which moves rightwards or leftwards with respective probabilities $\rho/(1+\rho)$ and $1/(1+\rho)$; when at 0, the particle moves to 1 at the next step. Show that the walk has a stationary distribution if and only if $\rho < 1$, and in this case the unique such distribution $\boldsymbol{\pi}$ is given by $\pi_0 = \frac{1}{2}(1-\rho)$, $\pi_n = \frac{1}{2}(1-\rho^2)\rho^{n-1}$ for $n \ge 1$.

2. Suppose now that the random walker of Exercise (11.2.1) delays its steps in the following way. When at the point n, it waits a random length of time having the exponential distribution with parameter θ_n before moving to its next position; different 'holding times' are independent of each other and of further information concerning the steps of the walk. Show that, subject to reasonable assumptions on the θ_n, the ensuing continuous-time process settles into an equilibrium distribution $\boldsymbol{\nu}$ given by $\nu_n = C\pi_n/\theta_n$ for some appropriate constant C.

By applying this result to the case when $\theta_0 = \lambda$, $\theta_n = \lambda+\mu$ for $n \ge 1$, deduce that the equilibrium distribution of the M(λ)/M(μ)/1 queue is $\nu_n = (1-\rho)\rho^n$, $n \ge 0$, where $\rho = \lambda/\mu < 1$.

3. Waiting time. Consider a M(λ)/M(μ)/1 queue with $\rho = \lambda/\mu$ satisfying $\rho < 1$, and suppose that the number $Q(0)$ of people in the queue at time 0 has the stationary distribution $\pi_n = (1-\rho)\rho^n$, $n \ge 0$. Let W be the time spent by a typical new arrival before he begins his service. Show that the distribution of W is given by $\mathbb{P}(W \le x) = 1-\rho e^{-x(\mu-\lambda)}$ for $x \ge 0$, and note that $\mathbb{P}(W = 0) = 1-\rho$.

4. A box contains i red balls and j lemon balls, and they are drawn at random without replacement. Each time a red (respectively lemon) ball is drawn, a particle doing a walk on $\{0, 1, 2, \dots\}$ moves one step to the right (respectively left); the origin is a retaining barrier, so that leftwards steps from the origin are suppressed. Let $\pi(n; i, j)$ be the probability that the particle ends at position n, having started at the origin. Write down a set of difference equations for the $\pi(n; i, j)$, and deduce that

$$\pi(n; i, j) = A(n; i, j) - A(n+1; i, j) \quad \text{for } i \le j+n$$

where $A(n; i, j) = \binom{i}{n}/\binom{j+n}{n}$.

5. Let Q be a M(λ)/M(μ)/1 queue with $Q(0) = 0$. Show that $p_n(t) = \mathbb{P}(Q(t) = n)$ satisfies

$$p_n(t) = \sum_{i,j \ge 0} \pi(n; i, j) \left(\frac{(\lambda t)^i e^{-\lambda t}}{i!}\right)\left(\frac{(\mu t)^j e^{-\mu t}}{j!}\right)$$

where the $\pi(n; i, j)$ are given in Exercise (11.2.4).

6. Let $Q(t)$ be the length of an M(λ)/M(μ)/1 queue at time t, and let $Z = \{Z_n\}$ be the jump chain of Q. Explain how the stationary distribution of Q may be derived from that of Z, and vice versa.

7. **Tandem queues.** Two queues have one server each, and all service times are independent and exponentially distributed, with parameter μ_i for queue i. Customers arrive at the first queue at the instants of a Poisson process of rate λ ($< \min\{\mu_1, \mu_2\}$), and on completing service immediately enter the second queue. The queues are in equilibrium. Show that:

(a) the output of the first queue is a Poisson process with intensity λ, and that its departures before time t are independent of the length of this queue at time t (this is known as *Burke's theorem*),

(b) the waiting times of a given customer in the two queues are not independent.

11.3 Exercises. M/G/1

1. Consider M(λ)/D(d)/1 where $\rho = \lambda d < 1$. Show that the mean queue length at moments of departure in equilibrium is $\frac{1}{2}\rho(2-\rho)/(1-\rho)$.

2. Consider M(λ)/M(μ)/1, and show that the moment generating function of a typical busy period is given by

$$M_B(s) = \frac{(\lambda + \mu - s) - \sqrt{(\lambda + \mu - s)^2 - 4\lambda\mu}}{2\lambda}$$

for all sufficiently small but positive values of s.

3. (a) Show that, for a M/G/1 queue, the sequence of times at which the server passes from being busy to being free constitutes a renewal process.

(b) When the above queue is in equilibrium, what is the moment generating function of the total time that an arriving customer spends in the queue including service?

4. **Loss system.** Consider the M(λ)/G/1 queue with no waiting room. Customers who arrive while the server is busy are lost. Show that the long run proportion of arrivals lost is $1/(1+\rho^{-1})$ where $\rho = \lambda\mathbb{E}(S)$ is the traffic intensity.

11.4 Exercises. G/M/1

1. Consider G/M(μ)/1, and let $\alpha_j = \mathbb{E}((\mu X)^j e^{-\mu X}/j!)$ where X is a typical interarrival time. Suppose the traffic intensity ρ is less than 1. Show that the equilibrium distribution $\boldsymbol{\pi}$ of the imbedded chain at moments of arrivals satisfies

$$\pi_n = \sum_{i=0}^{\infty} \alpha_i \pi_{n+i-1} \qquad \text{for } n \geq 1.$$

Look for a solution of the form $\pi_n = \theta^n$ for some θ, and deduce that the unique stationary distribution is given by $\pi_j = (1-\eta)\eta^j$ for $j \geq 0$, where η is the smallest positive root of the equation $s = M_X(\mu(s-1))$.

2. Consider a G/M(μ)/1 queue in equilibrium. Let η be the smallest positive root of the equation $x = M_X(\mu(x-1))$ where M_X is the moment generating function of an interarrival time. Show that the mean number of customers ahead of a new arrival is $\eta(1-\eta)^{-1}$, and the mean waiting time is $\eta\{\mu(1-\eta)\}^{-1}$.

3. Consider D(1)/M(μ)/1 where $\mu > 1$. Show that the continuous-time queue length $Q(t)$ does not converge in distribution as $t \to \infty$, even though the imbedded chain at the times of arrivals is ergodic.

11.5 Exercises. G/G/1

1. Show that, for a G/G/1 queue, the starting times of the busy periods of the server constitute a renewal process.

2. Consider a G/M(μ)/1 queue in equilibrium, together with the dual (unstable) M(μ)/G/1 queue. Show that the idle periods of the latter queue are exponentially distributed. Use the theory of duality of queues to deduce for the former queue that: (a) the waiting-time distribution is a mixture of an exponential distribution and an atom at zero, and (b) the equilibrium queue length is geometric.

3. Consider G/M(μ)/1, and let G be the distribution function of $S - X$ where S and X are typical (independent) service and interarrival times. Show that the *Wiener–Hopf equation*

$$F(x) = \int_{-\infty}^{x} F(x-y)\,dG(y), \qquad x \ge 0,$$

for the limiting waiting-time distribution F is satisfied by $F(x) = 1 - \eta e^{-\mu(1-\eta)x}$, $x \ge 0$. Here, η is the smallest positive root of the equation $x = M_X(\mu(x-1))$, where M_X is the moment generating function of X.

11.6 Exercise. Heavy traffic

1. Consider the M(λ)/M(μ)/1 queue with $\rho = \lambda/\mu < 1$. Let Q_ρ be a random variable with the equilibrium queue distribution, and show that $(1-\rho)Q_\rho$ converges in distribution as $\rho \uparrow 1$, the limit distribution being exponential with parameter 1.

11.7 Exercises. Networks of queues

1. Consider an open migration process with c stations, in which individuals arrive at station j at rate ν_j, individuals move from i to j at rate $\lambda_{ij}\phi_i(n_i)$, and individuals depart from i at rate $\mu_i\phi_i(n_i)$, where n_i denotes the number of individuals currently at station i. Show when $\phi_i(n_i) = n_i$ for all i that the system behaves as though the customers move independently through the network. Identify the explicit form of the stationary distribution, subject to an assumption of irreducibility, and explain a connection with the Bartlett theorem of Problem (8.10.6).

2. Let Q be an M(λ)/M(μ)/s queue where $\lambda < s\mu$, and assume Q is in equilibrium. Show that the process of departures is a Poisson process with intensity λ, and that departures up to time t are independent of the value of $Q(t)$.

3. Customers arrive in the manner of a Poisson process with intensity λ in a shop having two servers. The service times of these servers are independent and exponentially distributed with respective parameters μ_1 and μ_2. Arriving customers form a single queue, and the person at the head of the queue moves to the first free server. When both servers are free, the next arrival is allocated a server chosen according to one of the following rules:

(a) each server is equally likely to be chosen,

(b) the server who has been free longer is chosen.

Assume that $\lambda < \mu_1 + \mu_2$, and the process is in equilibrium. Show in each case that the process of departures from the shop is a Poisson process, and that departures prior to time t are independent of the number of people in the shop at time t.

4. Difficult customers. Consider an M(λ)/M(μ)/1 queue modified so that on completion of service the customer leaves with probability δ, or rejoins the queue with probability $1-\delta$. Find the distribution of the total time a customer spends being served. Hence show that equilibrium is possible if $\lambda < \delta\mu$, and find the stationary distribution. Show that, in equilibrium, the departure process is Poisson, but if the rejoining customer goes to the end of the queue, the composite arrival process is not Poisson.

5. Consider an open migration process in equilibrium. If there is no path by which an individual at station k can reach station j, show that the stream of individuals moving directly from station j to station k forms a Poisson process.

6. Show that an open migration process is non-explosive.

7. Let X be an irreducible, continuous-time Markov chain with generator $\mathbf{G}$ on the state space $T = S \cup \{\infty\}$, where S is countable and non-empty. Show that a distribution $\boldsymbol{\pi}$ on T satisfies $\boldsymbol{\pi}\mathbf{G} = \mathbf{0}$ if and only if, for $j \in S$, $\sum_{i \in T} \pi_i g_{ij} = 0$.

11.8 Problems

1. Finite waiting room. Consider M(λ)/M(μ)/k with the constraint that arriving customers who see N customers in the line ahead of them leave and never return. Find the stationary distribution of queue length for the cases $k = 1$ and $k = 2$.

2. Baulking. Consider M(λ)/M(μ)/1 with the constraint that if an arriving customer sees n customers in the line ahead of him, he joins the queue with probability $p(n)$ and otherwise leaves in disgust.

(a) Find the stationary distribution of queue length if $p(n) = (n+1)^{-1}$.

(b) Find the stationary distribution $\boldsymbol{\pi}$ of queue length if $p(n) = 2^{-n}$, and show that the probability that an arriving customer joins the queue (in equilibrium) is $\mu(1 - \pi_0)/\lambda$.

3. Series. In a Moscow supermarket customers queue at the cash desk to pay for the goods they want; then they proceed to a second line where they wait for the goods in question. If customers arrive in the shop like a Poisson process with parameter λ and all service times are independent and exponentially distributed, parameter μ_1 at the first desk and μ_2 at the second, find the stationary distributions of queue lengths, when they exist, and show that, at any given time, the two queue lengths are independent in equilibrium.

4. Batch (or bulk) service. Consider M/G/1, with the modification that the server may serve up to m customers simultaneously. If the queue length is less than m at the beginning of a service period then she serves everybody waiting at that time. Find a formula which is satisfied by the probability generating function of the stationary distribution of queue length at the times of departures, and evaluate this generating function explicitly in the case when $m = 2$ and service times are exponentially distributed.

5. Consider M(λ)/M(μ)/1 where $\lambda < \mu$. Find the moment generating function of the length B of a typical busy period, and show that $\mathbb{E}(B) = (\mu - \lambda)^{-1}$ and $\text{var}(B) = (\lambda + \mu)/(\mu - \lambda)^3$. Show that the density function of B is

$$f_B(x) = \frac{\sqrt{\mu/\lambda}}{x} e^{-(\lambda+\mu)x} I_1\big(2x\sqrt{\lambda\mu}\big) \quad \text{for } x > 0$$

where I_1 is a modified Bessel function.

6. Consider M(λ)/G/1 in equilibrium. Obtain an expression for the mean queue length at departure times. Show that the mean waiting time in equilibrium of an arriving customer is $\frac{1}{2}\lambda\mathbb{E}(S^2)/(1-\rho)$ where S is a typical service time and $\rho = \lambda\mathbb{E}(S)$.

Amongst all possible service-time distributions with given mean, find the one for which the mean waiting time is a minimum.

7. Let W_t be the time which a customer would have to wait in a M(λ)/G/1 queue if he were to arrive at time t. Show that the distribution function $F(x; t) = \mathbb{P}(W_t \le x)$ satisfies

$$\frac{\partial F}{\partial t} = \frac{\partial F}{\partial x} - \lambda F + \lambda \mathbb{P}(W_t + S \le x)$$

where S is a typical service time, independent of W_t.

Suppose that $F(x, t) \to H(x)$ for all x as $t \to \infty$, where H is a distribution function satisfying $0 = h - \lambda H + \lambda \mathbb{P}(U + S \le x)$ for $x > 0$, where U is independent of S with distribution function H, and h is the density function of H on $(0, \infty)$. Show that the moment generating function M_U of U satisfies

$$M_U(\theta) = \frac{(1-\rho)\theta}{\lambda + \theta - \lambda M_S(\theta)}$$

where ρ is the traffic intensity. You may assume that $\mathbb{P}(S = 0) = 0$.

8. Consider a G/G/1 queue in which the service times are constantly equal to 2, whilst the interarrival times take either of the values 1 and 4 with equal probability $\frac{1}{2}$. Find the limiting waiting time distribution.

9. Consider an extremely idealized model of a telephone exchange having infinitely many channels available. Calls arrive in the manner of a Poisson process with intensity λ, and each requires one channel for a length of time having the exponential distribution with parameter μ, independently of the arrival process and of the duration of other calls. Let $Q(t)$ be the number of calls being handled at time t, and suppose that $Q(0) = I$.

Determine the probability generating function of $Q(t)$, and deduce $\mathbb{E}(Q(t))$, $\mathbb{P}(Q(t) = 0)$, and the limiting distribution of $Q(t)$ as $t \to \infty$.

Assuming the queue is in equilibrium, find the proportion of time that no channels are occupied, and the mean length of an idle period. Deduce that the mean length of a busy period is $(e^{\lambda/\mu} - 1)/\lambda$.

10. Customers arrive in a shop in the manner of a Poisson process with intensity λ, where $0 < \lambda < 1$. They are served one by one in the order of their arrival, and each requires a service time of unit length. Let $Q(t)$ be the number in the queue at time t. By comparing $Q(t)$ with $Q(t + 1)$, determine the limiting distribution of $Q(t)$ as $t \to \infty$ (you may assume that the quantities in question converge). Hence show that the mean queue length in equilibrium is $\lambda(1 - \frac{1}{2}\lambda)/(1 - \lambda)$.

Let W be the waiting time of a newly arrived customer when the queue is in equilibrium. Deduce from the results above that $\mathbb{E}(W) = \frac{1}{2}\lambda/(1 - \lambda)$.

11. Consider M(λ)/D(1)/1, and suppose that the queue is empty at time 0. Let T be the earliest time at which a customer departs leaving the queue empty. Show that the moment generating function M_T of T satisfies

$$\log\left(1 - \frac{s}{\lambda}\right) + \log M_T(s) = (s - \lambda)\big(1 - M_T(s)\big),$$

and deduce the mean value of T, distinguishing between the cases $\lambda < 1$ and $\lambda \ge 1$.

12. Suppose $\lambda < \mu$, and consider a M(λ)/M(μ)/1 queue Q in equilibrium.

(a) Show that Q is a reversible Markov chain.

(b) Deduce the equilibrium distributions of queue length and waiting time.

(c) Show that the times of departures of customers form a Poisson process, and that $Q(t)$ is independent of the times of departures prior to t.

(d) Consider a sequence of K single-server queues such that customers arrive at the first in the manner of a Poisson process, and (for each j) on completing service in the jth queue each customer moves to the $(j + 1)$th. Service times in the jth queue are exponentially distributed with parameter μ_j, with as much independence as usual. Determine the (joint) equilibrium distribution of the queue lengths, when $\lambda < \mu_j$ for all j.

13. Consider the queue M(λ)/M(μ)/k, where $k \geq 1$. Show that a stationary distribution $\boldsymbol{\pi}$ exists if and only if $\lambda < k\mu$, and calculate it in this case.

Suppose that the cost of operating this system in equilibrium is

$$Ak + B\sum_{n=k}^{\infty}(n-k+1)\pi_n,$$

the positive constants A and B representing respectively the costs of employing a server and of the dissatisfaction of delayed customers.

Show that, for fixed μ, there is a unique value λ^* in the interval $(0, \mu)$ such that it is cheaper to have $k = 1$ than $k = 2$ if and only if $\lambda < \lambda^*$.

14. Customers arrive in a shop in the manner of a Poisson process with intensity λ. They form a single queue. There are two servers, labelled 1 and 2, server i requiring an exponentially distributed time with parameter μ_i to serve any given customer. The customer at the head of the queue is served by the first idle server; when both are idle, an arriving customer is equally likely to choose either.

(a) Show that the queue length settles into equilibrium if and only if $\lambda < \mu_1 + \mu_2$.

(b) Show that, when in equilibrium, the queue length is a time-reversible Markov chain.

(c) Deduce the equilibrium distribution of queue length.

(d) Generalize your conclusions to queues with many servers.

15. Consider the D(1)/M(μ)/1 queue where $\mu > 1$, and let Q_n be the number of people in the queue just before the nth arrival. Let Q_μ be a random variable having as distribution the stationary distribution of the Markov chain $\{Q_n\}$. Show that $(1-\mu^{-1})Q_\mu$ converges in distribution as $\mu \downarrow 1$, the limit distribution being exponential with parameter 2.

16. Kendall's taxicabs. Taxis arrive at a stand in the manner of a Poisson process with intensity τ, and passengers arrive in the manner of an (independent) Poisson process with intensity π. If there are no waiting passengers, the taxis wait until passengers arrive, and then move off with the passengers, one to each taxi. If there is no taxi, passengers wait until they arrive. Suppose that initially there are neither taxis nor passengers at the stand. Show that the probability that n passengers are waiting at time t is $(\pi/\tau)^{\frac{1}{2}n}e^{-(\pi+\tau)t}I_n(2t\sqrt{\pi\tau})$, where $I_n(x)$ is the modified Bessel function, i.e. the coefficient of z^n in the power series expansion of $\exp\{\frac{1}{2}x(z+z^{-1})\}$.

17. Machines arrive for repair as a Poisson process with intensity λ. Each repair involves two stages, the ith machine to arrive being under repair for a time $X_i + Y_i$, where the pairs (X_i, Y_i), $i = 1, 2, \dots$, are independent with a common joint distribution. Let $U(t)$ and $V(t)$ be the numbers of machines in the X-stage and Y-stage of repair at time t. Show that $U(t)$ and $V(t)$ are independent Poisson random variables.

18. Ruin. An insurance company pays independent and identically distributed claims $\{K_n : n \geq 1\}$ at the instants of a Poisson process with intensity λ, where $\lambda\mathbb{E}(K_1) < 1$. Premiums are received at constant rate 1. Show that the maximum deficit M the company will ever accumulate has moment generating function

$$\mathbb{E}(e^{\theta M}) = \frac{(1-\rho)\theta}{\lambda+\theta-\lambda\mathbb{E}(e^{\theta K})}.$$

19. (a) **Erlang's loss formula.** Consider M(λ)/M(μ)/s with baulking, in which a customer departs immediately if, on arrival, he sees all the servers occupied ahead of him. Show that, in equilibrium, the probability that all servers are occupied is

$$\pi_s = \frac{\rho^s/s!}{\sum_{j=0}^{s}\rho^j/j!}, \qquad \text{where } \rho = \lambda/\mu.$$

(b) Consider an M(λ)/M(μ)/∞ queue with channels (servers) numbered $1, 2, \ldots$. On arrival, a customer will choose the lowest numbered channel that is free, and be served by that channel. Show in the notation of part (a) that the fraction p_c of time that channel c is busy is $p_c = \rho(\pi_{c-1} - \pi_c)$ for $c \geq 2$, and $p_1 = \pi_1$.

20. For an M(λ)/M(μ)/1 queue with $\lambda < \mu$, when in equilibrium, show that the expected time until the queue is first empty is $\lambda(\mu - \lambda)^{-2}$.

21. Consider an M(λ)/G/∞ queue. Use the renewal–reward theorem to show that the expected duration of a busy period is $(e^{\rho} - 1)/\lambda$ where $\rho = \lambda\mathbb{E}(S)$.

12

Martingales

12.1 Exercises. Introduction

1. (a) If $(Y, \mathcal{F})$ is a martingale, show that $\mathbb{E}(Y_n) = \mathbb{E}(Y_0)$ for all n.

(b) If $(Y, \mathcal{F})$ is a submartingale (respectively supermartingale) with finite means, show that $\mathbb{E}(Y_n) \geq \mathbb{E}(Y_0)$ (respectively $\mathbb{E}(Y_n) \leq \mathbb{E}(Y_0)$).

2. Let $(Y, \mathcal{F})$ be a martingale, and show that $\mathbb{E}(Y_{n+m} \mid \mathcal{F}_n) = Y_n$ for all $n, m \geq 0$.

3. Let Z_n be the size of the nth generation of a branching process with $Z_0 = 1$, having mean family size μ and extinction probability η. Show that $Z_n\mu^{-n}$ and η^{Z_n} define martingales.

4. Let $\{S_n : n \geq 0\}$ be a simple symmetric random walk on the integers with $S_0 = k$. Show that S_n and $S_n^2 - n$ are martingales. Making assumptions similar to those of de Moivre (see Example (12.1.4)), find the probability of ruin and the expected duration of the game for the gambler's ruin problem.

5. Let $(Y, \mathcal{F})$ be a martingale with the property that $\mathbb{E}(Y_n^2) < \infty$ for all n. Show that, for $i \leq j \leq k$, $\mathbb{E}\{(Y_k - Y_j)Y_i\} = 0$, and $\mathbb{E}\{(Y_k - Y_j)^2 \mid \mathcal{F}_i\} = \mathbb{E}(Y_k^2 \mid \mathcal{F}_i) - \mathbb{E}(Y_j^2 \mid \mathcal{F}_i)$. Suppose there exists K such that $\mathbb{E}(Y_n^2) \leq K$ for all n. Show that the sequence $\{Y_n\}$ converges in mean square as $n \to \infty$.

6. Let Y be a martingale and let u be a convex function mapping $\mathbb{R}$ to $\mathbb{R}$. Show that $\{u(Y_n) : n \geq 0\}$ is a submartingale provided that $\mathbb{E}(u(Y_n)^+) < \infty$ for all n.

Show that $|Y_n|$, Y_n^2, and Y_n^+ constitute submartingales whenever the appropriate moment conditions are satisfied.

7. Let Y be a submartingale and let u be a convex non-decreasing function mapping $\mathbb{R}$ to $\mathbb{R}$. Show that $\{u(Y_n) : n \geq 0\}$ is a submartingale provided that $\mathbb{E}(u(Y_n)^+) < \infty$ for all n.

Show that (subject to a moment condition) Y_n^+ constitutes a submartingale, but that $|Y_n|$ and Y_n^2 need not constitute submartingales.

8. Let X be a discrete-time Markov chain with countable state space S and transition matrix $\mathbf{P}$. Suppose that $\psi : S \to \mathbb{R}$ is bounded and satisfies $\sum_{j \in S} p_{ij}\psi(j) \leq \lambda\psi(i)$ for some $\lambda > 0$ and all $i \in S$. Show that $\lambda^{-n}\psi(X_n)$ constitutes a supermartingale.

9. Let $G_n(s)$ be the probability generating function of the size Z_n of the nth generation of a branching process, where $Z_0 = 1$ and $\text{var}(Z_1) > 0$. Let H_n be the inverse function of the function G_n, viewed as a function on the interval $[0, 1]$, and show that $M_n = \{H_n(s)\}^{Z_n}$ defines a martingale with respect to the sequence Z.

12.2 Exercises. Martingale differences and Hoeffding's inequality

1. Knapsack problem. It is required to pack a knapsack to maximum benefit. Suppose you have n objects, the ith object having volume V_i and worth W_i, where $V_1, V_2, \dots, V_n, W_1, W_2, \dots, W_n$ are independent non-negative random variables with finite means, and $W_i \le M$ for all i and some fixed M. Your knapsack has volume c, and you wish to maximize the total worth of the objects packed in it. That is, you wish to find the vector $z_1, z_2, \dots, z_n$ of 0's and 1's such that $\sum_1^n z_i V_i \le c$ and which maximizes $\sum_1^n z_i W_i$. Let Z be the maximal possible worth of the knapsack's contents, and show that $\mathbb{P}(|Z - \mathbb{E}Z| \ge x) \le 2\exp\{-x^2/(2nM^2)\}$ for $x > 0$.

2. Graph colouring. Given n vertices $v_1, v_2, \dots, v_n$, for each $1 \le i < j \le n$ we place an edge between v_i and v_j with probability p; different pairs are joined independently of each other. We call v_i and v_j *neighbours* if they are joined by an edge. The *chromatic number* χ of the ensuing graph is the minimal number of pencils of different colours which are required in order that each vertex may be coloured differently from each of its neighbours. Show that $\mathbb{P}(|\chi - \mathbb{E}\chi| \ge x) \le 2\exp\{-\frac{1}{2}x^2/n\}$ for $x > 0$.

3. Maurer's inequality.

(a) Let X and Y be random variables such that $X \le b < \infty$ a.s., $\mathbb{E}(X \mid Y) = 0$, and $\mathbb{E}(X^2 \mid Y) \le \sigma^2 < \infty$. Show that

$$\mathbb{E}(e^{\theta X} \mid Y) \le \exp\left\{\tfrac{1}{2}\theta^2(b^2 + \sigma^2)\right\}, \qquad \theta \ge 0.$$

(b) Let $(M, \mathcal{F})$ be a martingale with $M_0 = 0$, having differences $D_r = M_r - M_{r-1}$, and suppose that $D_n \le b_n < \infty$ a.s., and $\mathbb{E}(D_n^2 \mid M_{n-1}) \le \sigma_n^2 < \infty$ for $n \ge 1$. Show that

$$\mathbb{P}(M_n \ge t) \le \exp\left\{-\frac{t^2}{2\sum_{r=1}^n (b_r^2 + \sigma_r^2)}\right\}, \qquad t \ge 0.$$

4. Quadratic variation. Let $(M, \mathcal{F})$ be a martingale with $M_0 = 0$, having differences $D_r = M_r - M_{r-1}$. The process given by $Q_n = \sum_{r=1}^n D_r^2$ is called the *optional quadratic variation* of M, while $V_n = \sum_{r=1}^n \mathbb{E}(D_r^2 \mid \mathcal{F}_{r-1})$ is called the *predictable quadratic variation* of M. Show that $X_n = M_n^2 - Q_n$ and $Y_n = M_n^2 - V_n$ define martingales with respect to $\mathcal{F}$.

12.3 Exercises. Crossings and convergence

1. Give a reasonable definition of a *downcrossing* of the interval $[a, b]$ by the random sequence $Y_0, Y_1, \dots$.

(a) Show that the number of downcrossings differs from the number of upcrossings by at most 1.

(b) If $(Y, \mathcal{F})$ is a submartingale, show that the number $D_n(a, b; Y)$ of downcrossings of $[a, b]$ by Y up to time n satisfies

$$\mathbb{E}D_n(a, b; Y) \le \frac{\mathbb{E}\{(Y_n - b)^+\}}{b - a}.$$

2. Let $(Y, \mathcal{F})$ be a supermartingale with finite means, and let $U_n(a, b; Y)$ be the number of upcrossings of the interval $[a, b]$ up to time n. Show that

$$\mathbb{E}U_n(a, b; Y) \le \frac{\mathbb{E}\{(Y_n - a)^-\}}{b - a}.$$

Deduce that $\mathbb{E}U_n(a, b; Y) \le a/(b-a)$ if Y is non-negative and $a \ge 0$.

3. Let X be a Markov chain with countable state space S and transition matrix $\mathbf{P}$. Suppose that X is irreducible and recurrent, and that $\psi : S \to S$ is a bounded function satisfying $\sum_{j\in S} p_{ij}\psi(j) \le \psi(i)$ for $i \in S$. Show that ψ is a constant function.

4. Let $Z_1, Z_2, \dots$ be independent random variables such that:

$$Z_n = \begin{cases} a_n & \text{with probability } \frac{1}{2}n^{-2}, \\ 0 & \text{with probability } 1 - n^{-2}, \\ -a_n & \text{with probability } \frac{1}{2}n^{-2}, \end{cases}$$

where $a_1 = 2$ and $a_n = 4\sum_{j=1}^{n-1} a_j$. Show that $Y_n = \sum_{j=1}^{n} Z_j$ defines a martingale. Show that $Y = \lim Y_n$ exists almost surely, but that there exists no M such that $\mathbb{E}|Y_n| \le M$ for all n.

5. Random adding martingale. Let $x_1, x_2, \dots, x_r \in \mathbb{R}$, and let the sequence $\{X_n : n \ge 1\}$ of random variables be given as follows. We set

$$X_n = \begin{cases} x_n & \text{if } 1 \le n \le r, \\ X_{U(n)} + X_{V(n)} & \text{if } n > r, \end{cases}$$

where $U(n)$ and $V(n)$ are uniformly distributed on $\{1, 2, \dots, n-1\}$, and the random variables $\{U(n), V(n) : n > r\}$ are independent.

Show that

$$M_n = \frac{1}{n(n+1)}\sum_{k=1}^{n} X_k, \qquad n = r, r+1, \dots,$$

is a martingale with respect to the sequence $\{X_n\}$.

Enthusiasts seeking a challenge are invited to show that the M converges almost surely and in mean square to a non-degenerate limit.

6. Pólya's urn revisited. Let R_n and B_n be the numbers of red and blue balls, respectively, in an urn at the nth stage, and assume $R_0 = B_0 = 1$. At each stage, a ball is drawn and returned together with a fresh ball of the other colour. Show that $M_n = (B_n - R_n)(B_n + R_n - 1)$ defines a martingale. Does it converge almost surely?

12.4 Exercises. Stopping times

1. If T_1 and T_2 are stopping times with respect to a filtration $\mathcal{F}$, show that $T_1 + T_2$, $\max\{T_1, T_2\}$, and $\min\{T_1, T_2\}$ are stopping times also.

2. Let $X_1, X_2, \dots$ be a sequence of non-negative independent random variables and let $N(t) = \max\{n : X_1 + X_2 + \cdots + X_n \le t\}$. Show that $N(t) + 1$ is a stopping time with respect to a suitable filtration to be specified.

3. Let $(Y, \mathcal{F})$ be a submartingale and $x > 0$. Show that

$$\mathbb{P}\left(\max_{0\le m\le n} Y_m \ge x\right) \le \frac{1}{x}\mathbb{E}(Y_n^+).$$

4. Let $(Y, \mathcal{F})$ be a non-negative supermartingale and $x > 0$. Show that

$$\mathbb{P}\left(\max_{0\le m\le n} Y_m \ge x\right) \le \frac{1}{x}\mathbb{E}(Y_0).$$

5. Let $(Y, \mathcal{F})$ be a submartingale and let S and T be stopping times satisfying $0 \le S \le T \le N$ for some deterministic N. Show that $\mathbb{E}Y_0 \le \mathbb{E}Y_S \le \mathbb{E}Y_T \le \mathbb{E}Y_N$.

6. Let $\{S_n\}$ be a simple random walk with $S_0 = 0$ such that $0 < p = \mathbb{P}(S_1 = 1) < \frac{1}{2}$. Use de Moivre's martingale to show that $\mathbb{E}(\sup_m S_m) \le p/(1-2p)$. Show further that this inequality may be replaced by an equality.

7. Let $\mathcal{F}$ be a filtration. For any stopping time T with respect to $\mathcal{F}$, denote by $\mathcal{F}_T$ the collection of all events A such that, for all n, $A \cap \{T \le n\} \in \mathcal{F}_n$. Let S and T be stopping times.

(a) Show that $\mathcal{F}_T$ is a σ-field, and that T is measurable with respect to this σ-field.

(b) If $A \in \mathcal{F}_S$, show that $A \cap \{S \le T\} \in \mathcal{F}_T$.

(c) Let S and T satisfy $S \le T$. Show that $\mathcal{F}_S \subseteq \mathcal{F}_T$.

8. Stopping an exchangeable sequence.

(a) Let T be a stopping time for an exchangeable sequence $X_1, X_2, \dots, X_n$. Show that, if $\mathbb{P}(T \le n-r) = 1$, then the random vector $(X_{T+1}, X_{T+2}, \dots, X_{T+r})$ has the same distribution as $(X_1, X_2, \dots, X_r)$.

(b) An urn contains v violet balls and w white balls, which are drawn at random without replacement. Let $m < w$, and let T be the number of the draw at which the mth white ball is drawn. What is the probability that the next ball is white?

12.5 Exercises. Optional stopping

1. Let $(Y, \mathcal{F})$ be a martingale and T a stopping time such that $\mathbb{P}(T < \infty) = 1$. Show that $\mathbb{E}(Y_T) = \mathbb{E}(Y_0)$ if either of the following holds:

(a) $\mathbb{E}(\sup_n |Y_{T\wedge n}|) < \infty$, (b) $\mathbb{E}(|Y_{T\wedge n}|^{1+\delta}) \le c$ for some $c, \delta > 0$ and all n.

2. Let $(Y, \mathcal{F})$ be a martingale. Show that $(Y_{T\wedge n}, \mathcal{F}_n)$ is a uniformly integrable martingale for any finite stopping time T such that either:

(a) $\mathbb{E}|Y_T| < \infty$ and $\mathbb{E}(|Y_n| I_{\{T>n\}}) \to 0$ as $n \to \infty$, or

(b) $\{Y_n\}$ is uniformly integrable.

3. Let $(Y, \mathcal{F})$ be a uniformly integrable martingale, and let S and T be finite stopping times satisfying $S \le T$. Prove that $Y_T = \mathbb{E}(Y_\infty \mid \mathcal{F}_T)$ and that $Y_S = \mathbb{E}(Y_T \mid \mathcal{F}_S)$, where Y_∞ is the almost sure limit as $n \to \infty$ of Y_n.

4. Let $\{S_n : n \ge 0\}$ be a simple symmetric random walk with $0 < S_0 < N$ and with absorbing barriers at 0 and N. Use the optional stopping theorem to show that the mean time until absorption is $\mathbb{E}\{S_0(N - S_0)\}$.

5. Let $\{S_n : n \ge 0\}$ be a simple symmetric random walk with $S_0 = 0$. Show that

$$Y_n = \frac{\cos\{\lambda[S_n - \frac{1}{2}(b-a)]\}}{(\cos\lambda)^n}$$

constitutes a martingale if $\cos\lambda \ne 0$.

Let a and b be positive integers. Show that the time T until absorption at one of two absorbing barriers at $-a$ and b satisfies

$$\mathbb{E}\big(\{\cos\lambda\}^{-T}\big) = \frac{\cos\{\frac{1}{2}\lambda(b-a)\}}{\cos\{\frac{1}{2}\lambda(b+a)\}}, \qquad 0 < \lambda < \frac{\pi}{b+a}.$$

6. Let $\{S_n : n \ge 0\}$ be a simple symmetric random walk on the positive and negative integers, with $S_0 = 0$. For each of the three following random variables, determine whether or not it is a stopping time and find its mean:

$$U = \min\{n \geq 5 : S_n = S_{n-5} + 5\}, \quad V = U - 5, \quad W = \min\{n : S_n = 1\}.$$

7. Let $S_n = a + \sum_{r=1}^{n} X_r$ be a simple symmetric random walk. The walk stops at the earliest time T when it reaches either of the two positions 0 or K where $0 < a < K$. Show that $M_n = \sum_{r=0}^{n} S_r - \frac{1}{3}S_n^3$ is a martingale and deduce that $\mathbb{E}\big(\sum_{r=0}^{T} S_r\big) = \frac{1}{3}(K^2 - a^2)a + a$.

8. Gambler's ruin. Let X_i be independent random variables each equally likely to take the values ± 1, and let $T = \min\{n : S_n \in \{-a, b\}\}$. Verify the conditions of the optional stopping theorem (12.5.1) for the martingale $S_n^2 - n$ and the stopping time T.

9. Family planning, Problem (3.11.30) revisited. Children are either female or male. Their sexes are independent random variables, being female with probability q or male with probability $p = 1 - q$. A woman ceases childbearing at stage T, and we write G_n and B_n for the numbers of girls and boys born to her up to and including stage n. Assume that T is a finite stopping time for the sequence $\{(G_n, B_n) : n \geq 1\}$. Show that, no matter the stopping rule that yields T, we have $\mathbb{E}(G_T)/\mathbb{E}(B_T) = q/p$. What can be said about $\mathbb{E}(G_T/B_T)$?

10. Let $X = \{X_n : n \geq 0\}$ be a Markov chain with state space $\{0, 1, \dots, b\}$, such that $i \to 0, b$ for $i \in \{1, 2, \dots, b-1\}$. If X is also a martingale, show that 0 and b are absorbing, and that, given X_0, the probability of absorption at b is X_0/b.

12.6 Exercise. The maximal inequality

1. Martingale laws of large numbers. Let $M_n = \sum_{r=1}^{n} D_r$ be a zero-mean martingale with difference sequence $\{D_r : r \geq 1\}$, such that $\sigma_r^2 = \text{var}(D_r) < \infty$. Show the following.

(a) We have that $\mathbb{E}(D_r D_s) = 0$ for $r \neq s$.

(b) If $n^{-2}\sum_{r=1}^{n} \sigma_r^2 \to 0$ as $n \to \infty$, then D satisfies the weak law of large numbers in that $n^{-1}M_n \xrightarrow{\text{P}} 0$.

(c) If $\sum_r \sigma_r^2/r^2 < \infty$, then D satisfies the strong law of large numbers in that $n^{-1}M_n \xrightarrow{\text{a.s.}} 0$.

12.7 Exercises. Backward martingales and continuous-time martingales

1. Let X be a continuous-time Markov chain with finite state space S and generator $\mathbf{G}$. Let $\eta = \{\eta(i) : i \in S\}$ be a root of the equation $\mathbf{G}\eta' = \mathbf{0}$. Show that $\eta(X(t))$ constitutes a martingale with respect to $\mathcal{F}_t = \sigma(\{X(u) : u \leq t\})$.

2. Let N be a Poisson process with intensity λ and $N(0) = 0$, and let $T_a = \min\{t : N(t) = a\}$, where a is a positive integer. Assuming that $\mathbb{E}\{\exp(\psi T_a)\} < \infty$ for sufficiently small positive ψ, use the optional stopping theorem to show that $\text{var}(T_a) = a\lambda^{-2}$. Show further that T_a has characteristic function $\phi_a(s) = [\lambda/(\lambda - is)]^a$.

3. Let $S_m = \sum_{r=1}^{m} X_r$, $m \leq n$, where the X_r are independent and identically distributed with finite mean. Denote by $U_1, U_2, \dots, U_n$ the order statistics of n independent variables which are uniformly distributed on $(0, t)$, and set $U_{n+1} = t$. Show that $R_m = S_m/U_{m+1}$, $0 \leq m \leq n$, is a backward martingale with respect to a suitable sequence of σ-fields, and deduce that

$$\mathbb{P}\big(R_m \geq 1 \text{ for some } m \leq n \mid S_n = y\big) \leq \min\{y/t, 1\}.$$

4. (a) Let W be a standard Wiener process. Show that the following are martingales:

$$\text{(i) } W(t), \quad \text{(ii) } W(t)^2 - t, \quad \text{(iii) } W(t)^3 - 3tW(t), \quad \text{(iv) } W(t)^4 - 6tW(t)^2 + 3t^2.$$

(b) Let T be the earliest time at which W exits the interval $(-a, a)$. Show that $\mathbb{E}(T^n) < \infty$ for $n \geq 1$. Use the above martingales to show that $\mathbb{E}(T^2) = \frac{5}{3}a^4$. [You may use appropriate optional stopping theorems and the dominated convergence theorem.]

5. Let X and Y be independent standard Wiener processes. Show that

$$M(t) = Y(t)X(t)^2 - \int_0^t Y(u)\,du$$

defines a martingale with respect to the natural filtration $\mathcal{F}_t = \sigma(\{X_v, Y_v : v \leq t\})$.

6. Brownian motion in a disc. Let $Z(t) = z + X(t) + iY(t)$, where X and Y are independent standard Wiener processes and $|z| < 1$. Show that the first passage time T to the unit circle has expected value $\mathbb{E}(T) = \frac{1}{2}(1 - |z|^2)$.

7. Kendall's taxicabs, Problem (11.8.16) revisited. Let X and Y be independent Poisson processes with rate λ, and let $Q(t) = X(t) - Y(t)$. For positive integers m, n, find:

(a) the probability that Q hits n before $-m$,

(b) the expected time for Q to hit either $-m$ or n.

8. Quadratic variation. Let N be a Poisson process with rate λ. Show the following.

(a) The predictable quadratic variation of N over the interval $[0, t]$ equals λt, which is to say that

$$\sum_{r=0}^{n-1} \mathbb{E}\Big(\big[N((r+1)t/n) - N(rt/n)\big]^2 \,\big|\, \mathcal{F}_{rt/n} \Big) \to \lambda t \qquad \text{as } n \to \infty,$$

where $\mathcal{F}_u = \sigma(\{N(s) : s \leq u\})$.

(b) The optional quadratic variation of $N(s) - \lambda s$ over $[0, t]$ equals $N(t)$. (See Exercise (12.2.4).)

(c) $M(t) = (N(t) - \lambda t)^2 - N(t)$ defines a martingale.

12.9 Problems

1. Let Z_n be the size of the nth generation of a branching process with immigration in which the mean family size is μ ($\neq 1$) and the mean number of immigrants per generation is m. Show that

$$Y_n = \mu^{-n} \left\{ Z_n - m \frac{1 - \mu^n}{1 - \mu} \right\}$$

defines a martingale.

2. In an age-dependent branching process, each individual gives birth to a random number of offspring at random times. At time 0, there exists a single progenitor who has N children at the subsequent times $B_1 \leq B_2 \leq \cdots \leq B_N$; his family may be described by the vector $(N, B_1, B_2, \ldots, B_N)$. Each subsequent member x of the population has a family described similarly by a vector $(N(x), B_1(x), \ldots, B_{N(x)}(x))$ having the same distribution as $(N, B_1, \ldots, B_N)$ and independent of all other individuals' families. The number $N(x)$ is the number of his offspring, and $B_i(x)$ is the time between the births of the parent and the ith offspring. Let $\{B_{n,r} : r \geq 1\}$ be the times of births of individuals in the nth generation. Let $M_n(\theta) = \sum_r e^{-\theta B_{n,r}}$, and show that $Y_n = M_n(\theta)/\mathbb{E}(M_1(\theta))^n$ defines a martingale with respect to $\mathcal{F}_n = \sigma(\{B_{m,r} : m \leq n,\ r \geq 1\})$, for any value of θ such that $\mathbb{E}M_1(\theta) < \infty$.

3. Let $(Y, \mathcal{F})$ be a martingale with $\mathbb{E}Y_n = 0$ and $\mathbb{E}(Y_n^2) < \infty$ for all n. Show that

$$\mathbb{P}\left(\max_{1\le k\le n} Y_k > x\right) \le \frac{\mathbb{E}(Y_n^2)}{\mathbb{E}(Y_n^2) + x^2}, \qquad x > 0.$$

4. Let $(Y, \mathcal{F})$ be a non-negative submartingale with $Y_0 = 0$, and let $\{c_n\}$ be a non-increasing sequence of positive numbers. Show that

$$\mathbb{P}\left(\max_{1\le k\le n} c_k Y_k \ge x\right) \le \frac{1}{x}\sum_{k=1}^{n} c_k \mathbb{E}(Y_k - Y_{k-1}), \qquad x > 0.$$

Such an inequality is sometimes named after subsets of Hájek, Rényi, and Chow. Deduce Kolmogorov's inequality for the sum of independent random variables. [Hint: Work with the martingale $Z_n = c_n Y_n - \sum_{k=1}^n c_k \mathbb{E}(X_k \mid \mathcal{F}_{k-1}) + \sum_{k=1}^n (c_{k-1} - c_k) Y_{k-1}$ where $X_k = Y_k - Y_{k-1}$.]

5. Suppose that the sequence $\{X_n : n \ge 1\}$ of random variables satisfies $\mathbb{E}(X_n \mid X_1, X_2, \dots, X_{n-1}) = 0$ for all n, and also $\sum_{k=1}^\infty \mathbb{E}(|X_k|^r)/k^r < \infty$ for some $r \in [1, 2]$. Let $S_n = \sum_{i=1}^n Z_i$ where $Z_i = X_i/i$, and show that

$$\mathbb{P}\left(\max_{1\le k\le n} |S_{m+k} - S_m| \ge x\right) \le \frac{1}{x^r}\mathbb{E}\big(|S_{m+n} - S_m|^r\big), \qquad x > 0.$$

Deduce that S_n converges a.s. as $n \to \infty$, and hence that $n^{-1}\sum_1^n X_k \xrightarrow{\text{a.s.}} 0$. [Hint: In the case $1 < r \le 2$, prove and use the fact that $h(u) = |u|^r$ satisfies $h(v) - h(u) \le (v-u)h'(u) + 2h((v-u)/2)$. Kronecker's lemma is useful for the last part.]

6. Let $X_1, X_2, \dots$ be independent random variables with

$$X_n = \begin{cases} 1 & \text{with probability } (2n)^{-1}, \\ 0 & \text{with probability } 1 - n^{-1}, \\ -1 & \text{with probability } (2n)^{-1}. \end{cases}$$

Let $Y_1 = X_1$ and for $n \ge 2$

$$Y_n = \begin{cases} X_n & \text{if } Y_{n-1} = 0, \\ nY_{n-1}|X_n| & \text{if } Y_{n-1} \ne 0. \end{cases}$$

Show that Y_n is a martingale with respect to $\mathcal{F}_n = \sigma(Y_1, Y_2, \dots, Y_n)$. Show that Y_n does not converge almost surely. Does Y_n converge in any way? Why does the martingale convergence theorem not apply?

7. Let $X_1, X_2, \dots$ be independent identically distributed random variables and suppose that $M(t) = \mathbb{E}(e^{tX_1})$ satisfies $M(t) = 1$ for some $t > 0$. Show that $\mathbb{P}(S_k \ge x \text{ for some } k) \le e^{-tx}$ for $x > 0$ and such a value of t, where $S_k = X_1 + X_2 + \cdots + X_k$.

8. Let Z_n be the size of the nth generation of a branching process with family-size probability generating function $G(s)$, and assume $Z_0 = 1$. Let ξ be the smallest positive root of $G(s) = s$. Use the martingale convergence theorem to show that, if $0 < \xi < 1$, then $\mathbb{P}(Z_n \to 0) = \xi$ and $\mathbb{P}(Z_n \to \infty) = 1 - \xi$.

9. Let $(Y, \mathcal{F})$ be a non-negative martingale, and let $Y_n^* = \max\{Y_k : 0 \le k \le n\}$. Show that

$$\mathbb{E}(Y_n^*) \le \frac{e}{e-1}\Big\{1 + \mathbb{E}\big(Y_n(\log Y_n)^+\big)\Big\}.$$

[Hint: $a\log^+ b \le a\log^+ a + b/e$ if $a, b \ge 0$, where $\log^+ x = \max\{0, \log x\}$.]

10. Let $X = \{X(t) : t \geq 0\}$ be a birth–death process with parameters λ_i, μ_i, where $\lambda_i = 0$ if and only if $i = 0$. Define $h(0) = 0$, $h(1) = 1$, and

$$h(j) = 1 + \sum_{i=1}^{j-1} \frac{\mu_1\mu_2 \cdots \mu_i}{\lambda_1\lambda_2 \cdots \lambda_i}, \qquad j \geq 2.$$

Show that $h(X(t))$ constitutes a martingale with respect to the filtration $\mathcal{F}_t = \sigma(\{X(u) : 0 \leq u \leq t\})$, whenever $\mathbb{E}h(X(t)) < \infty$ for all t. (You may assume that the forward equations are satisfied.)

Fix n, and let $m < n$; let $\pi(m)$ be the probability that the process is absorbed at 0 before it reaches size n, having started at size m. Show that $\pi(m) = 1 - \{h(m)/h(n)\}$.

11. Let $(Y, \mathcal{F})$ be a submartingale such that $\mathbb{E}(Y_n^+) \leq M$ for some M and all n.

(a) Show that $M_n = \lim_{m\to\infty} \mathbb{E}(Y_{n+m}^+ \mid \mathcal{F}_n)$ exists (almost surely) and defines a martingale with respect to $\mathcal{F}$.

(b) Show that Y_n may be expressed in the form $Y_n = X_n - Z_n$ where $(X, \mathcal{F})$ is a non-negative martingale, and $(Z, \mathcal{F})$ is a non-negative supermartingale. This representation of Y is sometimes termed the 'Krickeberg decomposition'.

(c) Let $(Y, \mathcal{F})$ be a martingale such that $\mathbb{E}|Y_n| \leq M$ for some M and all n. Show that Y may be expressed as the difference of two non-negative martingales.

12. Let $£Y_n$ be the assets of an insurance company after n years of trading. During each year it receives a total (fixed) income of $£P$ in premiums. During the nth year it pays out a total of $£C_n$ in claims. Thus $Y_{n+1} = Y_n + P - C_{n+1}$. Suppose that $C_1, C_2, \ldots$ are independent $N(\mu, \sigma^2)$ variables and show that the probability of ultimate bankruptcy satisfies

$$\mathbb{P}\big(Y_n \leq 0 \text{ for some } n\big) \leq \exp\left\{-\frac{2(P-\mu)Y_0}{\sigma^2}\right\}.$$

13. Pólya's urn. A bag contains red and blue balls, with initially r red and b blue where $rb > 0$. A ball is drawn from the bag, its colour noted, and then it is returned to the bag together with a new ball of the same colour. Let R_n be the number of red balls after n such operations.

(a) Show that $Y_n = R_n/(n + r + b)$ is a martingale which converges almost surely and in mean.

(b) Let T be the number of balls drawn until the first blue ball appears, and suppose that $r = b = 1$. Show that $\mathbb{E}\{(T+2)^{-1}\} = \frac{1}{4}$.

(c) Suppose $r = b = 1$, and show that $\mathbb{P}(Y_n \geq \frac{3}{4} \text{ for some } n) \leq \frac{2}{3}$.

14. Here is a modification of the last problem. Let $\{A_n : n \geq 1\}$ be a sequence of random variables, each being a non-negative integer. We are provided with the bag of Problem (12.9.13), and we add balls according to the following rules. At each stage a ball is drawn from the bag, and its colour noted; we assume that the distribution of this colour depends only on the current contents of the bag and not on any further information concerning the A_n. We return this ball together with A_n new balls of the same colour. Write R_n and B_n for the numbers of red and blue balls in the urn after n operations, and let $\mathcal{F}_n = \sigma(\{R_k, B_k : 0 \leq k \leq n\})$. Show that $Y_n = R_n/(R_n + B_n)$ defines a martingale. Suppose $R_0 = B_0 = 1$, let T be the number of balls drawn until the first blue ball appears, and show that

$$\mathbb{E}\left(\frac{1 + A_T}{2 + \sum_{i=1}^{T} A_i}\right) = \frac{1}{2},$$

so long as $\sum_n \big(2 + \sum_{i=1}^{n} A_i\big)^{-1} = \infty$ a.s.

15. Labouchere system. Here is a gambling system for playing a fair game. Choose a sequence $x_1, x_2, \dots, x_n$ of positive numbers.

Wager the sum of the first and last numbers on an evens bet. If you win, delete those two numbers; if you lose, append their sum as an extra term x_{n+1} $(= x_1 + x_n)$ at the right-hand end of the sequence.

You play iteratively according to the above rule. If the sequence ever contains one term only, you wager that amount on an evens bet. If you win, you delete the term, and if you lose you append it to the sequence to obtain two terms.

Show that, with probability 1, the game terminates with a profit of $\sum_1^n x_i$, and that the time until termination has finite mean.

This looks like another clever strategy. Show that the mean size of your maximum deficit is infinite. (When Henry Labouchere was sent down from Trinity College, Cambridge, in 1852, his gambling debts exceeded £6000.)

16. Here is a martingale approach to the question of determining the mean number of tosses of a coin before the first appearance of the sequence HHH. A large casino contains infinitely many gamblers $G_1, G_2, \dots$, each with an initial fortune of \$1. A croupier tosses a coin repeatedly. For each n, gambler G_n bets as follows. Just before the nth toss he stakes his \$1 on the event that the nth toss shows heads. The game is assumed fair, so that he receives a total of $\$p^{-1}$ if he wins, where p is the probability of heads. If he wins this gamble, then he *repeatedly* stakes his entire current fortune on heads, at the same odds as his first gamble. At the first subsequent tail he loses his fortune and leaves the casino, penniless. Let S_n be the casino's profit (losses count negative) after the nth toss. Show that S_n is a martingale. Let N be the number of tosses before the first appearance of HHH; show that N is a stopping time and hence find $\mathbb{E}(N)$.

Now adapt this scheme to calculate the mean time to the first appearance of the sequence HTH.

17. Let $\{(X_k, Y_k) : k \geq 1\}$ be a sequence of independent identically distributed random vectors such that each X_k and Y_k takes values in the set $\{-1, 0, 1, 2, \dots\}$. Suppose that $\mathbb{E}(X_1) = \mathbb{E}(Y_1) = 0$ and $\mathbb{E}(X_1 Y_1) = c$, and furthermore X_1 and Y_1 have finite non-zero variances. Let U_0 and V_0 be positive integers, and define $(U_{n+1}, V_{n+1}) = (U_n, V_n) + (X_{n+1}, Y_{n+1})$ for each $n \geq 0$. Let $T = \min\{n : U_n V_n = 0\}$ be the first hitting time by the random walk (U_n, V_n) of the axes of $\mathbb{R}^2$. Show that $\mathbb{E}(T) < \infty$ if and only if $c < 0$, and that $\mathbb{E}(T) = -\mathbb{E}(U_0 V_0)/c$ in this case. [Hint: You might show that $U_n V_n - cn$ is a martingale.]

18. The game 'Red Now' may be played by a single player with a well shuffled conventional pack of 52 playing cards. At times $n = 1, 2, \dots, 52$ the player turns over a new card and observes its colour. Just once in the game he must say, just before exposing a card, "Red Now". He wins the game if the next exposed card is red. Let R_n be the number of red cards remaining face down after the nth card has been turned over. Show that $X_n = R_n/(52 - n)$, $0 \leq n < 52$, defines a martingale. Show that there is no strategy for the player which results in a probability of winning different from $\frac{1}{2}$.

19. A businessman has a redundant piece of equipment which he advertises for sale, inviting "offers over £1000". He anticipates that, each week for the foreseeable future, he will be approached by one prospective purchaser, the offers made in week $0, 1, \dots$ being £$1000X_0$, £$1000X_1, \dots$, where $X_0, X_1, \dots$ are independent random variables with a common density function f and finite mean. Storage of the equipment costs £$1000c$ per week and the prevailing rate of interest is α (> 0) per week. Explain why a sensible strategy for the businessman is to sell in the week T, where T is a stopping time chosen so as to maximize

$$\mu(T) = \mathbb{E}\left\{(1+\alpha)^{-T} X_T - \sum_{n=1}^{T} (1+\alpha)^{-n} c\right\}.$$

Show that this problem is equivalent to maximizing $\mathbb{E}\{(1+\alpha)^{-T} Z_T\}$ where $Z_n = X_n + c/\alpha$.

Show that there exists a unique positive real number γ with the property that

$$\alpha\gamma = \int_{\gamma}^{\infty} \mathbb{P}(Z_n > y)\,dy,$$

and that, for this value of γ, the sequence $V_n = (1+\alpha)^{-n}\max\{Z_n, \gamma\}$ constitutes a supermartingale. Deduce that the optimal strategy for the businessman is to set a target price τ (which you should specify in terms of γ) and sell the first time he is offered at least this price.

In the case when $f(x) = 2x^{-3}$ for $x \geq 1$, and $c = \alpha = \frac{1}{90}$, find his target price and the expected number of weeks he will have to wait before selling.

20. Let Z be a branching process satisfying $Z_0 = 1$, $\mathbb{E}(Z_1) < 1$, and $\mathbb{P}(Z_1 \geq 2) > 0$. Show that $\mathbb{E}(\sup_n Z_n) \leq \eta/(\eta - 1)$, where η is the largest root of the equation $x = G(x)$ and G is the probability generating function of Z_1.

21. Matching. In a cloakroom there are K coats belonging to K people who make an attempt to leave by picking a coat at random. Those who pick their own coat leave, the rest return the coats and try again at random. Let N be the number of rounds of attempts until everyone has left. Show that $\mathbb{E}N = K$ and $\mathrm{var}(N) \leq K$.

22. Let W be a standard Wiener process, and define

$$M(t) = \int_0^t W(u)\,du - \tfrac{1}{3}W(t)^3.$$

Show that $M(t)$ is a martingale, and deduce that the expected area under the path of W until it first reaches one of the levels a (> 0) or b (< 0) is $-\frac{1}{3}ab(a+b)$.

23. Let $W = (W_1, W_2, \dots, W_d)$ be a d-dimensional Wiener process, the W_i being independent one-dimensional Wiener processes with $W_i(0) = 0$ and variance parameter $\sigma^2 = d^{-1}$. Let $R(t)^2 = W_1(t)^2 + W_2(t)^2 + \cdots + W_d(t)^2$, and show that $R(t)^2 - t$ is a martingale. Deduce that the mean time to hit the sphere of $\mathbb{R}^d$ with radius a is a^2.

24. Let W be a standard one-dimensional Wiener process, and let $a, b > 0$. Let T be the earliest time at which W visits either of the two points $-a$, b. Show that $\mathbb{P}(W(T) = b) = a/(a+b)$ and $\mathbb{E}(T) = ab$. In the case $a = b$, find $\mathbb{E}(e^{-sT})$ for $s > 0$.

25. Let (a_n) be a real sequence satisfying $a_n \in (0, 1)$, and let $\{U_n : n \geq 1\}$ be independent random variables with the uniform distribution on $(0, 1)$. Define

$$X_{n+1} = \begin{cases} (1 - a_n)X_n + a_n & \text{if } X_n > U_{n+1}, \\ (1 - a_n)X_n & \text{otherwise}, \end{cases}$$

where $X_0 = \rho \in (0, 1)$.

(a) Show that the sequence $X = \{X_n : n \geq 0\}$ is a martingale with respect to the filtration $\mathcal{F}_n = \sigma(X_0, X_1, \dots, X_n)$, and that X_n converges a.s. and in mean square to some X_∞.

(b) Show that the infinite sum $\sum_{n=1}^{\infty} \mathbb{E}\{(X_{n+1} - X_n)^2 \mid \mathcal{F}_n\}$ converges a.s. and in mean to some random variable A with $\mathbb{E}(A) = \mathbb{E}(X_\infty^2) - \rho^2$.

(c) Hence prove that $S = \sum_{n=0}^{\infty} a_n^2 X_n(1 - X_n)$ is a.s. finite.

(d) Deduce that, if $\sum_n a_n^2 = \infty$, then X_∞ takes only the values 0 and 1. In this case, what is $\mathbb{P}(X_\infty = 1)$?

26. Exponential inequality for Wiener process. Let W be a standard Wiener process, and show that

$$\mathbb{P}\left(\sup_{0\leq t\leq T} W(t) \geq x\right) \leq \exp\{-\tfrac{1}{2}x^2/T\}, \qquad x > 0.$$

You may assume a version of Doob's maximal inequality (12.6.1) for continuous parameter submartingales.

27. Insurance, Problem (8.10.7) revisited. An insurance company receives premiums (net of costs) at rate ρ per unit time. Claims $X_1, X_2, \dots$ are independent random variables with the exponential distribution with parameter μ, and they arrive at the times of a Poisson process of rate λ (assume the usual independence, as well as $\lambda, \mu, \rho > 0$). Let $Y(t)$ be the assets of the company at time t where $Y(0) = y > 0$. Show that

$$\mathbb{P}\big(Y(t) \leq 0 \text{ for some } t > 0\big) = \begin{cases} \left(1 - \dfrac{\theta}{\mu}\right) e^{-\theta y} & \text{if } \theta > 0, \\ 1 & \text{otherwise,} \end{cases}$$

where $\theta = \mu - (\lambda/\rho)$.

13

Diffusion processes

13.2 Exercise. Brownian motion

1. Total variation. Show that the total variation of a standard Wiener process over any non-trivial interval does not exist.
[The total variation $V_{a,b}(f)$ of a function $f : \mathbb{R} \to \mathbb{R}$ over the bounded interval $[a, b]$ is defined as

$$V_{a,b}(f) = \sup \sum |f(s_{r+1}) - f(s_r)|,$$

where the supremum is taken over all increasing sequences $s_0 = a, s_1, s_2, \ldots, s_n = b$. You may find it useful that the quadratic variation of the Wiener process exists and is non-zero almost surely.]

13.3 Exercises. Diffusion processes

1. Let $X = \{X(t) : t \geq 0\}$ be a simple birth–death process with parameters $\lambda_n = n\lambda$ and $\mu_n = n\mu$. Suggest a diffusion approximation to X.

2. Bartlett's equation. Let D be a diffusion with instantaneous mean and variance $a(t, x)$ and $b(t, x)$, and let $M(t, \theta) = \mathbb{E}(e^{\theta D(t)})$, the moment generating function of $D(t)$. Use the forward diffusion equation to derive *Bartlett's equation*:

$$\frac{\partial M}{\partial t} = \theta a\left(t, \frac{\partial}{\partial \theta}\right) M + \frac{1}{2}\theta^2 b\left(t, \frac{\partial}{\partial \theta}\right) M$$

where we interpret

$$g\left(t, \frac{\partial}{\partial \theta}\right) M = \sum_n \gamma_n(t) \frac{\partial^n M}{\partial \theta^n}$$

if $g(t, x) = \sum_{n=0}^{\infty} \gamma_n(t) x^n$.

3. Write down Bartlett's equation in the case of the Wiener process D having drift m and instantaneous variance 1, and solve it subject to the boundary condition $D(0) = 0$.

4. Write down Bartlett's equation in the case of an Ornstein–Uhlenbeck process D having instantaneous mean $a(t, x) = -x$ and variance $b(t, x) = 1$, and solve it subject to the boundary condition $D(0) = 0$.

5. Bessel process. Let $W_1(t)$, $W_2(t)$, $W_3(t)$ be independent Wiener processes. The positive $R = R(t)$ such that $R^2 = W_1^2 + W_2^2 + W_3^2$ is the three-dimensional *Bessel process*. Show that R is a Markov process. Is this result true in a general number n of dimensions?

6. Show that the transition density for the Bessel process defined in Exercise (13.3.5) is

$$\begin{aligned} f(t, y \mid s, x) &= \frac{\partial}{\partial y}\mathbb{P}\big(R(t) \le y \mid R(s) = x\big) \\ &= \frac{y/x}{\sqrt{2\pi(t-s)}}\left\{\exp\left(-\frac{(y-x)^2}{2(t-s)}\right) - \exp\left(-\frac{(y+x)^2}{2(t-s)}\right)\right\}. \end{aligned}$$

7. If W is a Wiener process and the function $g : \mathbb{R} \to \mathbb{R}$ is continuous and strictly monotone, show that $g(W)$ is a continuous Markov process.

8. Let W be a Wiener process. Which of the following define martingales?
(a) $e^{\sigma W(t)}$, (b) $cW(t/c^2)$, (c) $tW(t) - \int_0^t W(s)\,ds$.

9. Exponential martingale, geometric Brownian motion. Let W be a standard Wiener process and define $S(t) = e^{at+bW(t)}$. Show that:

(a) S is a Markov process,

(b) S is a martingale (with respect to the filtration generated by W) if and only if $a + \frac{1}{2}b^2 = 0$, and in this case $\mathbb{E}(S(t)) = 1$.

10. Find the transition density for the Markov process of Exercise (13.3.9a).

11. Verify that the transition density (13.3.4) of the standard Wiener process satisfies the Chapman–Kolmogorov equations in continuous time.

13.4 Exercises. First passage times

1. Let W be a standard Wiener process and let $X(t) = \exp\{i\theta W(t) + \frac{1}{2}\theta^2 t\}$ where $i = \sqrt{-1}$. Show that X is a martingale with respect to the filtration given by $\mathcal{F}_t = \sigma(\{W(u) : u \le t\})$.

2. Let T be the (random) time at which a standard Wiener process W hits the 'barrier' in space–time given by $y = at + b$ where $a \le 0$ and $b > 0$; that is, $T = \inf\{t : W(t) = at + b\}$. Use the result of Exercise (13.4.1) to show that the density function of T has Laplace transform given by

$$\mathbb{E}(e^{-\psi T}) = \exp\{|a|b - b\sqrt{a^2 + 2\psi}\}, \qquad \psi \ge 0.$$

You may assume that the conditions of the optional stopping theorem are satisfied.

3. Let W be a standard Wiener process, and let T be the time of the last zero of W prior to time t. Show that $\mathbb{P}(T \le u) = (2/\pi)\sin^{-1}\sqrt{u/t}$ for $0 \le u \le t$.

4. Let X and Y be independent standard Wiener processes, so that (X, Y) is a planar Wiener process. Show that the value of Y at the time of the first passage of X to the level $x > 0$ has the Cauchy distribution with density $f(y) = x/[\pi(x^2 + y^2)]$.

5. Inverse Gaussian density. Let $D(t) = W(t) - at$ where W is a standard Wiener process and $a \le 0$ is a *drift* parameter. Use the result of Exercise (13.4.2) to show that the density function of the first passage time by D to the point $b > 0$ is

$$f(x) = \frac{b}{\sqrt{2\pi x^3}}\exp\left\{-\frac{(ax+b)^2}{2x}\right\}, \qquad x > 0.$$

6. Show that the standard Wiener process $W(t)$ and its maximum process $M(t) = \sup\{W(s) : 0 \le s \le t\}$ satisfy

$$\mathbb{P}\big(M(t) \ge m,\ W(t) \le w\big) = 1 - \Phi\big((2m - w)/\sqrt{t}\big), \qquad m \ge 0, \quad 0 \le w \le m,$$

where Φ is the $N(0, 1)$ distribution function.

7. Find the density function of $\sqrt{T(x)}$ where $T(x)$ is the first passage time of a standard Wiener process to the point x. Show that it is the same as that of $1/|Z|$ for a suitable normally distributed random variable Z.

8. Continuation. Show that $T(x)$ has the same distribution as $T(1)x^2$, and deduce that the process $\{T(x) : x \geq 0\}$ is self-similar. Hence or otherwise, show that $T(x)$ has a stable distribution with exponent $\frac{1}{2}$. [See Feller 1971, p. 174.]

9. Feller's diffusion approximation to the branching process, Example (13.3.12) revisited. Let $X = \{X(t) : t \geq 0\}$ be the diffusion model of Example (13.3.12) with $X(0) = x > 0$, and with instantaneous mean and variance ax and bx where $a < 0$ and $b > 0$. Let $Y = \int_0^T X(u)\,du$ be the integrated size of the accumulated population up to the extinction time $T = \inf\{t \geq 0 : X(t) = 0\}$. Show that the moment generating function $m(x, \psi) = \mathbb{E}_x(e^{-\psi Y})$ satisfies

$$\frac{1}{2}b\frac{d^2m}{dx^2} + a\frac{dm}{dx} - \psi m = 0, \qquad \psi \geq 0.$$

Deduce that Y has a 'first-passage' distribution of the type of Exercise (13.4.2), and specify which.

13.5 Exercises. Barriers

1. Let D be a standard Wiener process with drift m starting from $D(0) = d > 0$, and suppose that there is a reflecting barrier at the origin. Show that the density function $f^{\mathrm{r}}(t, y)$ of $D(t)$ satisfies $f^{\mathrm{r}}(t, y) \to 0$ as $t \to \infty$ if $m \geq 0$, whereas $f^{\mathrm{r}}(t, y) \to 2|m|e^{-2|m|y}$ for $y > 0$, as $t \to \infty$ if $m < 0$.

2. Conditioned Wiener process. Let W be a standard Wiener process started at $W(0) = x \in (0, a)$, with absorbing barriers at 0 and a. Let C be the process W conditioned on the event that W is absorbed at a. Show that the instantaneous mean and variance of C, when $C(t) = c \in (0, a)$, are $1/c$ and 1, respectively.

3. Continuation. Let T be the first passage time by C to the point a. Show that $\mathbb{E}_x(T) = \frac{1}{3}(a^2 - x^2)$ for $x \in (0, a)$.

13.6 Exercises. Excursions and the Brownian bridge

1. Brownian meander. Let W be a standard Wiener process. Show that the conditional density function of $W(t)$, given that $W(u) > 0$ for $0 < u < t$, is $g(x) = (x/t)e^{-x^2/(2t)}$, $x > 0$.

2. Show that the autocovariance function of the Brownian bridge is $c(s, t) = \min\{s, t\} - st$, $0 \leq s, t \leq 1$.

3. Let W be a standard Wiener process, and let $\widehat{W}(t) = W(t) - tW(1)$. Show that $\{\widehat{W}(t) : 0 \leq t \leq 1\}$ is a Brownian bridge.

4. If W is a Wiener process with $W(0) = 0$, show that $\widetilde{W}(t) = (1 - t)W(t/(1 - t))$ for $0 \leq t < 1$, $\widetilde{W}(1) = 0$, defines a Brownian bridge.

Deduce, with help from Corollary (13.4.14), that the probability of the Brownian bridge ever rising above the height $m > 0$ is e^{-2m^2}.

5. Let $0 < s < t < 1$. Show that the probability that the Brownian bridge has no zeros in the interval (s, t) is $(2/\pi)\cos^{-1}\sqrt{(t - s)/[t(1 - s)]}$.

13.7 Exercises. Stochastic calculus

1. Doob's L_2 inequality. Let W be a standard Wiener process, and show that

$$\mathbb{E}\Big(\max_{0\le s\le t} |W_s|^2\Big) \le 4\mathbb{E}(W_t^2).$$

2. Let W be a standard Wiener process. Fix $t > 0$, $n \ge 1$, and let $\delta = t/n$. Show that $Z_n = \sum_{j=0}^{n-1}(W_{(j+1)\delta} - W_{j\delta})^2$ satisfies $Z_n \to t$ in mean square as $n \to \infty$.

3. Let W be a standard Wiener process. Fix $t > 0$, $n \ge 1$, and let $\delta = t/n$. Let $V_j = W_{j\delta}$ and $\Delta_j = V_{j+1} - V_j$. Evaluate the limits of the following as $n \to \infty$:
(a) $I_1(n) = \sum_j V_j\Delta_j$,
(b) $I_2(n) = \sum_j V_{j+1}\Delta_j$,
(c) $I_3(n) = \sum_j \frac{1}{2}(V_{j+1} + V_j)\Delta_j$,
(d) $I_4(n) = \sum_j W_{(j+\frac{1}{2})\delta}\Delta_j$.

4. Let W be a standard Wiener process. Show that $U(t) = e^{-\beta t}W(e^{2\beta t})$ defines a stationary Ornstein–Uhlenbeck process.

5. Let W be a standard Wiener process. Show that $U_t = W_t - \beta\int_0^t e^{-\beta(t-s)}W_s\,ds$ defines an Ornstein–Uhlenbeck process.

13.8 Exercises. The Itô integral

In the absence of any contrary indication, W denotes a standard Wiener process, and $\mathcal{F}_t$ is the smallest σ-field containing all null events with respect to which every member of $\{W_u : 0 \le u \le t\}$ is measurable.

1. (a) Verify directly that $\displaystyle\int_0^t s\,dW_s = tW_t - \int_0^t W_s\,ds$.

(b) Verify directly that $\displaystyle\int_0^t W_s^2\,dW_s = \tfrac{1}{3}W_t^3 - \int_0^t W_s\,ds$.

(c) Show that $\displaystyle\mathbb{E}\bigg(\Big[\int_0^t W_s\,dW_s\Big]^2\bigg) = \int_0^t \mathbb{E}(W_s^2)\,ds$.

2. Let $X_t = \int_0^t W_s\,ds$. Show that X is a Gaussian process, and find its autocovariance and autocorrelation function.

3. Let $(\Omega, \mathcal{F}, \mathbb{P})$ be a probability space, and suppose that $X_n \xrightarrow{\text{m.s.}} X$ as $n \to \infty$. If $\mathcal{G} \subseteq \mathcal{F}$, show that $\mathbb{E}(X_n \mid \mathcal{G}) \xrightarrow{\text{m.s.}} \mathbb{E}(X \mid \mathcal{G})$.

4. Let ψ_1 and ψ_2 be predictable step functions, and show that

$$\mathbb{E}\{I(\psi_1)I(\psi_2)\} = \mathbb{E}\Big(\int_0^\infty \psi_1(t)\psi_2(t)\,dt\Big),$$

whenever both sides exist.

5. Assuming that *Gaussian white noise* $G_t = dW_t/dt$ exists in sufficiently many senses to appear as an integrand, show by integrating the stochastic differential equation $dX_t = -\beta X_t\,dt + dW_t$ that

$$X_t = W_t - \beta \int_0^t e^{-\beta(t-s)} W_s\,ds,$$

if $X_0 = 0$.

6. Let ψ be an adapted process with $\|\psi\| < \infty$. Show that $\|I(\psi)\|_2 = \|\psi\|$.

13.9 Exercises. Itô's formula

In the absence of any contrary indication, W denotes a standard Wiener process, and $\mathcal{F}_t$ is the smallest σ-field containing all null events with respect to which every member of $\{W_u : 0 \le u \le t\}$ is measurable.

1. Let X and Y be independent standard Wiener processes. Show that, with $R_t^2 = X_t^2 + Y_t^2$,

$$Z_t = \int_0^t \frac{X_s}{R_s}\,dX_s + \int_0^t \frac{Y_s}{R_s}\,dY_s$$

is a Wiener process. [Hint: Use Theorem (13.8.11) and Example (12.7.10).] Hence show that the squared process R^2, called the *squared Bessel process*, satisfies

$$R_t^2 = 2\int_0^t R_s\,dW_s + 2t.$$

Generalize this conclusion to n dimensions.

2. Write down the SDE obtained via Itô's formula for the process $Y_t = W_t^4$, and deduce that $\mathbb{E}(W_t^4) = 3t^2$.

3. Show that $Y_t = tW_t$ is an Itô process, and write down the corresponding SDE.

4. Wiener process on a circle. Let $Y_t = e^{iW_t}$. Show that $Y = X_1 + iX_2$ is a process on the unit circle satisfying

$$dX_1 = -\tfrac{1}{2}X_1\,dt - X_2\,dW, \quad dX_2 = -\tfrac{1}{2}X_2\,dt + X_1\,dW.$$

5. Find the SDEs satisfied by the processes:

(a) $X_t = W_t/(1+t)$,

(b) $X_t = \sin W_t$,

(c) [Wiener process on an ellipse] $X_t = a\cos W_t$, $Y_t = b\sin W_t$, where $ab \neq 0$.

6. Find a random function $X(t)$ satisfying

$$dX(t) = \left(\tfrac{1}{2}X + \sqrt{1+X^2}\right)dt + \sqrt{1+X^2}\,dW(t),$$

with $X(0) = 0$.

13.10 Exercises. Option pricing

In the absence of any contrary indication, W denotes a standard Wiener process, and $\mathcal{F}_t$ is the smallest σ-field containing all null events with respect to which every member of $\{W_u : 0 \le u \le t\}$ is measurable. The process $S_t = \exp((\mu - \frac{1}{2}\sigma^2)t + \sigma W_t)$ is a geometric Brownian motion, and $r \ge 0$ is the interest rate.

1. (a) Let Z have the $N(\gamma, \tau^2)$ distribution. Show that

$$\mathbb{E}\big((ae^Z - K)^+\big) = ae^{\gamma+\frac{1}{2}\tau^2}\Phi\left(\frac{\log(a/K)+\gamma}{\tau} + \tau\right) - K\Phi\left(\frac{\log(a/K)+\gamma}{\tau}\right)$$

where Φ is the $N(0, 1)$ distribution function.

(b) Let $\mathbb{Q}$ be a probability measure under which σW is a Wiener process with drift $r - \mu$ and instantaneous variance σ^2. Show for $0 \le t \le T$ that

$$\mathbb{E}_{\mathbb{Q}}\big((S_T - K)^+ \bigm| \mathcal{F}_t\big) = S_t e^{r(T-t)}\Phi(d_1(t, S_t)) - K\Phi(d_2(t, S_t))$$

where

$$d_1(t, x) = \frac{\log(x/K) + (r + \frac{1}{2}\sigma^2)(T-t)}{\sigma\sqrt{T-t}}, \quad d_2(t, x) = d_1(t, x) - \sigma\sqrt{T-t}.$$

2. Consider a portfolio which, at time t, holds $\xi(t, S)$ units of stock and $\psi(t, S)$ units of bond, and assume these quantities depend only on the values of S_u for $0 \le u \le t$. Find the function ψ such that the portfolio is self-financing in the three cases:

(a) $\xi(t, S) = 1$ for all t, S,

(b) $\xi(t, S) = S_t$,

(c) $\xi(t, S) = \int_0^t S_v\, dv$.

3. Suppose the stock price S_t is itself a Wiener process and the interest rate r equals 0, so that a unit of bond has unit value for all time. In the notation of Exercise (13.10.2), which of the following define self-financing portfolios?

(a) $\xi(t, S) = \psi(t, S) = 1$ for all t, S,

(b) $\xi(t, S) = 2S_t,\ \psi(t, S) = -S_t^2 - t$,

(c) $\xi(t, S) = -t,\ \psi(t, S) = \int_0^t S_s\, ds$,

(d) $\xi(t, S) = \int_0^t S_s\, ds,\ \psi(t, S) = -\int_0^t S_s^2\, ds$.

4. An 'American call option' differs from a European call option in that it may be exercised by the buyer *at any time up to the expiry date*. Show that the value of the American call option is the same as that of the corresponding European call option, and that there is no advantage to the holder of such an option to exercise it strictly before its expiry date.

5. Show that the Black–Scholes value at time 0 of the European call option is an increasing function of the initial stock price, the exercise date, the interest rate, and the volatility, and is a decreasing function of the strike price.

13.11 Exercises. Passage probabilities and potentials

1. Let G be the closed sphere with radius ϵ and centre at the origin of $\mathbb{R}^d$ where $d \geq 3$. Let $\mathbf{W}$ be a d-dimensional Wiener process starting from $\mathbf{W}(0) = \mathbf{w} \notin G$. Show that the probability that $\mathbf{W}$ visits G is $(\epsilon/r)^{d-2}$, where $r = |\mathbf{w}|$.

2. Let G be an infinite connected graph with finite vertex degrees. Let Δ_n be the set of vertices x which are distance n from 0 (that is, the shortest path from x to 0 contains n edges), and let N_n be the total number of edges joining pairs x, y of vertices with $x \in \Delta_n$, $y \in \Delta_{n+1}$. Show that a random walk on G is recurrent if $\sum_i N_i^{-1} = \infty$.

3. Let G be a connected graph with finite vertex degrees, and let H be a connected subgraph of G. Show that a random walk on H is recurrent if a random walk on G is recurrent, but that the converse is not generally true.

13.12 Problems

1. Let W be a standard Wiener process, that is, a process with independent increments and continuous sample paths such that $W(s+t) - W(s)$ is $N(0, t)$ for $t > 0$. Let α be a positive constant. Show that:

(a) $\alpha W(t/\alpha^2)$ is a standard Wiener process,

(b) $W(t + \alpha) - W(\alpha)$ is a standard Wiener process,

(c) the process V, given by $V(t) = tW(1/t)$ for $t > 0$, $V(0) = 0$, is a standard Wiener process.

2. Let $X = \{X(t) : t \geq 0\}$ be a Gaussian process with continuous sample paths, zero means, and autocovariance function $c(s, t) = u(s)v(t)$ for $s \leq t$ where u and v are continuous functions. Suppose that the ratio $r(t) = u(t)/v(t)$ is continuous and strictly increasing with inverse function r^{-1}. Show that $W(t) = X(r^{-1}(t))/v(r^{-1}(t))$ is a standard Wiener process on a suitable interval of time.

If $c(s, t) = s(1 - t)$ for $s \leq t < 1$, express X in terms of W.

3. Let $\beta > 0$, and show that $U(t) = e^{-\beta t} W(e^{2\beta t} - 1)$ is an Ornstein–Uhlenbeck process if W is a standard Wiener process.

4. Let $V = \{V(t) : t \geq 0\}$ be an Ornstein–Uhlenbeck process with instantaneous mean $a(t, x) = -\beta x$ where $\beta > 0$, with instantaneous variance $b(t, x) = \sigma^2$, and with $U(0) = u$. Show that $V(t)$ is $N(ue^{-\beta t}, \sigma^2(1 - e^{-2\beta t})/(2\beta))$. Deduce that $V(t)$ is asymptotically $N(0, \frac{1}{2}\sigma^2/\beta)$ as $t \to \infty$, and show that V is strongly stationary if $V(0)$ is $N(0, \frac{1}{2}\sigma^2/\beta)$.

Show that such a process is the *only* stationary Gaussian Markov process with continuous autocovariance function, and find its spectral density function.

5. **Feller's diffusion approximation to the branching process.** Let $D = \{D(t) : t \geq 0\}$ be a diffusion process with instantaneous mean $a(t, x) = \alpha x$ and instantaneous variance $b(t, x) = \beta x$ where α and β are positive constants. Let $D(0) = d$. Show that the moment generating function of $D(t)$ is

$$M(t, \theta) = \exp\left\{\frac{2\alpha d\theta e^{\alpha t}}{\beta\theta(1 - e^{\alpha t}) + 2\alpha}\right\}.$$

Find the mean and variance of $D(t)$, and show that $\mathbb{P}(D(t) = 0) \to e^{-2d\alpha/\beta}$ as $t \to \infty$.

6. Let D be an Ornstein–Uhlenbeck process with $D(0) = 0$, and place reflecting barriers at $-c$ and d where $c, d > 0$. Find the limiting distribution of D as $t \to \infty$.

7. Let $X_0, X_1, \dots$ be independent $N(0, 1)$ variables, and show that

$$W(t) = \frac{t}{\sqrt{\pi}} X_0 + \sqrt{\frac{2}{\pi}} \sum_{k=1}^{\infty} \frac{\sin(kt)}{k} X_k$$

defines a standard Wiener process on $[0, \pi]$.

8. Let W be a standard Wiener process with $W(0) = 0$. Place absorbing barriers at $-b$ and b, where $b > 0$, and let W^{a} be W absorbed at these barriers. Show that $W^{\mathrm{a}}(t)$ has density function

$$f^{\mathrm{a}}(y, t) = \frac{1}{\sqrt{2\pi t}} \sum_{k=-\infty}^{\infty} (-1)^k \exp\left\{-\frac{(y - 2kb)^2}{2t}\right\}, \qquad -b < y < b,$$

which may also be expressed as

$$f^{\mathrm{a}}(y, t) = \sum_{n=1}^{\infty} a_n e^{-\lambda_n t} \sin\left(\frac{n\pi(y + b)}{2b}\right), \qquad -b < y < b,$$

where $a_n = b^{-1} \sin(\frac{1}{2}n\pi)$ and $\lambda_n = n^2\pi^2/(8b^2)$.

Hence calculate $\mathbb{P}(\sup_{0 \le s \le t} |W(s)| > b)$ for the unrestricted process W.

9. Let D be a Wiener process with drift m, and suppose that $D(0) = 0$. Place absorbing barriers at the points $x = -a$ and $x = b$ where a and b are positive real numbers. Show that the probability p_a that the process is absorbed at $-a$ is given by

$$p_a = \frac{e^{2mb} - 1}{e^{2m(a+b)} - 1}.$$

10. Let W be a standard Wiener process and let $F(u, v)$ be the event that W has no zero in the interval (u, v).

(a) If $ab > 0$, show that $\mathbb{P}\big(F(0, t) \mid W(0) = a,\ W(t) = b\big) = 1 - e^{-2ab/t}$.

(b) If $W(0) = 0$ and $0 < t_0 \le t_1 \le t_2$, show that

$$\mathbb{P}\big(F(t_0, t_2) \mid F(t_0, t_1)\big) = \frac{\sin^{-1}\sqrt{t_0/t_2}}{\sin^{-1}\sqrt{t_0/t_1}}.$$

(c) Deduce that, if $W(0) = 0$ and $0 < t_1 \le t_2$, then $\mathbb{P}(F(0, t_2) \mid F(0, t_1)) = \sqrt{t_1/t_2}$.

11. Let W be a standard Wiener process. Show that

$$\mathbb{P}\left(\sup_{0 \le s \le t} |W(s)| \ge w\right) \le 2\mathbb{P}(|W(t)| \ge w) \le \frac{2t}{w^2} \qquad \text{for } w > 0.$$

Set $t = 2^n$ and $w = 2^{2n/3}$ and use the Borel–Cantelli lemma to show that $t^{-1}W(t) \to 0$ a.s. as $t \to \infty$.

12. Let $\mathbf{W}$ be a two-dimensional Wiener process with $\mathbf{W}(0) = \mathbf{w}$, and let F be the unit circle. What is the probability that $\mathbf{W}$ visits the upper semicircle G of F before it visits the lower semicircle H?

13. Let W_1 and W_2 be independent standard Wiener processes; the pair $\mathbf{W}(t) = (W_1(t), W_2(t))$ represents the position of a particle which is experiencing Brownian motion in the plane. Let l be some straight line in $\mathbb{R}^2$, and let P be the point on l which is closest to the origin O. Draw a diagram. Show that

(a) the particle visits l, with probability one,

(b) if the particle hits l for the first time at the point R, then the distance PR (measured as positive or negative as appropriate) has the Cauchy density function $f(x) = d/\{\pi(d^2+x^2)\}$, $-\infty < x < \infty$, where d is the distance OP,

(c) the angle $\widehat{POR}$ is uniformly distributed on $[-\frac{1}{2}\pi, \frac{1}{2}\pi]$.

14. Lévy's conformal invariance property. Let $\phi(x+iy) = u(x,y) + iv(x,y)$ be an analytic function on the complex plane with real part $u(x,y)$ and imaginary part $v(x,y)$, and assume that

$$\left(\frac{\partial u}{\partial x}\right)^2 + \left(\frac{\partial u}{\partial y}\right)^2 = 1.$$

Let (W_1, W_2) be the planar Wiener process of Problem (13.12.13) above. Show that the pair $u(W_1, W_2)$, $v(W_1, W_2)$ is also a planar Wiener process.

15. Let $M(t) = \max_{0 \le s \le t} W(s)$, where W is a standard Wiener process. Show that $M(t) - W(t)$ has the same distribution as $M(t)$.

16. Let W be a standard Wiener process, $u \in \mathbb{R}$, and let $Z = \{t : W(t) = u\}$. Show that Z is a null set (i.e. has Lebesgue measure zero) with probability one.

17. Let $M(t) = \max_{0 \le s \le t} W(s)$, where W is a standard Wiener process. Show that $M(t)$ is attained at exactly one point in $[0, t]$, with probability one.

18. Sparre Andersen theorem. Let $s_0 = 0$ and $s_m = \sum_{j=1}^{m} x_j$, where $(x_j : 1 \le j \le n)$ is a given sequence of real numbers. Of the $n!$ permutations of $(x_j : 1 \le j \le n)$, let A_r be the number of permutations in which exactly r values of $(s_m : 0 \le m \le n)$ are strictly positive, and let B_r be the number of permutations in which the maximum of $(s_m : 0 \le m \le n)$ first occurs at the rth place. Show that $A_r = B_r$ for $0 \le r \le n$. [Hint: Use induction on n.]

19. Arc sine laws. For the standard Wiener process W, let A be the amount of time u during the time interval $[0, t]$ for which $W(u) > 0$; let L be the time of the last visit to the origin before t; and let R be the time when W attains its maximum in $[0, t]$. Show that A, L, and R have the same distribution function $F(x) = (2/\pi)\sin^{-1}\sqrt{x/t}$ for $0 \le x \le t$. [Hint: Use the results of Problems (13.12.15)–(13.12.18).]

20. Let W be a standard Wiener process, and let U_x be the amount of time spent below the level x (≥ 0) during the time interval $(0, 1)$, that is, $U_x = \int_0^1 I_{\{W(t)<x\}}\,dt$. Show that U_x has density function

$$f_{U_x}(u) = \frac{1}{\pi\sqrt{u(1-u)}}\exp\left(-\frac{x^2}{2u}\right), \qquad 0 < u < 1.$$

Show also that

$$V_x = \begin{cases} \sup\{t \le 1 : W_t = x\} & \text{if this set is non-empty,} \\ 1 & \text{otherwise,} \end{cases}$$

has the same distribution as U_x.

21. Tanaka's example. Let $\text{sign}(x) = 1$ if $x > 0$ and $\text{sign}(x) = -1$ otherwise.

(a) Show that $V_t = \int_0^t \text{sign}(W_s)\,dW_s$ defines a standard Wiener process, where W is itself such a process.

(b) Deduce that V is the solution of the SDE $dX = \text{sign}(X_t)\,dW$ with $W(0) = 0$.

(c) Use the result of Problem (13.12.16) with $u = 0$ to show that $-V$ is also a solution of this SDE.

(d) Use the results of Example (12.7.10) and Theorem (13.8.11) to deduce that any solution of the SDE has the fdds of the Wiener process. [You have shown that solutions of the SDE are not unique in a pathwise sense, but do have unique fdds.]

22. After the level of an industrial process has been set at its desired value, it wanders in a random fashion. To counteract this the process is periodically reset to this desired value, at times $0, T, 2T, \dots$. If W_t is the deviation from the desired level, t units of time after a reset, then $\{W_t : 0 \le t < T\}$ can be modelled by a standard Wiener process. The behaviour of the process after a reset is independent of its behaviour before the reset. While W_t is outside the range $(-a, a)$ the output from the process is unsatisfactory and a cost is incurred at rate C per unit time. The cost of each reset is R. Show that the period T which minimises the long-run average cost per unit time is T^*, where

$$R = C \int_0^{T^*} \frac{a}{\sqrt{(2\pi t)}} \exp\left(-\frac{a^2}{2t}\right) dt.$$

23. An economy is governed by the Black–Scholes model in which the stock price behaves as a geometric Brownian motion with volatility σ, and there is a constant interest rate r. An investor likes to have a constant proportion γ ($\in (0, 1)$) of the current value of her self-financing portfolio in stock and the remainder in the bond. Show that the value function of her portfolio has the form $V_t = f(t) S_t^{\gamma}$ where $f(t) = c \exp\{(1-\gamma)(\frac{1}{2}\gamma\sigma^2 + r)t\}$ for some constant c depending on her initial wealth.

24. Let $u(t, x)$ be twice continuously differentiable in x and once in t, for $x \in \mathbb{R}$ and $t \in [0, T]$. Let W be the standard Wiener process. Show that u is a solution of the heat equation

$$\frac{\partial u}{\partial t} = \frac{1}{2} \frac{\partial^2 u}{\partial x^2}$$

if and only if the process $U_t = u(T - t, W_t)$, $0 \le t \le T$, has zero drift.

25. Walk on spheres.

(a) Let W be the standard Wiener process in $d = 3$ dimensions starting at the origin 0, and let T be its first passage time to the sphere with radius r and centre at 0. Show that $W(T)$ is independent of T.

(b) Let C be an open convex region containing 0, with smooth boundary B. For $x \in C$, let $M(x)$ be the largest sphere with centre x that is inscribable in $B \cup C$, with radius $r(x)$. Define the process $\{R_n : n \ge 0\}$ thus. We set $R_0 = 0$; given $R_1, R_2, \dots, R_n$ for $n \ge 0$, R_{n+1} is uniformly distributed on the sphere $M_n := M(R_n)$. Now define the increasing sequence $\{T_n : n \ge 0\}$ by: $T_0 = 0$, T_1 is the first passage time of W to $M(0)$; T_2 is the first subsequent time at which W hits M_1, and so on.

(i) Show that the sequences (R_n) and $(W(T_n))$ have the same distributions.

(ii) Deduce that $R_\infty = \lim_{n\to\infty} R_n$ has the same distribution as $W(T_B)$, where T_B is the first passage time of W to B.

(iii) Let d be the diameter of the smallest sphere that contains $B \cup C$. For $0 < a < d/2$, let $S(a) = \inf\{n : r(R_n) \le a\}$ be the least n at which R_n is within distance a of the boundary B. Show that $S(a)$ is a.s. finite, and moreover that $\mathbb{E}(S(a)) \le d/a$.

1

Events and their probabilities

1.2 Solutions. Events as sets

1. (a) Let $a \in (\bigcup A_i)^c$. Then $a \notin \bigcup A_i$, so that $a \in A_i^c$ for all i. Hence $(\bigcup A_i)^c \subseteq \bigcap A_i^c$. Conversely, if $a \in \bigcap A_i^c$, then $a \notin A_i$ for every i. Hence $a \notin \bigcup A_i$, and so $\bigcap A_i^c \subseteq (\bigcup A_i)^c$. The first De Morgan law follows.

(b) Applying part (a) to the family $\{A_i^c : i \in I\}$, we obtain that $\left(\bigcup_i A_i^c\right)^c = \bigcap_i (A_i^c)^c = \bigcap_i A_i$. Taking the complement of each side yields the second law.

2. Clearly

(i) $A \cap B = (A^c \cup B^c)^c$,

(ii) $A \setminus B = A \cap B^c = (A^c \cup B)^c$,

(iii) $A \triangle B = (A \setminus B) \cup (B \setminus A) = (A^c \cup B)^c \cup (A \cup B^c)^c$.

Now $\mathcal{F}$ is closed under the operations of countable unions and complements, and therefore each of these sets lies in $\mathcal{F}$.

3. Let us number the players $1, 2, \dots, 2^n$ in the order in which they appear in the initial table of draws. The set of victors in the first round is a point in the space $V_n = \{1, 2\} \times \{3, 4\} \times \cdots \times \{2^n - 1, 2^n\}$. Renumbering these victors in the same way as done for the initial draw, the set of second-round victors can be thought of as a point in the space V_{n-1}, and so on. The sample space of all possible outcomes of the tournament may therefore be taken to be $V_n \times V_{n-1} \times \cdots \times V_1$, a set containing $2^{2^{n-1}} 2^{2^{n-2}} \cdots 2^1 = 2^{2^n - 1}$ points.

Should we be interested in the ultimate winner only, we may take as sample space the set $\{1, 2, \dots, 2^n\}$ of all possible winners.

4. We must check that $\mathcal{G}$ satisfies the definition of a σ-field:

(a) $\varnothing \in \mathcal{F}$, and therefore $\varnothing = \varnothing \cap B \in \mathcal{G}$,

(b) if $A_1, A_2, \dots \in \mathcal{F}$, then $\bigcup_i (A_i \cap B) = \left(\bigcup_i A_i\right) \cap B \in \mathcal{G}$,

(c) if $A \in \mathcal{F}$, then $A^c \in \mathcal{F}$ so that $B \setminus (A \cap B) = A^c \cap B \in \mathcal{G}$.

Note that $\mathcal{G}$ is a σ-field *of subsets of* B but not a σ-field of subsets of Ω, since $C \in \mathcal{G}$ does not imply that $C^c = \Omega \setminus C \in \mathcal{G}$.

5. (a), (b), and (d) are identically true; (c) is true if and only if $A \subseteq C$.

1.3 Solutions. Probability

1. (i) We have (using the fact that $\mathbb{P}$ is a non-decreasing set function) that

$$\mathbb{P}(A \cap B) = \mathbb{P}(A) + \mathbb{P}(B) - \mathbb{P}(A \cup B) \geq \mathbb{P}(A) + \mathbb{P}(B) - 1 = \tfrac{1}{12}.$$

Also, since $A \cap B \subseteq A$ and $A \cap B \subseteq B$, $\mathbb{P}(A \cap B) \le \min\{\mathbb{P}(A), \mathbb{P}(B)\} = \frac{1}{3}$.

These bounds are attained in the following example. Pick a number at random from $\{1, 2, \dots, 12\}$. Taking $A = \{1, 2, \dots, 9\}$ and $B = \{9, 10, 11, 12\}$, we find that $A \cap B = \{9\}$, and so $\mathbb{P}(A) = \frac{3}{4}$, $\mathbb{P}(B) = \frac{1}{3}$, $\mathbb{P}(A \cap B) = \frac{1}{12}$. To attain the upper bound for $\mathbb{P}(A \cap B)$, take $A = \{1, 2, \dots, 9\}$ and $B = \{1, 2, 3, 4\}$.

(ii) Likewise we have in this case $\mathbb{P}(A \cup B) \le \min\{\mathbb{P}(A) + \mathbb{P}(B), 1\} = 1$, and $\mathbb{P}(A \cup B) \ge \max\{\mathbb{P}(A), \mathbb{P}(B)\} = \frac{3}{4}$. These bounds are attained in the examples above.

2. (i) We have (using the continuity property of $\mathbb{P}$) that

$$\mathbb{P}(\text{no head ever}) = \lim_{n\to\infty} \mathbb{P}(\text{no head in first } n \text{ tosses}) = \lim_{n\to\infty} 2^{-n} = 0,$$

so that $\mathbb{P}(\text{some head turns up}) = 1 - \mathbb{P}(\text{no head ever}) = 1$.

(ii) Given a fixed sequence s of heads and tails of length k, we consider the sequence of tosses arranged in disjoint groups of consecutive outcomes, each group being of length k. There is probability 2^{-k} that any given one of these is s, independently of the others. The event {one of the first n such groups is s} is a subset of the event {s occurs in the first nk tosses}. Hence (using the general properties of probability measures) we have that

$$\begin{aligned}
\mathbb{P}(s \text{ turns up eventually}) &= \lim_{n\to\infty} \mathbb{P}(s \text{ occurs in the first } nk \text{ tosses}) \\
&\ge \lim_{n\to\infty} \mathbb{P}(s \text{ occurs as one of the first } n \text{ groups}) \\
&= 1 - \lim_{n\to\infty} \mathbb{P}(\text{none of the first } n \text{ groups is } s) \\
&= 1 - \lim_{n\to\infty} (1 - 2^{-k})^n = 1.
\end{aligned}$$

3. Lay out the saucers in order, say as RRWWSS. The cups may be arranged in 6! ways, but since each pair of a given colour may be switched without changing the appearance, there are $6! \div (2!)^3 = 90$ distinct arrangements. By assumption these are equally likely. In how many such arrangements is no cup on a saucer of the same colour? The only acceptable arrangements in which cups of the same colour are paired off are WWSSRR and SSRRWW; by inspection, there are a further eight arrangements in which the first pair of cups is either SW or WS, the second pair is either RS or SR, and the third either RW or WR. Hence the required probability is $10/90 = \frac{1}{9}$.

4. We prove this by induction on n, considering first the case $n = 2$. Certainly $B = (A \cap B) \cup (B \setminus A)$ is a union of disjoint sets, so that $\mathbb{P}(B) = \mathbb{P}(A \cap B) + \mathbb{P}(B \setminus A)$. Similarly $A \cup B = A \cup (B \setminus A)$, and so

$$\mathbb{P}(A \cup B) = \mathbb{P}(A) + \mathbb{P}(B \setminus A) = \mathbb{P}(A) + \big\{\mathbb{P}(B) - \mathbb{P}(A \cap B)\big\}.$$

Hence the result is true for $n = 2$. Let $m \ge 2$ and suppose that the result is true for $n \le m$. Then it is true for pairs of events, so that

$$\begin{aligned}
\mathbb{P}\left(\bigcup_1^{m+1} A_i\right) &= \mathbb{P}\left(\bigcup_1^{m} A_i\right) + \mathbb{P}(A_{m+1}) - \mathbb{P}\left\{\left(\bigcup_1^{m} A_i\right) \cap A_{m+1}\right\} \\
&= \mathbb{P}\left(\bigcup_1^{m} A_i\right) + \mathbb{P}(A_{m+1}) - \mathbb{P}\left\{\bigcup_1^{m} (A_i \cap A_{m+1})\right\}.
\end{aligned}$$

Using the induction hypothesis, we may expand the two relevant terms on the right-hand side to obtain the result.

Let A_1, A_2, and A_3 be the respective events that you fail to obtain the ultimate, penultimate, and ante-penultimate Vice-Chancellors. Then the required probability is, by symmetry,

$$1-\mathbb{P}\left(\bigcup_1^3 A_i\right) = 1-3\mathbb{P}(A_1)+3\mathbb{P}(A_1\cap A_2)-\mathbb{P}(A_1\cap A_2\cap A_3)$$
$$= 1-3(\tfrac{4}{5})^6+3(\tfrac{3}{5})^6-(\tfrac{2}{5})^6.$$

5. By the continuity of $\mathbb{P}$, Exercise (1.2.1), and Problem (1.8.11),

$$\mathbb{P}\left(\bigcap_{r=1}^{\infty} A_r\right) = \lim_{n\to\infty}\mathbb{P}\left(\bigcap_{r=1}^{n} A_r\right) = \lim_{n\to\infty}\left[1-\mathbb{P}\left(\left(\bigcap_{r=1}^{n} A_r\right)^{\mathrm{c}}\right)\right]$$
$$= 1-\lim_{n\to\infty}\mathbb{P}\left(\bigcup_{r=1}^{n} A_r^{\mathrm{c}}\right) \ge 1-\lim_{n\to\infty}\sum_{r=1}^{n}\mathbb{P}(A_r^{\mathrm{c}}) = 1.$$

6. We have that $1=\mathbb{P}\left(\bigcup_1^n A_r\right)=\sum_r\mathbb{P}(A_r)-\sum_{r<s}\mathbb{P}(A_r\cap A_s)=np-\frac{1}{2}n(n-1)q$. Hence $p\ge n^{-1}$, and $\frac{1}{2}n(n-1)q=np-1\le n-1$.

7. Since at least one of the A_r occurs,

$$1=\mathbb{P}\left(\bigcup_1^n A_r\right)=\sum_r\mathbb{P}(A_r)-\sum_{r<s}\mathbb{P}(A_r\cap A_s)+\sum_{r<s<t}\mathbb{P}(A_r\cap A_s\cap A_t)$$
$$= np-\binom{n}{2}q+\binom{n}{3}x.$$

Since at least two of the events occur with probability $\frac{1}{2}$,

$$\tfrac{1}{2}=\mathbb{P}\left(\bigcup_{r<s}(A_r\cap A_s)\right)=\sum_{r<s}\mathbb{P}(A_r\cap A_s)-\tfrac{1}{2}\sum_{\substack{r<s\\ t<u\\ (r,s)\neq(t,u)}}\mathbb{P}(A_r\cap A_s\cap A_t\cap A_u)+\cdots.$$

By a careful consideration of the first three terms in the latter series, we find that

$$\frac{1}{2}=\binom{n}{2}q-3\binom{n}{3}x+\binom{n}{3}x.$$

Hence $\frac{3}{2}=np-\binom{n}{3}x$, so that $p\ge 3/(2n)$. Also, $\binom{n}{2}q=2np-\frac{5}{2}$, whence $q\le 4/n$.

1.4 Solutions. Conditional probability

1. By the definition of conditional probability,

$$\mathbb{P}(A\mid B)=\frac{\mathbb{P}(A\cap B)}{\mathbb{P}(B)}=\frac{\mathbb{P}(B\cap A)}{\mathbb{P}(A)}\frac{\mathbb{P}(A)}{\mathbb{P}(B)}=\mathbb{P}(B\mid A)\frac{\mathbb{P}(A)}{\mathbb{P}(B)}$$

if $\mathbb{P}(A)\mathbb{P}(B) \neq 0$. Hence

$$\frac{\mathbb{P}(A \mid B)}{\mathbb{P}(A)} = \frac{\mathbb{P}(B \mid A)}{\mathbb{P}(B)},$$

whence the last part is immediate.

2. Set $A_0 = \Omega$ for notational convenience. Expand each term on the right-hand side to obtain

$$\prod_{r=1}^{n} \mathbb{P}\Bigl(A_r \Bigm| \bigcap_{k=1}^{r-1} A_k\Bigr) = \prod_{r=1}^{n} \frac{\mathbb{P}(\bigcap_1^r A_k)}{\mathbb{P}(\bigcap_1^{r-1} A_k)} = \mathbb{P}\Bigl(\bigcap_1^n A_k\Bigr).$$

3. Let M be the event that the first coin is double-headed, R the event that it is double-tailed, and N the event that it is normal. Let H_{l}^i be the event that the lower face is a head on the ith toss, T_{u}^i the event that the upper face is a tail on the ith toss, and so on. Then, using conditional probability *ad nauseam*, we find:

(i) $$\mathbb{P}(H_{\mathrm{l}}^1) = \tfrac{2}{5}\mathbb{P}(H_{\mathrm{l}}^1 \mid M) + \tfrac{1}{5}\mathbb{P}(H_{\mathrm{l}}^1 \mid R) + \tfrac{2}{5}\mathbb{P}(H_{\mathrm{l}}^1 \mid N) = \tfrac{2}{5} + 0 + \tfrac{2}{5}\cdot\tfrac{1}{2} = \tfrac{3}{5}.$$

(ii) $$\mathbb{P}(H_{\mathrm{l}}^1 \mid H_{\mathrm{u}}^1) = \frac{\mathbb{P}(H_{\mathrm{l}}^1 \cap H_{\mathrm{u}}^1)}{\mathbb{P}(H_{\mathrm{u}}^1)} = \frac{\mathbb{P}(M)}{\mathbb{P}(H_{\mathrm{u}}^1)} = \tfrac{2}{5}\big/\tfrac{3}{5} = \tfrac{2}{3}.$$

(iii) $$\begin{aligned}\mathbb{P}(H_{\mathrm{l}}^2 \mid H_{\mathrm{u}}^1) &= 1\cdot\mathbb{P}(M \mid H_{\mathrm{u}}^1) + \tfrac{1}{2}\mathbb{P}(N \mid H_{\mathrm{u}}^1)\\ &= \mathbb{P}(H_{\mathrm{l}}^1 \mid H_{\mathrm{u}}^1) + \tfrac{1}{2}\bigl(1 - \mathbb{P}(H_{\mathrm{l}}^1 \mid H_{\mathrm{u}}^1)\bigr) = \tfrac{2}{3} + \tfrac{1}{2}\cdot\tfrac{1}{3} = \tfrac{5}{6}.\end{aligned}$$

(iv)

$$\mathbb{P}(H_{\mathrm{l}}^2 \mid H_{\mathrm{u}}^1 \cap H_{\mathrm{u}}^2) = \frac{\mathbb{P}(H_{\mathrm{l}}^2 \cap H_{\mathrm{u}}^1 \cap H_{\mathrm{u}}^2)}{\mathbb{P}(H_{\mathrm{u}}^1 \cap H_{\mathrm{u}}^2)} = \frac{\mathbb{P}(M)}{1\cdot\mathbb{P}(M) + \frac{1}{4}\cdot\mathbb{P}(N)} = \frac{\frac{2}{5}}{\frac{2}{5} + \frac{1}{10}} = \frac{4}{5}.$$

(v) From (iv), the probability that he discards a double-headed coin is $\frac{4}{5}$, the probability that he discards a normal coin is $\frac{1}{5}$. (There is of course no chance of it being double-tailed.) Hence, by conditioning on the discard,

$$\mathbb{P}(H_{\mathrm{u}}^3) = \tfrac{4}{5}\mathbb{P}(H_{\mathrm{u}}^3 \mid M) + \tfrac{1}{5}\mathbb{P}(H_{\mathrm{u}}^3 \mid N) = \tfrac{4}{5}\bigl(\tfrac{1}{4} + \tfrac{1}{2}\cdot\tfrac{1}{2}\bigr) + \tfrac{1}{5}\bigl(\tfrac{1}{2} + \tfrac{1}{2}\cdot\tfrac{1}{4}\bigr) = \tfrac{21}{40}.$$

4. The final calculation of $\frac{2}{3}$ refers not to a *single* draw of one ball from an urn containing three, but rather to a composite experiment comprising more than one stage (in this case, two stages). While it is true that {two black, one white} is the only fixed collection of balls for which a random choice is black with probability $\frac{2}{3}$, the composition of the urn is *not determined* prior to the final draw.

After all, if Carroll's argument were correct then it would apply also in the situation when the urn originally contains just one ball, either black or white. The final probability is now $\frac{3}{4}$, implying that the original ball was one half black and one half white! Carroll was himself aware of the fallacy in this argument.

5. (a) One cannot compute probabilities without knowing the rules governing the conditional probabilities. We assume that each of the 6 orderings of the car and goats are equally likely. Let C_i be the event that the ith door conceals the car, G the event that you see a goat, and B the event that you see Bill.

(i) We have that

$$\mathbb{P}(C_3 \mid G) = \frac{\mathbb{P}(C_3 \cap G \mid C_1)\mathbb{P}(C_1) + \mathbb{P}(C_3 \cap G \mid C_1^{\mathrm{c}})\mathbb{P}(C_1^{\mathrm{c}})}{\mathbb{P}(G \mid C_1)\mathbb{P}(C_1) + \mathbb{P}(G \mid C_1^{\mathrm{c}})\mathbb{P}(C_1^{\mathrm{c}})} = \frac{0\cdot\frac{1}{3} + 1\cdot\frac{2}{3}}{1\cdot\frac{1}{3} + 1\cdot\frac{2}{3}} = \frac{2}{3}.$$

(ii) Use a similar formula, and note that $\mathbb{P}(C_3 \cap B \mid C_1^c) = \mathbb{P}(B \mid C_1^c) = \frac{1}{2}$.

(iii) This time, $\mathbb{P}(C_3 \cap G \mid C_1^c) = \mathbb{P}(G \mid C_1^c) = \frac{1}{2}$.

(b) Let $\alpha \in [\frac{1}{2}, \frac{2}{3}]$, and suppose the presenter possesses a coin which falls with heads upwards with probability $\beta = 6\alpha - 3$. He flips the coin before the show, and adopts strategy (i) if and only if the coin shows heads, and otherwise strategy (iii). The probability in question is now $\frac{2}{3}\beta + \frac{1}{2}(1-\beta) = \alpha$. You never lose by swapping, but whether you gain depends on the presenter's protocol.

(c) Let D denote the first door chosen, and consider the following protocols:

(iv) If D conceals a goat, open it. Otherwise open one of the other two doors at random. In this case $p = 0$.

(v) If D conceals the car, open it. Otherwise open the unique remaining door which conceals a goat. In this case $p = 1$.

As in part (b), a randomized algorithm provides the protocol necessary for the last part.

6. This is immediate by the definition of conditional probability.

7. Let C_i be the colour of the ith ball picked, and use the obvious notation.

(a) Since each urn contains the same number $n-1$ of balls, the second ball picked is equally likely to be any of the $n(n-1)$ available. One half of these balls are magenta, whence $\mathbb{P}(C_2 = \mathrm{M}) = \frac{1}{2}$.

(b) By conditioning on the choice of urn,

$$\mathbb{P}(C_2 = \mathrm{M} \mid C_1 = \mathrm{M}) = \frac{\mathbb{P}(C_1, C_2 = \mathrm{M})}{\mathbb{P}(C_1 = \mathrm{M})} = \sum_{r=1}^{n} \frac{(n-r)(n-r-1)}{n(n-1)(n-2)} \Big/ \frac{1}{2} = \frac{2}{3}.$$

8. With R denoting red-haired and S not red-haired, in the obvious notation,

$$\begin{aligned}\mathbb{P}(\mathrm{BB} \mid \text{at least one } \mathrm{B_R}) &= \frac{\mathbb{P}(\mathrm{B_R B_R} \cup \mathrm{B_R B_S} \cup \mathrm{B_S B_R})}{\mathbb{P}(\mathrm{B_R B_R} \cup \mathrm{B_R B_S} \cup \mathrm{B_S B_R} \cup \mathrm{G B_R} \cup \mathrm{B_R G})} \\ &= \frac{\frac{1}{4}r^2 + \frac{1}{2}r(1-r)}{\frac{1}{4}r^2 + \frac{1}{2}r(1-r) + \frac{1}{2}r} = \frac{2-r}{4-r}.\end{aligned}$$

Assuming children have no preference for day of the week, and sex and weekday are independent, the answer is the above with $r = \frac{1}{7}$.

1.5 Solutions. Independence

1. Clearly

$$\begin{aligned}\mathbb{P}(A^c \cap B) &= \mathbb{P}(B \setminus \{A \cap B\}) = \mathbb{P}(B) - \mathbb{P}(A \cap B) \\ &= \mathbb{P}(B) - \mathbb{P}(A)\mathbb{P}(B) = \mathbb{P}(A^c)\mathbb{P}(B).\end{aligned}$$

For the final part, apply the first part to the pair B, A^c.

2. Suppose $i < j$ and $m < n$. If $j < m$, then A_{ij} and A_{mn} are determined by distinct independent rolls, and are therefore independent. For the case $j = m$ we have that

$$\begin{aligned}\mathbb{P}(A_{ij} \cap A_{jn}) &= \mathbb{P}(i\text{th, } j\text{th, and } n\text{th rolls show same number}) \\ &= \sum_{r=1}^{6} \tfrac{1}{6}\mathbb{P}\big(j\text{th and } n\text{th rolls both show } r \,\big|\, i\text{th shows } r\big) = \tfrac{1}{36} = \mathbb{P}(A_{ij})\mathbb{P}(A_{jn}),\end{aligned}$$

as required. However, if $i \neq j \neq k$, $\mathbb{P}(A_{ij} \cap A_{jk} \cap A_{ik}) = \frac{1}{36} \neq \frac{1}{216} = \mathbb{P}(A_{ij})\mathbb{P}(A_{jk})\mathbb{P}(A_{ik})$.

3. That (a) implies (b) is trivial. Suppose then that (b) holds. Consider the outcomes numbered $i_1, i_2, \dots, i_m$, and let $u_j \in \{\text{H}, \text{T}\}$ for $1 \le j \le m$. Let S_j be the set of all sequences of length $M = \max\{i_j : 1 \le j \le m\}$ showing u_j in the i_jth position. Clearly $|S_j| = 2^{M-1}$ and $\left|\bigcap_j S_j\right| = 2^{M-m}$. Therefore,

$$\mathbb{P}(S_j) = \frac{2^{M-1}}{2^M} = \frac{1}{2}, \qquad \mathbb{P}\left(\bigcap_j S_j\right) = \frac{2^{M-m}}{2^M} = \frac{1}{2^m},$$

so that $\mathbb{P}\left(\bigcap_j S_j\right) = \prod_j \mathbb{P}(S_j)$.

4. Suppose $|A| = a$, $|B| = b$, $|A \cap B| = c$, and A and B are independent. Then $\mathbb{P}(A \cap B) = \mathbb{P}(A)\mathbb{P}(B)$, which is to say that $c/p = (a/p) \cdot (b/p)$, and hence $ab = pc$. If $ab \neq 0$ then $p \mid ab$ (i.e. p divides ab). However, p is prime, and hence either $p \mid a$ or $p \mid b$. Therefore, either $A = \Omega$ or $B = \Omega$ (or both).

5. (a) Flip two coins; let A be the event that the first shows H, let B be the event that the second shows H, and let C be the event that they show the same. Then A and B are independent, but not conditionally independent given C.

(b) Roll two dice; let A be the event that the smaller is 3, let B be the event that the larger is 6, and let C be the event that the smaller score is no more than 3, and the larger is 4 or more. Then A and B are conditionally independent given C, but not independent.

(c) The definitions are equivalent if $\mathbb{P}(C) = 1$.

6. $(\frac{9}{10})^7 < \frac{1}{2}$.

7. (a) $\mathbb{P}(A \cap B) = \frac{1}{8} = \frac{1}{4} \cdot \frac{1}{2} = \mathbb{P}(A)\mathbb{P}(B)$, and $\mathbb{P}(B \cap C) = \frac{3}{8} = \frac{1}{2} \cdot \frac{3}{4} = \mathbb{P}(B)\mathbb{P}(C)$.

(b) $\mathbb{P}(A \cap C) = 0 \neq \mathbb{P}(A)\mathbb{P}(C)$.

(c) Only in the trivial cases when children are either almost surely boys or almost surely girls.

(d) No.

8. No. $\mathbb{P}(\text{all alike}) = \frac{1}{4}$.

9. $\mathbb{P}(\text{1st shows } r \text{ and sum is 7}) = \frac{1}{36} = \frac{1}{6} \cdot \frac{1}{6} = \mathbb{P}(\text{1st shows } r)\mathbb{P}(\text{sum is 7})$.

10. We have $\mathbb{P}(A_1) = \frac{1}{9}$ and $\mathbb{P}(A_2) = \mathbb{P}(A_3) = \frac{1}{2}$. Furthermore,

$$\mathbb{P}(A_1 \cap A_2 \cap A_3) = \mathbb{P}(X = 3,\ Y = 6) = \frac{1}{36},$$

and the equality follows. The events are not independent since they are not pairwise independent, for example,

$$\mathbb{P}(A_2 \cap A_3) = \frac{1}{6} \neq \frac{1}{4} = \mathbb{P}(A_2)\mathbb{P}(A_3).$$

1.7 Solutions. Worked examples

1. Write EF for the event that there is an open road from E to F, and EF^{c} for the complement of this event; write E $\leftrightarrow$ F if there is an open route from E to F, and E $\not\leftrightarrow$ F if there is none. Now $\{\text{A} \leftrightarrow \text{C}\} = \text{AB} \cap \text{BC}$, so that

$$\mathbb{P}(\text{AB} \mid \text{A} \not\leftrightarrow \text{C}) = \frac{\mathbb{P}(\text{AB}, \text{A} \not\leftrightarrow \text{C})}{\mathbb{P}(\text{A} \not\leftrightarrow \text{C})} = \frac{\mathbb{P}(\text{AB}, \text{B} \not\leftrightarrow \text{C})}{1 - \mathbb{P}(\text{A} \leftrightarrow \text{C})} = \frac{(1-p^2)p^2}{1-(1-p^2)^2}.$$

By a similar calculation (or otherwise) in the second case, one obtains the same answer:

$$\mathbb{P}(\text{AB} \mid \text{A} \not\leftrightarrow \text{C}) = \frac{(1-p^2)p^3}{1-(1-p^2)^2 p - (1-p)} = \frac{(1-p^2)p^2}{1-(1-p^2)^2}.$$

2. Let A be the event of exactly one ace, and KK be the event of exactly two kings. Then $\mathbb{P}(\text{A} \mid \text{KK}) = \mathbb{P}(\text{A} \cap \text{KK})/\mathbb{P}(\text{KK})$. Now, by counting acceptable combinations,

$$\mathbb{P}(\text{A} \cap \text{KK}) = \binom{4}{1}\binom{4}{2}\binom{44}{10}\bigg/\binom{52}{13}, \quad \mathbb{P}(\text{KK}) = \binom{4}{2}\binom{48}{11}\bigg/\binom{52}{13},$$

so the required probability is

$$\binom{4}{1}\binom{4}{2}\binom{44}{10}\bigg/\binom{4}{2}\binom{48}{11} = \frac{7\cdot 11\cdot 37}{3\cdot 46\cdot 47} \simeq 0.44.$$

3. *First method*: Suppose that the coin is being tossed by a special machine which is not switched off when the walker is absorbed. If the machine ever produces N heads in succession, then either the game finishes at this point or it is already over. From Exercise (1.3.2), such a sequence of N heads must (with probability one) occur sooner or later.

Alternative method: Write down the difference equations for p_k, the probability the game finishes at 0 having started at k, and for $\widehat{p}_k$, the corresponding probability that the game finishes at N; actually these two difference equations are the same, but the respective boundary conditions are different. Solve these equations and add their solutions to obtain the total 1.

4. It is a tricky question. One of the present authors is in agreement, since if $\mathbb{P}(A \mid C) > \mathbb{P}(B \mid C)$ and $\mathbb{P}(A \mid C^c) > \mathbb{P}(B \mid C^c)$ then

$$\begin{aligned}\mathbb{P}(A) &= \mathbb{P}(A \mid C)\mathbb{P}(C) + \mathbb{P}(A \mid C^c)\mathbb{P}(C^c)\\ &> \mathbb{P}(B \mid C)\mathbb{P}(C) + \mathbb{P}(B \mid C^c)\mathbb{P}(C^c) = \mathbb{P}(B).\end{aligned}$$

The other author is more suspicious of the question, and points out that there is a difficulty arising from the use of the word 'you'. In Example (1.7.10), Simpson's paradox, whilst drug I is preferable to drug II for both males and females, it is drug II that wins overall.

5. Let L_k be the label of the kth card. Then, using symmetry,

$$\mathbb{P}\big(L_k = m \,\big|\, L_k > L_r \text{ for } 1 \le r < k\big) = \frac{\mathbb{P}(L_k = m)}{\mathbb{P}(L_k > L_r \text{ for } 1 \le r < k)} = \frac{1}{m}\bigg/\frac{1}{k} = k/m.$$

6. There are $\binom{2b}{m}$ equally likely ways of assigning the men to the pairs. The number of assignations with no male pair is $2^m\binom{b}{m}$. The answer is $2^m\binom{b}{m}/\binom{2b}{m}$.

1.8 Solutions to problems

1. (a) *Method I*: There are 36 equally likely outcomes, and just 10 of these contain exactly one six. The answer is therefore $\frac{10}{36} = \frac{5}{18}$.

Method II: Since the throws have independent outcomes,

$$\mathbb{P}(\text{first is 6, second is not 6}) = \mathbb{P}(\text{first is 6})\mathbb{P}(\text{second is not 6}) = \tfrac{1}{6}\cdot\tfrac{5}{6} = \tfrac{5}{36}.$$

There is an equal probability of the event {first is not 6, second is 6}.

(b) A die shows an odd number with probability $\frac{1}{2}$; by independence, $\mathbb{P}(\text{both odd}) = \frac{1}{2} \cdot \frac{1}{2} = \frac{1}{4}$.

(c) Write S for the sum, and $\{i, j\}$ for the event that the first is i and the second j. Then $\mathbb{P}(S = 4) = \mathbb{P}(1, 3) + \mathbb{P}(2, 2) + \mathbb{P}(3, 1) = \frac{3}{36}$.

(d) Similarly

$$\begin{aligned}\mathbb{P}(S \text{ divisible by } 3) &= \mathbb{P}(S = 3) + \mathbb{P}(S = 6) + \mathbb{P}(S = 9) + \mathbb{P}(S = 12) \\ &= \{\mathbb{P}(1, 2) + \mathbb{P}(2, 1)\} \\ &\qquad + \{\mathbb{P}(1, 5) + \mathbb{P}(2, 4) + \mathbb{P}(3, 3) + \mathbb{P}(4, 2) + \mathbb{P}(5, 1)\} \\ &\qquad + \{\mathbb{P}(3, 6) + \mathbb{P}(4, 5) + \mathbb{P}(5, 4) + \mathbb{P}(6, 3)\} + \mathbb{P}(6, 6) \\ &= \tfrac{12}{36} = \tfrac{1}{3}.\end{aligned}$$

2. (a) By independence, $\mathbb{P}(n - 1 \text{ tails, followed by a head}) = 2^{-n}$.

(b) If n is odd, $\mathbb{P}(\# \text{ heads} = \# \text{ tails}) = 0$; $\#A$ denotes the cardinality of the set A. If n is even, there are $\binom{n}{n/2}$ sequences of outcomes with $\frac{1}{2}n$ heads and $\frac{1}{2}n$ tails. Any given sequence of heads and tails has probability 2^{-n}; therefore $\mathbb{P}(\# \text{ heads} = \# \text{ tails}) = 2^{-n}\binom{n}{n/2}$.

(c) There are $\binom{n}{2}$ sequences containing 2 heads and $n - 2$ tails. Each sequence has probability 2^{-n}, and therefore $\mathbb{P}(\text{exactly two heads}) = \binom{n}{2}2^{-n}$.

(d) Clearly

$$\mathbb{P}(\text{at least 2 heads}) = 1 - \mathbb{P}(\text{no heads}) - \mathbb{P}(\text{exactly one head}) = 1 - 2^{-n} - \binom{n}{1}2^{-n}.$$

3. (a) Recall De Morgan's Law (Exercise (1.2.1)): $\bigcap_i A_i = \left(\bigcup_i A_i^{\mathrm{c}}\right)^{\mathrm{c}}$, which lies in $\mathcal{F}$ since it is the complement of a countable union of complements of sets in $\mathcal{F}$.

(b) $\mathcal{H}$ is a σ-field because:

(i) $\varnothing \in \mathcal{F}$ and $\varnothing \in \mathcal{G}$; therefore $\varnothing \in \mathcal{H}$.

(ii) If $A_1, A_2, \dots$ is a sequence of sets belonging to both $\mathcal{F}$ and $\mathcal{G}$, then their union lies in both $\mathcal{F}$ and $\mathcal{G}$, which is to say that $\mathcal{H}$ is closed under the operation of taking countable unions.

(iii) Likewise A^{c} is in $\mathcal{H}$ if A is in both $\mathcal{F}$ and $\mathcal{G}$.

(c) We display an example. Let

$$\Omega = \{a, b, c\}, \quad \mathcal{F} = \{\{a\}, \{b, c\}, \varnothing, \Omega\}, \quad \mathcal{G} = \{\{a, b\}, \{c\}, \varnothing, \Omega\}.$$

Then $\mathcal{H} = \mathcal{F} \cup \mathcal{G}$ is given by $\mathcal{H} = \{\{a\}, \{c\}, \{a, b\}, \{b, c\}, \varnothing, \Omega\}$. Note that $\{a\} \in \mathcal{H}$ and $\{c\} \in \mathcal{H}$, but the union $\{a, c\}$ is not in $\mathcal{H}$, which is therefore not a σ-field.

4. In each case $\mathcal{F}$ may be taken to be the set of all subsets of Ω, and the probability of any member of $\mathcal{F}$ is the sum of the probabilities of the elements therein.

(a) $\Omega = \{\mathrm{H}, \mathrm{T}\}^3$, the set of all triples of heads (H) and tails (T). With the usual assumption of independence, the probability of any given triple containing h heads and $t = 3 - h$ tails is $p^h(1 - p)^t$, where p is the probability of heads on each throw.

(b) In the obvious notation, $\Omega = \{\mathrm{U}, \mathrm{V}\}^2 = \{\mathrm{UU}, \mathrm{VV}, \mathrm{UV}, \mathrm{VU}\}$. Also $\mathbb{P}(\mathrm{UU}) = \mathbb{P}(\mathrm{VV}) = \frac{2}{4} \cdot \frac{1}{3}$ and $\mathbb{P}(\mathrm{UV}) = \mathbb{P}(\mathrm{VU}) = \frac{2}{4} \cdot \frac{2}{3}$.

(c) Ω is the set of finite sequences of tails followed by a head, $\{\mathrm{T}^n\mathrm{H} : n \geq 0\}$, together with the infinite sequence T^∞ of tails. Now, $\mathbb{P}(\mathrm{T}^n\mathrm{H}) = (1 - p)^n p$, and $\mathbb{P}(\mathrm{T}^\infty) = \lim_{n\to\infty}(1 - p)^n = 0$ if $p \neq 0$.

5. As usual, $\mathbb{P}(A \,\triangle\, B) = \mathbb{P}\big((A \cup B) \setminus \mathbb{P}(A \cap B)\big) = \mathbb{P}(A \cup B) - \mathbb{P}(A \cap B)$.

6. Clearly, by Exercise (1.4.2),

$$\begin{aligned}\mathbb{P}(A \cup B \cup C) &= \mathbb{P}\big((A^c \cap B^c \cap C^c)^c\big) = 1 - \mathbb{P}(A^c \cap B^c \cap C^c)\\ &= 1 - \mathbb{P}(A^c \mid B^c \cap C^c)\mathbb{P}(B^c \mid C^c)\mathbb{P}(C^c).\end{aligned}$$

7. (a) If A is independent of itself, then $\mathbb{P}(A) = \mathbb{P}(A \cap A) = \mathbb{P}(A)^2$, so that $\mathbb{P}(A) = 0$ or 1.

(b) If $\mathbb{P}(A) = 0$ then $0 = \mathbb{P}(A \cap B) = \mathbb{P}(A)\mathbb{P}(B)$ for all B. If $\mathbb{P}(A) = 1$ then $\mathbb{P}(A \cap B) = \mathbb{P}(B)$, so that $\mathbb{P}(A \cap B) = \mathbb{P}(A)\mathbb{P}(B)$.

8. $\Omega \cup \varnothing = \Omega$ and $\Omega \cap \varnothing = \varnothing$, and therefore $1 = \mathbb{P}(\Omega \cup \varnothing) = \mathbb{P}(\Omega) + \mathbb{P}(\varnothing) = 1 + \mathbb{P}(\varnothing)$, implying that $\mathbb{P}(\varnothing) = 0$.

9. (i) $\mathbb{Q}(\varnothing) = \mathbb{P}(\varnothing \mid B) = 0$. Also $\mathbb{Q}(\Omega) = \mathbb{P}(\Omega \mid B) = \mathbb{P}(B)/\mathbb{P}(B) = 1$.
(ii) Let $A_1, A_2, \dots$ be disjoint members of $\mathcal{F}$. Then $\{A_i \cap B : i \geq 1\}$ are disjoint members of $\mathcal{F}$, implying that

$$\mathbb{Q}\left(\bigcup_1^\infty A_i\right) = \mathbb{P}\left(\bigcup_1^\infty A_i \,\middle|\, B\right) = \frac{\mathbb{P}(\bigcup_1^\infty (A_i \cap B))}{\mathbb{P}(B)} = \sum_1^\infty \frac{\mathbb{P}(A_i \cap B)}{\mathbb{P}(B)} = \sum_1^\infty \mathbb{Q}(A_i).$$

Finally, since $\mathbb{Q}$ is a probability measure,

$$\mathbb{Q}(A \mid C) = \frac{\mathbb{Q}(A \cap C)}{\mathbb{Q}(C)} = \frac{\mathbb{P}(A \cap C \mid B)}{\mathbb{P}(C \mid B)} = \frac{\mathbb{P}(A \cap B \cap C)}{\mathbb{P}(B \cap C)} = \mathbb{P}(A \mid B \cap C).$$

The order of the conditioning (C before B, or *vice versa*) is thus irrelevant.

10. As usual,

$$\mathbb{P}(A) = \mathbb{P}\left(\bigcup_1^\infty (A \cap B_j)\right) = \sum_1^\infty \mathbb{P}(A \cap B_j) = \sum_1^\infty \mathbb{P}(A \mid B_j)\mathbb{P}(B_j).$$

11. The first inequality is trivially true if $n = 1$. Let $m \geq 1$ and assume that the inequality holds for $n \leq m$. Then

$$\begin{aligned}\mathbb{P}\left(\bigcup_1^{m+1} A_i\right) &= \mathbb{P}\left(\bigcup_1^m A_i\right) + \mathbb{P}(A_{m+1}) - \mathbb{P}\left(\bigcup_1^m (A_i \cap A_{m+1})\right)\\ &\leq \mathbb{P}\left(\bigcup_1^m A_i\right) + \mathbb{P}(A_{m+1}) \leq \sum_1^{m+1} \mathbb{P}(A_i),\end{aligned}$$

by the hypothesis. The result follows by induction. Secondly, by the first part,

$$\mathbb{P}\left(\bigcap_1^n A_i\right) = \mathbb{P}\left(\left(\bigcup_1^n A_i^c\right)^c\right) = 1 - \mathbb{P}\left(\bigcup_1^n A_i^c\right) \geq 1 - \sum_1^n \mathbb{P}(A_i^c).$$

12. We have that

$$\begin{aligned}\mathbb{P}\Big(\bigcap_1^n A_i\Big) &= \mathbb{P}\Big(\Big(\bigcup_1^n A_i^c\Big)^c\Big) = 1 - \mathbb{P}\Big(\bigcup_1^n A_i^c\Big)\\ &= 1 - \sum_i \mathbb{P}(A_i^c) + \sum_{i<j} \mathbb{P}(A_i^c \cap A_j^c) - \cdots + (-1)^n \mathbb{P}\Big(\bigcap_1^n A_i^c\Big) \quad \text{by Exercise (1.3.4)}\\ &= 1 - n + \sum_i \mathbb{P}(A_i) + \binom{n}{2} - \sum_{i<j} \mathbb{P}(A_i \cup A_j) - \binom{n}{3} + \cdots\\ &\qquad + (-1)^n \binom{n}{n} - (-1)^n \mathbb{P}\Big(\bigcup_1^n A_i\Big) \quad \text{using De Morgan's laws again}\\ &= (1-1)^n + \sum_i \mathbb{P}(A_i) - \cdots - (-1)^n \mathbb{P}\Big(\bigcup_1^n A_i\Big) \quad \text{by the binomial theorem.}\end{aligned}$$

13. Clearly,

$$\mathbb{P}(N_k) = \sum_{\substack{S \subseteq \{1,2,\ldots,n\}\\ |S|=k}} \mathbb{P}\Big(\bigcap_{i \in S} A_i \bigcap_{j \notin S} A_j^c\Big).$$

For any such given S, we write $A_S = \bigcap_{i \in S} A_i$. Then

$$\mathbb{P}\Big(\bigcap_{i \in S} A_i \bigcap_{j \notin S} A_j^c\Big) = \mathbb{P}(A_S) - \sum_{j \notin S} \mathbb{P}(A_{S \cup \{j\}}) + \sum_{\substack{j<k\\ j,k \notin S}} \mathbb{P}(A_{S \cup \{j,k\}}) - \cdots$$

by Exercise (1.3.4). Hence

$$\mathbb{P}(N_k) = \sum_{|S|=k} \mathbb{P}(A_S) - \sum_{|S|=k+1} \binom{k+1}{k} \mathbb{P}(A_S) + \cdots + (-1)^{n-k} \binom{n}{k} \mathbb{P}(A_1 \cap \cdots \cap A_n)$$

where a typical summation is over all subsets S of $\{1, 2, \ldots, n\}$ having the required cardinality.

Let A_i be the event that a copy of the ith bust is obtained. Then, by symmetry,

$$\mathbb{P}(N_3) = \binom{5}{3}\alpha_3 - \binom{5}{4}\binom{4}{3}\alpha_4 + \binom{5}{3}\alpha_5$$

where α_j is the probability that the j most recent Vice-Chancellors are obtained. Now α_3 is given in Exercise (1.3.4), and α_4 and α_5 may be calculated similarly.

14. Assuming the conditional probabilities are defined,

$$\mathbb{P}(A_j \mid B) = \frac{\mathbb{P}(A_j \cap B)}{\mathbb{P}(B)} = \frac{\mathbb{P}(B \mid A_j)\mathbb{P}(A_j)}{\mathbb{P}\big(B \cap (\bigcup_1^n A_i)\big)} = \frac{\mathbb{P}(B \mid A_j)\mathbb{P}(A_j)}{\sum_{i=1}^n \mathbb{P}(B \mid A_i)\mathbb{P}(A_i)}.$$

15. (a) We have that

$$\begin{aligned}\mathbb{P}(N = 2 \mid S = 4) &= \frac{\mathbb{P}(\{N=2\} \cap \{S=4\})}{\mathbb{P}(S=4)} = \frac{\mathbb{P}(S=4 \mid N=2)\mathbb{P}(N=2)}{\sum_i \mathbb{P}(S=4 \mid N=i)\mathbb{P}(N=i)}\\ &= \frac{\frac{1}{12}\cdot\frac{1}{4}}{\frac{1}{6}\cdot\frac{1}{2} + \frac{1}{12}\cdot\frac{1}{4} + \frac{3}{216}\cdot\frac{1}{8} + \frac{1}{6^4}\cdot\frac{1}{16}}.\end{aligned}$$

(b) Secondly,

$$\mathbb{P}(S=4 \mid N \text{ even}) = \frac{\mathbb{P}(S=4 \mid N=2)\frac{1}{4} + \mathbb{P}(S=4 \mid N=4)\frac{1}{16}}{\mathbb{P}(N \text{ even})}$$
$$= \frac{\frac{1}{12}\cdot\frac{1}{4} + \frac{1}{6^4}\cdot\frac{1}{16}}{\frac{1}{4}+\frac{1}{16}+\cdots} = \frac{4^2 3^3 + 1}{4^4 3^3}.$$

(c) Writing D for the number shown by the first die,

$$\mathbb{P}(N=2 \mid S=4, D=1) = \frac{\mathbb{P}(N=2, S=4, D=1)}{\mathbb{P}(S=4, D=1)} = \frac{\frac{1}{6}\cdot\frac{1}{6}\cdot\frac{1}{4}}{\frac{1}{6}\cdot\frac{1}{6}\cdot\frac{1}{4}+\frac{1}{6}\cdot\frac{2}{36}\cdot\frac{1}{8}+\frac{1}{6^4}\cdot\frac{1}{16}}.$$

(d) Writing M for the maximum number shown, if $1 \le r \le 6$,

$$\mathbb{P}(M \le r) = \sum_{j=1}^{\infty} \mathbb{P}(M \le r \mid N=j)2^{-j} = \sum_{j=1}^{\infty}\left(\frac{r}{6}\right)^j \frac{1}{2^j} = \frac{r}{12}\left(1-\frac{r}{12}\right)^{-1} = \frac{r}{12-r}.$$

Finally, $\mathbb{P}(M=r) = \mathbb{P}(M \le r) - \mathbb{P}(M \le r-1)$.

16. (a) $\omega \in B$ if and only if, for all n, $\omega \in \bigcup_{i=n}^{\infty} A_i$, which is to say that ω belongs to infinitely many of the A_n.

(b) $\omega \in C$ if and only if, for some n, $\omega \in \bigcap_{i=n}^{\infty} A_i$, which is to say that ω belongs to all but a finite number of the A_n.

(c) It suffices to note that B is a countable intersection of countable unions of events, and is therefore an event.

(d) We have that

$$C_n = \bigcap_{i=n}^{\infty} A_i \subseteq A_n \subseteq \bigcup_{i=n}^{\infty} A_i = B_n,$$

and therefore $\mathbb{P}(C_n) \le \mathbb{P}(A_n) \le \mathbb{P}(B_n)$. By the continuity of probability measures (1.3.5), if $C_n \to C$ then $\mathbb{P}(C_n) \to \mathbb{P}(C)$, and if $B_n \to B$ then $\mathbb{P}(B_n) \to \mathbb{P}(B)$. If $B = C = A$ then

$$\mathbb{P}(A) = \mathbb{P}(C) \le \lim_{n\to\infty} \mathbb{P}(A_n) \le \mathbb{P}(C) = \mathbb{P}(A).$$

17. If B_n and C_n are independent for all n then, using the fact that $C_n \subseteq B_n$,

$$\mathbb{P}(B_n)\mathbb{P}(C_n) = \mathbb{P}(B_n \cap C_n) = \mathbb{P}(C_n) \to \mathbb{P}(C) \qquad \text{as } n \to \infty,$$

and also $\mathbb{P}(B_n)\mathbb{P}(C_n) \to \mathbb{P}(B)\mathbb{P}(C)$ as $n \to \infty$, so that $\mathbb{P}(C) = \mathbb{P}(B)\mathbb{P}(C)$, whence either $\mathbb{P}(C) = 0$ or $\mathbb{P}(B) = 1$ or both. In any case $\mathbb{P}(B \cap C) = \mathbb{P}(B)\mathbb{P}(C)$.

If $A_n \to A$ then $A = B = C$ so that $\mathbb{P}(A)$ equals 0 or 1.

18. It is standard (Theorem (1.3.5)) that $\mathbb{P}$ is continuous if it is countably additive. Suppose then that $\mathbb{P}$ is finitely additive and continuous. Let $A_1, A_2, \ldots$ be disjoint events. Then $\bigcup_1^{\infty} A_i = \lim_{n\to\infty} \bigcup_1^n A_i$, so that, by continuity and finite-additivity,

$$\mathbb{P}\left(\bigcup_1^{\infty} A_i\right) = \lim_{n\to\infty} \mathbb{P}\left(\bigcup_1^n A_i\right) = \lim_{n\to\infty} \sum_1^n \mathbb{P}(A_i) = \sum_1^{\infty} \mathbb{P}(A_i).$$

19. The network of friendship is best represented as a square with diagonals, with the corners labelled A, B, C, and D. Draw a diagram. Each link of the network is absent with probability p. We write EF for the event that a typical link EF is present, and EF^c for its complement. We write A $\leftrightarrow$ D for the event that A is connected to D by present links.

(d)
$$\begin{aligned}\mathbb{P}(\mathrm{A}\leftrightarrow\mathrm{D}\mid\mathrm{AD^c}) &= \mathbb{P}(\mathrm{A}\leftrightarrow\mathrm{D}\mid\mathrm{AD^c}\cap\mathrm{BC^c})p+\mathbb{P}(\mathrm{A}\leftrightarrow\mathrm{D}\mid\mathrm{AD^c}\cap\mathrm{BC})(1-p)\\ &= \{1-(1-(1-p)^2)^2\}p+(1-p^2)^2(1-p).\end{aligned}$$

(c)
$$\begin{aligned}\mathbb{P}(\mathrm{A}\leftrightarrow\mathrm{D}\mid\mathrm{BC^c}) &= \mathbb{P}(\mathrm{A}\leftrightarrow\mathrm{D}\mid\mathrm{AD^c}\cap\mathrm{BC^c})p+\mathbb{P}(\mathrm{A}\leftrightarrow\mathrm{D}\mid\mathrm{BC^c}\cap\mathrm{AD})(1-p)\\ &= \{1-(1-(1-p)^2)^2\}p+(1-p).\end{aligned}$$

(b)
$$\begin{aligned}\mathbb{P}(\mathrm{A}\leftrightarrow\mathrm{D}\mid\mathrm{AB^c}) &= \mathbb{P}(\mathrm{A}\leftrightarrow\mathrm{D}\mid\mathrm{AB^c}\cap\mathrm{AD^c})p+\mathbb{P}(\mathrm{A}\leftrightarrow\mathrm{D}\mid\mathrm{AB^c}\cap\mathrm{AD})(1-p)\\ &= (1-p)\{1-p(1-(1-p)^2)\}p+(1-p).\end{aligned}$$

(a)
$$\begin{aligned}\mathbb{P}(\mathrm{A}\leftrightarrow\mathrm{D}) &= \mathbb{P}(\mathrm{A}\leftrightarrow\mathrm{D}\mid\mathrm{AD^c})p+\mathbb{P}(\mathrm{A}\leftrightarrow\mathrm{D}\mid\mathrm{AD})(1-p)\\ &= \{1-(1-(1-p)^2)^2\}p^2+(1-p^2)^2p(1-p)+(1-p).\end{aligned}$$

20. We condition on the result of the first toss. If this is a head, then we require an odd number of heads in the next $n-1$ tosses. Similarly, if the first toss is a tail, we require an even number of heads in the next $n-1$ tosses. Hence

$$p_n = p(1-p_{n-1})+(1-p)p_{n-1} = (1-2p)p_{n-1}+p$$

with $p_0 = 1$. As an alternative to induction, we may seek a solution of the form $p_n = A+B\lambda^n$. Substitute this into the above equation to obtain

$$A+B\lambda^n = (1-2p)A+(1-2p)B\lambda^{n-1}+p$$

and $A+B=1$. Hence $A=\frac{1}{2}$, $B=\frac{1}{2}$, $\lambda = 1-2p$.

21. Let $A = \{$run of r heads precedes run of s tails$\}$, $B = \{$first toss is a head$\}$, and $C = \{$first s tosses are tails$\}$. Then

$$\mathbb{P}(A\mid B^c) = \mathbb{P}(A\mid B^c\cap C)\mathbb{P}(C\mid B^c)+\mathbb{P}(A\mid B^c\cap C^c)\mathbb{P}(C^c\mid B^c) = 0+\mathbb{P}(A\mid B)(1-q^{s-1}),$$

where $p = 1-q$ is the probability of heads on any single toss. Similarly $\mathbb{P}(A\mid B) = p^{r-1}+\mathbb{P}(A\mid B^c)(1-p^{r-1})$. We solve for $\mathbb{P}(A\mid B)$ and $\mathbb{P}(A\mid B^c)$, and use the fact that $\mathbb{P}(A) = \mathbb{P}(A\mid B)p+\mathbb{P}(A\mid B^c)q$, to obtain

$$\mathbb{P}(A) = \frac{p^{r-1}(1-q^s)}{p^{r-1}+q^{s-1}-p^{r-1}q^{s-1}}.$$

22. (a) (i) Since every cherry has the same chance to be this cherry, notwithstanding the fact that five are now in the pig, the probability that the cherry in question contains a stone is $\frac{5}{20}=\frac{1}{4}$.
(ii) Think about it the other way round. *First* a random stone is removed, and *then* the pig chooses his fruit. This does not change the relevant probabilities. Let C be the event that the removed cherry contains a stone, and let P be the event that the pig gets at least one stone. Then $\mathbb{P}(P\mid C)$ is the probability that out of 19 cherries, 15 of which are stoned, the pig gets a stone. Therefore

$$\mathbb{P}(P\mid C) = 1-\mathbb{P}\big(\text{pig chooses only stoned cherries}\,\big|\,C\big) = 1-\tfrac{15}{19}\cdot\tfrac{14}{18}\cdot\tfrac{13}{17}\cdot\tfrac{12}{16}\cdot\tfrac{11}{15}.$$

(b) Yes, by symmetry, so long as the 'inadvertently' withheld tickets were withheld independently of their numbers.

23. Label the seats $1, 2, \dots, 2n$ clockwise. For the sake of definiteness, we dictate that seat 1 be occupied by a woman; this determines the sex of the occupant of every other seat. For $1 \le k \le 2n$, let A_k be the event that seats $k, k+1$ are occupied by one of the couples (we identify seat $2n+1$ with seat 1). The required probability is

$$\mathbb{P}\left(\bigcap_1^{2n} A_i^{\mathrm{c}}\right) = 1 - \mathbb{P}\left(\bigcup_1^{2n} A_i\right) = 1 - \sum_i \mathbb{P}(A_i) + \sum_{i<j} \mathbb{P}(A_i \cap A_j) - \cdots.$$

Now, $\mathbb{P}(A_i) = n(n-1)!^2/n!^2$, since there are n couples who may occupy seats i and $i+1$, $(n-1)!$ ways of distributing the remaining $n-1$ women, and $(n-1)!$ ways of distributing the remaining $n-1$ men. Similarly, if $1 \le i < j \le 2n$, then

$$\mathbb{P}(A_i \cap A_j) = \begin{cases} n(n-1)\dfrac{(n-2)!^2}{n!^2} & \text{if } |i-j| \ne 1 \\ 0 & \text{if } |i-j| = 1, \end{cases}$$

subject to $\mathbb{P}(A_1 \cap A_{2n}) = 0$. In general,

$$\mathbb{P}(A_{i_1} \cap A_{i_2} \cap \cdots \cap A_{i_k}) = \frac{n!}{(n-k)!} \frac{(n-k)!^2}{n!^2} = \frac{(n-k)!}{n!}$$

if $i_1 < i_2 < \cdots < i_k$ and $i_{j+1} - i_j \ge 2$ for $1 \le j < k$, and $2n + i_1 - i_k \ge 2$; otherwise this probability is 0. Hence

$$\mathbb{P}\left(\bigcap_1^{2n} A_i^{\mathrm{c}}\right) = \sum_{k=0}^{n} (-1)^k \frac{(n-k)!}{n!} S_{k,n}$$

where $S_{k,n}$ is the number of ways of choosing k non-overlapping pairs of adjacent seats.

Finally, we calculate $S_{k,n}$. Consider first the number $N_{k,m}$ of ways of picking k non-overlapping pairs of adjacent seats from a line (rather than a circle) of m seats labelled $1, 2, \dots, m$. There is a one–one correspondence between the set of such arrangements and the set of $(m-k)$–vectors containing k 1's and $(m-2k)$ 0's. To see this, take such an arrangement of seats, and count 0 for an unchosen seat and 1 for a chosen pair of seats; the result is such a vector. Conversely take such a vector, read its elements in order, and construct the arrangement of seats in which each 0 corresponds to an unchosen seat and each 1 corresponds to a chosen pair. It follows that $N_{k,m} = \binom{m-k}{k}$.

Turning to $S_{k,n}$, either the pair $2n, 1$ is chosen or it is not. If it is chosen, we require another $k-1$ pairs out of a line of $2n-2$ seats. If it is not chosen, we require k pairs out of a line of $2n$ seats. Therefore

$$S_{k,n} = N_{k-1,2n-2} + N_{k,2n} = \binom{2n-k-1}{k-1} + \binom{2n-k}{k} = \binom{2n-k}{k} \frac{2n}{2n-k}.$$

24. Think about the experiment as laying down the $b+r$ balls from left to right in a random order. The number of possible orderings equals the number of ways of placing the blue balls, namely $\binom{b+r}{b}$. The number of ways of placing the balls so that the first k are blue, and the next red, is the number of ways of placing the red balls so that the first is in position $k+1$ and the remainder are amongst the $r+b-k-1$ places to the right, namely $\binom{r+b-k-1}{r-1}$. The required result follows.

The probability that the last ball is red is $r/(r+b)$, the same as the chance of being red for the ball in any other given position in the ordering.

25. We argue by induction on the total number of balls in the urn. Let p_{ac} be the probability that the last ball is azure, and suppose that $p_{ac} = \frac{1}{2}$ whenever $a, c \ge 1$, $a + c \le k$. Let α and σ be such that

$\alpha, \sigma \geq 1$, $\alpha + \sigma = k + 1$. Let A_i be the event that i azure balls are drawn before the first carmine ball, and let C_j be the event that j carmine balls are drawn before the first azure ball. We have, by taking conditional probabilities and using the induction hypothesis, that

$$\begin{aligned} p_{\alpha\sigma} &= \sum_{i=1}^{\alpha} p_{\alpha-i,\sigma}\mathbb{P}(A_i) + \sum_{j=1}^{\sigma} p_{\alpha,\sigma-j}\mathbb{P}(C_j) \\ &= p_{0,\sigma}\mathbb{P}(A_\alpha) + p_{\alpha,0}\mathbb{P}(C_\sigma) + \tfrac{1}{2}\sum_{i=1}^{\alpha-1}\mathbb{P}(A_i) + \tfrac{1}{2}\sum_{j=1}^{\sigma-1}\mathbb{P}(C_j). \end{aligned}$$

Now $p_{0,\sigma} = 0$ and $p_{\alpha,0} = 1$. Also, by an easy calculation,

$$\mathbb{P}(A_\alpha) = \frac{\alpha}{\alpha+\sigma} \cdot \frac{\alpha-1}{\alpha+\sigma-1} \cdots \frac{1}{\sigma+1} = \frac{\alpha!\,\sigma!}{(\alpha+\sigma)!} = \mathbb{P}(C_\sigma).$$

It follows from the above two equations that

$$p_{\alpha\sigma} = \tfrac{1}{2}\left(\sum_{i=1}^{\alpha}\mathbb{P}(A_i) + \sum_{j=1}^{\sigma}\mathbb{P}(C_j)\right) + \tfrac{1}{2}\big(\mathbb{P}(C_\sigma) - \mathbb{P}(A_\alpha)\big) = \tfrac{1}{2}.$$

26. (a) If she says the ace of hearts is present, then this imparts no information about the other card, which is equally likely to be any of the three other possibilities.

(b) In the given protocol, interchange hearts and diamonds.

27. Writing A if A tells the truth, and $\mathrm{A^c}$ otherwise, etc., the only outcomes consistent with D telling the truth are ABCD, $\mathrm{AB^cC^cD}$, $\mathrm{A^cBC^cD}$, and $\mathrm{A^cB^cCD}$, with a total probability of $\frac{13}{81}$. Likewise, the only outcomes consistent with D lying are $\mathrm{A^cB^cC^cD^c}$, $\mathrm{A^cBCD^c}$, $\mathrm{AB^cCD^c}$, and $\mathrm{ABC^cD^c}$, with a total probability of $\frac{28}{81}$. Writing S for the given statement, we have that

$$\mathbb{P}(\mathrm{D} \mid \mathrm{S}) = \frac{\mathbb{P}(\mathrm{D} \cap \mathrm{S})}{\mathbb{P}(\mathrm{D} \cap \mathrm{S}) + \mathbb{P}(\mathrm{D^c} \cap \mathrm{S})} = \frac{\frac{13}{81}}{\frac{13}{81} + \frac{28}{81}} = \frac{13}{41}.$$

Eddington himself thought the answer to be $\frac{25}{71}$; hence the 'controversy'. He argued that a truthful denial leaves things unresolved, so that if, for example, B truthfully denies that C contradicts D, then we cannot deduce that C supports D. He deduced that the only sequences which are inconsistent with the given statement are $\mathrm{AB^cCD}$ and $\mathrm{AB^cC^cD^c}$, and therefore

$$\mathbb{P}(\mathrm{D} \mid \mathrm{S}) = \frac{\frac{25}{81}}{\frac{25}{81} + \frac{46}{81}} = \frac{25}{71}.$$

Which side are *you* on?

28. Let B_r be the event that the rth vertex of a randomly selected cube is blue, and note that $\mathbb{P}(B_r) = \frac{1}{10}$. By Boole's inequality,

$$\mathbb{P}\left(\bigcup_{r=1}^{8} B_r\right) \leq \sum_{r=1}^{8}\mathbb{P}(B_r) = \tfrac{8}{10} < 1,$$

so at least 20 per cent of such cubes have only red vertices.

29. (a) $\mathbb{P}(B \mid A) = \mathbb{P}(A \cap B)/\mathbb{P}(A) = \mathbb{P}(A \mid B)\mathbb{P}(B)/\mathbb{P}(A) > \mathbb{P}(B)$.
(b) $\mathbb{P}(A \mid B^c) = \mathbb{P}(A \cap B^c)/\mathbb{P}(B^c) = \{\mathbb{P}(A) - \mathbb{P}(A \cap B)\}/\mathbb{P}(B^c) < \mathbb{P}(A)$.
(c) No. Consider the case $A \cap C = \varnothing$.

30. (a) The number of possible combinations of birthdays of m people is 365^m; the number of combinations of different birthdays is $365!/(365-m)!$. Use your calculator for the final part.
(b) The answer is

$$p_1 = 365\binom{364}{n-2}\frac{m!}{2}\frac{1}{365^m}.$$

(c) In the same manner as on Earth,

$$p_0 = \frac{M}{M}\cdot\frac{M-1}{M}\cdots\frac{M-m+1}{M},$$

so that, as $M \to \infty$,

$$\log p_0 = \sum_{k=1}^{m-1}\log\left(1-\frac{k}{M}\right)$$
$$= -\sum_{k=1}^{m-1}\frac{k}{M} + \mathrm{O}(M^{-2}) = -\frac{m(m-1)}{2M} + \mathrm{O}(M^{-2}).$$

31. (a) $\binom{n-r+1}{r}\bigg/\binom{n}{r}$.

(b) $(r-1)\binom{n-r+1}{r-1}\bigg/\binom{n}{r}$.

(c) $\dfrac{1}{r!}$.

(d) $1\bigg/\binom{n}{r}$.

(e) $\binom{r}{k}\binom{n-r}{r-k}\bigg/\binom{n}{r}$.

32. In the obvious notation, $\mathbb{P}(w\mathrm{S}, x\mathrm{H}, y\mathrm{D}, z\mathrm{C}) = \binom{13}{w}\binom{13}{x}\binom{13}{y}\binom{13}{z}\big/\binom{52}{13}$. Now use your calculator. Turning to the 'shape vector' (w, x, y, z) with $w \ge x \ge y \ge z$,

$$\mathbb{P}(w, x, y, z) = \begin{cases} 4\mathbb{P}(w\mathrm{S}, x\mathrm{H}, y\mathrm{D}, z\mathrm{C}) & \text{if } w \ne x = y = z, \\ 12\mathbb{P}(w\mathrm{S}, x\mathrm{H}, y\mathrm{D}, z\mathrm{C}) & \text{if } w = x \ne y \ne z, \end{cases}$$

on counting the disjoint ways of obtaining the shapes in question.

33. Use your calculator, and divide each of the following by $\binom{52}{5}$.

$$\binom{13}{1}\binom{4}{3}\binom{12}{3}\binom{4}{1}^3, \quad \binom{13}{2}\binom{4}{2}^2\binom{11}{1}\binom{4}{1}, \quad \binom{13}{1}\binom{4}{3}\binom{12}{2}\binom{4}{1}^2,$$
$$10\binom{4}{1}^5 - 10\binom{4}{1}, \quad \binom{4}{1}\binom{13}{5}, \quad \binom{13}{1}\binom{4}{3}\binom{12}{1}\binom{4}{2},$$
$$\binom{13}{1}\binom{4}{4}\binom{12}{1}\binom{4}{1}, \quad 10\binom{4}{1}.$$

34. Divide each of the following by 6^5.

$$\frac{6!\,5!}{3!\,(2!)^2},\quad \frac{6!\,5!}{3!\,(2!)^3},\quad \frac{6!\,5!}{2!\,(3!)^2},$$
$$6!,\quad \frac{6!\,5!}{4!\,3!\,2!},\quad \frac{6!\,5!}{(4!)^2},$$
$$\frac{6!\,5!}{(5!)^2}.$$

35. Let S_r denote the event that you receive r similar answers, and T the event that they are correct. Denote the event that your interlocutor is a tourist by V. Then $T \cap V^c = \varnothing$, and

$$\mathbb{P}(T \mid S_r) = \frac{\mathbb{P}(T \cap V \cap S_r)}{\mathbb{P}(S_r)} = \frac{\mathbb{P}(T \cap S_r \mid V)\mathbb{P}(V)}{\mathbb{P}(S_r)}.$$

Hence:

(a) $\mathbb{P}(T \mid S_1) = \frac{3}{4} \times \frac{2}{3}/1 = \frac{1}{2}$.

(b) $\mathbb{P}(T \mid S_2) = (\frac{3}{4})^2 \cdot \frac{2}{3}/[\{(\frac{3}{4})^2 + (\frac{1}{4})^2\}\frac{2}{3} + \frac{1}{3}] = \frac{1}{2}$.

(c) $\mathbb{P}(T \mid S_3) = (\frac{3}{4})^3 \cdot \frac{2}{3}/[\{(\frac{3}{4})^3 + (\frac{1}{4})^3\}\frac{2}{3} + \frac{1}{3}] = \frac{9}{20}$.

(d) $\mathbb{P}(T \mid S_4) = (\frac{3}{4})^4 \cdot \frac{2}{3}/[\{(\frac{3}{4})^4 + (\frac{1}{4})^4\}\frac{2}{3} + \frac{1}{3}] = \frac{27}{70}$.

(e) If the last answer differs, then the speaker is surely a tourist, so the required probability is

$$\frac{(\frac{3}{4})^3 \cdot \frac{1}{4}}{(\frac{3}{4})^3 \times \frac{1}{4} + (\frac{1}{4})^3 . \frac{3}{4}} = \frac{9}{10}.$$

36. Let E (respectively W) denote the event that the answer East (respectively West) is given.

(a) Using conditional probability,

$$\mathbb{P}(\text{East correct} \mid \text{E}) = \frac{\epsilon\mathbb{P}(\text{E} \mid \text{East correct})}{\mathbb{P}(\text{E})} = \frac{\epsilon \cdot \frac{2}{3} . \frac{3}{4}}{\frac{1}{2}\epsilon + (\frac{2}{3} . \frac{1}{4} + \frac{1}{3})(1+\epsilon)} = \epsilon,$$

$$\mathbb{P}(\text{East correct} \mid \text{W}) = \frac{\epsilon(\frac{2}{3} . \frac{1}{4} + \frac{1}{3})}{\epsilon(\frac{1}{6} + \frac{1}{3}) + \frac{2}{3} . \frac{3}{4}(1-\epsilon)} = \epsilon.$$

(b) Likewise, one obtains for the answer EE,

$$\frac{\epsilon \cdot \frac{2}{3}(\frac{3}{4})^2}{\epsilon \cdot \frac{2}{3}(\frac{3}{4})^2 + (1-\epsilon)\big(\frac{2}{3}(\frac{1}{4})^2 + \frac{1}{3}\big)} = \epsilon,$$

and for the answer WW,

$$\frac{\epsilon\big(\frac{2}{3}(\frac{1}{4})^2 + \frac{1}{3}\big)}{\epsilon \cdot \frac{3}{8} + (1-\epsilon)\frac{3}{8}} = \epsilon.$$

(c) Similarly for EEE,

$$\epsilon(\tfrac{2}{3})(\tfrac{3}{4})^2(\{\epsilon(\tfrac{2}{3})(\tfrac{3}{4})^3 + (1-\epsilon)(\tfrac{2}{3}(\tfrac{1}{4})3 + \tfrac{1}{3})\} = \frac{9\epsilon}{11 - 2\epsilon}$$

and for WWW,

$$\frac{\epsilon\{(\frac{2}{3})(\frac{1}{4})^3 + \frac{1}{3}\}}{\epsilon\big[(\frac{2}{3})(\frac{1}{4})^3 + \frac{1}{3}\big] + (1-\epsilon)\frac{2}{3}(\frac{3}{4})^3} = \frac{11\epsilon}{9+2\epsilon}.$$

Then for $\epsilon = \frac{9}{20}$, the first is $\frac{81}{202}$; the second is $\frac{1}{2}$, as you would expect if you look at Problem (1.8.35).

37. Use induction. The inductive step employs Boole's inequality and the fact that

$$\mathbb{P}\Big(\bigcup_{r=1}^{n+1} A_r\Big) = \mathbb{P}(A_{n+1}) + \mathbb{P}\Big(\bigcup_{r=1}^{n} A_r\Big) - \mathbb{P}\Big(\bigcup_{r=1}^{n}(A_r \cap A_{n+1})\Big).$$

38. We propose to prove by induction that

$$\mathbb{P}\Big(\bigcup_{r=1}^{n} A_r\Big) \le \sum_{r=1}^{n}\mathbb{P}(A_r) - \sum_{2\le r\le n}\mathbb{P}(A_r \cap A_1).$$

There is nothing special about the choice of A_1 in this inequality, which will therefore hold with any suffix k playing the role of the suffix 1. Kounias's inequality is then implied.

The above inequality holds trivially when $n = 1$. Assume that it holds for some value of n (≥ 1). We have that

$$\begin{aligned}\mathbb{P}\Big(\bigcup_{r=1}^{n+1} A_r\Big) &= \mathbb{P}\Big(\bigcup_{r=1}^{n} A_r\Big) + \mathbb{P}(A_{n+1}) - \mathbb{P}\Big(A_{n+1} \cap \bigcup_{r=1}^{n} A_r\Big)\\ &\le \sum_{r=1}^{n}\mathbb{P}(A_r) - \sum_{2\le r\le n}\mathbb{P}(A_r\cap A_1) + \mathbb{P}(A_{n+1}) - \mathbb{P}\Big(A_{n+1}\cap\bigcup_{r=1}^{n} A_r\Big)\\ &\le \sum_{r=1}^{n+1}\mathbb{P}(A_r) - \sum_{2\le r\le n+1}\mathbb{P}(A_r \cap A_1)\end{aligned}$$

since $\mathbb{P}(A_{n+1}\cap A_1) \le \mathbb{P}\big(A_{n+1}\cap \bigcup_{r=1}^{n} A_r\big)$.

39. Let $n \ge 2$. We label the passengers $1, 2, \dots, n$ in order, and we label the seats in such a way that the seat assignment of passenger i is also labelled i. Write F for the event that the last passenger finds his or her assigned seat to be free. Let K (≥ 2) be the seat taken by passenger 1, so that $\mathbb{P}(F) = (n-1)^{-1}\sum_{k=2}^{n}\alpha_k$ where $\alpha_k = \mathbb{P}(F \mid K = k)$. Note that $\alpha_n = 0$. Passengers $2, 3, \dots, K-1$ occupy their correct seats. Passenger K either occupies seat 1, in which case all subsequent passengers take their correct seats, or s/he occupies some seat L satisfying $L > K$. In the latter case, passengers $K+1, K+2, \dots, L-1$ are correctly seated. We obtain thus that

$$\alpha_k = \frac{1}{n-k+1}(1 + \alpha_{k+1} + \alpha_{k+2} + \cdots + \alpha_n), \qquad 2 \le k < n.$$

Therefore $\alpha_k = \frac{1}{2}$ for $2 \le k < n$, by induction, and so $\mathbb{P}(F) = \frac{1}{2}(n-2)/(n-1)$.

If passenger 1 sits in seat K where K is chosen uniformly at random from $\{1, 2, \dots, n\}$, then

$$\mathbb{P}(F) = \frac{1}{n}\sum_{k=1}^{n}\alpha_k = \frac{1}{n} + \frac{n-1}{n}\cdot\frac{n-2}{2(n-1)} = \frac{1}{2}.$$

Here is the smart solution to the last part. When the nth passenger chooses a seat, there is only one available, and this must be either seat 1 or seat n (since seat $k \in \{2, 3, \dots, n-1\}$, if free, would have been claimed earlier by passenger k). Each of these two possibilities $1, n$ has probability $\frac{1}{2}$, since no earlier decision has recognised any difference between them.

40. Placing a ship uniformly at random amounts to picking a great circle at random and then randomly selecting the centre of one of the two corresponding hemispheres as the location of the ship. Let R_i be the ith of the $n^2 - n + 2$ regions of the hint, and let A_i be the event that R_i is controlled by no ship. Since R_i must lie in the uncontrolled hemisphere of every ship, we have $\mathbb{P}(A_i) = 2^{-n}$. Now, for $i \neq j$, there exists a great circle that separates R_i and R_j, so that the events A_i are disjoint. Therefore,

$$\mathbb{P}\left(\bigcup_i A_i\right) = (n^2 - n + 2)2^{-n},$$

and the claim follows.

41. (a) Given $X = k$, you belong to A with probability k/n. Therefore, the probability that your team is A and has size k is $(k/n)/(n-1)$, whence the answer is $2k/[(n(n-1)]$.

(b) Let T_k be the event that your team has size k, and C the event that you are its captain. Since there are two captains, $\mathbb{P}(C) = 2/n$. Then,

$$\mathbb{P}(T_k \mid C) = \frac{\mathbb{P}(T_k \cap C)}{\mathbb{P}(C)} = \frac{1}{k} \cdot \frac{2k}{n(n-1)} \frac{1}{\mathbb{P}(C)} = \frac{1}{n-1},$$

the uniform distribution on $\{1, 2, \dots, n-1\}$.

42. Let n be odd (a similar argument applies if n is even). For A to win on the nth toss, her series of $\frac{1}{2}(n+1)$ flips must end in a head, preceded by a (possibly empty) run of tails of some length in $\{1, 2, \dots, \frac{1}{2}(n-1)\}$, preceded by a (possibly empty) run of heads. There are $\frac{1}{2}(n-1)$ such sequences each with probability $(\frac{1}{2})^{(n+1)/2}$. In addition, B needs not to win before the nth toss, which happens if and only if he obtained a run of tails, followed by a run of heads (either run may be empty). There are $\frac{1}{2}(n+1)$ such sequences, each with probability $(\frac{1}{2})^{(n-1)/2}$. The answer is the product, namely $(n-1)(n+1)(\frac{1}{2})^{n+2}$.

Set $n = 2k + 1$. The chance that A wins overall is

$$\sum_{k=1}^{\infty} 2k(2k+2)(\tfrac{1}{2})^{2k+3} = \tfrac{1}{8} \sum_{k=1}^{\infty} k(k+1)(\tfrac{1}{4})^{k-1} = \tfrac{16}{27}.$$

You may find it useful that, for $|z| < 1$,

$$\sum_{k=1}^{\infty} k(k+1)z^{k-1} = \frac{d^2}{dz^2}\left(\frac{z^2}{1-z}\right).$$

43. Let N_m denote the event in question, and let B_m be the event that the mth flip completes an earliest run of length k. Then, $\mathbb{P}(B_m) = (1-p)p^k \mathbb{P}(N_{m-k-1})$ for $m \geq k+1$. Since N_{m-1} is the disjoint union $N_{m-1} = N_m \cup B_m$,

$$\mathbb{P}(N_{m-1}) = \mathbb{P}(N_m) + \mathbb{P}(B_m),$$

and the result follows.

Set $k = 2$. The auxiliary equation is $x^3 - x^2 + p^2 q = 0$ where $p + q = 1$. It has $x - p$ as a factor, and hence the roots are $x = p, x_-, x_+$ where

$$x_\pm = \tfrac{1}{2}\left(q \pm \sqrt{q^2 + 4pq}\right).$$

By the boundary conditions $\rho_0 = \rho_1 = 1$, $\rho_2 = 1 - p^2$, we have $\rho_m = A + Bx_-^m + Cx_+^m$ where

$$A = 0, \quad B = \frac{1 - x_+}{x_- - x_+}, \quad C = \frac{1 - x_-}{x_+ - x_-}.$$

In particular, if $p = \frac{1}{2}$, we have $x_\pm = \frac{1}{4}(1 \pm \sqrt{5})$, and

$$\rho_m = \tfrac{1}{10}(5 - 3\sqrt{5})x_-^m + \tfrac{1}{10}(5 + 3\sqrt{5})x_+^m, \qquad m \geq 1.$$

2

Random variables and their distributions

2.1 Solutions. Random variables

1. (a) If $a > 0$, $x \in \mathbb{R}$, then $\{\omega : aX(\omega) \le x\} = \{\omega : X(\omega) \le x/a\} \in \mathcal{F}$ since X is a random variable. If $a < 0$,

$$\{\omega : aX(\omega) \le x\} = \{\omega : X(\omega) \ge x/a\} = \left\{\bigcup_{n\ge1}\left\{\omega : X(\omega) \le \frac{x}{a} - \frac{1}{n}\right\}\right\}^{c}$$

which lies in $\mathcal{F}$ since it is the complement of a countable union of members of $\mathcal{F}$. If $a = 0$,

$$\{\omega : aX(\omega) \le x\} = \begin{cases} \varnothing & \text{if } x < 0, \\ \Omega & \text{if } x \ge 0; \end{cases}$$

in either case, the event lies in $\mathcal{F}$.
(b) For $\omega \in \Omega$, $X(\omega) - X(\omega) = 0$, so that $X - X$ is the zero random variable (that this *is* a random variable follows from part (i) with $a = 0$). Similarly $X(\omega) + X(\omega) = 2X(\omega)$.

2. Set $Y = aX + b$. We have that

$$\mathbb{P}(Y \le y) = \begin{cases} \mathbb{P}\big(X \le (y-b)/a\big) = F\big((y-b)/a\big) & \text{if } a > 0, \\ \mathbb{P}\big(X \ge (y-b)/a\big) = 1 - \lim_{x\uparrow(y-b)/a} F(x) & \text{if } a < 0. \end{cases}$$

Finally, if $a = 0$, then $Y = b$, so that $\mathbb{P}(Y \le y)$ equals 0 if $b > y$ and 1 if $b \le y$.

3. Assume that any specified sequence of heads and tails with length n has probability 2^{-n}. There are exactly $\binom{n}{k}$ such sequences with k heads.

If heads occurs with probability p then, assuming the independence of outcomes, the probability of any given sequence of k heads and $n-k$ tails is $p^k(1-p)^{n-k}$. The answer is therefore $\binom{n}{k}p^k(1-p)^{n-k}$.

4. Write $H = \lambda F + (1-\lambda)G$. Then $\lim_{x\to-\infty} H(x) = 0$, $\lim_{x\to\infty} H(x) = 1$, and clearly H is non-decreasing and right-continuous. Therefore H is a distribution function.

5. The function $g(F(x))$ is a distribution function whenever g is continuous and non-decreasing on $[0, 1]$, with $g(0) = 0$, $g(1) = 1$. This is easy to check in each special case.

6. Let U be uniform on the finite set S, and let $\varnothing \ne R \subseteq S$. For $u \in R$, we have

$$\mathbb{P}(U = u \mid U \in R) = \frac{\mathbb{P}(U = s)}{\mathbb{P}(U \in R)} = \frac{1/s}{r/s} = \frac{1}{r},$$

where $r = |R|$ and $s = |S|$.

2.2 Solutions. The law of averages

1. Let p be the potentially embarrassed fraction of the population, and suppose that each sampled individual would truthfully answer "yes" with probability p independently of all other individuals. In the modified procedure, the chance that someone says yes is $p + \frac{1}{2}(1-p) = \frac{1}{2}(1+p)$. If the proportion of yes's is now ϕ, then $2\phi - 1$ is a decent estimate of p.

The advantage of the given procedure is that it allows individuals to answer "yes" without their being identified with certainty as having the embarrassing property.

2. Clearly $H_n + T_n = n$, so that $(H_n - T_n)/n = (2H_n/n) - 1$. Therefore

$$\mathbb{P}\left(2p - 1 - \epsilon \le \frac{1}{n}(H_n - T_n) \le 2p - 1 + \epsilon\right) = \mathbb{P}\left(\left|\frac{1}{n}H_n - p\right| \le \frac{\epsilon}{2}\right) \to 1$$

as $n \to \infty$, by the law of large numbers (2.2.1).

3. Let $I_n(x)$ be the indicator function of the event $\{X_n \le x\}$. By the law of averages, $n^{-1}\sum_{r=1}^{n} I_r(x)$ converges in the sense of (2.2.1) and (2.2.6) to $\mathbb{P}(X_n \le x) = F(x)$.

2.3 Solutions. Discrete and continuous variables

1. With $\delta = \sup_m |a_m - a_{m-1}|$, we have that

$$|F(x) - G(x)| \le |F(a_m) - F(a_{m-1})| \le |F(x+\delta) - F(x-\delta)|$$

for $x \in [a_{m-1}, a_m)$. Hence $G(x)$ approaches $F(x)$ for any x at which F is continuous.

2. For y lying in the range of g, $\{Y \le y\} = \{X \le g^{-1}(y)\} \in \mathcal{F}$.

3. Certainly Y is a random variable, using the result of the previous Exercise (2.3.2). Also

$$\mathbb{P}(Y \le y) = \mathbb{P}\big(F^{-1}(X) \le y\big) = \mathbb{P}\big(X \le F(y)\big) = F(y)$$

as required. If F is discontinuous then $F^{-1}(x)$ is not defined for all x, so that Y is not well defined. If F is non-decreasing and continuous, but not strictly increasing, then $F^{-1}(x)$ is not always defined uniquely. Such difficulties may be circumvented by defining $F^{-1}(x) = \inf\{y : F(y) \ge x\}$.

4. The function $\lambda f + (1-\lambda)g$ is non-negative and integrable over $\mathbb{R}$ to 1. Finally, fg is not necessarily a density, though it may be: e.g., if $f = g = 1$, $0 \le x \le 1$ then $f(x)g(x) = 1$, $0 \le x \le 1$.

5. (a) If $d > 1$, then $\int_1^\infty cx^{-d}\,dx = c/(d-1)$. Therefore f is a density function if $c = d-1$, and $F(x) = 1 - x^{-(d-1)}$ when this holds. If $d \le 1$, then f has infinite integral and cannot therefore be a density function.
(b) By differentiating $F(x) = e^x/(1+e^x)$, we see that F is the distribution function, and $c = 1$.

2.4 Solutions. Worked examples

1. (a) If $y \ge 0$,

$$\mathbb{P}(X^2 \le y) = \mathbb{P}(X \le \sqrt{y}) - \mathbb{P}(X < -\sqrt{y}) = F(\sqrt{y}) - F(-\sqrt{y}).$$

(b) We must assume that $X \ge 0$. If $y \ge 0$,

$$\mathbb{P}(\sqrt{X} \le y) = \mathbb{P}(0 \le X \le y^2) = F(y^2).$$

(c) If $-1 \le y \le 1$,

$$\begin{aligned}\mathbb{P}(\sin X \le y) &= \sum_{n=-\infty}^{\infty} \mathbb{P}\big((2n+1)\pi - \sin^{-1} y \le X \le (2n+2)\pi + \sin^{-1} y\big) \\ &= \sum_{n=-\infty}^{\infty} \Big\{F\big((2n+2)\pi + \sin^{-1} y\big) - F\big((2n+1)\pi - \sin^{-1} y\big)\Big\}.\end{aligned}$$

(d) $\mathbb{P}(G^{-1}(X) \le y) = \mathbb{P}(X \le G(y)) = F(G(y))$.
(e) If $0 \le y \le 1$, then $\mathbb{P}(F(X) \le y) = \mathbb{P}(X \le F^{-1}(y)) = F(F^{-1}(y)) = y$. There is a small difficulty if F is not *strictly* increasing, but this is overcome by defining $F^{-1}(y) = \sup\{x : F(x) = y\}$.
(f) $\mathbb{P}(G^{-1}(F(X)) \le y) = \mathbb{P}(F(X) \le G(y)) = G(y)$.

2. It is the case that, for $x \in \mathbb{R}$, $F_Y(x)$ and $F_Z(x)$ approach $F(x)$ as $a \to -\infty$, $b \to \infty$.

2.5 Solutions. Random vectors

1. Write $f_{xw} = \mathbb{P}(X = x, W = w)$. Then $f_{00} = f_{21} = \frac{1}{4}$, $f_{10} = \frac{1}{2}$, and $f_{xw} = 0$ for other pairs x, w.

2. (a) We have that

$$f_{X,Y}(x, y) = \begin{cases} p & \text{if } (x, y) = (1, 0), \\ 1 - p & \text{if } (x, y) = (0, 1), \\ 0 & \text{otherwise.} \end{cases}$$

(b) Secondly,

$$f_{X,Z}(x, z) = \begin{cases} 1 - p & \text{if } (x, z) = (0, 0), \\ p & \text{if } (x, z) = (1, 0), \\ 0 & \text{otherwise.} \end{cases}$$

3. Differentiating gives $f_{X,Y}(x, y) = e^{-x}/\{\pi(1 + y^2)\}$, $x \ge 0$, $y \in \mathbb{R}$.

4. Let $A = \{X \le b,\ c < Y \le d\}$, $B = \{a < X \le b,\ Y \le d\}$. Clearly

$$\mathbb{P}(A) = F(b, d) - F(b, c), \quad \mathbb{P}(B) = F(b, d) - F(a, d), \quad \mathbb{P}(A \cup B) = F(b, d) - F(a, c);$$

now $\mathbb{P}(A \cap B) = \mathbb{P}(A) + \mathbb{P}(B) - \mathbb{P}(A \cup B)$, which gives the answer. Draw a map of $\mathbb{R}^2$ and plot the regions of values of (X, Y) involved.

5. The given expression equals

$$\mathbb{P}(X = x,\ Y \le y) - \mathbb{P}(X = x,\ Y \le y - 1) = \mathbb{P}(X = x,\ Y = y).$$

Secondly, for $1 \le x \le y \le 6$,

$$f(x, y) = \begin{cases} \left(\dfrac{y - x + 1}{6}\right)^r - 2\left(\dfrac{y - x}{6}\right)^r + \left(\dfrac{y - x - 1}{6}\right)^r & \text{if } x < y, \\ \left(\dfrac{1}{6}\right)^r & \text{if } x = y. \end{cases}$$

6. No, because F is twice differentiable with $\partial^2 F/\partial x \partial y < 0$.

2.7 Solutions to problems

1. By the independence of the tosses,

$$\mathbb{P}(X > m) = \mathbb{P}(\text{first } m \text{ tosses are tails}) = (1-p)^m.$$

Hence

$$\mathbb{P}(X \le x) = \begin{cases} 1-(1-p)^{\lfloor x \rfloor} & \text{if } x \ge 0, \\ 0 & \text{if } x < 0. \end{cases}$$

Remember that $\lfloor x \rfloor$ denotes the integer part of x.

2. (a) If X takes values $\{x_i : i \ge 1\}$ then $X = \sum_{i=1}^{\infty} x_i I_{A_i}$ where $A_i = \{X = x_i\}$.

(b) Partition the real line into intervals of the form $[k2^{-m}, (k+1)2^{-m})$, $-\infty < k < \infty$, and define $X_m = \sum_{k=-\infty}^{\infty} k2^{-m} I_{k,m}$ where $I_{k,m}$ is the indicator function of the event $\{k2^{-m} \le X < (k+1)2^{-m}\}$. Clearly X_m is a random variable, and $X_m(\omega) \uparrow X(\omega)$ as $m \to \infty$ for all ω.

(c) Suppose $\{X_m\}$ is a sequence of random variables such that $X_m(\omega) \uparrow X(\omega)$ for all ω. Then $\{X \le x\} = \bigcap_m \{X_m \le x\}$, which is a countable intersection of events and therefore lies in $\mathcal{F}$.

3. (a) We have that

$$\{X + Y \le x\} = \bigcap_{n=1}^{\infty} \bigcup_{r \in \mathbb{Q}^+} \left(\{X \le r\} \cap \{Y \le x - r + n^{-1}\}\right)$$

where $\mathbb{Q}^+$ is the set of positive rationals.

In the second case, if XY is a positive function, then $XY = \exp\{\log X + \log Y\}$; now use Exercise (2.3.2) and the above. For the general case, note first that $|Z|$ is a random variable whenever Z is a random variable, since $\{|Z| \le a\} = \{Z \le a\} \setminus \{Z < -a\}$ for $a \ge 0$. Now, if $a \ge 0$, then $\{XY \le a\} = \{XY < 0\} \cup \{|XY| \le a\}$ and

$$\{XY < 0\} = \big(\{X < 0\} \cap \{Y > 0\}\big) \cup \big(\{X > 0\} \cap \{Y < 0\}\big).$$

Similar relations are valid if $a < 0$.

Finally $\{\min\{X, Y\} > x\} = \{X > x\} \cap \{Y > y\}$, the intersection of events.

(b) It is enough to check that $\alpha X + \beta Y$ is a random variable whenever $\alpha, \beta \in \mathbb{R}$ and X, Y are random variables. This follows from the argument above.

If Ω is finite, we may take as basis the set $\{I_A : A \in \mathcal{F}\}$ of all indicator functions of events.

4. (a) $F(\frac{3}{2}) - F(\frac{1}{2}) = \frac{1}{2}$.

(b) $F(2) - F(1) = \frac{1}{2}$.

(c) $\mathbb{P}(X^2 \le X) = \mathbb{P}(X \le 1) = \frac{1}{2}$.

(d) $\mathbb{P}(X \le 2X^2) = \mathbb{P}(X \ge \frac{1}{2}) = \frac{3}{4}$.

(e) $\mathbb{P}(X + X^2 \le \frac{3}{4}) = \mathbb{P}(X \le \frac{1}{2}) = \frac{1}{4}$.

(f) $\mathbb{P}(\sqrt{X} \le z) = \mathbb{P}(X \le z^2) = \frac{1}{2}z^2$ if $0 \le z \le \sqrt{2}$.

5. $\mathbb{P}(X = -1) = 1 - p$, $\mathbb{P}(X = 0) = 0$, $\mathbb{P}(X \ge 1) = \frac{1}{2}p$.

6. There are 6 intervals of 5 minutes, preceding the arrival times of buses. Each such interval has probability $\frac{5}{60} = \frac{1}{12}$, so the answer is $6 \cdot \frac{1}{12} = \frac{1}{2}$.

7. Let T and B be the numbers of people on given typical flights of EasyPeasy Airlines and Rye-Loaf Airways. From Exercise (2.1.3),

$$\mathbb{P}(T=k) = \binom{10}{k}\left(\frac{9}{10}\right)^k\left(\frac{1}{10}\right)^{10-k}, \quad \mathbb{P}(B=k) = \binom{20}{k}\left(\frac{9}{10}\right)^k\left(\frac{1}{10}\right)^{20-k}.$$

Now

$$\mathbb{P}(\text{EPA overbooked}) = \mathbb{P}(T=10) = (\tfrac{9}{10})^{10},$$
$$\mathbb{P}(\text{RLA overbooked}) = \mathbb{P}(B \ge 19) = 20(\tfrac{9}{10})^{19}(\tfrac{1}{10}) + (\tfrac{9}{10})^{20},$$

of which the latter is the larger.

8. Assuming the coins are fair, the chance of getting at least five heads is $(\frac{1}{2})^6 + 6(\frac{1}{2})^6 = \frac{7}{64}$.

9. (a) We have that

$$\mathbb{P}(X^+ \le x) = \begin{cases} 0 & \text{if } x < 0, \\ F(x) & \text{if } x \ge 0. \end{cases}$$

(b) Secondly,

$$\mathbb{P}(X^- \le x) = \begin{cases} 0 & \text{if } x < 0, \\ 1 - \lim_{y \uparrow -x} F(y) & \text{if } x \ge 0. \end{cases}$$

(c) $\mathbb{P}(|X| \le x) = \mathbb{P}(-x \le X \le x)$ if $x \ge 0$. Therefore

$$\mathbb{P}(|X| \le x) = \begin{cases} 0 & \text{if } x < 0, \\ F(x) - \lim_{y \uparrow -x} F(y) & \text{if } x \ge 0. \end{cases}$$

(d) $\mathbb{P}(-X \le x) = 1 - \lim_{y \uparrow -x} F(y)$.

10. By the continuity of probability measures (1.3.5),

$$\mathbb{P}(X = x_0) = \lim_{y \uparrow x_0} \mathbb{P}(y < X \le x_0) = F(x_0) - \lim_{y \uparrow x_0} F(y) = F(x_0) - F(x_0-),$$

using general properties of F. The result follows.

11. (a) Define $m = \sup\{x : F(x) < \frac{1}{2}\}$. Then $F(y) < \frac{1}{2}$ for $y < m$, and $F(m) \ge \frac{1}{2}$ (if $F(m) < \frac{1}{2}$ then $F(m') < \frac{1}{2}$ for some $m' > m$, by the right-continuity of F, a contradiction). Hence m is a median, and is smallest with this property.

A similar argument may be used to show that $M = \sup\{x : F(x) \le \frac{1}{2}\}$ is a median, and is largest with this property. The set of medians is then the closed interval $[m, M]$.

(b) Since F is continuous, $\lim_{y \uparrow m} F(y) = F(m) \le \frac{1}{2} \le F(m)$.

12. (a) Let the dice show X and Y. Write $S = X + Y$ and $f_i = \mathbb{P}(X = i)$, $g_i = \mathbb{P}(Y = i)$. Assume that $\mathbb{P}(S = 2) = \mathbb{P}(S = 7) = \mathbb{P}(S = 12) = \frac{1}{11}$. Now

$$\mathbb{P}(S=2) = \mathbb{P}(X=1)\mathbb{P}(Y=1) = f_1 g_1,$$
$$\mathbb{P}(S=12) = \mathbb{P}(X=6)\mathbb{P}(Y=6) = f_6 g_6,$$
$$\mathbb{P}(S=7) \ge \mathbb{P}(X=1)\mathbb{P}(Y=6) + \mathbb{P}(X=6)\mathbb{P}(Y=1) = f_1 g_6 + f_6 g_1.$$

It follows that $f_1 g_1 = f_6 g_6$, and also

$$\frac{1}{11} = \mathbb{P}(S=7) \ge f_1 g_1\left(\frac{g_6}{g_1} + \frac{f_6}{f_1}\right) = \frac{1}{11}\left(x + \frac{1}{x}\right)$$

where $x = g_6/g_1$. However $x + x^{-1} > 1$ for all $x > 0$, a contradiction.

(b) Let F and L be the two numbers shown, and check that $\mathbb{P}(F + L = k \bmod 6 \mid L = l) = \frac{1}{6}$ for $1 \le l, k \le 6$.

13. (a) Clearly d_{L} satisfies (i). As for (ii), suppose that $d_{\mathrm{L}}(F, G) = 0$. Then

$$F(x) \le \lim_{\epsilon \downarrow 0}\{G(x + \epsilon) + \epsilon\} = G(x)$$

and

$$F(y) \ge \lim_{\epsilon \downarrow 0}\{G(y - \epsilon) - \epsilon\} = G(y-).$$

Now $G(y-) \ge G(x)$ if $y > x$; taking the limit as $y \downarrow x$ we obtain

$$F(x) \ge \lim_{y \downarrow x} G(y-) \ge G(x),$$

implying that $F(x) = G(x)$ for all x.

Finally, if $F(x) \le G(x + \epsilon) + \epsilon$ and $G(x) \le H(x + \delta) + \delta$ for all x and some $\epsilon, \delta > 0$, then $F(x) \le H(x + \delta + \epsilon) + \epsilon + \delta$ for all x. A similar lower bound for $F(x)$ is valid, implying that $d_{\mathrm{L}}(F, H) \le d_{\mathrm{L}}(F, G) + d_{\mathrm{L}}(G, H)$.

(b) Clearly d_{TV} satisfies (i), and $d_{\mathrm{TV}}(X, Y) = 0$ if and only if $\mathbb{P}(X = Y) = 1$. By the usual triangle inequality,

$$\big|\mathbb{P}(X = k) - \mathbb{P}(Z = k)\big| \le \big|\mathbb{P}(X = k) - \mathbb{P}(Y = k)\big| + \big|\mathbb{P}(Y = k) - \mathbb{P}(Z = k)\big|,$$

and (iii) follows by summing over k.

We have that

$$\begin{aligned} 2\big|\mathbb{P}(X \in A) - \mathbb{P}(Y \in A)\big| &= \big|\big(\mathbb{P}(X \in A) - \mathbb{P}(Y \in A)\big) - \big(\mathbb{P}(X \in A^{\mathrm{c}}) - \mathbb{P}(Y \in A^{\mathrm{c}})\big)\big| \\ &= \left|\sum_k \big(\mathbb{P}(X = k) - \mathbb{P}(Y = k)\big) J_A(k)\right| \end{aligned}$$

where $J_A(k)$ equals 1 if $k \in A$ and equals -1 if $k \in A^{\mathrm{c}}$. Therefore,

$$2\big|\mathbb{P}(X \in A) - \mathbb{P}(Y \in A)\big| \le \sum_k \big|\mathbb{P}(X = k) - \mathbb{P}(Y = k)\big| \cdot |J_A(k)| = d_{\mathrm{TV}}(X, Y).$$

Equality holds if $A = \{k : \mathbb{P}(X = k) > \mathbb{P}(Y = k)\}$.

14. (a) Note that

$$\frac{\partial^2 F}{\partial x \partial y} = -e^{-x-y} < 0, \qquad x, y > 0,$$

so that F is not a joint distribution function.

(b) In this case

$$\frac{\partial^2 F}{\partial x \partial y} = \begin{cases} e^{-y} & \text{if } 0 \le x \le y, \\ 0 & \text{if } 0 \le y \le x, \end{cases}$$

and in addition

$$\int_0^\infty \int_0^\infty \frac{\partial^2 F}{\partial x \partial y}\, dx\, dy = 1.$$

Hence F is a joint distribution function, and easy substitutions reveal the marginals:

$$F_X(x) = \lim_{y\to\infty} F(x, y) = 1 - e^{-x}, \quad x \geq 0,$$

$$F_Y(y) = \lim_{x\to\infty} F(x, y) = 1 - e^{-y} - ye^{-y}, \quad y \geq 0.$$

15. Suppose that, for some $i \neq j$, we have $p_i < p_j$ and B_i is to the left of B_j. Write m for the position of B_i and r for the position of B_j, and consider the effect of interchanging B_i and B_j. For $k \leq m$ and $k > r$, $\mathbb{P}(T \geq k)$ is unchanged by the move. For $m < k \leq r$, $\mathbb{P}(T \geq k)$ is decreased by an amount $p_j - p_i$, since this is the increased probability that the search is successful at the mth position. Therefore the interchange of B_i and B_j is desirable.

It follows that the only ordering in which $\mathbb{P}(T \geq k)$ can be reduced for no k is that ordering in which the books appear in decreasing order of probability. In the event of ties, it is of no importance how the tied books are placed.

16. Intuitively, it may seem better to go *first* since the first person has greater choice. This conclusion is in fact false. Denote the coins by C_1, C_2, C_3 in order, and suppose you go *second*. If your opponent chooses C_1 then you choose C_3, because $\mathbb{P}(C_3 \text{ beats } C_1) = \frac{2}{5} + \frac{2}{5}\cdot\frac{3}{5} = \frac{16}{25} > \frac{1}{2}$. Likewise $\mathbb{P}(C_1 \text{ beats } C_2) = \mathbb{P}(C_2 \text{ beats } C_3) = \frac{3}{5} > \frac{1}{2}$. Whichever coin your opponent picks, you can arrange to have a better than evens chance of winning.

17. Various difficulties arise in sequential decision theory, even in simple problems such as this one. The following simple argument yields the optimal policy. Suppose that you have made a unsuccessful searches "ahead" and b unsuccessful searches "behind" (if any of these searches were successful, then there is no further problem). Let A be the event that the correct direction is ahead. Then

$$\begin{aligned}\mathbb{P}(A \mid \text{current knowledge}) &= \frac{\mathbb{P}(\text{current knowledge} \mid A)\mathbb{P}(A)}{\mathbb{P}(\text{current knowledge})} \\ &= \frac{(1-p)^a\alpha}{(1-p)^a\alpha + (1-p)^b(1-\alpha)},\end{aligned}$$

which exceeds $\frac{1}{2}$ if and only if $(1-p)^a\alpha > (1-p)^b(1-\alpha)$. The optimal policy is to compare $(1-p)^a\alpha$ with $(1-p)^b(1-\alpha)$. You search ahead if the former is larger and behind otherwise; in the event of a tie, do either.

18. (a) There are $\binom{64}{8}$ possible layouts, of which 8+8+2 are linear. The answer is $18/\binom{64}{8}$.

(b) Each row and column must contain exactly one pawn. There are 8 possible positions in the first row. Having chosen which of these is occupied, there are 7 positions in the second row which are admissible, 6 in the third, and so one. The answer is $8!/\binom{64}{8}$.

19. (a) The density function is $f(x) = F'(x) = 2xe^{-x^2}$, $x \geq 0$.

(b) The density function is $f(x) = F'(x) = x^2e^{-1/x}$, $x > 0$.

(c) The density function is $f(x) = F'(x) = 2(e^x + e^{-x})^{-2}$, $x \in \mathbb{R}$.

(d) This is not a distribution function because $F'(1) < 0$.

20. We have that

$$\mathbb{P}(U = V) = \int_{\{(u,v):u=v\}} f_{U,V}(u, v)\,du\,dv = 0.$$

The random variables X, Y are continuous but not jointly continuous: there exists no integrable function $f : [0,1]^2 \to \mathbb{R}$ such that

$$\mathbb{P}(X \leq x, Y \leq y) = \int_{u=0}^{x}\int_{v=0}^{y} f(u, v)\,du\,dv, \quad 0 \leq x, y \leq 1.$$

21. We have that

$$\mathbb{P}(Y_1 = u,\ Y_2 = v) = \mathbb{P}\left(X \le \frac{1}{u + \frac{1}{v+1}}\right) - \mathbb{P}\left(X \le \frac{1}{u + \frac{1}{v}}\right) = \frac{v+1}{u(v+1)+1} - \frac{v}{uv+1}.$$

Likewise, $\mathbb{P}(Y_1 = u) = \dfrac{1}{u(u+1)}$.

22. (a) Let $v \in V$, and $a \in \mathbb{F}$ with $a \neq 0$. Then

$$\mathbb{P}(v + aX_i = y) = \mathbb{P}\big(X_i = (y - v)/a\big) = \frac{1}{|V|} = \frac{1}{q^n},$$

the uniform distribution on V. The claim follows.
(b) Let N_m be the event that $X_1, X_2, \dots, X_m$ are linearly dependent. Then

$$\begin{aligned}\mathbb{P}(N_m) &= \mathbb{P}\big(a_1X_1 + \cdots + a_mX_m = 0 \text{ for some } (a_1, \dots, a_m) \neq \mathbf{0}\big)\\ &\le \sum_{\mathbf{a}\neq\mathbf{0}} \mathbb{P}\Big(\sum_{i=1}^{m} a_iX_i = 0\Big) = \sum_{\mathbf{a}\neq\mathbf{0}} \frac{1}{q^n} = \frac{q^m - 1}{q^n} < q^{-(n-m)}.\end{aligned}$$

Let S_m be the linear span of $X_1, \dots, X_m$. By conditional probability,

$$\mathbb{P}(N_m) = \mathbb{P}(N_{m-1}) + \mathbb{P}(X_m \in S_{m-1} \mid N_{m-1}^{\mathrm{c}})\mathbb{P}(N_{m-1}^{\mathrm{c}}).$$

Off the event N_{m-1}, $|S_{m-1}| = q^{m-1}$, so that

$$\mathbb{P}(N_m) = \mathbb{P}(N_{m-1}) + \frac{q^{m-1}}{q^n}(1 - \mathbb{P}(N_{m-1})) \ge \frac{q^{m-1}}{q^n}.$$

23. (a, b) A convex function u satisfies: u is absolutely continuous on closed intervals; its right and left derivatives exist everywhere and are equal except on a countable set; those derivatives are non-decreasing wherever they exist, and the left derivative is no greater than the right derivative everywhere,
(c) Check the definition.

3

Discrete random variables

3.1 Solutions. Probability mass functions

1. (a) $C^{-1} = \sum_1^\infty 2^{-k} = 1$.
(b) $C^{-1} = \sum_1^\infty 2^{-k}/k = \log 2$.
(c) $C^{-1} = \sum_1^\infty k^{-2} = \pi^2/6$.
(d) $C^{-1} = \sum_1^\infty 2^k/k! = e^2 - 1$.

2. (i) $\frac{1}{2}$; $1 - (2\log 2)^{-1}$; $1 - 6\pi^{-2}$; $(e^2 - 3)/(e^2 - 1)$.
(ii) 1; 1; 1; 1 and 2.
(iii) It is the case that $\mathbb{P}(X \text{ even}) = \sum_{k=1}^\infty \mathbb{P}(X = 2k)$, and the answers are therefore (a) $\frac{1}{3}$, (b) $1 - (\log 3)/(\log 4)$, (c) $\frac{1}{4}$. (d) We have that

$$\sum_{k=1}^\infty \frac{2^{2k}}{(2k)!} = \sum_{i=0}^\infty \frac{2^i + (-2)^i}{2(i!)} - 1 = \tfrac{1}{2}(e^2 + e^{-2}) - 1,$$

so the answer is $\frac{1}{2}(1 - e^{-2})$.

3. The number X of heads on the second round is the same as if we toss all the coins twice and count the number which show heads on both occasions. Each coin shows heads twice with probability p^2, so $\mathbb{P}(X = k) = \binom{n}{k} p^{2k}(1 - p^2)^{n-k}$.

4. Let D_k be the number of digits (to base 10) in the integer k. Then

$$\mathbb{P}(N = k) = \mathbb{P}(N = k \mid T = D_k)\mathbb{P}(T = D_k) = \frac{1}{|S_{D_k}|} 2^{-D_k}.$$

5. (a) The assertion follows for the binomial distribution because $k(n-k) \le (n-k+1)(k+1)$. The Poisson case is trivial.

(b) This follows from the fact that $k^8 \ge (k^2 - 1)^4$.

(c) The geometric mass function $f(k) = qp^k$, $k \ge 0$.

3.2 Solutions. Independence

1. We have that

$$\mathbb{P}(X = 1, Z = 1) = \mathbb{P}(X = 1, Y = 1) = \tfrac{1}{4} = \mathbb{P}(X = 1)\mathbb{P}(Z = 1).$$

This, together with three similar equations, shows that X and Z are independent. Likewise, Y and Z are independent. However

$$\mathbb{P}(X = 1, Y = 1, Z = -1) = 0 \neq \tfrac{1}{8} = \mathbb{P}(X = 1)\mathbb{P}(Y = 1)\mathbb{P}(Z = -1),$$

so that X, Y, and Z are not independent.

2. (a) If $x \geq 1$,

$$\begin{aligned}\mathbb{P}\big(\min\{X, Y\} \leq x\big) &= 1 - \mathbb{P}(X > x, Y > x) = 1 - \mathbb{P}(X > x)\mathbb{P}(Y > x)\\ &= 1 - 2^{-x} \cdot 2^{-x} = 1 - 4^{-x}.\end{aligned}$$

(b) $\mathbb{P}(Y > X) = \mathbb{P}(Y < X)$ by symmetry. Also

$$\mathbb{P}(Y > X) + \mathbb{P}(Y < X) + \mathbb{P}(Y = X) = 1.$$

Since

$$\mathbb{P}(Y = X) = \sum_x \mathbb{P}(Y = X = x) = \sum_x 2^{-x} \cdot 2^{-x} = \tfrac{1}{3},$$

we have that $\mathbb{P}(Y > X) = \frac{1}{3}$.

(c) $\frac{1}{3}$ by part (b).

(d)

$$\begin{aligned}\mathbb{P}(X \geq kY) &= \sum_{y=1}^{\infty} \mathbb{P}\big(X \geq kY, Y = y\big)\\ &= \sum_{y=1}^{\infty} \mathbb{P}\big(X \geq ky, Y = y\big) = \sum_{y=1}^{\infty} \mathbb{P}\big(X \geq ky)\mathbb{P}(Y = y)\\ &= \sum_{y=1}^{\infty}\sum_{x=0}^{\infty} 2^{-ky-x}2^{-y} = \frac{2}{2^{k+1}-1}.\end{aligned}$$

(e)

$$\begin{aligned}\mathbb{P}(X \text{ divides } Y) &= \sum_{k=1}^{\infty} \mathbb{P}(Y = kX) = \sum_{k=1}^{\infty}\sum_{x=1}^{\infty} \mathbb{P}(Y = kx, X = x)\\ &= \sum_{k=1}^{\infty}\sum_{x=1}^{\infty} 2^{-kx}2^{-x} = \sum_{k=1}^{\infty} \frac{1}{2^{k+1}-1}.\end{aligned}$$

(f) Let $r = m/n$ where m and n are coprime. Then

$$\mathbb{P}(X = rY) = \sum_{k=1}^{\infty} \mathbb{P}(X = km, Y = kn) = \sum_{k=1}^{\infty} 2^{-km}2^{-kn} = \frac{1}{2^{m+n}-1}.$$

3. (a) We have that

$$\begin{aligned}\mathbb{P}(X_1 < X_2 < X_3) &= \sum_{i<j<k} (1 - p_1)(1 - p_2)(1 - p_3)p_1^{i-1}p_2^{j-1}p_3^{k-1}\\ &= \sum_{i<j} (1 - p_1)(1 - p_2)p_1^{i-1}p_2^{j-1}p_3^{j}\\ &= \sum_i \frac{(1 - p_1)(1 - p_2)p_1^{i-1}(p_2p_3)^i p_3}{1 - p_2p_3}\\ &= \frac{(1 - p_1)(1 - p_2)p_2p_3^2}{(1 - p_2p_3)(1 - p_1p_2p_3)}.\end{aligned}$$

(b)
$$\mathbb{P}(X_1 \le X_2 \le X_3) = \sum_{i \le j \le k} (1-p_1)(1-p_2)(1-p_3)p_1^{i-1}p_2^{j-1}p_3^{k-1}$$
$$= \frac{(1-p_1)(1-p_2)}{(1-p_2p_3)(1-p_1p_2p_3)}.$$

4. (a) Either substitute $p_1 = p_2 = p_3 = \frac{5}{6}$ in the result of Exercise (3.2.3b), or argue as follows, with the obvious notation. The event $\{A < B < C\}$ occurs only if one of the following occurs on the first round:

(i) A and B both rolled 6,

(ii) A rolled 6, B and C did not,

(iii) none rolled 6.

Hence, using conditional probabilities,

$$\mathbb{P}(A < B < C) = \left(\tfrac{1}{6}\right)^2 + \tfrac{1}{6}\left(\tfrac{5}{6}\right)^2\mathbb{P}(B < C) + \left(\tfrac{5}{6}\right)^3\mathbb{P}(A < B < C),$$

In calculating $\mathbb{P}(B < C)$ we may ignore A's rolls, and an argument similar to the above tells us that

$$\mathbb{P}(B < C) = \left(\tfrac{5}{6}\right)^2\mathbb{P}(B < C) + \tfrac{1}{6}.$$

Hence $\mathbb{P}(B < C) = \frac{6}{11}$, yielding $\mathbb{P}(A < B < C) = \frac{216}{1001}$.

(b) One may argue as above. Alternatively, let N be the total number of rolls before the first 6 appears. The probability that A rolls the first 6 is

$$\mathbb{P}\big(N \in \{1,4,7,\dots\}\big) = \sum_{k=1,4,7,\dots} \left(\tfrac{5}{6}\right)^{k-1}\tfrac{1}{6} = \tfrac{36}{91}.$$

Once A has thrown the first 6, the game restarts with the players rolling in order BCABCA.... Hence the probability that B rolls the next 6 is $\frac{36}{91}$ also, and similarly for the probability that C throws the third 6. The answer is therefore $\left(\frac{36}{91}\right)^3$.

5. The vector $(-X_r : 1 \le r \le n)$ has the same joint distribution as $(X_r : 1 \le r \le n)$, and the claim follows.

Let $X + 2$ and $Y + 2$ have joint mass function f, where $f_{i,j}$ is the (i, j)th entry in the matrix

$$\begin{pmatrix} \frac{1}{6} & \frac{1}{12} & \frac{1}{12} \\ 0 & \frac{1}{6} & \frac{1}{6} \\ \frac{1}{6} & \frac{1}{12} & \frac{1}{12} \end{pmatrix}, \qquad 1 \le i, j \le 3.$$

Then

$$\mathbb{P}(X = -1) = \mathbb{P}(X = 1) = \mathbb{P}(Y = -1) = \mathbb{P}(Y = 1) = \tfrac{1}{3}, \quad \mathbb{P}(X = 0) = \mathbb{P}(Y = 0) = \tfrac{1}{3},$$
$$\mathbb{P}(X + Y = -2) = \tfrac{1}{6} \ne \tfrac{1}{12} = \mathbb{P}(X + Y = 2).$$

3.3 Solutions. Expectation

1. (a) No!
(b) Let X have mass function: $f(-1) = \frac{1}{9}$, $f(\frac{1}{2}) = \frac{4}{9}$, $f(2) = \frac{4}{9}$. Then

$$\mathbb{E}(X) = -\tfrac{1}{9} + \tfrac{2}{9} + \tfrac{8}{9} = 1 = -\tfrac{1}{9} + \tfrac{8}{9} + \tfrac{2}{9} = \mathbb{E}(1/X).$$

2. (a) If you have already j distinct types of object, the probability that the next packet contains a different type is $(c-j)/c$, and the probability that it does not is j/c. Hence the number of days required has the geometric distribution with parameter $(c-j)/c$; this distribution has mean $c/(c-j)$.
(b) The time required to collect all the types is the sum of the successive times to collect each new type. The mean is therefore

$$\sum_{j=0}^{c-1} \frac{c}{c-j} = c\sum_{k=1}^{c} \frac{1}{k}.$$

3. (a) Let I_{ij} be the indicator function of the event that players i and j throw the same number. Then

$$\mathbb{E}(I_{ij}) = \mathbb{P}(I_{ij} = 1) = \sum_{i=1}^{6} \left(\tfrac{1}{6}\right)^2 = \tfrac{1}{6}, \qquad i \neq j.$$

The total score of the group is $S = \sum_{i<j} I_{ij}$, so

$$\mathbb{E}(S) = \sum_{i<j} \mathbb{E}(I_{ij}) = \frac{1}{6}\binom{n}{2}.$$

We claim that the family $\{I_{ij} : i < j\}$ is pairwise independent. The crucial calculation for this is as follows: if $i < j < k$ then

$$\mathbb{E}(I_{ij}I_{jk}) = \mathbb{P}(i, j, \text{and } k \text{ throw same number}) = \sum_{r=1}^{6} \left(\tfrac{1}{6}\right)^3 = \tfrac{1}{36} = \mathbb{E}(I_{ij})\mathbb{E}(I_{jk}).$$

Hence

$$\mathrm{var}(S) = \mathrm{var}\left(\sum_{i<j} I_{ij}\right) = \sum_{i<j} \mathrm{var}(I_{ij}) = \binom{n}{2}\mathrm{var}(I_{12})$$

by symmetry. But $\mathrm{var}(I_{12}) = \frac{1}{6}\left(1 - \frac{1}{6}\right)$.
(b) Let X_{ij} be the common score of players i and j, so that $X_{ij} = 0$ if their scores are different. This time the total score is $S = \sum_{i<j} X_{ij}$, and

$$\mathbb{E}(S) = \binom{n}{2}\mathbb{E}(X_{12}) = \binom{n}{2}\frac{1}{6}\cdot\frac{7}{2} = \frac{7}{12}\binom{n}{2}.$$

The X_{ij} are *not* pairwise independent, and you have to slog it out thus:

$$\begin{aligned}\mathrm{var}(S) &= \mathbb{E}\left\{\left(\sum_{i<j} X_{ij}\right)^2\right\} - \mathbb{E}(S)^2 \\ &= \binom{n}{2}\mathbb{E}(X_{12}^2) + 6\binom{n}{3}\mathbb{E}(X_{12}X_{23}) + \left\{\binom{n}{2}^2 - \binom{n}{2} - 6\binom{n}{3}\right\}\mathbb{E}(X_{12})^2 - \left(\frac{7}{12}\right)^2\binom{n}{2}^2 \\ &= \frac{35}{16}\binom{n}{2} + \frac{35}{72}\binom{n}{3}.\end{aligned}$$

4. The expected reward is $\sum_{k=1}^{\infty} 2^{-k} \cdot 2^k = \infty$. If your utility function is u, then your 'fair' entrance fee is $\sum_{k=1}^{\infty} 2^{-k} u(2^k)$. For example, if $u(k) = c(1 - k^{-\alpha})$ for $k \geq 1$, where $c, \alpha > 0$, then the fair fee is

$$c \sum_{k=1}^{\infty} 2^{-k}(1 - 2^{-\alpha k}) = c\left(1 - \frac{1}{2^{\alpha+1} - 1}\right).$$

This fee is certainly not 'fair' for the person offering the wager, unless possibly he is a noted philanthropist.

5. We have that $\mathbb{E}(X^\alpha) = \sum_{x=1}^{\infty} x^\alpha / \{x(x+1)\}$, which is finite if and only if $\alpha < 1$.

6. Clearly

$$\operatorname{var}(a + X) = \mathbb{E}\big(\{(a + X) - \mathbb{E}(a + X)\}^2\big) = \mathbb{E}\big(\{X - \mathbb{E}(X)\}^2\big) = \operatorname{var}(X).$$

7. For each r, bet $\{1 + \pi(r)\}^{-1}$ on horse r. If the rth horse wins, your payoff is $\{\pi(r) + 1\}\{1 + \pi(r)\}^{-1} = 1$, which is in excess of your total stake $\sum_k \{\pi(k) + 1\}^{-1}$.

8. We may assume that: (a) after any given roll of the die, your decision whether or not to stop depends only on the value V of the current roll; (b) if it is optimal to stop for $V = r$, then it is also optimal to stop when $V > r$.

Consider the strategy: stop the first time that the die shows r or greater. Let $S(r)$ be the expected score achieved by following this stategy. By elementary calculations,

$$S(6) = 6 \cdot \mathbb{P}(6 \text{ appears before } 1) + 1 \cdot \mathbb{P}(1 \text{ appears before } 6) = \tfrac{7}{2},$$

and similarly $S(5) = 4$, $S(4) = 4$, $S(3) = \frac{19}{5}$, $S(2) = \frac{7}{2}$. The optimal strategy is therefore to stop at the first throw showing 4, 5, or 6. Similar arguments may be used to show that 'stop at 5 or 6' is the rule to maximize the expected squared score.

9. Proceeding as in Exercise (3.3.8), we find the expected returns for the same strategies to be:

$$S(6) = \tfrac{7}{2} - 3c, \quad S(5) = 4 - 2c, \quad S(4) = 4 - \tfrac{3}{2}c, \quad S(3) = \tfrac{19}{5} - \tfrac{6}{5}c, \quad S(2) = \tfrac{7}{2} - c.$$

If $c = \frac{1}{3}$, it is best to stop when the score is at least 4; if $c = 1$, you should stop when the score is at least 3. The respective expected scores are $\frac{7}{2}$ and $\frac{13}{5}$.

10. (a) Evidently,

$$\mathbb{E}d_Y = \frac{1}{m}\sum_v d_v = \frac{2|E|}{m},$$

and

$$\mathbb{P}(Z = z) = \sum_{v \sim z} \mathbb{P}(Y = v)\frac{1}{d_v} = \frac{1}{m}\sum_{v \sim z}\frac{1}{d_v}$$

where the sum is over all neighbours v of z. Therefore,

$$\mathbb{E}d_Z = \frac{1}{m}\sum_{\langle v,z\rangle \in E}' \frac{d_z}{d_v} = \frac{1}{m}\sum_{\langle v,z\rangle \in E}'' \frac{1}{2}\left(\frac{d_z}{d_v} + \frac{d_v}{d_z}\right),$$

where $\sum'$ (respectively, $\sum''$) denotes the sum over unordered pairs v, z (respectively, ordered pairs v, z). The result follows by the fact that, for $a > 0$, we have $a + a^{-1} \geq 2$.

(b) You should expect your friends to have more friends than you, on average.

11. Possible values of G are $x, -x+y, -x-y+z, -x-y-z$, with respective probabilities $\frac{1}{2}, \frac{1}{4}, \frac{1}{8}, \frac{1}{8}$. Therefore, $\mathbb{E}G = 0$, $\text{var}(G) = x^2 + \frac{1}{2}y^2 + \frac{1}{4}z^2$, and $\mathbb{P}(G < 0) = \frac{1}{2}$.

Suppose Savage orders his stakes according to the permutation abc of xyz. Then

$$\mathbb{P}(G < 0) = \tfrac{1}{4}I(-a+b<0) + \tfrac{1}{8}I(-a-b+c<0) + \tfrac{1}{8},$$

where $I(Z)$ is the indicator function of Z. We see that $\mathbb{P}(G < 0)$ can only be a minimum if $b > a$, which we assume henceforth. There are now three possibilities:

$$\begin{array}{llll} abc = zyx, & \text{var}(G) = z^2 + \frac{1}{2}y^2 + \frac{1}{4}x^2, & -a-b+c = x-y-z, & \mathbb{P}(G<0) \le \frac{1}{4}, \\ abc = zxy, & \text{var}(G) = z^2 + \frac{1}{2}x^2 + \frac{1}{4}y^2, & -a-b+c = y-x-z < 0, & \mathbb{P}(G<0) = \frac{1}{4}, \\ abc = yxz, & \text{var}(G) = y^2 + \frac{1}{2}x^2 + \frac{1}{4}z^2, & -a-b+c = z-x-y < 0, & \mathbb{P}(G<0) = \frac{1}{4}. \end{array}$$

The variance is a minimum in the first case, which is also the only case for which it is possible (depending on the values of x, y, z) that $\mathbb{P}(G < 0) < \frac{1}{4}$. The answer is $abc = zyx$: 'least first, greatest last'.

12. (a) Let w be Rth in the lexicographic ordering of the n words, so that R is uniform on $\{1, 2, \dots, n\}$. Then

$$c_n = n - 1 + \mathbb{E}(c_{R-1} + c_{n-R}) = n - 1 + \frac{2}{n}\sum_{j=1}^{n-1} c_j.$$

By subtracting successive equations,

$$\frac{c_{n+1}}{n+2} = \frac{2n}{(n+1)(n+2)} + \frac{c_n}{n+1},$$

so that

$$c_{n+1} = 2(n+2)\sum_{j=1}^{n} \frac{j}{(j+1)(j+2)}.$$

(b) Express the summand using partial fractions, and use the fact that $\sum_{r=1}^{n} r^{-1} = \log n + \gamma + \mathrm{O}(n^{-1})$ as $n \to \infty$.

(c) Since $\sum_n \mathbb{P}(N = n) = 1$, we have $A = 4$. Now, by part (a),

$$\mathbb{E}(c_N) = A\sum_{n=2}^{\infty} \frac{2(n+1)}{(n-1)n(n+1)} \sum_{j=1}^{n-1} \frac{j}{(j+1)(j+2)}.$$

We interchange the order of summation, calculate the inner sum over n, and then the sum over j, to find $\mathbb{E}(c_N) = A = 4$.

3.4 Solutions. Indicators and matching

1. (a) Let I_j be the indicator function of the event that the outcome of the $(j+1)$th toss is different from the outcome of the jth toss. The number R of distinct runs is given by $R = 1 + \sum_{j=1}^{n-1} I_j$. Hence

$$\mathbb{E}(R) = 1 + (n-1)\mathbb{E}(I_1) = 1 + (n-1)2pq,$$

where $q = 1 - p$. Now remark that I_j and I_k are independent if $|j - k| > 1$, so that

$$\mathbb{E}\{(R-1)^2\} = \mathbb{E}\left\{\left(\sum_{j=1}^{n-1} I_j\right)^2\right\} = (n-1)\mathbb{E}(I_1) + 2(n-2)\mathbb{E}(I_1 I_2) + \left\{(n-1)^2 - (n-1) - 2(n-2)\right\}\mathbb{E}(I_1)^2.$$

Now $\mathbb{E}(I_1) = 2pq$ and $\mathbb{E}(I_1 I_2) = p^2 q + pq^2 = pq$, and therefore

$$\begin{aligned}\operatorname{var}(R) = \operatorname{var}(R-1) &= (n-1)\mathbb{E}(I_1) + 2(n-2)\mathbb{E}(I_1 I_2) - \left\{(n-1) + 2(n-2)\right\}\mathbb{E}(I_1)^2 \\ &= 2pq\big(2n - 3 - 2pq(3n-5)\big).\end{aligned}$$

(b) Let I_r be the indicator function of the event that a run of heads begins at the rth position. Then

$$\mathbb{E}I_r = \mathbb{P}(I_r = 1) = \begin{cases} \dfrac{h}{h+t} & \text{if } r = 1, \\[2ex] \dfrac{ht}{(h+t)(h+t-1)} & \text{if } r > 1. \end{cases}$$

Hence, the number R of head runs has mean

$$\mathbb{E}R = \sum_{r=1}^{h+t} \mathbb{E}I_r = \frac{h}{h+t} + \sum_{r=2}^{h+t} \frac{ht}{(h+t)(h+t-1)} = \frac{h(t+1)}{h+t}.$$

The computation of variance is slightly complex but the answer is fairly sweet. Since $I_r I_{r+1} = 0$, we have

$$\mathbb{E}(R^2) = \mathbb{E}\left(\left[\sum_{r=1}^{h+t} I_r\right]^2\right) = \mathbb{E}R + \sum_{|i-j|\ge 2} \mathbb{E}(I_i I_j).$$

Now, if $|i - j| \ge 2$,

$$\mathbb{E}(I_i I_j) = \begin{cases} \dfrac{h}{h+t}\cdot\dfrac{t}{h+t-1}\cdot\dfrac{h-1}{h+t-2} & \text{if } \min\{i,j\} = 1, \\[2ex] \dfrac{h}{h+t}\cdot\dfrac{t}{h+t-1}\cdot\dfrac{h-1}{h+t-2}\cdot\dfrac{t-1}{h+t-3} & \text{otherwise.} \end{cases}$$

It follows after a calculation that

$$\operatorname{var}(R) = \frac{ht(h-1)(t+1)}{(h+t)^2(h+t-1)}.$$

2. The required total is $T = \sum_{i=1}^{k} X_i$, where X_i is the number shown on the ith ball. Hence $\mathbb{E}(T) = k\mathbb{E}(X_1) = \frac{1}{2}k(n+1)$. Now calculate, boringly,

$$\begin{aligned}\mathbb{E}\left\{\left(\sum_{i=1}^{k} X_i\right)^2\right\} &= k\mathbb{E}(X_1^2) + k(k-1)\mathbb{E}(X_1 X_2) \\ &= \frac{k}{n}\sum_{1}^{n} j^2 + \frac{k(k-1)}{n(n-1)} 2\sum_{i>j} ij \\ &= \frac{k}{n}\left\{\tfrac{1}{3}n(n+1)(n+2) - \tfrac{1}{2}n(n+1)\right\} \\ &\qquad + \frac{k(k-1)}{n(n-1)}\sum_{j=1}^{n} j\{n(n+1) - j(j+1)\} \\ &= \tfrac{1}{6}k(n+1)(2n+1) + \tfrac{1}{12}k(k-1)(3n+2)(n+1).\end{aligned}$$

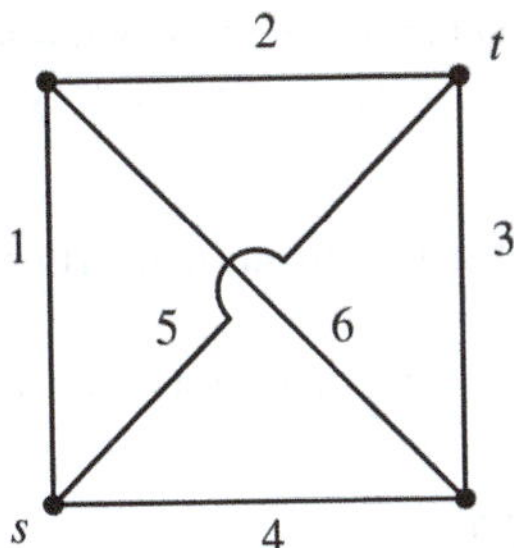

Figure 3.1. The network with source s and sink t.

Hence

$$\operatorname{var}(T) = k(n+1)\Big\{\tfrac{1}{6}(2n+1) + \tfrac{1}{12}(k-1)(3n+2) - \tfrac{1}{4}k(n+1)\Big\} = \tfrac{1}{12}(n+1)k(n-k).$$

3. Each couple survives with probability

$$\binom{2n-2}{m}\Big/\binom{2n}{m} = \left(1-\frac{m}{2n}\right)\left(1-\frac{m}{2n-1}\right),$$

so the required mean is

$$n\left(1-\frac{m}{2n}\right)\left(1-\frac{m}{2n-1}\right).$$

4. Any given red ball is in urn R after stage k if and only if it has been selected an even number of times. The probability of this is

$$\sum_{m \text{ even}} \binom{k}{m}\left(\frac{1}{n}\right)^m\left(1-\frac{1}{n}\right)^{k-m} = \frac{1}{2}\left\{\left[\left(1-\frac{1}{n}\right)+\frac{1}{n}\right]^k + \left[\left(1-\frac{1}{n}\right)-\frac{1}{n}\right]^k\right\}$$
$$= \frac{1}{2}\left\{1+\left(1-\frac{2}{n}\right)^k\right\},$$

and the mean number of such red balls is n times this probability.

5. Label the edges and vertices as in Figure 3.1. The structure function is

$$\zeta(X) = X_5 + (1-X_5)\Big\{(1-X_1)X_4\big[X_3+(1-X_3)X_2X_6\big]$$
$$+X_1\Big[X_2+(1-X_2)\Big(X_3\langle X_6+X_4(1-X_6)\rangle\Big)\Big]\Big\}.$$

For the reliability, see Problem (1.8.19a).

6. The structure function is $I_{\{S\geq k\}}$, the indicator function of $\{S \geq k\}$ where $S = \sum_{c=1}^n X_c$. The reliability is therefore $\sum_{i=k}^n \binom{n}{i} p^i (1-p)^{n-i}$.

7. Independently colour each vertex livid or bronze with probability $\frac{1}{2}$ each, and let L be the random set of livid vertices. Then $\mathbb{E}N_L = \frac{1}{2}|E|$. There must exist one or more possible values of N_L which are at least as large as its mean.

8. Let I_r be the indicator function that the rth pair have opposite polarity, so that $X = 1+\sum_{r=1}^{n-1} I_r$. We have that $\mathbb{P}(I_r = 1) = \frac{1}{2}$ and $\mathbb{P}(I_r = I_{r+1} = 1) = \frac{1}{4}$, whence $\mathbb{E}X = \frac{1}{2}(n+1)$ and $\operatorname{var} X = \frac{1}{4}(n-1)$.

9. (a) Let A_i be the event that the integer i remains in the ith position. Then

$$\begin{aligned}\mathbb{P}\Big(\bigcup_{r=1}^{n} A_r\Big) &= \sum_r \mathbb{P}(A_r) - \sum_{r<s} \mathbb{P}(A_r \cap A_s) + \cdots + (-1)^{n-1}\mathbb{P}\Big(\bigcap_r A_r\Big)\\ &= n\cdot\frac{1}{n} - \binom{n}{2}\frac{1}{n(n-1)} + \cdots + (-1)^{n-1}\frac{1}{n!}.\end{aligned}$$

Therefore the number M of matches satisfies

$$\mathbb{P}(M=0) = \frac{1}{2!} - \frac{1}{3!} + \cdots + (-1)^n \frac{1}{n!}.$$

Now

$$\begin{aligned}\mathbb{P}(M=r) &= \binom{n}{r}\mathbb{P}(r \text{ given numbers match, and the remaining } n-r \text{ are deranged})\\ &= \frac{n!}{r!(n-r)!}\,\frac{(n-r)!}{n!}\left(\frac{1}{2!} - \frac{1}{3!} + \cdots + (-1)^{n-r}\frac{1}{(n-r)!}\right).\end{aligned}$$

(b) We have that

$$\begin{aligned}d_{n+1} &= \sum_{r=2}^{n+1} \#\{\text{derangements with 1 in the } r\text{th place}\}\\ &= n\big[\#\{\text{derangements which swap 1 with 2}\}\\ &\qquad + \#\{\text{derangements in which 1 is in the 2nd place and 2 is not in the 1st place}\}\big]\\ &= nd_{n-1} + nd_n,\end{aligned}$$

where $\#A$ denotes the cardinality of the set A. By rearrangement, $d_{n+1} - (n+1)d_n = -(d_n - nd_{n-1})$. Set $u_n = d_n - nd_{n-1}$ and note that $u_2 = 1$, to obtain $u_n = (-1)^n$, $n \ge 2$, and hence

$$d_n = \frac{n!}{2!} - \frac{n!}{3!} + \cdots + (-1)^n \frac{n!}{n!}.$$

Now divide by $n!$ to obtain the results above.
(c) The answer is m/n by symmetry.

10. Let $I_{r,s}$ be the indicator function of the event that students r and s share the same birthday. Then

$$\mathbb{E}B = \mathbb{E}\sum_r \sum_{s>r} I_{r,s} = \binom{n}{2}\frac{1}{365}.$$

Furthermore,

$$\begin{aligned}\operatorname{var}(B) = \mathbb{E}(B^2) - \mathbb{E}(B)^2 &= \binom{n}{2}\mathbb{E}(I_{1,2}^2) + n(n+1)\mathbb{E}(I_{1,2}I_{2,3})\\ &\qquad + \left\{\binom{n}{2}^2 - \binom{n}{2} - n(n+1)\right\}\mathbb{E}(I_{1,2}I_{3,4}) - \binom{n}{2}^2\frac{1}{365^2}\\ &= \binom{n}{2}\frac{364}{365^2}.\end{aligned}$$

11. Choose the array A such that each disk touches its 6 neighbouring disks. Now cast A onto the plane randomly, and let U be the event that the origin 0 is not covered by a disk. By an easy calculation,

$$\mathbb{P}(U) = \frac{\frac{\sqrt{3}}{4} - \frac{\pi}{8}}{\frac{\sqrt{3}}{4}} = 1 - \tfrac{1}{2}\pi/\sqrt{3} \simeq 0.093,$$

the proportion of the plane not covered by the disks. Hence, the probability that one or more of ten chosen points is not covered is no greater than

$$\sum_{r=1}^{10} \mathbb{P}(U) \simeq 0.93.$$

The result follows since this is strictly less than 1.

12. By conditioning on the weather on day n,

$$w_{n+1} = pw_n + (1-p)(1-w_n).$$

Hence $w_n = \frac{1}{2} + \frac{1}{2}(2p-1)^n$ for $n \geq 0$. The mean number of wet days in the next week is $\sum_{r=1}^{7} w_r$.

13. Let I_i be the indicator function that the ith ball chosen is green. Then $\mathbb{E}I_i = g/b$ and $\mathbb{E}(I_1 I_2) = g(g-1)/[b(b-1)]$. Therefore, the number G of green balls in the sample satisfies

$$\mathbb{E}G = n\mathbb{E}I_1 = ng/b, \quad \mathbb{E}(G^2) = \mathbb{E}\left(\left[\sum_{i=1}^{n} I_i\right]^2\right) = n\mathbb{E}I_1 + n(n-1)\mathbb{E}(I_1 I_2),$$

whence

$$\operatorname{var} G = \mathbb{E}(G^2) - \mathbb{E}(G)^2 = \frac{ng(b-n)(b-g)}{b^2(b-1)}.$$

14. Let I_r be the indicator function that the rth couple sits together. First, we seat the men at random in the odd-numbered chairs. Since each man has two empty adjacent chairs, $\mathbb{E}I_r = \mathbb{P}(I_r = 1) = 2/n$. We have for $r \neq s$ that

$$\begin{aligned}\mathbb{P}(I_r = I_s = 1) = \mathbb{P}(I_s = 1 \mid I_r = 1)\mathbb{P}(I_r = 1) &= \left(\frac{1}{n-1} \cdot \frac{1}{n-1} + \frac{n-2}{n-1} \cdot \frac{2}{n-1}\right)\frac{2}{n} \\ &= \frac{2n-3}{(n-1)^2} \cdot \frac{2}{n}.\end{aligned}$$

Now, $\mathbb{E}X = n\mathbb{E}I_1 = 2$, and

$$\begin{aligned}\operatorname{var} X &= \mathbb{E}\left(\left[\sum_{r=1}^{n} I_r\right]^2\right) - 4 \\ &= \sum_{r=1}^{n} \mathbb{P}(I_r = 1) + \sum_{r \neq s} \mathbb{P}(I_r = I_s = 1) - 4 = 2 - \frac{2}{n-1}.\end{aligned}$$

3.5 Solutions. Examples of discrete variables

1. There are $n!/(n_1!n_2!\cdots n_t!)$ sequences of outcomes in which the ith possible outcome occurs n_i times for each i. The probability of any such sequence is $p_1^{n_1}p_2^{n_2}\cdots p_t^{n_t}$, and the result follows.

2. The total number H of heads satisfies

$$\mathbb{P}(H=x)=\sum_{n=x}^{\infty}\mathbb{P}(H=x\mid N=n)\mathbb{P}(N=n)=\sum_{n=x}^{\infty}\binom{n}{x}p^x(1-p)^{n-x}\frac{\lambda^n e^{-\lambda}}{n!}$$
$$=\frac{(\lambda p)^x e^{-\lambda p}}{x!}\sum_{n=x}^{\infty}\frac{\{\lambda(1-p)\}^{n-x}e^{-\lambda(1-p)}}{(n-x)!}.$$

The last summation equals 1, since it is the sum of the values of the Poisson mass function with parameter $\lambda(1-p)$.

3. $dp_n/d\lambda=p_{n-1}-p_n$ where $p_{-1}=0$. Hence $(d/d\lambda)\mathbb{P}(X\le n)=p_n(\lambda)$.

4. The probability of a marked animal in the nth place is a/b. Conditional on this event, the chance of $n-1$ preceding places containing $m-1$ marked and $n-m$ unmarked animals is

$$\binom{a-1}{m-1}\binom{b-a}{n-m}\bigg/\binom{b-1}{n-1},$$

as required. Now let X_j be the number of unmarked animals between the $j-1$th and jth marked animals, if all were caught. By symmetry, $\mathbb{E}X_j=(b-a)/(a+1)$, whence $\mathbb{E}X=m(\mathbb{E}X_1+1)=m(b+1)/(a+1)$.

5. By conditional probabilities, integration by parts, and Exercise (3.5.3),

$$\mathbb{P}(Y\le n)=\sum_{r=0}^{n}\int_0^\infty p_r(\lambda)f(\lambda)\,d\lambda$$
$$=\sum_{r=0}^{n}\Big[p_r(\lambda)F(\lambda)\Big]_0^\infty-\int_0^\infty\left(\frac{d}{d\lambda}\mathbb{P}(Y\le n)\right)F(\lambda)\,d\lambda$$
$$=0+\int_0^\infty p_n(\lambda)F(\lambda)\,d\lambda.$$

The second part is similar.

3.6 Solutions. Dependence

1. Remembering Problem (2.7.3b), it suffices to show that $\mathrm{var}(aX+bY)<\infty$ if $a,b\in\mathbb{R}$ and $\mathrm{var}(X),\mathrm{var}(Y)<\infty$. Now,

$$\mathrm{var}(aX+bY)=\mathbb{E}\left(\{aX+bY-\mathbb{E}(aX+bY)\}^2\right)$$
$$=a^2\,\mathrm{var}(X)+2ab\,\mathrm{cov}(X,Y)+b^2\,\mathrm{var}(Y)$$
$$\le a^2\,\mathrm{var}(X)+2ab\sqrt{\mathrm{var}(X)\,\mathrm{var}(Y)}+b^2\,\mathrm{var}(Y)$$
$$=\left(a\sqrt{\mathrm{var}(X)}+b\sqrt{\mathrm{var}(Y)}\right)^2$$

where we have used the Cauchy–Schwarz inequality (3.6.9) applied to $X-\mathbb{E}(X)$, $Y-\mathbb{E}(Y)$.

2. Let N_i be the number of times the ith outcome occurs. Then N_i has the binomial distribution with parameters n and p_i.

3. For $x = 1, 2, \ldots,$

$$\begin{aligned}\mathbb{P}(X = x) &= \sum_{y=1}^{\infty} \mathbb{P}(X = x, Y = y)\\ &= \sum_{y=1}^{\infty} \frac{C}{2}\left\{\frac{1}{(x+y-1)(x+y)} - \frac{1}{(x+y)(x+y+1)}\right\}\\ &= \frac{C}{2x(x+1)} = \frac{C}{2}\left(\frac{1}{x} - \frac{1}{x+1}\right),\end{aligned}$$

and hence $C = 2$. Clearly Y has the same mass function. Finally $\mathbb{E}(X) = \sum_{x=1}^{\infty}(x+1)^{-1} = \infty$, so the covariance does not exist.

4. $\text{Max}\{u, v\} = \frac{1}{2}(u+v) + \frac{1}{2}|u - v|$, and therefore

$$\begin{aligned}\mathbb{E}\big(\max\{X^2, Y^2\}\big) &= \tfrac{1}{2}\mathbb{E}(X^2 + Y^2) + \tfrac{1}{2}\mathbb{E}\big|(X-Y)(X+Y)\big|\\ &\le 1 + \tfrac{1}{2}\sqrt{\mathbb{E}\big((X-Y)^2\big)\mathbb{E}\big((X+Y)^2\big)}\\ &= 1 + \tfrac{1}{2}\sqrt{(2-2\rho)(2+2\rho)} = 1 + \sqrt{1-\rho^2},\end{aligned}$$

where we have used the Cauchy–Schwarz inequality.

5. (a) $\log y \le y - 1$ with equality if and only if $y = 1$. Therefore,

$$\mathbb{E}\left(\log \frac{f_Y(X)}{f_X(X)}\right) \le \mathbb{E}\left[\frac{f_Y(X)}{f_X(X)} - 1\right] = 0,$$

with equality if and only if $f_Y = f_X$.
(b) This holds likewise, with equality if and only if $f(x, y) = f_X(x) f_Y(y)$ for all x, y, which is to say that X and Y are independent.

6. (a) $a + b + c = \mathbb{E}\{I_{\{X>Y\}} + I_{\{Y>Z\}} + I_{\{Z>X\}}\} \le 2$, whence $\min\{a, b, c\} \le \frac{2}{3}$. Equality is attained, for example, if the vector (X, Y, Z) takes only three values with probabilities $f(2, 1, 3) = f(3, 2, 1) = f(1, 3, 2) = \frac{1}{3}$.
(b) $\mathbb{P}(X < Y) = \mathbb{P}(Y < X)$, etc.
(c) We have that $c = a = p$ and $b = 1 - p^2$. Also $\sup \min\{p, (1-p^2)\} = \frac{1}{2}(\sqrt{5} - 1)$.

7. We have for $1 \le x \le 9$ that

$$f_X(x) = \sum_{y=0}^{9} \log\left(1 + \frac{1}{10x + y}\right) = \log \prod_{y=0}^{9}\left(1 + \frac{1}{10x+y}\right) = \log\left(1 + \frac{1}{x}\right).$$

By calculation, $\mathbb{E}X \simeq 3.44$.

8. (i) $f_X(j) = c\sum_{k=0}^{\infty}\left\{\frac{j}{j!}a^j\frac{a^k}{k!} + \frac{k}{k!}a^k\frac{a^j}{j!}\right\} = c\frac{e^a(j+a)a^j}{j!}$.
(ii) $1 = \sum_j f_X(j) = 2ace^{2a}$, whence $c = e^{-2a}/(2a)$.

(iii) $f_{X+Y}(r) = \sum_{j=0}^{r} \frac{cra^r}{j!\,(r-j)!} = \frac{cra^r 2^r}{r!}$, $r \geq 0$.

(iv) $\mathbb{E}(X+Y-1) = \sum_{r=1}^{\infty} \frac{cr(r-1)(2a)^r}{r!} = 2a$. Now $\mathbb{E}(X) = \mathbb{E}(Y)$, and therefore $\mathbb{E}(X) = a + \frac{1}{2}$.

9. We have that $\text{cov}(U, V) = \text{var}\, Y > 0$ and $\text{cov}(V, W) = \text{var}\, Z > 0$. However, $\text{cov}(U, W) = -\text{var}\, X < 0$.

10. Let (U, V) and (X, Y) be independent, identically distributed random vectors. By independence and the given identity,

$$\mathbb{E}(V^2)\mathbb{E}(X^2) + \mathbb{E}(U^2)\mathbb{E}(Y^2) - 2\mathbb{E}(UV)\mathbb{E}(XY) \geq 0,$$

with equality if and only if $\mathbb{P}(VX = UY) = 1$. Hence, $\mathbb{E}(X^2)\mathbb{E}(Y^2) \geq \mathbb{E}(XY)^2$ with equality if and only if $\mathbb{P}(aX = bY) = 1$ for some real a, b, not both being zero.

11. Without loss of generality let $\mathbb{E}X = 0$. The fundamental inequality is $t - X \leq (t - X)I_{\{X<t\}}$, where I_A is the indicator function of A (just check the two cases $X < t$ and $X \geq t$). Take expectations and use the Cauchy–Schwarz inequality to obtain

$$t = \mathbb{E}(t - X) \leq \mathbb{E}\big[(t - X)I_{\{X<t\}}\big] \leq \sqrt{\mathbb{E}((t-X)^2)\mathbb{P}(X < t)}.$$

If $t > 0$, then

$$t^2 \leq \mathbb{E}\big((t-X)^2\big)\mathbb{P}(X < t) = (t^2 + \text{var}\, X)\mathbb{P}(X < t).$$

3.7 Solutions. Conditional distributions and conditional expectation

1. (a) We have that

$$\begin{aligned}
\mathbb{E}(aY + bZ \mid X = x) &= \sum_{y,z} (ay + bz)\mathbb{P}(Y = y, Z = z \mid X = x) \\
&= a \sum_{y,z} y\mathbb{P}(Y = y, Z = z \mid X = x) + b \sum_{y,z} z\mathbb{P}(Y = y, Z = z \mid X = x) \\
&= a \sum_{y} y\mathbb{P}(Y = y \mid X = x) + b \sum_{z} z\mathbb{P}(Z = z \mid X = x).
\end{aligned}$$

Parts (b)–(e) are verified by similar trivial calculations. Turning to (f),

$$\begin{aligned}
\mathbb{E}\{\mathbb{E}(Y \mid X, Z) \,|\, X = x\} &= \sum_{z} \left\{ \sum_{y} y\mathbb{P}(Y = y \mid X = x, Z = z)\mathbb{P}(X = x, Z = z \mid X = x) \right\} \\
&= \sum_{z} \sum_{y} y \frac{\mathbb{P}(Y = y, X = x, Z = z)}{\mathbb{P}(X = x, Z = z)} \cdot \frac{\mathbb{P}(X = x, Z = z)}{\mathbb{P}(X = x)} \\
&= \sum_{y} y\mathbb{P}(Y = y \mid X = x) = \mathbb{E}(Y \mid X = x) \\
&= \mathbb{E}\Big\{\mathbb{E}(Y \mid X) \,\Big|\, X = x, Z = z\Big\}, \quad \text{by part (e).}
\end{aligned}$$

2. If ϕ and ψ are two such functions then $\mathbb{E}\big((\phi(X) - \psi(X))g(X)\big) = 0$ for any suitable g. Setting $g(X) = I_{\{X=x\}}$ for any $x \in \mathbb{R}$ such that $\mathbb{P}(X = x) > 0$, we obtain $\phi(x) = \psi(x)$. Therefore $\mathbb{P}(\phi(X) = \psi(X)) = 1$.

3. We do not seriously expect you to want to do this one. However, if you insist, the method is to check in each case that both sides satisfy the appropriate definition, and then to appeal to uniqueness, deducing that the sides are almost surely equal (see Williams 1991, p. 88).

4. The natural definition is given by

$$\operatorname{var}(Y \mid X = x) = \mathbb{E}\big(\{Y - \mathbb{E}(Y \mid X = x)\}^2 \,\big|\, X = x\big).$$

Now,

$$\begin{aligned}\operatorname{var}(Y) = \mathbb{E}(\{Y - \mathbb{E}Y\}^2) &= \mathbb{E}\Big[\mathbb{E}\big(\{Y - \mathbb{E}(Y \mid X) + \mathbb{E}(Y \mid X) - \mathbb{E}Y\}^2 \,\big|\, X\big)\Big] \\ &= \mathbb{E}\big(\operatorname{var}(Y \mid X)\big) + \operatorname{var}\big(\mathbb{E}(Y \mid X)\big)\end{aligned}$$

since the mean of $\mathbb{E}(Y \mid X)$ is $\mathbb{E}Y$, and the cross product is, by Exercise (3.7.1e),

$$\begin{aligned}&2\mathbb{E}\Big[\mathbb{E}\big(\{Y - \mathbb{E}(Y \mid X)\}\{\mathbb{E}(Y \mid X) - \mathbb{E}Y\} \,\big|\, X\big)\Big] \\ &\qquad = 2\mathbb{E}\Big[\{\mathbb{E}(Y \mid X) - \mathbb{E}Y\}\mathbb{E}\{Y - \mathbb{E}(Y \mid X) \,\big|\, X\}\Big] = 0\end{aligned}$$

since $\mathbb{E}\{Y - \mathbb{E}(Y \mid X) \,\big|\, X\} = \mathbb{E}(Y \mid X) - \mathbb{E}(Y \mid X) = 0$.

5. We have that

$$\mathbb{E}(T - t \mid T > t) = \sum_{r=0}^{\infty} \mathbb{P}(T > t + r \mid T > t) = \sum_{r=0}^{\infty} \frac{\mathbb{P}(T > t + r)}{\mathbb{P}(T > t)}.$$

(a)
$$\mathbb{E}(T - t \mid T > t) = \sum_{r=0}^{N-t} \frac{N - t - r}{N - t} = \tfrac{1}{2}(N - t + 1).$$

(b)
$$\mathbb{E}(T - t \mid T > t) = \sum_{r=0}^{\infty} \frac{2^{-(t+r)}}{2^{-t}} = 2.$$

6. Clearly

$$\mathbb{E}(S \mid N = n) = \mathbb{E}\Big(\sum_{i=1}^{n} X_i\Big) = \mu n,$$

and hence $\mathbb{E}(S \mid N) = \mu N$. It follows that $\mathbb{E}(S) = \mathbb{E}\{\mathbb{E}(S \mid N)\} = \mathbb{E}(\mu N)$.

By Exercise (3.7.4),

$$\operatorname{var} S = \mathbb{E}(\operatorname{var}(S \mid N)) + \operatorname{var}(\mathbb{E}(S \mid N)) = \mathbb{E}(N\sigma^2) + \operatorname{var}(\mu N) = \sigma^2 \mathbb{E}N + \mu^2 \operatorname{var} N.$$

7. A robot passed is in fact faulty with probability $\pi = \{\phi(1 - \delta)\}/(1 - \phi\delta)$. Thus the number of faulty passed robots, given Y, is $\text{bin}(n - Y, \pi)$, with mean $(n - Y)\{\phi(1 - \delta)\}/(1 - \phi\delta)$. Hence

$$\mathbb{E}(X \mid Y) = Y + \frac{(n - Y)\phi(1 - \delta)}{1 - \phi\delta}.$$

8. (a) Let m be the family size, ϕ_r the indicator that the rth child is female, and μ_r the indicator of a male. The numbers G, B of girls and boys satisfy

$$G = \sum_{r=1}^{m} \phi_r, \quad B = \sum_{r=1}^{m} \mu_r, \quad \mathbb{E}(G) = \tfrac{1}{2}m = \mathbb{E}(B).$$

(It will be shown later that the result remains true for random m under reasonable conditions.) We have not used the property of independence.
(b) With M the event that the selected child is male,

$$\mathbb{E}(G \mid M) = \mathbb{E}\left(\sum_{r=1}^{m-1} \phi_r\right) = \tfrac{1}{2}(m-1) = \mathbb{E}(B).$$

The independence is necessary for this argument.

9. Conditional expectation is defined in terms of the conditional distribution, so the first step is not justified. Even if this step were accepted, the second equality is generally false.

10. By conditioning on X_{n-1},

$$\mathbb{E}X_n = \mathbb{E}\big[\mathbb{E}(X_n \mid X_{n-1})\big] = \mathbb{E}\big[p(X_{n-1}+1) + (1-p)(X_{n-1}+1+\widehat{X}_n)\big]$$

where $\widehat{X}_n$ has the same distribution as X_n. Hence $\mathbb{E}X_n = (1+\mathbb{E}X_{n-1})/p$. Solve this subject to $\mathbb{E}X_1 = p^{-1}$.

11. In the jointly discrete case,

$$\begin{aligned}\operatorname{cov}(X, Y \mid Z) &= \mathbb{E}\Big(\big(X - \mathbb{E}(X \mid Z)\big)\big(Y - \mathbb{E}(Y \mid Z)\big) \,\Big|\, Z\Big) \\ &= \sum_{x,y} \big(x - \mathbb{E}(X \mid Z)\big)\big(y - \mathbb{E}(Y \mid Z)\big) f_{X,Y|Z}(x, y \mid Z),\end{aligned}$$

where $f_{X,Y|Z}$ is the conditional joint mass function of X, Y given Z. For the next part, write down $\operatorname{cov}(X, Y)$, and use conditional expectation and the pull-through property of Exercise (3.7.1e), to obtain the claim.

12. Let B be the event that the first ball drawn is blue, and suppose B occurs. Between two consecutively chosen blue balls, and after the last blue ball, there is a run of red balls. Since there are r red balls in all, by symmetry, the first red run has mean length r/b, making $2 + (r/b)$ when including the first two blue balls. Writing D for the mean number required, we have shown that

$$\mathbb{E}(D \mid B) = \frac{r}{b} + 2, \qquad \mathbb{E}(D \mid B^{\mathrm{c}}) = \frac{b}{r} + 2,$$

so that, by conditional probability, $\mathbb{E}D = \mathbb{E}(D \mid B)\mathbb{P}(B) + \mathbb{E}(D \mid B^{\mathrm{c}})\mathbb{P}(B^{\mathrm{c}}) = 3$.

13. (a) By symmetry, $\mathbb{E}X = \frac{1}{2}n$. Furthermore,

$$\begin{aligned}\mathbb{E}(X(X-1)) &= \frac{1}{n+1}\sum_{k=1}^{n} k(k-1) \\ &= \frac{1}{n+1}\cdot\frac{1}{3}\sum_{k=1}^{n}\{(k+1)k(k-1) - k(k-1)(k-2)\} = \frac{1}{3}n(n-1).\end{aligned}$$

Therefore, $\operatorname{var}(X) = \frac{1}{3}n(n-1) + \frac{1}{2}n - \frac{1}{4}n^2 = \frac{1}{12}n^2 + \frac{1}{6}n$.
(b) Since Y has a conditional binomial distribution $\operatorname{bin}(n, X/n)$, we have $\mathbb{E}Y = \mathbb{E}(\mathbb{E}(Y \mid X)) = \mathbb{E}X = \frac{1}{2}n$. Similarly,

$$\begin{aligned}\mathbb{E}(Y^2) &= \mathbb{E}(\mathbb{E}(Y^2 \mid X)) \\ &= \mathbb{E}\left(n\frac{X}{n}\left(1 - \frac{X}{n}\right) + X^2\right) = \mathbb{E}X + \frac{n-1}{n}\mathbb{E}(X^2) = \frac{1}{6}(2n^2 + 2n - 1).\end{aligned}$$

and the formula for var Y follows in the usual way.

We have $\mathbb{E}(X+Y) = \mathbb{E}X + \mathbb{E}Y = n$. Now,

$$\mathbb{E}(XY) = \mathbb{E}\big(\mathbb{E}(XY \mid X)\big) = \mathbb{E}\big(X\mathbb{E}(Y \mid X)\big) = \mathbb{E}(X^2),$$

so that

$$\mathbb{E}\big((X+Y)^2\big) = 3\mathbb{E}(X^2) + \mathbb{E}(Y^2) = \tfrac{1}{6}(8n^2 + 5n - 1),$$

whence $\operatorname{var}(X+Y) = \mathbb{E}((X+Y)^2) - n^2 = \frac{1}{6}(2n^2 + 5n - 1)$.

14. This is just summation. First,

$$\mathbb{P}(X = x) = \sum_{y=x}^{\infty} \frac{\lambda^y e^{-2\lambda}}{x!\,(y-x)!} = \frac{\lambda^x e^{-2\lambda}}{x!} \sum_{y=x}^{\infty} \frac{\lambda^{y-x}}{(y-x)!} = \frac{\lambda^x}{x!} e^{-\lambda},$$

$$\mathbb{P}(Y = y) = \sum_{x=0}^{y} \frac{\lambda^y e^{-2\lambda}}{x!\,(y-x)!} = \frac{\lambda^y e^{-2\lambda}}{y!} \sum_{x=0}^{y} \binom{y}{x} = \frac{(2\lambda)^y}{y!} e^{-2\lambda}.$$

The conditional mass function is

$$f_{X|Y}(x \mid y) = \frac{f(x,y)}{f_Y(y)} = \binom{y}{x}\left(\frac{1}{2}\right)^y, \qquad x = 0, 1, 2, \dots, y,$$

the $\text{bin}(y, \frac{1}{2})$ distribution.

3.8 Solutions. Sums of random variables

1. By the convolution theorem,

$$\begin{aligned}
\mathbb{P}(X+Y=z) &= \sum_k \mathbb{P}(X=k)\mathbb{P}(Y=z-k) \\
&= \begin{cases} \dfrac{k+1}{(m+1)(n+1)} & \text{if } 0 \le z \le m \wedge n, \\[2mm] \dfrac{(m \wedge n)+1}{(m+1)(n+1)} & \text{if } m \wedge n < z < m \vee n, \\[2mm] \dfrac{m+n+1-k}{(m+1)(n+1)} & \text{if } m \vee n \le z \le m+n, \end{cases}
\end{aligned}$$

where $m \wedge n = \min\{m, n\}$ and $m \vee n = \max\{m, n\}$.

2. If $z \ge 2$,

$$\mathbb{P}(X+Y=z) = \sum_{k=1}^{\infty} \mathbb{P}(X=k, Y=z-k) = \frac{C}{z(z+1)}.$$

Also, if $z \geq 0$,

$$\begin{aligned}\mathbb{P}(X - Y = z) &= \sum_{k=1}^{\infty} \mathbb{P}(X = k + z, Y = k)\\ &= C \sum_{k=1}^{\infty} \frac{1}{(2k + z - 1)(2k + z)(2k + z + 1)}\\ &= \tfrac{1}{2} C \sum_{k=1}^{\infty} \left\{ \frac{1}{(2k + z - 1)(2k + z)} - \frac{1}{(2k + z)(2k + z + 1)} \right\}\\ &= \tfrac{1}{2} C \sum_{r=1}^{\infty} \frac{(-1)^{r+1}}{(r + z)(r + z + 1)}.\end{aligned}$$

By symmetry, if $z \leq 0$, $\mathbb{P}(X - Y = z) = \mathbb{P}(X - Y = -z) = \mathbb{P}(X - Y = |z|)$.

3. $\displaystyle\sum_{r=1}^{z-1} \alpha(1 - \alpha)^{r-1} \beta(1 - \beta)^{z-r-1} = \frac{\alpha\beta\{(1 - \beta)^{z-1} - (1 - \alpha)^{z-1}\}}{\alpha - \beta}.$

4. Repeatedly flip a coin that shows heads with probability p. Let X_r be the number of flips after the $r - 1$th head up to, and including, the rth. Then X_r is geometric with parameter p. The number of flips Z to obtain n heads is negative binomial, and $Z = \sum_{r=1}^{n} X_r$ by construction.

5. Sam. Let X_n be the number of sixes shown by $6n$ dice, so that $X_{n+1} = X_n + Y$ where Y has the same distribution as X_1 and is independent of X_n. Then,

$$\begin{aligned}\mathbb{P}(X_{n+1} \geq n + 1) &= \sum_{r=0}^{6} \mathbb{P}(X_n \geq n + 1 - r)\mathbb{P}(Y = r)\\ &= \mathbb{P}(X_n \geq n) + \sum_{r=0}^{6} \big[\mathbb{P}(X_n \geq n + 1 - r) - \mathbb{P}(X_n \geq n)\big]\mathbb{P}(Y = r).\end{aligned}$$

We set $g(k) = \mathbb{P}(X_n = k)$ and use the fact, easily proved, that $g(n) \geq g(n - 1) \geq \cdots \geq g(n - 5)$ to find that the last sum is no bigger than

$$g(n) \sum_{r=0}^{6} (r - 1)\mathbb{P}(Y = r) = g(n)\big(\mathbb{E}(Y) - 1\big).$$

The claim follows since $\mathbb{E}(Y) = 1$.

6. (i) $\displaystyle \text{LHS} = \sum_{n=0}^{\infty} n g(n) e^{-\lambda} \lambda^n / n! = \lambda \sum_{n=1}^{\infty} \frac{g(n) e^{-\lambda}}{(n - 1)!} \lambda^{n-1} = \text{RHS}.$

(ii) Conditioning on N and X_N,

$$\begin{aligned}\text{LHS} &= \mathbb{E}\big(\mathbb{E}(S g(S) \mid N)\big) = \mathbb{E}\big\{N \mathbb{E}(X_N g(S) \,\big|\, N)\big\}\\ &= \sum_{n=0}^{\infty} \frac{e^{-\lambda}\lambda^n}{n!} n \sum_{x=0}^{\infty} x \mathbb{E}\left(g\left(\sum_{r=1}^{n-1} X_r + x \right) \right) \mathbb{P}(X_0 = x)\\ &= \lambda \sum_{x=0}^{\infty} x \mathbb{E}\big(g(S + x)\big)\mathbb{P}(X_0 = x) = \text{RHS}.\end{aligned}$$

7. See Exercise (3.7.6).

8. By conditional probability,

$$\begin{aligned}\mathbb{P}(X+Y=n+1,\ X=k) &= \mathbb{P}(X+Y=n+1 \mid X=k)\mathbb{P}(X=k)\\ &= \frac{p\lambda\mu^n}{q}\left(\frac{q}{\mu}\right)^k, \qquad 1 \le k \le n.\end{aligned}$$

Therefore,

$$\mathbb{P}(X+Y=n+1) = \sum_{k=1}^{n} \mathbb{P}(X+Y=n+1,\ X=k) = p\lambda\frac{\mu^n - q^n}{\mu - q}, \qquad n \ge 1,$$

$$\mathbb{P}(X=k \mid X+Y=n+1) = \frac{(\mu-q)q^{k-1}\mu^{n-k}}{\mu^n - q^n}, \qquad 1 \le k \le n.$$

When $q = \mu$, either recompute (using the fact that the result of the first calculation does not depend on k), or use L'Hôpital's rule. The outcome is that, conditional on $X+Y=n+1$, X is uniform on $\{1, 2, \dots, n\}$, and moreover $\mathbb{P}(X+Y=n+1) = np^2q^{n-1}$ for $n \ge 1$.

3.9 Solutions. Simple random walk

1. (a) Consider an infinite sequence of tosses of a coin, any one of which turns up heads with probability p. With probability one there will appear a run of N heads sooner or later. If the coin tosses are 'driving' the random walk, then absorption occurs no later than this run, so that ultimate absorption is (almost surely) certain. Let S be the number of tosses before the first run of N heads. Certainly $\mathbb{P}(S > Nr) \le (1-p^N)^r$, since Nr tosses may be divided into r blocks of N tosses, each of which is such a run with probability p^N. Hence $\mathbb{P}(S = s) \le (1-p^N)^{\lfloor s/N \rfloor}$, and in particular $\mathbb{E}(S^k) < \infty$ for all $k \ge 1$. By the above argument, $\mathbb{E}(T^k) < \infty$ also.

2. If $S_0 = k$ then the first step X_1 satisfies

$$\mathbb{P}(X_1 = 1 \mid W) = \frac{\mathbb{P}(X_1 = 1)\mathbb{P}(W \mid X_1 = 1)}{\mathbb{P}(W)} = \frac{pp_{k+1}}{p_k}.$$

Let T be the duration of the walk. Then

$$\begin{aligned}J_k &= \mathbb{E}(T \mid S_0 = k, W)\\ &= \mathbb{E}(T \mid S_0 = k, W, X_1 = 1)\mathbb{P}(X_1 = 1 \mid S_0 = k, W)\\ &\qquad + \mathbb{E}(T \mid S_0 = k, W, X_1 = -1)\mathbb{P}(X_1 = -1 \mid S_0 = k, W)\\ &= \left(1 + J_{k+1}\right)\frac{p_{k+1}p}{p_k} + \left(1 + J_{k-1}\right)\left(1 - \frac{p_{k+1}p}{p_k}\right)\\ &= 1 + \frac{pp_{k+1}J_{k+1}}{p_k} + \frac{\left(p_k - pp_{k+1}\right)J_{k-1}}{p_k},\end{aligned}$$

as required.

Certainly $J_0 = 0$. If $p = \frac{1}{2}$ then $p_k = 1 - (k/N)$, so the difference equation becomes

$$(N-k-1)J_{k+1} - 2(N-k)J_k + (N-k+1)J_{k-1} = 2(k-N)$$

for $1 \le k \le N-1$ (where J_N is interpreted as 0). Setting $u_k = (N-k)J_k$, we obtain

$$u_{k+1} - 2u_k + u_{k-1} = 2(k-N),$$

with general solution $u_k = A + Bk - \frac{1}{3}(N-k)^3$ for constants A and B. Now $u_0 = u_N = 0$, and therefore $A = \frac{1}{3}N^3$, $B = -\frac{1}{3}N^2$, implying that $J_k = \frac{1}{3}\{N^2 - (N-k)^2\}$, $0 \le k < N$.

3. The recurrence relation may be established as in Exercise (3.9.2). Set $u_k = (\rho^k - \rho^N)J_k$ and use the fact that $p_k = (\rho^k - \rho^N)/(1-\rho^N)$ where $\rho = q/p$, to obtain

$$pu_{k+1} - (1-r)u_k + qu_{k-1} = \rho^N - \rho^k.$$

The solution is

$$u_k = A + B\rho^k + \frac{k(\rho^k + \rho^N)}{p-q},$$

for constants A and B. The boundary conditions, $u_0 = u_N = 0$, yield the answer.

4. Conditioning in the obvious way on the result of the first toss, we obtain

$$p_{mn} = pp_{m-1,n} + (1-p)p_{m,n-1}, \qquad \text{if } m, n \ge 1.$$

The boundary conditions are $p_{m0} = 0$, $p_{0n} = 1$, if $m, n \ge 1$.

5. Let Y be the number of negative steps of the walk up to absorption. Then $\mathbb{E}(X+Y) = D_k$ and

$$X - Y = \begin{cases} N-k & \text{if the walk is absorbed at } N, \\ -k & \text{if the walk is absorbed at } 0. \end{cases}$$

Hence $\mathbb{E}(X-Y) = (N-k)(1-p_k) - kp_k$, and solving for $\mathbb{E}X$ gives the result.

6. (a) As in Example (3.9.6), $m_k := \mathbb{E}D_k$ satisfies $m_k = k(a-k)$. By conditioning on the first step, $v_k := \text{var}\, D_k$ satisfies

$$\begin{aligned} v_k &= \tfrac{1}{2}\mathbb{E}\left(\{D_{k+1} - m_{k+1} + m_{k+1} - m_k + 1\}^2\right) + \tfrac{1}{2}\mathbb{E}\left(\{D_{k-1} - m_{k-1} + m_{k-1} - m_k + 1\}^2\right) \\ &= \tfrac{1}{2}v_{k+1} + \tfrac{1}{2}v_{k-1} + 2(2k-a)^2, \end{aligned}$$

when $0 < k < a$, by expanding and simplifying. Either solve this difference equation subject to the boundary conditions $v_0 = v_a = 0$, or insert the given solutions into the equation and check the boundary conditions.

(b) Let $p < \frac{1}{2}$. The non-zero steps of the walk execute a symmetric walk, whose duration d_k has mean and variance given in part (a). The blocks of consecutive zero-steps have independent lengths, which are independent of the symmetric walk. Each block length has a geometric distribution with mean $1/(2p)$ and variance $(1-2p)/(2p)^2$. The result now follows by Exercise (3.8.7).

7. Condition on the first step and use the result of Example (1.7.4) to find that

$$r_k = \frac{1}{2}\cdot\frac{a-k-1}{a-k} + \frac{1}{2}\cdot\frac{k-1}{k} = 1 - \frac{a}{2k(a-k)}.$$

The number R_k of returns to k before absorption is geometrically distributed on $\{0, 1, 2, \dots\}$ with mean $r_k/(1-r_k) = [2k(a-k)/a] - 1$. Let $x < k$. The mean number of visits to x is the probability of ever visiting x multiplied by $1 + R_x$, that is,

$$\frac{a-k}{a-x}\left\{1 + \frac{2x(a-x)}{a} - 1\right\} = \frac{2x(a-k)}{a}.$$

The case $x > k$ follows by symmetry. Curiously, if the barrier at a is reflecting,

$$v_x = \begin{cases} 2x & \text{if } x \le k, \\ 2k & \text{if } x \ge k. \end{cases}$$

3.10 Solutions. Random walk: counting sample paths

1. Conditioning on the first step X_1,

$$\mathbb{P}(T = 2n) = \tfrac{1}{2}\mathbb{P}(T = 2n \mid X_1 = 1) + \tfrac{1}{2}\mathbb{P}(T = 2n \mid X_1 = -1)$$
$$= \tfrac{1}{2} f_{-1}(2n-1) + \tfrac{1}{2} f_1(2n-1)$$

where $f_b(m)$ is the probability that the first passage to b of a symmetric walk, starting from 0, takes place at time m. From the hitting time theorem (3.10.14),

$$f_1(2n-1) = f_{-1}(2n-1) = \frac{1}{2n-1}\mathbb{P}(S_{2n-1} = 1) = \frac{1}{2n-1}\binom{2n-1}{n}2^{-(2n-1)},$$

which therefore is the value of $\mathbb{P}(T = 2n)$.

For the last part, note first that $\sum_1^\infty \mathbb{P}(T = 2n) = 1$, which is to say that $\mathbb{P}(T < \infty) = 1$; either appeal to your favourite method in order to see this, or observe that $\mathbb{P}(T = 2n)$ is the coefficient of s^{2n} in the expansion of $F(s) = 1 - \sqrt{1 - s^2}$. The required result is easily obtained by expanding the binomial coefficient using Stirling's formula.

2. By equation (3.10.13) of PRP, for $r \geq 0$,

$$\begin{aligned}\mathbb{P}(M_n = r) &= \mathbb{P}(M_n \geq r) - \mathbb{P}(M_n \geq r+1)\\ &= 2\mathbb{P}(S_n \geq r+1) + \mathbb{P}(S_n = r) - 2\mathbb{P}(S_n \geq r+2) - \mathbb{P}(S_n = r+1)\\ &= \mathbb{P}(S_n = r) + \mathbb{P}(S_n = r+1)\\ &= \max\{\mathbb{P}(S_n = r), \mathbb{P}(S_n = r+1)\}\end{aligned}$$

since only one of these two terms is non-zero.

3. By considering the random walk reversed, we see that the probability of a first visit to S_{2n} at time $2k$ is the same as the probability of a last visit to S_0 at time $2n - 2k$. The result is then immediate from the arc sine law (3.10.19) for the last visit to the origin.

4. Let $A_{\pm a}$ be the event that the walk is absorbed at $\pm a$, and T_t the event that absorption takes place at time t. A path is absorbed at a at time t if and only (i) if it does not visit $-a$ and (ii) it comprises $\frac{1}{2}(t-a)$ leftward steps and $\frac{1}{2}(t+a)$ rightward steps. Reversing the directions of all steps gives a path that is absorbed at $-a$ at time t. By summing over such paths,

$$\frac{\mathbb{P}(A_a \cap T_t)}{\mathbb{P}(A_{-a} \cap T_t)} = \left(\frac{p}{q}\right)^a,$$

where $p + q = 1$. Therefore,

$$\mathbb{P}(A_a \mid T_t) = \frac{(p/q)^a}{1 + (p/q)^a},$$

which is independent of t.

5. Write $\mathbb{P}_k$ for $\mathbb{P}$ conditioned on $S_0 = k$. We claim that, for $n \geq 1$,

$$(*) \qquad \mathbb{P}_k(T = n) = \frac{k}{n}\mathbb{P}_k(S_n = 0), \qquad k \geq 0,$$

and we prove this by induction on n. The equality is evidently true when $k = 0$, for all $n \geq 1$. When $n = 1$, both sides equal 0 when $k \neq 1$, and both sides equal $\mathbb{P}(X_1 = -1)$ when $k = 1$. Thus (*) holds when $n = 1$.

Let $n \geq 2$, and suppose (*) holds for $n-1$. Let $k \geq 1$. For $s \geq -1$,

$$\mathbb{P}_k(T = n \mid X_1 = s) = \mathbb{P}_{k+s}(T = n-1) = \frac{k+s}{n-1}\mathbb{P}_{k+s}(S_{n-1} = 0),$$

by the induction hypothesis, whence

$$\mathbb{P}_k(T = n) = \sum_{s=-1}^{\infty} \frac{k+s}{n-1}\mathbb{P}_{k+s}(S_{n-1} = 0)\mathbb{P}(X_1 = s).$$

Since $\mathbb{P}(A \mid B)\mathbb{P}(B) = \mathbb{P}(B \mid A)\mathbb{P}(A)$,

$$\begin{aligned}\sum_{s=-1}^{\infty} (k+s)\mathbb{P}_{k+s}(S_{n-1} = 0)\mathbb{P}(X_1 = s) &= \sum_{s=-1}^{\infty} (k+s)\mathbb{P}_k(X_1 = s \mid S_n = 0)\mathbb{P}_k(S_n = 0) \\ &= \mathbb{P}(S_n = 0)\sum_{s=-1}^{\infty} (k+s)\mathbb{P}_k(X_1 = s \mid S_n = 0) \\ &= \mathbb{P}(S_n = 0)\big[k + \mathbb{E}_k(X_1 \mid S_n = 0)\big].\end{aligned}$$

By symmetry, $\mathbb{E}_k(X_1 \mid S_n = 0) = -k/n$, so that

$$\mathbb{P}_k(T = n) = \frac{k}{n}\mathbb{P}(S_n = 0)$$

as required. [See also Theorem (5.3.7). This proof is due to van der Hofstad and Keane 2008, who explain also that the hitting time theorem is equivalent to the ballot theorem (3.10.6).]

3.11 Solutions to problems

1. (a) Clearly, for all $a, b \in \mathbb{R}$,

$$\begin{aligned}\mathbb{P}\big(g(X) = a, h(Y) = b\big) &= \sum_{\substack{x,y:\\ g(x)=a, h(y)=b}} \mathbb{P}(X = x, Y = y) \\ &= \sum_{\substack{x,y:\\ g(x)=a, h(y)=b}} \mathbb{P}(X = x)\mathbb{P}(Y = y) \\ &= \sum_{x:g(x)=a} \mathbb{P}(X = x) \sum_{y:h(y)=b} \mathbb{P}(Y = y) \\ &= \mathbb{P}(g(X) = a)\mathbb{P}(h(Y) = b).\end{aligned}$$

(b) See the definition (3.2.1) of independence.
(c) The only remaining part which requires proof is that X and Y are independent if $f_{X,Y}(x, y) = g(x)h(y)$ for all $x, y \in \mathbb{R}$. Suppose then that this holds. Then

$$f_X(x) = \sum_y f_{X,Y}(x, y) = g(x)\sum_y h(y), \quad f_Y(y) = \sum_x f_{X,Y}(x, y) = h(y)\sum_x g(x).$$

Now

$$1 = \sum_x f_X(x) = \sum_x g(x)\sum_y h(y),$$

so that

$$f_X(x)f_Y(y) = g(x)h(y)\sum_x g(x)\sum_y h(y) = g(x)h(y) = f_{X,Y}(x,y).$$

2. If $\mathbb{E}(X^2) = \sum_x x^2\mathbb{P}(X = x) = 0$ then $\mathbb{P}(X = x) = 0$ for $x \neq 0$. Hence $\mathbb{P}(X = 0) = 1$. Therefore, if $\text{var}(X) = 0$, it follows that $\mathbb{P}(X - \mathbb{E}X = 0) = 1$.

3. (a)

$$\mathbb{E}\big(g(X)\big) = \sum_y y\mathbb{P}\big(g(X) = y\big) = \sum_y \sum_{x:g(x)=y} y\mathbb{P}(X = x) = \sum_x g(x)\mathbb{P}(X = x)$$

as required.

(b)

$$\begin{aligned}\mathbb{E}\big(g(X)h(Y)\big) &= \sum_{x,y} g(x)h(y)f_{X,Y}(x,y) && \text{by Lemma(3.6.6)}\\ &= \sum_{x,y} g(x)h(y)f_X(x)f_Y(y) && \text{by independence}\\ &= \sum_x g(x)f_X(x)\sum_y h(y)f_Y(y) = \mathbb{E}(g(X))\mathbb{E}(h(Y)).\end{aligned}$$

4. (a) Clearly $f_X(i) = f_Y(i) = \frac{1}{3}$ for $i = 1, 2, 3$.

(b) $(X+Y)(\omega_1) = 3$, $(X+Y)(\omega_2) = 5$, $(X+Y)(\omega_3) = 4$, and therefore $f_{X+Y}(i) = \frac{1}{3}$ for $i = 3, 4, 5$.

(c) $(XY)(\omega_1) = 2$, $(XY)(\omega_2) = 6$, $(XY)(\omega_3) = 3$, and therefore $f_{XY}(i) = \frac{1}{3}$ for $i = 2, 3, 6$.

(d) Similarly $f_{X/Y}(i) = \frac{1}{3}$ for $i = \frac{1}{2}, \frac{2}{3}, 3$.

(e)

$$f_{Y|Z}(2 \mid 2) = \frac{\mathbb{P}(Y = 2, Z = 2)}{\mathbb{P}(Z = 2)} = \frac{\mathbb{P}(\omega_1)}{\mathbb{P}(\omega_1 \cup \omega_2)} = \frac{1}{2},$$

and similarly $f_{Y|Z}(3 \mid 2) = \frac{1}{2}$, $f_{Y|Z}(1 \mid 1) = 1$, and other values are 0.

(f) Likewise $f_{Z|Y}(2 \mid 2) = f_{Z|Y}(2 \mid 3) = f_{Z|Y}(1 \mid 1) = 1$.

5. (a) $\displaystyle\sum_{n=1}^{\infty} \frac{k}{n(n+1)} = k\sum_{n=1}^{\infty}\left\{\frac{1}{n} - \frac{1}{n+1}\right\} = k$, and therefore $k = 1$.

(b) $\sum_{n=1}^{\infty} kn^{\alpha} = k\zeta(-\alpha)$ where ζ is the Riemann zeta function, and we require $\alpha < -1$ for the sum to converge. In this case $k = \zeta(-\alpha)^{-1}$.

6. (a) We have that

$$\begin{aligned}\mathbb{P}(X + Y = n) &= \sum_{k=0}^{n} \mathbb{P}(X = n-k)\mathbb{P}(Y = k) = \sum_{k=0}^{n} \frac{e^{-\lambda}\lambda^{n-k}}{(n-k)!}\cdot\frac{e^{-\mu}\mu^k}{k!}\\ &= \frac{e^{-\lambda-\mu}}{n!}\sum_{k=0}^{n}\binom{n}{k}\lambda^{n-k}\mu^k = \frac{e^{-\lambda-\mu}(\lambda+\mu)^n}{n!}.\end{aligned}$$

(b)

$$\begin{aligned}\mathbb{P}(X = k \mid X + Y = n) &= \frac{\mathbb{P}(X = k, X + Y = n)}{\mathbb{P}(X + Y = n)}\\ &= \frac{\mathbb{P}(X = k)\mathbb{P}(Y = n-k)}{\mathbb{P}(X + Y = n)} = \binom{n}{k}\frac{\lambda^k\mu^{n-k}}{(\lambda+\mu)^n}.\end{aligned}$$

Hence the conditional distribution is $\text{bin}(n, \lambda/(\lambda+\mu))$.

7. (i) We have that

$$\begin{aligned}\mathbb{P}(X = n+k \mid X > n) &= \frac{\mathbb{P}(X = n+k,\, X > n)}{\mathbb{P}(X > n)} \\ &= \frac{p(1-p)^{n+k-1}}{\sum_{j=n+1}^{\infty} p(1-p)^{j-1}} = p(1-p)^{k-1} = \mathbb{P}(X = k).\end{aligned}$$

(ii) Many random variables of interest are 'waiting times', i.e. the time one must wait before the occurrence of some event A of interest. If such a time is geometric, the lack-of-memory property states that, given that A has not occurred by time n, the time to wait for A starting from n has the same distribution as it did to start with. With sufficient imagination this can be interpreted as a failure of memory by the process giving rise to A.

(iii) No. This is because, by the above, any such process satisfies $G(k+n) = G(k)G(n)$ where $G(n) = \mathbb{P}(X > n)$. Hence $G(k+1) = G(1)^{k+1}$ and X is geometric.

8. Clearly,

$$\begin{aligned}\mathbb{P}(X+Y = k) &= \sum_{j=0}^{k} \mathbb{P}(X = k-j,\, Y = j) \\ &= \sum_{j=0}^{k} \binom{m}{k-j} p^{k-j} q^{m-k+j} \binom{n}{j} p^j q^{n-j} \\ &= p^k q^{m+n-k} \sum_{j=0}^{k} \binom{m}{k-j}\binom{n}{j} = p^k q^{m+n-k}\binom{m+n}{k}\end{aligned}$$

which is $\text{bin}(m+n, p)$.

9. Turning immediately to the second request, by the binomial theorem,

$$\tfrac{1}{2}(x+y)^n + \tfrac{1}{2}(y-x)^n = \tfrac{1}{2}\sum_k \binom{n}{k} y^{n-k}\{x^k + (-x)^k\} = \sum_{k \text{ even}} \binom{n}{k} x^k y^{n-k}$$

as required. Now,

$$\begin{aligned}\mathbb{P}(N \text{ even}) &= \sum_{k \text{ even}} \binom{n}{k} p^k (1-p)^{n-k} \\ &= \tfrac{1}{2}\{(p+1-p)^n + (1-p-p)^n\} = \tfrac{1}{2}\{1 + (1-2p)^n\}\end{aligned}$$

in agreement with Problem (1.8.20).

10. There are $\binom{b}{k}$ ways of choosing k blue balls, and $\binom{N-b}{n-k}$ ways of choosing $n-k$ red balls. The total number of ways of choosing n balls is $\binom{N}{n}$, and the claim follows. Finally,

$$\begin{aligned}
\mathbb{P}(B=k) &= \binom{n}{k}\frac{b!}{(b-k)!}\cdot\frac{(N-b)!}{(N-b-n+k)!}\cdot\frac{(N-n)!}{N!}\\
&= \binom{n}{k}\left\{\frac{b}{N}\cdot\frac{b-1}{N}\cdots\frac{b-k+1}{N}\right\}\\
&\qquad\times\left\{\frac{N-b}{N}\cdots\frac{N-b-n+k+1}{N}\right\}\left\{\frac{N}{N}\cdots\frac{N-n+1}{N}\right\}^{-1}\\
&\to \binom{n}{k}p^k(1-p)^{n-k} \qquad \text{as } N\to\infty.
\end{aligned}$$

11. Using the result of Problem (3.11.8),

$$\begin{aligned}
\mathbb{P}(X=k \mid X+Y=N) &= \frac{\mathbb{P}(X=k)\mathbb{P}(Y=N-k)}{\mathbb{P}(X+Y=N)}\\
&= \frac{\binom{n}{k}p^kq^{n-k}\binom{n}{N-k}p^{N-k}q^{n-N+k}}{\binom{2n}{N}p^Nq^{2n-N}} = \frac{\binom{n}{k}\binom{n}{N-k}}{\binom{2n}{N}}.
\end{aligned}$$

12. (a) $\mathbb{E}(X)=c+d$, $\mathbb{E}(Y)=b+d$, and $\mathbb{E}(XY)=d$, so $\operatorname{cov}(X,Y)=d-(c+d)(b+d)$, and X and Y are uncorrelated if and only if this equals 0.
(b) For independence, we require $f(i,j)=\mathbb{P}(X=i)\mathbb{P}(Y=j)$ for all i,j, which is to say that

$$a=(a+b)(a+c),\quad b=(a+b)(b+d),\quad c=(c+d)(a+c),\quad d=(b+d)(c+d).$$

Now $a+b+c+d=1$, and with a little work one sees that any one of these relations implies the other three. Therefore X and Y are independent if and only if $d=(b+d)(c+d)$, the same condition as for uncorrelatedness.

13. (a) We have that

$$\mathbb{E}(X)=\sum_{m=0}^{\infty}m\mathbb{P}(X=m)=\sum_{m=0}^{\infty}\sum_{n=0}^{m-1}\mathbb{P}(X=m)=\sum_{n=0}^{\infty}\sum_{m=n+1}^{\infty}\mathbb{P}(X=m)=\sum_{n=0}^{\infty}\mathbb{P}(X>n).$$

(b) *First method.* Let N be the number of balls drawn. Then, by (a),

$$\begin{aligned}
\mathbb{E}(N) &= \sum_{n=0}^{r}\mathbb{P}(N>n)=\sum_{n=0}^{r}\mathbb{P}(\text{first } n \text{ balls are red})\\
&= \sum_{n=0}^{r}\frac{r}{b+r}\frac{r-1}{b+r-1}\cdots\frac{r-n+1}{b+r-n+1}=\sum_{n=0}^{r}\frac{r!}{(b+r)!}\frac{(b+r-n)!}{(r-n)!}\\
&= \frac{r!\,b!}{(b+r)!}\sum_{n=0}^{r}\binom{n+b}{b}=\frac{b+r+1}{b+1},
\end{aligned}$$

where we have used the combinatorial identity $\sum_{n=0}^{r}\binom{n+b}{b}=\binom{r+b+1}{b+1}$. To see this, either use the simple identity $\binom{x}{r-1}+\binom{x}{r}=\binom{x+1}{r}$ repeatedly, or argue as follows. Changing the order of summation, we find that

$$\sum_{r=0}^{\infty} x^r \sum_{n=0}^{r} \binom{n+b}{b} = \frac{1}{1-x} \sum_{n=0}^{\infty} x^n \binom{n+b}{b}$$
$$= (1-x)^{-(b+2)} = \sum_{r=0}^{\infty} x^r \binom{b+r+1}{b+1}$$

by the (negative) binomial theorem. Equating coefficients of x^r, we obtain the required identity.

Second method. Writing $m(b, r)$ for the mean in question, and conditioning on the colour of the first ball, we find that

$$m(b,r) = \frac{b}{b+r} + \{1 + m(b, r-1)\} \frac{r}{b+r}.$$

With appropriate boundary conditions and a little effort, one may obtain the result.

Third method. Withdraw all the balls, and let N_i be the number of red balls drawn between the ith and $(i+1)$th blue ball ($N_0 = N$, and N_b is defined analogously). Think of a possible 'colour sequence' as comprising r reds, split by b blues into $b+1$ red sequences. There is a one–one correspondence between the set of such sequences with $N_0 = i$, $N_m = j$ (for given i, j, m) and the set of such sequences with $N_0 = j$, $N_m = i$; just interchange the '0th' red run with the mth red run. In particular $\mathbb{E}(N_0) = \mathbb{E}(N_m)$ for all m. Now $N_0 + N_1 + \cdots + N_b = r$, so that $\mathbb{E}(N_m) = r/(b+1)$, whence the claim is immediate.

(c) We use the notation just introduced. In addition, let B_r be the number of blue balls remaining after the removal of the last red ball. The length of the last 'colour run' is $N_b + B_r$, only one of which is non-zero. The answer is therefore $r/(b+1) + b/(r+1)$, by the argument of the third solution to part (b).

(d) The first equation follows by part (a) with $\mathbb{P}(U \geq r) = \mathbb{P}(X \geq r)\mathbb{P}(Y \geq r)$, and the second since $\mathbb{P}(V \geq r) = \mathbb{P}(\{X \geq r\} \cup \{Y \geq r\})$, the probability of an independent union. The third holds by the fact that $UV = XY$.

(e) We have that

$$\sum_{r=0}^{\infty} 2r\mathbb{P}(X > r) = \mathbb{E}\left(\sum_{r=1}^{\infty} 2r I_{\{X>r\}}\right) = \mathbb{E}\left(\sum_{r=1}^{X-1} 2r\right) = \mathbb{E}(X(X-1)).$$

Similarly,

$$\sum_{r=1}^{\infty} 3r(r+1)\mathbb{P}(X > r) = \mathbb{E}(X(X^2-1)).$$

14. We have that $\mathbb{E}(X_k) = p_k$ and $\text{var}(X_k) = p_k(1-p_k)$, and the claims follow in the usual way, the first by the linearity of $\mathbb{E}$ and the second by the independence of the X_k; see Theorems (3.3.8) and (3.3.11).

Let $s = \sum_k p_k$, and let Z be a random variable taking each of the values $p_1, p_2, \ldots, p_n$ with equal probability n^{-1}. Now $\mathbb{E}(Z^2) - \mathbb{E}(Z)^2 = \text{var}(Z) \geq 0$, so that

$$\sum_k \frac{1}{n} p_k^2 \geq \left(\sum_k \frac{1}{n} p_k\right)^2 = \frac{s^2}{n^2}$$

with equality if and only if Z is (almost surely) constant, which is to say that $p_1 = p_2 = \cdots = p_n$. Hence

$$\text{var}(Y) = \sum_k p_k - \sum_k p_k^2 \leq s - \frac{s^2}{n}$$

with equality if and only if $p_1 = p_2 = \cdots = p_n$. Essentially the same route may be followed using a Lagrange multiplier.

This conclusion is not contrary to informed intuition, but experience shows it to be contrary to much uninformed intuition.

15. A matrix $\mathbf{V}$ has zero determinant if and only if it is singular, which is to say if and only if there is a non-zero vector $\mathbf{x}$ such that $\mathbf{x}\mathbf{V}\mathbf{x}' = 0$. However,

$$\mathbf{x}\mathbf{V}(\mathbf{X})\mathbf{x}' = \mathbb{E}\left\{\left(\sum_k x_k(X_k - \mathbb{E}X_k)\right)^2\right\}.$$

Hence, by the result of Problem (3.11.2), $\sum_k x_k(X_k - \mathbb{E}X_k)$ is constant with probability one, and the result follows.

16. The random variables $X + Y$ and $|X - Y|$ are uncorrelated since

$$\begin{aligned}\operatorname{cov}\big(X+Y, |X-Y|\big) &= \mathbb{E}\{(X+Y)|X-Y|\} - \mathbb{E}(X+Y)\mathbb{E}(|X-Y|)\\ &= \tfrac{1}{4} + \tfrac{1}{4} - 1\cdot\tfrac{1}{2} = 0.\end{aligned}$$

However,

$$\tfrac{1}{4} = \mathbb{P}(X+Y=0, |X-Y|=0) \neq \mathbb{P}(X+Y=0)\mathbb{P}(|X-Y|=0) = \tfrac{1}{4}\cdot\tfrac{1}{2} = \tfrac{1}{8},$$

so that $X + Y$ and $|X - Y|$ are dependent.

17. Let I_k be the indicator function of the event that there is a match in the kth place. Then $\mathbb{P}(I_k = 1) = n^{-1}$, and for $k \neq j$,

$$\mathbb{P}(I_k = 1, I_j = 1) = \mathbb{P}(I_j = 1 \mid I_k = 1)\mathbb{P}(I_k = 1) = \frac{1}{n(n-1)}.$$

Now $X = \sum_{k=1}^n I_k$, so that $\mathbb{E}(X) = \sum_{k=1}^n n^{-1} = 1$ and

$$\begin{aligned}\operatorname{var}(X) &= \mathbb{E}(X^2) - (\mathbb{E}X)^2 = \mathbb{E}\left(\sum_1^n I_k\right)^2 - 1\\ &= \sum_1^n \mathbb{E}(I_k)^2 + \sum_{j\neq k}\mathbb{E}(I_jI_k) - 1 = 1 + 2\binom{n}{2}\frac{1}{n(n-1)} - 1 = 1.\end{aligned}$$

We have by the usual (mis)matching argument of Example (3.4.3) that

$$\mathbb{P}(X = r) = \frac{1}{r!}\sum_{i=0}^{n-r}\frac{(-1)^i}{i!}, \qquad 0 \le r \le n-2,$$

which tends to $e^{-1}/r!$ as $n \to \infty$.

18. (a) Let $Y_1, Y_2, \ldots, Y_n$ be Bernoulli with parameter p_2, and $Z_1, Z_2, \ldots, Z_n$ Bernoulli with parameter p_1/p_2, and suppose the usual independence. Define $A_i = Y_iZ_i$, a Bernoulli random variable that has parameter $\mathbb{P}(A_i = 1) = \mathbb{P}(Y_i = 1)\mathbb{P}(Z_i = 1) = p_1$. Now $(A_1, A_2, \ldots, A_n) \le (Y_1, Y_2, \ldots, Y_n)$ so that $f(\mathbf{A}) \le f(\mathbf{Y})$. Hence $e(p_1) = \mathbb{E}(f(\mathbf{A})) \le \mathbb{E}(f(\mathbf{Y})) = e(p_2)$.

(b) Suppose first that $n = 1$, and let X and X' be independent Bernoulli variables with parameter p. We claim that

$$\{f(X) - f(X')\}\{g(X) - g(X')\} \ge 0;$$

to see this consider the three cases $X = X'$, $X < X'$, $X > X'$ separately, using the fact that f and g are increasing. Taking expectations, we obtain

$$\mathbb{E}(\{f(X) - f(X')\}\{g(X) - g(X')\}) \geq 0,$$

which may be expanded to find that

$$\begin{aligned} 0 &\leq \mathbb{E}\big(f(X)g(X)\big) - \mathbb{E}\big(f(X')g(X)\big) - \mathbb{E}\big(f(X)g(X')\big) + \mathbb{E}\big(f(X')g(X')\big) \\ &= 2\Big\{\mathbb{E}\big(f(X)g(X)\big) - \mathbb{E}(f(X))\mathbb{E}(g(X))\Big\} \end{aligned}$$

by the properties of X and X'.

Suppose that the result is valid for all n satisfying $n < k$ where $k \geq 2$. Now

$$(*) \qquad \mathbb{E}\big(f(\mathbf{X})g(\mathbf{X})\big) = \mathbb{E}\left\{\mathbb{E}\big(f(\mathbf{X})g(\mathbf{X}) \mid X_1, X_2, \ldots, X_{k-1}\big)\right\};$$

here, the conditional expectation given $X_1, X_2, \ldots, X_{k-1}$ is defined in very much the same way as in Definition (3.7.3), with broadly similar properties, in particular Theorem (3.7.4); see also Exercises (3.7.1, 3). If $X_1, X_2, \ldots, X_{k-1}$ are given, then $f(\mathbf{X})$ and $g(\mathbf{X})$ may be thought of as increasing functions of the single remaining variable X_k, and therefore

$$\mathbb{E}\big(f(\mathbf{X})g(\mathbf{X}) \mid X_1, X_2, \ldots, X_{k-1}\big) \geq \mathbb{E}\big(f(\mathbf{X}) \mid X_1, X_2, \ldots, X_{k-1}\big)\mathbb{E}\big(g(\mathbf{X}) \mid X_1, X_2, \ldots, X_{k-1}\big)$$

by the induction hypothesis. Furthermore

$$f'(\mathbf{X}) = \mathbb{E}\big(f(\mathbf{X}) \mid X_1, X_2, \ldots, X_{k-1}\big), \quad g'(\mathbf{X}) = \mathbb{E}\big(g(\mathbf{X}) \mid X_1, X_2, \ldots, X_{k-1}\big),$$

are increasing functions of the $k-1$ variables $X_1, X_2, \ldots, X_{k-1}$, implying by the induction hypothesis that $\mathbb{E}(f'(\mathbf{X})g'(\mathbf{X})) \geq \mathbb{E}(f'(\mathbf{X}))\mathbb{E}(g'(\mathbf{X}))$. We substitute this into $(*)$ to obtain

$$\mathbb{E}\big(f(\mathbf{X})g(\mathbf{X})\big) \geq \mathbb{E}(f'(\mathbf{X}))\mathbb{E}(g'(\mathbf{X})) = \mathbb{E}(f(\mathbf{X}))\mathbb{E}(g(\mathbf{X}))$$

by the definition of f' and g'.

19. Certainly $R(p) = \mathbb{E}(I_A) = \sum_\omega I_A(\omega)\mathbb{P}(\omega)$ and $\mathbb{P}(\omega) = p^{N(\omega)}q^{m-N(\omega)}$ where $p + q = 1$. Differentiating, we obtain

$$\begin{aligned} R'(p) &= \sum_\omega I_A(\omega)p^{N(\omega)}q^{m-N(\omega)}\left(\frac{N(\omega)}{p} - \frac{m - N(\omega)}{q}\right) \\ &= \frac{1}{pq}\sum_\omega I_A(\omega)\mathbb{P}(\omega)\big(N(\omega) - mp\big) \\ &= \frac{1}{pq}\mathbb{E}\big(I_A(N - mp)\big) = \frac{1}{pq}\{\mathbb{E}(I_A N) - \mathbb{E}(I_A)\mathbb{E}(N)\} = \frac{1}{pq}\operatorname{cov}(I_A, N). \end{aligned}$$

Applying the Cauchy–Schwarz inequality (3.6.9) to the latter covariance, we find that $R'(p) \leq (pq)^{-1}\sqrt{\operatorname{var}(I_A)\operatorname{var}(N)}$. However I_A is Bernoulli with parameter $R(p)$, so that $\operatorname{var}(I_A) = R(p)(1 - R(p))$, and finally N is $\operatorname{bin}(m, p)$ so that $\operatorname{var}(N) = mp(1 - p)$, whence the upper bound for $R'(p)$ follows.

As for the lower bound, use the general fact that $\operatorname{cov}(X+Y, Z) = \operatorname{cov}(X, Z) + \operatorname{cov}(Y, Z)$ to deduce that $\operatorname{cov}(I_A, N) = \operatorname{cov}(I_A, I_A) + \operatorname{cov}(I_A, N - I_A)$. Now I_A and $N - I_A$ are increasing functions of ω, in the sense of Problem (3.11.18); you should check this. Hence $\operatorname{cov}(I_A, N) \geq \operatorname{var}(I_A) + 0$ by the result of that problem. The lower bound for $R'(p)$ follows.

20. (a) Let each edge be *blue* with probability p_1 and *yellow* with probability p_2; assume these two events are independent of each other and of the colourings of all other edges. Call an edge *green* if it is both blue and yellow, so that each edge is *green* with probability $p_1 p_2$. If there is a working *green* connection from source to sink, then there is also a blue connection and a yellow connection. Thus

$$\begin{aligned}\mathbb{P}(\text{green connection}) &\le \mathbb{P}(\text{blue connection, and yellow connection})\\ &= \mathbb{P}(\text{blue connection})\mathbb{P}(\text{yellow connection})\end{aligned}$$

so that $R(p_1 p_2) \le R(p_1)R(p_2)$.

(b) This is somewhat harder, and may be proved by induction on the number n of edges of G. If $n = 1$ then a consideration of the two possible cases yields that either $R(p) = 1$ for all p, or $R(p) = p$ for all p. In either case the required inequality holds.

Suppose then that the inequality is valid whenever $n < k$ where $k \ge 2$, and consider the case when G has k edges. Let e be an edge of G and write $\omega(e)$ for the state of e; $\omega(e) = 1$ if e is working, and $\omega(e) = 0$ otherwise. Writing A for the event that there is a working connection from source to sink, we have that

$$\begin{aligned}R(p^\gamma) &= \mathbb{P}_{p^\gamma}(A \mid \omega(e) = 1)p^\gamma + \mathbb{P}_{p^\gamma}(A \mid \omega(e) = 0)(1 - p^\gamma)\\ &\le \mathbb{P}_p(A \mid \omega(e) = 1)^\gamma p^\gamma + \mathbb{P}_p(A \mid \omega(e) = 0)^\gamma(1 - p^\gamma)\end{aligned}$$

where $\mathbb{P}_\alpha$ is the appropriate probability measure when each edge is working with probability α. The inequality here is valid since, if $\omega(e)$ is given, then the network G is effectively reduced in size by one edge; the induction hypothesis is then utilized for the case $n = k - 1$. It is a minor chore to check that

$$x^\gamma p^\gamma + y^\gamma(1 - p)^\gamma \le \{xp + y(1 - p)\}^\gamma \qquad \text{if } x \ge y \ge 0;$$

to see this, check that equality holds when $x = y \ge 0$ and that the derivative of the left-hand side with respect to x is at most the corresponding derivative of the right-hand side when $x, y \ge 0$. Apply the latter inequality with $x = \mathbb{P}_p(A \mid \omega(e) = 1)$ and $y = \mathbb{P}_p(A \mid \omega(e) = 0)$ to obtain

$$R(p^\gamma) \le \big\{\mathbb{P}_p(A \mid \omega(e) = 1)p + \mathbb{P}_p(A \mid \omega(e) = 0)(1 - p)\big\}^\gamma = R(p)^\gamma.$$

21. (a) The number X of such extraordinary individuals has the $\text{bin}(10^7, 10^{-7})$ distribution. Hence $\mathbb{E}X = 1$ and

$$\begin{aligned}\mathbb{P}(X > 1 \mid X \ge 1) &= \frac{\mathbb{P}(X > 1)}{\mathbb{P}(X > 0)} = \frac{1 - \mathbb{P}(X = 0) - \mathbb{P}(X = 1)}{1 - \mathbb{P}(X = 0)}\\ &= \frac{1 - (1 - 10^{-7})^{10^7} - 10^7 \cdot 10^{-7}(1 - 10^{-7})^{10^7 - 1}}{1 - (1 - 10^{-7})^{10^7}}\\ &\simeq \frac{1 - 2e^{-1}}{1 - e^{-1}} \simeq 0.4.\end{aligned}$$

(Shades of (3.5.4) here: X is approximately Poisson distributed with parameter 1.)

(b) Likewise

$$\mathbb{P}(X > 2 \mid X \ge 2) \simeq \frac{1 - 2e^{-1} - \frac{1}{2}e^{-1}}{1 - 2e^{-1}} \simeq 0.3.$$

(c) Provided $m \ll N = 10^7$,

$$\mathbb{P}(X = m) = \frac{N!}{m!\,(N - m)!}\left(\frac{1}{N}\right)^m \left(1 - \frac{1}{N}\right)^{N - m} \simeq \frac{e^{-1}}{m!},$$

the Poisson distribution. Assume that "reasonably confident that n is all" means that $\mathbb{P}(X > n \mid X \geq n) \leq r$ for some suitable small number r. Assuming the Poisson approximation, $\mathbb{P}(X > n) \leq r\mathbb{P}(X \geq n)$ if and only if

$$e^{-1} \sum_{k=n+1}^{\infty} \frac{1}{k!} \leq re^{-1} \sum_{k=n}^{\infty} \frac{1}{k!}.$$

For any given r, the smallest acceptable value of n may be determined numerically. If r is small, then very roughly $n \simeq 1/r$ will do (e.g., if $r = 0.05$ then $n \simeq 20$).

(d) No level p of improbability is sufficiently small for one to be sure that the person is specified uniquely. If $p = 10^{-7}\alpha$, then X is $\text{bin}(10^7, 10^{-7}\alpha)$, which is approximately Poisson with parameter α. Therefore, in this case,

$$\mathbb{P}(X > 1 \mid X \geq 1) \simeq \frac{1 - e^{-\alpha} - \alpha e^{-\alpha}}{1 - e^{-\alpha}} = \rho, \qquad \text{say.}$$

An acceptable value of ρ for a very petty offence might be $\rho \simeq 0.05$, in which case $\alpha \simeq 0.1$ and so $p = 10^{-8}$ might be an acceptable level of improbability. For a capital offence, one would normally require a much smaller value of ρ. We note that the rules of evidence do not allow an overt discussion along these lines in a court of law in the United Kingdom.

22. The number G of girls has the binomial distribution $\text{bin}(2n, p)$. Hence

$$\begin{aligned}\mathbb{P}(G \geq 2n - G) = \mathbb{P}(G \geq n) &= \sum_{k=n}^{2n} \binom{2n}{k} p^k q^{2n-k} \\ &\leq \binom{2n}{n} \sum_{k=n}^{\infty} p^k q^{2n-k} = \binom{2n}{n} p^n q^n \frac{q}{q-p},\end{aligned}$$

where we have used the fact that $\binom{2n}{k} \leq \binom{2n}{n}$ for all k.

With $p = 0.485$ and $n = 10^4$, we have using Stirling's formula (Exercise (3.10.1)) that

$$\begin{aligned}\binom{2n}{n} p^n q^n \frac{q}{q-p} &\leq \frac{1}{\sqrt{(n\pi)}} \{(1 - 0.03)(1 + 0.03)\}^n \frac{0.515}{0.03} \\ &= \frac{0.515}{3\sqrt{\pi}} \left(1 - \frac{9}{10^4}\right)^{10^4} \leq 1.23 \times 10^{-5}.\end{aligned}$$

It follows that the probability that boys outnumber girls for 82 successive years is at least $(1 - 1.23 \times 10^{-5})^{82} \geq 0.99899$.

23. Let M be the number of such visits. If $k \neq 0$, then $M \geq 1$ if and only if the particle hits 0 before it hits N, an event with probability $1 - kN^{-1}$ by equation (1.7.7). Having hit 0, the chance of another visit to 0 before hitting N is $1 - N^{-1}$, since the particle at 0 moves immediately to 1 whence there is probability $1 - N^{-1}$ of another visit to 0 before visiting N. Hence

$$\mathbb{P}(M \geq r \mid S_0 = k) = \left(1 - \frac{k}{N}\right)\left(1 - \frac{1}{N}\right)^{r-1}, \qquad r \geq 1,$$

so that

$$\begin{aligned}\mathbb{P}(M = j \mid S_0 = k) &= \mathbb{P}(M \geq j \mid S_0 = k) - \mathbb{P}(M \geq j + 1 \mid S_0 = 0) \\ &= \left(1 - \frac{k}{N}\right)\left(1 - \frac{1}{N}\right)^{j-1} \frac{1}{N}, \qquad j \geq 1.\end{aligned}$$

24. Either read the solution to Exercise (3.9.4), or the following two related solutions neither of which uses difference equations.

First method. Let T_k be the event that A wins and exactly k tails appear. Then $k < n$ so that $\mathbb{P}(\text{A wins}) = \sum_{k=0}^{n-1} \mathbb{P}(T_k)$. However $\mathbb{P}(T_k)$ is the probability that $m+k$ tosses yield m heads, k tails, and the last toss is heads. Hence

$$\mathbb{P}(T_k) = \binom{m+k-1}{m-1} p^m q^k,$$

whence the result follows.

Second method. Suppose the coin is tossed $m+n-1$ times. If the number of heads is m or more, then A must have won; conversely if the number of heads is $m-1$ or less, then the number of tails is n or more, so that B has won. The number of heads is $\text{bin}(m+n-1, p)$ so that

$$\mathbb{P}(\text{A wins}) = \sum_{k=m}^{m+n-1} \binom{m+n-1}{k} p^k q^{m+n-1-k}.$$

25. The chance of winning, having started from k, is

$$\frac{1-(q/p)^k}{1-(q/p)^N} \quad \text{which may be written as} \quad \frac{1-(q/p)^{\frac{1}{2}k}}{1-(q/p)^{\frac{1}{2}N}} \cdot \frac{1+(q/p)^{\frac{1}{2}k}}{1+(q/p)^{\frac{1}{2}N}};$$

see Example (3.9.6). If k and N are even, doubling the stake is equivalent to playing the original game with initial fortune $\frac{1}{2}k$ and the price of the Jaguar set at $\frac{1}{2}N$. The probability of winning is now

$$\frac{1-(q/p)^{\frac{1}{2}k}}{1-(q/p)^{\frac{1}{2}N}},$$

which is larger than before, since the final term in the above display is greater than 1 (when $p < \frac{1}{2}$).

If $p = \frac{1}{2}$, doubling the stake makes no difference to the chance of winning. If $p > \frac{1}{2}$, it is better to decrease the stake.

26. This is equivalent to taking the limit as $N \to \infty$ in the previous Problem (3.11.25). In the limit when $p \neq \frac{1}{2}$, the probability of ultimate bankruptcy is

$$\lim_{N\to\infty} \frac{(q/p)^k - (q/p)^N}{1-(q/p)^N} = \begin{cases} (q/p)^k & \text{if } p > \frac{1}{2}, \\ 1 & \text{if } p < \frac{1}{2}, \end{cases}$$

where $p + q = 1$. If $p = \frac{1}{2}$, the corresponding limit is $\lim_{N\to\infty}(1 - k/N) = 1$.

27. Using the technique of reversal, we have that

$$\begin{aligned}
\mathbb{P}(R_n = R_{n-1} + 1) &= \mathbb{P}(S_{n-1} \neq S_n, S_{n-2} \neq S_n, \dots, S_0 \neq S_n) \\
&= \mathbb{P}(X_n \neq 0, X_{n-1} + X_n \neq 0, \dots, X_1 + \dots + X_n \neq 0) \\
&= \mathbb{P}(X_1 \neq 0, X_2 + X_1 \neq 0, \dots, X_n + \dots + X_1 \neq 0) \\
&= \mathbb{P}(S_1 \neq 0, S_2 \neq 0, \dots, S_n \neq 0) = \mathbb{P}(S_1 S_2 \cdots S_n \neq 0).
\end{aligned}$$

It follows that $\mathbb{E}(R_n) = \mathbb{E}(R_{n-1}) + \mathbb{P}(S_1 S_2 \cdots S_n \neq 0)$ for $n \geq 1$, whence

$$\frac{1}{n}\mathbb{E}(R_n) = \frac{1}{n}\left\{1 + \sum_{m=1}^{n} \mathbb{P}(S_1 S_2 \cdots S_m \neq 0)\right\} \to \mathbb{P}(S_k \neq 0 \text{ for all } k \geq 1)$$

since $\mathbb{P}(S_1 S_2 \cdots S_m \neq 0) \downarrow \mathbb{P}(S_k \neq 0 \text{ for all } k \geq 1)$ as $m \to \infty$.

There are various ways of showing that the last probability equals $|p - q|$, and here is one. Suppose $p > q$. If $X_1 = 1$, the probability of never subsequently hitting the origin equals $1 - (q/p)$, by the calculation in the solution to Problem (3.11.26) above. If $X_1 = -1$, the probability of staying away from the origin subsequently is 0. Hence the answer is $p(1 - (q/p)) + q \cdot 0 = p - q$.

If $q > p$, the same argument yields $q - p$, and if $p = q = \frac{1}{2}$ the answer is 0.

28. Consider first the event that M_{2n} is first attained at time $2k$. This event occurs if and only if: (i) the walk makes a first passage to S_{2k} (> 0) at time $2k$, and (ii) the walk thereafter does not exceed S_{2k}. These two events are independent. The chance of (i) is, by reversal and symmetry,

$$\begin{aligned}
&\mathbb{P}(S_{2k-1} < S_{2k}, S_{2k-2} < S_{2k}, \dots, S_0 < S_{2k}) \\
&\quad = \mathbb{P}(X_{2k} > 0, X_{2k-1} + X_{2k} > 0, \dots, X_1 + \cdots + X_{2k} > 0) \\
&\quad = \mathbb{P}(X_1 > 0, X_1 + X_2 > 0, \dots, X_1 + \cdots + X_{2k} > 0) \\
&\quad = \mathbb{P}(S_i > 0 \text{ for } 1 \leq i \leq 2k) = \tfrac{1}{2}\mathbb{P}(S_i \neq 0 \text{ for } 1 \leq i \leq 2k) \\
&\quad = \tfrac{1}{2}\mathbb{P}(S_{2k} = 0) \qquad \text{by equation (3.10.23).}
\end{aligned}$$

As for the second event, we may translate S_{2k} to the origin to obtain the probability of (ii):

$$\mathbb{P}(S_{2k+1} \leq S_{2k}, \dots, S_{2n} \leq S_{2k}) = \mathbb{P}(M_{2n-2k} = 0) = \mathbb{P}(S_{2n-2k} = 0),$$

where we have used the result of Exercise (3.10.2). The answer is therefore as given.

The probabilities of (i) and (ii) are unchanged in the case $i = 2k + 1$; the basic reason for this is that S_{2r} is even, and S_{2r+1} odd, for all r.

29. Let $u_k = \mathbb{P}(S_k = 0)$, $f_k = \mathbb{P}(S_k = 0, S_i \neq 0 \text{ for } 1 \leq i < k)$, and use conditional probability (or recall from equation (3.10.25)) to obtain

$$(*) \qquad u_{2n} = \sum_{k=1}^{n} u_{2n-2k} f_{2k}.$$

Now $N_1 = 2$, and therefore it suffices to prove that $\mathbb{E}(N_n) = \mathbb{E}(N_{n-1})$ for $n \geq 2$. Let N'_{n-1} be the number of points visited by the walk $S_1, S_2, \dots, S_n$ exactly once (we have removed S_0). Then

$$N_n = \begin{cases} N'_{n-1} + 1 & \text{if } S_k \neq S_0 \text{ for } 1 \leq k \leq n, \\ N'_{n-1} - 1 & \text{if } S_k = S_0 \text{ for exactly one } k \text{ in } \{1, 2, \dots, n\}, \\ N'_{n-1} & \text{otherwise.} \end{cases}$$

Hence, writing $\alpha_n = \mathbb{P}(S_k \neq 0 \text{ for } 1 \leq k \leq n)$,

$$\begin{aligned}
\mathbb{E}(N_n) &= \mathbb{E}(N'_{n-1}) + \alpha_n - \mathbb{P}(S_k = S_0 \text{ exactly once}) \\
&= \mathbb{E}(N_{n-1}) + \alpha_n - \left\{ f_2 \alpha_{n-2} + f_4 \alpha_{n-4} + \cdots + f_{2\lfloor n/2 \rfloor} \right\}
\end{aligned}$$

where $\lfloor x \rfloor$ is the integer part of x. Now $\alpha_{2m} = \alpha_{2m+1} = u_{2m}$ by equation (3.10.23). If $n = 2k$ is even, then

$$\mathbb{E}(N_{2k}) - \mathbb{E}(N_{2k-1}) = u_{2k} - \{ f_2 u_{2k-2} + \cdots + f_{2k} \} = 0 \qquad \text{by } (*).$$

If $n = 2k + 1$ is odd, then

$$\mathbb{E}(N_{2k+1}) - \mathbb{E}(N_{2k}) = u_{2k} - \left\{ f_2 u_{2k-2} + \cdots + f_{2k} \right\} = 0 \qquad \text{by } (*).$$

In either case the claim is proved.

30. (a) Not much.

(b) The rhyme may be interpreted in any of several ways. Interpreting it as meaning that families stop at their first son, we may represent the sample space of a typical family as $\{B, GB, G^2B, \dots\}$, with $\mathbb{P}(G^nB) = 2^{-(n+1)}$. The mean number of girls is $\sum_{n=1}^{\infty} n\mathbb{P}(G^nB) = \sum_{n=1}^{\infty} n2^{-(n+1)} = 1$; there is exactly one boy.

The empirical sex ratio for large populations will be near to 1:1, by the law of large numbers. However the variance of the number of girls in a typical family is var(#girls) $= 2$, whilst var(#boys) $= 0$; $\#A$ denotes the cardinality of A. Considerable variation from 1:1 is therefore possible in smaller populations, but in either direction. In a large number of small populations, the number of large predominantly female families would be balanced by a large number of male singletons.

31. Any positive integer m has a unique factorization in the form $m = \prod_i p_i^{m(i)}$ for non-negative integers $m(1), m(2), \dots$. Hence,

$$\mathbb{P}(M = m) = \prod_i \mathbb{P}\big(N(i) = m(i)\big) = \prod_i \left(1 - \frac{1}{p_i^{\beta}}\right) \frac{1}{p_i^{\beta m(i)}} = C\left(\prod_i p_i^{-m(i)}\right)^{\beta} = \frac{C}{m^{\beta}}$$

where $C = \prod_i (1 - p_i^{-\beta})$. Now $\sum_m \mathbb{P}(M = m) = 1$, so that $C^{-1} = \sum_m m^{-\beta}$.

32. Number the plates $0, 1, 2, \dots, N$ where 0 is the starting plate, fix k satisfying $0 < k \le N$, and let A_k be the event that plate number k is the last to be visited. In order to calculate $\mathbb{P}(A_k)$, we cut the table open at k, and bend its outside edge into a line segment, along which the plate numbers read $k, k+1, \dots, N, 0, 1, \dots, k$ in order. It is convenient to relabel the plates as $-(N+1-k), -(N-k), \dots, -1, 0, 1, \dots, k$. Now A_k occurs if and only if a symmetric random walk, starting from 0, visits both $-(N-k)$ and $k-1$ *before* it visits either $-(N+1-k)$ or k. Suppose it visits $-(N-k)$ before it visits $k-1$. The (conditional) probability that it subsequently visits $k-1$ before visiting $-(N+1-k)$ is the same as the probability that a symmetric random walk, starting from 1, hits N before it hits 0, a probability of N^{-1} by (1.7.7). The same argument applies if the cake visits $k-1$ before it visits $-(N-k)$. Therefore $\mathbb{P}(A_k) = N^{-1}$.

33. With j denoting the jth best vertex, the walk has transition probabilities $p_{jk} = (j-1)^{-1}$ for $1 \le k < j$. By conditional expectation,

$$r_j = 1 + \frac{1}{j-1}\sum_{k=1}^{j-1} r_k, \qquad r_1 = 0.$$

Induction now supplies the result. Since $r_j \sim \log j$ for large j, the worst-case expectation is about $\log \binom{n}{m}$.

34. Let p_n denote the required probability. If (m_r, m_{r+1}) is first pair to make a dimer, then m_1 is ultimately uncombined with probability p_{r-1}. By conditioning on the first pair, we find that $p_n = (p_1 + p_2 + \cdots + p_{n-2})/(n-1)$, giving $n(p_{n+1} - p_n) = -(p_n - p_{n-1})$. Therefore, $n!(p_{n+1} - p_n) = (-1)^{n-1}(p_2 - p_1) = (-1)^n$, and the claim follows by summing.

Finally,

$$\mathbb{E}U_n = \sum_{r=1}^{n} \mathbb{P}(m_r \text{ is uncombined}) = p_n + p_1 p_{n-1} + \cdots + p_{n-1}p_1 + p_n,$$

since the rth molecule may be thought of as an end molecule of two sequences of length r and $n-r+1$. Now $p_n \to e^{-1}$ as $n \to \infty$, and it is an easy exercise of analysis to obtain that $n^{-1}\mathbb{E}U_n \to e^{-2}$.

35. First,

$$\lambda^k = \left(\sum_i p_i\right)^k = \sum_{r_1, r_2, \dots, r_k} p_{r_1} p_{r_2} \cdots p_{r_k} \ge k! \sum_{\{r_1, \dots, r_k\}} p_{r_1} p_{r_2} \cdots p_{r_k},$$

where the last summation is over all subsets $\{r_1, \dots, r_k\}$ of k distinct elements of $\{1, 2, \dots, n\}$. Secondly,

$$\begin{aligned} \lambda^k &\le k! \sum_{\{r_1, \dots, r_k\}} p_{r_1} p_{r_2} \cdots p_{r_k} + \binom{k}{2} \sum_i p_i^2 \sum_{r_1, \dots, r_{k-2}} p_{r_1} p_{r_2} \cdots p_{r_{k-2}} \\ &\le k! \sum_{r_1, \dots, r_k} p_{r_1} p_{r_2} \cdots p_{r_k} + \binom{k}{2} \max_i p_i \left(\sum_j p_j\right)^{k-1}. \end{aligned}$$

Hence

$$(*) \qquad \sum_{\{r_1, \dots, r_k\}} p_{r_1} p_{r_2} \cdots p_{r_k} = \frac{\lambda^k}{k!}\left\{1 + \mathrm{O}\left(\frac{k^2}{\lambda} \max_i p_i\right)\right\}.$$

By Taylor's theorem applied to the function $\log(1-x)$, there exist θ_r satisfying $0 < \theta_r < \{2(1-c)^2)\}^{-1}$ such that

$$(**) \qquad \prod_{r=1}^{n} (1 - p_r) = \prod_r \exp\{-p_r - \theta_r p_r^2\} = \exp\left\{-\lambda - \lambda \mathrm{O}\left(\max_i p_i\right)\right\}.$$

Finally,

$$\mathbb{P}(X = k) = \left(\prod_r (1 - p_r)\right) \sum_{\{r_1, \dots, r_k\}} \frac{p_{r_1} \cdots p_{r_k}}{(1 - p_{r_1}) \cdots (1 - p_{r_k})}.$$

The claim follows from $(*)$ and $(**)$.

36. It is elementary that

$$\mathbb{E}(\overline{Y}) = \frac{1}{n} \sum_{r=1}^{N} \mathbb{E}(X_r) = \frac{1}{n} \sum_{r=1}^{N} x_r \cdot \frac{n}{N} = \mu.$$

We write $\overline{Y} - \mathbb{E}(\overline{Y})$ as the mixture of indicator variables thus:

$$\overline{Y} - \mathbb{E}(\overline{Y}) = \sum_{r=1}^{N} \frac{x_r}{n} \left(I_r - \frac{n}{N}\right).$$

It follows from the fact

$$\mathbb{E}(I_i I_j) = \frac{n}{N} \cdot \frac{n-1}{N-1}, \qquad i \ne j,$$

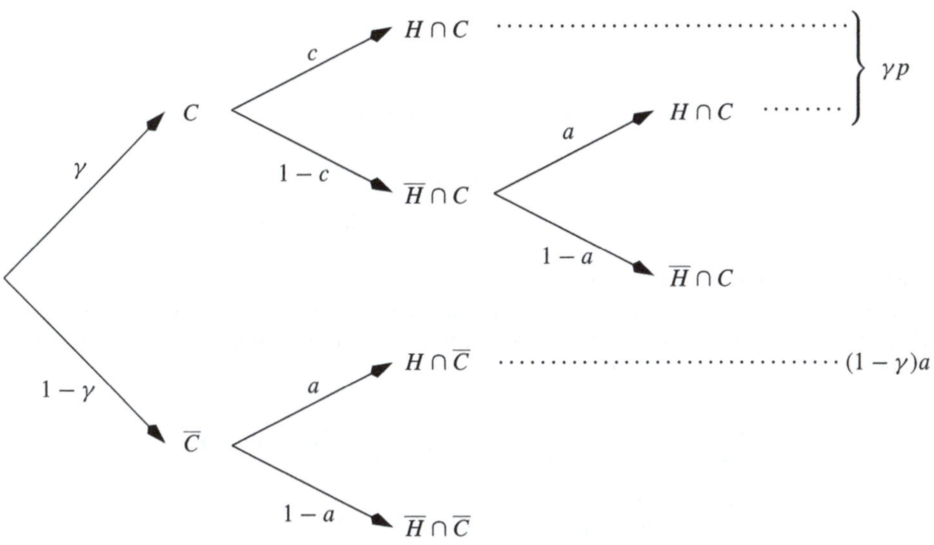

Figure 3.2. The tree of possibility and probability in Problem (3.11.37). The presence of the disease is denoted by C, and hospitalization by H; their negations are denoted by $\overline{C}$ and $\overline{H}$.

that

$$\begin{aligned}
\operatorname{var}(\overline{Y}) &= \sum_{r=1}^{N} \frac{x_r^2}{n^2} \mathbb{E}\left\{\left(I_r - \frac{n}{N}\right)^2\right\} + \sum_{i \neq j} \frac{x_i x_j}{n^2} \mathbb{E}\left\{\left(I_i - \frac{n}{N}\right)\left(I_j - \frac{n}{N}\right)\right\} \\
&= \sum_{r=1}^{N} \frac{x_r^2}{n^2} \frac{n}{N}\left(1 - \frac{n}{N}\right) + \sum_{i \neq j} \frac{x_i x_j}{n^2} \left\{\frac{n}{N}\frac{n-1}{N-1} - \frac{n^2}{N^2}\right\} \\
&= \sum_{r=1}^{N} x_r^2 \frac{N-n}{N^2 n} - \sum_{i \neq j} x_i x_j \frac{N-n}{n(N-1)N^2} \\
&= \frac{N-n}{Nn(N-1)} \left\{\sum_{r=1}^{N} x_r^2 - \frac{1}{N}\sum_{r=1}^{N} x_r^2 - \frac{1}{N}\sum_{i \neq j} x_i x_j\right\} \\
&= \frac{N-n}{n(N-1)} \frac{1}{N} \left\{\sum_{r=1}^{N} x_r^2 - N\overline{x}^2\right\} = \frac{N-n}{n(N-1)} \frac{1}{N} \sum_{r=1}^{N} (x_r - \overline{x})^2.
\end{aligned}$$

37. The tree in Figure 3.2 illustrates the possibilities and probabilities. If G contains n individuals, X is $\text{bin}(n, \gamma p + (1-\gamma)a)$ and Y is $\text{bin}(n, \gamma p)$. It is not difficult to see that $\operatorname{cov}(X, Y) = n\gamma p(1-\nu)$ where $\nu = \gamma p + (1-\gamma)a$. Also, $\operatorname{var}(Y) = n\gamma p(1-\gamma p)$ and $\operatorname{var}(X) = nk(1-\nu)$. The result follows from the definition of correlation.

38. (a) This is an extension of Exercise (3.5.2). With $\mathbb{P}_n$ denoting the probability measure conditional on $N = n$, we have that

$$\mathbb{P}_n(X_i = r_i \text{ for } 1 \leq i \leq k) = \frac{n!}{r_1! r_2! \cdots r_k! s!} f(1)^{r_1} f(2)^{r_2} \cdots f(k)^{r_k} (1 - F(k))^s,$$

where $s = n - \sum_{i=1}^{k} r_i$. Therefore,

$$\mathbb{P}(X_i = r_i \text{ for } 1 \le i \le k) = \sum_{n=0}^{\infty} \mathbb{P}_n(X_i = r_i \text{ for } 1 \le i \le k)\mathbb{P}(N = n)$$
$$= \prod_{i=1}^{k} \left\{ \frac{\nu^{r_i} f(i)^{r_i} e^{-\nu f(i)}}{r_i!} \right\} \sum_{s=0}^{\infty} \frac{\nu^s (1 - F(k))^s}{s!} e^{-\nu(1-F(k))}.$$

The final sum is a Poisson sum, and equals 1.
(b) We use an argument relevant to Wald's equation. The event $\{T \le n-1\}$ depends only on the random variables $X_1, X_2, \dots, X_{n-1}$, and these are independent of X_n. It follows that X_n is independent of the event $\{T \ge n\} = \{T \le n - 1\}^{\mathrm{c}}$. Hence,

$$\mathbb{E}(S) = \sum_{i=1}^{\infty} \mathbb{E}(X_i I_{\{T \ge i\}}) = \sum_{i=1}^{\infty} \mathbb{E}(X_i)\mathbb{E}(I_{\{T \ge i\}}) = \sum_{i=1}^{\infty} \mathbb{E}(X_i)\mathbb{P}(T \ge i)$$
$$= \sum_{i=1}^{\infty} \nu f(i) \sum_{t=i}^{\infty} \mathbb{P}(T = t) = \nu \sum_{t=1}^{\infty} \mathbb{P}(T = t) \sum_{i=1}^{t} f(i)$$
$$= \nu \sum_{t=1}^{\infty} \mathbb{P}(T = t) F(t) = \mathbb{E}(F(T)).$$

39. (a) Place an absorbing barrier at $a + 1$, and let p_a be the probability that the particle is absorbed at 0. By conditioning on the first step, we obtain that

$$p_n = \frac{1}{n+2}(p_0 + p_1 + p_2 + \cdots + p_{n+1}), \qquad 1 \le n \le a.$$

The boundary conditions are $p_0 = 1$, $p_{a+1} = 0$. It follows that $p_{n+1} - p_n = (n + 1)(p_n - p_{n-1})$ for $2 \le n \le a$, and in addition $p_2 - p_1 = p_1 - 1$. By iteration,

$$p_{n+1} - p_n = \tfrac{1}{2}(n + 1)!\,(p_2 - p_1) = \tfrac{1}{2}(n + 1)!\,(p_1 - 1).$$

Setting $n = a$ we obtain that $-p_a = \frac{1}{2}(a + 1)!\,(p_1 - 1)$. By summing over $2 \le n < a$,

$$p_a - p_1 = (p_1 - 1) + \tfrac{1}{2}(p_1 - 1) \sum_{j=3}^{a} j!,$$

and we eliminate p_1 to conclude that

$$p_a = \frac{(a + 1)!}{4 + 3! + 4! + \cdots + (a + 1)!}.$$

It is now easy to see that, for given r, $p_r = p_r(a) \to 1$ as $a \to \infty$, so that ultimate absorption at 0 is (almost) certain, irrespective of the starting point. The limit is justified by the continuity of probability measures, Theorem (1.3.5).
(b) Let λ_r be the probability that the last step is from 1 to 0, having started at r. Then

$$(*) \qquad \lambda_1 = \tfrac{1}{3}(1 + \lambda_1 + \lambda_2),$$
$$(**) \qquad (r + 2)\lambda_r = \lambda_1 + \lambda_2 + \cdots + \lambda_{r+1}, \qquad r \ge 2.$$

It follows that

$$\lambda_r - \lambda_{r-1} = \frac{1}{r+1}(\lambda_{r+1} - \lambda_r), \qquad r \geq 3,$$

whence

$$\lambda_r - \lambda_{r-1} = \frac{1}{(r+1)(r+2)\cdots(m+1)}(\lambda_{m+1} - \lambda_m), \qquad 3 \leq r < m.$$

Letting $m \to \infty$, we deduce that $\lambda_r = \lambda_{r-1}$ for $r \geq 3$, so that $\lambda_r = \lambda_2$ for $r \geq 2$. From (**) with $r = 2$, we have $\lambda_2 = \frac{1}{2}\lambda_1$, and, from (*), $\lambda_1 = \frac{2}{3}$.

(c) There are a couple of ways of doing this, of which one follows. As in part (a), we introduce a second absorbing barrier at $a+1$, and shall later allow $a \to \infty$. Let $\nu_r = \nu_r(a)$ be the mean duration having started at $r \in \{1, 2, \dots, a\}$. The recurrence relation is

$$(*) \qquad \nu_r = 1 + \frac{1}{r+2}(\nu_1 + \nu_2 + \cdots + \nu_{r+1}), \qquad 1 \leq r \leq a,$$

with boundary condition $\nu_{a+1} = 0$. Arguing as above, $d_{r+1} := (\nu_{r+1} - \nu_r)/(r+1)!$ satisfies

$$d_{r+1} - d_r = -\frac{1}{(r+1)!}, \qquad 2 \leq r \leq a,$$

which we sum to obtain

$$d_{k+1} - d_2 = -\sum_{r=3}^{k+1} \frac{1}{r!}, \qquad 2 \leq k \leq a,$$

and again to obtain

$$0 - \nu_2 = \sum_{k=2}^{a} (k+1)! \left(d_2 - \sum_{r=3}^{k+1} \frac{1}{r!} \right).$$

By (*) with $r = 1$, we have $\nu_2 = 2\nu_1 - 3$ and $d_2 = \frac{1}{2}(\nu_1 - 3)$, which we substitute into the above equation and solve for $\nu_1 = \nu_1(a)$, thus obtaining

$$\nu_1(a) = \frac{3 + \frac{3}{2}\sum_k (k+1)! + \sum_k (k+1)!\, e_k}{2 + \frac{1}{2}\sum_k (k+1)!},$$

where $\sum_k$ is over $2 \leq k \leq a$ and

$$e_k = \sum_{r=3}^{k+1} \frac{1}{r!} \to e - \frac{5}{2} \qquad \text{as } k \to \infty.$$

Divide through by $\sum_k (k+1)!$ and let $a \to \infty$ to deduce (via a little real analysis) that $\nu_1(a) \to 3 + 2e - 5 = 2(e-1)$, which is the required answer.

The final limit may be justified using the monotone convergence theorem of Section 5.6. A shorter and more satisfying solution using the theory of Markov chains may be found at Exercise (6.4.16).

40. We label the vertices $1, 2, \dots, n$, and we let π be a random permutation of this set. Let K be the set of vertices v with the property that $\pi(w) > \pi(v)$ for all neighbours w of v. It is not difficult to see that K is an independent set, whence $\alpha(G) \geq |K|$. Therefore, $\alpha(G) \geq \mathbb{E}|K| = \sum_v \mathbb{P}(v \in K)$. For any vertex v, a random permutation π is equally likely to assign any given ordering to the set comprising v and its neighbours. Also, $v \in K$ if and only if v is the earliest element in this ordering, whence $\mathbb{P}(v \in K) = 1/(d_v + 1)$. The result follows.

41. If she bets S, the gain is $-cS + (1+r)IS - S$, where I is the indicator function of a win. The mean gain is therefore $(-c + p(1+r) - 1)S = gS$. With $S = fF$, we have the new fortune $F' = F + (-c + (1+r)I - 1)fF$ with mean $\mathbb{E}F' = F(1+gf)$.

Under the given conditions, we have $0 < g < 1$. In part (a), the mean return is largest when f is a maximum, which is to say $f_a = 1$. In part (b), she seeks to maximize

$$\mathbb{E}\left\{\log\left[1 + f((1+r)I - (1+c))\right]\right\} = p\log\bigl(1 + f(r-c)\bigr) + q\log\bigl(1 - f(1+c)\bigr).$$

The minimum is achieved when the derivative is zero, which occurs at

$$f_b = \frac{g}{(r-c)(1+c)} < 1.$$

This gives a slower rate of growth than $f_a = 1$ but with a diminished risk of bankruptcy.

42. By conditional expectation, $m_n := \mathbb{E}(X_n)$ satisfies

$$m_{n+1} = \frac{2}{n}\sum_{r=1}^{n} m_r, \qquad n \geq r,$$

which implies by subtraction that $nm_{n+1} - (n-1)m_n = 2m_n$. Therefore,

$$m_n = \frac{n}{n-1}m_{n-1} = \frac{n}{n-1}\cdot\frac{n-1}{n-2}\cdots\frac{r+2}{r+1}m_{r+1} = \frac{2n}{r(r+1)}\sum_{k=1}^{r} x_k.$$

43. It is immediate that $X_1 = 1$, $X_2 = 0$, and X_3 takes values in $\{-1, 0, 1\}$ with respective probabilities $\frac{1}{4}, \frac{1}{2}, \frac{1}{4}$. By symmetry, $\mathbb{E}X_n = 0$ for $n \geq 2$, and indeed

$$\mathbb{E}(X_n \mid X_1, X_2, \ldots, X_{n-1}) = 0, \qquad n \geq 2.$$

With

$$a_n = \mathbb{E}\left(\left(\sum_{r=1}^{n} X_r\right)^2\right), \qquad v_n = \mathbb{E}(X_n^2),$$

we have $a_1 = a_2 = 1$ and $v_1 = 1$, $v_2 = 0$, $v_3 = \frac{1}{2}$. Since $X_{n+1} = X_U - X_V$ in the obvious notation, we have by conditional expectation given $X_1, X_2, \ldots, X_n$ that

$$(*) \qquad v_{n+1} = \mathbb{E}(X_U^2) + \mathbb{E}(X_V^2) - 2\mathbb{E}(X_U X_V) = \frac{2}{n}\sum_{r=1}^{n} v_r - \frac{2}{n^2}a_n, \qquad n \geq 1.$$

Also, by iteration at the last step,

$$a_{n+1} = \mathbb{E}(X_{n+1}^2) + 2\mathbb{E}\left(X_{n+1}\sum_{r=1}^{n} X_r\right) + a_n = v_{n+1} + 0 + a_n = \sum_{r=1}^{n+1} v_r, \qquad n \geq 1.$$

On eliminating a_n in (*) and subtracting successive equations, we obtain

$$\frac{n^2}{2(n-1)}v_{n+1} - \frac{(n-1)^2}{2(n-2)}v_n = v_n, \qquad n \geq 3,$$

which yields, on iteration,

$$v_{n+1} = (n-1)\left(1-\frac{3}{n^2}\right)\cdots\left(1-\frac{3}{3^2}\right)v_3 = -(n-1)\prod_{k=1}^{n}\left(1-\frac{3}{k^2}\right), \qquad n \ge 3.$$

Hence,

$$\frac{1}{n}\operatorname{var}(X_n) \to -\prod_{k=1}^{\infty}\left(1-\frac{3}{k^2}\right).$$

Further details of random adding may be found in Clifford and Stirzaker 2019. See also Exercise (12.3.5).

44. Assume 'at random' means uniformly at random. Given that the ring is in box j, the mean number of boxes opened is $1 + B_j/n$ where

$$B_j = \begin{cases} (b_{n-1} + \cdots + b_{n-j+1}) + (b_j + \cdots + b_{n-1}) & \text{if } 1 < j < n, \\ b_1 + b_2 + \cdots + b_{n-1} & \text{if } j = 1, n. \end{cases}$$

which we average over j to obtain

$$b_n = 1 + \frac{2}{n^2}\sum_{k=1}^{n-1} kb_k, \qquad n \ge 1.$$

Therefore, $x_k = kb_k$ satisfies $nx_n = n^2 + 2\sum_{k=1}^{n-1} x_k$, whence

$$nx_n - (n-1)x_{n-1} = 2n - 1 + 2x_{n-1}, \qquad n \ge 1,$$

where $x_0 = 0$. Set $y_n = x_n/(n+1)$ to find that

$$y_n - y_{n-1} = -\frac{1}{n} + \frac{3}{n+1}, \qquad n \ge 1.$$

Iterate this, and substitute back for b_n to obtain

$$b_n = \frac{3}{n} - \frac{3(n+1)}{n} + \frac{2(n+1)}{n}\sum_{k=1}^{n}\frac{1}{k} \sim 2\log n \qquad \text{as } n \to \infty.$$

45. This time, $s_r := b_{2^r}$ satisfies

$$s_r = 1 + \left(1 - \frac{1}{n}\right)s_{r-1}$$

with solution

$$s_r = \frac{r2^r}{2^r - 1} - 1, \qquad r \ge 1.$$

The answer is asymptotic to $\log n/\log 2$: 'bisecting search' is more efficient on average than 'random search'.

46. (a) Let B_i be the event that the word is in box i. Given B_i, the mean number of attempts before success is $1/c_i$, so that the unconditional mean is $\mu(c) = \sum_i p_i/c_i$. Using a Lagrange multiplier, $\mu(c)$ is a minimum for given (p_i) if $c_i \propto \sqrt{p_i}$, and the minimal mean is $\left(\sum_i \sqrt{p_i}\right)^2$.

(b) Rank the p_i in decreasing order as $p_{(i)}$ (thus, $p_{(1)}$ is the largest and $p_{(n)}$ the smallest), and search the boxes in that order. The mean number of days taken is $\sum_i i p_{(i)}$. This may be seen to be a minimum by considering the effect of interchanging the order of any two boxes.

(c) On every search, inspect a box having probability $p_{(1)}$. The mean is $1/p_{(1)}$. Any other strategy has larger mean.

47. Let G_r (respectively, J_s) be the event that Gwen wins r games (respectively, John wins s games), and let G (respectively, J) be the event that Gwen (respectively, John) wins the match. Then, for $r \geq 1$,

$$\begin{aligned}\mathbb{P}(G_r \cap J) &= \mathbb{P}\big(\text{Gwen wins } r \text{ of first } n+r \text{ games, John wins the } (n+r+1)\text{th game}\big)\\ &= \gamma^r \delta^{n+1}\binom{n+r}{r} = \left[\gamma^{r-1}\delta^{n+1}\binom{n+r-1}{r-1}\right] \times \left(\frac{\gamma(n+r)}{r}\right)\\ &= \mathbb{P}(G_{r-1} \cap J) \times \left(\frac{\gamma(n+r)}{r}\right),\end{aligned}$$

whence $rf_r = (n+r)\gamma f_{r-1}$.

With N denoting the total number of games played, and $g_r = \mathbb{P}(G \cap J_r)$,

$$(*) \qquad \begin{aligned}T_n &= \mathbb{E}(N \mid G)\mathbb{P}(G) + \mathbb{E}(N \mid J)\mathbb{P}(J)\\ &= P_n \sum_{r=0}^{n}(n+r+1)g_r + Q_n \sum_{r=0}^{n}(n+r+1)f_r.\end{aligned}$$

Now,

$$\begin{aligned}\mu_f := \sum_{r=1}^{n} rf_r &= \sum_{r=1}^{n}(n+r)\gamma f_{r-1}\\ &= \sum_{j=0}^{n}\gamma j f_j - \gamma n f_n + \gamma(n+1)(1-f_n)\\ &= \gamma\mu_f + \gamma(n+1) - \gamma(2n+1)f_n,\end{aligned}$$

with a similar expression for μ_g. Plug these expressions into (*) and hold your breath to obtain the answer. For the last part you will need the more refined form of Stirling's formula:

$$n! \sim n^{n+\frac{1}{2}}e^{-n+O(n^{-1/2})}\sqrt{2\pi}.$$

48. Both sides count the number of possible colourings, the left according to the colours of the individual components, and the right according to the colour classes. Multiply through by $e^{-1}/c!$ and sum over c to obtain

$$\begin{aligned}e^{-1}\sum_{c=0}^{\infty}\frac{c^n}{c!} &= e^{-1}\sum_{c=0}^{\infty}\sum_{k=1}^{n} S(n,k)\frac{1}{(c-k)!}\\ &= e^{-1}\sum_{k=1}^{n} S(n,k)\sum_{s=0}^{\infty}\frac{1}{s!} = \sum_{k=1}^{n} S(n,k).\end{aligned}$$

49. For any of the rules, at the point when the winner is decided (that is, either Bertha or Harold have won n games exactly), Bertha will have served no more than n games, and Harold no more than $n-1$

games. One may now allow them to continue to play and accrue points until Bertha and Harold have served exactly n and $n-1$ games, respectively. The order of the additional games is immaterial, and they do not change the winner. Therefore, the probability of Bertha winning is the same as before.

50. (a) The entropy $H(X)$ depends only on the *values* taken by $f_X(\cdot)$.
(b) Take logarithms and then expectations of

$$\frac{f_X(x)}{f_{X|Y}(x \mid y)} = \frac{f_X(x) f_Y(y)}{f_{X,Y}(x,y)}.$$

(c) Use the fact that $f_{X,Y}(x,y) = f_X(x) f_Y(y)$.
(d) A Bernoulli variable with parameter p has entropy $H := -p \log p - (1-p)\log(1-p) > 0$, whence the bin$(n, p)$ distribution has entropy nH, which increases with n.
(e) The required entropy is

$$H(p) = -\sum_{k=0}^{\infty} p(1-p)^k \{\log p + k \log(1-p)\} = \log p - \frac{1-p}{p} \log(1-p).$$

The derivative is negative.

51. (a) This follows by the definition of entropy.
(b) Use the result of Problem (3.11.51b). Strict monotonicity follows by the fact (see Exercise (3.6.5)) that $I(N; K) \geq 0$, with equality if and only if N and K are independent (which they clearly are not).

52. (a) Let $p_1, p_2, \dots, p_r$ be distinct primes. An integer n is divisible by every p_i if and only if it is divisible by the product $\pi := p_1 p_2 \cdots p_r$. Therefore,

$$\mathbb{P}\left(\bigcap_{i=1}^{r} E_{p_i}\right) = \sum_{n=1}^{\infty} \frac{(n\pi)^{-\beta}}{\zeta(\beta)} = \pi^{-\beta} = \prod_{i=1}^{r} (p_i)^{-\beta} = \prod_{i=1}^{r} \mathbb{P}(E_{p_i}).$$

(b) Every integer except 1 is divisible by some prime. Therefore,

$$\mathbb{P}(X=1) = \mathbb{P}\left(\bigcap_{p \in \Pi} E_p^{\mathrm{c}}\right) = \prod_{p \in \Pi} \left(1 - \frac{1}{p^\beta}\right),$$

where Π denotes the primes. However, $\mathbb{P}(X=1) = 1/\zeta(\beta)$. There was surreptitious use of the continuity of probability measures here, see Theorem (1.3.5).
(c) By a similar argument to the above, the events $E_{p^2} = \{p^2 \mid X\}$ are independent, and $\mathbb{P}(E_{p^2}) = p^{-2\beta}$. The probability that X is square-free is

$$\mathbb{P}\left(\bigcap_{p \in \Pi} E_{p^2}^{\mathrm{c}}\right) = \prod_{p \in \Pi} \left(1 - \frac{1}{p^2}\right).$$

(d) We have $H = m$ if and only if $X = mx$ and $Y = my$ for some coprime x, y. Therefore,

$$\mathbb{P}(H = m) = \sum_{\gcd\{x,y\}=1} \frac{(mx)^{-\beta}}{\zeta(\beta)} \cdot \frac{(my)^{-\beta}}{\zeta(\beta)} = m^{-2\beta} \mathbb{P}(H=1).$$

By summing over m, we obtain $\mathbb{P}(H=1) = 1/\zeta(m^{2\beta})$, and the claim follows.

53. We have $\mathbb{P}(S \ge 3) = 1$ and, for $3 \le r \le n$,

$$\mathbb{P}(S \ge r+1 \mid S \ge r) = \frac{n-r+1}{n-1}.$$

Therefore,

$$\mathbb{P}(S \ge r+1) = \frac{(n-1)!}{(n-r)!\,(n-1)^{r-1}}, \qquad r \ge 3.$$

The first two moments satisfy, as $n \to \infty$,

$$\mathbb{E}(S) = \sum_{r=0}^{\infty} \mathbb{P}(S > r) \sim \sqrt{\pi n/2}, \qquad \mathbb{E}(S^2) = 2\sum_{r=1}^{\infty} r\mathbb{P}(S \ge r) + \mathbb{E}(S) \sim 2n.$$

54. (a) For a given card, let p_r be the probability it is in its correct place after r shuffles. If it is, then it remains there after one transposition with probability $\binom{n-1}{2}/\binom{n}{2} = (n-2)/n$; if it is not, then it returns there with probability $1/\binom{n}{2}$. Therefore,

$$p_r = \frac{n-2}{n} p_{r-1} + \frac{2}{n(n-1)}(1 - p_{r-1}), \qquad r \ge 1,$$

whence

$$p_r = \frac{1}{n} + \frac{n-1}{n}\left(\frac{n-3}{n-1}\right)^r, \qquad r \ge 1,$$

noting that $p_0 = 1$.
(b) The required expectation is

$$\mathbb{E}(C_r) = np_r = 1 + (n-1)\left(\frac{n-3}{n-1}\right)^r, \qquad r \ge 1.$$

(c) This is approximately 2 if

$$(n-1)\left(\frac{n-3}{n-1}\right)^r \approx 1.$$

The left side may be approximated by $ne^{-2r/n}$, which is roughly 1 if $r \approx \frac{1}{2}n \log n$.

55. Let S_j be the set of vertices at distance j from 0. Let m_j be the mean number of steps to pass from a given vertex in S_j to 0. By conditioning on the first step, we have

$$\begin{aligned} m_j &= 1 + \frac{j}{d} m_{j-1} + \frac{d-j}{d} m_{j+1}, \qquad 1 \le j < d, \\ m_d &= 1 + m_{d-1}, \end{aligned}$$

subject to $m_0 = 0$. This recurrence relation may be expressed in the form

$$\binom{d-1}{j}(m_{j+1} - m_j) - \binom{d-1}{j-1}(m_j - m_{j-1}) = -\binom{d}{j},$$

which, on summing, yields

$$\binom{d-1}{j}(m_{j+1} - m_j) = m_1 - \sum_{r=1}^{j} \binom{d}{r}.$$

Set $j = d-1$ and use the fact that $m_d - m_{d-1} = 1$ to find that

$$m_1 = 1 + \sum_{r=1}^{d-1} \binom{d}{r} = 2^d - 1.$$

Hence,

$$m_{j+1} - m_j = \left\{ \sum_{r=j+1}^{d} \binom{d}{r} \right\} \bigg/ \binom{d-1}{j} = 2^d \mathbb{P}(S > j) \bigg/ \binom{d-1}{j},$$

where S has the bin$(d, \frac{1}{2})$ distribution. Therefore, $m_d = m_1 + \Sigma_d$ where

$$0 \le \Sigma_d = \sum_{j=1}^{d-1} (m_{j+1} - m_j) \le 2^d \sum_{j=1}^{d-1} \mathbb{P}(S > j) \bigg/ \binom{d-1}{j}.$$

Split the sum into three terms: (i) $j = 1$, (ii) $2 \le j \le \frac{3}{4}d$, (iii) $j > \frac{3}{4}d$. The first term is no larger than $2^d/(d-1)$. The second term is no larger than $2^d d / \binom{d-1}{2}$. The third is (by Bernstein's inequality (2.2.4) or similar) no larger than $2^d e^{-\alpha d}$ for some $\alpha > 0$. In conclusion, $m_d = (1 + o(1))2^d$. Improved asymptotics can be calculated.

56. (a) Let $m \ge 2$, and $\beta_k = \mathbb{P}(m \text{ is hit} \mid X_0 = k)$. By conditioning on the first step,

$$(k) \qquad \beta_k = \frac{1}{n-k+1}(\beta_{k+1} + \beta_{k+2} + \cdots + \beta_{m-1} + 1), \qquad 1 \le k < m.$$

Setting $k = m-1$, we obtain $\beta_{m-1} = 1/(n-m+2)$. On subtracting (k) from $(k+1)$, we obtain $\beta_{k+1} = \beta_k$ for $1 \le k < m-1$, so that $\beta_1 = \beta_{m-1} = 1/(n-m+2)$ as required. The case $m = 1$ follows by the fact that absorption occurs at either 1 or n.

(b) We use the notation in the solution to Problem (1.8.39). Set $X_0 = 1$. Passenger 1 selects some seat X_1. If $X_1 = 1$ then all passengers sit in their assigned seats. If $X_1 > 1$, passengers $2, 3, \dots, X_1 - 1$ occupy their correct seats, and passenger X_1 sits in some seat X_2 satisfying either $X_2 = 1$ or $X_2 > X_1$. This process is iterated, and the resulting sequence $X = (X_r : r \ge 0)$ is a walk of the type studied in part (a). Passenger $m \ge 2$ finds his/her seat occupied if and only if m is hit by X, and this occurs with probability $1/(n-m+2)$.

Here is the quick solution to part (b). When passenger m chooses a seat, seats $2, 3, \dots, m-1$ are already taken (since, if seat $r \in \{2, 3, \dots, m-1\}$ were free, it would have been claimed earlier by passenger r, which is a contradiction), and one further seat also. The choices made so far include no information about the label of this further seat which, by symmetry, is equally likely to be any of the $n - m + 2$ seats labelled $1, m, m+1, \dots, n$. Therefore, it is seat m with probability $1/(n-m+2)$.

57. (a) Let I be the indicator function of the event $\{X > a\mathbb{E}(X)\}$, where $a \in [0, 1]$. Then,

$$\begin{aligned} \mathbb{E}(X) &= \mathbb{E}(XI) + \mathbb{E}(X(1-I)) \le \mathbb{E}(XI) + a\mathbb{E}(X) \\ &\le \sqrt{\mathbb{E}(X^2)\mathbb{E}(I)} + a\mathbb{E}(X), \end{aligned}$$

by the Cauchy–Schwarz inequality at the second step. The claim follows by rearrangement.

(b) Set $a = 0$ to obtain

$$\mathbb{P}(X = 0) = 1 - \mathbb{P}(X > 0) \le 1 - \frac{(\mathbb{E}X)^2}{\mathbb{E}(X^2)} = \frac{\operatorname{var}(X)}{\mathbb{E}(X^2)}.$$

Recall that $\operatorname{var}(X) = \mathbb{E}(X^2) - (\mathbb{E}X)^2 \ge 0$.

(c) We have that $\mathbb{E}X = \sum_{r=1}^{n} \mathbb{P}(A_r)$, so that

$$\begin{aligned}\operatorname{var}(X) &= \mathbb{E}(X^2) - \mathbb{E}(X)^2 \\ &= \sum_{r=1}^{n} \mathbb{P}(A_r) + \sum\nolimits^{*} \mathbb{P}(A_r \cap A_s) + \sum\nolimits^{+} \mathbb{P}(A_r \cap A_s) - \mathbb{E}(X)^2 \\ &\le \mathbb{E}X + \sum\nolimits^{*} \mathbb{P}(A_r \cap A_s),\end{aligned}$$

where $\sum^{+}$ is the sum over unordered pairs r, s for which A_r and A_s are independent.

4

Continuous random variables

4.1 Solutions. Probability density functions

1. (a) $\{x(1-x)\}^{-\frac{1}{2}}$ is the derivative of $\sin^{-1}(2x-1)$, and therefore $C = \pi^{-1}$.
(b) $C = 1$, since

$$\int_{-\infty}^{\infty} \exp(-x - e^{-x})\,dx = \lim_{K\to\infty} \big[\exp(-e^{-x})\big]_{-K}^{K} = 1.$$

(c) Substitute $v = (1+x^2)^{-1}$ to obtain

$$\int_{-\infty}^{\infty} \frac{dx}{(1+x^2)^m} = \int_0^1 v^{m-\frac{3}{2}}(1-v)^{-\frac{1}{2}}\,dv = B(\tfrac{1}{2}, m-\tfrac{1}{2})$$

where $B(\cdot,\cdot)$ is a beta function; see paragraph (4.4.8) and Exercise (4.4.2). Hence, if $m > \frac{1}{2}$,

$$C^{-1} = B(\tfrac{1}{2}, m-\tfrac{1}{2}) = \frac{\Gamma(\frac{1}{2})\Gamma(m-\frac{1}{2})}{\Gamma(m)}.$$

2. (i) The distribution function F_Y of Y is

$$F_Y(y) = \mathbb{P}(Y \le y) = \mathbb{P}(aX \le y) = \mathbb{P}(X \le y/a) = F_X(y/a).$$

So, differentiating, $f_Y(y) = a^{-1} f_X(y/a)$.
(ii) Certainly

$$F_{-X}(x) = \mathbb{P}(-X \le x) = \mathbb{P}(X \ge -x) = 1 - \mathbb{P}(X \le -x)$$

since $\mathbb{P}(X = -x) = 0$. Hence $f_{-X}(x) = f_X(-x)$. If X and $-X$ have the same distribution function then $f_{-X}(x) = f_X(x)$, whence the claim follows. Conversely, if $f_X(-x) = f_X(x)$ for all x, then, by substituting $u = -x$,

$$\mathbb{P}(-X \le y) = \mathbb{P}(X \ge -y) = \int_{-y}^{\infty} f_X(x)\,dx = \int_{-\infty}^{y} f_X(-u)\,du = \int_{-\infty}^{y} f_X(u)\,du = \mathbb{P}(X \le y),$$

whence X and $-X$ have the same distribution function.

3. Since $\alpha \ge 0$, $f \ge 0$, and $g \ge 0$, it follows that $\alpha f + (1-\alpha)g \ge 0$. Also

$$\int_{\mathbb{R}} \{\alpha f + (1-\alpha)g\}\,dx = \alpha\int_{\mathbb{R}} f\,dx + (1-\alpha)\int_{\mathbb{R}} g\,dx = \alpha + 1 - \alpha = 1.$$

If X is a random variable with density f, and Y a random variable with density g, then $\alpha f+(1-\alpha)g$ is the density of a random variable Z which takes the value X with probability α and Y otherwise.

Some minor technicalities are necessary in order to find an appropriate probability space for such a Z. If X and Y are defined on the probability space $(\Omega, \mathcal{F}, \mathbb{P})$, it is necessary to define the product space $(\Omega, \mathcal{F}, \mathbb{P}) \times (\Sigma, \mathcal{G}, \mathbb{Q})$ where $\Sigma = \{0, 1\}$, $\mathcal{G}$ is the set of all subsets of Σ, and $\mathbb{Q}(0) = \alpha$, $\mathbb{Q}(1) = 1 - \alpha$. For $\omega \times \sigma \in \Omega \times \Sigma$, we define

$$Z(\omega \times \sigma) = \begin{cases} X(\omega) & \text{if } \sigma = 0, \\ Y(\omega) & \text{if } \sigma = 1. \end{cases}$$

4. (a) By definition, $r(x) = \lim_{h \downarrow 0} \dfrac{1}{h} \dfrac{F(x+h) - F(x)}{1 - F(x)} = \dfrac{f(x)}{1 - F(x)}$.

(b) We have that

$$\frac{d}{dx}\left\{\frac{H(x)}{x}\right\} = \frac{d}{dx}\left\{\frac{1}{x}\int_0^x r(y)\,dy\right\} = \frac{r(x)}{x} - \frac{1}{x^2}\int_0^x r(y)\,dy = \frac{1}{x^2}\int_0^x [r(x) - r(y)]\,dy,$$

which is non-negative if r is non-increasing.

(c) $H(x)/x$ is non-decreasing if and only if, for $0 \le \alpha \le 1$,

$$\frac{1}{\alpha x} H(\alpha x) \le \frac{1}{x} H(x) \qquad \text{for all } x \ge 0,$$

which is to say that $-\alpha^{-1}\log[1 - F(\alpha x)] \le -\log[1 - F(x)]$. We exponentiate to obtain the claim.

(d) Likewise, if $H(x)/x$ is non-decreasing then $H(\alpha t) \le \alpha H(t)$ for $0 \le \alpha \le 1$ and $t \ge 0$, whence $H(\alpha t) + H(t - \alpha t) \le H(t)$ as required.

4.2 Solutions. Independence

1. Let N be the required number. Then $\mathbb{P}(N = n) = F(K)^{n-1}[1 - F(K)]$ for $n \ge 1$, the geometric distribution with mean $[1 - F(K)]^{-1}$.

2. (i) $\text{Max}\{X, Y\} \le v$ if and only if $X \le v$ and $Y \le v$. Hence, by independence,

$$\mathbb{P}\big(\max\{X, Y\} \le v\big) = \mathbb{P}(X \le v, Y \le v) = \mathbb{P}(X \le v)\mathbb{P}(Y \le v) = F(v)^2.$$

Differentiate to obtain the density function of $V = \max\{X, Y\}$.

(ii) Similarly $\min\{X, Y\} > u$ if and only if $X > u$ and $Y > u$. Hence

$$\mathbb{P}(U \le u) = 1 - \mathbb{P}(U > u) = 1 - \mathbb{P}(X > u)\mathbb{P}(Y > u) = 1 - [1 - F(u)]^2,$$

giving $f_U(u) = 2f(u)[1 - F(u)]$.

3. The 24 permutations of the order statistics are equally likely by symmetry, and thus have equal probability. Hence $\mathbb{P}(X_1 < X_2 < X_3 < X_4) = \frac{1}{24}$, and $\mathbb{P}(X_1 > X_2 < X_3 < X_4) = \frac{3}{24}$, by enumerating the possibilities.

4. $\mathbb{P}(Y(y) > k) = F(y)^k$ for $k \ge 1$. Hence $\mathbb{E}Y(y) = F(y)/[1 - F(y)] \to \infty$ as $y \to \infty$. Therefore,

$$\begin{aligned} \mathbb{P}\big(Y(y) > \mathbb{E}Y(y)\big) &= \{1 - [1 - F(y)]\}^{\lfloor F(y)/[1-F(y)] \rfloor} \\ &\sim \exp\left\{1 - [1 - F(y)]\left\lfloor \frac{F(y)}{1 - F(y)} \right\rfloor\right\} \to e^{-1} \quad \text{as } y \to \infty. \end{aligned}$$

5. By independence, for $i \neq r$,

$$\mathbb{P}\big(\{X_i \le X_r\} \cup \{Y_i \le Y_r\}\big) = 1 - \mathbb{P}\big(\{X_i > X_r\} \cap \{Y_i > Y_r\}\big) = 1 - \tfrac{1}{2} \cdot \tfrac{1}{2} = \tfrac{3}{4}.$$

Therefore, $\mathbb{P}(P_i \text{ is peripheral}) = (\frac{3}{4})^{n-1}$. The number of peripheral points is $\sum_{i=1}^{n} I_i$ where I_i is the indicator function that P_i is peripheral. The answer is $n\mathbb{E}(I_1) = n(\frac{3}{4})^{n-1}$.

6. (a) The required probability is the area of the region of $[0, 1]^2$, in the u/v-plane, lying below the curve $v = u^2$ and above the line $v = x$. Draw a picture.

(b) By part (a) with $x = 0$, we have $\mathbb{P}(V < U^2) = \frac{1}{3}$, whence

$$\mathbb{P}(V > x \mid U^2 > V) = \frac{\mathbb{P}(x < V < U^2)}{\mathbb{P}(V < U^2)} = 1 - 3x + 2x^{3/2}.$$

The required conditional density function is

$$f(x) = \frac{d}{dx}\big(1 - \mathbb{P}(V > x \mid U^2 > V)\big) = 3(1 - x^{1/2}), \qquad x \in (0, 1).$$

(c) The quadratic has two distinct real roots if and only if its discriminant is strictly positive, which is to say that $U^2 - V > 0$. As above, this has probability $\frac{1}{3}$.

(d) The roots are $R_\pm = -U \pm \sqrt{U^2 - V}$. When $U^2 - V > 0$, we have $R_\pm < 1$. Furthermore, $R_+ > R_- > -1$ if and only if $V > 2U - 1$. Therefore, the required probability is $\mathbb{P}(V > 2U - 1 \mid U^2 - V > 0)$. The area of the region $\{(u, v) : v > 2u - 1,\ u^2 - v > 0\} \cap [0, 1]^2$ is $\frac{1}{12}$, and the final answer is $\frac{1}{12}/\frac{1}{3} = \frac{1}{4}$.

7. (a) Consider the points taken in clockwise order. There are n choices for the point to be regarded as 'first', and for each choice the required probability is $(\frac{1}{2})^{n-1}$. By the partition theorem, the answer is $n(\frac{1}{2})^{n-1}$.

(b) This is a variant of Flash's Problem (1.8.40). In the language of that problem, the key extra observation is that n spaceships fail to control the entire surface of the sphere if and only if they lie within some open hemisphere.

Argue as follows to see the last claim. A hemisphere H has an equator, and there is a corresponding pole S not lying in H (S is the centre of the complement of H). If the ships lie within some open hemisphere H, then some neighbourhood of the pole S is uncontrolled. Conversely, if some neighbourhood of a point P is uncontrolled, then the ships lie in the open hemisphere H with pole P.

4.3 Solutions. Expectation

1. (a) $\mathbb{E}(X^\alpha) = \int_0^\infty x^\alpha e^{-x}\, dx < \infty$ if and only if $\alpha > -1$.

(b) In this case

$$\mathbb{E}(|X|^\alpha) = \int_{-\infty}^{\infty} \frac{C|x|^\alpha}{(1 + x^2)^m}\, dx < \infty$$

if and only if $-1 < \alpha < 2m - 1$.

2. We have that

$$1 = \mathbb{E}\left(\frac{\sum_1^n X_i}{S_n}\right) = \sum_{i=1}^{n} \mathbb{E}(X_i / S_n).$$

By symmetry, $\mathbb{E}(X_i/S_n) = \mathbb{E}(X_1/S_n)$ for all i, and hence $1 = n\mathbb{E}(X_1/S_n)$. Therefore

$$\mathbb{E}(S_m/S_n) = \sum_{i=1}^{m} \mathbb{E}(X_i/S_n) = m\mathbb{E}(X_1/S_n) = m/n.$$

3. Either integrate by parts or use Fubini's theorem:

$$\begin{aligned} r\int_0^\infty x^{r-1}\mathbb{P}(X > x)\,dx &= r\int_0^\infty x^{r-1}\left\{\int_{y=x}^\infty f(y)\,dy\right\}dx \\ &= \int_{y=0}^\infty f(y)\left\{\int_{x=0}^{y} rx^{r-1}\,dx\right\}dy = \int_0^\infty y^r f(y)\,dy. \end{aligned}$$

An alternative proof is as follows. Let I_x be the indicator of the event that $X > x$, so that $\int_0^\infty I_x\,dx = X$. Taking expectations, and taking a minor liberty with the integral which may be made rigorous, we obtain $\mathbb{E}X = \int_0^\infty \mathbb{E}(I_x)\,dx$. A similar argument may be used for the more general case.

4. We may suppose without loss of generality that $\mu = 0$ and $\sigma = 1$. Assume further that $m > 1$. In this case, at least half the probability mass lies to the right of 1, whence $\mathbb{E}(XI_{\{X\ge m\}}) \ge \frac{1}{2}$. Now $0 = \mathbb{E}(X) = \mathbb{E}\{X[I_{\{X\ge m\}} + I_{\{X<m\}}]\}$, implying that $\mathbb{E}(XI_{\{X<m\}}) \le -\frac{1}{2}$. Likewise,

$$\mathbb{E}(X^2 I_{\{X\ge m\}}) \ge \tfrac{1}{2}, \quad \mathbb{E}(X^2 I_{\{X<m\}}) \le \tfrac{1}{2}.$$

By the definition of the median, and the fact that X is continuous,

$$\mathbb{E}(X \mid X < m) \le -1, \quad \mathbb{E}(X^2 \mid X < m) \le 1.$$

It follows that $\operatorname{var}(X \mid X < m) \le 0$, which implies in turn that, conditional on $\{X < m\}$, X is almost surely concentrated at a single value. This contradicts the continuity of X, and we deduce that $m \le 1$. The possibility $m < -1$ may be ruled out similarly, or by considering the random variable $-X$.

5. (a) It is a standard to write $X = X^+ - X^-$ where $X^+ = \max\{X, 0\}$ and $X^- = -\min\{X, 0\}$. Now X^+ and X^- are non-negative, and so, by Lemma (4.3.4),

$$\begin{aligned} \mu = \mathbb{E}(X) = \mathbb{E}(X^+) - \mathbb{E}(X^-) &= \int_0^\infty \mathbb{P}(X > x)\,dx - \int_0^\infty \mathbb{P}(X < -x)\,dx \\ &= \int_0^\infty [1 - F(x)]\,dx - \int_0^\infty F(-x)\,dx = \int_0^\infty [1 - F(x)]\,dx - \int_{-\infty}^0 F(x)\,dx. \end{aligned}$$

It is a triviality that

$$\mu = \int_0^\mu F(x)\,dx + \int_0^\mu [1 - F(x)]\,dx$$

and the equation follows with $a = \mu$. It is easy to see that it cannot hold with any other value of a, since both sides are monotonic functions of a.

(b) We have that

$$\mathbb{E}|X - a| = \int_a^\infty \mathbb{P}(X > x)\,dx + \int_{-\infty}^a \mathbb{P}(X < x)\,dx,$$

so that

$$\frac{d}{da}\mathbb{E}|X - a| = \mathbb{P}(X < a) - \mathbb{P}(X > a).$$

This is strictly positive when a is greater than all medians, negative when a is less than all medians, and zero otherwise.

6. Let $x \in \mathbb{R}$, and let I_j be the indicator function that $X_j > x$. Then $I_j I_k$ is the indicator function that $\min\{X_j, X_k\} > x$, and similarly for other products of the I_j. Now,

$$\begin{aligned} I\big(\max X_j > x\big) &= 1 - \prod_{j=1}^{n}(1 - I_j) \\ &= \sum_j I_j - \sum_{j<k} I\big(\min\{X_j, X_k\} > x\big) + \cdots + (-1)^{n+1} I\big(\min X_j > x\big). \end{aligned}$$

Integrate over x to obtain the result.

7. Let $X = X^+ - X^-$ where $X^+ = \max\{X, 0\}$ and $X^- = -\min\{X, 0\}$. Since $\mathbb{E}(X^r) < \infty$, we have $0 \le \mathbb{E}((X^r)^+), \mathbb{E}((X^r)^-) < \infty$. Note that

$$(X^r)^+ = \begin{cases} (X^+)^r & \text{if } r \text{ is odd,} \\ |X|^r & \text{if } r \text{ is even.} \end{cases}$$

Let r be odd (a similar argument holds if r is even). By Exercise (4.3.3),

$$\mathbb{E}\big((X^r)^+\big) = \int_0^\infty r x^{r-1} \mathbb{P}(X > x)\, dx < \infty.$$

For the second part, since X has a density function f, X^+ has density function f on $(0, \infty)$ together with an atom at 0. Therefore, for $K > 0$, $Y = (X^r)^+$ satisfies

$$\begin{aligned} \mathbb{E}(Y) &= \int_0^\infty x^r f(x)\, dx \\ &= \int_0^K x^r f(x)\, dx + \int_K^\infty x^r f(x)\, dx \ge \int_0^K x^r f(x)\, dx + K^r \int_K^\infty f(x)\, dx \\ &= \int_0^K x^r f(x)\, dx + K^r \mathbb{P}(X > K). \end{aligned}$$

Let $K \to \infty$ and use the fact that $\mathbb{E}(Y) < \infty$.

8. For the first part, take $g(x) = x$ and $h(x) = f(x)/b$. For the second, take $g(x) = x^n$.

9. Use the expansion

$$\mathbb{E}\big[(X-a)^2\big] = \mathbb{E}\big[\big((X - \mathbb{E}X) + (\mathbb{E}X - a)\big)^2\big] = \operatorname{var}(X) + [\mathbb{E}X - a]^2.$$

10. There is probability $(r^2)^n$ that all n arrows hit within the circle with radius r, whence

$$f_R(r) = \frac{d}{dr} r^{2n} = 2nr^{2n-1}, \qquad r \in [0, 1].$$

The mean area of the circle in question is

$$\mathbb{E}(\pi R^2) = \pi \int_0^1 r^2 2nr^{2n-1}\, dr = \frac{\pi n}{n+1}.$$

Conditional on R, the other $n-1$ arrows are distributed on a disk with radius R. By the above, the area A in question has conditional expectation $\mathbb{E}(A \mid R) = R^2 \pi (n-1)/n$, so that

$$\mathbb{E}(A) = \mathbb{E}(\mathbb{E}(A \mid R)) = \frac{n-1}{n} \mathbb{E}(\pi R^2) = \frac{\pi(n-1)}{n+1}.$$

11. Suppose first that $M = 0$. In the notation of Khinchin's representation, we have $\mathbb{E}X = \frac{1}{2}\mathbb{E}Y$, and

$$\sigma^2 = \mathbb{E}(U^2)\mathbb{E}(Y^2) - (\mathbb{E}U\mathbb{E}Y)^2 = \tfrac{1}{3}\mathbb{E}(Y^2) - \tfrac{1}{4}\mathbb{E}(Y)^2.$$

On eliminating $\mathbb{E}Y$, we obtain $\mathbb{E}(X)^2 + \mathrm{var}(Y) = 3\sigma^2$. The result follows on considering the random variable $X - M$. Equality holds in the given inequality if and only if Y is almost surely constant, so X is uniform.

Finally, noting Exercise (4.3.4), we have by the triangle inequality that $|M - m| < (1 + \sqrt{3})\sigma$, though in fact it can be shown that $|M - m| \le \sigma\sqrt{3}$ and this bound is sharp.

12. We use the Khinchin representation of unimodality from Exercise (4.3.11). We may think of the velocity V as comprising two independent elements: a speed $|V|$, and a random direction. Let W be the distance to the endpoint of the interval in the direction of travel, and note that W is uniformly distributed on $[0, 1]$, and is independent of $|V|$. Now, $T = W/|V|$, and the claim follows by Khinchin's representation.

13. (a) As in the solution to Exercise (4.3.12), and after a change of variables, the relationship between the density function f of T and the density function g of the speed is found to be

(*) $$f(t) = \int_0^{1/t} xg(x)\,dx, \qquad t > 0.$$

(b) With g as given, we have that $f(t) = 1 - (1 + 1/t)e^{-1/t}$ for $t > 0$.

4.4 Solutions. Examples of continuous variables

1. (i) Integrating by parts,

$$\Gamma(t) = \int_0^\infty x^{t-1}e^{-x}\,dx = (t-1)\int_0^\infty x^{t-2}e^{-x}\,dx = (t-1)\Gamma(t-1).$$

If n is an integer, then it follows that $\Gamma(n) = (n-1)\Gamma(n-1) = \cdots = (n-1)!\,\Gamma(1)$ where $\Gamma(1) = 1$.
(ii) We have, using the substitution $u^2 = x$, that

$$\begin{aligned}\Gamma(\tfrac{1}{2})^2 &= \left\{\int_0^\infty x^{-\frac{1}{2}}e^{-x}\,dx\right\}^2 = \left\{\int_0^\infty 2e^{-u^2}\,du\right\}^2 \\ &= 4\int_0^\infty e^{-u^2}\,du\int_0^\infty e^{-v^2}\,dv = 4\int_{r=0}^\infty\int_{\theta=0}^{\pi/2} e^{-r^2}r\,dr\,d\theta = \pi\end{aligned}$$

as required. For integral n,

$$\Gamma(n+\tfrac{1}{2}) = (n-\tfrac{1}{2})\Gamma(n-\tfrac{1}{2}) = \cdots = (n-\tfrac{1}{2})(n-\tfrac{3}{2})\cdots\tfrac{1}{2}\Gamma(\tfrac{1}{2}) = \frac{(2n)!}{4^n n!}\sqrt{\pi}.$$

2. By the definition of the gamma function,

$$\Gamma(a)\Gamma(b) = \int_0^\infty x^{a-1}e^{-x}\,dx\int_0^\infty y^{b-1}e^{-y}\,dy = \int_0^\infty\int_0^\infty e^{-(x+y)}x^{a-1}y^{b-1}\,dx\,dy.$$

Now set $u = x + y$, $v = x/(x + y)$, obtaining

$$\begin{aligned}&\int_{u=0}^\infty\int_{v=0}^1 e^{-u}u^{a+b-1}v^{a-1}(1-v)^{b-1}\,dv\,du \\ &\qquad = \int_0^\infty u^{a+b-1}e^{-u}\,du\int_0^1 v^{a-1}(1-v)^{b-1}\,dv = \Gamma(a+b)B(a,b).\end{aligned}$$

3. If g is strictly decreasing then $\mathbb{P}(g(X) \le y) = \mathbb{P}(X \ge g^{-1}(y)) = 1 - g^{-1}(y)$ so long as $0 \le g^{-1}(y) \le 1$. Therefore $\mathbb{P}(g(X) \le y) = 1 - e^{-y}$, $y \ge 0$, if and only if $g^{-1}(y) = e^{-y}$, which is to say that $g(x) = -\log x$ for $0 < x < 1$.

4. We have that

$$\mathbb{P}(X \le x) = \int_{-\infty}^{x} \frac{1}{\pi(1+u^2)}\,du = \frac{1}{2} + \frac{1}{\pi}\tan^{-1} x.$$

Also,

$$\mathbb{E}(|X|^\alpha) = \int_{-\infty}^{\infty} \frac{|x|^\alpha}{\pi(1+x^2)}\,dx$$

is finite if and only if $|\alpha| < 1$.

5. Writing Φ for the $N(0,1)$ distribution function, $\mathbb{P}(Y \le y) = \mathbb{P}(X \le \log y) = \Phi(\log y)$. Hence

$$f_Y(y) = \frac{1}{y} f_X(\log y) = \frac{1}{y\sqrt{2\pi}} e^{-\frac{1}{2}(\log y)^2}, \qquad 0 < y < \infty.$$

6. Integrating by parts,

$$\begin{aligned}\text{LHS} &= \int_{-\infty}^{\infty} g(x)\left\{(x-\mu)\frac{1}{\sigma}\phi\left(\frac{x-\mu}{\sigma}\right)\right\} dx \\ &= -\left[g(x)\sigma\phi\left(\frac{x-\mu}{\sigma}\right)\right]_{-\infty}^{\infty} + \int_{-\infty}^{\infty} g'(x)\sigma\phi\left(\frac{x-\mu}{\sigma}\right) dx = \text{RHS}.\end{aligned}$$

7. (a) $r(x) = \alpha\beta x^{\beta-1}$.
(b) $r(x) = \lambda$.
(c) $r(x) = \dfrac{\lambda\alpha e^{-\lambda x} + \mu(1-\alpha)e^{-\mu x}}{\alpha e^{-\lambda x} + (1-\alpha)e^{-\mu x}}$, which approaches $\min\{\lambda, \mu\}$ as $x \to \infty$.

8. (a) Clearly $\phi' = -x\phi$. Using this identity and integrating by parts repeatedly,

$$\begin{aligned}1 - \Phi(x) &= \int_x^{\infty} \phi(u)\,du = -\int_x^{\infty} \frac{\phi'(u)}{u}\,du = \frac{\phi(x)}{x} + \int_x^{\infty} \frac{\phi'(u)}{u^3}\,du \\ &= \frac{\phi(x)}{x} - \frac{\phi(x)}{x^3} - \int_x^{\infty} \frac{3\phi'(u)}{u^5}\,du = \frac{\phi(x)}{x} - \frac{\phi(x)}{x^3} + \frac{3\phi(x)}{x^5} - \int_x^{\infty} \frac{15\phi(u)}{u^6}\,du.\end{aligned}$$

(b) Let $c \in \mathbb{R}$, $d = (c-\mu)/\sigma$, and also $x \in \mathbb{R}$, $Z = (X-\mu)/\sigma$, $z = (x-\mu)/\sigma$. Then,

$$\mathbb{P}(X > x \mid X > c) = \mathbb{P}(Z > z \mid Z > d),$$

so that the conditional density function of X given that $X > c$ is

$$g(x) = \frac{1}{\sigma} \cdot \frac{\phi(z)}{1 - \Phi(d)}, \qquad x > c.$$

Therefore,

$$\mathbb{E}(X \mid X > c) = \int_c^{\infty} x g(x)\,dx = \int_c^{\infty} \frac{x\phi(z)}{\sigma(1-\Phi(d))}\,dx = \int_d^{\infty} \frac{(z\sigma+\mu)\phi(z)}{\sigma(1-\Phi(d))}\,\sigma dz,$$

by the substitution $x = z\sigma + \mu$. Hence,

$$\mathbb{E}(X \mid X > c) = \frac{\sigma}{1-\Phi(d)} \int_d^{\infty} z\phi(z)\,dz + \mu,$$

and the answer follows by integration. The second formula follows similarly.

9. We have that

$$\mathbb{P}(U \le V \le W) = \int_0^\infty \lambda e^{-\lambda u}\mathbb{P}(u \le V \le W)\,du = \int_0^\infty \lambda e^{-\lambda u}\int_u^\infty \mu e^{-\mu v}\mathbb{P}(v \le W)\,dv\,du,$$

where $\mathbb{P}(v \le W) = e^{-\nu v}$. Insert and integrate.

10. By the tail integral formula,

$$\mathbb{E}\big(\min\{U, X\}\big) = \int_0^\infty \mathbb{P}(\min\{U, X\} > t)\,dt = \int_0^1 (1-t)e^{-\lambda t}\,dt = \frac{1}{\lambda}e^{-\lambda} - \frac{1}{\lambda^2}(1-e^{-\lambda}).$$

11. (a) We have

$$F_Y(y) = \mathbb{P}(aX \le y - yX) = \mathbb{P}\left(X \le \frac{y}{a+y}\right) = \frac{y}{a+y}, \qquad y > 0.$$

(b) By the definition of S,

$$\mathbb{P}(S \ge s) = (1 - F_Y(s))^n = \left(\frac{a}{a+s}\right)^n, \qquad s > 0,$$

and the Pareto density follows on differentiating. The mean is

$$\mathbb{E}S = \int_0^\infty \mathbb{P}(S > s)\,ds = \frac{a}{n-1}.$$

(c) In the usual way,

$$\mathbb{P}(T \le t) = \mathbb{P}(Y_1 \le t)^n = \left(\frac{y}{a+y}\right)^n, \qquad t > 0.$$

12. The Cauchy distribution function is

$$F(x) = \frac{1}{2} + \frac{1}{\pi}\tan^{-1}x, \qquad x \in \mathbb{R}.$$

For $y \in (0, 1)$,

$$\begin{aligned}\mathbb{P}(Y \le y) &= \mathbb{P}\left(1 + X^2 \ge \frac{1}{y}\right) = 2\mathbb{P}\left(X \ge \sqrt{\frac{1-y}{y}}\right)\\ &= 2\left(1 - F\big(\sqrt{(1-y)/y}\big)\right) = 1 - \frac{2}{\pi}\tan^{-1}\sqrt{\frac{1-y}{y}}\\ &= 1 - \frac{2}{\pi}\cos^{-1}\sqrt{y} = \frac{2}{\pi}\sin^{-1}\sqrt{y}.\end{aligned}$$

13. (a) The answers are $(c-b)^n/(2c)^n$ and $(c+b)^n/(2c)^n$, respectively.
(b) We have that

$$\begin{aligned}\mathbb{P}(Z \le z) &= \sum_{i=1}^{2n+1} \mathbb{P}(Z \le z,\ Z = X_i) = (2n+1)\mathbb{P}(Z \le z,\ Z = X_1)\\ &= (2n+1)\int_{-c}^{z} f_X(x)\mathbb{P}\big(n \text{ of the } X_i \text{ are less than } x, \text{ and } n \text{ are greater}\big)\,dx\\ &= (2n+1)\int_{-c}^{z} \frac{1}{2c}\binom{2n}{n}\frac{(c^2-x^2)^n}{(2c)^{2n}}\,dx.\end{aligned}$$

Differentiate to obtain the answer.
(c) Since f_Z is a density function,

$$c_n := \int_{-c}^{c} (c^2 - z^2)^n\,dz = \frac{(n!)^2}{(2n+1)!}(2c)^{2n+1}.$$

(d) Since Z and $-Z$ have the same distribution, $\mathbb{E}Z = 0$. Therefore,

$$\begin{aligned}\operatorname{var}(Z) = \mathbb{E}(Z^2) &= \int_{-c}^{c} z^2 f_Z(z)\,dz\\ &= \frac{1}{c_n}\left\{c^2\int_{-c}^{c}(c^2-z^2)^n\,dz - \int_{-c}^{c}(c^2-z^2)^{n+1}\,dz\right\}\\ &= \frac{1}{c_n}(c^2 c_n - c_{n+1}) = \frac{c^2}{2n+3}.\end{aligned}$$

14. We have for $y > 0$ that

$$F_Y(y) = \mathbb{P}(Y \le y) = \mathbb{P}\left(X \le \frac{y}{1+y}\right) = F_X(y/(1+y)).$$

Differentiate to get the density function f_Y. By the definition of Y,

$$\mathbb{E}(Y^n) = \frac{1}{B(a,b)}\int_0^\infty \left(\frac{x}{1-x}\right)^n x^{a-1}(1-x)^{b-1}\,dx = \frac{B(a+n,b-n)}{B(a,b)},$$

as required.

15. With f denoting the density function of Z,

$$\begin{aligned}\mathbb{E}|Z-t| &= \int_0^\infty |z-t|f(z)\,dz = \int_0^t -(z-t)f(z)\,dz + \int_t^\infty (z-t)f(z)\,dz\\ &= \int_0^\infty (z-t)f(z)\,dz - 2\int_0^t (z-t)f(z)\,dz.\end{aligned}$$

The penultimate integral equals 0 since $\mathbb{E}Z = t$. The other term is

$$-2\int_0^t \frac{z^t e^{-z}}{\Gamma(t)}\,dz + 2\int_0^t \frac{tz^{t-1}e^{-z}}{\Gamma(t)}\,dz = 2\frac{t^t e^{-t}}{\Gamma(t)},$$

on integrating the first by parts.

4.5 Solutions. Dependence

1. (i) As the product of non-negative continuous functions, f is non- negative and continuous. Also

$$g(x) = \tfrac{1}{2}e^{-|x|} \int_{-\infty}^{\infty} \frac{1}{\sqrt{2\pi x^{-2}}} e^{-\frac{1}{2}x^2y^2}\, dy = \tfrac{1}{2}e^{-|x|}$$

if $x \neq 0$, since the integrand is the $N(0, x^{-2})$ density function. It is easily seen that $g(0) = 0$, so that g is discontinuous, while

$$\int_{-\infty}^{\infty} g(x)\, dx = \int_{-\infty}^{\infty} \tfrac{1}{2}e^{-|x|}\, dx = 1.$$

(ii) Clearly $f_Q \geq 0$ and

$$\int_{-\infty}^{\infty}\int_{-\infty}^{\infty} f_Q(x, y)\, dx\, dy = \sum_{n=1}^{\infty} \left(\tfrac{1}{2}\right)^n \cdot 1 = 1.$$

Also f_Q is the uniform limit of continuous functions on any subset of $\mathbb{R}^2$ of the form $[-M, M] \times \mathbb{R}$; hence f_Q is continuous. Hence f_Q is a continuous density function. On the other hand

$$\int_{-\infty}^{\infty} f_Q(x, y)\, dy = \sum_{n=1}^{\infty} \left(\tfrac{1}{2}\right)^n g(x - q_n),$$

where g is discontinuous at 0.

(iii) Take Q to be the set of the rationals, in some order.

2. We may assume that the centre of the rod is uniformly positioned in a square of size $a \times b$, while the acute angle between the rod and a line of the first grid is uniform on $[0, \frac{1}{2}\pi]$. If the latter angle is θ then, with the aid of a diagram, one finds that there is no intersection if and only if the centre of the rod lies within a certain inner rectangle of size $(a - r\cos\theta) \times (b - r\sin\theta)$. Hence the probability of an intersection is

$$\frac{2}{\pi ab}\int_0^{\pi/2} \{ab - (a - r\cos\theta)(b - r\sin\theta)\}\, d\theta = \frac{2r}{\pi ab}(a + b - \tfrac{1}{2}r).$$

3. (a) Let I be the indicator of the event that the first needle intersects a line, and let J be the indicator that the second needle intersects a line. By the result of Exercise (4.5.2), $\mathbb{E}(I) = \mathbb{E}(J) = 2/\pi$; hence $Z = I + J$ satisfies $\mathbb{E}(\frac{1}{2}Z) = 2/\pi$.

We have that

$$\begin{aligned}\operatorname{var}(\tfrac{1}{2}Z) &= \tfrac{1}{4}\{\mathbb{E}(I^2) + \mathbb{E}(J^2) + 2\mathbb{E}(IJ)\} - \mathbb{E}(\tfrac{1}{2}Z)^2\\ &= \tfrac{1}{4}\{\mathbb{E}(I) + \mathbb{E}(J) + 2\mathbb{E}(IJ)\} - \frac{4}{\pi^2} = \frac{1}{\pi} - \frac{4}{\pi^2} + \frac{1}{2}\mathbb{E}(IJ).\end{aligned}$$

In the notation of (4.5.8), if $0 < \theta < \frac{1}{2}\pi$, then two intersections occur if $z < \frac{1}{2}\min\{\sin\theta, \cos\theta\}$ or $1 - z < \frac{1}{2}\min\{\sin\theta, \cos\theta\}$. With a similar component when $\frac{1}{2}\pi \leq \theta < \pi$, we find that

$$\begin{aligned}\mathbb{E}(IJ) = \mathbb{P}(\text{two intersections}) &= \frac{4}{\pi}\iint_R dz\, d\theta\\ &= \frac{4}{\pi}\int_0^{\pi/2} \tfrac{1}{2}\min\{\sin\theta, \cos\theta\}\, d\theta = \frac{4}{\pi}\int_0^{\pi/4} \sin\theta\, d\theta = \frac{4}{\pi}\left(1 - \frac{1}{\sqrt{2}}\right),\end{aligned}$$

where

$$R = \{(z, \theta) : 0 < z < \tfrac{1}{2}\min\{\sin\theta, \cos\theta\},\ 0 < \theta < \tfrac{1}{2}\pi\}.$$

Hence,

$$\operatorname{var}(\tfrac{1}{2}Z) = \frac{1}{\pi} - \frac{4}{\pi^2} + \frac{1}{\pi}(2 - \sqrt{2}) = \frac{3 - \sqrt{2}}{\pi} - \frac{4}{\pi^2}.$$

(b) For Buffon's needle, the variance of the number of intersections is $(2/\pi) - (2/\pi)^2$ which exceeds $\operatorname{var}(\frac{1}{2}Z)$. You should therefore use Buffon's cross.

(c) Let G be the number of intersections of the unit needle with the two orthogonal unit grids. Trigonometric calculations similar to those of part (a) yield $\mathbb{P}(G = 2) = 1/\pi$ and $\mathbb{P}(G = 1) = 2/\pi$. Hence the estimator $\frac{1}{2}G$ has mean $2/\pi$ and variance $3/(2\pi) - 4/\pi^2$, which is less than that for Buffon's cross thrown onto parallel lines. You would prefer to use the needle on the orthogonal grids.

4. (i) $F_U(u) = 1 - (1-u)(1-u)$ if $0 < u < 1$, and so $\mathbb{E}(U) = \int_0^1 2u(1-u)\,du = \frac{1}{3}$. (Alternatively, place three points independently at random on the circumference of a circle of circumference 1. Measure the distances X and Y from the first point to the other two, along the circumference clockwise. Clearly X and Y are independent and uniform on $[0, 1]$. Hence by circular symmetry, $\mathbb{E}(U) = \mathbb{E}(V - U) = \mathbb{E}(1 - V) = \frac{1}{3}$.)

(ii) Clearly $UV = XY$, so that $\mathbb{E}(UV) = \mathbb{E}(X)\mathbb{E}(Y) = \frac{1}{4}$. Hence

$$\operatorname{cov}(U, V) = \mathbb{E}(UV) - \mathbb{E}(U)\mathbb{E}(V) = \tfrac{1}{4} - \tfrac{1}{3}(1 - \tfrac{1}{3}) = \tfrac{1}{36},$$

since $\mathbb{E}(V) = 1 - \mathbb{E}(U)$ by 'symmetry'.

5. (a) If X and Y are independent then, by (4.5.6) and independence,

$$\begin{aligned}\mathbb{E}(g(X)h(Y)) &= \iint g(x)h(y)f_{X,Y}(x, y)\,dx\,dy \\ &= \int g(x)f_X(x)\,dx \int h(y)f_Y(y)\,dy = \mathbb{E}(g(X))\mathbb{E}(h(Y)).\end{aligned}$$

By independence

$$\mathbb{E}(e^{\frac{1}{2}(X+Y)}) = \mathbb{E}(e^{\frac{1}{2}X})^2 = \left\{\int_0^\infty e^{\frac{1}{2}x}e^{-x}\,dx\right\}^2 = 4.$$

(b) We may assume that X and Y have zero mean. Since they are independent,

$$\mathbb{E}\big((X - Y)^2\big) = \mathbb{E}(X^2) - 2\mathbb{E}(XY) + \mathbb{E}(Y^2) = 2\mathbb{E}(X^2) = 2\operatorname{var}(X).$$

6. If O is the centre of the circle, take the radius OA as origin of coordinates. That is, A $= (1, 0)$, B $= (1, \Theta)$, C $= (1, \Phi)$, in polar coordinates, where we choose the labels in such a way that $0 \le \Theta \le \Phi$. The pair Θ, Φ has joint density function $f(\theta, \phi) = (2\pi^2)^{-1}$ for $0 < \theta < \phi < 2\pi$.

The three angles of ABC are $\frac{1}{2}\Theta$, $\frac{1}{2}(\Phi - \Theta)$, $\pi - \frac{1}{2}\Phi$. You should plot in the θ/ϕ-plane the set of pairs (θ, ϕ) such that $0 < \theta < \phi < 2\pi$ and such that at least one of the three angles exceeds $x\pi$. Then integrate f over this region to obtain the result. The shape of the region depends on whether or not $x < \frac{1}{2}$. The density function g of the largest angle is given by differentiation:

$$g(x) = \begin{cases} 6(3x - 1) & \text{if } \frac{1}{3} \le x \le \frac{1}{2}, \\ 6(1 - x) & \text{if } \frac{1}{2} \le x \le 1. \end{cases}$$

The expectation is found to be $\frac{11}{18}\pi$.

7. We have that $\mathbb{E}(\overline{X}) = \mu$, and therefore $\mathbb{E}(X_r - \overline{X}) = 0$. Furthermore,

$$\mathbb{E}\{\overline{X}(X_r - \overline{X})\} = \frac{1}{n}\mathbb{E}\Big(\sum_s X_r X_s\Big) - \mathbb{E}(\overline{X}^2) = \frac{1}{n}\{\sigma^2 + n\mu^2\} - \big(\operatorname{var}(\overline{X}) + \mathbb{E}(\overline{X})^2\big)$$
$$= \frac{1}{n}\{\sigma^2 + n\mu^2\} - \left(\frac{\sigma^2}{n} + \mu^2\right) = 0.$$

8. The condition is that $\mathbb{E}(Y)\operatorname{var}(X) + \mathbb{E}(X)\operatorname{var}(Y) = 0$.

9. If X and Y are positive, then S positive entails T positive, which displays the dependence. Finally, $S^2 = X$ and $T^2 = Y$.

10. By symmetry, $\mathbb{P}(X > Y) = \mathbb{P}(Y > X)$, and by continuity, $\mathbb{P}(X = Y) = 0$. Therefore, $\mathbb{P}(X > Y) = \frac{1}{2}$.

For any three unequal numbers a, b, c, there are six distinct orderings by size, and for the continuous random variables X, Y, Z, by symmetry each has probability $\frac{1}{6}$. There are three such orderings consistent with $X > Y$, namely, ZXY, XZY, XYZ; of these, two are consistent with $Z > Y$, namely, ZXY, XZY. Therefore, $\mathbb{P}(Z > Y \mid X > Y) = \frac{2}{3}$.

11. Multiply up and take expectations. It follows that

$$2\operatorname{cov}(X, Y) = \mathbb{E}\iint \{I(U \le x) - I(X \le x)\}\{I(V \le y) - I(Y \le y)\}\,dx\,dy$$
$$= \iint 2\{F(x, y) - F_X(x)F_Y(y)\}\,dx\,dy,$$

where we have used Fubini's theorem to pass the expectation into the integral. The above integral representation, which may not be totally obvious, is seen by examining the range of values of x and y where the integrand is non-zero.

12. Let F and f denote the joint distribution and density functions of X, Y. Then $Z = \max\{X, Y\}$ has distribution function $F(z, z)$, with expectation

$$\mathbb{E}Z = \int_{-\infty}^{\infty} z\{F_x(z, z) + F_y(z, z)\}\,dz,$$

where (with a similar expression for $F_y(x, y)$)

$$F_x(x, y) := \frac{\partial F(x, y)}{\partial x} = \int_{-\infty}^{y} f(x, v)\,dv$$
$$= \int_{-\infty}^{y} \frac{1}{2\pi\sqrt{1-\rho^2}} \exp\left\{-\frac{1}{2(1-\rho^2)}\left[(1-\rho^2)x^2 + (v - \rho x)^2\right]\right\} dv$$
$$= \phi(x)\Phi\left(\frac{y - \rho x}{\sqrt{1-\rho^2}}\right),$$

and ϕ, Φ are the $N(0, 1)$ density and distribution functions. By integration by parts,

$$\mathbb{E}Z = 2\int_{-\infty}^{\infty} z\phi(z)\Phi\left(\frac{z(1-\rho)}{\sqrt{1-\rho^2}}\right) dz$$
$$= \sqrt{2/\pi}\int_{-\infty}^{\infty} e^{-\frac{1}{2}z^2}\sqrt{\frac{1-\rho}{1+\rho}}\,\phi\left(z\sqrt{\frac{1-\rho}{1+\rho}}\right) dz = \sqrt{\frac{1-\rho}{\pi}}.$$

13. For $n \in \{0, 1, 2 \dots\}$ and $y \in (0, 1)$,

$$\mathbb{P}(\{N = n\} \cap \{M < y\}) = \mathbb{P}(n < X < n + y) = e^{-\lambda n} - e^{-\lambda(n+y)} = e^{-\lambda n}(1 - e^{-\lambda y}).$$

Since this factorizes, M and N are independent with

$$\mathbb{P}(N = n) = e^{-\lambda n}(1 - e^{-\lambda}), \qquad f_M(y) = \frac{\lambda e^{-\lambda y}}{1 - e^{-\lambda}}.$$

14. The expectation equals

$$I := \frac{1}{2\pi} \iint \exp\left\{-\tfrac{1}{2}\left[(1 - 2a)x^2 - 4bxy + (1 - 2c)y^2\right]\right\} dx\, dy.$$

Recalling the bivariate normal density function of Example (4.5.9), we see that $I = \sigma\tau\sqrt{1 - \rho^2}$ where

$$\sigma^2(1 - \rho^2) = \frac{1}{1 - 2a}, \quad \tau^2(1 - \rho^2) = \frac{1}{1 - 2c}, \quad \sigma\tau(1 - \rho^2) = \frac{\rho}{2b}.$$

Hence

$$\rho^2 = \frac{4b^2}{(1 - 2a)(1 - 2c)}, \quad \sigma\tau\sqrt{1 - \rho^2} = \frac{1}{\sqrt{(1 - 2a)(1 - 2c) - 4b^2}}.$$

15. Without loss of generality, we may consider the *standard* bivariate normal density function, and we calculate

$$\begin{aligned} 1 + \phi^2 &= \frac{1}{2\pi(1 - \rho^2)} \iint \exp\left\{\frac{1}{2}x^2 + \frac{1}{2}y^2 - \frac{1}{1 - \rho^2}(x^2 - 2\rho xy + y^2)\right\} dx\, dy \\ &= \frac{1}{1 - \rho^2}, \end{aligned}$$

either by considering the bivariate normal density with $\sigma^2 = \tau^2 = (1 + \rho^2)/(1 - \rho^2)$, having correlation $2\rho/(1 + \rho^2)$, or by routine integration.

For the second part,

$$\begin{aligned} I &= \mathbb{E} \log\left\{\frac{f(X, Y)}{g(x)h(y)}\right\} \\ &= -\frac{1}{2}\log(1 - \rho^2) + \mathbb{E}\left\{\frac{X^2}{2\sigma^2} + \frac{Y^2}{2\tau^2} - \frac{1}{2(1 - \rho^2)}\left(\frac{X^2}{\sigma^2} - \frac{2\rho XY}{\sigma\tau} + \frac{Y^2}{2\tau^2}\right)\right\} \\ &= -\tfrac{1}{2}\log(1 - \rho^2). \end{aligned}$$

Note that

$$I = \begin{cases} 0 & \text{if } \rho = 0, \\ \infty & \text{if } \rho = 1. \end{cases}$$

16. By symmetry, $\mathbb{E}(\cos(n\pi X)) = 0$ for $n \geq 1$. Furthermore,

$$\mathbb{E}\{\cos(m\pi X)\cos(n\pi X)\} = \tfrac{1}{2}\mathbb{E}\{\cos((m + n)\pi X) + \cos((m - n)\pi X)\} = 0,$$

for $m \neq n$. Therefore, they are uncorrelated. On the other hand, $\cos(2\pi X) = 2\cos^2(\pi X) - 1$, so that

$$\mathbb{P}(\{|\cos(\pi X)| < 1/\sqrt{2}\} \cap \{\cos(2\pi X) > 0\}) = 0,$$

whereas

$$\mathbb{P}(\{|\cos(\pi X)| < 1/\sqrt{2}\})\mathbb{P}(\{\cos(2\pi X) > 0\}) \neq 0.$$

17. Using indicator functions,

$$\begin{aligned}
\tfrac{1}{2}\mathbb{E}|X-Y| &= \tfrac{1}{2}\mathbb{E}\{(X-Y)I(X>Y)+(Y-X)I(X<Y)\} \\
&= \mathbb{E}\{(Y-X)I(X<Y)\} && \text{by symmetry} \\
&= \mathbb{E}\int_{X\wedge Y}^{Y} dy = \mathbb{E}\int_0^\infty I(y<Y)\,dy - \mathbb{E}\int_0^\infty I(y<X)I(y<Y)\,dy \\
&= \int_0^\infty (1-F(y))\,dy - \int_0^\infty (1-F(y))^2\,dy && \text{by independence} \\
&= \mathbb{E}X - \int_0^\infty (1-F(y))^2\,dy,
\end{aligned}$$

where $x \wedge y = \min\{x, y\}$. The above uses the non-negativity of X, Y. The second form for the MAD follows by making the change of variables $u = F(x)$, $v = F(y)$ in the expression

$$\mathbb{E}|X-Y| = \iint_{\mathbb{R}^2} |x-y|\,dF(x)\,dF(y).$$

Routine integrations yield the required MADs: (a) $\frac{1}{3}$, (b) λ, (c) $2a/[(a-1)(2a-1)]$. For (d), note that $X-Y$ is $N(0,2)$, so that $\mathbb{E}|X-Y| = \sqrt{2/\pi}$.

18. By symmetry in $x = y = 0$, the joint density function of $|X|, |Y|$ is

$$f(x,y) = \frac{2}{\pi}e^{-\frac{1}{2}(x^2+y^2)}, \qquad x, y > 0,$$

and by symmetry in the line $x = y$, that of U, V is

$$g(u,v) = \frac{4}{\pi}e^{-\frac{1}{2}(u^2+v^2)}, \qquad 0 < u < v < \infty.$$

Therefore, using polar coordinates,

$$\begin{aligned}
\mathbb{E}(U/V) &= \frac{4}{\pi}\iint_{0<u<v<\infty} \frac{u}{v}e^{-\frac{1}{2}(u^2+v^2)}\,du\,dv \\
&= \frac{4}{\pi}\int_0^\infty re^{-\frac{1}{2}r^2}\,dr\int_{\pi/4}^{\pi/2}\cot\theta\,d\theta = \frac{4}{\pi}\cdot 1\cdot\log\sqrt{2}.
\end{aligned}$$

19. Three lengths can form a triangle if and only if the sum of any two is no greater than the third. The triangle Δ is defined by the inequalities $x+y \geq 1$ and $x, y \leq 1$, and any such pair (x, y) satisfies the given condition.

For $(x, y) \in \Delta$, the angle $\theta = \widehat{ABC}$ is given by the cosine formula by

$$\cos\theta = \frac{x^2+(2-x-y)^2-y^2}{2x(2-x-y)},$$

and θ is obtuse if and only if this is strictly negative. This occurs if and only if $f(x) < y$ where

$$f(x) = \frac{x^2-2x+2}{2-x}.$$

By drawing a picture, or otherwise, we see that

$$\mathbb{P}(f(X) < Y) = 2\int_0^1 [1 - f(x)]\, dx = 3 - 4\log 2.$$

20. Since f cannot be factorized as a function of x multiplied by a function of y, X and Y are not independent. Alternatively, you may calculate the marginal density functions by integration. By routine integrations, $\mathbb{E}X = \mathbb{E}Y = \frac{1}{2}$ and $\mathbb{E}(XY) = \frac{1}{4}$. Therefore, $\text{cov}(X, Y) = 0$.

4.6 Solutions. Conditional distributions and conditional expectation

1. The point is picked according to the uniform distribution on the surface of the unit sphere, which is to say that, for any suitable subset C of the surface, the probability the point lies in C is the surface integral $\int_C (4\pi)^{-1}\, dS$. Changing to polar coordinates, $x = \cos\theta\cos\phi$, $y = \sin\theta\cos\phi$, $z = \sin\phi$, subject to $x^2+y^2+z^2 = 1$, this surface integral becomes $(4\pi)^{-1}\int_C |\cos\phi|\, d\theta\, d\phi$, whence the joint density function of Θ and Φ is

$$f(\theta,\phi) = \frac{1}{4\pi}|\cos\phi|, \qquad |\phi| \le \tfrac{1}{2}\pi,\ 0 \le \theta < 2\pi.$$

The marginals are then $f_\Theta(\theta) = (2\pi)^{-1}$, $f_\Phi(\phi) = \frac{1}{2}|\cos\phi|$, and the conditional density functions are

$$f_{\Theta|\Phi}(\theta \mid \phi) = \frac{1}{2\pi}, \qquad f_{\Phi|\Theta}(\phi \mid \theta) = \tfrac{1}{2}|\cos\phi|,$$

for appropriate θ and ϕ. Thus Θ and Φ are independent. The fact that the conditional density functions are different from each other is sometimes referred to as 'Borel's paradox'.

2. We have that

$$\psi(x) = \int_{-\infty}^{\infty} y\frac{f_{X,Y}(x,y)}{f_X(x)}\, dy$$

and therefore

$$\begin{aligned}\mathbb{E}\big(\psi(X)g(X)\big) &= \int_{-\infty}^{\infty}\int_{-\infty}^{\infty} y\frac{f_{X,Y}(x,y)}{f_X(x)} g(x) f_X(x)\, dx\, dy\\ &= \int_{-\infty}^{\infty}\int_{-\infty}^{\infty} \{y\, g(x)\} f_{X,Y}(x,y)\, dx\, dy = \mathbb{E}(Yg(X)).\end{aligned}$$

3. Take Y to be a random variable with mean ∞, say $f_Y(y) = y^{-2}$ for $1 \le y < \infty$, and let $X = Y$. Then $\mathbb{E}(Y \mid X) = X$ which is (almost surely) finite.

4. (a) We have that

$$f_X(x) = \int_x^{\infty} \lambda^2 e^{-\lambda y}\, dy = \lambda e^{-\lambda x}, \qquad 0 \le x < \infty,$$

so that $f_{Y|X}(y \mid x) = \lambda e^{\lambda(x-y)}$, for $0 \le x \le y < \infty$.

(b) Similarly,

$$f_X(x) = \int_0^{\infty} x e^{-x(y+1)}\, dy = e^{-x}, \qquad 0 \le x < \infty,$$

so that $f_{Y|X}(y \mid x) = xe^{-xy}$, for $0 \le y < \infty$.

5. We have that

$$\mathbb{P}(Y=k)=\int_0^1 \mathbb{P}(Y=k\mid X=x)f_X(x)\,dx=\int_0^1\binom{n}{k}x^k(1-x)^{n-k}\frac{x^{a-1}(1-x)^{b-1}}{B(a,b)}\,dx$$
$$=\binom{n}{k}\frac{B(a+k,n-k+b)}{B(a,b)}.$$

In the special case $a=b=1$, this yields

$$\mathbb{P}(Y=k)=\binom{n}{k}\frac{\Gamma(k+1)\Gamma(n-k+1)}{\Gamma(n+2)}=\frac{1}{n+1},\quad 0\le k\le n,$$

whence Y is uniformly distributed.

We have in general that

$$\mathbb{E}(Y)=\int_0^1 \mathbb{E}(Y\mid X=x)f_X(x)\,dx=\frac{na}{a+b},$$

and, by a similar computation of $\mathbb{E}(Y^2)$,

$$\operatorname{var}(Y)=\frac{nab(a+b+n)}{(a+b)^2(a+b+1)}.$$

6. By conditioning on X_1,

$$G_n(x)=\mathbb{P}(N>n)=\int_0^x G_{n-1}(x-u)\,du=\int_0^x G_{n-1}(v)\,dv.$$

Now $G_0(v)=1$ for all $v\in(0,1]$, and the result follows by induction. Now,

$$\mathbb{E}N=\sum_{n=0}^{\infty}\mathbb{P}(N>n)=e^x.$$

More generally,

$$G_N(s)=\sum_{n=1}^{\infty}s^n\mathbb{P}(N=n)=\sum_{n=1}^{\infty}s^n\left(\frac{x^{n-1}}{(n-1)!}-\frac{x^n}{n!}\right)=(s-1)e^{sx}+1,$$

whence $\operatorname{var}(N)=G_N''(1)+G_N'(1)-G_N'(1)^2=2xe^x+e^x-e^{2x}$.

7. We may assume without loss of generality that $\mathbb{E}X=\mathbb{E}Y=0$. By the Cauchy–Schwarz inequality,

$$\mathbb{E}(XY)^2=\mathbb{E}\big(X\mathbb{E}(Y\mid X)\big)^2\le\mathbb{E}(X^2)\mathbb{E}\big(\mathbb{E}(Y\mid X)^2\big).$$

Hence,

$$\mathbb{E}\big(\operatorname{var}(Y\mid X)\big)=\mathbb{E}(Y^2)-\mathbb{E}\big(\mathbb{E}(Y\mid X)^2\big)\le\mathbb{E}Y^2-\frac{\mathbb{E}(XY)^2}{\mathbb{E}(X^2)}=(1-\rho^2)\operatorname{var}(Y).$$

8. One way, as essentially followed in the solution to Exercise (4.4.9), is to evaluate

$$\int_0^\infty\int_x^\infty\int_y^\infty \lambda\mu\nu e^{-\lambda x-\mu y-\nu z}\,dx\,dy\,dz.$$

Another way is to observe that $\min\{Y, Z\}$ is exponentially distributed with parameter $\mu + \nu$, whence $\mathbb{P}(X < \min\{Y, Z\}) = \lambda/(\lambda + \mu + \nu)$. Similarly, $\mathbb{P}(Y < Z) = \mu/(\mu + \nu)$, and the product of these two terms is the required answer.

9. By integration, for $x, y > 0$,

$$f_Y(y) = \int_0^y f(x, y)\,dx = \tfrac{1}{6}cy^3e^{-y}, \quad f_X(x) = \int_x^\infty f(x, y)\,dy = cxe^{-x},$$

whence $c = 1$. It is simple to check the values of $f_{X|Y}(x \mid y) = f(x, y)/f_Y(y)$ and $f_{Y|X}(y \mid x)$, and then deduce by integration that $\mathbb{E}(X \mid Y = y) = \frac{1}{2}Y$ and $\mathbb{E}(Y \mid X = x) = x + 2$.

10. We have that $N > n$ if and only if X_0 is largest of $\{X_0, X_1, \dots, X_n\}$, an event having probability $1/(n + 1)$. Therefore, $\mathbb{P}(N = n) = 1/\{n(n + 1)\}$ for $n \geq 1$. Next, on the event $\{N = n\}$, X_n is the largest, whence

$$\mathbb{P}(X_N \leq x) = \sum_{n=1}^\infty \frac{F(x)^{n+1}}{n(n+1)} = \sum_{n=1}^\infty \frac{F(x)^{n+1}}{n} - \sum_{n=1}^\infty \frac{F(x)^{n+1}}{n+1} + F(x),$$

as required. Finally,

$$\mathbb{P}(M = m) = \mathbb{P}(X_0 \geq X_1 \geq \cdots \geq X_{m-1}) - \mathbb{P}(X_0 \geq X_1 \geq \cdots \geq X_m) = \frac{1}{m!} - \frac{1}{(m+1)!}.$$

11. We argue naively.
(a) The joint density function of U and X is uniform on the region $x \in [0, 1]$, $x - u \in [0, 1]$, and U has a triangular distribution in that $f_U(u) = 1 - |u|$ for $u \in [-1, 1]$. Therefore,

$$f(x \mid D) = \left.\frac{f_{U,X}(u, x)}{f_U(u)}\right|_{u=0} = 1.$$

(b) The joint density function of V and X is $f_{V,X}(v, x) = x$ on the region $x \in [0, 1]$, $xv \in [0, 1]$, and

$$f_V(v) = \begin{cases} \dfrac{1}{2} & \text{if } 0 \leq v \leq 1, \\[2mm] \dfrac{1}{2v^2} & \text{if } v \geq 1. \end{cases}$$

Therefore,

$$f(x \mid D) = \left.\frac{f_{V,X}(v, x)}{f_V(v)}\right|_{v=1} = 2x.$$

(c) This is an example of Borel's paradox: when conditioning on an event of probability 0, the answer depends on how the conditioning is done.

12. (a) By conditional probability,

$$\mathbb{P}(X < Y) = \int_{-\infty}^\infty \mathbb{P}(X < Y \mid Y = y) f_Y(y)\,dy = \int_{-\infty}^\infty F_X(y) f_Y(y)\,dy.$$

(b) Suppose A changes their number if and only if $U_A < a$, and B likewise when $U_B < b$, for predetermined values a, b. (The optimal strategies are in fact of this form.) The new numbers V_A, V_B have densities

$$f_A(v) = \begin{cases} a & \text{if } 0 < v < a, \\ 1 + a & \text{if } a \leq v < 1, \end{cases} \qquad f_B(v) = \begin{cases} b & \text{if } 0 < v < b, \\ 1 + b & \text{if } b \leq v < 1. \end{cases}$$

By part (a), assuming $a \le b$,

$$\begin{aligned}\mathbb{P}(B \text{ wins}) &= \int_0^a bau\,du + \int_a^b b\big(a^2 + (u-a)(1+a)\big)\,du + \int_b^1 (1+b)\big(a^2 + (u-a)(1+a)\big)\,du \\ &= \tfrac{1}{2} + \tfrac{1}{2}(a-b)(b+ab-1).\end{aligned}$$

If A chooses $a = \frac{1}{2}(\sqrt{5}-1)$, then

$$\mathbb{P}(B \text{ wins}) \begin{cases} < \frac{1}{2} & \text{if } b > a, \\ = \frac{1}{2} & \text{if } b = a. \end{cases}$$

A similar calculation and conclusion holds when $b \le a$. Player B calculates likewise.

13. (a) For $r \ge 2$, by symmetry each element in $\{X_1, X_2, \dots, X_r\}$ is equally likely to be the maximum of the set. Therefore, $\mathbb{P}(A_r) = 1/r$. Conditional on A_r, the relative ordering of $X_1, X_2, \dots, X_{r-1}$ has the same distribution as before, and the independence follows.

(b) By independence,

$$\operatorname{var}(R_n) = \operatorname{var}\left(\sum_{r=1}^n I_{A_r}\right) = \sum_{r=1}^n \operatorname{var}(I_{A_r}) = \sum_{r=1}^n \frac{1}{r}\left(1 - \frac{1}{r}\right).$$

(c) The index T of the first record time after 1 satisfies

$$\mathbb{P}(T > n) = \mathbb{P}(X_1 > X_r \text{ for } r = 2, 3, \dots, n) = \frac{1}{n},$$

whence $\mathbb{E}T = \infty$.

14. Let X be the length of the shorter piece of the first break, so $f_X(x) = 2$ for $x \in [0, \frac{1}{2}]$. The other two pieces have lengths $U(1-X)$ and $(1-U)(1-X)$ where U is uniformly distributed on $[0, 1]$. These three lengths can form a triangle (as in Exercise (4.6.19)) if and only if

$$U(1-X) \le X + (1-U)(1-X), \qquad (1-U)(1-X) \le X + U(1-X),$$

noting that we have already that $X < 1 - X$. Conditional on $X = x$, the above occurs if and only if $1 - 2x \le 2U(1-x) \le 1$, which has probability

$$\mathbb{P}\left(\frac{1-2x}{2(1-x)} \le U \le \frac{1}{2(1-x)}\right) = \frac{1}{2(1-x)} - \frac{1-2x}{2(1-x)} = \frac{x}{1-x}.$$

The unconditional probability is

$$\int_0^{\frac{1}{2}} \frac{x}{1-x} f_X(x)\,dx = \int_0^{\frac{1}{2}} \frac{2x}{1-x}\,dx = 2\log 2 - 1.$$

15. We have that

$$\mathbb{E}(U_{(1)}) = \int_0^1 \mathbb{P}(U_{(1)} > x)\,dx = \int_0^1 (1-x)^n\,dx = \frac{1}{n+1}.$$

By symmetry, $\mathbb{E}(U_{(n)}) = 1 - \mathbb{E}(U_{(1)})$. Conditional on $U_1, U_2, \dots, U_n$, the given event occurs with probability $1 - (U_{(n)} - U_{(1)})$, so that the unconditional probability is

$$1 - \mathbb{E}(U_{(n)}) + \mathbb{E}(U_{(1)}) = \frac{2}{n+1}.$$

4.7 Solutions. Functions of random variables

1. We observe that, if $0 \le u \le 1$,

$$\begin{aligned}\mathbb{P}(XY \le u) &= \mathbb{P}(XY \le u, Y \le u) + \mathbb{P}(XY \le u, Y > u) = \mathbb{P}(Y \le u) + \mathbb{P}(X \le u/Y,\ Y > u)\\ &= u + \int_u^1 \frac{u}{y}\,dy = u(1 - \log u).\end{aligned}$$

By the independence of XY and Z,

$$\mathbb{P}(XY \le u, Z^2 \le v) = \mathbb{P}(XY \le u)\mathbb{P}(Z \le \sqrt{v}) = u\sqrt{v}(1 - \log u), \quad 0 < u, v < 1.$$

Differentiate to obtain the joint density function

$$g(u, v) = \frac{\log(1/u)}{2\sqrt{v}}, \qquad 0 \le u, v \le 1.$$

Hence

$$\mathbb{P}(XY \le Z^2) = \iint_{0 \le u \le v \le 1} \frac{\log(1/u)}{2\sqrt{v}}\,du\,dv = \tfrac{5}{9}.$$

Arguing more directly,

$$\mathbb{P}(XY \le Z^2) = \iiint_{\substack{0 \le x,y,z \le 1\\ xy \le z^2}} dx\,dy\,dz = \tfrac{5}{9}.$$

2. The transformation $x = uv$, $y = u - uv$ has Jacobian

$$J = \begin{vmatrix} v & u \\ 1 - v & -u \end{vmatrix} = -u.$$

Hence $|J| = |u|$, and therefore $f_{U,V}(u, v) = ue^{-u}$, for $0 \le u < \infty$, $0 \le v \le 1$. Hence U and V are independent, and $f_V(v) = 1$ on $[0, 1]$ as required.

3. Arguing directly,

$$\mathbb{P}(\sin X \le y) = \mathbb{P}(X \le \sin^{-1} y) = \frac{2}{\pi}\sin^{-1} y, \qquad 0 \le y \le 1,$$

so that $f_Y(y) = 2/\big(\pi\sqrt{1 - y^2}\big)$, for $0 \le y \le 1$. Alternatively, make a one-dimensional change of variables.

4. (a) $\mathbb{P}(\sin^{-1} X \le y) = \mathbb{P}(X \le \sin y) = \sin y$, for $0 \le y \le \frac{1}{2}\pi$. Hence $f_Y(y) = \cos y$, for $0 \le y \le \frac{1}{2}\pi$.

(b) Similarly, $\mathbb{P}(\sin^{-1} X \le y) = \frac{1}{2}(1 + \sin y)$, for $-\frac{1}{2}\pi \le y \le \frac{1}{2}\pi$, so that $f_Y(y) = \frac{1}{2}\cos y$, for $-\frac{1}{2}\pi \le y \le \frac{1}{2}\pi$.

5. Consider the mapping $w = x$, $z = (y - \rho x)/\sqrt{1 - \rho^2}$ with inverse $x = w$, $y = \rho w + z\sqrt{1 - \rho^2}$ and Jacobian

$$J = \begin{vmatrix} 1 & 0 \\ \rho & \sqrt{1 - \rho^2} \end{vmatrix} = \sqrt{1 - \rho^2}.$$

The mapping is one–one, and therefore W $(= X)$ and Z satisfy

$$f_{W,Z}(w, z) = \frac{\sqrt{1-\rho^2}}{2\pi\sqrt{1-\rho^2}} \exp\left\{-\frac{1}{2(1-\rho^2)}(1-\rho^2)(w^2+z^2)\right\} = \frac{1}{2\pi}e^{-\frac{1}{2}(w^2+z^2)},$$

implying that W and Z are independent $N(0, 1)$ variables. Now

$$\{X > 0, Y > 0\} = \left\{W > 0, Z > -W\rho \Big/ \sqrt{1-\rho^2}\right\},$$

and therefore, moving to polar coordinates,

$$\mathbb{P}(X > 0, Y > 0) = \int_{\theta=\alpha}^{\frac{1}{2}\pi}\int_{r=0}^{\infty} \frac{1}{2\pi}e^{-\frac{1}{2}r^2} r\,dr\,d\theta = \int_{\alpha}^{\frac{1}{2}\pi}\frac{1}{2\pi}\,d\theta$$

where $\alpha = -\tan^{-1}\big(\rho/\sqrt{1-\rho^2}\big) = -\sin^{-1}\rho$.

6. We confine ourselves to the more interesting case when $\rho \neq 1$. Writing $X = U$, $Y = \rho U + \sqrt{1-\rho^2}V$, we have that U and V are independent $N(0, 1)$ variables. It is easy to check that $Y > X$ if and only if $(1-\rho)U < \sqrt{1-\rho^2}V$. Turning to polar coordinates,

$$\mathbb{E}(\max\{X, Y\}) = \int_0^\infty \frac{re^{-\frac{1}{2}r^2}}{2\pi}\left[\int_\psi^{\psi+\pi}\left\{\rho r\cos\theta + r\sqrt{1-\rho^2}\sin\theta\right\}d\theta + \int_{\psi-\pi}^{\psi} r\cos\theta\,d\theta\right]dr$$

where $\tan\psi = \sqrt{(1-\rho)/(1+\rho)}$. Some algebra yields the result. For the second part,

$$\mathbb{E}(\max\{X, Y\}^2) = \mathbb{E}(X^2 I_{\{X>Y\}}) + \mathbb{E}(Y^2 I_{\{Y>X\}}) = \mathbb{E}(X^2 I_{\{X<Y\}}) + \mathbb{E}(Y^2 I_{\{Y<X\}}),$$

by the symmetry of the marginals of X and Y. Adding, we obtain $2\mathbb{E}(\max\{X, Y\}^2) = \mathbb{E}(X^2) + \mathbb{E}(Y^2) = 2$.

7. We have that

$$\mathbb{P}(X < Y,\ Z > z) = \mathbb{P}(z < X < Y) = \frac{\lambda}{\lambda+\mu}e^{-(\lambda+\mu)z} = \mathbb{P}(X < Y)\mathbb{P}(Z > z).$$

(a) $\mathbb{P}(X = Z) = \mathbb{P}(X < Y) = \dfrac{\lambda}{\lambda+\mu}$.

(b) By conditioning on Y,

$$\mathbb{P}\big((X-Y)^+ = 0\big) = \mathbb{P}(X \le Y) = \frac{\lambda}{\lambda+\mu}, \quad \mathbb{P}\big((X-Y)^+ > w\big) = \frac{\mu}{\lambda+\mu}e^{-\lambda w} \quad \text{for } w > 0.$$

By conditioning on X,

$$\begin{aligned}\mathbb{P}(V > v) = \mathbb{P}(|X-Y| > v) &= \int_0^\infty \mathbb{P}(Y > v+x) f_X(x)\,dx + \int_v^\infty \mathbb{P}(Y < x-v) f_X(x)\,dx \\ &= \frac{\mu e^{-\lambda v} + \lambda e^{-\mu v}}{\lambda+\mu}, \qquad v > 0.\end{aligned}$$

(c) By conditioning on X, the required probability is found to be

$$\int_0^t \lambda e^{-\lambda x}\int_{t-x}^\infty \mu e^{-\mu y}\,dy\,dx = \frac{\lambda}{\mu-\lambda}\{e^{-\lambda t} - e^{-\mu t}\}.$$

8. Either make a change of variables and find the Jacobian, or argue directly. With the convention that $\sqrt{r^2-u^2}=0$ when $r^2-u^2<0$, we have that

$$F(r,x)=\mathbb{P}(R\le r,\ X\le x)=\frac{2}{\pi}\int_{-r}^{x}\sqrt{r^2-u^2}\,du,$$

$$f(r,x)=\frac{\partial^2 F}{\partial r\,\partial x}=\frac{2r}{\pi\sqrt{r^2-x^2}},\qquad |x|<r<1.$$

9. As in the previous exercise,

$$\mathbb{P}(R\le r,\ Z\le z)=\frac{3}{4\pi}\int_{-r}^{z}\pi(r^2-w^2)\,dw.$$

Hence $f(r,z)=\frac{3}{2}r$ for $|z|<r<1$. This question may be solved in spherical polars also.

10. The transformation $s=x+y$, $r=x/(x+y)$, has inverse $x=rs$, $y=(1-r)s$ and Jacobian $J=s$. Therefore,

$$\begin{aligned}f_R(r)&=\int_0^\infty f_{R,S}(r,s)\,ds=\int_0^\infty f_{X,Y}\big(rs,(1-r)s\big)s\,ds\\&=\int_0^\infty \lambda e^{-\lambda rs}\mu e^{-\mu(1-r)s}s\,ds=\frac{\lambda\mu}{\{\lambda r+\mu(1-r)\}^2},\qquad 0\le r\le 1.\end{aligned}$$

11. We have that

$$\mathbb{P}(Y\le y)=\mathbb{P}\left(X^2\ge\frac{a}{y}-1\right)=2\mathbb{P}\left(X\le-\sqrt{\frac{a}{y}-1}\right),$$

whence

$$f_Y(y)=2f_X\big(-\sqrt{(a/y)-1}\big)=\frac{1}{\pi\sqrt{y(a-y)}},\qquad 0\le y\le a.$$

12. Using the result of Example (4.6.7), and integrating by parts, we obtain

$$\begin{aligned}\mathbb{P}(X>a,\ Y>b)&=\int_a^\infty \phi(x)\left\{1-\Phi\left(\frac{b-\rho x}{\sqrt{1-\rho^2}}\right)\right\}dx\\&=[1-\Phi(a)][1-\Phi(c)]+\int_a^\infty[1-\Phi(x)]\phi\left(\frac{b-\rho x}{\sqrt{1-\rho^2}}\right)\frac{\rho}{\sqrt{1-\rho^2}}\,dx.\end{aligned}$$

Since $[1-\Phi(x)]/\phi(x)$ is decreasing, the last term on the right is no greater than

$$\frac{1-\Phi(a)}{\phi(a)}\int_a^\infty \phi(x)\phi\left(\frac{b-\rho x}{\sqrt{1-\rho^2}}\right)\frac{\rho}{\sqrt{1-\rho^2}}\,dx,$$

which yields the upper bound after an integration.

13. The random variable Y is symmetric and, for $a>0$,

$$\mathbb{P}(Y>a)=\mathbb{P}(0<X<a^{-1})=\int_0^{a^{-1}}\frac{du}{\pi(1+u^2)}=\int_\infty^a\frac{-v^{-2}\,dv}{\pi(1+v^{-2})},$$

by the transformation $v = 1/u$. For another example, consider the density function

$$f(x) = \begin{cases} \frac{1}{2}x^{-2} & \text{if } x > 1, \\ \frac{1}{2} & \text{if } 0 \le x \le 1. \end{cases}$$

14. (a) The transformation $w = x + y$, $z = x/(x+y)$ is one–one from $(0,\infty)^2$ to $(0,\infty) \times (0,1)$, with inverse $x = wz$, $y = (1-z)w$, and Jacobian $J = w$, whence

$$\begin{aligned} f(w,z) &= w \cdot \frac{\lambda(\lambda wz)^{\alpha-1} e^{-\lambda wz}}{\Gamma(\alpha)} \cdot \frac{\lambda(\lambda(1-z)w)^{\beta-1} e^{-\lambda(1-z)w}}{\Gamma(\beta)} \\ &= \frac{\lambda(\lambda w)^{\alpha+\beta-1} e^{-\lambda w}}{\Gamma(\alpha+\beta)} \cdot \frac{z^{\alpha-1}(1-z)^{\beta-1}}{B(\alpha,\beta)}, \qquad w > 0,\ 0 < z < 1. \end{aligned}$$

Hence W and Z are independent, and Z is beta distributed with parameters α and β.
(b) We have $R = Z/(1-Z)$, so that

$$\mathbb{P}(R \le r) = \mathbb{P}\left(Z \le \frac{r}{r+1}\right), \quad f_R(r) = f_Z\big(r/(r+1)\big)\frac{d}{dr}\left(\frac{r}{r+1}\right),$$

and the claim follows.

15. Differentiate the given equation to obtain

$$f_Y(y)(1 - F_X(y)) = \lambda f_X(y)(1 - F_Y(y)).$$

As in Exercise (4.6.12),

$$\mathbb{P}(X > Y) = \int_0^\infty f_Y(y)(1 - F_X(y))\,dy, \qquad \mathbb{P}(Y > X) = \int_0^\infty f_X(y)(1 - F_Y(y))\,dy.$$

16. Let $U = Y$, $V = XY$, and consider the invertible transformation $u = y$, $v = xy$, acting on $(0,\infty) \times (-1,1)$. Its inverse has Jacobian $J = 1/u$, whence

$$f_{U,V}(u,v) = \frac{1}{\pi} \cdot \frac{u e^{-\frac{1}{2}u^2}}{\sqrt{u^2 - v^2}}, \qquad u > 0,\ |v| < u.$$

Thus,

$$f_V(v) = \int_{|v|}^\infty f_{U,V}(u,v)\,du = \frac{1}{\sqrt{2\pi}} e^{-\frac{1}{2}v^2}.$$

You may care to review (4.4.4) and Problem (4.14.1).

17. (a) This follows from the fact that, for measurable $A \subseteq S$,

$$\mathbb{P}(U \in A \mid U \in S) = \frac{\mathbb{P}(U \in A)}{\mathbb{P}(U \in S)} = \frac{|A|}{|S|}.$$

(b) Since U is uniformly distributed, $\mathbb{P}(U_r = 1) = \frac{1}{2}$. Let $s > r$. By part (a), $\mathbb{P}(U_s = 1 \mid U_r = 1) = \frac{1}{2}$, whence

$$\mathbb{P}(U_r = 1, U_s = 1) = \tfrac{1}{2} \cdot \tfrac{1}{2} = \mathbb{P}(U_r = 1)\mathbb{P}(U_s = 1).$$

Now, V has distribution function $\mathbb{P}(V \le v) = v^2$ for $v \in [0, 1]$. Therefore,

$$\begin{aligned}\mathbb{P}(V_1 = 1) &= \mathbb{P}(V > \tfrac{1}{2}) = \tfrac{3}{4},\\ \mathbb{P}(V_2 = 1) &= \mathbb{P}\big(V \in (\tfrac{1}{4}, \tfrac{1}{2}) \cup (\tfrac{3}{4}, 1)\big) = \tfrac{5}{8},\\ \mathbb{P}(V_1 = 1,\ V_2 = 1) &= \mathbb{P}\big(V \in (\tfrac{3}{4}, 1)\big) = \tfrac{7}{16},\end{aligned}$$

whence $\operatorname{cov}(V_1, V_2) = \frac{7}{16} - \frac{5}{8}\cdot\frac{3}{4} = -\frac{1}{32}$. Finally, as $n \to \infty$,

$$\mathbb{P}(V_n = 1) = \sum_{k=1}^{2^{n-1}} \mathbb{P}\left(\frac{2k-1}{2^n} < \sqrt{U} < \frac{2k}{2^n}\right) = \sum_{k=1}^{2^{n-1}} \frac{4k-1}{2^{2n}} = \frac{2^n+1}{2^{n+1}} \to \frac{1}{2}.$$

18. (a) The joint density function of (X, Y) is $(2\pi)^{-1}e^{-\frac{1}{2}(x^2+y^2)}$, which has spherical symmetry. Therefore, $\Theta = \tan^{-1}(Y/X)$ is uniformly distributed on $(0, 2\pi)$, and the claim follows. See also Problem (4.14.16).

(b) Let U, V be independent $N(0, 1)$ variables, and let

$$X = U, \qquad Y = \rho U + \sqrt{1-\rho^2}V,$$

as in Example (4.7.12). Then $Y/X = \rho + \sqrt{1-\rho^2}(V/U)$ has, by part (a), the rescaled and shifted Cauchy density function given in the question.

19. Represent X and Y in terms of the independent $N(0, 1)$ variables U, V of the last solution. Then,

$$\mathbb{E}(Y \mid X > x) = \rho\mathbb{E}(U \mid U > x) = \rho \cdot \frac{1}{\Phi(-x)} \int_x^\infty u\phi(u)\,du = \frac{\rho\phi(x)}{\Phi(-x)}.$$

20. Once again, we represent X, Y in terms of the independent $N(0, 1)$ variables U, V of the solution to Exercise (4.7.18). The Jacobian of the corresponding transformation $(u, v) \mapsto (x, y)$ is $\sqrt{1-\rho^2}$, and

$$\mathbb{P}\big(X^2 + Y^2 - 2\rho XY \le c\big) = \mathbb{P}\left(U^2 + V^2 \le \frac{c}{1-\rho^2}\right) = 1 - \exp\left\{-\frac{1}{2}\frac{c}{1-\rho^2}\right\}, \quad c > 0,$$

after changing to polar coordinates. Using the same transformation, the area of the ellipse $x^2 + y^2 - 2\rho xy \le c$ is found to be

$$A = \iint_{R_c} \sqrt{1-\rho^2}\,du\,dv = \frac{\pi c}{\sqrt{1-\rho^2}},$$

where $R_c = \big\{(u, v) : u^2 + v^2 \le c/(1-\rho^2)\big\}$.

21. The closest integer C is odd if, for some $n \ge 0$, we have $\frac{1}{2}(4n+1) < R < \frac{1}{2}(4n+3)$, which is to say that $(X, Y) \in T$ where $T := \bigcup_n T_n$ and

$$T_n = \big\{(x, y) : (4n+1)y < 2x < (4n+3)y\big\} \cap [0, 1]^2.$$

Draw a picture of the regions T_n, and note that $T_2, T_3, \dots$ are triangles. The required probability is the sum of the areas of the T_n, namely,

$$\tfrac{1}{2}(\tfrac{1}{2} + \tfrac{1}{3}) + \tfrac{1}{2}(\tfrac{2}{5} - \tfrac{2}{7}) + \tfrac{1}{2}(\tfrac{2}{9} - \tfrac{2}{11}) + \cdots = -\tfrac{1}{4} + \big[1 - \tfrac{1}{3} + \tfrac{1}{5} - \tfrac{1}{7} + \tfrac{1}{9} - \tfrac{1}{11} + \cdots\big],$$

which is well known to equal $\frac{1}{4}(\pi - 1)$.

22. For $0 \le c \le \zeta$ and $C = -2\log(c/\zeta)$,

$$\mathbb{P}\big(f(X,Y) \le c\big) = \mathbb{P}\left(X^2 + \frac{(Y-\rho X)^2}{1-\rho^2} \ge C\right) = \mathbb{P}\big(U^2 + V^2 \ge C\big)$$

where U, V are independent $N(0,1)$ variables (recall Exercise (4.7.18)). The last equals

$$\int_{R_c} \frac{1}{2\pi} e^{-\frac{1}{2}(u^2+v^2)}\,du\,dv = \int_{\sqrt{C}}^{\infty} re^{-\frac{1}{2}r^2}\,dr = e^{-\frac{1}{2}C} = \frac{c}{\zeta},$$

where $R_c = \{(u,v) : u^2 + v^2 \ge C\}$.

23. Let T_i be the time of arrival of person i, so that the T_i are independent and uniformly distributed on $[0,1]$. Let $T_{(k)}$ be the kth of the T_i written in increasing order; the sequence $\mathbf{T} = (T_{(1)}, T_{(2)}, \dots, T_{(n)})$ is the sequence of 'order statistics' of the T_i. Since the T_i are continuous, the vector $\mathbf{T}$ has a joint density function $f_{\mathbf{T}}$ having as support the subset S of $[0,1]^n$ containing all increasing sequences. It is left as an exercise (see Problem (4.14.21)) to show that

$$(*) \qquad f_{\mathbf{T}}(\mathbf{t}) = n!, \qquad \mathbf{t} \in S.$$

We note for future use that

$$(**) \qquad \mathbb{P}\big(0 < T_{(1)} < T_{(2)} < \cdots < T_{(n)} < a\big) = a^n, \qquad a \in (0,1),$$

since this is the probability that every T_i is less than a.

Let $0 < \delta < 1/(n-1)$, and let R be the set of increasing sequences $\mathbf{t} = (t_1, t_2, \dots, t_n) \in S$ such that $0 < t_1 < t_2 - \delta < t_3 - 2\delta < \cdots < t_n - (n-1)\delta$. We are asked to find the probability $\mathbb{P}(\mathbf{T} \in R)$. Towards this end, we perform a change of variables.

Define the map $h : S \to hS$ by $h(\mathbf{t}) = \mathbf{u}$ where $u_k = t_k - (k-1)\delta$ for $k = 1, 2, \dots, n$ and $\mathbf{t} = (t_1, t_2, \dots, t_n) \in S$. The mapping h is invertible on its range hS, but we must take care over this range. We shall not need to specify hS entirely; it will suffice to note that $R' \subseteq hS$ where R' is the subset of S containing all increasing sequences $\mathbf{u}$ satisfying $0 < u_1 < u_2 < \cdots < u_n < 1 - (n-1)\delta$. Note that $R = h^{-1}R'$.

The random vector $\mathbf{U} = (U_1, U_2, \dots, U_n) = h(\mathbf{T})$ has a density function $f_{\mathbf{U}}$ that may be obtained from $f_{\mathbf{T}}$ by the Jacobian formula. By the definition of h, the Jacobian satisfies $J = 1$, and it follows that $f_{\mathbf{U}}(\mathbf{u}) = f_{\mathbf{T}}(h^{-1}(\mathbf{u})) = n!$ for $\mathbf{u} \in hS$, by (*).

By integrating $f_{\mathbf{U}}$ over R', and using the fact that $R = h^{-1}R'$, we deduce that

$$\begin{aligned}
\mathbb{P}(\mathbf{T} \in R) = \mathbb{P}(\mathbf{U} \in R') &= \int_{R'} n!\,d\mathbf{u} \\
&= \mathbb{P}\big(0 < T_{(1)} < T_{(2)} < \cdots < T_{(n)} < 1 - (n-1)\delta\big) \\
&= \big(1 - (n-1)\delta\big)^n \qquad \text{by } (**).
\end{aligned}$$

24. The mapping $T : (x,y) \mapsto (w,z)$ given by $w = xy$, $z = y/x$ is invertible, and its inverse has Jacobian $J = 1/(2z)$. Therefore,

$$f_{W,Z}(w,z) = \frac{1}{2z}, \qquad (w,z) \in T\big([0,1]^2\big).$$

After some thought, one sees that the domain $T([0,1]^2)$ of $f_{W,Z}$ is the set of (w,z) such that $z \ge 0$ and $w \le \min\{z, 1/z\}$. Hence, W has density

$$f_W(w) = \int_w^{1/w} \frac{1}{2z}\,dz = -\log w, \qquad 0 < w < 1,$$

and likewise integrating out w gives

$$f_Z(z) = \begin{cases} \dfrac{1}{2} & \text{if } 0 < z < 1, \\ \dfrac{1}{2z^2} & \text{if } z > 1. \end{cases}$$

25. (a) By making the change of variable $Y = (X-\mu)/\sigma$, we may assume that $\mu = 0$ and $\sigma = 1$. The $N(0,1)$ density function ϕ satisfies $\phi'(x) = -x\phi(x)$. Integrate by parts to obtain

$$\mathbb{E}\{Xg(X)\} = \int_{-\infty}^{\infty} g(x)[x\phi(x)]\,dx = \int_{-\infty}^{\infty} g'(x)\phi(x)\,dx = \mathbb{E}(g'(X)).$$

(b) We may assume that X, Y have the standard bivariate normal distribution with correlation ρ. Express X, Y in terms of the independent $N(0,1)$ variables U, V as in the solution to Exercise (4.7.18). Then

$$\mathbb{E}(Yg(X)) = \rho\mathbb{E}(Ug(U)), \qquad \mathbb{E}(g'(X))\operatorname{cov}(X,Y) = \rho\mathbb{E}(g'(U)),$$

and the result follows by part (a) applied to U.

(c) Let h be a bounded function. Solve the differential equation $g'(x) - xg(x) = h(x)$, using an integrating factor, to find that

$$g(x) = e^{\frac{1}{2}x^2}\int_{-\infty}^{x} h(y)e^{-\frac{1}{2}y^2}\,dy.$$

Let $y \in \mathbb{R}$ and let $h(x) = I_{\{x \le y\}} - \Phi(y)$ where Φ is the $N(0,1)$ distribution function. By assumption, with g given above,

$$0 = \mathbb{E}\big(g'(X) - Xg(X)\big) = \mathbb{E}(h(X)) = \mathbb{P}(X \le y) - \Phi(y), \qquad y \in \mathbb{R}.$$

26. The Cauchy–Schwarz inequality for integrable functions $f, g : [0, K] \to \mathbb{R}$ states that

$$\left[\int_0^K f(x)g(x)\,dx\right]^2 \le \int_0^K f(x)^2\,dx \int_0^K g(x)^2\,dx.$$

You can derive this from the more familiar Cauchy–Schwarz inequality of (4.5.12).

By Exercise (4.3.9) and the above inequality applied to g and $f \equiv 1$,

$$\begin{aligned}
\operatorname{var} G(X) &\le \mathbb{E}\left\{\big(G(X) - G(0)\big)^2\right\} = \mathbb{E}\left\{\left(\int_0^X g(y)\,dy\right)^2\right\} \\
&\le \mathbb{E}\left\{\int_0^X 1\,dy \int_0^X g(y)^2\,dy\right\} = \mathbb{E}\left\{X\int_0^X g(y)^2\,dy\right\} \\
&= -\int_{y=-\infty}^{0}\int_{x=-\infty}^{y} g(y)^2 x\phi(x)\,dx\,dy + \int_{y=0}^{\infty}\int_{x=y}^{\infty} g(y)^2 x\phi(x)\,dx\,dy,
\end{aligned}$$

by Fubini's theorem, where ϕ is the $N(0,1)$ density function. Substituting for ϕ and integrating over x, we obtain $\int_{-\infty}^{\infty} g(y)^2\phi(y)\,dy = \mathbb{E}(g(X)^2)$.

For the lower bound, with $\mu = \mathbb{E}G(X)$,

$$\begin{aligned}\operatorname{var} G(X) &= \mathbb{E}\left\{(G(X)-\mu)^2\right\} = \mathbb{E}(X^2)\mathbb{E}\left\{(G(X)-\mu)^2\right\} \quad \text{since } \mathbb{E}(X^2)=1\\ &\ge \left\{\mathbb{E}\big(X(G(X)-\mu)\big)\right\}^2 \qquad \text{by the Cauchy–Schwarz inequality}\\ &= \left\{\mathbb{E}g(X)\right\}^2,\end{aligned}$$

by Stein's identity of Exercise (4.7.25).

27. Let $u = x$, $v = x + y$, whose inverse map has Jacobian 1, whence

$$f_{U,V}(u,v) = \tfrac{2}{3}ve^{-u}, \qquad 0 < u < v < u+1.$$

Finally,

$$f_V(v) = \int_{\max\{0,v-1\}}^{v} \tfrac{2}{3}ve^{-u}\,du = \begin{cases} \tfrac{2}{3}ve^{-v}(e-1) & \text{if } v \ge 1,\\ \tfrac{2}{3}v(1-e^{-v}) & \text{if } v < 1.\end{cases}$$

4.8 Solutions. Sums of random variables

1. By the convolution formula (4.8.2), $Z = X + Y$ has density function

$$f_Z(z) = \int_0^z \lambda\mu e^{-\lambda x}e^{-\mu(z-x)}\,dx = \frac{\lambda\mu}{\mu-\lambda}\left(e^{-\lambda z} - e^{-\mu z}\right), \qquad z \ge 0,$$

if $\lambda \ne \mu$. What happens if $\lambda = \mu$? (Z has a gamma distribution in this case.)

2. Using the convolution formula (4.8.2), $W = \alpha X + \beta Y$ has density function

$$f_W(w) = \int_{-\infty}^{\infty} \frac{1}{\pi\alpha(1+(x/\alpha)^2)}\cdot\frac{1}{\pi\beta(1+\{(w-x)/\beta\}^2)}\,dx,$$

which equals the limit of a complex integral:

$$\lim_{R\to\infty}\int_D \frac{\alpha\beta}{\pi^2}\cdot\frac{1}{z^2+\alpha^2}\cdot\frac{1}{(z-w)^2+\beta^2}\,dz$$

where D is the semicircle in the upper complex plane with diameter $[-R, R]$ on the real axis. Evaluating the residues at $z = i\alpha$ and $z = w + i\beta$ yields

$$\begin{aligned} f_W(w) &= \frac{\alpha\beta 2\pi i}{\pi^2}\left\{\frac{1}{2i\alpha}\cdot\frac{1}{(i\alpha-w)^2+\beta^2} + \frac{1}{2i\beta}\cdot\frac{1}{(w+i\beta)^2+\alpha^2}\right\}\\ &= \frac{1}{\pi(\alpha+\beta)}\cdot\frac{1}{1+\{w/(\alpha+\beta)\}^2}, \qquad -\infty < w < \infty\end{aligned}$$

after some manipulation. Hence W has a Cauchy distribution also.

3. Using the convolution formula (4.8.2),

$$f_Z(z) = \int_0^z \tfrac{1}{2}ze^{-z}\,dy = \tfrac{1}{2}z^2e^{-z}, \qquad z \ge 0.$$

4. Let f_n be the density function of S_n. By convolution,

$$f_2(x) = \lambda_1 e^{-\lambda_1 x} \cdot \frac{\lambda_2}{\lambda_2 - \lambda_1} + \lambda_2 e^{-\lambda_2 x} \cdot \frac{\lambda_1}{\lambda_1 - \lambda_2} = \sum_{r=1}^{2} \lambda_r e^{-\lambda_r x} \prod_{\substack{s=1\\ s\neq r}}^{n} \frac{\lambda_s}{\lambda_s - \lambda_r}.$$

This leads to the guess that

$$(*) \qquad f_n(x) = \sum_{r=1}^{n} \lambda_r e^{-\lambda_r x} \prod_{\substack{s=1\\ s\neq r}}^{n} \frac{\lambda_s}{\lambda_s - \lambda_r}, \qquad n \geq 2,$$

which may be proved by induction as follows. Assume that $(*)$ holds for $n \leq N$. Then

$$\begin{aligned} f_{N+1}(x) &= \int_0^x \sum_{r=1}^{N} \lambda_r e^{-\lambda_r (x-y)} \lambda_{N+1} e^{-\lambda_{N+1} y} \prod_{\substack{s=1\\ s\neq r}}^{N} \frac{\lambda_s}{\lambda_s - \lambda_r}\, dy \\ &= \sum_{r=1}^{N} \lambda_r e^{-\lambda_r x} \prod_{\substack{s=1\\ s\neq r}}^{N+1} \frac{\lambda_s}{\lambda_s - \lambda_r} + A e^{-\lambda_{N+1} x}, \end{aligned}$$

for some constant A. We integrate over x to find that

$$1 = \sum_{r=1}^{N} \prod_{\substack{s=1\\ s\neq r}}^{N+1} \frac{\lambda_s}{\lambda_s - \lambda_r} + \frac{A}{\lambda_{N+1}},$$

and $(*)$ follows with $n = N + 1$ on solving for A.

5. (a) The density function of $X + Y$ is, by convolution,

$$f_2(x) = \begin{cases} x & \text{if } 0 \leq x \leq 1, \\ 2 - x & \text{if } 1 \leq x \leq 2. \end{cases}$$

Therefore, for $1 \leq x \leq 2$,

$$f_3(x) = \int_0^1 f_2(x-y)\, dy = \int_{x-1}^{1} (x-y)\, dy + \int_0^{x-1} (2 - x + y)\, dy = \tfrac{3}{4} - (x - \tfrac{3}{2})^2.$$

Likewise,

$$f_3(x) = \begin{cases} \tfrac{1}{2}x^2 & \text{if } 0 \leq x \leq 1, \\ \tfrac{1}{2}(3-x)^2 & \text{if } 2 \leq x \leq 3. \end{cases}$$

(b) This holds by a simple induction.

(c) The vector X fails to form a polygon if and only if there exists $j \in \{1, 2, \dots, n\}$ such that $X_j > S_n - X_j$. Since X_j and $1 - X_j$ have the same distribution, the last inequality has the same probability p_j as that of the event $S_n < 1$. By part (b), $\mathbb{P}(S_n < 1) = \int_0^1 f_n(x)\, dx = 1/n!$. Therefore, the answer is $1 - np_j = 1 - 1/(n-1)!$.

6. The covariance satisfies $\text{cov}(U, V) = \mathbb{E}(X^2 - Y^2) = 0$, as required. If X and Y are symmetric random variables taking values ± 1, then

$$\mathbb{P}(U = 2, V = 2) = 0 \quad \text{but} \quad \mathbb{P}(U = 2)\mathbb{P}(V = 2) > 0.$$

If X and Y are independent $N(0,1)$ variables, $f_{U,V}(u,v) = (4\pi)^{-1}e^{-\frac{1}{4}(u^2+v^2)}$, which factorizes as a function of u multiplied by a function of v.

7. From the representation $X = \sigma\rho U + \sigma\sqrt{1-\rho^2}V$, $Y = \tau U$, where U and V are independent $N(0,1)$, we learn that

$$\mathbb{E}(X \mid Y = y) = \mathbb{E}(\sigma\rho U \mid U = y/\tau) = \frac{\sigma\rho y}{\tau}.$$

Similarly,

$$\mathbb{E}(X^2 \mid Y = y) = \mathbb{E}\big((\sigma\rho U)^2 + \sigma^2(1-\rho^2)V^2 \,\big|\, U = y/\tau\big) = \left(\frac{\sigma\rho y}{\tau}\right)^2 + \sigma^2(1-\rho^2)$$

whence $\mathrm{var}(X \mid Y) = \sigma^2(1-\rho^2)$. For parts (c) and (d), simply calculate that $\mathrm{cov}(X, X+Y) = \sigma^2 + \rho\sigma\tau$, $\mathrm{var}(X+Y) = \sigma^2 + 2\rho\sigma\tau + \tau^2$, and

$$1 - \rho(X, X+Y)^2 = \frac{\tau^2(1-\rho^2)}{\sigma^2 + 2\rho\sigma\tau + \tau^2}.$$

8. First recall that $\mathbb{P}(|X| \le y) = 2\Phi(y) - 1$. We shall use the fact that $U = (X+Y)/\sqrt{2}$, $V = (X-Y)/\sqrt{2}$ are independent and $N(0,1)$ distributed. Let Δ be the triangle of $\mathbb{R}^2$ with vertices $(0,0)$, $(0,Z)$, $(Z,0)$. Then

$$\begin{aligned}\mathbb{P}(Z \le z \mid X > 0,\ Y > 0) &= 4\mathbb{P}\big((X,Y) \in \Delta\big) = \mathbb{P}\big(|U| \le z/\sqrt{2}, |V| \le z/\sqrt{2}\big) \quad \text{by symmetry}\\ &= 2\big\{2\Phi(z/\sqrt{2}) - 1\big\}^2,\end{aligned}$$

whence the conditional density function is

$$f(z) = 2\sqrt{2}\big\{2\Phi(z/\sqrt{2}) - 1\big\}\phi(z/\sqrt{2}).$$

Finally,

$$\begin{aligned}\mathbb{E}(Z \mid X > 0,\ Y > 0) &= 2\mathbb{E}(X \mid X > 0,\ Y > 0)\\ &= 2\mathbb{E}(X \mid X > 0) = 4\mathbb{E}(XI_{\{X>0\}}) = 4\int_0^\infty \frac{x}{\sqrt{2\pi}} e^{-\frac{1}{2}x^2}\,dx.\end{aligned}$$

9. By a simple application of the Jacobian method, the joint density function of U, V is

$$f_{U,V}(u,v) = \frac{1}{2\pi(a^2+b^2)} \exp\left\{-\frac{1}{2}\frac{u^2+v^2}{a^2+b^2}\right\},$$

whence U and V are independent with the $N(0, a^2+b^2)$ distribution.

10. For $a > 0$ and a random variable Z, define

$$Z_a = \begin{cases} a & \text{if } Z > a,\\ Z & \text{if } |Z| \le a,\\ -a & \text{if } Z < -a.\end{cases}$$

Now, $X_a + Y_a = (X_a+Y_a)^+ - (X_a+Y_a)^-$, and $(X_a+Y_a)^\pm \to (X+Y)^\pm$ as $a \to \infty$, where the convergence is monotone in each case. Hence,

$$\begin{aligned}\mathbb{E}X_a + \mathbb{E}Y_a = \mathbb{E}(X_a+Y_a) &= \mathbb{E}\big((X_a+Y_a)^+\big) - \mathbb{E}\big((X_a+Y_a)^-\big)\\ &\to \mathbb{E}(X+Y) \qquad \text{as } a \to \infty,\end{aligned}$$

whenever the last expectation exists. Finally, note that the limit of the left side, above, depends on the marginal distribution functions only. [This potentially baffling question is of interest because of examples that show the falseness of the equivalent statement for sums of three random variables.]

11. (a) By the result of Exercise (3.7.4) and symmetry,

$$\mathbb{E}\big(\text{var}(U \mid S)\big) = \text{var}(U) - \text{var}\big(\mathbb{E}(U \mid S)\big) = \tfrac{1}{12} - \text{var}(\tfrac{1}{3}S) = \tfrac{1}{12} - \tfrac{1}{9} \cdot \tfrac{3}{12} = \tfrac{1}{18}.$$

(b) This is an unpleasant calculation using the result of Exercise (4.8.5a), above, and the details are left to the assiduous reader. There are three cases, depending on whether $s < 1$, $1 < s < 2$, or $s > 2$.

4.9 Solutions. Multivariate normal distribution

1. Since $\mathbf{V}$ is symmetric, there exists a non-singular matrix $\mathbf{M}$ such that $\mathbf{M}' = \mathbf{M}^{-1}$ and $\mathbf{V} = \mathbf{M}\boldsymbol{\Lambda}\mathbf{M}^{-1}$, where $\boldsymbol{\Lambda}$ is the diagonal matrix with diagonal entries the eigenvalues $\lambda_1, \lambda_2, \dots, \lambda_n$ of $\mathbf{V}$. Let $\boldsymbol{\Lambda}^{\frac{1}{2}}$ be the diagonal matrix with diagonal entries $\sqrt{\lambda_1}, \sqrt{\lambda_2}, \dots, \sqrt{\lambda_n}$; $\boldsymbol{\Lambda}^{\frac{1}{2}}$ is well defined since $\mathbf{V}$ is non-negative definite. Writing $\mathbf{W} = \mathbf{M}\boldsymbol{\Lambda}^{\frac{1}{2}}\mathbf{M}'$, we have that $\mathbf{W} = \mathbf{W}'$ and also

$$\mathbf{W}^2 = (\mathbf{M}\boldsymbol{\Lambda}^{\frac{1}{2}}\mathbf{M}^{-1})(\mathbf{M}\boldsymbol{\Lambda}^{\frac{1}{2}}\mathbf{M}^{-1}) = \mathbf{M}\boldsymbol{\Lambda}\mathbf{M}^{-1} = \mathbf{V}$$

as required. Clearly $\mathbf{W}$ is non-singular if and only if $\boldsymbol{\Lambda}^{\frac{1}{2}}$ is non-singular. This happens if and only if $\lambda_i > 0$ for all i, which is to say that $\mathbf{V}$ is positive definite.

2. By Theorem (4.9.6), $\mathbf{Y}$ has the multivariate normal distribution with mean vector $\mathbf{0}$ and covariance matrix

$$\mathbf{W}^{-1}\mathbf{V}\mathbf{W}^{-1} = \mathbf{W}^{-1}(\mathbf{W}^2)\mathbf{W}^{-1} = \mathbf{I}.$$

3. Clearly $Y = (\mathbf{X} - \boldsymbol{\mu})\mathbf{a}' + \boldsymbol{\mu}\mathbf{a}'$ where $\mathbf{a} = (a_1, a_2, \dots, a_n)$. Using Theorem (4.9.6) as in the previous solution, $(\mathbf{X} - \boldsymbol{\mu})\mathbf{a}'$ is univariate normal with mean $\mathbf{0}$ and variance $\mathbf{a}\mathbf{V}\mathbf{a}'$. Hence Y is normal with mean $\boldsymbol{\mu}\mathbf{a}'$ and variance $\mathbf{a}\mathbf{V}\mathbf{a}'$.

4. Make the transformation $u = x + y$, $v = x - y$, with inverse $x = \frac{1}{2}(u + v)$, $y = \frac{1}{2}(u - v)$, so that $|J| = \frac{1}{2}$. The exponent of the bivariate normal density function is

$$-\frac{1}{2(1-\rho^2)}(x^2 - 2\rho xy + y^2) = -\frac{1}{4(1-\rho^2)}\big\{u^2(1-\rho) + v^2(1+\rho)\big\},$$

and therefore $U = X + Y$, $V = X - Y$ have joint density

$$f(u, v) = \frac{1}{4\pi\sqrt{1-\rho^2}} \exp\left\{-\frac{u^2}{4(1+\rho)} - \frac{v^2}{4(1-\rho)}\right\},$$

whence U and V are independent with respective distributions $N(0, 2(1+\rho))$ and $N(0, 2(1-\rho))$.

5. That Y is $N(0, 1)$ follows by showing that $\mathbb{P}(Y \le y) = \mathbb{P}(X \le y)$ for each of the cases $y \le -a$, $|y| < a$, $y \ge a$.

Secondly,

$$\rho(a) = \mathbb{E}(XY) = \int_{-a}^{a} x^2\phi(x)\,dx - \int_{-\infty}^{-a} x^2\phi(x)\,dx - \int_{a}^{\infty} x^2\phi(x)\,dx = 1 - 4\int_{a}^{\infty} x^2\phi(x)\,dx.$$

The answer to the final part is *no*; X and Y are $N(0,1)$ variables, but the pair (X,Y) is *not* bivariate normal. One way of seeing this is as follows. There exists a root a of the equation $\rho(a)=0$. With this value of a, *if* the pair X, Y is bivariate normal, then X and Y are independent. This conclusion is manifestly false: in particular, we have that $\mathbb{P}(X>a,\ Y>a)\neq\mathbb{P}(X>a)\mathbb{P}(Y>a)$.

6. Recall from Exercise (4.8.7) that for any pair of centred normal random variables

$$\mathbb{E}(X\mid Y)=\frac{\operatorname{cov}(X,Y)}{\operatorname{var}Y}Y,\qquad \operatorname{var}(X\mid Y)=\{1-\rho(X,Y)^2\}\operatorname{var}X.$$

The first claim follows immediately. Likewise,

$$\operatorname{var}(X_j\mid X_k)=\{1-\rho(X_j,X_k)^2\}\operatorname{var}X_j=\left\{1-\frac{\sum_r c_{jr}c_{kr}}{\sqrt{\sum_r c_{jr}^2\sum_r c_{kr}^2}}\right\}\sum_r c_{jr}^2.$$

7. As in the above exercise, we calculate $a=\mathbb{E}(X_1\mid\sum_1^n X_r)$ and $b=\operatorname{var}(X_1\mid\sum_1^n X_r)$ using the facts that $\operatorname{var}X_1=v_{11}$, $\operatorname{var}\big(\sum_1^n X_i\big)=\sum_{ij}v_{ij}$, and $\operatorname{cov}\big(X_1,\sum_1^n X_r\big)=\sum_r v_{1r}$.

8. Let $p=\mathbb{P}(X>0,\ Y>0,\ Z>0)=\mathbb{P}(X<0,\ Y<0,\ Z<0)$. Then

$$\begin{aligned}1-p&=\mathbb{P}\big(\{X>0\}\cup\{Y>0\}\cup\{Z>0\}\big)\\&=\mathbb{P}(X>0)+\mathbb{P}(Y>0)+\mathbb{P}(Z>0)+p\\&\qquad-\mathbb{P}(X>0,\ Y>0)-\mathbb{P}(Y>0,\ Z>0)-\mathbb{P}(X>0,\ Z>0)\\&=\frac{3}{2}+p-\left[\frac{3}{4}+\frac{1}{2\pi}\{\sin^{-1}\rho_1+\sin^{-1}\rho_2+\sin^{-1}\rho_3\}\right].\end{aligned}$$

9. Let U, V, W be independent $N(0,1)$ variables, and represent X, Y, Z as $X=U$, $Y=\rho_1 U+\sqrt{1-\rho_1^2}V$,

$$Z=\rho_3 U+\frac{\rho_2-\rho_1\rho_3}{\sqrt{1-\rho_1^2}}V+\sqrt{\frac{1-\rho_1^2-\rho_2^2-\rho_3^2+2\rho_1\rho_2\rho_3}{(1-\rho_1^2)}}W.$$

We have that $U=X$, $V=(Y-\rho_1X)/\sqrt{1-\rho_1^2}$, and $\mathbb{E}(Z\mid X,Y)$ follows immediately, as does the conditional variance.

10. Yes. Since U, V have a bivariate normal distribution, they are independent if and only if they are uncorrelated. Now, $\mathbb{E}(U)=\mathbb{E}(V)=0$ and

$$\mathbb{E}(UV)=2(\cos^2\theta-\sin^2\theta)\mathbb{E}(XY)=2\rho\cos(2\theta),$$

which equals zero if and only if $\theta=\frac{1}{4}\pi,\frac{3}{4}\pi$.

11. The vector (X,Y) takes values in the set $L^+\cup L^-$ where L^+ (respectively, L^-) is the line $y=x$ (respectively, $y=-x$). Since this has zero Lebesgue measure, the vector has a singular distribution. If (X,Y) has a singular bivariate normal distribution then all linear combinations are jointly normal. However, $X+Y$ has an atom of size $\frac{1}{2}$ at zero, and otherwise is continuously distributed.

4.10 Solutions. Distributions arising from the normal distribution

1. *First method.* We have from paragraph (4.4.6) that the $\chi^2(m)$ density function is

$$f_m(x) = \frac{1}{\Gamma(m/2)} 2^{-m/2} x^{\frac{1}{2}m-1} e^{-\frac{1}{2}x}, \qquad x \ge 0.$$

The density function of $Z = X_1 + X_2$ is, by the convolution formula,

$$\begin{aligned} g(z) &= c\int_0^z x^{\frac{1}{2}m-1} e^{-\frac{1}{2}x} (z-x)^{\frac{1}{2}n-1} e^{-\frac{1}{2}(z-x)}\,dx \\ &= cz^{\frac{1}{2}(m+n)-1} e^{-\frac{1}{2}z} \int_0^1 u^{\frac{1}{2}m-1}(1-u)^{\frac{1}{2}n-1}\,du \end{aligned}$$

by the substitution $u = x/z$, where c is a constant. Hence $g(z) = c'z^{\frac{1}{2}(m+n)-1}e^{-\frac{1}{2}z}$ for $z \ge 0$, for an appropriate constant c', as required.

Second method. If m and n are integral, the following argument is neat. Let $Z_1, Z_2, \dots, Z_{m+n}$ be independent $N(0, 1)$ variables. Then X_1 has the same distribution as $Z_1^2 + Z_2^2 + \cdots + Z_m^2$, and X_2 the same distribution as $Z_{m+1}^2 + Z_{m+2}^2 + \cdots + Z_{m+n}^2$ (see Problem (4.14.12)). Hence $X_1 + X_2$ has the same distribution as $Z_1^2 + \cdots + Z_{m+n}^2$, i.e. the $\chi^2(m+n)$ distribution.

2. (i) The $t(r)$ distribution is symmetric with finite mean, and hence this mean is 0.

(ii) Here is one way. Let U and V be independent $\chi^2(r)$ and $\chi^2(s)$ variables (respectively). Then

$$\mathbb{E}\left(\frac{U/r}{V/s}\right) = \frac{s}{r}\mathbb{E}(U)\mathbb{E}(V^{-1})$$

by independence. Now U is $\Gamma(\frac{1}{2}, \frac{1}{2}r)$ and V is $\Gamma(\frac{1}{2}, \frac{1}{2}s)$, so that $\mathbb{E}(U) = r$ and

$$\mathbb{E}(V^{-1}) = \int_0^\infty \frac{1}{v}\frac{2^{-s/2}}{\Gamma(\frac{1}{2}s)} v^{\frac{1}{2}s-1}e^{-\frac{1}{2}v}\,dv = \frac{\Gamma(\frac{1}{2}s-1)}{2\Gamma(\frac{1}{2}s)}\int_0^\infty \frac{2^{-\frac{1}{2}(s-2)}}{\Gamma(\frac{1}{2}s-1)} v^{\frac{1}{2}s-2}e^{-\frac{1}{2}v}\,dv = \frac{1}{s-2}$$

if $s > 2$, since the integrand is a density function. Hence

$$\mathbb{E}\left(\frac{U/r}{V/s}\right) = \frac{s}{s-2} \qquad \text{if } s > 2.$$

(iii) If $s \le 2$ then $\mathbb{E}(V^{-1}) = \infty$.

3. Substitute $r = 1$ into the $t(r)$ density function.

4. *First method.* Find the density function of X/Y, using a change of variables. The answer is $F(2, 2)$.

Second method. X and Y are independent $\chi^2(2)$ variables (just check the density functions), and hence X/Y is $F(2, 2)$.

5. (i) The vector $(\overline{X}, X_1 - \overline{X}, X_2 - \overline{X}, \dots, X_n - \overline{X})$ has, by Theorem (4.9.6), a multivariate normal distribution. We have as in Exercise (4.5.7) that $\text{cov}(\overline{X}, X_r - \overline{X}) = 0$ for all r, which implies that $\overline{X}$ is independent of each X_r. Using the form of the multivariate normal density function, it follows that $\overline{X}$ is independent of the family $\{X_r - \overline{X} : 1 \le r \le n\}$, and hence of any function of these variables. Now $S^2 = (n-1)^{-1}\sum_r (X_r - \overline{X})^2$ is such a function.

(ii) The key fact is that two random variables U, V with a bivariate normal distribution are independent if and only if $\text{cov}(U, V) = 0$. First consider the case $n = 2$, for which $\overline{X} = \frac{1}{2}(X_1 + X_2)$ and $S^2 = \frac{1}{2}(X_1 - X_2)^2$. Since $\text{cov}(X_1 + X_2, X_1 - X_2) = 0$, and $X_1 + X_2$, $X_1 - X_2$ have a bivariate normal distribution, they are independent. Therefore, $\overline{X}$ and S^2 are independent when $n = 2$. Suppose the claim holds for $n \le N$ where $N \ge 2$, and let $n = N + 1$. It may be checked that, in the natural notation,

$$(*) \quad \overline{X}_{N+1} = \frac{1}{N+1}(X_{N+1} + N\overline{X}_N), \quad NS_{N+1}^2 = (N-1)S_N^2 + \frac{N}{N+1}(X_{N+1} - \overline{X}_N)^2.$$

Since $\overline{X}_N$ and $X_{N+1} - \overline{X}_N$ have a bivariate normal distribution with zero covariance, they are independent. The induction step follows by (*).

The same approach may be used to prove Theorem (4.10.1).

6. The choice of fixed vector is immaterial, since the joint distribution of the X_j is spherically symmetric, and we therefore take this vector to be $(0, 0, \dots, 0, 1)$. We make the change of variables $U^2 = Q^2 + X_n^2$, $\tan\Psi = Q/X_n$, where $Q^2 = \sum_{r=1}^{n-1} X_r^2$ and $Q \ge 0$. Since Q has the $\chi^2(n-1)$ distribution, and is independent of X_n, the pair Q, X_n has joint density function

$$f(q, x) = \frac{e^{-\frac{1}{2}x^2}}{\sqrt{2\pi}} \cdot \frac{\frac{1}{2}(\frac{1}{2}q)^{\frac{1}{2}(n-3)} e^{-\frac{1}{2}x}}{\Gamma(\frac{1}{2}(n-1))}, \qquad x \in \mathbb{R},\ q > 0.$$

The theory is now slightly easier than the practice. We solve for U, Ψ, find the Jacobian, and deduce the joint density function $f_{U,\Psi}(u, \psi)$ of U, Ψ. We now integrate over u, and choose the constant so that the total integral is 1.

Here is a second solution (for a third see Exercise (4.10.7)). Recall that X_n^2 has the $\chi^2(1)$ distribution, and Q^2 the $\chi^2(n-1)$ distribution. The conditional density function of $S := X_n^2$ given $T + S := Q^2 + X_n^2 = 1$ is proportional to

$$g(s) = s^{-\frac{1}{2}}(1-s)^{\frac{1}{2}(n-1)-1} e^{-\frac{1}{2}s} e^{-\frac{1}{2}(1-s)} = s^{-\frac{1}{2}}(1-s)^{\frac{1}{2}(n-3)}.$$

Since this is a beta density, the constant of proportionality is $C = 1/B(\frac{1}{2}, \frac{1}{2}(n-1))$, as in Example (4.4.8). As above, $X_n = \cos\Psi$. By the change of variable formula, the conditional density function of Ψ, given $Q^2 + X_n^2 = 1$, is

$$\begin{aligned} Cg(\cos^2\psi)\cos\psi\sin\psi &= \frac{C}{\cos\psi}(\sin^2\psi)^{\frac{1}{2}(n-3)}\cos\psi\sin\psi \\ &= C(\sin\psi)^{n-2}, \qquad \psi \in [0, \pi]. \end{aligned}$$

It is a minor complication that the mapping $(\cos\psi)^2 \mapsto x^2$ is two-to-one from $[0, \pi]$ to $[0, \infty)$, and this accounts for the dis/appearance of a factor 2 in the calculation.

7. The joint density function of $Y = (Y_1, Y_2, \dots, Y_n)$ has spherical symmetry on its support, the $(n-1)$-sphere S_{n-1}. Therefore, Y is uniformly distributed on S_{n-1}. This uniform density may be expressed in spherical polar coordinates as $(\sin\phi_1)^{n-2}(\sin\phi_2)^{n-3}\cdots\sin\phi_{n-2}$. We integrate out all variables except the polar angle ϕ_1, thereby obtaining that the density function of Ψ, as in Exercise (4.10.6), is proportional to $(\sin\phi)^{n-2}$. The constant of proportionality may be obtained by direct integration.

4.11 Solutions. Sampling from a distribution

1. Uniform on the set $\{1, 2, \dots, n\}$.

2. The result holds trivially when $n = 2$, and more generally by induction on n.

3. We may assume without loss of generality that $\lambda = 1$ (since Z/λ is $\Gamma(\lambda, t)$ if Z is $\Gamma(1, t)$). Let U, V be independent random variables which are uniformly distributed on $[0, 1]$. We set $X = -t \log V$ and note that X has the exponential distribution with parameter $1/t$. It is easy to check that

$$\frac{1}{\Gamma(t)} x^{t-1} e^{-x} \le c f_X(x) \qquad \text{for } x > 0,$$

where $c = t^t e^{-t+1}/\Gamma(t)$. Also, conditional on the event A that

$$U \le \frac{X^{t-1} e^{-t}}{\Gamma(t)} t e^{-X/t},$$

X has the required gamma distribution. This observation may be used as a basis for sampling using the rejection method. We note that $A = \{\log U \le (n-1)\big(\log(X/n) - (X/n) + 1\big)\}$. We have that $\mathbb{P}(A) = 1/c$, and therefore there is a mean number c of attempts before a sample of size 1 is obtained.

4. Use your answer to Exercise (4.11.3) to sample X from $\Gamma(1, \alpha)$ and Y from $\Gamma(1, \beta)$. By Exercise (4.7.14), $Z = X/(X+Y)$ has the required distribution.

5. (a) This is the beta distribution with parameters 2, 2. Use the result of Exercise (4.11.4).

(b) The required $\Gamma(1, 2)$ variables may be more easily obtained and used by forming $X = -\log(U_1 U_2)$ and $Y - \log(U_3 U_4)$ where $\{U_i : 1 \le i \le 4\}$ are independent and uniform on $[0, 1]$.

(c) Let U_1, U_2, U_3 be as in (b) above, and let Z be the second order statistic $U_{(2)}$. That is, Z is the middle of the three values taken by the U_i; see Problem (4.14.21). The random variable Z has the required distribution.

(d) As a slight variant, take $Z = \max\{U_1, U_2\}$ conditional on the event $\{Z \le U_3\}$.

(e) Finally, let $X = \sqrt{U_1}/(\sqrt{U_1} + \sqrt{U_2})$, $Y = \sqrt{U_1} + \sqrt{U_2}$. The distribution of X, conditional on the event $\{Y \le 1\}$, is as required.

6. We use induction. The result is obvious when $n = 2$. Let $n \ge 3$ and let $\mathbf{p} = (p_1, p_2, \dots, p_n)$ be a probability vector. Since $\mathbf{p}$ sums to 1, its minimum entry $p_{(1)}$ and maximum entry $p_{(n)}$ must satisfy

$$p_{(1)} \le \frac{1}{n} < \frac{1}{n-1}, \qquad p_{(1)} + p_{(n)} \ge p_{(1)} + \frac{1 - p_{(1)}}{n-1} = \frac{1 + (n-2)p_{(1)}}{n-1} \ge \frac{1}{n-1}.$$

We relabel the entries of the vector $\mathbf{p}$ such that $p_1 = p_{(1)}$ and $p_2 = p_{(n)}$, and set $\mathbf{v}_1 = \big((n-1)p_1, 1 - (n-1)p_1, 0, \dots, 0\big)$. Then

$$\mathbf{p} = \frac{1}{n-1}\mathbf{v}_1 + \frac{n-2}{n-1}\mathbf{p}_{n-1} \quad \text{where} \quad \mathbf{p}_{n-1} = \frac{n-1}{n-2}\left(0, p_1 + p_2 - \frac{1}{n-1}, p_3, \dots, p_n\right),$$

is a probability vector with at most $n-1$ non-zero entries. The induction step is complete.

It is a consequence that sampling from a discrete distribution may be achieved by sampling from a collection of Bernoulli random variables.

7. It is an elementary exercise to show that $\mathbb{P}(R^2 \le 1) = \frac{1}{4}\pi$, and that, conditional on this event, the vector (T_1, T_2) is uniformly distributed on the unit disk. Assume henceforth that $R^2 \le 1$, and write (R, Θ) for the point (T_1, T_2) expressed in polar coordinates. We have that R and Θ are independent with joint density function $f_{R,\Theta}(r, \theta) = r/\pi$, $0 \le r \le 1$, $0 \le \theta < 2\pi$. Let (Q, Ψ) be the polar coordinates of (X, Y), and note that $\Psi = \Theta$ and $e^{-\frac{1}{2}Q^2} = R^2$. The random variables Q and Ψ are independent, and, by a change of variables, Q has density function $f_Q(q) = qe^{-\frac{1}{2}q^2}$, $q > 0$. We recognize the distribution of (Q, Ψ) as that of the polar coordinates of (X, Y) where X and Y are independent $N(0, 1)$ variables. [Alternatively, the last step may be achieved by a two-dimensional change of variables.]

8. We have that

$$\mathbb{P}(X = k) = \mathbb{P}\left(\left\lfloor \frac{\log U}{\log q} \right\rfloor = k - 1\right) = \mathbb{P}(q^k < U \le q^{k-1}) = q^{k-1}(1 - q), \quad k \ge 1.$$

9. The polar coordinates (R, Θ) of (X, Y) have joint density function

$$f_{R,\Theta}(r, \theta) = \frac{2r}{\pi}, \qquad 0 \le r \le 1,\ -\tfrac{1}{2}\pi \le \theta \le \tfrac{1}{2}\pi.$$

Make a change of variables to find that $Y/X = \tan\Theta$ has the Cauchy distribution.

10. By the definition of Z,

$$\begin{aligned}\mathbb{P}(Z = m) &= h(m) \prod_{r=0}^{m-1} \big(1 - h(r)\big) \\ &= \mathbb{P}(X > 0)\mathbb{P}(X > 1 \mid X > 0) \cdots \mathbb{P}(X = m \mid X > m - 1) = \mathbb{P}(X = m).\end{aligned}$$

11. Suppose g is increasing, so that $h(\mathbf{x}) = -g(\mathbf{1} - \mathbf{x})$ is increasing also. By the FKG inequality of Problem (3.11.18b), $\kappa = \operatorname{cov}(g(\mathbf{U}), -g(1 - \mathbf{U})) \ge 0$, yielding the result.

Estimating I by the average $(2n)^{-1}\sum_{r=1}^{2n} g(\mathbf{U}_r)$ of $2n$ random vectors $\mathbf{U}_r$ requires a sample of size $2n$ and yields an estimate having some variance $2n\sigma^2$. If we estimate I by the average $(2n)^{-1}\{\sum_{r=1}^{n} g(\mathbf{U}_r) + g(1 - \mathbf{U}_r)\}$, we require a sample of size only n, and we obtain an estimate with the smaller variance $2n(\sigma^2 - \kappa)$.

12. (a) By the law of the unconscious statistician,

$$\mathbb{E}\left[\frac{g(Y)f_X(Y)}{f_Y(Y)}\right] = \int \frac{g(y)f_X(y)}{f_Y(y)} f_Y(y)\,dy = I.$$

(b) This is immediate from the fact that the variance of a sum of independent variables is the sum of their variances; see Theorem (3.3.11b).

This an application of the strong law of large numbers, Theorem (7.5.1).

13. (a) If U is uniform on $[0, 1]$, then $X = \sin(\frac{1}{2}\pi U)$ has the required distribution. This is an example of the inverse transform method.

(b) If U is uniform on $[0, 1]$, then $1 - U^2$ has density function $g(x) = \{2\sqrt{1-x}\}^{-1}$, $0 \le x \le 1$. Now $g(x) \ge (\pi/4)f(x)$, which fact may be used as a basis for the rejection method.

14. A random vector (A, B, C) is uniform on the unit sphere if: (i) C is uniform on $[-1, 1]$ (recall Archimedes's cylinder theorem), and (ii) given $C = c$, (A, B) is uniform on the circle with radius $\sqrt{1-c^2}$. We may write $(X, Y) = (R\cos\Theta, R\sin\Theta)$ where R, Θ are independent, $1 - 2R^2$ is uniform on $[-1, 1]$, and $(r\cos\Theta, r\sin\Theta)$ is uniform on the circle of radius r. Choose r such that $r^2 = 1 - (1 - 2R^2)^2$, to find that

$$\begin{aligned}(U, V, W) &= \left(2R\cos\Theta\sqrt{1-R^2},\, 2R\sin\Theta\sqrt{1-R^2},\, 1 - 2R^2\right) \\ &= \left(\sqrt{1-(1-2R^2)^2}\cos\Theta,\, \sqrt{1-(1-2R^2)^2}\sin\Theta,\, 1 - 2R^2\right)\end{aligned}$$

is uniformly distributed on the unit sphere in $\mathbb{R}^3$.

15. With A the event that $U < f_1(Y)/f_Y(Y)$, and

$$I = \int_{-\infty}^{\infty} f_2(x)\,dx = 1 - \int_{-\infty}^{\infty} f_1(x)\,dx,$$

we have

$$\begin{aligned}\mathbb{P}(Z \le z) &= \mathbb{P}(Y \le z,\ A) + \mathbb{P}(A^c)\int_{-\infty}^{z} \frac{1}{I} f_2(x)\,dx \\ &= \int_{-\infty}^{z} \frac{f_1(y)}{f_Y(y)} \cdot f_Y(y)\,dy + \int_{-\infty}^{\infty}\left(1 - \frac{f_1(y)}{f_Y(y)}\right) f_Y(y)\,dy \int_{-\infty}^{z} \frac{1}{I} f_2(x)\,dx \\ &= \int_{-\infty}^{z} \big(f_1(x) + f_2(x)\big)\,dx.\end{aligned}$$

16. (a) We have $\mathbb{P}(G(U) \le x) = \mathbb{P}(U \le G^{-1}(x)) = G^{-1}(x)$, and the required quantile function is the inverse of G^{-1}.
(b) By Theorem (4.11.1), $F^{-1}(U)$ has the same distribution as X, so that, for $v \in (0,1)$,

$$\begin{aligned}H_X(v) &= \mathbb{E}\big(X \mid F(X) > v\big) = \mathbb{E}\big(X \mid X > F^{-1}(v)\big) \\ &= \frac{1}{1-v}\int_{F^{-1}(v)}^{\infty} x f(x)\,dx \qquad \text{since } \mathbb{P}(F(X) > v) = 1 - v \\ &= \frac{1}{1-v}\int_{v}^{\infty} F^{-1}(w)\,dw \qquad \text{on substituting } w = F(x).\end{aligned}$$

[The Hardy–Littlewood transform is of interest as being the random variable that is least in stochastic order among all random variables that are greater than X in stop-loss order; see Exercise (4.12.7).]

17. Certainly $G(x, u)$ is non-decreasing in x and u, and for any $y \in (0,1)$ there exists (x_y, u_y) such that $G(x_y, u_y) = y$. Hence,

$$\begin{aligned}\mathbb{P}(G(X,U) \le y) &= \mathbb{P}\Big(\{X < x_y\} \cup \big(\{X = x_y\} \cap \{U \le u_y\}\big)\Big) \\ &= \mathbb{P}(X < x_y) + u_y \mathbb{P}(X = x_y) \\ &= G(x_y, u_y) = y,\end{aligned}$$

as required for the uniformity of $G(X, U)$. Finally, $F(X-) \le W \le F(X)$, and $X = F^{-1}(W)$ follows by the definition $F^{-1}(w) = \inf\{x : F(x) \ge w\}$,

18. Let $U_i = F_i(X_i)$, and let G be the joint distribution function of $(U_1, U_2, \dots, U_n)$. Then

$$\begin{aligned}\mathbb{P}(X_i \le x_i,\ i = 1, 2, \dots, n) &= \mathbb{P}\big(F_i(X_i) \le F_i(x_i),\ i = 1, 2, \dots, n\big) \\ &= G(F_1(x_1), F_2(x_2), , \dots, F_n(x_n)).\end{aligned}$$

4.12 Solutions. Coupling and Poisson approximation

1. Suppose that $\mathbb{E}(u(X)) \geq \mathbb{E}(u(Y))$ for all increasing functions u. Let $c \in \mathbb{R}$ and set $u = I_c$ where

$$I_c(x) = \begin{cases} 1 & \text{if } x > c, \\ 0 & \text{if } x \leq c, \end{cases}$$

to find that $\mathbb{P}(X > c) = \mathbb{E}(I_c(X)) \geq \mathbb{E}(I_c(Y)) = \mathbb{P}(Y > c)$.

Conversely, suppose that $X \geq_{\text{st}} Y$. We may assume by Theorem (4.12.3) that X and Y are defined on the same sample space, and that $\mathbb{P}(X \geq Y) = 1$. Let u be an increasing function. Then $\mathbb{P}(u(X) \geq u(Y)) \geq \mathbb{P}(X \geq Y) = 1$, whence $\mathbb{E}(u(X) - u(Y)) \geq 0$ whenever this expectation exists.

2. Let $\alpha = \mu/\lambda$, and let $\{I_r : r \geq 1\}$ be independent Bernoulli random variables with parameter α. Then $Z = \sum_{r=1}^{X} I_r$ has the Poisson distribution with parameter $\lambda\alpha = \mu$, and $Z \leq X$.

3. Use the argument in the solution to Problem (2.7.13).

4. For any $A \subseteq \mathbb{R}$,

$$\begin{aligned} \mathbb{P}(X \neq Y) \geq \mathbb{P}(X \in A,\ Y \in A^{\text{c}}) &= \mathbb{P}(X \in A) - \mathbb{P}(X \in A,\ Y \in A) \\ &\geq \mathbb{P}(X \in A) - \mathbb{P}(Y \in A), \end{aligned}$$

and similarly with X and Y interchanged. Hence,

$$\mathbb{P}(X \neq Y) \geq \sup_{A \subseteq \mathbb{R}} \big|\mathbb{P}(X \in A) - \mathbb{P}(Y \in A)\big| = \tfrac{1}{2} d_{\text{TV}}(X, Y).$$

5. For any positive x and y, we have that $(y - x)^+ + x \wedge y = y$, where $x \wedge y = \min\{x, y\}$. It follows that

$$\sum_k \{f_X(k) - f_Y(k)\}^+ = \sum_k \{f_Y(k) - f_X(k)\}^+ = 1 - \sum_k f_X(k) \wedge f_Y(k),$$

and by the definition of $d_{\text{TV}}(X, Y)$ that the common value in this display equals $\frac{1}{2} d_{\text{TV}}(X, Y) = \delta$. Let U be a Bernoulli variable with parameter $1 - \delta$, and let V, W, Z be independent integer-valued variables with

$$\begin{aligned} \mathbb{P}(V = k) &= \{f_X(k) - f_Y(k)\}^+/\delta, \\ \mathbb{P}(W = k) &= \{f_Y(k) - f_X(k)\}^+/\delta, \\ \mathbb{P}(Z = k) &= f_X(k) \wedge f_Y(k)/(1 - \delta). \end{aligned}$$

Then $X' = UZ + (1 - U)V$ and $Y' = UZ + (1 - U)W$ have the required marginals, and $\mathbb{P}(X' = Y') = \mathbb{P}(U = 1) = 1 - \delta$. See also Problem (7.11.16d).

6. Evidently $d_{\text{TV}}(X, Y) = |p - q|$, and we may assume without loss of generality that $p \geq q$. We have from Exercise (4.12.4) that $\mathbb{P}(X = Y) \leq 1 - (p - q)$. Let U and Z be independent Bernoulli variables with respective parameters $1 - p + q$ and $q/(1 - p + q)$. The pair $X' = U(Z - 1) + 1$, $Y' = UZ$ has the same marginal distributions as the pair X, Y, and $\mathbb{P}(X' = Y') = \mathbb{P}(U = 1) = 1 - p + q$ as required.

In order to achieve the minimum, we set $X'' = 1 - X'$ and $Y'' = Y'$, so that $\mathbb{P}(X'' = Y'') = 1 - \mathbb{P}(X' = Y') = p - q$.

7. Choosing $u(x) = (x - a)^+$ shows that increasing convex order (see the next exercise) implies increasing stop-loss order. Conversely, suppose $X \leq_{\text{sl}} Y$. Let c be an increasing, convex function that is bounded from below, with right derivative c' and left limit $c(-\infty) = \lim_{u\to-\infty} c(u)$. Such convex functions on $\mathbb{R}$ are quite well behaved: (i) they are continuous, (ii) differentiable except possibly on some countable set, (iii) twice differentiable almost everywhere with non-decreasing second derivative, and (iv) $|u|c'(u) \to 0$ as $u \to -\infty$. We assume for simplicity that c is twice differentiable on $\mathbb{R}$. By integration by parts,

$$\int_{-\infty}^{\infty} (x-s)^+ \, dc'(s) = c(x) - c(-\infty), \qquad x \in \mathbb{R}.$$

Therefore, by Fubini's theorem,

$$\begin{aligned}
\mathbb{E}(c(X)) &= \int_{-\infty}^{\infty} c(x)\, dF_X(x) \\
&= \int_{-\infty}^{\infty} \left[c(-\infty) + \int_{-\infty}^{\infty} (x-s)^+ \, dc'(s) \right] dF_X(x) \\
&= c(-\infty) + \int_{-\infty}^{\infty} \left[\int_{-\infty}^{\infty} (x-s)^+ \, dF_X(x) \right] dc'(s) \\
&\le c(-\infty) + \int_{-\infty}^{\infty} \left[\int_{-\infty}^{\infty} (x-s)^+ \, dF_Y(x) \right] dc'(s) = \mathbb{E}(c(Y)),
\end{aligned}$$

where we have used the assumption $X \le_{\text{sl}} Y$ at the last stage.†

8. (a) Let $X \le_{\text{cx}} Y$, and choose $u_1(x) = x$, $u_2(x) = -x$, and $u_3(x) = (x - \mathbb{E}X)^2$.

(b) The 'only if' holds by part (a), and choosing $u(x) = (x-a)^+$. Assume that $\mathbb{E}X = \mathbb{E}Y$ and $X \le_{\text{sl}} Y$, and let u be a convex function. Since u is convex, it can be expressed as $u = c + d$ where c (respectively, d) is convex and increasing (respectively, decreasing). By Exercise (4.12.7), $\mathbb{E}(c(X)) \le \mathbb{E}(c(Y))$, and thus it suffices to show that $\mathbb{E}(d(X)) \le \mathbb{E}(d(Y))$. This is implied (as in Exercise (4.11.7) applied to $-X$ and $-Y$) by $\mathbb{E}((a-X)^+) \le \mathbb{E}((a-Y)^+)$ for $a \in \mathbb{R}$, which we prove next.

We have $X - a = (X-a)^+ - (X-a)^-$, so that $\mathbb{E}((a-X)^+) = \mathbb{E}((X-a)^+) + a - \mathbb{E}X$. Now, $\mathbb{E}X = \mathbb{E}Y$ and $X \le_{\text{sl}} Y$, whence $\mathbb{E}((a-X)^+) \le \mathbb{E}((a-Y)^+)$ as required.

4.13 Solutions. Geometrical probability

1. The angular coordinates Ψ and Σ of A and B have joint density $f(\psi, \sigma) = (2\pi)^{-2}$. We make the change of variables from $(p, \theta) \mapsto (\psi, \sigma)$ by $p = \cos\{\frac{1}{2}(\sigma - \psi)\}$, $\theta = \frac{1}{2}(\pi + \sigma + \psi)$, with inverse

$$\psi = \theta - \tfrac{1}{2}\pi - \cos^{-1} p, \quad \sigma = \theta - \tfrac{1}{2}\pi + \cos^{-1} p,$$

and Jacobian $|J| = 2/\sqrt{1-p^2}$.

2. Let A be the left shaded region and B the right shaded region in the figure. Writing λ for the random line, by Example (4.13.2),

$$\begin{aligned}
\mathbb{P}(\lambda \text{ meets both } S_1 \text{ and } S_2) &= \mathbb{P}(\lambda \text{ meets both } A \text{ and } B) \\
&= \mathbb{P}(\lambda \text{ meets } A) + \mathbb{P}(\lambda \text{ meets } B) - \mathbb{P}(\lambda \text{ meets either } A \text{ or } B) \\
&\propto b(A) + b(B) - b(H) = b(X) - b(H),
\end{aligned}$$

whence $\mathbb{P}(\lambda \text{ meets } S_2 \mid \lambda \text{ meets } S_1) = [b(X) - b(H)]/b(S_1)$.

†Think of dF_X as $f_X(x)\,dx$ if you are unfamiliar with this form of integral, to be defined in Section 5.6.

The case when $S_2 \subseteq S_1$ is treated in Example (4.13.2). When $S_1 \cap S_2 \neq \varnothing$ and $S_1 \triangle S_2 \neq \varnothing$, the argument above shows the answer to be $[b(S_1) + b(S_2) - b(H)]/b(S_1)$.

3. With $|I|$ the length of the intercept I of λ_1 with S_2, we have that $\mathbb{P}(\lambda_2 \text{ meets } I) = 2|I|/b(S_1)$, by the Buffon needle calculation (4.13.2). The required probability is

$$\frac{1}{2}\int_0^{2\pi}\int_{-\infty}^{\infty}\frac{2|I|}{b(S_1)}\cdot\frac{dp\,d\theta}{b(S_1)} = \int_0^{2\pi}\frac{|S_2|}{b(S_1)^2}\,d\theta = \frac{2\pi|S_2|}{b(S_1)^2}.$$

4. If the two points are denoted $\mathrm{P} = (X_1, Y_1)$, and $\mathrm{Q} = (X_2, Y_2)$, then

$$\mathbb{E}(Z^2) = \mathbb{E}(|\mathrm{PQ}|^2) = 2\mathbb{E}\big((X_1 - X_2)^2\big) = 4\operatorname{var}(X_1) = \frac{8}{\pi a^2}\int_{-a}^{a} x^2\sqrt{a^2 - x^2}\,dx = a^2.$$

We use Crofton's method in order to calculate $\mathbb{E}(Z)$. Consider a disc D of radius x surrounded by an annulus A of width h. We set $\lambda(x) = \mathbb{E}(Z \mid \mathrm{P}, \mathrm{Q} \in D)$, and find that

$$\lambda(x+h) = \lambda(x)\left(1 - \frac{4h}{x} - \mathrm{o}(h)\right) + 2\mathbb{E}(Z \mid \mathrm{P} \in D,\ \mathrm{Q} \in A)\left(\frac{2h}{x} + \mathrm{o}(h)\right).$$

Now

$$\mathbb{E}(Z \mid \mathrm{P} \in D,\ \mathrm{Q} \in A) = \frac{2}{\pi x^2}\int_0^{\frac{1}{2}\pi}\int_0^{2x\cos\theta} r^2\,dr\,d\theta + \mathrm{o}(1) = \frac{32x}{9\pi},$$

whence

$$\frac{d\lambda}{dx} = -\frac{4\lambda}{x} + \frac{128}{9\pi},$$

which is easily integrated subject to $\lambda(0) = 0$ to give the result.

5. (i) We may assume without loss of generality that the sphere has radius 1. The length $X = |\mathrm{AO}|$ has density function $f(x) = 3x^2$ for $0 \le x \le 1$. The triangle includes an obtuse angle if B lies either in the hemisphere opposite to A, or in the sphere with centre $\frac{1}{2}X$ and radius $\frac{1}{2}X$, or in the segment cut off by the plane through A perpendicular to AO. Hence,

$$\begin{aligned}\mathbb{P}(\text{obtuse}) &= \tfrac{1}{2} + \mathbb{E}\big((\tfrac{1}{2}X)^3\big) + (\tfrac{4}{3}\pi)^{-1}\mathbb{E}\left(\int_X^1 \pi(1-y^2)\,dy\right)\\ &= \tfrac{1}{2} + \tfrac{1}{16} + (\tfrac{4}{3})^{-1}\mathbb{E}(\tfrac{2}{3} - X + \tfrac{1}{3}X^3) = \tfrac{5}{8}.\end{aligned}$$

(ii) In the case of the circle, X has density function $2x$ for $0 \le x \le 1$, and similar calculations yield

$$\mathbb{P}(\text{obtuse}) = \frac{1}{2} + \frac{1}{8} + \frac{1}{\pi}\mathbb{E}\big(\cos^{-1}X - X\sqrt{1 - X^2}\big) = \frac{3}{4}.$$

6. Choose the x-axis along AB. With $\mathrm{P} = (X, Y)$ and $\mathrm{G} = (\gamma_1, \gamma_2)$,

$$\mathbb{E}|\mathrm{ABP}| = \tfrac{1}{2}|\mathrm{AB}|\,\mathbb{E}(Y) = \tfrac{1}{2}|\mathrm{AB}|\gamma_2 = |\mathrm{ABG}|.$$

7. We use Exercise (4.13.6). First fix P, and then Q, to find that

$$\mathbb{E}|\mathrm{APQ}| = \mathbb{E}\big[\mathbb{E}(|\mathrm{APQ}| \mid \mathrm{P})\big] = \mathbb{E}|\mathrm{APG}_2| = |\mathrm{AG}_1\mathrm{G}_2|.$$

With $b = |\mathrm{AB}|$ and h the height of the triangle ABC on the base AB, we have that $|\mathrm{G_1G_2}| = \frac{1}{3}b$ and the height of the triangle $\mathrm{AG_1G_2}$ is $\frac{2}{3}h$. Hence,

$$\mathbb{E}|\mathrm{APQ}| = \tfrac{1}{2} \cdot \tfrac{1}{3}b \cdot \tfrac{2}{3}h = \tfrac{2}{9}|\mathrm{ABC}|.$$

8. Let the scale factor for the random triangle be X, where $X \in (0, 1)$. For a triangle with scale factor x, any given vertex can lie anywhere in a certain triangle having area $(1-x)^2|\mathrm{ABC}|$. Picking one at random from all possible such triangles amounts to supposing that X has density function $f(x) = 3(1-x)^2$, $0 \le x \le 1$. Hence the mean area is

$$|\mathrm{ABC}|\,\mathbb{E}(X^2) = |\mathrm{ABC}| \int_0^1 3x^2(1-x)^2\,dx = \tfrac{1}{10}|\mathrm{ABC}|.$$

9. We have by conditioning that, for $0 \le z \le a$,

$$\begin{aligned} F(z, a+da) &= F(z,a)\left(\frac{a}{a+da}\right)^2 + \mathbb{P}(X \ge a-z)\cdot\frac{2a\,da}{(a+da)^2} + \mathrm{o}(da) \\ &= F(z,a)\left(1 - \frac{2\,da}{a}\right) + \frac{z}{a}\cdot\frac{2da}{a} + \mathrm{o}(da), \end{aligned}$$

and the equation follows by taking the limit as $da \downarrow 0$. The boundary condition may be taken to be $F(a, a) = 1$, and we deduce that

$$F(z,a) = \frac{2z}{a} - \left(\frac{z}{a}\right)^2, \qquad 0 \le z \le a.$$

Likewise, by use of conditional expectation,

$$m_r(a+da) = m_r(a)\left(1 - \frac{2\,da}{a}\right) + \mathbb{E}\big((a-X)^r\big)\cdot\frac{2\,da}{a} + \mathrm{o}(da).$$

Now, $\mathbb{E}((a-X)^r) = a^r/(r+1)$, yielding the required equation. The boundary condition is $m_r(0) = 0$, and therefore

$$m_r(a) = \frac{2a^r}{(r+1)(r+2)}.$$

10. If n great circles meet each other, not more than two at any given point, then there are $2\binom{n}{2}$ intersections. It follows that there are $4\binom{n}{2}$ segments between vertices, and Euler's formula gives the number of regions as $n(n-1)+2$. We may think of the plane as obtained by taking the limit as $R \to \infty$ and 'stretching out' the sphere. Each segment is a side of two polygons, so the average number of sides satisfies

$$\frac{4n(n-1)}{2+n(n-1)} \to 4 \qquad \text{as } n \to \infty.$$

11. By making an affine transformation, we may without loss of generality assume the triangle has vertices $\mathrm{A} = (0,1)$, $\mathrm{B} = (0,0)$, $\mathrm{C} = (1,0)$. With $\mathrm{P} = (X, Y)$, we have that

$$\mathrm{L} = \left(\frac{X}{1-Y}, 0\right), \quad \mathrm{M} = \left(\frac{X}{X+Y}, \frac{Y}{X+Y}\right), \quad \mathrm{N} = \left(0, \frac{Y}{1-X}\right).$$

Hence,

$$\mathbb{E}|\mathrm{BLN}| = 2\int_{\mathrm{ABC}} \frac{xy}{2(1-x)(1-y)}\,dx\,dy = \int_0^1 \left(-x - \frac{x}{1-x}\log x\right)dx = \frac{\pi^2}{6} - \frac{3}{2},$$

and likewise $\mathbb{E}|\text{CLM}| = \mathbb{E}|\text{ANM}| = \frac{1}{6}\pi^2 - \frac{3}{2}$. It follows that

$$\mathbb{E}|\text{LMN}| = \tfrac{1}{2}(10 - \pi^2) = (10 - \pi^2)|\text{ABC}|.$$

12. Let the points be P, Q, R, S. By Example (4.13.6),

$$\mathbb{P}\big(\text{one lies inside the triangle formed by the other three}\big) = 4\mathbb{P}(\text{S} \in \text{PQR}) = 4 \cdot \tfrac{1}{12}.$$

13. We use Crofton's method. Let $m(a)$ be the mean area, and condition on whether points do or do not fall in the annulus with internal and external radii a, $a+h$. Then

$$m(a+h) = m(a)\left(\frac{a}{a+h}\right)^6 + \left[\frac{6h}{a} + \mathrm{o}(h)\right]\widehat{m}(a),$$

where $\widehat{m}(a)$ is the mean area of a triangle having one vertex P on the boundary of the circle. Using polar coordinates with P as origin,

$$\begin{aligned}\pi^2 a^4 \widehat{m}(a) &= \frac{1}{2}\int_0^{\pi}\int_0^{\pi}\int_0^{2a\cos\theta}\int_0^{2a\cos\psi} r_1^2 r_2^2\, dr_1\, dr_2\, \sin|\theta - \psi|\, d\theta\, d\psi \\ &= \frac{32a^6}{9}\int_0^{\pi}\int_0^{\pi} \sin^3\theta \sin^3\psi \sin|\theta-\psi|\, d\theta\, d\psi = \frac{35a^6\pi}{36}.\end{aligned}$$

Letting $h \downarrow 0$ above, we obtain

$$\frac{dm}{da} = -\frac{6m}{a} + \frac{6}{a}\cdot\frac{35a^2}{36\pi},$$

whence $m(a) = (35a^2)/(48\pi)$.

14. Let a be the radius of C, and let R be the distance of A from the centre. Conditional on R, the required probability is $(a-R)^2/a^2$, whence the answer is $\mathbb{E}((a-R)^2/a^2) = \int_0^1 (1-r)^2 2r\, dr = \frac{1}{6}$.

15. Let a be the radius of C, and let R be the distance of A from the centre. As in Exercise (4.13.14), the answer is $\mathbb{E}((a-R)^3/a^3) = \int_0^1 (1-r)^3 3r^2\, dr = \frac{1}{20}$.

16. We assume without loss of generality that one of the points is at the north pole. If the other is at (X, Y, Z), then $D^2 = X^2 + Y^2 + (1-Z)^2 = 2 - 2Z$. By Archimedes's cylinder theorem, Z is uniformly distributed on $[-1, 1]$. Therefore, for $d \in (0, 1)$,

$$\mathbb{P}(D \le d) = \mathbb{P}(D^2 \le d^2) = \mathbb{P}(Z \ge 1 - \tfrac{1}{2}d^2) = \tfrac{1}{4}d^2.$$

17. The length of the diameter is 2, and of the semicircle π. Therefore,

$$\mathbb{P}(D) = 2\cdot\frac{2\pi}{(2+\pi)^2}, \qquad \mathbb{P}(N) = \left(\frac{\pi}{2+\pi}\right)^2.$$

Let Θ be the angle between a given radius and the point on the semicircle, and let Y be the distance from the centre to the point on the diameter. The joint density function of the pair Θ, Y is

$$f(\theta, y) = \frac{1}{2\pi}, \qquad \theta \in (0, \pi),\ y \in (-1, 1).$$

The area of the triangle is $A = \frac{1}{2}|Y|\sin\Theta$, so that

$$\mathbb{E}(A \mid D) = \frac{1}{2\pi}\int_{\theta=0}^{\pi} d\theta \int_{y=-1}^{1} \frac{1}{2}|y|\sin\theta\, dy = \frac{1}{2\pi}\cdot\frac{1}{2}\big[-\cos\theta\big]_0^{\pi} = \frac{1}{2\pi}.$$

Conditional on N, the chord between the radii to the chosen points subtends an angle Θ at the centre with density function $g(\theta) = 2(\pi - \theta)/\pi^2$, for $\theta \in (0, \pi)$. (You should check this.) Therefore,

$$\mathbb{E}(A \mid N) = \int_0^\pi \frac{1}{2}\sin\theta \cdot \frac{2(\pi-\theta)}{\pi^2}\, d\theta = \frac{1}{\pi}.$$

In conclusion,

$$\mathbb{E}(A) = \mathbb{E}(A \mid D)\mathbb{P}(D) + \mathbb{E}(A \mid N)\mathbb{P}(N) = \frac{1}{2+\pi}.$$

18. Rotating the wheel clockwise from an arbitrary marked point, let $P_1, P_2, \dots, P_n$ be the initial points, in order, of the inspected arcs. Let A_i be the arc of length x starting at P_i, let I_i be the event that $P_k \notin A_i$ for all $k \neq i$, and let N be the number of events I_i that occur. The required probability is $\mathbb{P}(N = 0)$.

By the rotation-invariance of the distribution of the set $\{P_i : i = 1, 2, \dots, n\}$, we have that $\mathbb{P}(I_i) = (1-x)^{n-1}$. Likewise, $\mathbb{P}(I_i \cap I_j) = (1-2x)^{n-2}$ for $i \neq j$, and so on. By the inclusion–exclusion formula (3.4.2),

$$\mathbb{P}(N = 0) = 1 - \mathbb{P}(N \geq 1) = 1 - \binom{n}{1}(1-x)^{n-1} + \binom{n}{2}(1-2x)^{n-1} - \cdots .$$

where the series terminates as stated.

4.14 Solutions to problems

1. (a) We have that

$$\iint_{\mathbb{R}^2} e^{-(x^2+y^2)}\, dx\, dy = \iint_{\mathbb{R}^2} e^{-r^2} r\, dr\, d\theta = \pi.$$

Secondly, $f \geq 0$, and it is easily seen that $\int_{-\infty}^{\infty} f(x)\, dx = 1$ using the substitution $y = (x - \mu)/(\sigma\sqrt{2})$. Alternatively, note that the first integral is the volume generated by the surface of revolution from $z = e^{-t^2}$, which equals

$$\int_0^1 \pi t(z)^2\, dz = -\int_0^1 \pi \log z\, dz = \pi.$$

(b) The mean is $\int_{-\infty}^{\infty} x(2\pi)^{-\frac{1}{2}} e^{-\frac{1}{2}x^2}\, dx$, which equals 0 since $xe^{-\frac{1}{2}x^2}$ is an odd integrable function. The variance is $\int_{-\infty}^{\infty} x^2(2\pi)^{-\frac{1}{2}} e^{-\frac{1}{2}x^2}\, dx$, easily integrated by parts to obtain 1.
(c) Note that

$$\frac{d}{dy}\left\{y^{-1}e^{-\frac{1}{2}y^2}\right\} = -(1+y^{-2})e^{-\frac{1}{2}y^2},$$
$$\frac{d}{dy}\left\{(y^{-1} - y^{-3})e^{-\frac{1}{2}y^2}\right\} = -(1-3y^{-4})e^{-\frac{1}{2}y^2},$$

and also $1 - 3y^{-4} < 1 < 1 + y^{-2}$. Multiply throughout these inequalities by $e^{-\frac{1}{2}y^2}/\sqrt{2\pi}$, and integrate over $[x, \infty)$, to obtain the required inequalities. More extensive inequalities may be found in Exercise (4.4.8).

(d) The required probability is $\alpha(x) = [1 - \Phi(x + a/x)]/[1 - \Phi(x)]$. By (c),

$$\alpha(x) = (1 + o(1))\frac{e^{-\frac{1}{2}(x+a/x)^2}}{e^{-\frac{1}{2}x^2}} \to e^{-a} \qquad \text{as } x \to \infty.$$

2. Clearly $f \ge 0$ if and only if $0 \le \alpha < \beta \le 1$. Also

$$C^{-1} = \int_\alpha^\beta (x - x^2)\,dx = \tfrac{1}{2}(\beta^2 - \alpha^2) - \tfrac{1}{3}(\beta^3 - \alpha^3).$$

3. The A_i partition the sample space, and $i - 1 \le X(\omega) < i$ if $\omega \in A_i$. Taking expectations and using the fact that $\mathbb{E}(I_i) = \mathbb{P}(A_i)$, we find that $S \le \mathbb{E}(X) \le 1 + S$ where

$$S = \sum_{i=2}^{\infty}(i-1)\mathbb{P}(A_i) = \sum_{i=2}^{\infty}\sum_{j=1}^{i-1} 1 \cdot \mathbb{P}(A_i) = \sum_{j=1}^{\infty}\sum_{i=j+1}^{\infty}\mathbb{P}(A_i) = \sum_{j=1}^{\infty}\mathbb{P}(X \ge j).$$

4. (a) (i) Let $F^{-1}(y) = \sup\{x : F(x) = y\}$, so that

$$\mathbb{P}(F(X) \le y) = \mathbb{P}(X \le F^{-1}(y)) = F(F^{-1}(y)) = y, \qquad 0 \le y \le 1.$$

(ii) $\mathbb{P}(-\log F(X) \le y) = \mathbb{P}(F(X) \ge e^{-y}) = 1 - e^{-y}$ if $y \ge 0$.
(b) Draw a picture. With $D = \mathrm{PR}$,

$$\mathbb{P}(D \le d) = \mathbb{P}(\tan \widehat{\mathrm{PQR}} \le d) = \mathbb{P}(\widehat{\mathrm{PQR}} \le \tan^{-1} d) = \frac{1}{\pi}\left(\frac{\pi}{2} + \tan^{-1} d\right).$$

Differentiate to obtain the result.
(c) By symmetry,

$$\mathbb{P}(X + Y > 0 \mid X = x) = \mathbb{P}(Y > -x) = \mathbb{P}(Y < x) = F(x),$$

so that $\mathbb{P}(X + Y > 0 \mid X) = F(X)$, which is uniformly distributed on $(0, 1)$, with mean $\frac{1}{2}$.

5. (a) Suppose that g is monotone and satisfies $g(s + t) = g(s)g(t)$ for $s, t \ge 0$. For an integer m, $g(m) = g(1)g(m-1) = \cdots = g(1)^m$. For rational $x = m/n$, $g(x)^n = g(m) = g(1)^m$ so that $g(x) = g(1)^x$; all such powers are interpreted as $\exp\{x \log g(1)\}$. Finally, if x is irrational, and g is monotone non-increasing (say), then $g(u) \le g(x) \le g(v)$ for all rationals u and v satisfying $v \le x \le u$. Hence $g(1)^u \le g(x) \le g(1)^v$. Take the limits as $u \downarrow x$ and $v \uparrow x$ through the rationals to obtain $g(x) = e^{\mu x}$ where $\mu = \log g(1)$.
(b) Clearly

$$\mathbb{P}(X > s + x \mid X > s) = \frac{\mathbb{P}(X > s + x)}{\mathbb{P}(X > s)} = \frac{e^{-\lambda(s+x)}}{e^{-\lambda s}} = e^{-\lambda x}$$

if $x, s \ge 0$, where λ is the parameter of the distribution.

Suppose that the non-negative random variable X has the lack-of-memory property. Then $G(x) = \mathbb{P}(X > x)$ is monotone and satisfies $G(s + x) = G(s)G(x)$ for $s, x \ge 0$. Hence $G(s) = e^{-\lambda s}$ for some λ; certainly $\lambda > 0$ since $G(s) \le 1$ for all s.
(c) The total cost is

$$C = \begin{cases} h & \text{if } M \le h, \\ h + a + b(M - h) & \text{if } M > h. \end{cases}$$

By the lack-of-memory property of the exponential distribution, $\mathbb{E}(M-h \mid M>h)=1$, whence

$$\mathbb{E}C = h\mathbb{P}(M \le h) + (h+a+b)\mathbb{P}(M>h) = h + (a+b)e^{-h}.$$

This is a minimum at $h=\log(a+b)$ when $a+b \ge 1$, and at $h=0$ when $a+b<1$. The latter is evidence against building the barrier when $a+b<1$.

6. If X and Y are independent, we may take $g=f_X$ and $h=f_Y$. Suppose conversely that $f(x,y)=g(x)h(y)$. Then

$$f_X(x) = g(x)\int_{-\infty}^{\infty} h(y)\,dy, \quad f_Y(y) = h(y)\int_{-\infty}^{\infty} g(x)\,dx$$

and

$$1 = \int_{-\infty}^{\infty} f_Y(y)\,dy = \int_{-\infty}^{\infty} g(x)\,dx \int_{-\infty}^{\infty} h(y)\,dy.$$

Hence $f_X(x)f_Y(y) = g(x)h(y) = f(x,y)$, so that X and Y are independent.

The given f cannot be factorized as a product since the domain depends on whether or not $x<y$.

7. They are not independent since $\mathbb{P}(Y<1,\ X>1)=0$ whereas $\mathbb{P}(Y<1)>0$ and $\mathbb{P}(X>1)>0$. As for the marginals,

$$f_X(x) = \int_x^{\infty} 2e^{-x-y}\,dy = 2e^{-2x}, \qquad f_Y(y) = \int_0^y 2e^{-x-y}\,dx = 2e^{-y}(1-e^{-y}),$$

for $x, y \ge 0$. Finally,

$$\mathbb{E}(XY) = \int_{x=0}^{\infty}\int_{y=x}^{\infty} xy2e^{-x-y}\,dx\,dy = 1$$

and $\mathbb{E}(X)=\frac{1}{2}$, $\mathbb{E}(Y)=\frac{3}{2}$, implying that $\text{cov}(X,Y)=\frac{1}{4}$.

8. As in Example (4.13.1), the desired property holds if and only if the length X of the chord satisfies $X \le \sqrt{3}$. Writing R for the distance from P to the centre O, and Θ for the acute angle between the chord and the line OP, we have that $X = 2\sqrt{1-R^2\sin^2\Theta}$, and therefore $\mathbb{P}(X \le \sqrt{3}) = \mathbb{P}(R\sin\Theta \ge \frac{1}{2})$. The answer is therefore

$$\mathbb{P}\left(R \ge \frac{1}{2\sin\Theta}\right) = \frac{2}{\pi}\int_0^{\frac{1}{2}\pi} \mathbb{P}\left(R \ge \frac{1}{2\sin\theta}\right)d\theta,$$

which equals $\frac{2}{3} - \sqrt{3}/(2\pi)$ in case (a) and $\frac{2}{3} + \pi^{-1}\log\tan(\pi/12)$ in case (b).

9. Evidently,

$$\mathbb{E}(U) = \mathbb{P}(Y \le g(X)) = \iint_{\substack{0\le x,y\le 1\\ y\le g(x)}} dx\,dy = \int_0^1 g(x)\,dx,$$

$$\mathbb{E}(V) = \mathbb{E}(g(X)) = \int_0^1 g(x)\,dx,$$

$$\mathbb{E}(W) = \tfrac{1}{2}\int_0^1 \{g(x)+g(1-x)\}\,dx = \int_0^1 g(x)\,dx.$$

Secondly,

$$\mathbb{E}(U^2) = \mathbb{E}(U) = J, \qquad \mathbb{E}(V^2) = \int_0^1 g(x)^2\,dx \le J \quad \text{since } g \le 1,$$

$$\begin{aligned}\mathbb{E}(W^2) &= \tfrac{1}{4}\left\{2\int_0^1 g(x)^2\,dx + 2\int_0^1 g(x)g(1-x)\,dx\right\}\\ &= \mathbb{E}(V^2) - \tfrac{1}{2}\int_0^1 g(x)\{g(x)-g(1-x)\}\,dx\\ &= \mathbb{E}(V^2) - \tfrac{1}{4}\int_0^1 \{g(x)-g(1-x)\}^2\,dx \le \mathbb{E}(V^2).\end{aligned}$$

Hence $\text{var}(W) \le \text{var}(V) \le \text{var}(U)$.

10. Clearly the claim is true for $n = 1$, since the $\Gamma(\lambda, 1)$ distribution *is* the exponential distribution. Suppose it is true for $n \le k$ where $k \ge 1$, and consider the case $n = k+1$. Writing f_n for the density function of S_n, we have by the convolution formula (4.8.2) that

$$f_{k+1}(x) = \int_0^x f_k(y) f_1(x-y)\,dy = \int_0^x \frac{\lambda^k}{\Gamma(k)} y^{k-1} e^{-\lambda y} \lambda e^{-\lambda(x-y)}\,dy = \frac{\lambda^{k+1}e^{-\lambda x}}{\Gamma(k)} \int_0^x y^{k-1}\,dy,$$

which is easily seen to be the $\Gamma(\lambda, k+1)$ density function.

11. (a) Let $Z_1, Z_2, \dots, Z_{m+n}$ be independent exponential variables with parameter λ. Then, by Problem (4.14.10), $X' = Z_1 + \cdots + Z_m$ is $\Gamma(\lambda, m)$, $Y' = Z_{m+1} + \cdots + Z_{m+n}$ is $\Gamma(\lambda, n)$, and $X' + Y'$ is $\Gamma(\lambda, m+n)$. The pair (X, Y) has the same joint distribution as the pair (X', Y'), and therefore $X + Y$ has the same distribution as $X' + Y'$, i.e. $\Gamma(\lambda, m+n)$.

(b) Using the transformation $u = x + y$, $v = x/(x+y)$, with inverse $x = uv$, $y = u(1-v)$, and Jacobian

$$J = \begin{vmatrix} v & u \\ 1-v & -u \end{vmatrix} = -u,$$

we find that $U = X + Y$, $V = X/(X+Y)$ have joint density function

$$\begin{aligned}f_{U,V}(u,v) = f_{X,Y}\big(uv, u(1-v)\big)|u| &= \frac{\lambda^{m+n}}{\Gamma(m)\Gamma(n)}(uv)^{m-1}\{u(1-v)\}^{n-1}e^{-\lambda u}u\\ &= \left\{\frac{\lambda^{m+n}}{\Gamma(m+n)}u^{m+n-1}e^{-\lambda u}\right\}\left\{\frac{v^{m-1}(1-v)^{n-1}}{B(m,n)}\right\}\end{aligned}$$

for $u \ge 0$, $0 \le v \le 1$. Hence U and V are independent, U being $\Gamma(\lambda, m+n)$, and V having the beta distribution with parameters m and n.

(c) Integrating by parts,

$$\begin{aligned}\mathbb{P}(X > t) &= \int_t^\infty \frac{\lambda^m}{(m-1)!}x^{m-1}e^{-\lambda x}\,dx\\ &= \left[-\frac{\lambda^{m-1}}{(m-1)!}x^{m-1}e^{-\lambda x}\right]_t^\infty + \int_t^\infty \frac{\lambda^{m-1}}{(m-2)!}x^{m-2}e^{-\lambda x}\,dx\\ &= e^{-\lambda t}\frac{(\lambda t)^{m-1}}{(m-1)!} + \mathbb{P}(X' > t)\end{aligned}$$

where X' is $\Gamma(\lambda, m-1)$. Hence, by induction,

$$\mathbb{P}(X > t) = \sum_{k=0}^{m-1} e^{-\lambda t}\frac{(\lambda t)^k}{k!} = \mathbb{P}(Z < m).$$

(d) This may be achieved by the usual change of variables technique. Alternatively, reflect that, using the notation and result of part (b), the invertible mapping $u = x + y$, $v = x/(x + y)$ maps a pair X, Y of independent ($\Gamma(\lambda, m)$ and $\Gamma(\lambda, n)$) variables to a pair U, V of independent ($\Gamma(\lambda, m + n)$ and $B(m, n)$) variables. Now $UV = X$, so that (figuratively)

$$\text{“}\Gamma(\lambda, m+n) \times B(m, n) = \Gamma(\lambda, m)\text{”}.$$

Replace n by $n - m$ to obtain the required conclusion.

12. (a) $Z = X_1^2$ satisfies

$$f_Z(z) = \frac{d}{dz}\mathbb{P}(X_1^2 \le z) = \frac{d}{dz}\left\{2\int_0^{\sqrt{z}} \frac{1}{\sqrt{2\pi}} e^{-\frac{1}{2}u^2}\, du\right\} = \frac{1}{\sqrt{2\pi}} z^{-\frac{1}{2}} e^{-\frac{1}{2}z}, \qquad z \ge 0,$$

the $\Gamma(\frac{1}{2}, \frac{1}{2})$ or $\chi^2(1)$ density function.

(b) If $z \ge 0$, $Z = X_1^2 + X_2^2$ satisfies

$$\mathbb{P}(Z \le z) = \mathbb{P}(X_1^2 + X_2^2 \le z) = \iint_{x^2+y^2 \le z} \frac{1}{2\pi} e^{-\frac{1}{2}(x^2+y^2)}\, dx\, dy$$

$$= \int_{r=0}^{\sqrt{z}} \int_{\theta=0}^{2\pi} \frac{1}{2\pi} e^{-\frac{1}{2}r^2} r\, dr\, d\theta = 1 - e^{-\frac{1}{2}z},$$

the $\chi^2(2)$ distribution function.

(c) One way is to work in n-dimensional polar coordinates! Alternatively use induction. It suffices to show that if X and Y are independent, X being $\chi^2(n)$ and Y being $\chi^2(2)$ where $n \ge 1$, then $Z = X + Y$ is $\chi^2(n + 2)$. However, by the convolution formula (4.8.2),

$$f_Z(z) = \int_0^z \left\{\frac{2^{-n/2}}{\Gamma(n/2)} x^{\frac{1}{2}n-1} e^{-\frac{1}{2}x}\right\}\left\{\tfrac{1}{2} e^{-\frac{1}{2}(z-x)}\right\} dx = ce^{-\frac{1}{2}z} z^{\frac{1}{2}n}, \qquad z \ge 0,$$

for some constant c. This is the $\chi^2(n + 2)$ density function as required.

13. Concentrate on where x occurs in $f_{X|Y}(x \mid y)$; any multiplicative constant can be sorted out later:

$$f_{X|Y}(x \mid y) = \frac{f_{X,Y}(x, y)}{f_Y(y)} = c_1(y) \exp\left\{-\frac{1}{2(1-\rho^2)}\left(\frac{x^2}{\sigma_1^2} - \frac{2x\mu_1}{\sigma_1^2} - \frac{2\rho x(y - \mu_2)}{\sigma_1\sigma_2}\right)\right\}$$

by Example (4.5.9), where $c_1(y)$ depends on y only. Hence

$$f_{X|Y}(x \mid y) = c_2(y) \exp\left\{-\frac{\left[x - \mu_1 - \rho\sigma_1(y - \mu_2)/\sigma_2\right]^2}{2(1-\rho^2)\sigma_1^2}\right\}, \qquad x \in \mathbb{R},$$

for some $c_2(y)$. This is the normal density function with mean $\mu_1 + \rho\sigma_1(y - \mu_2)/\sigma_2$ and variance $\sigma_1^2(1 - \rho^2)$. See also Exercise (4.8.7).

14. Set $u = y/x$, $v = x$, with inverse $x = v$, $y = uv$, and $|J| = |v|$. Hence the pair $U = Y/X$, $V = X$ has joint density $f_{U,V}(u, v) = f_{X,Y}(v, uv)|v|$ for $-\infty < u, v < \infty$. Therefore $f_U(u) = \int_{-\infty}^{\infty} f(v, uv)|v|\, dv$.

15. By the result of Problem (4.14.14), $U = Y/X$ has density function

$$f_U(u) = \int_{-\infty}^{\infty} f(y)f(uy)|y|\, dy,$$

and therefore it suffices to show that U has the Cauchy distribution if and only if $Z = \tan^{-1} U$ is uniform on $(-\frac{1}{2}\pi, \frac{1}{2}\pi)$. Clearly

$$\mathbb{P}(Z \le \theta) = \mathbb{P}(U \le \tan\theta), \quad -\tfrac{1}{2}\pi < \theta < \tfrac{1}{2}\pi,$$

whence $f_Z(\theta) = f_U(\tan\theta)\sec^2\theta$. Therefore $f_Z(\theta) = \pi^{-1}$ (for $|\theta| < \frac{1}{2}\pi$) if and only if

$$f_U(u) = \frac{1}{\pi(1+u^2)}, \qquad -\infty < u < \infty.$$

When f is the $N(0, 1)$ density,

$$\int_{-\infty}^{\infty} f(x)f(xy)|x|\,dx = 2\int_0^{\infty} \frac{1}{2\pi} e^{-\frac{1}{2}x^2(1+y^2)} x\,dx,$$

which is easily integrated directly to obtain the Cauchy density. In the second case, we have the following integral:

$$\int_{-\infty}^{\infty} \frac{a^2|x|}{(1+x^4)(1+x^4y^4)}\,dx.$$

In this case, make the substitution $z = x^2$ and expand as partial fractions.

16. The transformation $x = r\cos\theta$, $y = r\sin\theta$ has Jacobian $J = r$, so that

$$f_{R,\Theta}(r,\theta) = \frac{1}{2\pi} r e^{-\frac{1}{2}r^2}, \qquad r > 0, \quad 0 \le \theta < 2\pi.$$

Therefore R and Θ are independent, Θ being uniform on $[0, 2\pi)$, and R^2 having distribution function

$$\mathbb{P}(R^2 \le a) = \int_0^{\sqrt{a}} r e^{-\frac{1}{2}r^2}\,dr = 1 - e^{-\frac{1}{2}a};$$

this is the exponential distribution with parameter $\frac{1}{2}$ (otherwise known as $\Gamma(\frac{1}{2}, 1)$ or $\chi^2(2)$). The density function of R is $f_R(r) = re^{-\frac{1}{2}r^2}$ for $r > 0$.

Now, by symmetry,

$$\mathbb{E}\left(\frac{X^2}{R^2}\right) = \frac{1}{2}\mathbb{E}\left(\frac{X^2+Y^2}{R^2}\right) = \frac{1}{2}.$$

In the first octant, i.e. in $\{(x, y) : 0 \le y \le x\}$, we have $\min\{x, y\} = y$, $\max\{x, y\} = x$. The joint density $f_{X,Y}$ is invariant under rotations, and hence the expectation in question is

$$8\int_{0\le y\le x} \frac{y}{x} f_{X,Y}(x,y)\,dx\,dy = 8\int_{\theta=0}^{\pi/4}\int_{r=0}^{\infty} \frac{\tan\theta}{2\pi} re^{-\frac{1}{2}r^2}\,dr\,d\theta = \frac{2}{\pi}\log 2.$$

17. (i) Using independence,

$$\mathbb{P}(U \le u) = 1 - \mathbb{P}(X > u, Y > u) = 1 - \big(1 - F_X(u)\big)\big(1 - F_Y(u)\big).$$

Similarly

$$\mathbb{P}(V \le v) = \mathbb{P}(X \le v, Y \le v) = F_X(v)F_Y(v).$$

(ii) (a) By (i), $\mathbb{P}(U \le u) = 1 - e^{-2u}$ for $u \ge 0$.

(b) Also, $Z = X + \frac{1}{2}Y$ satisfies

$$\mathbb{P}(Z > v) = \int_0^\infty \mathbb{P}\big(Y > 2(v-x)\big) f_X(x)\,dx = \int_0^v e^{-2(v-x)} e^{-x}\,dx + \int_v^\infty e^{-x}\,dx$$
$$= e^{-2v}(e^v - 1) + e^{-v} = 1 - (1 - e^{-v})^2 = \mathbb{P}(V > v).$$

Therefore $\mathbb{E}(V) = \mathbb{E}(X) + \frac{1}{2}\mathbb{E}(Y) = \frac{3}{2}$, and $\text{var}(V) = \text{var}(X) + \frac{1}{4}\,\text{var}(Y) = \frac{5}{4}$ by independence.

18. (a) We have that

$$\mathbb{P}(X \le Y) = \int_0^\infty \mathbb{P}(X \le y)\mu e^{-\mu y}\,dy = \int_0^\infty (1 - e^{-\lambda y})\mu e^{-\mu y}\,dy = \frac{\lambda}{\mu + \lambda}.$$

(b) Clearly, for $w > 0$,

$$\mathbb{P}(U \le u, W > w) = \mathbb{P}(U \le u, W > w, X \le Y) + \mathbb{P}(U \le u, W > w, X > Y).$$

Now

$$\mathbb{P}(U \le u, W > w, X \le Y) = \mathbb{P}(X \le u, Y > X + w) = \int_0^u \lambda e^{-\lambda x} e^{-\mu(x+w)}\,dx$$
$$= \frac{\lambda}{\lambda + \mu} e^{-\mu w}(1 - e^{-(\lambda+\mu)u})$$

and similarly

$$\mathbb{P}(U \le u, W > w, X > Y) = \frac{\mu}{\lambda + \mu} e^{-\lambda w}(1 - e^{-(\lambda+\mu)u}).$$

Hence, for $0 \le u \le u + w < \infty$,

$$\mathbb{P}(U \le u, W > w) = (1 - e^{-(\lambda+\mu)u})\left(\frac{\lambda}{\lambda + \mu} e^{-\mu w} + \frac{\mu}{\lambda + \mu} e^{-\lambda w}\right),$$

an expression which factorizes into the product of a function of u with a function of w. Hence U and W are independent.

19. $U = X + Y$, $V = X$ have joint density function $f_Y(u - v) f_X(v)$, $0 \le v \le u$. Hence

$$f_{V|U}(v \mid u) = \frac{f_{U,V}(u, v)}{f_U(u)} = \frac{f_Y(u - v) f_X(v)}{\int_0^u f_Y(u - y) f_X(y)\,dy}.$$

(a) We are given that $f_{V|U}(v \mid u) = u^{-1}$ for $0 \le v \le u$; then

$$f_Y(u - v) f_X(v) = \frac{1}{u}\int_0^u f_Y(u - y) f_X(y)\,dy$$

is a function of u alone, implying that

$$\begin{aligned} f_Y(u - v) f_X(v) &= f_Y(u) f_X(0) && \text{by setting } v = 0 \\ &= f_Y(0) f_X(u) && \text{by setting } v = u. \end{aligned}$$

In particular $f_Y(u)$ and $f_X(u)$ differ only by a multiplicative constant; they are both density functions, implying that this constant is 1, and $f_X = f_Y$. Substituting this throughout the above display, we

find that $g(x) = f_X(x)/f_X(0)$ satisfies $g(0) = 1$, g is continuous, and $g(u - v)g(v) = g(u)$ for $0 \le v \le u$. From the hint, $g(x) = e^{-\lambda x}$ for $x \ge 0$ for some $\lambda > 0$ (remember that g is integrable).
(b) Arguing similarly, we find that

$$f_Y(u - v)f_X(v) = \frac{c}{u^{\alpha+\beta-1}} v^{\alpha-1}(u - v)^{\beta-1} \int_0^u f_Y(u - y)f_X(y)\,dy$$

for $0 \le v \le u$ and some constant c. Setting $f_X(v) = \chi(v)v^{\alpha-1}$, $f_Y(y) = \eta(y)y^{\beta-1}$, we obtain $\eta(u - v)\chi(v) = h(u)$ for $0 \le v \le u$, and for some function h. Arguing as before, we find that η and χ are proportional to negative exponential functions, so that X and Y have gamma distributions.

20. We are given that U is uniform on [0, 1], so that $0 \le X, Y \le \frac{1}{2}$ almost surely. For $0 < \epsilon < \frac{1}{4}$,

$$\epsilon = \mathbb{P}(X + Y < \epsilon) \le \mathbb{P}(X < \epsilon, Y < \epsilon) = \mathbb{P}(X < \epsilon)^2,$$

and similarly

$$\epsilon = \mathbb{P}(X + Y > 1 - \epsilon) \le \mathbb{P}(X > \tfrac{1}{2} - \epsilon, Y > \tfrac{1}{2} - \epsilon) = \mathbb{P}(X > \tfrac{1}{2} - \epsilon)^2,$$

implying that $\mathbb{P}(X < \epsilon) \ge \sqrt{\epsilon}$ and $\mathbb{P}(X > \frac{1}{2} - \epsilon) \ge \sqrt{\epsilon}$. Now

$$\begin{aligned}2\epsilon = \mathbb{P}(\tfrac{1}{2} - \epsilon < X + Y < \tfrac{1}{2} + \epsilon) &\ge \mathbb{P}(X > \tfrac{1}{2} - \epsilon, Y < \epsilon) + \mathbb{P}(X < \epsilon, Y > \tfrac{1}{2} - \epsilon)\\ &= 2\mathbb{P}(X > \tfrac{1}{2} - \epsilon)\mathbb{P}(X < \epsilon) \ge 2(\sqrt{\epsilon})^2.\end{aligned}$$

Therefore all the above inequalities are in fact equalities, implying that $\mathbb{P}(X < \epsilon) = \mathbb{P}(X > \frac{1}{2} - \epsilon) = \sqrt{\epsilon}$ if $0 < \epsilon < \frac{1}{4}$. Hence a contradiction:

$$\tfrac{1}{8} = \mathbb{P}(X + Y < \tfrac{1}{8}) = \mathbb{P}(X, Y < \tfrac{1}{8}) - \mathbb{P}(X, Y < \tfrac{1}{8},\ X + Y \ge \tfrac{1}{8}) < \mathbb{P}(X < \tfrac{1}{8}, Y < \tfrac{1}{8}) = \tfrac{1}{8}.$$

21. Evidently

$$\mathbb{P}(X_{(1)} \le y_1, \dots, X_{(n)} \le y_n) = \sum_{\pi} \mathbb{P}\left(X_{\pi_1} \le y_1, \dots, X_{\pi_n} \le y_n, X_{\pi_1} < \cdots < X_{\pi_n}\right)$$

where the sum is over all permutations $\pi = (\pi_1, \pi_2, \dots, \pi_n)$ of $(1, 2, \dots, n)$. By symmetry, each term in the sum is equal, whence the sum equals

$$n!\,\mathbb{P}(X_1 \le y_1, \dots, X_n \le y_n, X_1 < X_2 < \cdots < X_n).$$

The integral form is then immediate. The joint density function is, by its definition, the integrand.

22. (a) In the notation of Problem (4.14.21), the joint density function of $X_{(2)}, \dots, X_{(n)}$ is

$$\begin{aligned}g_2(y_2, \dots, y_n) &= \int_{-\infty}^{y_2} g(y_1, \dots, y_n)\,dy_1\\ &= n!\,L(y_2, \dots, y_n)F(y_2)f(y_2)f(y_3)\cdots f(y_n)\end{aligned}$$

where F is the common distribution function of the X_i. Similarly $X_{(3)}, \dots, X_{(n)}$ have joint density

$$g_3(y_3, \dots, y_n) = \tfrac{1}{2}n!\,L(y_3, \dots, y_n)F(y_3)^2 f(y_3)\cdots f(y_n),$$

and by iteration, $X_{(k)}, \dots, X_{(n)}$ have joint density

$$g_k(y_k, \dots, y_n) = \frac{n!}{(k-1)!} L(y_k, \dots, y_n) F(y_k)^{k-1} f(y_k) \cdots f(y_n).$$

We now integrate over $y_n, y_{n-1}, \dots, y_{k+1}$ in turn, arriving at

$$f_{X_{(k)}}(y_k) = \frac{n!}{(k-1)!\,(n-k)!} F(y_k)^{k-1} \{1 - F(y_k)\}^{n-k} f(y_k).$$

(b) It is neater to argue directly. Fix x, and let I_r be the indicator function of the event $\{X_r \le x\}$, and let $S = I_1 + I_2 + \cdots + I_n$. Then S is distributed as $\text{bin}(n, F(x))$, and

$$\mathbb{P}(X_{(k)} \le x) = \mathbb{P}(S \ge k) = \sum_{l=k}^{n} \binom{n}{l} F(x)^l (1 - F(x))^{n-l}.$$

Differentiate to obtain, with $F = F(x)$,

$$\begin{aligned} f_{X_{(k)}}(x) &= \sum_{l=k}^{n} \binom{n}{l} f(x) \left\{ l F^{l-1}(1-F)^{n-l} - (n-l) F^l (1-F)^{n-l-1} \right\} \\ &= k \binom{n}{k} f(x) F^{k-1} (1-F)^{n-k} \end{aligned}$$

by successive cancellation of the terms in the series.

23. Using the result of Problem (4.14.21), the joint density function is $g(\mathbf{y}) = n!\,L(\mathbf{y}) T^{-n}$ for $0 \le y_i \le T$, $1 \le i \le n$, where $\mathbf{y} = (y_1, y_2, \dots, y_n)$.

24. (a) We make use of Problems (4.14.22)–(4.14.23). The density function of $X_{(k)}$ is $f_k(x) = k\binom{n}{k} x^{k-1}(1-x)^{n-k}$ for $0 \le x \le 1$, so that the density function of $nX_{(k)}$ is

$$\frac{1}{n} f_k(x/n) = \frac{k}{k!} \frac{n(n-1)\cdots(n-k+1)}{n^k} x^{k-1} \left(1 - \frac{x}{n}\right)^{n-k} \to \frac{1}{(k-1)!} x^{k-1} e^{-x}$$

as $n \to \infty$. The limit is the $\Gamma(1, k)$ density function.

(b) For an increasing sequence $x_{(1)}, x_{(2)}, \dots, x_{(n)}$ in $[0, 1]$, we define the sequence $u_n = -n \log x_{(n)}$, $u_k = -k \log(x_{(k)}/x_{(k+1)})$ for $1 \le k < n$. This mapping has inverse

$$x_{(n)} = e^{-u_n/n}, \quad x_{(k)} = x_{(k+1)} e^{-u_k/k} = \exp\left\{ -\sum_{i=k}^{n} i^{-1} u_i \right\},$$

with Jacobian $J = (-1)^n e^{-u_1-u_2-\cdots-u_n}/n!$. Applying the mapping to the sequence $X_{(1)}, X_{(2)}, \dots, X_{(n)}$ we obtain a family $U_1, U_2, \dots, U_n$ of random variables with joint density $g(u_1, u_2, \dots, u_n) = e^{-u_1-u_2-\cdots-u_n}$ for $u_i \ge 0$, $1 \le i \le n$, yielding that the U_i are independent and exponentially distributed, with parameter 1. Finally $\log X_{(k)} = -\sum_{i=k}^{n} i^{-1} U_i$.

(c) In the notation of part (b), $Z_k = \exp(-U_k)$ for $1 \le k \le n$, a collection of independent variables. Finally, U_k is exponential with parameter 1, and therefore

$$\mathbb{P}(Z_k \le z) = \mathbb{P}(U_k \ge -\log z) = e^{\log z} = z, \qquad 0 < z \le 1.$$

25. (i) (X_1, X_2, X_3) is uniformly distributed over the unit cube of $\mathbb{R}^3$, and the answer is therefore the volume of that set of points (x_1, x_2, x_3) of the cube which allow a triangle to be formed. A triangle

is impossible if $x_1 \geq x_2 + x_3$, or $x_2 \geq x_1 + x_3$, or $x_3 \geq x_1 + x_2$. This defines three regions of the cube which overlap only at the origin. Each of these regions is a tetrahedron; for example, the region $x_3 \geq x_1 + x_2$ is an isosceles tetrahedron with vertices $(0, 0, 0)$, $(1, 0, 1)$, $(0, 1, 1)$, $(0, 0, 1)$, with volume $\frac{1}{6}$. Hence the required probability is $1 - 3 \cdot \frac{1}{6} = \frac{1}{2}$.
(ii) The rods of length $x_1, x_2, \dots, x_n$ fail to form a polygon if either $x_1 \geq x_2 + x_3 + \cdots + x_n$ or any of the other $n - 1$ corresponding inequalities hold. We therefore require the volume of the n-dimensional hypercube with n corners removed. The inequality $x_1 \geq x_2 + x_3 + \cdots + x_n$ corresponds to the convex hull of the points $(0, 0, \dots, 0)$, $(1, 0, \dots, 0)$, $(1, 1, 0, \dots, 0)$, $(1, 0, 1, 0, \dots, 0)$, $\dots$, $(1, 0, \dots, 0, 1)$. Mapping $x_1 \mapsto 1 - x_1$, we see that this has the same volume V_n as has the convex hull of the origin $\mathbf{0}$ together with the n unit vectors $\mathbf{e}_1, \mathbf{e}_2, \dots, \mathbf{e}_n$. Clearly $V_2 = \frac{1}{2}$, and we claim that $V_n = 1/n!$. Suppose this holds for $n < k$, and consider the case $n = k$. Then

$$V_k = \int_0^1 dx_1\, V_{k-1}(\mathbf{0}, x_1\mathbf{e}_2, \dots, x_1\mathbf{e}_k)$$

where $V_{k-1}(\mathbf{0}, x_1\mathbf{e}_2, \dots, x_1\mathbf{e}_k)$ is the $(k-1)$-dimensional volume of the convex hull of $\mathbf{0}, x_1\mathbf{e}_2, \dots, x_1\mathbf{e}_k$. Now

$$V_{k-1}(\mathbf{0}, x_1\mathbf{e}_2, \dots, x_1\mathbf{e}_k) = x_1^{k-1} V_{k-1} = \frac{x_1^{k-1}}{(k-1)!},$$

so that

$$V_k = \int_0^1 \frac{x_1^{k-1}}{(k-1)!}\, dx_1 = \frac{1}{k!}.$$

The required probability is therefore $1 - n/(n!) = 1 - \{(n-1)!\}^{-1}$.

26. (i) The lengths of the pieces are $U = \min\{X_1, X_2\}$, $V = |X_1 - X_2|$, $W = 1 - U - V$, and we require that $U < V + W$, etc, as in the solution to Problem (4.14.25). In terms of the X_i we require

$$\begin{aligned} \text{either}: \quad & X_1 < \tfrac{1}{2}, \quad |X_1 - X_2| < \tfrac{1}{2}, \quad 1 - X_2 < \tfrac{1}{2}, \\ \text{or}: \quad & X_2 < \tfrac{1}{2}, \quad |X_1 - X_2| < \tfrac{1}{2}, \quad 1 - X_1 < \tfrac{1}{2}. \end{aligned}$$

Plot the corresponding region of $\mathbb{R}^2$. One then sees that the area of the region is $\frac{1}{4}$, which is therefore the probability in question.
(ii) The pieces may form a polygon if no piece is as long as the sum of the lengths of the others. Since the total length is 1, this requires that each piece has length less than $\frac{1}{2}$. Neglecting certain null events, this fails to occur if and only if the disjoint union of events $A_0 \cup A_1 \cup \cdots \cup A_n$ occurs, where

$$A_0 = \{\text{no break in } (0, \tfrac{1}{2}]\}, \quad A_k = \{\text{no break in } (X_k, X_k + \tfrac{1}{2}]\} \text{ for } 1 \leq k \leq n;$$

remember that there is a permanent break at 1. Now $\mathbb{P}(A_0) = (\frac{1}{2})^n$, and for $k \geq 1$,

$$\mathbb{P}(A_k) = \int_0^1 \mathbb{P}(A_k \mid X_k = x)\, dx = \int_0^{\frac{1}{2}} (\tfrac{1}{2})^{n-1}\, dx = (\tfrac{1}{2})^n;$$

Hence $\mathbb{P}(A_0 \cup A_1 \cup \cdots \cup A_n) = (n+1)2^{-n}$ whence the required probability is $1 - (n+1)2^{-n}$.

27. (a) The function $g(t) = (t^p/p) + (t^{-q}/q)$, for $t > 0$, has a unique minimum at $t = 1$, and hence $g(t) \geq g(1) = 1$ for $t > 0$. Substitute $t = x^{1/q}y^{-1/p}$ where

$$x = \frac{|X|}{\{\mathbb{E}|X^p|\}^{1/p}}, \quad y = \frac{|Y|}{\{\mathbb{E}|Y^q|\}^{1/q}},$$

(we may as well suppose that $\mathbb{P}(XY = 0) \neq 1$) to find that

$$\frac{|X|^p}{p\mathbb{E}|X^p|} + \frac{|Y|^q}{q\mathbb{E}|Y^q|} \geq \frac{|XY|}{\{\mathbb{E}|X^p|\}^{1/p}\{\mathbb{E}|Y^q|\}^{1/q}}.$$

Hölder's inequality follows by taking expectations.

(b) We have, with $Z = |X + Y|$,

$$\begin{aligned}\mathbb{E}(Z^p) &= \mathbb{E}(Z \cdot Z^{p-1}) \leq \mathbb{E}(|X|Z^{p-1}) + \mathbb{E}(|Y|Z^{p-1}) \\ &\leq \{\mathbb{E}|X^p|\}^{1/p}\{\mathbb{E}(Z^p)\}^{1/q} + \{\mathbb{E}|Y^p|\}^{1/p}\{\mathbb{E}(Z^p)\}^{1/q}\end{aligned}$$

by Hölder's inequality, where $p^{-1} + q^{-1} = 1$. Divide by $\{\mathbb{E}(Z^p)\}^{1/q}$ to get the result.

28. Apply the Cauchy–Schwarz inequality to $|Z|^{\frac{1}{2}(b-a)}$ and $|Z|^{\frac{1}{2}(b+a)}$, where $0 \leq a \leq b$, to obtain $\{\mathbb{E}|Z^b|\}^2 \leq \mathbb{E}|Z^{b-a}|\,\mathbb{E}|Z^{b+a}|$. Now take logarithms: $2g(b) \leq g(b-a) + g(b+a)$ for $0 \leq a \leq b$. Also $g(p) \to g(0) = 1$ as $p \downarrow 0$ (by dominated convergence). These two properties of g imply that g is convex on intervals of the form $[0, M)$ if it is finite on this interval. The reference to dominated convergence may be avoided by using Hölder instead of Cauchy–Schwarz.

By convexity, $g(x)/x$ is non-decreasing in x, and therefore $g(r)/r \geq g(s)/s$ if $0 < s \leq r$.

For the second part, let

$$\lambda = \int \cdots \int \left\{ \prod_{k=1}^{n} |x_k^{r_k}| - \prod_{k=1}^{n} |x_k^{s}| \right\} dF(x_1)\, dF(x_2) \cdots dF(x_n).$$

The integral† is unchanged by any permutation of the variables $x_1, x_2, \ldots, x_n$, and therefore it may replaced by its arithmetic mean over all such permutations, denoted $\langle \cdot \rangle$. That is

$$\lambda = \int \cdots \int \left\{ \langle |x_1^{r_1}|, \ldots, |x_n^{r_n}| \rangle - \prod_{k=1}^{n} |x_k^{s}| \right\} dF(x_1)\, dF(x_2) \cdots dF(x_n).$$

We now replace the arithmetic mean by the geometric mean, namely $\{|x_1| \cdots |x_n|\}^s$, to obtain the given inequality by the arithmetic/geometric mean inequality,

29. Assume that X, Y, Z are jointly continuously distributed with joint density function f. Then

$$\mathbb{E}(X \mid Y = y, Z = z) = \int x f_{X|Y,Z}(x \mid y, z)\, dx = \int x \frac{f(x, y, z)}{f_{Y,Z}(y, z)}\, dx.$$

Hence

$$\begin{aligned}\mathbb{E}\{\mathbb{E}(X \mid Y, Z) \mid Y = y\} &= \int \mathbb{E}(X \mid Y = y, Z = z) f_{Z|Y}(z \mid y)\, dz \\ &= \iint x \frac{f(x, y, z)}{f_{Y,Z}(y, z)} \frac{f_{Y,Z}(y, z)}{f_Y(y)}\, dx\, dz \\ &= \iint x \frac{f(x, y, z)}{f_Y(y)}\, dx\, dz = \mathbb{E}(X \mid Y = y).\end{aligned}$$

Alternatively, use the general properties of conditional expectation as laid down in Section 4.6.

---

†Think of dF_X as $f_X(x)\, dx$ if you are unfamiliar with this form of integral, to be defined in Section 5.6.

30. The first car to arrive in a car park of length $x+1$ effectively divides it into two disjoint parks of length Y and $x-Y$, where Y is the position of the car's leftmost point. Now Y is uniform on $[0,x]$, and the formula follows by conditioning on Y. Laplace transforms are the key to exploring the asymptotic behaviour of $m(x)/x$ as $x\to\infty$.

31. (a) If the needle were of length d, the answer would be $2/\pi$ as before. Think about the new needle as being obtained from a needle of length d by dividing it into two parts, an 'active' part of length L, and a 'passive' part of length $d-L$, and then counting only intersections involving the active part. The chance of an 'active intersection' is now $(2/\pi)(L/d)=2L/(\pi d)$.

(b) As in part (a), the angle between the line and the needle is independent of the distance between the line and the needle's centre, each having the same distribution as before. The answer is therefore unchanged.

(c) The following argument lacks a little rigour, which may be supplied as a consequence of the statement that S has finite length. For $\epsilon>0$, let $x_1,x_2,\dots,x_n$ be points on S, taken in order along S, such that x_0 and x_n are the endpoints of S, and $|x_{i+1}-x_i|<\epsilon$ for $0\le i<n$; $|x-y|$ denotes the Euclidean distance from x to y. Let J_i be the straight line segment joining x_i to x_{i+1}, and let I_i be the indicator function of $\{J_i\cap\lambda\ne\varnothing\}$. If ϵ is sufficiently small, the total number of intersections between $J_0\cup J_1\cup\cdots\cup J_{n-1}$ and S has mean

$$\sum_{i=0}^{n-1}\mathbb{E}(I_i)=\frac{2}{\pi d}\sum_{i=0}^{n-1}|x_{i+1}-x_i|$$

by part (b). In the limit as $\epsilon\downarrow 0$, we have that $\sum_i\mathbb{E}(I_i)$ approaches the required mean, while

$$\frac{2}{\pi d}\sum_{i=0}^{n-1}|x_{i+1}-x_i|\to\frac{2L(S)}{\pi d}.$$

32. (i) Fix Cartesian axes within the gut in question. Taking one end of the needle as the origin, the other end is uniformly distributed on the unit sphere of $\mathbb{R}^3$. With the X-ray plate parallel to the x/y-plane, the projected length V of the needle satisfies $V\ge v$ if and only if $|Z|\le\sqrt{1-v^2}$, where Z is the (random) z-coordinate of the 'loose' end of the needle. Hence, for $0\le v\le 1$,

$$\mathbb{P}(V\ge v)=\mathbb{P}\big(-\sqrt{1-v^2}\le Z\le\sqrt{1-v^2}\big)=\frac{4\pi\sqrt{1-v^2}}{4\pi}=\sqrt{1-v^2},$$

since $4\pi\sqrt{1-v^2}$ is the surface area of that part of the unit sphere satisfying $|z|\le\sqrt{1-v^2}$ (use Archimedes's theorem of the circumscribing cylinder, or calculus). Therefore V has density function $f_V(v)=v/\sqrt{1-v^2}$ for $0\le v\le 1$.

(ii) Draw a picture, if you can. The lengths of the projections are determined by the angle Θ between the plane of the cross and the X-ray plate, together with the angle Ψ of rotation of the cross about an axis normal to its arms. Assume that Θ and Ψ are independent and uniform on $[0,\frac{1}{2}\pi]$. If the axis system has been chosen with enough care, we find that the lengths A, B of the projections of the arms are given by

$$A=\sqrt{\cos^2\Psi+\sin^2\Psi\cos^2\Theta},\qquad B=\sqrt{\sin^2\Psi+\cos^2\Psi\cos^2\Theta},$$

with inverse

$$\Theta=\cos^{-1}\sqrt{A^2+B^2-1},\qquad \Psi=\tan^{-1}\sqrt{\frac{1-A^2}{1-B^2}}.$$

Some slog is now required to calculate the Jacobian J of this mapping, and the answer will be $f_{A,B}(a,b) = 4|J|\pi^{-2}$ for $0 < a, b < 1$, $a^2 + b^2 > 1$.

33. The order statistics of the X_i have joint density function

$$h(x_1, x_2, \ldots, x_n) = \lambda^n n! \exp\left(-\sum_{i=1}^{n} x_i\right)$$

on the set I of increasing sequences of positive reals. Define the one–one mapping from I onto $(0, \infty)^n$ by

$$y_1 = nx_1, \qquad y_r = (n+1-r)(x_r - x_{r-1}) \quad \text{for } 1 < r \le n,$$

with inverse $x_r = \sum_{k=1}^{r} y_k/(n-k+1)$ for $r \ge 1$. The Jacobian is $(n!)^{-1}$, whence the joint density function of $Y_1, Y_2, \ldots, Y_n$ is

$$\frac{1}{n!}\lambda^n n! \exp\left(-\sum_{i=1}^{n} x_i(\mathbf{y})\right) = \lambda^n \exp\left(-\sum_{k=1}^{n} y_k\right).$$

34. Recall Problem (4.14.4). First, $Z_i = F(X_i)$, $1 \le i \le n$, is a sequence of independent variables with the uniform distribution on $[0, 1]$. Secondly, a variable U has the latter distribution if and only if $-\log U$ has the exponential distribution with parameter 1.

It follows that $L_i = -\log F(X_i)$, $1 \le i \le n$, is a sequence of independent variables with the exponential distribution. The order statistics $L_{(1)}, \ldots, L_{(n)}$ are *in order* $-\log F(X_{(n)}), \ldots, -\log F(X_{(1)})$, since the function $-\log F(\cdot)$ is non-increasing. Applying the result of Problem (4.14.33), $E_1 = -n\log F(X_{(n)})$ and

$$E_r = -(n+1-r)\left\{\log F(X_{(n+1-r)}) - \log F(X_{(n+2-r)})\right\}, \quad 1 < r \le n,$$

are independent with the exponential distribution. Therefore $\exp(-E_r)$, $1 \le r \le n$, are independent with the uniform distribution.

35. One may be said to be in state j if the first $j-1$ prizes have been rejected and the jth prize has just been viewed, and it is the best so far. There are two possible decisions at this stage: either accept the jth prize, or reject it and continue. The mean return of the first decision equals the probability j/n that the overall best prize lies in the first j, and the mean return of the second is the maximal probability $f(j)$ that one may obtain the best prize having rejected the first j. Thus the maximal mean return $V(j)$ in state j satisfies

$$V(j) = \max\{j/n, f(j)\}.$$

Now j/n increases with j, and $f(j)$ decreases with j (since a possible stategy is to reject the $(j+1)$th prize also). Therefore there exists J such that $j/n \le f(j)$ if and only if $j \le J$. This confirms the optimal stategy as having the following form: reject the first J prizes out of hand, and accept the subsequent prize which is the best of those viewed so far. If there is no such prize, we pick the last prize presented.

Let Π_J be the probability of achieving the best prize by following the above strategy. Let A_k be the event that you pick the kth prize, and B the event that the prize picked is the best. Then,

$$\Pi_J = \sum_{k=J+1}^{n} \mathbb{P}(B \mid A_k)\mathbb{P}(A_k) = \sum_{k=J+1}^{n} \left(\frac{k}{n}\right)\left(\frac{J}{k-1}\cdot\frac{1}{k}\right) = \frac{J}{n}\sum_{k=J+1}^{n}\frac{1}{k-1},$$

and we choose the integer J which maximizes this expresion.

When n is large, we have the asymptotic relation $\Pi_J \simeq (J/n)\log(n/J)$. The maximum of the function $h_n(x) = (x/n)\log(n/x)$ occurs at $x = n/e$, and we deduce that $J \simeq n/e$. [A version of this problem was posed by Cayley (1875). Our solution is due to Lindley (1961).]

36. The joint density function of (X, Y, Z) is

$$f(x,y,z) = \frac{1}{(2\pi\sigma^2)^{3/2}} \exp\{-\tfrac{1}{2}(r^2 - 2\lambda x - 2\mu y - 2\nu z + \lambda^2 + \mu^2 + \nu^2)\}$$

where $r^2 = x^2 + y^2 + z^2$. The conditional density of X, Y, Z given $R = r$ is therefore proportional to $\exp\{\lambda x + \mu y + \nu z\}$. Now choosing spherical polar coordinates with axis in the direction (λ, μ, ν), we obtain a density function proportional to $\exp(a\cos\theta)\sin\theta$, where $a = r\sqrt{\lambda^2 + \mu^2 + \nu^2}$. The constant is chosen in such a way that the total probability is unity.

37. (a) $\phi'(x) = -x\phi(x)$, so $H_1(x) = x$. Differentiate the equation for H_n to obtain $H_{n+1}(x) = xH_n(x) - H_n'(x)$, and use induction to deduce that H_n is a polynomial of degree n as required. Integrating by parts gives, when $m \le n$,

$$\begin{aligned}\int_{-\infty}^{\infty} H_m(x)H_n(x)\phi(x)\,dx &= (-1)^n \int_{-\infty}^{\infty} H_m(x)\phi^{(n)}(x)\,dx \\ &= (-1)^{n-1}\int_{-\infty}^{\infty} H_m'(x)\phi^{(n-1)}(x)\,dx \\ &= \cdots = (-1)^{n-m}\int_{-\infty}^{\infty} H_m^{(m)}(x)\phi^{(n-m)}(x)\,dx,\end{aligned}$$

and the claim follows by the fact that $H_m^{(m)}(x) = m!$.

(b) By Taylor's theorem and the first part,

$$\phi(x)\sum_{n=0}^{\infty}\frac{t^n}{n!}H_n(x) = \sum_{n=0}^{\infty}\frac{(-t)^n}{n!}\phi^{(n)}(x) = \phi(x-t),$$

whence

$$\sum_{n=0}^{\infty}\frac{t^n}{n!}H_n(x) = \exp\{-\tfrac{1}{2}(x-t)^2 + \tfrac{1}{2}x^2\} = \exp(xt - \tfrac{1}{2}t^2).$$

38. The polynomials of Problem (4.14.37) are orthogonal, and there are unique expansions (subject to mild conditions) of the form $u(x) = \sum_{r=0}^{\infty} a_r H_r(x)$ and $v(x) = \sum_{r=0}^{\infty} b_r H_r(x)$. Without loss of generality, we may assume that $\mathbb{E}(U) = \mathbb{E}(V) = 0$, whence, by Problem (4.14.37a), $a_0 = b_0 = 0$. By (4.14.37a) again,

$$\operatorname{var}(U) = \mathbb{E}(u(X)^2) = \sum_{r=1}^{\infty} a_r^2 r!, \quad \operatorname{var}(V) = \sum_{r=1}^{\infty} b_r^2 r!.$$

By (4.14.37b) above,

$$\mathbb{E}\left(\sum_{m=0}^{\infty}\frac{H_m(X)s^m}{m!}\sum_{n=0}^{\infty}\frac{H_n(Y)t^n}{n!}\right) = \mathbb{E}\big(\exp\{sX - \tfrac{1}{2}s^2 + tY - \tfrac{1}{2}t^2\}\big) = e^{st\rho}.$$

By considering the coefficient of $s^m t^n$,

$$\mathbb{E}\big(H_m(X)H_n(Y)\big) = \begin{cases} \rho^n n! & \text{if } m = n, \\ 0 & \text{if } m \ne n, \end{cases}$$

and so

$$\operatorname{cov}(U,V)=\mathbb{E}\Bigl(\sum_{m=1}^{\infty}a_m H_m(X)\sum_{n=1}^{\infty}b_n H_n(Y)\Bigr)=\sum_{n=1}^{\infty}a_n b_n \rho^n n!,$$

$$|\rho(U,V)|=\frac{|\rho|\,|\sum_{n=1}^{\infty}a_n b_n\rho^{n-1}n!|}{\sqrt{\sum_{n=1}^{\infty}a_n^2 n!\sum_{n=1}^{\infty}b_n^2 n!}}\le|\rho|\frac{\sum_{n=1}^{\infty}|a_n b_n|n!}{\sqrt{\sum_{n=1}^{\infty}a_n^2 n!\sum_{n=1}^{\infty}b_n^2 n!}}\le|\rho|,$$

where we have used the Cauchy–Schwarz inequality at the last stage.

39. (a) Let $Y_r = X_{(r)} - X_{(r-1)}$ with the convention that $X_{(0)} = 0$ and $X_{(n+1)} = 1$. By Problem (4.14.21) and a change of variables, we may see that $Y_1, Y_2, \dots, Y_{n+1}$ have the distribution of a point chosen uniformly at random in the simplex of non-negative vectors $\mathbf{y} = (y_1, y_2, \dots, y_{n+1})$ with sum 1. [This may also be derived using a Poisson process representation and Theorem (6.12.7).] Consequently, the Y_j are identically distributed, and their joint distribution is invariant under permutations of the indices of the Y_j. Now $\sum_{r=1}^{n+1} Y_r = 1$ and, by taking expectations, $(n+1)\mathbb{E}(Y_1) = 1$, whence $\mathbb{E}(X_{(r)}) = r\mathbb{E}(Y_1) = r/(n+1)$.
(b) We have that

$$\mathbb{E}(Y_1^2)=\int_0^1 x^2 n(1-x)^{n-1}\,dx=\frac{2}{(n+1)(n+2)},$$

$$1=\mathbb{E}\Bigl[\Bigl(\sum_{r=1}^{n+1}Y_r\Bigr)^2\Bigr]=(n+1)\mathbb{E}(Y_1^2)+n(n+1)\mathbb{E}(Y_1Y_2),$$

implying that

$$\mathbb{E}(Y_1Y_2)=\frac{1}{(n+1)(n+2)},$$

and also

$$\mathbb{E}(X_{(r)}X_{(s)})=r\mathbb{E}(Y_1^2)+r(s-1)\mathbb{E}(Y_1Y_2)=\frac{r(s+1)}{(n+1)(n+2)}.$$

The required covariance follows.

40. (a) By paragraph (4.4.6), X^2 is $\Gamma(\frac12,\frac12)$ and Y^2+Z^2 is $\Gamma(\frac12,1)$. Now use the results of Exercise (4.7.14).
(b) Since the distribution of X^2/R^2 is independent of the value of $R^2 = X^2+Y^2+Z^2$, it is valid also if the three points are picked independently and uniformly within the sphere.

41. (a) By symmetry, $\phi(x) = \phi(-x)$ and $\Phi(x) = 1-\Phi(-x)$. Therefore, $g(x) = \phi(x) + \phi(x)\bigl(\Phi(\lambda x)-\Phi(-\lambda x)\bigr)$. The second term is an odd function since $\Phi(\lambda x)-\Phi(-\lambda x)$ is odd, and it therefore integrates to 0 over $\mathbb{R}$. Since $g \ge 0$, g is a density function.

Both $|X|$ and $|Y|$ have density function $2\phi(x)$ for $x > 0$. The joint density of X, $|Y|$, when independent, is $2\phi(x)\phi(y)$, for $x \in \mathbb{R}$, $y \in (0,\infty)$. Set $u = (x+\lambda|y|)/\sqrt{1+\lambda^2}$, $v = |y|$, whose inverse has Jacobian $J = \sqrt{1+\lambda^2}$, to obtain

$$f_{U,V}(u,v)=\frac{2\sqrt{1+\lambda^2}}{2\pi}\exp\Bigl\{-\tfrac12\bigl(u\sqrt{1+\lambda^2}-\lambda v\bigr)^2-\tfrac12 v^2\Bigr\},\qquad u\in\mathbb{R},\ v\in(0,\infty).$$

Integrate out v, and make a change of variables, to obtain $f_U(u) = 2\phi(u)\Phi(\lambda u) = g(u)$.
(b) Part (i) is as above. Alternatively, since X_1 and λX_2 are independent and symmetric,

$$\tfrac12=\mathbb{P}(X_1<\lambda X_2)=\int_{-\infty}^{\infty}F_1(\lambda y)f_2(y)\,dy=\int_{-\infty}^{\infty}\tfrac12 g_2(y)\,dy,$$

wth a similar argument for g_1. Finally,

$$\begin{aligned}\mathbb{P}(X_2 < z \mid X_1 < \lambda X_2) &= 2\mathbb{P}(X_2 < z,\ X_1 < \lambda X_2)\\ &= 2\int_{-\infty}^{z} F_1(\lambda y) f_2(y)\,dy = \int_{-\infty}^{z} g_2(y)\,dy.\end{aligned}$$

42. The required probability equals

$$\begin{aligned}\mathbb{P}\Big(\{X_3 - \tfrac{1}{2}(X_1+X_2)\}^2 + \{Y_3 - \tfrac{1}{2}(Y_1+Y_2)\}^2 \le \tfrac{1}{4}(X_1-X_2)^2 + \tfrac{1}{4}(Y_1-Y_2)^2\Big)&\\ = \mathbb{P}(U_1^2 + U_2^2 \le V_1^2 + V_2^2)&\end{aligned}$$

where U_1, U_2 are $N(0, \frac{3}{2})$, V_1, V_2 are $N(0, \frac{1}{2})$, and U_1, U_2, V_1, V_2 are independent. The answer is therefore

$$\begin{aligned}p &= \mathbb{P}\big(\tfrac{3}{2}(N_1^2 + N_2^2) \le \tfrac{1}{2}(N_3^2 + N_4^2)\big) \quad \text{where the } N_i \text{ are independent } N(0,1)\\ &= \mathbb{P}(K_1 \le \tfrac{1}{3}K_2) \qquad \text{where the } K_i \text{ are independent chi-squared } \chi^2(2)\\ &= \mathbb{P}\left(\frac{K_1}{K_1+K_2} \le \frac{1}{4}\right) = \mathbb{P}(B \le \tfrac{1}{4}) = \tfrac{1}{4}\end{aligned}$$

where we have used the result of Exercise (4.7.14), and B is a beta-distributed random variable with parameters 1, 1.

43. The argument of Problem (4.14.42) leads to the expression

$$\begin{aligned}\mathbb{P}(U_1^2 + U_2^2 + U_3^2 \le V_1^2 + V_2^2 + V_3^2) &= \mathbb{P}(K_1 \le \tfrac{1}{3}K_2) \quad \text{where the } K_i \text{ are } \chi^2(3)\\ &= \mathbb{P}(B \le \tfrac{1}{4}) = \frac{1}{3} - \frac{\sqrt{3}}{4\pi},\end{aligned}$$

where B is beta-distributed with parameters $\frac{3}{2}, \frac{3}{2}$.

44. (a) Simply expand thus: $\mathbb{E}[(X-\mu)^3] = \mathbb{E}[X^3 - 3X^2\mu + 3X\mu^2 - \mu^3]$ where $\mu = \mathbb{E}(X)$.

(b) $\mathrm{var}(S_n) = n\sigma^2$ and $\mathbb{E}[(S_n - n\mu)^3] = n\mathbb{E}[(X_1-\mu)^3]$ plus terms which equal zero because $\mathbb{E}(X_1 - \mu) = 0$.

(c) If Y is Bernoulli with parameter p, then $\mathrm{skw}(Y) = (1-2p)/\sqrt{pq}$, and the claim follows by (b).

(d) $m_1 = \lambda$, $m_2 = \lambda + \lambda^2$, $m_3 = \lambda^3 + 3\lambda^2 + \lambda$, and the claim follows by (a).

(e) Since λX is $\Gamma(1, t)$, we may as well assume that $\lambda = 1$. It is immediate that $\mathbb{E}(X^n) = \Gamma(t+n)/\Gamma(t)$, whence

$$\mathrm{skw}(X) = \frac{t(t+1)(t+2) - 3t\cdot t(t+1) + 2t^3}{t^{3/2}} = \frac{2}{\sqrt{t}}.$$

45. We find as above that $\mathrm{kur}(X) = (m_4 - 4m_3m_1 + 6m_2m_1^2 - 3m_1^4)/\sigma^4$ where $m_k = \mathbb{E}(X^k)$.

(a) $m_4 = 3\sigma^4$ for the $N(0, \sigma^2)$ distribution, whence $\mathrm{kur}(X) = 3\sigma^4/\sigma^4$.

(b) $m_r = r!/\lambda^r$, and the result follows.

(c) In this case, $m_4 = \lambda^4 + 6\lambda^3 + 7\lambda^2 + \lambda$, $m_3 = \lambda^3 + 3\lambda^2 + \lambda$, $m_2 = \lambda^2 + \lambda$, and $m_1 = \lambda$.

(d) $(\mathrm{var}\, S_n)^2 = n^2\sigma^4$ and $\mathbb{E}[(S_n - nm_1)^4] = n\mathbb{E}[(X_1 - m_1)^4] + 3n(n-1)\sigma^4$.

46. We have as $n \to \infty$ that

$$\mathbb{P}(X_{(n)} \le x + \log n) = \{1 - e^{-(x+\log n)}\}^n = \left(1 - \frac{e^{-x}}{n}\right)^n \to e^{-e^{-x}}, \qquad -\infty < x < \infty.$$

Write $Y_n = X_{(n)} - \log n$. By Lemma (4.3.4) and a non-rigorous interchange of a limit and an integral sign,

$$\mathbb{E}(Y_n) = \mathbb{E}(Y_n^+) - \mathbb{E}(Y_n^-) = \int_0^\infty \mathbb{P}(Y_n > x)\,dx - \int_0^\infty \mathbb{P}(Y_n < -x)\,dx$$
$$\to \int_0^\infty \{1 - e^{-e^{-x}} - e^{-e^x}\}\,dx.$$

On the other hand, by the lack-of-memory property, $\mathbb{E}(X_{(1)}) = n^{-1}$, $\mathbb{E}(X_{(2)}) = n^{-1} + (n-1)^{-1}$, and so on, whence

$$\mathbb{E}(Y_n) = \mathbb{E}(X_{(n)} - \log n) = \frac{1}{n} + \frac{1}{n-1} + \cdots + 1 - \log n \to \gamma.$$

47. By the argument presented in Section 4.11, conditional on acceptance, X has density function f_S. You might use this method when f_S is itself particularly tiresome or expensive to calculate. If $a(x)$ and $b(x)$ are easy to calculate and are close to f_S, much computational effort may be saved.

48. $M = \max\{U_1, U_2, \ldots, U_Y\}$ satisfies

$$\mathbb{P}(M \le t) = \mathbb{E}(t^Y) = \frac{e^t - 1}{e - 1}.$$

Thus,

$$\mathbb{P}(Z \ge z) = \mathbb{P}(X \ge \lfloor z \rfloor + 2) + \mathbb{P}\big(X = \lfloor z \rfloor + 1,\ Y \le \lfloor z \rfloor + 1 - z\big)$$
$$= \frac{(e-1)e^{-\lfloor z \rfloor - 2}}{1 - e^{-1}} + (e-1)e^{-\lfloor z \rfloor - 1} \cdot \frac{e^{\lfloor z \rfloor + 1 - z} - 1}{e - 1} = e^{-z}.$$

49. (a) Y has density function e^{-y} for $y > 0$, and X has density function $f_X(x) = \alpha e^{-\alpha x}$ for $x > 0$. Now $Y \ge \frac{1}{2}(X - \alpha)^2$ if and only if

$$V \alpha e^{-\alpha X} \frac{e^{\frac{1}{2}\alpha^2}}{\alpha} \sqrt{\frac{2}{\pi}} \le \sqrt{\frac{2}{\pi}} e^{-\frac{1}{2}X^2},$$

which is to say that $aVf_X(X) \le f(X)$, where $a = \alpha^{-1} e^{\frac{1}{2}\alpha^2} \sqrt{2/\pi}$. Recalling the argument of Example (4.11.5), we conclude that, conditional on this event occurring, X has density function f.

(b) The number of rejections is geometrically distributed with mean a^{-1}, so the optimal value of α is that which minimizes $\alpha e^{-\frac{1}{2}\alpha^2} \sqrt{\pi/2}$, that is, $\alpha = 1$.

(c) Setting

$$Z = \begin{Bmatrix} +X & \text{with probability } \frac{1}{2} \\ -X & \text{with probability } \frac{1}{2} \end{Bmatrix} \quad \text{conditional on } Y > \tfrac{1}{2}(X - \alpha)^2,$$

we obtain a random variable Z with the $N(0, 1)$ distribution.

50. (a) $\mathbb{E}(X) = \displaystyle\int_0^1 \sqrt{1 - u^2}\,du = \pi/4$.

(b) $\mathbb{E}(Y) = \dfrac{2}{\pi} \displaystyle\int_0^{\frac{1}{2}\pi} \sin\theta\,d\theta = 2/\pi$.

51. You are asked to calculate the mean distance of a randomly chosen pebble from the nearest collection point. Running through the cases, where we suppose the circle has radius a and we write P for the position of the pebble,

(i) $$\mathbb{E}|\mathrm{OP}| = \frac{1}{\pi a^2}\int_0^{2\pi}\int_0^a r^2\,dr\,d\theta = \frac{2a}{3}.$$

(ii) $$\mathbb{E}|\mathrm{AP}| = \frac{2}{\pi a^2}\int_0^{\frac{1}{2}\pi}\int_0^{2a\cos\theta} r^2\,dr\,d\theta = \frac{32a}{9\pi}.$$

(iii) $$\begin{aligned}\mathbb{E}(|\mathrm{AP}| \wedge |\mathrm{BP}|) &= \frac{4}{\pi a^2}\left[\int_0^{\frac{1}{4}\pi}\int_0^{a\sec\theta} r^2\,dr\,d\theta + \int_{\frac{1}{4}\pi}^{\frac{1}{2}\pi}\int_0^{2a\cos\theta} r^2\,dr\,d\theta\right] \\ &= \frac{4a}{3\pi}\left\{\frac{16}{3} - \frac{17}{6}\sqrt{2} + \frac{1}{2}\log(1+\sqrt{2})\right\} \simeq \frac{2a}{3}\times 1.13.\end{aligned}$$

(iv)
$$\mathbb{E}(|\mathrm{AP}| \wedge |\mathrm{BP}| \wedge |\mathrm{CP}|) = \frac{6}{\pi a^2}\left\{\int_0^{\frac{1}{3}\pi}\int_0^{x} r^2\,dr\,d\theta + \int_{\frac{1}{3}\pi}^{\frac{1}{2}\pi}\int_0^{2a\cos\theta} r^2\,dr\,d\theta\right\}$$

where $x = a\sin(\frac{1}{3}\pi)\operatorname{cosec}(\frac{2}{3}\pi - \theta)$

$$\begin{aligned}&= \frac{2a}{\pi}\left\{\int_0^{\frac{1}{3}\pi}\frac{1}{8}3\sqrt{3}\operatorname{cosec}^3\left(\frac{\pi}{3}+\theta\right)d\theta + \int_{\frac{1}{3}\pi}^{\frac{1}{2}\pi} 8\cos^3\theta\,d\theta\right\} \\ &= \frac{2a}{\pi}\left\{\frac{16}{3} - \frac{11}{4}\sqrt{3} + \frac{3\sqrt{3}}{16}\log\frac{3}{2}\right\} \simeq \frac{2a}{3}\times 0.67.\end{aligned}$$

52. By Problem (4.14.4), the displacement of R relative to P is the sum of two independent Cauchy random variables. By Exercise (4.8.2), this sum has also a Cauchy distribution, and inverting the transformation shows that Θ is uniformly distributed.

53. We may assume without loss of generality that R has length 1. Note that Δ occurs if and only if the sum of any two parts exceeds the length of the third part.

(a) If the breaks are at X, Y, where $0 < X < Y < 1$, then Δ occurs if and only if $2Y > 1$, and $2(Y - X) < 1$ and $2X < 1$. These inequalities are satisfied with probability $\frac{1}{4}$.

(b) The length X of the shorter piece has density function $f_X(x) = 2$ for $0 \le x \le \frac{1}{2}$. The other pieces are of length $(1-X)Y$ and $(1-X)(1-Y)$, where Y is uniform on $(0, 1)$. The event Δ occurs if and only if $2Y < 2XY + 1$ and $X + Y - XY > \frac{1}{2}$, and this has probability

$$2\int_0^{\frac{1}{2}}\left\{\frac{1}{2(1-x)} - \frac{1-2x}{2(1-x)}\right\}dx = \log(4/e).$$

(c) The three lengths are X, $\frac{1}{2}(1-X)$, $\frac{1}{2}(1-X)$, where X is uniform on $(0, 1)$. The event Δ occurs if and only if $X < \frac{1}{2}$.

(d) This triangle is obtuse if and only if

$$\frac{X/2}{(1-X)} > \frac{1}{\sqrt{2}},$$

which is to say that $X > \sqrt{2} - 1$. Hence,

$$\mathbb{P}(\text{obtuse} \mid \Delta) = \frac{\mathbb{P}(\sqrt{2} - 1 < X < \frac{1}{2})}{\mathbb{P}(X < \frac{1}{2})} = 3 - 2\sqrt{2}.$$

54. The shorter piece has density function $f_X(x) = 2$ for $0 \le x \le \frac{1}{2}$. Hence,

$$\mathbb{P}(R \le r) = \mathbb{P}\left(\frac{X}{1-X} \le r\right) = \frac{2r}{1+r},$$

with density function $f_R(r) = 2/(1-r)^2$ for $0 \le r \le 1$. Therefore,

$$\mathbb{E}(R) = \int_0^1 \mathbb{P}(R > r)\,dr = \int_0^1 \frac{1-r}{1+r}\,dr = 2\log 2 - 1,$$
$$\mathbb{E}(R^2) = \int_0^1 2r\mathbb{P}(R > r)\,dr = \int_0^1 \frac{2r(1-r)}{1+r}\,dr = 3 - 4\log 2,$$

and $\text{var}(R) = 2 - (2\log 2)^2$.

55. With an obvious notation,

$$\mathbb{E}(R^2) = \mathbb{E}[(X_1 - X_2)^2] + \mathbb{E}[(Y_1 - Y_2)^2] = 4\mathbb{E}(X_1^2) - 4\{\mathbb{E}(X_1)\}^2 = 4\cdot\tfrac{1}{3}a^2 - 4(\tfrac{1}{2}a)^2 = \tfrac{1}{3}a^2.$$

By a natural re-scaling, we may assume that $a = 1$. Now, $X_1 - X_2$ and $Y_1 - Y_2$ have the same triangular density symmetric on $(-1, 1)$, whence $(X_1 - X_2)^2$ and $(Y_1 - Y_2)^2$ have distribution function $F(z) = 2\sqrt{z} - z$ and density function $f_Z(z) = z^{-\frac{1}{2}} - 1$, for $0 \le z \le 1$. Therefore R^2 has the density f given by

$$f(r) = \begin{cases} \displaystyle\int_0^r \left(\frac{1}{\sqrt{z}} - 1\right)\left(\frac{1}{\sqrt{r-z}} - 1\right) dz & \text{if } 0 \le r \le 1, \\ \displaystyle\int_{r-1}^1 \left(\frac{1}{\sqrt{z}} - 1\right)\left(\frac{1}{\sqrt{r-z}} - 1\right) dz & \text{if } 1 \le r \le 2. \end{cases}$$

The claim follows since

$$\int_a^b \frac{1}{\sqrt{z}}\frac{1}{\sqrt{r-z}}\,dz = 2\left(\sin^{-1}\sqrt{\frac{b}{r}} - \sin^{-1}\sqrt{\frac{a}{r}}\right) \quad \text{for } 0 \le a \le b \le 1.$$

56. We use an argument similar to that used for Buffon's needle. Dropping the paper at random amounts to dropping the lattice at random on the paper. The mean number of points of the lattice in a small element of area dA is dA. By the additivity of expectations, the mean number of points on the paper is A. There must therefore exist a position for the paper in which it covers at least $\lceil A \rceil$ points.

57. Consider a small element of surface dS. Positioning the rock at random amounts to shining light at this element from a randomly chosen direction. On averaging over all possible directions, we see that the mean area of the shadow cast by dS is proportional to the area of dS. We now integrate over the surface of the rock, and use the additivity of expectation, to deduce that the area A of the random shadow satisfies $\mathbb{E}(A) = cS$ for some constant c which is independent of the shape of the rock. By considering the special case of the sphere, we find $c = \frac{1}{4}$. It follows that at least one orientation of the rock gives a shadow of area at least $\frac{1}{4}S$.

58. (a) We have from Problem (4.14.11b) that $Y_r = X_r/(X_1 + \cdots + X_r)$ is independent of $X_1 + \cdots + X_r$, and therefore of the variables $X_{r+1}, X_{r+2}, \cdots, X_{k+1}, X_1 + \cdots + X_{k+1}$. Therefore Y_r is independent of $\{Y_{r+s} : s \ge 1\}$, and the claim follows.

(b) Let $S = X_1 + \cdots + X_{k+1}$. The inverse transformation $x_1 = z_1 s$, $x_2 = z_2 s, \ldots, x_k = z_k s$, $x_{k+1} = s - z_1 s - z_2 s - \cdots - z_k s$ has Jacobian

$$J = \begin{vmatrix} s & 0 & 0 & \cdots & z_1 \\ 0 & s & 0 & \cdots & z_2 \\ \vdots & \vdots & \vdots & \ddots & \vdots \\ 0 & 0 & 0 & \cdots & z_k \\ -s & -s & -s & \cdots & 1 - z_1 - \cdots - z_k \end{vmatrix} = s^k.$$

The joint density function of $X_1, X_2, \ldots, X_k, S$ is therefore (with $\sigma = \sum_{r=1}^{k+1} \beta_r$),

$$\left\{ \prod_{r=1}^{k} \frac{\lambda^{\beta_r} (z_r s)^{\beta_r - 1} e^{-\lambda z_r s}}{\Gamma(\beta_r)} \right\} \cdot \frac{\lambda^{\beta_{k+1}} \{s(1 - z_1 - \cdots - z_k)\}^{\beta_{k+1} - 1} e^{-\lambda s(1 - z_1 - \cdots - z_k)}}{\Gamma(\beta_{k+1})}$$

$$= f(\lambda, \boldsymbol{\beta}, s) \left(\prod_{r=1}^{k} z_r^{\beta_r - 1} \right) (1 - z_1 - \cdots - z_k)^{\beta_{k+1} - 1},$$

where f is a function of the given variables. The result follows by integrating over s.

59. Let $\mathbf{C} = (c_{rs})$ be an orthogonal $n \times n$ matrix with $c_{ni} = 1/\sqrt{n}$ for $1 \leq i \leq n$. Let $Y_{ir} = \sum_{s=1}^{n} X_{is} c_{rs}$, and note that the vectors $\mathbf{Y}_r = (Y_{1r}, Y_{2r}, \ldots, Y_{nr})$, $1 \leq r \leq n$, are multivariate normal. Clearly $\mathbb{E} Y_{ir} = 0$, and

$$\mathbb{E}(Y_{ir} Y_{js}) = \sum_{t,u} c_{rt} c_{su} \mathbb{E}(X_{it} X_{ju}) = \sum_{t,u} c_{rt} c_{su} \delta_{tu} v_{ij} = \sum_{t} c_{rt} c_{st} v_{ij} = \delta_{rs} v_{ij},$$

where δ_{tu} is the Kronecker delta, since $\mathbf{C}$ is orthogonal. It follows that the set of vectors $\mathbf{Y}_r$ has the same joint distribution as the set of $\mathbf{X}_r$. Since $\mathbf{C}$ is orthogonal, $X_{ir} = \sum_{s=1}^{n} c_{sr} Y_{is}$, and therefore

$$S_{ij} = \sum_{r,s,t} c_{sr} c_{tr} Y_{is} Y_{jt} - \frac{1}{n} \sum_r X_{ir} \sum_r X_{jr} = \sum_{s,t} \delta_{st} Y_{is} Y_{jt} - \frac{1}{\sqrt{n}} \sum_r X_{ir} \frac{1}{\sqrt{n}} \sum_r X_{jr}$$

$$= \sum_s Y_{is} Y_{js} - Y_{in} Y_{jn} = \sum_{s=1}^{n-1} Y_{is} Y_{js}.$$

This has the same distribution as T_{ij} because the $\mathbf{Y}_r$ and the $\mathbf{X}_r$ are identically distributed.

60. We sketch this. Let $\mathbb{E}|\mathrm{PQR}| = m(a)$, and use Crofton's method. A point randomly dropped in $S(a + da)$ lies in $S(a)$ with probability

$$\left(\frac{a}{a + da} \right)^2 = 1 - \frac{2da}{a} + o(da).$$

Hence

$$\frac{dm}{da} = -\frac{6m}{a} + \frac{6m_b}{a},$$

where $m_b(a)$ is the conditional mean of $|\mathrm{PQR}|$ given that P is constrained to lie on the boundary of $S(a)$. Let $b(x)$ be the conditional mean of $|\mathrm{PQR}|$ given that P lies a distance x down one vertical edge.

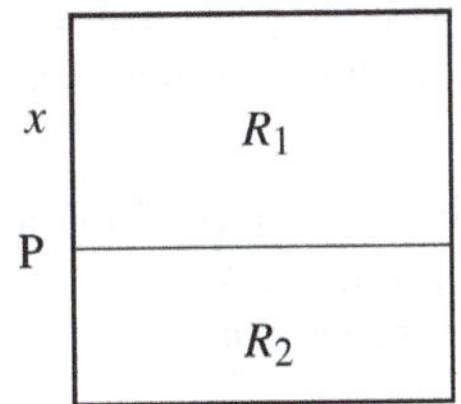

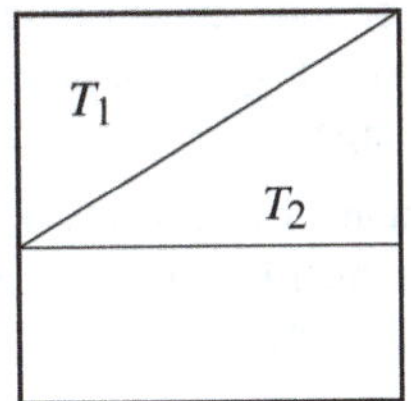

By conditioning on whether Q and R lie above or beneath P we find, in an obvious notation, that

$$b(x) = \left(\frac{x}{a}\right)^2 m_{R_1} + \left(\frac{a-x}{a}\right)^2 m_{R_2} + \frac{2x(a-x)}{a^2} m_{R_1,R_2}.$$

By Exercise (4.13.6) (see also Exercise (4.13.7)), $m_{R_1,R_2} = \frac{1}{2}(\frac{1}{2}a)(\frac{1}{2}a) = \frac{1}{8}a^2$. In order to find m_{R_1}, we condition on whether Q and R lie in the triangles T_1 or T_2, and use an obvious notation.

Recalling Example (4.13.6), we have that $m_{T_1} = m_{T_2} = \frac{4}{27} \cdot \frac{1}{2}ax$. Next, arguing as we did in that example,

$$m_{T_1,T_2} = \tfrac{1}{2} \cdot \tfrac{4}{9}\{ax - \tfrac{1}{4}ax - \tfrac{1}{4}ax - \tfrac{1}{8}ax\}.$$

Hence, by conditional expectation,

$$m_{R_1} = \tfrac{1}{4} \cdot \tfrac{4}{27} \cdot \tfrac{1}{2}ax + \tfrac{1}{4} \cdot \tfrac{4}{27} \cdot \tfrac{1}{2}ax + \tfrac{1}{2} \cdot \tfrac{4}{9} \cdot \tfrac{3}{8}ax = \tfrac{13}{108}ax.$$

We replace x by $a - x$ to find m_{R_2}, whence in total

$$b(x) = \left(\frac{x}{a}\right)^2 \frac{13ax}{108} + \left(\frac{a-x}{a}\right)^2 \frac{13a(a-x)}{108} + \frac{2x(a-x)}{a^2} \cdot \frac{a^2}{8} = \frac{13}{108}a^2 - \frac{12ax}{108} + \frac{12x^2}{108}.$$

Since the height of P is uniformly distributed on $[0, a]$, we have that

$$m_b(a) = \frac{1}{a}\int_0^a b(x)\,dx = \frac{11a^2}{108}.$$

We substitute this into the differential equation to obtain the solution $m(a) = \frac{11}{144}a^2$.

Turning to the last part, by making an affine transformation, we may without loss of generality take the parallelogram to be a square. The points form a convex quadrilateral when no point lies inside the triangle formed by the other three, and the required probability is therefore $1 - 4m(a)/a^2 = 1 - \frac{44}{144} = \frac{25}{36}$.

61. Choose four points P, Q, R, S uniformly at random inside C, and let T be the event that their convex hull is a triangle. By considering which of the four points lies in the interior of the convex hull of the other three, we see that $\mathbb{P}(T) = 4\mathbb{P}(\text{S} \in \text{PQR}) = 4\mathbb{E}|\text{PQR}|/|C|$. Having chosen P, Q, R, the four points form a triangle if and only if S lies in either the triangle PQR or the shaded region A. Thus, $\mathbb{P}(T) = \{|A| + \mathbb{E}|\text{PQR}|\}/|C|$, and the claim follows on solving for $\mathbb{P}(T)$.

62. Since $\mathbf{X}$ has zero means and covariance matrix $\mathbf{I}$, we have that $\mathbb{E}(\mathbf{Z}) = \boldsymbol{\mu} + \mathbb{E}(\mathbf{X})\mathbf{L} = \boldsymbol{\mu}$, and the covariance matrix of $\mathbf{Z}$ is $\mathbb{E}(\mathbf{L}'\mathbf{X}'\mathbf{X}\mathbf{L}) = \mathbf{L}'\mathbf{I}\mathbf{L} = \mathbf{V}$.

63. Let $\mathbf{D} = (d_{ij}) = \mathbf{AB} - \mathbf{C}$. The claim is trivial if $\mathbf{D} = \mathbf{0}$, and so we assume the converse. Choose i, k such that $d_{ik} \neq 0$, and write $y_i = \sum_{j=1}^n d_{ij}x_j = S + d_{ik}x_k$. Now $\mathbb{P}(y_i = 0) = \mathbb{E}\big(\mathbb{P}(x_k = -S/d_{ik} \mid S)\big)$. For any given S, there is probability at least $\frac{1}{2}$ that $x_k \neq -S/d_{ik}$, and the second claim follows.

Let $\mathbf{x}_1, \mathbf{x}_2, \dots, \mathbf{x}_m$ be independent random vectors with the given distribution. If $\mathbf{D} \neq \mathbf{0}$, the probability that $\mathbf{D}\mathbf{x}_s = \mathbf{0}$ for $1 \le s \le m$ is at most $(\frac{1}{2})^m$, which may be made as small as required by choosing m sufficiently large.

64. (a) Let N_i be the number of packets opened until the appearance of the first object of type i. Then N_i has a geometric distribution with parameter $1/n$, so that $\mathbb{E}N_i = 1$ and $\mathbb{E}(N_i^2) = 2n^2 - n$. Furthermore, for $i \neq j$, $\min\{N_i, N_j\}$ is geometric with parameter $2/n$, so

$$(*) \qquad \mathbb{E}\big(\min\{N_i, N_j\}\big) = \frac{n}{2}, \qquad \mathbb{E}\big(\min\{N_i, N_j\}^2\big) = 2\left(\frac{n}{2}\right)^2 - \frac{n}{2},$$

and more generally. By the result of Exercise (4.3.6),

$$\begin{aligned}
\mathbb{E}(T_n) &= \sum_{i=1}^{n} \mathbb{E}(N_i) - \sum_{i<j} \mathbb{E}\big(\min\{N_i, N_j\}\big) + \cdots \\
&= \binom{n}{1} \cdot n - \binom{n}{2} \cdot \frac{n}{2} + \binom{n}{3} \cdot \frac{n}{3} - \cdots + (-1)^{n+1}\binom{n}{n}\frac{n}{n} \\
&= n \sum_{i=1}^{n} \frac{(-1)^{i+1}}{i}\binom{n}{i}.
\end{aligned}$$

Another way of regarding T_n is as $\sum_{i=1}^{n} R_i$, where R_i is the number of packets sampled to obtain the ith new type after previously obtaining the $(i-1)$th new type. The R_i are independent and geometric, and R_i has mean $n/(n-i+1)$. Therefore,

$$\mathbb{E}(T_n) = \sum_{i=1}^{n} \frac{n}{n-i+1} = n \sum_{r=1}^{n} \frac{1}{r}.$$

(b) We have that

$$\sum_{r=1}^{n} \frac{(-1)^{r+1}}{r}\binom{n}{r} = \sum_{r=1}^{n} (-1)^{r+1}\binom{n}{r}\int_0^1 x^{r-1}\,dx = -\int_0^1 \frac{1}{x}\{(1-x)^n - 1\}\,dx.$$

Denote this sum as S_n, and note that

$$S_n - S_{n-1} = \int_0^1 (1-x)^{n-1}\,dx = \frac{1}{n},$$

and $S_1 = 1$. The identity follows.

(c) We have $T_n^2 = \max_i\{N_i^2\}$. By repeating the above calculations,

$$\begin{aligned}
\mathbb{E}(T_n^2) &= \binom{n}{1} \cdot (2n^2 - n) - \binom{n}{2}\left(2\left(\frac{n}{2}\right)^2 - \frac{n}{2}\right) + \cdots + (-1)^{n+1}\binom{n}{n}\left(2\left(\frac{n}{n}\right)^2 - \frac{n}{n}\right) \\
&= 2n^3 \sum_{r=1}^{n} \frac{(-1)^{r+1}}{r^2}\binom{n}{r} - n^2 \sum_{r=1}^{n} \frac{(-1)^{r+1}}{r}\binom{n}{r}.
\end{aligned}$$

However,

$$\begin{aligned}
\mathbb{E}\left\{\left(\sum_{r=1}^{n} R_r\right)^2\right\} &= \sum_{r=1}^{n} \operatorname{var}(R_r) + n^2\left(\sum_{r=1}^{n} \frac{1}{r}\right)^2 \\
&= \sum_{r=1}^{n} \frac{n(n-r)}{r^2} + n^2\left(\sum_{r=1}^{n} \frac{1}{r}\right).
\end{aligned}$$

The result follows from the above.

65. Let α (respectively, β, γ) be the event that A (respectively, B, C) is a vertex of the random triangle T. By the remark preceding Example (4.13.6), this ratio of means is invariant under affine transformations of the plane. By taking ABC to be equilateral, we have by symmetry that

$$\mathbb{E}\left(\frac{|T|}{|\mathrm{ABC}|}\right) = \mathbb{E}\left(\frac{|T|}{|\mathrm{ABC}|} \,\middle|\, \delta\right), \qquad \delta = \alpha, \beta, \gamma,$$

where $|Z|$ denotes the area of Z.

Now take ABC to be a right-angle isosceles triangle with short sides of length 1, with A the right-angled vertex, and condition on the event α. Consider the event $E(x, y) = \{X < x, Y < y\}$, where X (respectively, Y) is the distance between A and the point of intersection of PQ with the line AB (respectively, AC), and let $D(x, y)$ be the subset of $\mathbb{R}^4$ generated by P $= (u, v)$ and Q $= (w, z)$ (in the Cartesian plane with A as origin) on which the event $E(x, y)$ occurs. Let $T : (u, v, w, z) \to (u', v', w', z')$ be the linear transformation of $\mathbb{R}^4$ that takes $D(x, y)$ to $D(1, 1)$. Then, for some constant c,

$$\mathbb{P}(E(x, y)) = c \iint_{D(x,y)} du\, dv\, dw\, dz = cx^2y^2 \iint_{D(1,1)} du'\, dv'\, dw'\, dz' = x^2y^2\mathbb{P}(E(1, 1)).$$

Now,

$$\mathbb{P}(E(x, y) \mid \alpha) = \mathbb{P}\big(E(x, y) \,\big|\, E(1, 1)\big) = \frac{\mathbb{P}(E(x, y))}{\mathbb{P}(E(1, 1))} = x^2y^2, \qquad 0 < x, y, < 1.$$

Therefore,

$$\begin{aligned}\mathbb{E}\left(\frac{|T|}{|\mathrm{ABC}|} \,\middle|\, \alpha\right) &= \mathbb{E}\left(\frac{XY/2}{1/2} \,\middle|\, \alpha\right) = \mathbb{E}(XY \mid \alpha) \\ &= \iint_{(0,1)^2} xy \cdot 4xy\, dx\, dy = \frac{4}{9},\end{aligned}$$

and the result follows.

66. Let

$$d(u, v) = |u + v| - |u - v| = 2\min\big\{|u|, |v|\big\}\mathrm{sign}(uv), \qquad u, v \in \mathbb{R}.$$

By Lemma (4.3.4),

$$\mathbb{E}|X + Y| - \mathbb{E}|X - Y| = \mathbb{E}d(X, Y) = \int_0^\infty J(t)\, dt,$$

where

$$J(t) = I\big(\min\{|X|, |Y|\}\mathrm{sign}(XY) > t\big).$$

We have $d(X, Y) = 0$ when $XY = 0$. By considering the four disjoint events arising from the signs of non-zero X and Y, we have

$$\begin{aligned}\mathbb{E}d(X, Y) &= 2\mathbb{E}\int_0^\infty \big\{I(X > t, Y > t) + I(X < -t, Y < -t) - I(X > t, Y < -t) \\ &\qquad\qquad - I(X < -t, Y > t)\big\}\, dt \\ &= 2\int_0^\infty \big\{\mathbb{P}(X > t) - \mathbb{P}(X < -t)\big\}\big\{\mathbb{P}(Y > t) - \mathbb{P}(Y < -t)\big\}\, dt \\ &= 2\int_0^\infty \big\{\mathbb{P}(X > t) - \mathbb{P}(X < -t)\big\}^2\, dt \geq 0.\end{aligned}$$

Equality holds if and only $F(t) = 1 - F(-t)$ for all > 0. Arguably, $d(X, Y)$ is a better measure of departure from symmetry (for zero-mean distributions) than the 'skewness' of Problem (4.14.44), which can be zero for asymmetric distributions.

67. We have that

$$\mathbb{P}(X_1 \le x) = \mathbb{P}(U_1 \le x^n) = x^n = \mathbb{P}(U_{(n)} \le x), \qquad x \in (0,1),$$

in agreement with the density of the order statistics $U_{(1)}, U_{(2)}, \dots, U_{(n)}$. See Problems (4.14.21, 23). We proceed by observing the $U_{(k)}$ as k decreases, and the X_l as l increases. Conditional on $U_j = U_{(n)} = u$, the remaining variables $U_1, U_2, \dots, U_{j-1}, U_{j+1}, \dots, U_n$ are independent and uniformly distributed on $[0, u]$, whence $U_{(n-1)}$ has the conditional distribution function $F(x) = (x/u)^{n-1}$, $0 \le x \le u$. On the other hand, conditional on $X_1 = u$, X_2 has distribution function F. Therefore, (X_1, X_2) has the same (unconditional distribution) as $(U_{(n)}, U_{(n-1)})$. The full claim follows similarly.

In the change-of-variables method, we set $x_1 = u_1^{1/n}$, and so on. The Jacobian of this transformation is the determinant of a triangular matrix, which turns out to be $1/n!$ after a calculation. Therefore, the Jacobian of the inverse transformation is $n!$. It follows that the vector X has joint density function $n!$ on the set of increasing n-sequences from $[0, 1]$.

68. With f the density function of X,

$$\begin{aligned} V &= \int_a^\infty (x-a) f(x)\,dx \\ &= \int_a^\infty \frac{x-\mu}{\sigma\sqrt{2\pi}} \exp\left\{-\frac{(x-\mu)^2}{2\sigma^2}\right\} dx + \int_a^\infty \frac{\mu-a}{\sigma\sqrt{2\pi}} \exp\left\{-\frac{(x-\mu)^2}{2\sigma^2}\right\} dx, \end{aligned}$$

and the formula for V follows after integration. Note that $V = \sigma\phi(y)[1 - yM(y)]$, where $M(y)$ is Mills's ratio of Exercise (4.4.8).

The asymptotic for V follows from the fact that $1 - yM(y) \sim y^{-2}$ as $y \to \infty$.

69. (a) This holds since

$$\tfrac{4}{27}y^3 - (y-x)x^2 = \tfrac{4}{27}(y+3x)(y - \tfrac{3}{2}x)^2 \ge 0, \qquad x, y \ge 0.$$

(b) We have for $0 < x < y$ that, with y as given in the question,

$$\mathbb{P}(X > x) = (y-x)g(x) \le \frac{4g(x)}{9x^2} \int_0^y v^2\,dv.$$

Now, by the definition of y and the monotonicity of g,

$$\begin{aligned} g(x)\int_0^y v^2\,dv &= g(x)\int_0^x v^2\,dv + \int_x^y \big(g(x) - g(v)\big)v^2\,dv + \int_x^y g(v)v^2\,dv \\ &\le \int_0^x g(v)v^2\,dv + y^2 \int_x^y \big(g(x) - g(v)\big)\,dv + \int_x^y g(v)v^2\,dv \\ &= \int_0^y g(v)v^2\,dv + y^2\left((y-x)g(x) - \int_x^y g(v)\,dv\right) \\ &= \int_0^y g(v)v^2\,dv + y^2 \int_y^\infty g(v)\,dv \le \mathbb{E}(X^2). \end{aligned}$$

The definition of y is used at the last step.

(c) Let X have density function f with a unique mode at 0. The density function of $Y = |X|$ is $g(y) = f(y) + f(-y)$ for $y \geq 0$, having a unique mode at 0. Apply part (b) to Y. [This proof is due to H. Cramér.]

70. (a) Set

$$G_k(s) = \mathbb{E}\big(s^{I_{k+1}+I_{k+2}+\cdots+I_n}\big) = \prod_{i=k+1}^{n} (q_i + p_i s),$$

and calculate $G_k'(0) = \mathbb{P}(I_{k+1} + \cdots + I_n = 1)$.

(b) We have that

$$\begin{aligned}\sigma_{k+1} - \sigma_k &= (R_{k+2} - q_{k+1}R_{k+1}) \prod_{k+2}^{n} q_i \\ &= (p_{k+1}R_{k+2} - q_{k+1}r_{k+1}) \prod_{k+2}^{n} q_i \quad \begin{cases} > 0 & \text{if } R_{k+2} > 1, \\ < 0 & \text{if } R_{k+2} < 1, \end{cases}\end{aligned}$$

since $q_{k+1}r_{k+1} = p_{k+1}$ and $\prod_i q_i > 0$. The sequence R_k is strictly decreasing in k, whence σ_k has a mode at $s = \max\{1, \max\{k : R_{k+1} \geq 1\}\}$. If $R_{s+1} = 1$, there is a modal interval of length 2.

(c) The rule 'stop at the first success (strictly) after time k' succeeds if and only if $I_{k+1} + \cdots + I_n = 1$, which occurs with probability σ_k. This is a maximum at $k = \tau$, which is to say that we stop at time τ or later, with τ as given. By the independence of the trials, this conclusion is optimal amongst the class of rules of the form 'stop at the first success after time T' for a stopping time T. The optimal stopping rule is therefore as claimed.

(d) Let $L = \max\{i : p_i = 1\}$, so that $r_L = \infty$ and $\tau \geq L$. If we continue beyond time τ, there is strictly positive probability of failure, whereas if we stop at L, we succeed with probability one. Therefore, the optimal strategy is to stop at time τ.

(e) This follows by the above.

(f) We adopt the language of Problem (4.14.35). Let I_i be the indicator function of the event that the ith prize is the best prize so far. By symmetry, given $I_{i+1} = 1$, the indicators $I_1, I_2, \ldots, I_i$ have the same distribution as their unconditional distribution. Therefore, I_{i+1} and the vector $(I_1, I_2, \ldots, I_i)$ are independent for every i. It follows that the I_i are independent. We apply Bruss's odds rule, with $p_i = 1/i$ and odds ratio $r_k = 1/(k-1)$ for $k \geq 2$, to find that the optimal stopping rule is to inspect the first τ prizes and to take the next best thereafter, where τ is the greatest k such that

$$R_k = \frac{1}{k-1} + \frac{1}{k} + \cdots + \frac{1}{n} \geq 1.$$

The success probability is $\sigma_\tau = (\tau - 1)R_\tau/n$. We have as $n \to \infty$ that $\tau/n \to e^{-1}$, $R_\tau \downarrow 1$, so that $\sigma_\tau \to e^{-1}$.

71. Let S_r be the supremum of $\mathbb{E}(X_T)$ over all stopping strategies T that do not stop before examining X_r, and let

$$M_r = \max\{X_r, S_{r+1}\} = S_{r+1} + (X_r - S_{r+1})^+, \qquad 1 \leq r \leq n.$$

Then $S_{n+1} = 0$, and $S_r = \mathbb{E}(M_r)$ for $0 \leq r \leq n$. Now, $S_0 \geq S_1 \geq \cdots \geq S_n \geq S_{n+1} = 0$, and hence $M_r \leq S_1 + (X_r - S_{r+1})^+$. Therefore,

$$\max\{X_r : 0 \leq r \leq n\} \leq \max\{M_r : 0 \leq r \leq n\} \leq S_1 + \sum_{r=0}^{n} (X_r - S_{r+1})^+,$$

so that

$$\mathbb{E}\big(\max\{X_r : 0 \leq r \leq n\}\big) \leq S_1 + \sum_{r=0}^{n} (S_r - S_{r+1}) = S_1 + S_0 \leq 2S_0.$$

72. Identify the stick with the interval $[0, s]$, and let B the largest position of $n + 1$ breaks. Then B has density function $f_B(x) = (n + 1)x^n/s^{n+1}$ for $x \in (0, s)$, and $p_{n+1}(s, y) = 0$ if $B > s - y$. Conditional on $B = x$, the other n breaks are independent and uniformly distributed on $(0, x)$. Therefore,

$$p_{n+1}(s, y) = \int_0^{s-y} p_n(x, y) f_B(x)\, dx.$$

(i) Note that $p_0(s, y) = I(y < s)$, and $p_1(s, y) = (s - 2y)^+/s$. The formula holds by induction on n.

(ii) We have

$$\mathbb{E}S_1 = \frac{1}{s^n} \int_0^{s/(n+1)} \{(s - (n + 1)y)^+\}^n\, dy = \frac{s}{(n + 1)^2} = \frac{s(H_{n+1} - H_n)}{n + 1},$$

in the notation of part (iii).

(iii) Given $S_1 = s_1$, the remaining n lengths can be written as $S_r = s_1 + T_r$, where $T_1 < T_2 < \cdots < T_n$ are the order statistics of the lengths of fragments obtained by breaking a stick of length $L = s - (n + 1)s_1$ into n parts. Now,

$$\mathbb{E}L = s - (n + 1)\mathbb{E}S_1 = \frac{ns}{n + 1},$$

and so, by the argument so far,

$$\mathbb{E}T_1 = \frac{1}{n^2} \cdot \frac{ns}{n + 1} = \frac{s}{n(n + 1)},$$

and

$$\mathbb{E}S_2 = \frac{s}{(n + 1)^2} + \frac{s}{n(n + 1)} = \frac{s(H_{n+1} - H_{n-1})}{n + 1}.$$

The general result follows by iteration.

73. (a) Since f integrates to 1, we have $c = 2 + \alpha$. The joint distribution function is

$$F(x, y) = \int_{u=1-y}^{x} \int_{v=1-u}^{y} cu^\alpha\, dv\, du = \frac{2 + \alpha}{1 + \alpha}(y - 1)x^{1+\alpha} + x^{2+\alpha} + \frac{(1 - y)^{2+\alpha}}{1 + \alpha},$$

for $x, y \in (0, 1)$.

(b) Such a triangle exists if and only if all of the following hold,

$$X + Y > 2 - X - Y, \quad X + 2 - X - Y > Y, \quad Y + 2 - X - Y > X,$$

and this is seen immediately to occur with probability 1.

(c) The angle Θ opposite Y satisfies, by the cosine rule,

$$\cos \Theta = \frac{X^2 + (2 - X - Y)^2 - Y^2}{2X(2 - X - Y)},$$

which is negative if and only if

$$Y > g(X) := \frac{X^2 - 2X + 2}{2 - X}.$$

Now, $1 - X \le g(X) \le 1$, so

$$\mathbb{P}(\Theta \text{ is obtuse}) = \mathbb{P}(Y > g(X)) = \int_{x=0}^{1} \int_{y=g(x)}^{1} f(x, y)\, dy\, dx = c \int_0^1 \frac{x(1-x)}{2-x} x^\alpha\, dx.$$

When $\alpha = 0$, this equals $3 - 4\log 2$.

74. (a) The rate at which you pass (or are passed by) vehicles with speed x is proportional to $|x - v|$, the absolute difference of speeds. Thus, by so-called size-biasing, the density π of the speeds of such vehicles satisfies

$$\pi(x) \propto |x - v| f(x),$$

where the constant of proportionality is chosen such that π integrates to 1. The required expected value is the mean of this distribution. Size-bias will be discussed later, and here is a preview. Imagine the overtaking traffic as being distributed on the freeway behind your vehicle. In the next unit of time, a vehicle travelling at speed x will pass you if and only if it is within distance $x - v$. Thus the rate at which you are passed by vehicles at speed x is proportional to $x - v$. This is the size-biasing effect that accounts for the factor $|x - v|$ in π above.

(b) Let $\mu - a < v < \mu < \mu + b$, and let f be symmetric about μ in its interval of support $[\mu - a, \mu + a]$. Then $m(v) - \mu = \mathbb{E}\{(X - \mu)|X - v|\}/\mathbb{E}|X - v|$, and

$$\begin{aligned}
\mathbb{E}\{(X-\mu)|X-v|\} &= \int_{\mu-a}^{v} (x-\mu)(v-x) f(x)\, dx + \int_{v}^{\mu} (x-\mu)(x-v) f(x)\, dx \\
&\qquad + \int_{\mu-a}^{\mu} (\mu - x)(2\mu - x - v) f(x)\, dx \\
&= 2\int_{\mu-a}^{v} (\mu - x)(\mu - v) f(x)\, dx + 2\int_{v}^{\mu} (\mu - x)^2 f(x)\, dx > 0,
\end{aligned}$$

since the integrands are positive. In summary, $m(v) - \mu$ is continuous and positive on $[0, \mu]$, it equals 0 at $v = \mu$, and $m(0) = \mathbb{E}(X^2)/\mathbb{E}(X) \ge \mathbb{E}(X) > 0$. Therefore, $m(v) - \mu$ is maximized at some $v_{\max} \in [0, \mu)$.

(c) By the same argument, $m(v) - \mu < 0$ for $v > \mu$.

(d) This is left to the reader.

5

Generating functions and their applications

5.1 Solutions. Generating functions

1. (a) If $|s| < (1-p)^{-1}$,

$$G(s) = \sum_{m=0}^{\infty} s^m \binom{n+m-1}{m} p^n (1-p)^m = \left\{ \frac{p}{1-s(1-p)} \right\}^n.$$

Therefore the mean is $G'(1) = n(1-p)/p$. The variance is $G''(1)+G'(1)-G'(1)^2 = n(1-p)/p^2$.
(b) If $|s| < 1$,

$$G(s) = \sum_{m=1}^{\infty} s^m \left(\frac{1}{m} - \frac{1}{m+1} \right) = 1 + \left(\frac{1-s}{s} \right) \log(1-s).$$

Therefore $G'(1) = \infty$, and there exist no moments of order 1 or greater.
(c) If $p < |s| < p^{-1}$,

$$G(s) = \sum_{m=-\infty}^{\infty} s^m \left(\frac{1-p}{1+p} \right) p^{|m|} = \frac{1-p}{1+p} \left\{ 1 + \frac{sp}{1-sp} + \frac{p/s}{1-(p/s)} \right\}.$$

The mean is $G'(1) = 0$, and the variance is $G''(1) = 2p(1-p)^{-2}$.

2. (i) Either hack it out, or use indicator functions I_A thus:

$$T(s) = \sum_{n=0}^{\infty} s^n \mathbb{P}(X > n) = \mathbb{E}\left(\sum_{n=0}^{\infty} s^n I_{\{n<X\}} \right) = \mathbb{E}\left(\sum_{n=0}^{X-1} s^n \right) = \mathbb{E}\left(\frac{1-s^X}{1-s} \right) = \frac{1-G(s)}{1-s}.$$

(ii) It follows that

$$T(1) = \lim_{s\uparrow 1} \left\{ \frac{1-G(s)}{1-s} \right\} = \lim_{s\uparrow 1} \frac{G'(s)}{1} = G'(1) = \mathbb{E}(X)$$

by L'Hôpital's rule. Also,

$$\begin{aligned} T'(1) &= \lim_{s\uparrow 1} \left\{ \frac{-(1-s)G'(s) + 1 - G(s)}{(1-s)^2} \right\} \\ &= \tfrac{1}{2} G''(1) = \tfrac{1}{2} \{ \text{var}(X) - G'(1) + G'(1)^2 \} \end{aligned}$$

whence the claim is immediate.

3. (i) We have that $G_{X,Y}(s,t) = \mathbb{E}(s^X t^Y)$, whence $G_{X,Y}(s,1) = G_X(s)$ and $G_{X,Y}(1,t) = G_Y(t)$.
(ii) If $\mathbb{E}|XY| < \infty$ then

$$\mathbb{E}(XY) = \mathbb{E}\left(XYs^{X-1}t^{Y-1}\right)\Big|_{s=t=1} = \frac{\partial^2}{\partial s\,\partial t} G_{X,Y}(s,t)\bigg|_{s=t=1}.$$

4. We write $G(s,t)$ for the joint generating function.

(a)
$$\begin{aligned} G(s,t) &= \sum_{j=0}^{\infty}\sum_{k=0}^{j} s^j t^k (1-\alpha)(\beta-\alpha)\alpha^j \beta^{k-j-1} \\ &= \sum_{j=0}^{\infty}\left(\frac{\alpha s}{\beta}\right)^j \frac{(1-\alpha)(\beta-\alpha)}{\beta}\cdot\frac{1-(\beta t)^{j+1}}{1-\beta t} && \text{if } \beta|t|<1 \\ &= \frac{(1-\alpha)(\beta-\alpha)}{(1-\beta t)\beta}\left\{\frac{1}{1-(\alpha s/\beta)} - \frac{\beta t}{1-\alpha s t}\right\} && \text{if } \frac{\alpha}{\beta}|s|<1 \\ &= \frac{(1-\alpha)(\beta-\alpha)}{(1-\alpha s t)(\beta-\alpha s)} \end{aligned}$$

(the condition $\alpha|st| < 1$ is implied by the other two conditions on s and t). The marginal generating functions are

$$G(s,1) = \frac{(1-\alpha)(\beta-\alpha)}{(1-\alpha s)(\beta-\alpha s)}, \quad G(1,t) = \frac{1-\alpha}{1-\alpha t},$$

and the covariance is easily calculated by the conclusion of Exercise (5.1.3) as $\alpha(1-\alpha)^{-2}$.
(b) Arguing similarly, we obtain $G(s,t) = (e-1)/\{e(1-te^{s-2})\}$ if $|t|e^{s-2} < 1$, with marginals

$$G(s,1) = \frac{1-e^{-1}}{1-e^{s-2}}, \quad G(1,t) = \frac{1-e^{-1}}{1-te^{-1}},$$

and covariance $e(e-1)^{-2}$.
(c) Once again,

$$G(s,t) = \frac{\log\{1-tp(1-p+sp)\}}{\log(1-p)} \qquad \text{if } |tp(1-p+sp)| < 1.$$

The marginal generating functions are

$$G(s,1) = \frac{\log\{1-p+p^2(1-s)\}}{\log(1-p)}, \quad G(1,t) = \frac{\log(1-tp)}{\log(1-p)},$$

and the covariance is

$$-\frac{p^2\{p+\log(1-p)\}}{(1-p)^2\{\log(1-p)\}^2}.$$

5. (i) We have that

$$\mathbb{E}(x^H y^T) = \sum_{k=0}^{n} x^k y^{n-k}\binom{n}{k}p^k(1-p)^{n-k} = (px+qy)^n$$

where $p + q = 1$.

(ii) More generally, if each toss results in one of t possible outcomes, the ith of which has probability p_i, then the corresponding quantity is a function of t variables, $x_1, x_2, \dots, x_t$, and is found to be $(p_1 x_1 + p_2 x_2 + \cdots + p_t x_t)^n$.

6. We have that

$$\mathbb{E}(s^X) = \mathbb{E}\{\mathbb{E}(s^X \mid U)\} = \int_0^1 \{1 + u(s-1)\}^n \, du = \frac{1}{n+1} \cdot \frac{1 - s^{n+1}}{1 - s},$$

the probability generating function of the uniform distribution. See also Exercise (4.6.5).

Here is a less direct but curious solution. Let $U, V_1, V_2, \dots, V_n$ be independent random variables with the uniform distribution on $[0, 1]$. Conditional on U, the variables $I_i = I(V_i \le U)$ are independent Bernoulli variables with parameter U, with sum S which has the $\text{bin}(n, U)$ distribution. Now,

$$S = \sum_{i=1}^{n} I(V_{(i)} \le u),$$

where the $V_{(i)}$ are the order statistics of the V_i. Set $V_{(0)} = 0$ and $V_{(n+1)} = 1$. Then $S = k$ if and only if $V_{(k)} \le U < V_{(k+1)}$, which has probability $\mathbb{E}(V_{(k+1)} - V_{(k)}) = 1/(n+1)$.

7. We have that

$$\begin{aligned} G_{X,Y,Z}(x, y, z) = G(x, y, z, 1) &= \tfrac{1}{8}(xyz + xy + yz + zx + x + y + z + 1) \\ &= \tfrac{1}{2}(x+1)\tfrac{1}{2}(y+1)\tfrac{1}{2}(z+1) = G_X(x)G_Y(y)G_Z(z), \end{aligned}$$

whence X, Y, Z are independent. The same conclusion holds for any other set of exactly three random variables. However, $G(x, y, z, w) \ne G_X(x)G_Y(y)G_Z(z)G_W(w)$.

8. (a) We have by differentiating that $\mathbb{E}(X^{2n}) = 0$, whence $\mathbb{P}(X = 0) = 1$. This is not a moment generating function.

(b) This is a moment generating function if and only if $\sum_r p_r = 1$, in which case it is that of a random variable X with $\mathbb{P}(X = a_r) = p_r$.

9. The coefficients of s^n in both combinations of G_1, G_2 are non-negative and sum to 1. They are therefore probability generating functions, as is $G(\alpha s)/G(\alpha)$ for the same reasons.

10. Since Z is discrete and M continuous, one needs to be careful over the interpretation of certain notation. Given $Z = z$, M has distribution function $\mathbb{P}(M \le m) = F(m)^z$ and density function

$$f_{M|Z}(m \mid z) = zF'(m)F(m)^{z-1}.$$

Therefore,

$$f_M(m) = \sum_{z=1}^{\infty} zF'(m)F(m)^{z-1}\mathbb{P}(Z = z) = F'(m)G'(F(m)).$$

Now,

$$\begin{aligned} \mathbb{E}(Z \mid M = m) &= \sum_{z=1}^{\infty} z \frac{\mathbb{P}(Z = z) f_{M|Z}(m \mid z)}{f_M(m)} \\ &= \frac{1}{F'(m)G'(F(m))} \sum_{z=1}^{\infty} z^2 F'(m)F(m)^{z-1}\mathbb{P}(Z = z) \\ &= \frac{F(m)G''(F(m)) + G'(F(m))}{G'(F(m))}. \end{aligned}$$

11. We have

$$\mathbb{E}(s^Y) = \sum_{x=1}^{\infty} pq^{x-1}s^{x\wedge n} = \sum_{x=1}^{n} pq^{x-1}s^x + \sum_{x=n+1}^{\infty} pq^{x-1}s^n = \frac{ps + (1-s)(qs)^n}{1-qs},$$

where $a \wedge b = \min\{a, b\}$. Find the mean by differentiating at $s = 1$.

12. (a) We have that $\pi = \sum_r u^r \mathbb{P}(X_i = r) = G_X(u)$.
(b) The required probability is $\sum_n \pi^n \mathbb{P}(N = n) = G_N(\pi)$. On the other hand, it is also $G_T(u)$ where T is the total number of balls. Therefore, $G_T(u) = G_N(G_X(u))$.
(c) By differentiating at $u = 1$,

$$\begin{aligned}\mathbb{E}(T) &= G_T'(1) = G_N'(1)G_X'(1) = \mathbb{E}(N)\mathbb{E}(X),\\ \mathrm{var}(T) &= G_T''(1) + G_T'(1) - G_T'(1)^2 = \mathrm{var}(N)\mathbb{E}(X)^2 + \mathbb{E}(N)\,\mathrm{var}(X),\end{aligned}$$

after a calculation. This is perhaps a little less elegant than using conditional expectation.
(d) The indicator function that a ball is unmarked has probability generating function $G_B(s) = 1 - u + us$. By differentiating $G_U(s) = G_N(G_X(G_B(s)))$, we find that

$$\begin{aligned}\mathbb{E}(U) &= u\mathbb{E}(N)\mathbb{E}(X),\\ \mathrm{var}(U) &= u^2\,\mathrm{var}(N)\mathbb{E}(X)^2 + \mathbb{E}(N)\big(u^2\,\mathrm{var}(X) + u(1-u)\mathbb{E}X\big).\end{aligned}$$

5.2 Solutions. Some applications

1. Let $G(s) = \mathbb{E}(s^X)$ and $G_S(s) = \sum_{j=0}^{n} s^j S_j$. By the result of Exercise (5.1.2),

$$T(s) = \sum_{m=0}^{\infty} s^m \mathbb{P}(X \ge m) = 1 + s\sum_{k=0}^{\infty} s^k \mathbb{P}(X > k) = 1 + \frac{s(1-G(s))}{1-s} = \frac{1-sG(s)}{1-s}.$$

Now,

$$G_S(s) = \sum_{m=0}^{n} s^m \mathbb{E}\binom{X}{m} = \mathbb{E}\left\{\sum_{m=0}^{n} s^m \binom{X}{m}\right\} = \mathbb{E}\{(1+s)^X\} = G(1+s)$$

so that

$$\frac{T(s)-T(0)}{s} = \frac{G_S(s-1)-G_S(0)}{s-1}$$

where we have used the fact that $T(0) = G_S(0) = 1$. Therefore

$$\sum_{i=1}^{n} s^{i-1}\mathbb{P}(X \ge i) = \sum_{j=1}^{n} (s-1)^{j-1} S_j.$$

Equating coefficients of s^{i-1}, we obtain as required that

$$\mathbb{P}(X \ge i) = \sum_{j=i}^{n} S_j \binom{j-1}{i-1}(-1)^{j-i}, \qquad 1 \le i \le n.$$

Similarly,

$$\frac{G_S(s)-G_S(0)}{s} = \frac{T(1+s)-T(0)}{1+s}$$

whence the second formula follows.

2. Let A_i be the event that the ith person is chosen by nobody, and let X be the number of events $A_1, A_2, \dots, A_n$ which occur. Clearly

$$\mathbb{P}(A_{i_1} \cap A_{i_2} \cap \dots \cap A_{i_j}) = \left(\frac{n-j}{n-1}\right)^j \left(\frac{n-j-1}{n-1}\right)^{n-j}$$

if $i_1 \neq i_2 \neq \dots \neq i_j$, since this event requires each of $i_1, \dots, i_j$ to choose from a set of $n-j$ people, and each of the others to choose from a set of size $n-j-1$. Using Waring's Theorem (Problem (1.8.13) or equation (5.2.14)),

$$\mathbb{P}(X = k) = \sum_{j=k}^{n} (-1)^{j-k} \binom{j}{k} S_j$$

where

$$S_j = \binom{n}{j} \left(\frac{n-j}{n-1}\right)^j \left(\frac{n-j-1}{n-1}\right)^{n-j}.$$

Using the result of Exercise (5.2.1),

$$\mathbb{P}(X \geq k) = \sum_{j=k}^{n} (-1)^{j-k} \binom{j-1}{k-1} S_j, \qquad 1 \leq k \leq n,$$

while $\mathbb{P}(X \geq 0) = 1$.

3. (a)

$$\mathbb{E}(x^{X+Y}) = \mathbb{E}\{\mathbb{E}(x^{X+Y} \mid Y)\} = \mathbb{E}\{x^Y e^{Y(x-1)}\} = \mathbb{E}\{(xe^{x-1})^Y\} = \exp\{\mu(xe^{x-1} - 1)\}.$$

(b) The probability generating function of X_1 is

$$G(s) = \sum_{k=1}^{\infty} \frac{\{s(1-p)\}^k}{k \log(1/p)} = \frac{\log\{1 - s(1-p)\}}{\log p}.$$

Using the 'compounding theorem' (5.1.25),

$$G_Y(s) = G_N(G(s)) = e^{\mu(G(s)-1)} = \left(\frac{p}{1 - s(1-p)}\right)^{-\mu/\log p}.$$

4. Clearly,

$$\mathbb{E}\left(\frac{1}{1+X}\right) = \mathbb{E}\left(\int_0^1 t^X \, dt\right) = \int_0^1 \mathbb{E}(t^X) \, dt = \int_0^1 (q + pt)^n \, dt = \frac{1 - q^{n+1}}{p(n+1)}$$

where $q = 1 - p$. In the limit,

$$\mathbb{E}\left(\frac{1}{1+X}\right) = \frac{1 - (1 - \lambda/n)^{n+1}}{\lambda(n+1)/n} + o(1) \to \frac{1 - e^{-\lambda}}{\lambda},$$

the corresponding moment of the Poisson distribution with parameter λ.

5. Conditioning on the outcome of the first toss, we obtain $h_n = qh_{n-1} + p(1 - h_{n-1})$ for $n \geq 1$, where $q = 1 - p$ and $h_0 = 1$. Multiply throughout by s^n and sum to find that $H(s) = \sum_{n=0}^{\infty} s^n h_n$ satisfies $H(s) - 1 = (q - p)sH(s) + ps/(1 - s)$, and so

$$H(s) = \frac{1 - qs}{(1 - s)\{1 - (q - p)s\}} = \frac{1}{2}\left\{\frac{1}{1 - s} + \frac{1}{1 - (q - p)s}\right\}.$$

6. By considering the event that HTH does not appear in n tosses and then appears in the next three, we find that $\mathbb{P}(X > n)p^2q = \mathbb{P}(X = n + 1)pq + \mathbb{P}(X = n + 3)$. We multiply by s^{n+3} and sum over n to obtain

$$\frac{1 - \mathbb{E}(s^X)}{1 - s} p^2 q s^3 = pqs^2 \mathbb{E}(s^X) + \mathbb{E}(s^X),$$

which may be solved as required. Let Z be the time at which THT first appears, so $Y = \min\{X, Z\}$. By a similar argument,

$$\mathbb{P}(Y > n)p^2q = \mathbb{P}(X = Y = n + 1)pq + \mathbb{P}(X = Y = n + 3) + \mathbb{P}(Z = Y = n + 2)p,$$
$$\mathbb{P}(Y > n)q^2p = \mathbb{P}(Z = Y = n + 1)pq + \mathbb{P}(Z = Y = n + 3) + \mathbb{P}(X = Y = n + 2)q.$$

We multipy by s^{n+1}, sum over n, and use the fact that $\mathbb{P}(Y = n) = \mathbb{P}(X = Y = n) + \mathbb{P}(Z = Y = n)$.

7. Suppose there are $n + 1$ matching letter/envelope pairs, numbered accordingly. Imagine the envelopes lined up in order, and the letters dropped at random onto these envelopes. Assume that exactly $j + 1$ letters land on their correct envelopes. The removal of any one of these $j + 1$ letters, together with the corresponding envelope, results after re-ordering in a sequence of length n in which exactly j letters are correctly placed. It is not difficult to see that, for each resulting sequence of length n, there are exactly $j + 1$ originating sequences of length $n + 1$. The first result follows. We multiply by s^j and sum over j to obtain the second. It is evident that $G_1(s) = s$. Either use induction, or integrate repeatedly, to find that $G_n(s) = \sum_{r=0}^{n}(s - 1)^r/r!$.

8. We have for $|s| < \mu + 1$ that

$$\mathbb{E}(s^X) = \mathbb{E}\{\mathbb{E}(s^X \mid \Lambda)\} = \mathbb{E}(e^{\Lambda(s-1)}) = \frac{\mu}{\mu - (s - 1)} = \frac{\mu}{\mu + 1}\sum_{k=0}^{\infty}\left(\frac{s}{\mu + 1}\right)^k.$$

9. Since the waiting times for new objects are geometric and independent,

$$\mathbb{E}(s^T) = s\left(\frac{3s}{4 - s}\right)\left(\frac{s}{2 - s}\right)\left(\frac{s}{4 - 3s}\right).$$

Using partial fractions, the coefficient of s^k is $\frac{3}{32}\{\frac{1}{2}(\frac{1}{4})^{k-4} - 4(\frac{1}{2})^{k-4} + \frac{9}{2}(\frac{3}{4})^{k-4}\}$, for $k \geq 4$.

10. Given $B = b$, R has the bin(b, r) distribution, so that

$$G_R(s) = \mathbb{E}(s^R) = \mathbb{E}\big[(1 - r + rs)^B\big] = (1 - pr + prs)^m,$$

whence R has the bin(m, pr) distribution. Then $A = R + N$, so that

$$\begin{aligned} G_A(s) &= \mathbb{E}\big[s^R(1 - \alpha + \alpha s)^{m-R}\big] = (1 - \alpha + \alpha s)^m\left(1 - pr + \frac{prs}{1 - \alpha + \alpha s}\right)^m \\ &= \big\{(1 - pr)(1 - \alpha) + (\alpha + (1 - \alpha)pr)s\big\}^m, \end{aligned}$$

so that A has the bin$(m, \alpha + (1 - \alpha)pr)$ distribution.

The above operation maps the bin(m, p_0) distribution to the bin(m, p_1) distribution where $p_1 = \alpha + (1-\alpha)rp_0$. Iteration of the process gives a sequence (p_n) satisfying $p_n \to p_\infty$ where $p_\infty = \alpha + (1-\alpha)rp_\infty$, that is, $p_\infty = \alpha/[1 - r(1-\alpha)] \in (0, 1)$.

11. The solution of Exercise (5.2.7) gives the probability generating function of the number X_n of matches as $G_n(s) = \sum_{r=0}^{n}(s-1)^r/r!$. If each match is discarded with probability $1-t$, the remaining number Y_n has probability generating function $G_n(1-t+ts)$, and the answer follows. The second part holds by the exponential power series.

12. The proof is by induction on the number n of values taken by X with strictly positive probability. Suppose $x_1 < x_2 < \cdots < x_n$, and the random variable X is such that $\mathbb{P}(X = x_i) = p_i > 0$ for $i = 1, 2, \ldots, n$ and $\sum_i p_i = 1$. If $n = 1$, we take $Y = x_1$. Assume the claim holds for $n \le N$ and let $n = N+1$. Let Z be a a weighted sum of independent Bernoulli variables $B_2, \ldots, B_{N+1}$ with mass function

$$\mathbb{P}(Z = x_k) = \frac{p_k}{1-p_1}, \qquad k = 2, 3, \ldots, N+1.$$

Let B_1 be Bernoulli with parameter $1 - p_1$, independent of Z, and let

$$Y = \begin{cases} x_1 & \text{if } B_1 = 0, \\ Z & \text{if } B_1 = 1. \end{cases}$$

13. (a) We have $B(z) = (1+z+z^2+\cdots+z^m)A(z)$, and the coefficient of z^m is $a_m + a_{m-1} + \cdots + a_0$.
(b) By inspecting (5.2.14), the question is equivalent to the assertion that, for $j \ge i$,

$$\sum_{r=j}^{n}(-1)^{r-j}\binom{r}{r-i}S_r \ge 0.$$

On substituting from (5.2.13), this is equivalent to

$$\sum_{r=j}^{n}(-1)^{r-j}\binom{r}{r-i}\sum_{s=r}^{n}\binom{s}{r}\mathbb{P}(X = s) \ge 0.$$

The coefficient of $\mathbb{P}(X = s)$ is

$$\sum_{r=j}^{s}(-1)^{r-j}\binom{r}{r-i}\binom{s}{r} = \binom{s}{i}\left\{\binom{s-i}{s-j} - \binom{s-i}{s-j+1} + \cdots + (-1)^{s-j}\binom{s-i}{0}\right\},$$

which is the coefficient of z^{s-j} in $(1-z)^{s-i-1}(1-z^{s-j+1})\binom{s}{i}(-1)^{s-j}$. This is positive as required. We used part (a) to get the last equality. [The authors hope that the readers can follow the above better than the former can.]

5.3 Solutions. Random walk

1. Let A_k be the event that the walk ever reaches the point k. Then $A_k \supseteq A_{k+1}$ if $k \ge 0$, so that

$$\mathbb{P}(M \ge r) = \mathbb{P}(A_r) = \mathbb{P}(A_0)\prod_{k=0}^{r-1}\mathbb{P}(A_{k+1} \mid A_k) = (p/q)^r, \qquad r \ge 0,$$

since $\mathbb{P}(A_{k+1} \mid A_k) = \mathbb{P}(A_1 \mid A_0) = p/q$ for $k \ge 0$ by Corollary (5.3.6).

2. (a) We have by Theorem (5.3.1c) that

$$\sum_{k=1}^{\infty} s^{2k} 2k f_0(2k) = sF_0'(s) = \frac{s^2}{\sqrt{1-s^2}} = s^2 \mathbb{P}_0(s) = \sum_{k=1}^{\infty} s^{2k} p_0(2k-2),$$

and the claim follows by equating the coefficients of s^{2k}.
(b) It is the case that $\alpha_n = \mathbb{P}(S_1 S_2 \cdots S_{2n} \neq 0)$ satisfies

$$\alpha_n = \sum_{\substack{k=2n+2 \\ k \text{ even}}}^{\infty} f_0(k),$$

with the convention that $\alpha_0 = 1$. We have used the fact that ultimate return to 0 occurs with probability 1. This sequence has generating function given by

$$\begin{aligned}\sum_{n=0}^{\infty} s^{2n} \sum_{\substack{k=2n+2 \\ k \text{ even}}}^{\infty} f_0(k) &= \sum_{\substack{k=2 \\ k \text{ even}}}^{\infty} f_0(k) \sum_{n=0}^{\frac{1}{2}k-1} s^{2n} \\ &= \frac{1-F_0(s)}{1-s^2} = \frac{1}{\sqrt{1-s^2}} \quad \text{by Theorem (5.3.1c)} \\ &= \mathbb{P}_0(s) = \sum_{n=0}^{\infty} s^{2n} \mathbb{P}(S_{2n} = 0).\end{aligned}$$

Now equate the coefficients of s^{2n}. (Alternatively, use Exercise (5.1.2) to obtain the generating function of the α_n directly.)

3. Draw a diagram of the square with the letters ABCD in clockwise order. Clearly $p_{\text{AA}}(m) = 0$ if m is odd. The walk is at A after $2n$ steps if and only if the numbers of leftward and rightward steps are equal *and* the numbers of upward and downward steps are equal. The number of ways of choosing $2k$ horizontal steps out of $2n$ is $\binom{2n}{2k}$. Hence

$$p_{\text{AA}}(2n) = \sum_{k=0}^{n} \binom{2n}{2k} \alpha^{2k} \beta^{2n-2k} = \tfrac{1}{2}\{(\alpha+\beta)^{2n} + (\alpha-\beta)^{2n}\} = \tfrac{1}{2}\{1 + (\alpha-\beta)^{2n}\}$$

with generating function

$$G_{\text{A}}(s) = \sum_{n=0}^{\infty} s^{2n} p_{\text{AA}}(2n) = \frac{1}{2}\left\{\frac{1}{1-s^2} + \frac{1}{1-\{s(\alpha-\beta)\}^2}\right\}.$$

Writing $F_{\text{A}}(s)$ for the probability generating function of the time T of first return, we use the argument which leads to Theorem (5.3.1a) to find that $G_{\text{A}}(s) = 1 + F_{\text{A}}(s)G_{\text{A}}(s)$, and therefore $F_{\text{A}}(s) = 1 - G_{\text{A}}(s)^{-1}$.

4. Write (X_n, Y_n) for the position of the particle at time n. It is an elementary calculation to show that the relations $U_n = X_n + Y_n$, $V_n = X_n - Y_n$ define independent simple symmetric random walks U and V. Now $T = \min\{n : U_n = m\}$, and therefore $G_T(s) = \{s^{-1}(1-\sqrt{1-s^2})\}^m$ for $|s| \le 1$ by Theorem (5.3.5).

Now $X - Y = V_T$, so that

$$G_{X-Y}(s) = \mathbb{E}\{\mathbb{E}(s^{V_T} \mid T)\} = \mathbb{E}\left\{\left(\frac{s+s^{-1}}{2}\right)^T\right\} = G_T\left(\tfrac{1}{2}(s+s^{-1})\right)$$

where we have used the independence of U and V. This converges if and only if $|\frac{1}{2}(s+s^{-1})| \le 1$, which is to say that $s = \pm 1$. Note that $G_T(s)$ converges in a non-trivial region of the complex plane.

5. Let T be the time of the first return of the walk S to its starting point 0. During the time-interval $(0, T)$, the walk is equally likely to be to the left or to the right of 0, and therefore

$$L_{2n} = \begin{cases} TR + L' & \text{if } T \le 2n, \\ 2nR & \text{if } T > 2n, \end{cases}$$

where R is Bernoulli with parameter $\frac{1}{2}$, L' has the distribution of L_{2n-T}, and R and L' are independent. It follows that $G_{2n}(s) = \mathbb{E}(s^{L_{2n}})$ satisfies

$$G_{2n}(s) = \sum_{k=1}^{n} \tfrac{1}{2}(1+s^{2k})G_{2n-2k}(s)f(2k) + \sum_{k>n} \tfrac{1}{2}(1+s^{2n})f(2k)$$

where $f(2k) = \mathbb{P}(T = 2k)$. (Remember that L_{2n} and T are even numbers.) Let $H(s,t) = \sum_{n=0}^{\infty} t^{2n} G_{2n}(s)$. Multiply through the above equation by t^{2n} and sum over n, to find that

$$H(s,t) = \tfrac{1}{2}H(s,t)\{F(t) + F(st)\} + \tfrac{1}{2}\{J(t) + J(st)\}$$

where $F(x) = \sum_{k=0}^{\infty} x^{2k} f(2k)$ and

$$J(x) = \sum_{n=0}^{\infty} x^{2n} \sum_{k>n} f(2k) = \frac{1}{\sqrt{1-x^2}}, \qquad |x| < 1,$$

by the calculation in the solution to Exercise (5.3.2). Using the fact that $F(x) = 1 - \sqrt{1-x^2}$, we deduce that $H(s,t) = 1/\sqrt{(1-t^2)(1-s^2t^2)}$. The coefficient of $s^{2k}t^{2n}$ is

$$\begin{aligned} \mathbb{P}(L_{2n} = 2k) &= \binom{-\frac{1}{2}}{n-k}(-1)^{n-k} \cdot \binom{-\frac{1}{2}}{k}(-1)^k \\ &= \binom{2k}{k}\binom{2n-2k}{n-k}\left(\frac{1}{2}\right)^{2n} = \mathbb{P}(S_{2k} = 0)\mathbb{P}(S_{2n-2k} = 0). \end{aligned}$$

6. We show that all three terms have the same generating function, using various results established for simple symmetric random walk. First, in the usual notation,

$$\sum_{m=0}^{\infty} 4m\mathbb{P}(S_{2m} = 0)s^{2m} = 2sP_0'(s) = \frac{2s^2}{(1-s^2)^{3/2}}.$$

Secondly, by Exercise (5.3.2),

$$\mathbb{E}(T \wedge 2m) = 2m\mathbb{P}(T > 2m) + \sum_{k=1}^{m} 2kf_0(2k) = 2m\mathbb{P}(S_{2m} = 0) + \sum_{k=1}^{m} \mathbb{P}(S_{2k-2} = 0).$$

Hence,

$$\sum_{m=0}^{\infty} s^{2m}\mathbb{E}(T \wedge 2m) = \frac{s^2 P_0(s)}{1-s^2} + sP_0'(s) = \frac{2s^2}{(1-s^2)^{3/2}}.$$

Finally, using the hitting time theorem (3.10.14), (5.3.7), and some algebra at the last stage,

$$\begin{aligned}\sum_{m=0}^{\infty} s^{2m} 2\mathbb{E}|S_{2m}| &= 4\sum_{m=0}^{\infty} s^{2m}\sum_{k=1}^{m} 2k\mathbb{P}(S_{2m}=2k) = 4\sum_{m=0}^{\infty} s^{2m}\sum_{k=1}^{m} 2mf_{2k}(2m)\\ &= 4s\frac{d}{ds}\sum_{m=0}^{\infty} s^{2m}\sum_{k=1}^{m} f_{2k}(2m) = 4s\frac{d}{ds}\frac{F_1(s)^2}{1-F_1(s)^2} = \frac{2s^2}{(1-s^2)^{3/2}}.\end{aligned}$$

7. Let I_n be the indicator of the event $\{S_n = 0\}$, so that $S_{n+1} = S_n + X_{n+1} + I_n$. In equilibrium, $\mathbb{E}(S_0) = \mathbb{E}(S_0) + \mathbb{E}(X_1) + \mathbb{E}(I_0)$, which implies that $\mathbb{P}(S_0 = 0) = \mathbb{E}(I_0) = -\mathbb{E}(X_1)$ and entails $\mathbb{E}(X_1) \le 0$. Furthermore, it is impossible that $\mathbb{P}(S_0 = 0) = 0$ since this entails $\mathbb{P}(S_0 = a) = 0$ for all $a < \infty$. Hence $\mathbb{E}(X_1) < 0$ if S is in equilibrium. Next, in equilibrium,

$$\mathbb{E}(z^{S_0}) = \mathbb{E}(z^{S_{n+1}}) = \mathbb{E}\big(z^{S_n+X_{n+1}+I_n}\{(1-I_n)+I_n\}\big).$$

Now,

$$\begin{aligned}\mathbb{E}\{z^{S_n+X_{n+1}+I_n}(1-I_n)\} &= \mathbb{E}(z^{S_n} \mid S_n > 0)\mathbb{E}(z^{X_1})\mathbb{P}(S_n > 0)\\ \mathbb{E}(z^{S_n+X_{n+1}+I_n} I_n) &= z\mathbb{E}(z^{X_1})\mathbb{P}(S_n = 0).\end{aligned}$$

Hence

$$\mathbb{E}(z^{S_0}) = \mathbb{E}(z^{X_1})\big[\{\mathbb{E}(z^{S_0}) - \mathbb{P}(S_0 = 0)\} + z\mathbb{P}(S_0 = 0)\big]$$

which yields the appropriate choice for $\mathbb{E}(z^{S_0})$.

8. The hitting time theorem (3.10.14), (5.3.7), states that $\mathbb{P}(T_{0b} = n) = (|b|/n)\mathbb{P}(S_n = b)$, whence

$$\mathbb{E}(T_{0b} \mid T_{0b} < \infty) = \frac{b}{\mathbb{P}(T_{0b} < \infty)}\sum_n \mathbb{P}(S_n = b).$$

The walk is transient if and only if $p \ne \frac{1}{2}$, and therefore $\mathbb{E}(T_{0b} \mid T_{0b} < \infty) < \infty$ if and only if $p \ne \frac{1}{2}$. Suppose henceforth that $p \ne \frac{1}{2}$.

The required conditional mean may be found by conditioning on the first step, or alternatively as follows. Assume first that $p < q$, so that $\mathbb{P}(T_{0b} < \infty) = (p/q)^b$ by Corollary (5.3.6). Then $\sum_n \mathbb{P}(S_n = b)$ is the mean of the number N of visits of the walk to b. Now

$$\mathbb{P}(N = r) = \left(\frac{p}{q}\right)^b \rho^{r-1}(1-\rho), \qquad r \ge 1,$$

where $\rho = \mathbb{P}(S_n = 0 \text{ for some } n \ge 1) = 1 - |p-q|$. Therefore $\mathbb{E}(N) = (p/q)^b/|p-q|$ and

$$\mathbb{E}(T_{0b} \mid T_{0b} < \infty) = \frac{b}{(p/q)^b}\cdot\frac{(p/q)^b}{|p-q|}.$$

We have when $p > q$ that $\mathbb{P}(T_{0b} < \infty) = 1$, and $\mathbb{E}(T_{01}) = (p-q)^{-1}$. The result follows from the fact that $\mathbb{E}(T_{0b}) = b\mathbb{E}(T_{01})$.

9. By Corollary (5.3.6), adapted to the current setting, $\mathbb{P}_1(H) = \min\{1, q/p\} = q/p$, where $\mathbb{P}_i$ denotes probability conditional on starting at i. By the Markov property, $\mathbb{P}_i(H) = (q/p)^i$ for $i \geq 1$. Therefore, for $i \geq 1$,

$$\mathbb{P}_i(X_1 = 1 \mid H) = \frac{\mathbb{P}_i(H \mid X_1 = 1)\mathbb{P}(X_1 = 1)}{\mathbb{P}_i(H)} = \frac{(q/p)^{i+1} p}{(q/p)^i} = q.$$

By iteration, the walk behaves like a walk with p and q interchanged until H occurs.

5.4 Solutions. Branching processes

1. Clearly $\mathbb{E}(Z_n \mid Z_m) = Z_m \mu^{n-m}$ since, given Z_m, Z_n is the sum of the numbers of $(n-m)$th generation descendants of Z_m progenitors. Hence $\mathbb{E}(Z_m Z_n \mid Z_m) = Z_m^2 \mu^{n-m}$ and $\mathbb{E}(Z_m Z_n) = \mathbb{E}\{\mathbb{E}(Z_m Z_n \mid Z_m)\} = \mathbb{E}(Z_m^2)\mu^{n-m}$. Hence

$$\operatorname{cov}(Z_m, Z_n) = \mu^{n-m}\mathbb{E}(Z_m^2) - \mathbb{E}(Z_m)\mathbb{E}(Z_n) = \mu^{n-m}\operatorname{var}(Z_m),$$

and, by Lemma (5.4.2),

$$\rho(Z_m, Z_n) = \mu^{n-m}\sqrt{\frac{\operatorname{var} Z_m}{\operatorname{var} Z_n}} = \begin{cases} \sqrt{\mu^{n-m}(1-\mu^m)/(1-\mu^n)} & \text{if } \mu \neq 1, \\ \sqrt{m/n} & \text{if } \mu = 1. \end{cases}$$

2. Suppose $0 \leq r \leq n$, and let $\mathfrak{C}_i$ be the set of members of the nth generation that are descendants of the ith member of the r generation. Pick a (uniform) random element of the nth generation. Conditional on $Z_1, Z_2, \dots, Z_n$, it belongs to $\mathfrak{C}_i$ with probability $\Pi_i = |\mathfrak{C}_i|/Z_n$. By symmetry or otherwise, $\mathbb{E}(\Pi_i \mid Z_r)$ does not depend on i. Since $\sum_i \mathbb{E}(\Pi_i \mid Z_r) = 1$, we have $\mathbb{E}(\Pi_i \mid Z_r) = Z_r^{-1}$.

By a similar argument, conditional on Z_r, the chance that two independently chosen individuals from the nth generation have the same rth generation ancestor is $\sum_i \mathbb{E}(\Pi_i^2 \mid Z_r)$. By the Cauchy–Schwarz inequality, this is at least $\sum_i \mathbb{E}(\Pi_i \mid Z_r)^2 = Z_r^{-1}$, with equality (when $r \neq 0$) if and only if the process is deterministic.

In conclusion,

$$\mathbb{P}(L \geq r) \geq \mathbb{E}(Z_r^{-1}).$$

If $0 < \mathbb{P}(Z_1 = 0) < 1$, then almost the same argument proves that $\mathbb{P}(L \geq r \mid Z_n > 0) \geq \mathbb{E}(Z_r^{-1} \mid Z_n > 0)$.

3. The number Z_n of nth generation decendants satisfies

$$\mathbb{P}(Z_n = 0) = G_n(0) = \begin{cases} \dfrac{n}{n+1} & \text{if } p = q, \\[2ex] \dfrac{q(p^n - q^n)}{p^{n+1} - q^{n+1}} & \text{if } p \neq q, \end{cases}$$

whence, for $n \geq 1$,

$$\mathbb{P}(T = n) = \mathbb{P}(Z_n = 0) - \mathbb{P}(Z_{n-1} = 0) = \begin{cases} \dfrac{1}{n(n+1)} & \text{if } p = q, \\[2ex] \dfrac{p^{n-1}q^n(p-q)^2}{(p^n - q^n)(p^{n+1} - q^{n+1})} & \text{if } p \neq q. \end{cases}$$

It follows that $\mathbb{E}(T) < \infty$ if and only if $p < q$.

4. (a) As usual,

$$G_2(s) = G(G(s)) = 1 - \alpha\{\alpha(1-s)^\beta\}^\beta = 1 - \alpha^{1+\beta}(1-s)^{\beta^2}.$$

This suggests that $G_n(s) = 1 - \alpha^{1+\beta+\cdots+\beta^{n-1}}(1-s)^{\beta^n}$ for $n \ge 1$; this formula may be proved easily by induction, using the fact that $G_n(s) = G(G_{n-1}(s))$.

(b) As in the above part (a),

$$G_2(s) = f^{-1}(P(f(f^{-1}(P(f(s)))))) = f^{-1}(P(P(f(s)))) = f^{-1}(P_2(f(s)))$$

where $P_2(s) = P(P(s))$. Similarly $G_n(s) = f^{-1}(P_n(f(s)))$ for $n \ge 1$, where $P_n(s) = P(P_{n-1}(s))$.

(c) With $P(s) = \alpha s/\{1-(1-\alpha)s\}$ where $\alpha = \gamma^{-1}$, it is an easy exercise to prove, by induction, that $P_n(s) = \alpha^n s/\{1-(1-\alpha^n)s\}$ for $n \ge 1$, implying that

$$G_n(s) = P_n(s^m)^{1/m} = \left\{\frac{\alpha^n s^m}{1-(1-\alpha^n)s^m}\right\}^{1/m}.$$

5. Let Z_n be the number of members of the nth generation. The $(n+1)$th generation has size $C_{n+1} + I_{n+1}$ where C_{n+1} is the number of natural offspring of the previous generation, and I_{n+1} is the number of immigrants. Therefore by the independence,

$$\mathbb{E}(s^{Z_{n+1}} \mid Z_n) = \mathbb{E}(s^{C_{n+1}} \mid Z_n)H(s) = G(s)^{Z_n}H(s),$$

whence

$$G_{n+1}(s) = \mathbb{E}(s^{Z_{n+1}}) = \mathbb{E}\{G(s)^{Z_n}\}H(s) = G_n(G(s))H(s).$$

6. By Example (5.4.3),

$$\mathbb{E}(s^{Z_n}) = \frac{n-(n-1)s}{n+1-ns} = \frac{n-1}{n} + \frac{1}{n^2(1+n^{-1}-s)}, \qquad n \ge 0.$$

Differentiate and set $s = 0$ to find that

$$\mathbb{E}(V_1) = \sum_{n=0}^{\infty}\mathbb{P}(Z_n = 1) = \sum_{n=0}^{\infty}\frac{1}{(n+1)^2} = \tfrac{1}{6}\pi^2.$$

Similarly,

$$\mathbb{E}(V_2) = \sum_{n=0}^{\infty}\frac{n}{(n+1)^3} = \sum_{n=0}^{\infty}\frac{1}{(n+1)^2} - \sum_{n=0}^{\infty}\frac{1}{(n+1)^3} = \tfrac{1}{6}\pi^2 - \sum_{n=0}^{\infty}\frac{1}{(n+1)^3},$$

$$\mathbb{E}(V_3) = \sum_{n=0}^{\infty}\frac{n^2}{(n+1)^4} = \sum_{n=0}^{\infty}\frac{(n+1)^2 - 2(n+1) + 1}{(n+1)^4} = \tfrac{1}{6}\pi^2 + \tfrac{1}{90}\pi^4 - 2\sum_{n=0}^{\infty}\frac{1}{(n+1)^3}.$$

The conclusion is obtained by eliminating $\sum_n (n+1)^{-3}$.

7. The family-size generating function is $G(s) = (q+ps)^2$, and the extinction probability η is the smallest non-negative root of $s = G(s)$, that is, $\eta = \min\{1, (q/p)^2\}$. By an easy calculation,

$$\mathbb{P}(X_1 = 1 \mid T < \infty) = 2pq,$$

$$\mathbb{P}(X_1 = 2 \mid T < \infty) = \begin{cases} q^2 & \text{if } q < p, \\ p^2 & \text{if } q \ge p, \end{cases}$$

whence

$$\mathbb{P}(X_1 = 0 \mid T < \infty) = \begin{cases} q^2 & \text{if } q < p, \\ p^2 & \text{if } q \geq p. \end{cases}$$

Using conditional expectation, $\mu = \mathbb{E}(T \mid T < \infty)$ satisfies

$$\mu = 1 + 2pq\mu + \begin{cases} 2q^2\mu & \text{if } q < p, \\ 2p^2\mu & \text{if } q \geq p. \end{cases}$$

The result follows. By the above, the conditional generating function is

$$\begin{cases} (p+qs)^2 & \text{if } q < p, \\ (q+ps)^2 & \text{if } q \geq p. \end{cases}$$

8. This follows by the Paley–Zygmund inequality of Problem (3.11.56b), with Lemma (5.4.2).

5.5 Solutions. Age-dependent branching processes

1. (i) The usual renewal argument shows as in Theorem (5.5.1) that

$$G_t(s) = \int_0^t G(G_{t-u}(s)) f_T(u)\, du + \int_t^\infty s f_T(u)\, du.$$

Differentiate with respect to t, to obtain

$$\frac{\partial}{\partial t} G_t(s) = G(G_0(s)) f_T(t) + \int_0^t \frac{\partial}{\partial t}\{G(G_{t-u}(s))\} f_T(u)\, du - s f_T(t).$$

Now $G_0(s) = s$, and

$$\frac{\partial}{\partial t}\{G(G_{t-u}(s))\} = -\frac{\partial}{\partial u}\{G(G_{t-u}(s))\},$$

so that, using the fact that $f_T(u) = \lambda e^{-\lambda u}$ if $u \geq 0$,

$$\int_0^t \frac{\partial}{\partial t}\{G(G_{t-u}(s))\} f_T(u)\, du = -\Big[G(G_{t-u}(s)) f_T(u)\Big]_0^t - \lambda \int_0^t G(G_{t-u}(s)) f_T(u)\, du,$$

having integrated by parts. Hence

$$\begin{aligned} \frac{\partial}{\partial t} G_t(s) &= G(s)\lambda e^{-\lambda t} + \Big\{-G(s)\lambda e^{-\lambda t} + \lambda G(G_t(s))\Big\} - \lambda\Big\{G_t(s) - \int_t^\infty s f_T(u)\, du\Big\} - s\lambda e^{-\lambda t} \\ &= \lambda\{G(G_t(s)) - G_t(s)\}. \end{aligned}$$

(ii) Substitute $G(s) = s^2$ into the last equation to obtain

$$\frac{\partial G_t}{\partial t} = \lambda(G_t^2 - G_t)$$

with boundary condition $G_0(s) = s$. Integrate to obtain $\lambda t + c(s) = \log\{1 - G_t^{-1}\}$ for some function $c(s)$. Using the boundary condition at $t = 0$, we find that $c(s) = \log\{1 - G_0^{-1}\} = \log\{1 - s^{-1}\}$, and hence $G_t(s) = se^{-\lambda t}/\{1 - s(1 - e^{-\lambda t})\}$. Expand in powers of s to find that $Z(t)$ has the geometric distribution $\mathbb{P}(Z(t) = k) = (1 - e^{-\lambda t})^{k-1} e^{-\lambda t}$ for $k \geq 1$.

2. The equation becomes

$$\frac{\partial G_t}{\partial t} = \tfrac{1}{2}(1 + G_t^2) - G_t$$

with boundary condition $G_0(s) = s$. This differential equation is easily solved with the result

$$G_t(s) = \frac{2s + t(1-s)}{2 + t(1-s)} = \frac{4/t}{2 + t(1-s)} - \frac{2-t}{t}.$$

We pick out the coefficient of s^n to obtain

$$\mathbb{P}(Z(t) = n) = \frac{4}{t(2+t)}\left(\frac{t}{2+t}\right)^n, \quad n \ge 1,$$

and therefore

$$\mathbb{P}(Z(t) \ge k) = \sum_{n=k}^{\infty} \frac{4}{t(2+t)}\left(\frac{t}{2+t}\right)^n = \frac{2}{t}\left(\frac{t}{2+t}\right)^k, \qquad k \ge 1.$$

It follows that, for $x > 0$ and in the limit as $t \to \infty$,

$$\mathbb{P}(Z(t) \ge xt \mid Z(t) > 0) = \frac{\mathbb{P}(Z(t) \ge xt)}{\mathbb{P}(Z(t) \ge 1)} = \left(\frac{t}{2+t}\right)^{\lceil xt \rceil - 1} = \left(1 + \frac{2}{t}\right)^{1 - \lceil xt \rceil} \to e^{-2x}.$$

5.6 Solutions. Expectation revisited

1. (a) (i) Set $a = \mathbb{E}(X)$ to find that $u(X) \ge u(\mathbb{E}X) + \lambda(X - \mathbb{E}X)$ for some fixed λ. Take expectations to obtain the result.

(ii) Let u be strictly convex. If equality holds, then $u(X) = u(\mathbb{E}X) + \lambda(X - \mathbb{E}X)$ a.s. Since u is strictly convex, it must be the case that $\mathbb{P}(X = \mathbb{E}X) = 1$.

(b) Let g be a density function with the given first two moments, and let f be the $N(\mu, \sigma^2)$ density function. The key step is the fact that

$$\int_{\mathbb{R}} g(x) \log f(x)\, dx = \int_{\mathbb{R}} f(x) \log f(x)\, dx,$$

which holds since the left and right sides are the same linear function of the respective variances of g and f. Therefore,

$$\begin{aligned} H(g) - H(f) &= -\int_{\mathbb{R}} g(x) \log g(x)\, dx + \int_{\mathbb{R}} f(x) \log f(x)\, dx \\ &= -\int_{\mathbb{R}} g(x) \log g(x)\, dx + \int_{\mathbb{R}} g(x) \log f(x)\, dx \\ &= \int_{\mathbb{R}} g(x) \log\big(f(x)/g(x)\big)\, dx \le \log\left(\int_{\mathbb{R}} g(x)\big(f(x)/g(x)\big)\, dx\right) = 0, \end{aligned}$$

where we used Jensen's inequality at the last stage with the convex function $u(x) = -\log x$. Since u is strictly convex on $(0, \infty)$, equality holds if and only if $f(x)/g(x) = c$ for some constant c, which necessarily equals 1.

2. Certainly $Z_n = \sum_{i=1}^{n} X_i$ and $Z = \sum_{i=1}^{\infty} |X_i|$ are such that $|Z_n| \le Z$, and the result follows by dominated convergence.

3. Apply Fatou's lemma to the sequence $\{-X_n : n \ge 1\}$ to find that

$$\mathbb{E}\Big(\limsup_{n\to\infty} X_n\Big) = -\mathbb{E}\Big(\liminf_{n\to\infty} -X_n\Big) \ge -\liminf_{n\to\infty} \mathbb{E}(-X_n) = \limsup_{n\to\infty} \mathbb{E}(X_n).$$

4. Suppose that $\mathbb{E}|X^r| < \infty$ where $r > 0$. We have that, if $x > 0$,

$$x^r \mathbb{P}(|X| \ge x) \le \int_{[x,\infty)} u^r \, dF(u) \to 0 \quad \text{as } x \to \infty,$$

where F is the distribution function of $|X|$.

Conversely suppose that $x^r \mathbb{P}(|X| \ge x) \to 0$ where $r \ge 0$, and let $0 \le s < r$. Now $\mathbb{E}|X^s| = \lim_{M\to\infty} \int_0^M u^s \, dF(u)$ and, by integration by parts,

$$\int_0^M u^s \, dF(u) = \Big[-u^s(1 - F(u))\Big]_0^M + \int_0^M s u^{s-1}(1 - F(u)) \, du.$$

The first term on the right-hand side is negative. The integrand in the second term satisfies $su^{s-1}\mathbb{P}(|X| > u) \le su^{s-1} \cdot u^{-r}$ for all large u. Therefore the integral is bounded uniformly in M, as required.

5. Suppose first that, for all $\epsilon > 0$, there exists $\delta = \delta(\epsilon) > 0$, such that $\mathbb{E}(|X|I_A) < \epsilon$ for all A satisfying $\mathbb{P}(A) < \delta$. Fix $\epsilon > 0$, and find x (> 0) such that $\mathbb{P}(|X| > x) < \delta(\epsilon)$. Then, for $y > x$,

$$\int_{-y}^{y} |u| \, dF_X(u) \le \int_{-x}^{x} |u| \, dF_X(u) + \mathbb{E}\left(|X| I_{\{|X|>x\}}\right) \le \int_{-x}^{x} |u| \, dF_X(u) + \epsilon.$$

Hence $\int_{-y}^{y} |u| \, dF_X(u)$ converges as $y \to \infty$, whence $\mathbb{E}|X| < \infty$.

Conversely suppose that $\mathbb{E}|X| < \infty$. It follows that $\mathbb{E}\left(|X| I_{\{|X|>y\}}\right) \to 0$ as $y \to \infty$. Let $\epsilon > 0$, and find y such that $\mathbb{E}\left(|X| I_{\{|X|>y\}}\right) < \frac{1}{2}\epsilon$. For any event A, $I_A \le I_{A\cap B^c} + I_B$ where $B = \{|X| > y\}$. Hence

$$\mathbb{E}(|X|I_A) \le \mathbb{E}\left(|X| I_{A\cap B^c}\right) + \mathbb{E}\left(|X| I_B\right) \le y\mathbb{P}(A) + \tfrac{1}{2}\epsilon.$$

Writing $\delta = \epsilon/(2y)$, we have that $\mathbb{E}(|X|I_A) < \epsilon$ if $\mathbb{P}(A) < \delta$.

6. Let $S = \min\{X, Y\}$, so that $X+Y = M+S$. Then $(X+Y)^2 = (M+S)^2$ and $(X-Y)^2 = (M-S)^2$. Take expectations and add to obtain

$$\mathbb{E}(X^2) + \mathbb{E}(Y^2) = \mathbb{E}(M^2) + \mathbb{E}(S^2).$$

Also, $X + Y = M + S$ and $|X - Y| = M - S$. Take expectations, square, and add:

$$\begin{aligned}
\mathbb{E}(M)^2 + \mathbb{E}(S)^2 &= \mathbb{E}(X)\mathbb{E}(Y) + \tfrac{1}{2}\mathbb{E}(X)^2 + \tfrac{1}{2}\mathbb{E}(Y)^2 + \tfrac{1}{2}(\mathbb{E}|X - Y|)^2 \\
&\ge \mathbb{E}(X)\mathbb{E}(Y) + \tfrac{1}{2}\mathbb{E}(X)^2 + \tfrac{1}{2}\mathbb{E}(Y)^2 + \tfrac{1}{2}(\mathbb{E}X - \mathbb{E}Y)^2 \quad \text{by Jensen's inequality} \\
&= \mathbb{E}(X)^2 + \mathbb{E}(Y)^2.
\end{aligned}$$

Hence $\operatorname{var}(M) + \operatorname{var}(S) \le \operatorname{var}(X) + \operatorname{var}(Y)$.

7. Let $R = 1/S$ when $S > 0$, and $R = 0$ otherwise. Then

$$\begin{aligned}\mathbb{P}(S > 0) = \mathbb{E}(RS) &= \sum_{r=1}^{n} \mathbb{E}(RI_r)\\ &= \sum_{r=1}^{n} \Big\{\mathbb{E}(RI_r \mid I_r = 1)\mathbb{P}(I_r = 1) + \mathbb{E}(RI_r \mid I_r = 0)\mathbb{P}(I_r = 0)\Big\}\\ &\ge \sum_{r=1}^{n} \mathbb{E}(S^{-1} \mid I_r = 1)\mathbb{P}(I_r = 1)\\ &\ge \sum_{r=1}^{n} \frac{\mathbb{P}(I_r = 1)}{\mathbb{E}(S \mid I_r = 1)},\end{aligned}$$

by Jensen's inequality applied with the convex function $u(x) = 1/x$ for $x > 0$.

5.7 Solutions. Characteristic functions

1. Let X have the Cauchy distribution, with characteristic function $\phi(s) = e^{-|s|}$. Setting $Y = X$, we have that $\phi_{X+Y}(t) = \phi(2t) = e^{-2|t|} = \phi_X(t)\phi_Y(t)$. However, X and Y are certainly dependent.

2. (i) It is the case that $\text{Re}\{\phi(t)\} = \mathbb{E}(\cos tX)$, so that, in the obvious notation,

$$\begin{aligned}\text{Re}\{1 - \phi(2t)\} &= \int_{-\infty}^{\infty} \{1 - \cos(2tx)\}\, dF(x) = 2\int_{-\infty}^{\infty} \{1 - \cos(tx)\}\{1 + \cos(tx)\}\, dF(x)\\ &\le 4\int_{-\infty}^{\infty} \{1 - \cos(tx)\}\, dF(x) = 4\,\text{Re}\{1 - \phi(t)\}.\end{aligned}$$

(ii) Note first that, if X and Y are independent with common characteristic function ϕ, then $X - Y$ has characteristic function

$$\psi(t) = \mathbb{E}(e^{itX})\mathbb{E}(e^{-itY}) = \phi(t)\phi(-t) = \phi(t)\overline{\phi(t)} = |\phi(t)|^2.$$

Apply the result of part (i) to the function ψ to obtain that $1 - |\phi(2t)|^2 \le 4(1 - |\phi(t)|^2)$. However $|\phi(t)| \le 1$, so that

$$1 - |\phi(2t)| \le 1 - |\phi(2t)|^2 \le 4(1 - |\phi(t)|^2) \le 8(1 - |\phi(t)|).$$

3. (a) With $m_k = \mathbb{E}(X^k)$, we have that

$$\mathbb{E}(e^{\theta X}) = 1 + \sum_{k=1}^{\infty} \frac{1}{k!} m_k \theta^k = 1 + S(\theta),$$

say, and therefore, for sufficiently small values of θ,

$$K_X(\theta) = \sum_{r=1}^{\infty} \frac{(-1)^{r+1}}{r} S(\theta)^r.$$

Expand $S(\theta)^r$ in powers of θ, and equate the coefficients of $\theta, \theta^2, \theta^3$, in turn, to find that $k_1(X) = m_1$, $k_2(X) = m_2 - m_1^2$, $k_3(X) = m_3 - 3m_1m_2 + 2m_1^3$.

(b) If X and Y are independent, $K_{X+Y}(\theta) = \log\{\mathbb{E}(e^{\theta X})\mathbb{E}(e^{\theta Y})\} = K_X(\theta) + K_Y(\theta)$, whence the claim is immediate.

4. The $N(0,1)$ variable X has moment generating function $\mathbb{E}(e^{\theta X}) = e^{\frac{1}{2}\theta^2}$, so that $K_X(\theta) = \frac{1}{2}\theta^2$.

5. (a) Suppose X takes values in $L(a,b)$. Then

$$|\phi_X(2\pi/b)| = \left|\sum_x e^{2\pi i x/b}\mathbb{P}(X=x)\right| = |e^{2\pi i a/b}|\left|\sum_m e^{2\pi i m}\mathbb{P}(X=a+bm)\right| = 1$$

since only numbers of the form $x = a+bm$ make non-zero contributions to the sum.

Suppose in addition that X has span b, and that $|\phi_X(T)| = 1$ for some $T \in (0, 2\pi/b)$. Then $\phi_X(T) = e^{ic}$ for some $c \in \mathbb{R}$. Now

$$\mathbb{E}\big(\cos(TX-c)\big) = \tfrac{1}{2}\mathbb{E}\big(e^{iTX-ic} + e^{-iTX+ic}\big) = 1,$$

using the fact that $\mathbb{E}(e^{-iTX}) = \overline{\phi_X(T)} = e^{-ic}$. However $\cos x \le 1$ for all x, with equality if and only if x is a multiple of 2π. It follows that $TX - c$ is a multiple of 2π, with probability 1, and hence that X takes values in the set $L(c/T, 2\pi/T)$. However $2\pi/T > b$, which contradicts the maximality of the span b. We deduce that no such T exists.

(b) This follows by the argument above.

6. This is a form of the 'Riemann–Lebesgue lemma'. It is a standard result of analysis that, for $\epsilon > 0$, there exists a step function g_ϵ such that $\int_{-\infty}^{\infty} |f(x) - g_\epsilon(x)|\,dx < \epsilon$. Let $\phi_\epsilon(t) = \int_{-\infty}^{\infty} e^{itx} g_\epsilon(x)\,dx$. Then

$$|\phi_X(t) - \phi_\epsilon(t)| = \left|\int_{-\infty}^{\infty} e^{itx}\big(f(x) - g_\epsilon(x)\big)\,dx\right| \le \int_{-\infty}^{\infty} |f(x) - g_\epsilon(x)|\,dx < \epsilon.$$

If we can prove that, for each ϵ, $|\phi_\epsilon(t)| \to 0$ as $t \to \pm\infty$, then it will follow that $|\phi_X(t)| < 2\epsilon$ for all large t, and the claim then follows.

Now $g_\epsilon(x)$ is a finite linear combination of functions of the form $cI_A(x)$ for reals c and intervals A, that is $g_\epsilon(x) = \sum_{k=1}^{K} c_k I_{A_k}(x)$; elementary integration yields

$$\phi_\epsilon(t) = \sum_{k=1}^{K} c_k \frac{e^{itb_k} - e^{ita_k}}{it}$$

where a_k and b_k are the endpoints of A_k. Therefore

$$|\phi_\epsilon(t)| \le \frac{2}{t}\sum_{k=1}^{K} c_k \to 0, \qquad \text{as } t \to \pm\infty.$$

7. If X is $N(\mu, 1)$, then the moment generating function of X^2 is

$$M_{X^2}(s) = \mathbb{E}(e^{sX^2}) = \int_{-\infty}^{\infty} e^{sx^2}\frac{1}{\sqrt{2\pi}}e^{-\frac{1}{2}(x-\mu)^2}\,dx = \frac{1}{\sqrt{1-2s}}\exp\left(\frac{\mu^2 s}{1-2s}\right),$$

if $s < \frac{1}{2}$, by completing the square in the exponent. It follows that

$$M_Y(s) = \prod_{j=1}^{n}\left\{\frac{1}{\sqrt{1-2s}}\exp\left(\frac{\mu_j^2 s}{1-2s}\right)\right\} = \frac{1}{(1-2s)^{n/2}}\exp\left(\frac{s\theta}{1-2s}\right).$$

It is tempting to substitute $s = it$ to obtain the answer. This procedure may be justified in this case using the theory of analytic continuation.

8. (a) $T^2 = X^2/(Y/n)$, where X^2 is $\chi^2(1; \mu^2)$ by Exercise (5.7.7), and Y is $\chi^2(n)$. Hence T^2 is $F(1, n; \mu^2)$.

(b) F has the same distribution function as

$$Z = \frac{(A^2 + B)/m}{V/n}$$

where A, B, V are independent, A being $N(\sqrt{\theta}, 1)$, B being $\chi^2(m-1)$, and V being $\chi^2(n)$. Therefore

$$\begin{aligned}\mathbb{E}(Z) &= \frac{1}{m}\left\{\mathbb{E}(A^2)\mathbb{E}\left(\frac{n}{V}\right) + (m-1)\mathbb{E}\left(\frac{B/(m-1)}{V/n}\right)\right\}\\ &= \frac{1}{m}\left\{(1+\theta)\frac{n}{n-2} + (m-1)\frac{n}{n-2}\right\} = \frac{n(m+\theta)}{m(n-2)},\end{aligned}$$

where we have used the fact (see Exercise (4.10.2)) that the $F(r, s)$ distribution has mean $s/(s-2)$ if $s > 2$.

9. Let $\widetilde{X}$ be independent of X with the same distribution. Then $|\phi|^2$ is the characteristic function of $X - \widetilde{X}$ and, by the inversion theorem,

$$\frac{1}{2\pi}\int_{-\infty}^{\infty} |\phi(t)|^2 e^{-itx}\,dt = f_{X-\widetilde{X}}(x) = \int_{-\infty}^{\infty} f(y)f(x+y)\,dy.$$

Now set $x = 0$. We require that the density function of $X - \widetilde{X}$ be differentiable at 0.

10. By definition,

$$e^{-ity}\phi_X(y) = \int_{-\infty}^{\infty} e^{iy(x-t)} f_X(x)\,dx.$$

Now multiply by $f_Y(y)$, integrate over $y \in \mathbb{R}$, and change the order of integration with an appeal to Fubini's theorem.

11. (a) We adopt the usual convention that integrals of the form $\int_u^v g(y)\,dF(y)$ include any atom of the distribution function F at the upper endpoint v but not at the lower endpoint u. It is a consequence that F_τ is right-continuous, and it is immediate that F_τ increases from 0 to 1. Therefore F_τ is a distribution function. The corresponding moment generating function is

$$M_\tau(t) = \int_{-\infty}^{\infty} e^{tx}\,dF_\tau(x) = \frac{1}{M(t)}\int_{-\infty}^{\infty} e^{tx+\tau x}\,dF(x) = \frac{M(t+\tau)}{M(t)}.$$

(b) The required moment generating function is

$$\frac{M_{X+Y}(t+\tau)}{M_{X+Y}(t)} = \frac{M_X(t+\tau)M_Y(t+\tau)}{M_X(t)M_Y(t)},$$

the product of the moment generating functions of the individual tilted distributions.

12. Change variables to polar coordinates and integrate over $\theta \in [0, 2\pi]$ to obtain the desired marginal density function.

13. (a) We have

$$\begin{aligned}\mathbb{E}(e^{sX+tY}) &= \int_{x=0}^{\infty}\int_{y=x}^{\infty} 2e^{-(1-s)x}e^{-(1-t)y}\,dx\,dy\\ &= 2\int_0^{\infty} e^{-(1-s)x}\cdot\frac{e^{-(1-t)x}}{1-t}\,dx = \frac{2}{(1-t)(2-s-t)}, \qquad s, t < 1.\end{aligned}$$

The coefficients of s, t, and st in the series expansion of this yield

$$\text{cov}(X, Y) = \mathbb{E}(XY) - \mathbb{E}(X)\mathbb{E}(Y) = 1 - \tfrac{1}{2} \cdot \tfrac{3}{2} = \tfrac{1}{4}.$$

(b) This time,

$$\begin{aligned}\mathbb{E}(e^{sX+tY}) &= \frac{e^{\frac{1}{2}(s^2+t^2)}}{2\pi(2+c^2)} \iint_{\mathbb{R}^2} (c^2 + x^2 + y^2) \exp\left\{-\tfrac{1}{2}(x-s)^2 - \tfrac{1}{2}(y-t)^2\right\} dx\, dy \\ &= \frac{e^{\frac{1}{2}(s^2+t^2)}}{2\pi(2+c^2)} \int_{-\infty}^{\infty} e^{-\frac{1}{2}(x-s)^2} \left\{(c^2+x^2)\sqrt{2\pi} + (1+t^2)\sqrt{2\pi}\right\} dx \\ &= \frac{e^{\frac{1}{2}(s^2+t^2)}}{2+c^2} \{c^2 + 1 + s^2 + 1 + t^2\}.\end{aligned}$$

The coefficients of s, t, and st are zero, so $\text{cov}(X, Y) = 0$. However, the joint moment generating function does not factorize, so X and Y are not independent.

5.8 Solutions. Examples of characteristic functions

1. (i) We have that $\overline{\phi(t)} = \overline{\mathbb{E}(e^{itX})} = \mathbb{E}(e^{-itX}) = \phi_{-X}(t)$.
(ii) If X_1 and X_2 are independent random variables with common characteristic function ϕ, then $\phi_{X_1+X_2}(t) = \phi_{X_1}(t)\phi_{X_2}(t) = \phi(t)^2$.
(iii) Similarly, $\phi_{X_1-X_2}(t) = \phi_{X_1}(t)\phi_{-X_2}(t) = \phi(t)\overline{\phi(t)} = |\phi(t)|^2$.
(iv) Let X have characteristic function ϕ, and let Z be equal to X with probability $\frac{1}{2}$ and to $-X$ otherwise. The characteristic function of Z is given by

$$\phi_Z(t) = \tfrac{1}{2}\left(\mathbb{E}(e^{itX}) + \mathbb{E}(e^{-itX})\right) = \tfrac{1}{2}\left(\phi(t) + \overline{\phi(t)}\right) = \text{Re}(\phi)(t),$$

where we have used the argument of part (i) above.
(v) If X is Bernoulli with parameter $\frac{1}{3}$, then its characteristic function is $\phi(t) = \frac{2}{3} + \frac{1}{3}e^{it}$. Suppose Y is a random variable with characteristic function $\psi(t) = |\phi(t)|$. Then $\psi(t)^2 = \phi(t)\phi(-t)$. Written in terms of random variables this asserts that $Y_1 + Y_2$ has the same distribution as $X_1 - X_2$, where the Y_i are independent with characteristic function ψ, and the X_i are independent with characteristic function ϕ. Now $X_j \in \{0, 1\}$, so that $X_1 - X_2 \in \{-1, 0, 1\}$, and therefore $Y_j \in \{-\frac{1}{2}, \frac{1}{2}\}$. Write $\alpha = \mathbb{P}(Y_j = \frac{1}{2})$. Then

$$\begin{aligned}\mathbb{P}(Y_1 + Y_2 = 1) &= \alpha^2 = \mathbb{P}(X_1 - X_2 = 1) = \tfrac{2}{9}, \\ \mathbb{P}(Y_1 + Y_2 = -1) &= (1-\alpha)^2 = \mathbb{P}(X_1 - X_2 = -1) = \tfrac{2}{9},\end{aligned}$$

implying that $\alpha^2 = (1-\alpha)^2$ so that $\alpha = \frac{1}{2}$, contradicting the fact that $\alpha^2 = \frac{2}{9}$. We deduce that no such variable Y exists.

2. For $t \geq 0$,

$$\mathbb{P}(X \geq x) = \mathbb{P}(e^{tX} \geq e^{tx}) \leq e^{-tx}\mathbb{E}(e^{tX}).$$

Now minimize over $t \geq 0$. When X is $N(0, 1)$ and $x > 0$, we have $M(t) = e^{\frac{1}{2}t^2}$, and the infimum in the first part is achieved when $t = x$.

3. The moment generating function of Z is

$$M_Z(t) = \mathbb{E}\left\{\mathbb{E}(e^{tXY} \mid Y)\right\} = \mathbb{E}\{M_X(tY)\} = \mathbb{E}\left\{\left(\frac{\lambda}{\lambda - tY}\right)^m\right\}$$
$$= \int_0^1 \left(\frac{\lambda}{\lambda - ty}\right)^m \frac{y^{n-1}(1-y)^{m-n-1}}{B(n, m-n)}\, dy.$$

Substitute $v = 1/y$ and integrate by parts to obtain that

$$I_{mn} = \int_1^\infty \frac{(v-1)^{m-n-1}}{(\lambda v - t)^m}\, dv$$

satisfies

$$I_{mn} = \left[-\frac{1}{\lambda(m-1)} \frac{(v-1)^{m-n-1}}{(\lambda v - t)^{m-1}}\right]_1^\infty + \frac{m-n-1}{\lambda(m-1)} I_{m-1,n} = c(m, n, \lambda) I_{m-1,n}$$

for some $c(m, n, \lambda)$. We iterate this to obtain

$$I_{mn} = c' I_{n+1,n} = c' \int_1^\infty \frac{dv}{(\lambda v - t)^{n+1}} = \frac{c'}{n\lambda} \cdot \frac{1}{(\lambda - t)^n}$$

for some c' depending on m, n, λ. Therefore $M_Z(t) = c''(\lambda - t)^{-n}$ for some c'' depending on m, n, λ. However $M_Z(0) = 1$, and hence $c'' = \lambda^n$, giving that Z is $\Gamma(\lambda, n)$. Throughout these calculations we have assumed that t is sufficiently small and positive. Alternatively, we could have set $t = is$ and used characteristic functions. See also Problem (4.14.12).

4. We have that

$$\mathbb{E}(e^{itX^2}) = \int_{-\infty}^\infty e^{itx^2} \frac{1}{\sqrt{2\pi\sigma^2}} \exp\left(-\frac{(x-\mu)^2}{2\sigma^2}\right) dx$$
$$= \int_{-\infty}^\infty \frac{1}{\sqrt{2\pi\sigma^2}} \exp\left(-\frac{\left[x - \mu(1 - 2\sigma^2 it)^{-1}\right]^2}{2\sigma^2(1 - 2\sigma^2 it)^{-1}}\right) \exp\left(\frac{it\mu^2}{1 - 2\sigma^2 it}\right) dx$$
$$= \frac{1}{\sqrt{1 - 2\sigma^2 it}} \exp\left(\frac{it\mu^2}{1 - 2\sigma^2 it}\right).$$

The integral is evaluated by using Cauchy's theorem when integrating around a sector in the complex plane. It is highly suggestive to observe that the integrand differs only by a multiplicative constant from a hypothetical normal density function with (complex) mean $\mu(1 - 2\sigma^2 it)^{-1}$ and (complex) variance $\sigma^2(1 - 2\sigma^2 it)^{-1}$.

5. (a) Use the result of Exercise (5.8.4) with $\mu = 0$ and $\sigma^2 = 1$: $\phi_{X_1^2}(t) = (1 - 2it)^{-\frac{1}{2}}$, the characteristic function of the $\chi^2(1)$ distribution.

(b) From (a), the sum S has characteristic function $\phi_S(t) = (1 - 2it)^{-\frac{1}{2}n}$, the characteristic function of the $\chi^2(n)$ distribution.

(c) We have that

$$\mathbb{E}(e^{itX_1/X_2}) = \mathbb{E}\{\mathbb{E}(e^{itX_1/X_2} \mid X_2)\} = \mathbb{E}\left(\phi_{X_1}(t/X_2)\right) = \mathbb{E}\left(\exp\{-\tfrac{1}{2}t^2/X_2^2\}\right).$$

Now

$$\mathbb{E}(\exp\{-\tfrac{1}{2}t^2/X_2^2\}) = \int_{-\infty}^{\infty} \frac{1}{\sqrt{2\pi}} \exp\left(-\frac{t^2}{2x^2} - \frac{x^2}{2}\right) dx.$$

There are various ways of evaluating this integral. Using the result of Problem (5.12.18c), we find that the answer is $e^{-|t|}$, whence X_1/X_2 has the Cauchy distribution.

(d) We have that

$$\begin{aligned}\mathbb{E}(e^{itX_1X_2}) &= \mathbb{E}\{\mathbb{E}(e^{itX_1X_2} \mid X_2)\} = \mathbb{E}(\phi_{X_1}(tX_2)) = \mathbb{E}(e^{-\frac{1}{2}t^2X_2^2}) \\ &= \int_{-\infty}^{\infty} \frac{1}{\sqrt{2\pi}} \exp\{-\tfrac{1}{2}x^2(1+t^2)\}\, dx = \frac{1}{\sqrt{1+t^2}},\end{aligned}$$

on observing that the integrand differs from the $N(0, (1+t^2)^{-\frac{1}{2}})$ density function only by a multiplicative constant. Now, examination of a standard work of reference, such as Abramowitz and Stegun (1965, Section 9.6.21), reveals that

$$\int_0^{\infty} \frac{\cos(xt)}{\sqrt{1+t^2}}\, dt = K_0(x),$$

where $K_0(x)$ is the second kind of modified Bessel function. Hence the required density, by the inversion theorem, is $f(x) = K_0(|x|)/\pi$. Note that, for small x, $K_0(x) \sim -\log x$, and for large positive x, $K_0(x) \sim e^{-x}\sqrt{\pi x/2}$.

As a matter of interest, note that we may also invert the more general characteristic function $\phi(t) = (1-it)^{-\alpha}(1+it)^{-\beta}$. Setting $1 - it = -z/x$ in the integral gives

$$\begin{aligned}f(x) &= \frac{1}{2\pi}\int_{-\infty}^{\infty} \frac{e^{-itx}}{(1-it)^{\alpha}(1+it)^{\beta}}\, dt = \frac{e^{-x}x^{\alpha-1}}{2^{\beta}2\pi i}\int_{-x-ix\infty}^{-x+ix\infty} \frac{e^{-z}\, dz}{(-z)^{\alpha}\,(1+z/(2x))^{\beta}} \\ &= \frac{e^x(2x)^{\frac{1}{2}(\beta-\alpha)}}{\Gamma(\alpha)} W_{\frac{1}{2}(\alpha-\beta),\frac{1}{2}(1-\alpha-\beta)}(2x)\end{aligned}$$

where W is a confluent hypergeometric function. When $\alpha = \beta$ this becomes

$$f(x) = \frac{(x/2)^{\alpha-\frac{1}{2}}}{\Gamma(\alpha)\sqrt{\pi}} K_{\alpha-\frac{1}{2}}(x)$$

where K is a Bessel function of the second kind.

(e) Using (d), we find that the required characteristic function is $\phi_{X_1X_2}(t)\phi_{X_3X_4}(t) = (1+t^2)^{-1}$. In order to invert this, either use the inversion theorem for the Cauchy distribution to find the required density to be $f(x) = \frac{1}{2}e^{-|x|}$ for $-\infty < x < \infty$, or alternatively express $(1+t^2)^{-1}$ as partial fractions, $(1+t^2)^{-1} = \frac{1}{2}\{(1-it)^{-1} + (1+it)^{-1}\}$, and recall that $(1-it)^{-1}$ is the characteristic function of an exponential distribution.

6. The joint characteristic function of $\mathbf{X} = (X_1, X_2, \ldots, X_n)$ satisfies $\phi_{\mathbf{X}}(\mathbf{t}) = \mathbb{E}(e^{i\mathbf{t}\mathbf{X}'}) = \mathbb{E}(e^{iY})$ where $\mathbf{t} = (t_1, t_2, \ldots, t_n) \in \mathbb{R}^n$ and $Y = \mathbf{t}\mathbf{X}' = t_1X_1 + \cdots + t_nX_n$. Now Y is normal with mean and variance

$$\mathbb{E}(Y) = \sum_{j=1}^{n} t_j\mathbb{E}(X_j) = \mathbf{t}\boldsymbol{\mu}', \qquad \operatorname{var}(Y) = \sum_{j,k=1}^{n} t_jt_k\operatorname{cov}(X_j, X_k) = \mathbf{t}\mathbf{V}\mathbf{t}',$$

where $\boldsymbol{\mu}$ is the mean vector of $\mathbf{X}$, and $\mathbf{V}$ is the covariance matrix of $\mathbf{X}$. Therefore $\phi_{\mathbf{X}}(\mathbf{t}) = \phi_Y(1) = \exp(i\mathbf{t}\boldsymbol{\mu}' - \frac{1}{2}\mathbf{t}\mathbf{V}\mathbf{t}')$ by paragraph (5.8.5).

Let $\mathbf{Z} = \mathbf{X} - \boldsymbol{\mu}$. It is easy to check that the vector $\mathbf{Z}$ has joint characteristic function $\phi_{\mathbf{Z}}(\mathbf{t}) = e^{-\frac{1}{2}\mathbf{t}\mathbf{V}\mathbf{t}'}$, which we recognize by (5.8.6) as being that of the $N(\mathbf{0}, \mathbf{V})$ distribution.

7. We have that $\mathbb{E}(Z) = 0$, $\mathbb{E}(Z^2) = 1$, and $\mathbb{E}(e^{tZ}) = \mathbb{E}\{\mathbb{E}(e^{tZ} \mid U, V)\} = \mathbb{E}(e^{\frac{1}{2}t^2}) = e^{\frac{1}{2}t^2}$. If X and Y have the bivariate normal distribution with correlation ρ, then the random variable $Z = (UX + VY)/\sqrt{U^2 + 2\rho UV + V^2}$ is $N(0, 1)$.

8. By definition, $\mathbb{E}(e^{itX}) = \mathbb{E}(\cos(tX)) + i\mathbb{E}(\sin(tX))$. By integrating by parts,

$$\int_0^\infty \cos(tx)\lambda e^{-\lambda x}\,dx = \frac{\lambda^2}{\lambda^2 + t^2}, \quad \int_0^\infty \sin(tx)\lambda e^{-\lambda x}\,dx = \frac{\lambda t}{\lambda^2 + t^2},$$

and

$$\frac{\lambda^2 + i\lambda t}{\lambda^2 + t^2} = \frac{\lambda}{\lambda - it}.$$

9. (a) We have that $e^{-|x|} = e^{-x}I_{\{x \ge 0\}} + e^{x}I_{\{x < 0\}}$, whence the required characteristic function is

$$\phi(t) = \frac{1}{2}\left(\frac{1}{1 - it} + \frac{1}{1 + it}\right) = \frac{1}{1 + t^2}.$$

(b) By a similar argument applied to the $\Gamma(1, 2)$ distribution, we have in this case that

$$\phi(t) = \frac{1}{2}\left(\frac{1}{(1 - it)^2} + \frac{1}{(1 + it)^2}\right) = \frac{1 - t^2}{(1 + t^2)^2}.$$

10. Suppose X has moment generating function $M(t)$. The proposed equation gives

$$M(t) = \int_0^1 M(ut)^2\,du = \frac{1}{t}\int_0^t M(v)^2\,dv.$$

Differentiate to obtain $tM' + M = M^2$, with solution $M(t) = \lambda/(\lambda + t)$. Thus the exponential distribution has the stated property.

11. We have that

$$\phi_{X,Y}(s, t) = \mathbb{E}(e^{isX + itY}) = \phi_{sX + tY}(1).$$

Now $sX + tY$ is $N(0, s^2\sigma^2 + 2st\sigma\tau\rho + \tau^2)$ where $\sigma^2 = \text{var}(X)$, $\tau^2 = \text{var}(Y)$, $\rho = \text{corr}(X, Y)$, and therefore

$$\phi_{X,Y}(s, t) = \exp\{-\tfrac{1}{2}(s^2\sigma^2 + 2st\sigma\tau\rho + t^2\tau^2)\}.$$

The fact that $\phi_{X,Y}$ may be expressed in terms of the characteristic function of a single normal variable is sometimes referred to as the *Cramér–Wold device*.

12. By writing $Y_i = (X_i - \mu)/\sigma$, we may assume without loss of generality (if we like) that $\mu = 0$ and $\sigma = 1$. The first part is proved as in the solution to Exercise (4.10.5).

The second displayed equation is by expansion. The two terms on the left side are independent, and the squared normals have χ^2 distributions. Taking characteristic functions, the characteristic function of $T = (n - 1)S^2/\sigma^2$ satisfies

$$\phi_T(t) \cdot \frac{1}{(1 - 2it)^{1/2}} = \frac{1}{(1 - 2it)^{n/2}},$$

and hence T has the $\chi^2(n - 1)$ distribution.

13. The linear combination $Y = \theta_1 X_1 + \theta_2 X_2 + \cdots + \theta_n X_n$ has a zero-mean normal distribution, and so

$$\sum_{r=0}^{\infty} \frac{i^r}{r!} \mathbb{E}(Y^r) = \mathbb{E}(e^{iY}) = e^{-\frac{1}{2}\mathbb{E}(Y^2)} = \sum_{k=0}^{\infty} \frac{(-1)^k}{2^k k!} \mathbb{E}(Y^2)^k.$$

The coefficient of $\theta_1 \theta_2 \cdots \theta_n$ in $\mathbb{E}(Y^r)$ is $n!\,\mathbb{E}(X_1 X_2 \cdots X_n)$ when $n = r$, and 0 otherwise; while the coefficient of $\theta_1 \theta_2 \cdots \theta_{2m}$ in $\mathbb{E}(Y^2)^m$ is $2^m m! \sum_r \prod_{i_r < j_r} \mathbb{E}(X_{i_r} X_{j_r})$. Now equate these.

Thus, for example, $\mathbb{E}(X_1 X_2 X_3) = 0$, while

$$\mathbb{E}(X_1 X_2 X_3 X_4) = \mathbb{E}(X_1 X_2)\mathbb{E}(X_3 X_4) + \mathbb{E}(X_1 X_3)\mathbb{E}(X_2 X_4) + \mathbb{E}(X_1 X_4)\mathbb{E}(X_2 X_3).$$

14. The moment generating function of S is

$$M_S(t) = \prod_{r=1}^{n} M_{X_r}(t) = \prod_{r=1}^{n} \frac{\lambda_r}{\lambda_r - t},$$

by the independence of the X_r.

If the λ_r are distinct, this may be expanded thus using partial fractions:

$$M_S(t) = \sum_{r=1}^{n} \frac{\lambda_r}{\lambda_r - t} \prod_{\substack{s=1 \\ s \neq r}}^{n} \frac{\lambda_s}{\lambda_s - \lambda_r}.$$

By inverting this, we find that the density function of S is a weighted sum of the functions $e^{-\lambda_r x}$.

Setting $t = 0$ in the above yields the first identity. Differentiating and setting $t = 0$ yields the second.

15. This is essentially Stein's equation of Exercise (4.7.25). By the change of variables $x = y + \theta$, with ϕ the $N(0, 1)$ density function,

$$\begin{aligned}
\mathbb{E}(e^{\theta X} f(X)) &= \int e^{\theta x} f(x)\phi(x)\,dx = \int e^{\theta(y+\theta)} f(y+\theta)\phi(y+\theta)\,dy \\
&= \frac{1}{\sqrt{2\pi}} \int \exp\{\theta y + \theta^2 - \tfrac{1}{2}(y+\theta)^2\} f(y+\theta)\,dy \\
&= e^{\frac{1}{2}\theta^2} \int f(y+\theta)\phi(y)\,dy.
\end{aligned}$$

Assuming we may differentiate through the integral sign, do so with respect to θ and set $\theta = 0$ for the second part.

16. Write $e^{itx} = \cos(tx) + i\sin(tx)$, and use the fact that sine is an odd function, to find that

$$\phi(t) = \frac{1}{\sqrt{2\pi}} \int_{-\infty}^{\infty} \cos(tx) e^{-\frac{1}{2}x^2}\,dx.$$

Differentiate through the integral sign, and integrate by parts, to obtain

$$\begin{aligned}
\sqrt{2\pi}\phi'(t) &= -\int_{-\infty}^{\infty} \sin(tx) x e^{-\frac{1}{2}x^2}\,dx \\
&= -\int_{-\infty}^{\infty} t\cos(tx) e^{-\frac{1}{2}x^2}\,dx = -t\sqrt{2\pi}\phi(t).
\end{aligned}$$

The interchange of integral and derivative may be justified by the fact that $e^{-\frac{1}{2}x^2}$ is bounded and integrable. Now solve the differential equation $\phi'(t) = -t\phi(t)$ subject to $\phi(0) = 1$.

17. (a) We have

$$\phi_Y(t) = \int_0^\infty \frac{xe^{itx}}{\mu}\,dF_X(x) = \frac{1}{i}\frac{d}{dt}\int_0^\infty \frac{e^{itx}}{\mu}\,dF_X(x) = \frac{1}{i\mu}\phi_X'(t),$$

where (for enthusiasts) the integral and derivative may be interchanged by appeal to the dominated convergence theorem.

(b) Set $\phi_X(t) = \big(\lambda/(\lambda - it)\big)^r$ and use part (a).

(c) The given Poisson distribution has characteristic function $\phi_X(t) = \exp(-\lambda + \lambda e^{it})$, and $\mu = \lambda$. Therefore, $\phi_X'(t)/(i\mu) = e^{it}\phi_X(t)$. Conversely, if Y is distributed as $X + 1$, we have $i\mu e^{it}\phi_X(t) = \phi_X'(t)$, which we solve subject to $\phi_X(0) = 1$.

Finally, if X has the bin(n, p) distribution, then $Y - 1$ has the bin$(n-1, p)$ distribution.

5.9 Solutions. Inversion and continuity theorems

1. Clearly, for $0 \le y \le 1$, $\mathbb{P}(X_n \le ny) = n^{-1}\lfloor ny \rfloor \to y$ as $n \to \infty$.

2. (a) The derivative of F_n is $f_n(x) = 1 - \cos(2n\pi x)$, for $0 \le x \le 1$. It is easy to see that f_n is non-negative and $\int_0^1 f_n(x)\,dx = 1$. Therefore F_n is a distribution function with density function f_n.

(b) As $n \to \infty$,

$$\left|\frac{\sin(2n\pi x)}{2n\pi}\right| \le \frac{1}{2n\pi} \to 0,$$

and so $F_n(x) \to x$ for $0 \le x \le 1$. On the other hand, $\cos(2n\pi x)$ does not converge unless $x \in \{0, 1\}$, and therefore $f_n(x)$ does not converge on $(0, 1)$.

3. We may express N as the sum $N = T_1 + T_2 + \cdots + T_k$ of independent variables each having the geometric distribution $\mathbb{P}(T_j = r) = pq^{r-1}$ for $r \ge 1$, where $p + q = 1$. Therefore

$$\phi_N(t) = \phi_{T_1}(t)^k = \left\{\frac{pe^{it}}{1 - qe^{it}}\right\}^k,$$

implying that $Z = 2Np$ has characteristic function

$$\phi_Z(t) = \phi_N(2pt) = \left\{\frac{pe^{2pit}}{1 - (1-p)e^{2pit}}\right\}^k = \left\{\frac{p(1 + 2pit + o(p))}{p(1 - 2it + o(1))}\right\}^k \to (1 - 2it)^{-k}$$

as $p \downarrow 0$, the characteristic function of the $\Gamma(\frac{1}{2}, k)$ distribution. The result follows by the continuity theorem (5.9.5).

4. All you need to know is the fact, easily proved, that $\psi_m(t) = e^{itm}$ satisfies

$$\int_{-\pi}^{\pi} \psi_j(t)\psi_k(t)\,dt = \begin{cases} 2\pi & \text{if } j + k = 0, \\ 0 & \text{if } j + k \ne 0, \end{cases}$$

for integers j and k.

Now, $\phi(t) = \sum_{j=-\infty}^{\infty} e^{itj}\mathbb{P}(X = j)$, so that

$$\frac{1}{2\pi}\int_{-\pi}^{\pi} e^{-itk}\phi(t)\,dt = \frac{1}{2\pi}\sum_{j=-\infty}^{\infty}\mathbb{P}(X = j)\int_{-\pi}^{\pi}\psi_j(t)\psi_{-k}(t)\,dt = \frac{1}{2\pi}\cdot\mathbb{P}(X = k)\cdot 2\pi.$$

If X is arithmetic with span λ, then X/λ is integer valued, whence

$$\mathbb{P}(X = k\lambda) = \frac{\lambda}{2\pi}\int_{-\pi/\lambda}^{\pi/\lambda} e^{-itk\lambda}\phi_X(t)\,dt.$$

5. Let X be uniformly distributed on $[-a, a]$, Y be uniformly distributed on $[-b, b]$, and let X and Y be independent. Then X has characteristic function $\sin(at)/(at)$, and Y has characteristic function $\sin(bt)/(bt)$. We apply the inversion theorem (5.9.1) to the characteristic function of $X + Y$ to find that

$$\frac{1}{2\pi}\int_{-\infty}^{\infty}\phi_{X+Y}(t)\,dt = \frac{1}{2\pi}\int_{-\infty}^{\infty}\frac{\sin(at)\sin(bt)}{abt^2}\,dt = f_{X+Y}(0) = \frac{a\wedge b}{2ab}.$$

6. It is elementary that

$$\int_0^{\infty}\exp\{f_n(x)\}\,dx = \int_0^{\infty}x^n e^{-x}\,dx = \Gamma(n+1) = n!.$$

In addition, $a = n$, $f_n''(a) = -n^{-1}$, and

$$\int_0^{\infty}\exp\{f_n(a) + \tfrac{1}{2}(x-a)^2 f_n''(a)\}\,dx = n^n e^{-n}\int_0^{\infty}\exp\left\{-\frac{(x-n)^2}{2n}\right\}dx \sim n^n e^{-n}\sqrt{2\pi n},$$

and Stirling's formula follows.

7. The vector $\mathbf{X}$ has joint characteristic function $\phi(\mathbf{t}) = \exp(-\frac{1}{2}\mathbf{tVt}')$. By the multidimensional version of the inversion theorem (5.9.1), the joint density function of $\mathbf{X}$ is

$$f(\mathbf{x}) = \frac{1}{(2\pi)^n}\int_{\mathbb{R}^n}\exp(-i\mathbf{tx}' - \tfrac{1}{2}\mathbf{tVt}')\,d\mathbf{t}.$$

Therefore, if $i \neq j$,

$$\frac{\partial f}{\partial v_{ij}} = \frac{1}{(2\pi)^n}\int_{\mathbb{R}^n} t_i t_j \exp(-i\mathbf{tx}' - \tfrac{1}{2}\mathbf{tVt}')\,d\mathbf{t} = \frac{\partial^2 f}{\partial x_i \partial x_j},$$

and similarly when $i = j$. When $i \neq j$,

$$\begin{aligned}\frac{\partial}{\partial v_{ij}}\mathbb{P}\big(\max_k X_k \le u\big) &= \int_Q \frac{\partial f}{\partial v_{ij}}\,d\mathbf{x} \quad \text{where } Q = \{\mathbf{x} : x_k \le u \text{ for } k = 1, 2, \dots, n\}\\ &= \int_Q \frac{\partial^2 f}{\partial x_i \partial x_j}\,d\mathbf{x} = \int' f\Big|_{x_i=x_j=-\infty}^{x_i=x_j=u} d\mathbf{x}' \ge 0,\end{aligned}$$

where $\int' \cdot\, d\mathbf{x}'$ is an integral over the variables x_k for $k \neq i, j$.

Therefore, $\mathbb{P}(\max_k X_k \le u)$ increases in every parameter v_{ij}, and is therefore greater than its value when $v_{ij} = 0$ for $i \neq j$, namely $\prod_k \mathbb{P}(X_k \le u)$.

8. By a two-dimensional version of the inversion theorem (5.9.1) applied to $\mathbb{E}(e^{i\mathbf{tX}'})$, $\mathbf{t} = (t_1, t_2)$,

$$\begin{aligned}\frac{\partial}{\partial\rho}\mathbb{P}(X_1 > 0, X_2 > 0) &= \frac{\partial}{\partial\rho}\int_0^{\infty}\int_0^{\infty}\left\{\frac{1}{4\pi^2}\iint_{\mathbb{R}^2}\exp(-i\mathbf{tx}' - \tfrac{1}{2}\mathbf{tVt}')\,d\mathbf{t}\right\}d\mathbf{x}\\ &= \frac{\partial}{\partial\rho}\frac{1}{4\pi^2}\iint_{\mathbb{R}^2}\frac{\exp(-\frac{1}{2}\mathbf{tVt}')}{(it_1)(it_2)}\,d\mathbf{t}\\ &= \frac{1}{4\pi^2}\iint_{\mathbb{R}^2}\exp(-\tfrac{1}{2}\mathbf{tVt}')\,d\mathbf{t} = \frac{2\pi\sqrt{|\mathbf{V}^{-1}|}}{4\pi^2} = \frac{1}{2\pi\sqrt{1-\rho^2}}.\end{aligned}$$

We integrate with respect to ρ to find that, in agreement with Exercise (4.7.5),

$$\mathbb{P}(X_1 > 0,\ X_2 > 0) = \frac{1}{4} + \frac{1}{2\pi}\sin^{-1}\rho.$$

9. (a) We have that

$$\mathbb{E}(e^{itY_n}) = \phi(t/(cn))^n = \left\{1 - \frac{|t|}{n} + o(1/n)\right\}^n \to e^{-|t|} \quad \text{as } n \to \infty,$$

and the Cauchy limit holds by the continuity theorem.
(b) In the usual way,

$$\begin{aligned}
\mathbb{E}(e^{it/U}) &= \frac{1}{2}\int_{-1}^{1} e^{it/u}\,du = \frac{1}{2}\int_0^1 (e^{-it/u} + e^{it/u})\,dx = \int_0^1 \cos(t/u)\,du \\
&= \int_{|t|}^{\infty} \frac{|t|\cos x}{x^2}\,dx \quad \text{by the transformation } x = t/u, \\
&= 1 - |t|\int_{|t|}^{\infty} \frac{1-\cos x}{x^2}\,dx \\
&= 1 - \frac{1}{2}\pi|t| + |t|\int_0^{|t|} \frac{1-\cos x}{x^2}\,dx = 1 - c|t| + o(t),
\end{aligned}$$

where $c = \pi/2$. We have used the fact that $\int_0^\infty x^{-2}(1-\cos x)\,dx = \pi/2$, which may be proved in a number of ways including by a contour integral in the complex plane. By part (a), the limiting distribution of Y_n is Cauchy.

5.10 Solutions. Two limit theorems

1. (a) Let $\{X_i : i \ge 1\}$ be a collection of independent Bernoulli random variables with parameter $\frac{1}{2}$. Then $S_n = \sum_1^n X_i$ is binomially distributed as $\text{bin}(n, \frac{1}{2})$. Hence, by the central limit theorem,

$$2^{-n} \sum_{\substack{k:\\ |k-\frac{1}{2}n| \le \frac{1}{2}x\sqrt{n}}} \binom{n}{k} = \mathbb{P}\left(\frac{|S_n - \frac{1}{2}n|}{\frac{1}{2}\sqrt{n}} \le x\right) \to \Phi(x) - \Phi(-x) = \int_{-x}^{x} \frac{1}{\sqrt{2\pi}} e^{-\frac{1}{2}y^2}\,dy,$$

where Φ is the $N(0,1)$ distribution function.
(b) Let $\{X_i : i \ge 1\}$ be a collection of independent Poisson random variables, each with parameter 1. Then $S_n = \sum_1^n X_i$ is Poisson with parameter n, and by the central limit theorem

$$e^{-n} \sum_{\substack{k:\\ |k-n| \le x\sqrt{n}}} \frac{n^k}{k!} = \mathbb{P}\left(\frac{|S_n - n|}{\sqrt{n}} \le x\right) \to \Phi(x) - \Phi(-x), \quad \text{as above.}$$

2. A superficially plausible argument asserts that, if all babies look the same, then the number X of correct answers in n trials is a random variable with the $\text{bin}(n, \frac{1}{2})$ distribution. Then, for large n,

$$\mathbb{P}\left(\frac{X - \frac{1}{2}n}{\frac{1}{2}\sqrt{n}} > 3\right) \simeq 1 - \Phi(3) \simeq \frac{1}{1000}$$

by the central limit theorem. For the given values of n and X,

$$\frac{X - \frac{1}{2}n}{\frac{1}{2}\sqrt{n}} = \frac{910 - 750}{5\sqrt{15}} \simeq 8.$$

Now we might say that the event $\{X - \frac{1}{2}n > \frac{3}{2}\sqrt{n}\}$ is sufficiently unlikely that its occurrence casts doubt on the original supposition that babies look the same.

A statistician would level a good many objections at drawing such a clear cut decision from such murky data, but this is beyond our scope to elaborate.

3. Clearly

$$\phi_Y(t) = \mathbb{E}\{\mathbb{E}(e^{itY} \mid X)\} = \mathbb{E}\{\exp(X(e^{it} - 1))\} = \left(\frac{1}{1 - (e^{it} - 1)}\right)^s = \left(\frac{1}{2 - e^{it}}\right)^s.$$

It follows that

$$\mathbb{E}(Y) = \frac{1}{i}\phi_Y'(0) = s, \qquad \mathbb{E}(Y^2) = -\phi_Y''(0) = s^2 + 2s,$$

whence $\text{var}(Y) = 2s$. Therefore the characteristic function of the normalized variable $Z = (Y - \mathbb{E}Y)/\sqrt{\text{var}(Y)}$ is

$$\phi_Z(t) = e^{-it\sqrt{s/2}}\phi_Y(t/\sqrt{2s}).$$

Now,

$$\begin{aligned}\log\{\phi_Y(t/\sqrt{2s})\} &= -s\log(2 - e^{it/\sqrt{2s}}) = s(e^{it/\sqrt{2s}} - 1) + \tfrac{1}{2}s(e^{it/\sqrt{2s}} - 1)^2 + \mathrm{o}(1)\\ &= it\sqrt{\tfrac{1}{2}s} - \tfrac{1}{4}t^2 - \tfrac{1}{4}t^2 + \mathrm{o}(1),\end{aligned}$$

where the $\mathrm{o}(1)$ terms are as $s \to \infty$. Hence $\log\{\phi_Z(t)\} \to -\frac{1}{2}t^2$ as $s \to \infty$, and the result follows by the continuity theorem (5.9.5).

Let $P_1, P_2, \dots$ be an infinite sequence of independent Poisson variables with parameter 1. Then $S_n = P_1 + P_2 + \cdots + P_n$ is Poisson with parameter n. Now Y has the Poisson distribution with parameter X, and so Y is distributed as S_X. Also, X has the same distribution as the sum of s independent exponential variables, implying that $X \to \infty$ as $s \to \infty$, with probability 1. This suggests by the central limit theorem that S_X (and hence Y also) is approximately normal in the limit as $s \to \infty$. We have neglected the facts that s and X are not generally integer valued.

4. Since X_1 is non-arithmetic, there exist integers $n_1, n_2, \dots, n_k$ with greatest common divisor 1 and such that $\mathbb{P}(X_1 = n_i) > 0$ for $1 \le i \le k$. There exists N such that, for all $n \ge N$, there exist non-negative integers $\alpha_1, \alpha_2, \dots, \alpha_k$ such that $n = \alpha_1 n_1 + \cdots + \alpha_k n_k$. If x is a non-negative integer, write $N = \beta_1 n_1 + \cdots + \beta_k n_k$, $N + x = \gamma_1 n_1 + \cdots + \gamma_k n_k$ for non-negative integers $\beta_1, \dots, \beta_k, \gamma_1 \dots, \gamma_k$. Now $S_n = X_1 + \cdots + X_n$ is such that

$$\mathbb{P}(S_B = N) \ge \mathbb{P}(X_j = n_i \text{ for } B_{i-1} < j \le B_i,\ 1 \le i \le k) = \prod_{i=1}^{k}\mathbb{P}(X_1 = n_i)^{\beta_i} > 0$$

where $B_0 = 0$, $B_i = \beta_1 + \beta_2 + \cdots + \beta_i$, $B = B_k$. Similarly $\mathbb{P}(S_G = N + x) > 0$ where $G = \gamma_1 + \gamma_2 + \cdots + \gamma_k$. Therefore

$$\mathbb{P}(S_G - S_{G,B+G} = x) \ge \mathbb{P}(S_G = N + x)\mathbb{P}(S_B = N) > 0$$

where $S_{G,B+G} = \sum_{i=G+1}^{B+G} X_i$. Also, $\mathbb{P}(S_B - S_{B,B+G} = -x) > 0$ as required.

5. Let $X_1, X_2, \dots$ be independent integer-valued random variables with mean 0, variance 1, span 1, and common characteristic function ϕ. We are required to prove that $\sqrt{n}\mathbb{P}(U_n = x) \to e^{-\frac{1}{2}x^2}/\sqrt{2\pi}$ as $n \to \infty$ where

$$U_n = \frac{1}{\sqrt{n}} S_n = \frac{X_1 + X_2 + \cdots + X_n}{\sqrt{n}}$$

and x is any number of the form $k/\sqrt{n}$ for integral k. The case of general μ and σ^2 is easily derived from this.

By the result of Exercise (5.9.4), for any such x,

$$\mathbb{P}(U_n = x) = \frac{1}{2\pi\sqrt{n}} \int_{-\pi\sqrt{n}}^{\pi\sqrt{n}} e^{-itx} \phi_{U_n}(t)\, dt,$$

since U_n is arithmetic. Arguing as in the proof of the local limit theorem (6),

$$2\pi \left| \sqrt{n}\mathbb{P}(U_n = x) - f(x) \right| \le I_n + J_n$$

where f is the $N(0, 1)$ density function, and

$$I_n = \int_{-\pi\sqrt{n}}^{\pi\sqrt{n}} \left| e^{-itx}\left(\phi_{U_n}(t) - e^{-\frac{1}{2}t^2}\right) \right| dt, \qquad J_n = \int_{|t| > \pi\sqrt{n}} |e^{-itx} e^{-\frac{1}{2}t^2}|\, dt.$$

Now $J_n = 2\sqrt{2\pi}\left(1 - \Phi(\pi\sqrt{n})\right) \to 0$ as $n \to \infty$, where Φ is the $N(0, 1)$ distribution function. As for I_n, pick $\delta \in (0, \pi)$. Then

$$I_n \le \int_{-\delta\sqrt{n}}^{\delta\sqrt{n}} \left| \phi(t/\sqrt{n})^n - e^{-\frac{1}{2}t^2} \right| dt + \int_{\delta\sqrt{n} < |t| < \pi\sqrt{n}} \left\{ \left| \phi(t/\sqrt{n})^n \right| + e^{-\frac{1}{2}t^2} \right\} dt.$$

The final term involving $e^{-\frac{1}{2}t^2}$ is dealt with as was J_n. By Exercise (5.7.5a), there exists $\lambda \in (0, 1)$ such that $|\phi(t)| < \lambda$ if $\delta \le |t| \le \pi$. This implies that

$$\int_{\delta\sqrt{n} < |t| < \pi\sqrt{n}} \left| \phi(t/\sqrt{n})^n \right| dt \le (\pi - \delta)\lambda^n \sqrt{n} \to 0,$$

and it remains only to show that

$$\int_{-\delta\sqrt{n}}^{\delta\sqrt{n}} \left| \phi(t/\sqrt{n})^n - e^{-\frac{1}{2}t^2} \right| dt \to 0 \quad \text{as } n \to \infty.$$

The proof of this is considerably simpler if we make the extra (though unnecessary) assumption that $m_3 = \mathbb{E}|X_1^3| < \infty$, and we assume this henceforth. It is a consequence of Taylor's theorem (see Theorem (5.7.4)) that $\phi(t) = 1 - \frac{1}{2}t^2 - \frac{1}{6}it^3 m_3 + o(t^3)$ as $t \to 0$. It follows that $\phi(t) = e^{-\frac{1}{2}t^2 + t^3\theta(t)}$ for some finite $\theta(t)$. Now $|e^x - 1| \le |x|e^{|x|}$, and therefore

$$\begin{aligned} \left| \phi(t/\sqrt{n})^n - e^{-\frac{1}{2}t^2} \right| &= e^{-\frac{1}{2}t^2} \left| \exp\left(t^3 n^{-\frac{1}{2}} \theta(tn^{-\frac{1}{2}})\right) - 1 \right| \\ &\le \frac{|t^3\theta(tn^{-\frac{1}{2}})|}{\sqrt{n}} \exp\left(\frac{|t^3\theta(tn^{-\frac{1}{2}})|}{\sqrt{n}} - \frac{1}{2}t^2 \right). \end{aligned}$$

Let $K_\delta = \sup\{|\theta(u)| : |u| \le \delta\}$, noting that $K_\delta < \infty$, and pick δ sufficiently small that $0 < \delta < \pi$ and $\delta K_\delta < \frac{1}{4}$. For $|t| < \delta\sqrt{n}$,

$$\left|\phi(t/\sqrt{n})^n - e^{-\frac{1}{2}t^2}\right| \le K_\delta \frac{|t|^3}{\sqrt{n}} \exp(t^2\delta K_\delta - \tfrac{1}{2}t^2) \le K_\delta \frac{|t|^3}{\sqrt{n}} e^{-\frac{1}{4}t^2},$$

and therefore

$$\int_{-\delta\sqrt{n}}^{\delta\sqrt{n}} \left|\phi(t/\sqrt{n})^n - e^{-\frac{1}{2}t^2}\right| dt \le \frac{K_\delta}{\sqrt{n}} \int_{-\delta\sqrt{n}}^{\delta\sqrt{n}} |t|^3 e^{-\frac{1}{4}t^2}\, dt \to 0 \quad \text{as } n \to \infty$$

as required.

6. The second moment of the X_i is

$$2\int_0^{e^{-1}} \frac{x^2}{2x(\log x)^2}\, dx = \int_{-\infty}^{-1} \frac{e^{2u}}{u^2}\, du$$

(substitute $x = e^u$), a finite integral. Therefore the X's have finite mean and variance. The density function is symmetric about 0, and so the mean is 0.

By the convolution formula, if $0 < x < e^{-1}$,

$$f_2(x) = \int_{-e^{-1}}^{e^{-1}} f(y)f(x-y)\, dy \ge \int_0^x f(y)f(x-y)\, dy \ge f(x)\int_0^x f(y)\, dy,$$

since $f(x-y)$, viewed as a function of y, is increasing on $[0, x]$. Hence

$$f_2(x) \ge \frac{f(x)}{2\log|x|} = \frac{1}{4|x|(\log|x|)^3}$$

for $0 < x < e^{-1}$. Continuing this procedure, we obtain

$$f_n(x) \ge \frac{k_n}{|x|(\log|x|)^{n+1}}, \qquad 0 < x < e^{-1},$$

for some positive constant k_n. Therefore $f_n(x) \to \infty$ as $x \to 0$, and in particular the density function of $(X_1 + \cdots + X_n)/\sqrt{n}$ does not converge to the appropriate normal density at the origin.

7. We have for $s > 0$ that

$$\begin{aligned}
\phi(is) &= \frac{1}{\sqrt{2\pi}} \int_0^\infty \exp(-(2x)^{-1} - xs)x^{-3/2}\, dx \\
&= \frac{1}{\sqrt{2\pi}} \int_0^\infty \exp(-\tfrac{1}{2}y^2 - sy^{-2})2\, dy \quad \text{by substituting } x = y^{-2} \\
&= \exp(-\sqrt{2s}),
\end{aligned}$$

by the result of Problem (5.12.18c), or by consulting a table of integrals. The required conclusion follows by analytic continuation in the upper half-plane. See Moran (1968, p. 271).

8. (a) The sum $S_n = \sum_{r=1}^n X_r$ has characteristic function $\mathbb{E}(e^{itS_n}) = \phi(t)^n = \phi(tn^2)$, whence $U_n = S_n/n$ has characteristic function $\phi(tn) = \mathbb{E}(e^{itnX_1})$. Therefore,

$$\mathbb{P}(S_n < c) = \mathbb{P}(nX_1 < c) = \mathbb{P}\left(X_1 < \frac{c}{n}\right) \to 0 \qquad \text{as } n \to \infty.$$

(b) $\mathbb{E}(e^{itT_n}) = \phi(t) = \mathbb{E}(e^{itX_1})$.

9. (a) Yes, because X_n is the sum of independent identically distributed random variables with non-zero variance.

(b) It cannot in general obey what we have called the central limit theorem, because $\text{var}(X_n) = (n^2 - n)\,\text{var}(\Theta) + n\mathbb{E}(\Theta)(1 - \mathbb{E}(\Theta))$ and $n\,\text{var}(X_1) = n\mathbb{E}(\Theta)(1 - \mathbb{E}(\Theta))$ are different whenever $\text{var}(\Theta) \neq 0$. Indeed the right 'normalization' involves dividing by n rather than $\sqrt{n}$. It may be shown when $\text{var}(\Theta) \neq 0$ that the distribution of X_n/n converges to that of the random variable Θ.

10. Denote the rth vote by B_r and let the lead of A over B when c votes have been counted be $D_c = \sum_{r=1}^{c} B_r$. Let $E_v = D_{\lambda v}/\sqrt{\lambda v}$ and $F_v = D_v/\sqrt{v}$. By a suitable central limit theorem, the pair (E_v, F_v) converges in distribution to the standard bivariate normal distribution with correlation

$$\rho = \text{cov}(E_v, F_v) = \frac{1}{v\sqrt{\lambda}}\mathbb{E}(D_{\lambda v}D_v) = \frac{1}{v\sqrt{\lambda}}\mathbb{E}(D_{\lambda v}^2) = \sqrt{\lambda}.$$

The probability that A leads after λv and again after v votes is

$$\mathbb{P}(D_{\lambda v} > 0,\ D_v > 0) = \mathbb{P}(E_v > 0,\ F_v > 0) \to \frac{1}{4} + \frac{1}{2\pi}\sin^{-1}\sqrt{\lambda} \quad \text{as } v \to \infty,$$

by the result of Exercise (4.7.5). Adding the same probability for B gives the result.

11. Let $X_1, X_2, \ldots$ be independent random variables with the Poisson distribution with parameter 1, so that $S_n = X_1 + \cdots + X_n$ has the Poisson distribution with parameter n. In particular, $(S_n - n)/\sqrt{n}$ converges to the $N(0, 1)$ distribution as $n \to \infty$. By the local central limit theorem (5.10.9),

$$\sqrt{n}\mathbb{P}(S_n = n) \to \frac{1}{\sqrt{2\pi}} \qquad \text{as } n \to \infty.$$

and the claim follows since $\mathbb{P}(S_n = n) = n^n e^{-n}/n!$.

12. Let N_k be the number of households of size k. The probability a random individual is in a household of size x is

$$s(x) = \frac{xN_x}{\sum_k kN_k} = \frac{xg(x)}{\sum_k kg(k)}$$

where $g(x) = N_x/\sum_k N_k$ is the proportion of households of size x. If the N_k are large, then $g(x) \approx f(x)$, and $\sum_k kg(k) \approx \mu$.

13. The density function of L satisfies, for small dz,

$$\begin{aligned} f_L(z)dz &= \mathbb{P}\big(L \in (z, z+dz)\big) + \text{o}(dz) \\ &= \sum_{r=1}^{n} \mathbb{P}\left(L \in (z, z+dz),\ \frac{S_{r-1}}{S_n} \le U \le \frac{S_r}{S_n}\right) + \text{o}(dz) \\ &= \sum_{r=1}^{n} \mathbb{P}\left(\frac{X_r}{S_n} \in (z, z+dz),\ \frac{S_{r-1}}{S_n} \le U \le \frac{S_r}{S_n}\right) + \text{o}(dz) \\ &= \sum_{r=1}^{n} \mathbb{P}\left(\frac{X_r}{S_n} \in (z, z+dz)\right)\big(z + \text{O}(dz)\big) \quad \text{by conditioning} \\ &= \sum_{r=1}^{n} zf_Z(z)dz + \text{o}(dz) = nzf_Z(z)dz + \text{o}(dz), \end{aligned}$$

whence $f_L(z) = nzf_Z(z) = (z/\mathbb{E}Z)f_Z(z)$.

14. We have $A_n = X_n + 2X_{n-1} + \cdots + nX_1$, so that A_n is $N(0, \sigma(n)^2)$ where $\sigma(n)^2 = \sum_{j=1}^n j^2 \sim Cn^3$. Therefore, with B distributed as $N(0, 1)$,

$$\mathbb{P}(-1 < A_n < 1) = \mathbb{P}\left(-\frac{1}{\sigma(n)} < B < \frac{1}{\sigma(n)}\right) \le \frac{C'}{n^{3/2}},$$

for some $C' < \infty$. This is summable, and the result follows by the forthcoming Borel–Cantelli Lemma (7.3.10a).

15. Chance guessing corresponds to random variables X_H and X_S, respectively, with the binomial distributions with respective means $\mu_H \approx 4898$, $\mu_S \approx 4258$, and standard deviations $\sigma_H \approx 50$, $\sigma_S \approx 46$. Thus, $Y_H = (X_H - \mu_H)/\sigma_H \approx 6.4$ and $Y_S = (X_S - \mu_S)/\sigma_S \approx 16.4$ are realizations of approximately $N(0, 1)$ variables. With Z a standard normal variable, $\mathbb{P}(Z \ge 6.4) \approx 10^{-9}$ and $\mathbb{P}(Z \ge 16.4) \approx e^{-128}$. These "astronomically" small numbers are beyond our practical experience. Indeed the error in the binomial/normal approximation may be more significant.

5.11 Solutions. Large deviations

1. We may write $S_n = \sum_1^n X_i$ where the X_i have moment generating function $M(t) = \frac{1}{2}(e^t + e^{-t})$. Applying the large deviation theorem (5.11.4), we obtain that, for $0 < a < 1$, $\mathbb{P}(S_n > an)^{1/n} \to \inf_{t>0}\{g(t)\}$ where $g(t) = e^{-at}M(t)$. Now g has a minimum when $e^t = \sqrt{(1+a)/(1-a)}$, where it takes the value $1/\sqrt{(1+a)^{1+a}(1-a)^{1-a}}$ as required. If $a \ge 1$, then $\mathbb{P}(S_n > an) = 0$ for all n.

2. (i) Let Y_n have the binomial distribution with parameters n and $\frac{1}{2}$. Then $2Y_n - n$ has the same distribution as the random variable S_n in Exercise (5.11.1). Therefore, if $0 < a < 1$,

$$\mathbb{P}(Y_n - \tfrac{1}{2}n > \tfrac{1}{2}an)^{1/n} = \mathbb{P}(S_n > an)^{1/n} \to \frac{1}{\sqrt{(1+a)^{1+a}(1-a)^{1-a}}},$$

and similarly for $\mathbb{P}(Y_n - \frac{1}{2}n < -\frac{1}{2}an)$, by symmetry. Hence

$$T_n^{1/n} = \left\{2^n \mathbb{P}\left(|Y_n - \tfrac{1}{2}n| > \tfrac{1}{2}an\right)\right\}^{1/n} \to \frac{4}{\sqrt{(1+a)^{1+a}(1-a)^{1-a}}}.$$

(ii) This time let $S_n = X_1 + \cdots + X_n$, the sum of independent Poisson variables with parameter 1. Then $T_n = e^n \mathbb{P}(S_n > n(1+a))$. The moment generating function of $X_1 - 1$ is $M(t) = \exp(e^t - 1 - t)$, and the large deviation theorem gives that $T_n^{1/n} \to e \inf_{t>0}\{g(t)\}$ where $g(t) = e^{-at}M(t)$. Now $g'(t) = (e^t - a - 1)\exp(e^t - at - t - 1)$ whence g has a minimum at $t = \log(a+1)$. Therefore $T_n^{1/n} \to eg(\log(1+a)) = \{e/(a+1)\}^{a+1}$.

3. Suppose that $M(t) = \mathbb{E}(e^{tX})$ is finite on the interval $[-\delta, \delta]$. Now, for $a > 0$, $M(\delta) \ge e^{\delta a}\mathbb{P}(X > a)$, so that $\mathbb{P}(X > a) \le M(\delta)e^{-\delta a}$. Similarly, $\mathbb{P}(X < -a) \le M(-\delta)e^{-\delta a}$.

Suppose conversely that such λ, μ exist. Then

$$M(t) \le \mathbb{E}(e^{|tX|}) = \int_{[0,\infty)} e^{|t|x}\, dF(x)$$

where F is the distribution function of $|X|$. Integrate by parts to obtain

$$M(t) \le 1 + \left[-e^{|t|x}[1 - F(x)]\right]_0^\infty + \int_0^\infty |t|e^{|t|x}[1 - F(x)]\, dx$$

(the term '1' takes care of possible atoms at 0). However $1 - F(x) \le \mu e^{-\lambda x}$, so that $M(t) < \infty$ if $|t|$ is sufficiently small.

4. The characteristic function of S_n/n is $\{e^{-|t/n|}\}^n = e^{-|t|}$, and hence S_n/n is Cauchy. Hence

$$\mathbb{P}(S_n > an) = \int_a^\infty \frac{dx}{\pi(1+x^2)} = \frac{1}{\pi}\left(\frac{\pi}{2} - \tan^{-1} a\right).$$

5. We have

$$\mathbb{E}(e^{tX_r}) = 1 + p_r(e^t - 1) \le \exp\{p_r(e^t - 1)\} \qquad \text{for } t > 0.$$

By Markov's inequality, for $t \ge 0$,

$$\begin{aligned}\mathbb{P}(S > (1+\epsilon)\mu) &= \mathbb{P}(e^{tS} > e^{t(1+\epsilon)\mu}) \\ &\le \frac{\mathbb{E}(e^{tS})}{e^{t(1+\epsilon)\mu}} \le \exp\left\{\sum_{r=1}^n p_r\big[(e^t - 1) - t(1+\epsilon)\big]\right\}.\end{aligned}$$

The right side is minimized over $t > 0$ by choosing $t = \log(1+\epsilon)$, giving the required result.

5.12 Solutions to problems

1. The probability generating function of the sum is

$$\left\{\frac{1}{6}\sum_{i=1}^{6} s^i\right\}^{10} = \left(\frac{1}{6}s\right)^{10}\left\{\frac{1-s^6}{1-s}\right\}^{10} = \left(\frac{1}{6}s\right)^{10}(1 - 10s^6 + \cdots)(1 + 10s + \cdots).$$

The coefficient of s^{27} is

$$\left(\frac{1}{6}\right)^{10}\left\{\binom{10}{2}\binom{14}{5} - \binom{10}{1}\binom{20}{11} + \binom{26}{17}\right\}.$$

2. (a) The initial sequences T, HT, HHT, HHH induce a partition of the sample space. By conditioning on this initial sequence, we obtain $f(k) = qf(k-1) + pqf(k-2) + p^2qf(k-3)$ for $k > 3$, where $p + q = 1$. Also $f(1) = f(2) = 0$, $f(3) = p^3$. In principle, this difference equation may be solved in the usual way (see Appendix I). An alternative is to use generating functions. Set $G(s) = \sum_{k=1}^\infty s^k f(k)$, multiply throughout the difference equation by s^k and sum, to find that $G(s) = p^3s^3/\{1 - qs - pqs^2 - p^2qs^3\}$. To find the coefficient of s^k, factorize the denominator, expand in partial fractions, and use the binomial series.

Another equation for $f(k)$ is obtained by observing that $X = k$ if and only if $X > k - 4$ and the last four tosses were THHH. Hence

$$f(k) = qp^3\left(1 - \sum_{i=1}^{k-4} f(i)\right), \qquad k > 3.$$

Applying the first argument to the mean, we find that $\mu = \mathbb{E}(X)$ satisfies $\mu = q(1+\mu) + pq(2+\mu) + p^2q(3+\mu) + 3p^3$ and hence $\mu = (1 + p + p^2)/p^3$.

As for HTH, consider the event that HTH does not occur in n tosses, and in addition the next three tosses give HTH. The number Y until the first occurrence of HTH satisfies

$$\mathbb{P}(Y > n)p^2q = \mathbb{P}(Y = n+1)pq + \mathbb{P}(Y = n+3), \quad n \ge 2.$$

Sum over n to obtain $\mathbb{E}(Y) = (pq+1)/(p^2q)$.

(b) $G_N(s) = (q+ps)^n$, in the obvious notation.

(i) $\mathbb{P}(2 \text{ divides } N) = \frac{1}{2}\{G_N(1) + G_N(-1)\}$, since only the coefficients of the *even* powers of s contribute to this probability.

(ii) Let ω be a complex cube root of unity. Then the coefficient of $\mathbb{P}(X = k)$ in $\frac{1}{3}\{G_N(1) + G_N(\omega) + G_N(\omega^2)\}$ is

$$\begin{aligned} \tfrac{1}{3}\{1+\omega^3+\omega^6\} &= 1, \qquad \text{if } k = 3r, \\ \tfrac{1}{3}\{1+\omega+\omega^2\} &= 0, \qquad \text{if } k = 3r+1, \\ \tfrac{1}{3}\{1+\omega^2+\omega^4\} &= 0, \qquad \text{if } k = 3r+2, \end{aligned}$$

for integers r. Hence $\frac{1}{3}\{G_N(1) + G_N(\omega) + G_N(\omega^2)\} = \sum_{r=0}^{\lfloor \frac{1}{3}n \rfloor} \mathbb{P}(N = 3r)$, the probability that N is a multiple of 3. Generalize this conclusion.

3. We have that $T = k$ if no run of n heads appears in the first $k-n-1$ throws, then there is a tail, and then a run of n heads. Therefore $\mathbb{P}(T = k) = \mathbb{P}(T > k-n-1)qp^n$ for $k \geq n+1$ where $p+q = 1$. Finally $\mathbb{P}(T = n) = p^n$. Multiply by s^k and sum to obtain a formula for the probability generating function G of T:

$$\begin{aligned} G(s) - p^n s^n &= qp^n \sum_{k=n+1}^{\infty} s^k \sum_{j>k-n-1} \mathbb{P}(T = j) = qp^n \sum_{j=1}^{\infty} \mathbb{P}(T = j) \sum_{k=n+1}^{n+j} s^k \\ &= \frac{qp^n s^{n+1}}{1-s} \sum_{j=1}^{\infty} \mathbb{P}(T = j)(1-s^j) = \frac{qp^n s^{n+1}}{1-s}(1-G(s)). \end{aligned}$$

Therefore

$$G(s) = \frac{p^n s^n - p^{n+1} s^{n+1}}{1 - s + qp^n s^{n+1}}.$$

4. The required generating function is

$$G(s) = \sum_{k=r}^{\infty} s^k \binom{k-1}{r-1} p^r (1-p)^{k-r} = \left(\frac{ps}{1-qs} \right)^r$$

where $p+q = 1$. The mean is $G'(1) = r/p$ and the variance is $G''(1) + G'(1) - \{G'(1)\}^2 = rq/p^2$.

5. It is standard, see equation (5.3.3), that $p_0(2n) = \binom{2n}{n}(pq)^n$. Using Stirling's formula,

$$p_0(2n) \sim \frac{(2n)^{2n+\frac{1}{2}} e^{-2n} \sqrt{2\pi}}{\{n^{n+\frac{1}{2}} e^{-n} \sqrt{2\pi}\}^2} (pq)^n = \frac{(4pq)^n}{\sqrt{\pi n}}.$$

The generating function $F_0(s)$ for the first return time is given by $F_0(s) = 1 - P_0(s)^{-1}$ where $P_0(s) = \sum_n s^{2n} p_0(2n)$. Therefore the probability of ultimate return is $F_0(1) = 1 - \lambda^{-1}$ where, by Abel's theorem,

$$\lambda = \sum_n p_0(2n) \begin{cases} = \infty & \text{if } p = q = \frac{1}{2}, \\ < \infty & \text{if } p \neq q. \end{cases}$$

Hence $F_0(1) = 1$ if and only if $p = \frac{1}{2}$.

6. (a) $R_n = X_n^2 + Y_n^2$ satisfies

$$\begin{aligned}\mathbb{E}(R_{n+1} - R_n) &= \mathbb{E}\{(X_{n+1}^2 - X_n^2) + (Y_{n+1}^2 - Y_n^2)\}\\ &= 2\mathbb{E}(X_{n+1}^2 - X_n^2) = 2\mathbb{E}\{\mathbb{E}(X_{n+1}^2 - X_n^2 \mid X_n)\}\\ &= 2\mathbb{E}\{\tfrac{1}{4}[(X_n+1)^2 - X_n^2] + \tfrac{1}{4}[(X_n-1)^2 - X_n^2]\} = 1.\end{aligned}$$

Hence $R_n = n + R_0 = n$.
(b) The quick way is to argue as in the solution to Exercise (5.3.4). Let $U_n = X_n + Y_n$, $V_n = X_n - Y_n$. Then U and V are simple symmetric random walks, and furthermore they are independent. Therefore

$$p_0(2n) = \mathbb{P}(U_{2n} = 0, V_{2n} = 0) = \mathbb{P}(U_{2n} = 0)\mathbb{P}(V_{2n} = 0) = \left\{\left(\frac{1}{2}\right)^{2n}\binom{2n}{n}\right\}^2,$$

by (5.3.3). Using Stirling's formula, $p_0(2n) \sim (n\pi)^{-1}$, and therefore $\sum_n p_0(2n) = \infty$, implying that the chance of eventual return is 1.

A longer method is as follows. The walk is at the origin at time 0 if and only if it has taken equal numbers of leftward and rightward steps, and also equal numbers of upward and downward steps. Therefore

$$p_0(2n) = \left(\frac{1}{4}\right)^{2n}\sum_{m=0}^{n}\frac{(2n)!}{(m!)^2\{(n-m)!\}^2} = \left(\frac{1}{2}\right)^{4n}\binom{2n}{n}^2.$$

7. (a) Let e_{ij} be the probability the walk ever reaches j having started from i. Clearly $e_{a0} = e_{a,a-1}e_{a-1,a-2}\cdots e_{10}$, since a passage to 0 from a requires a passage to $a-1$, then a passage to $a-2$, and so on. By homogeneity, $e_{a0} = (e_{10})^a$.

By conditioning on the value of the first step, we find that $e_{10} = pe_{30} + qe_{00} = pe_{10}^3 + q$. The cubic equation $x = px^3 + q$ has roots $x = 1, c, d$, where

$$c = \frac{-p - \sqrt{p^2+4pq}}{2p}, \quad d = \frac{-p + \sqrt{p^2+4pq}}{2p}.$$

Now $|c| > 1$, and $|d| \ge 1$ if and only if $p^2 + 4pq \ge 9p^2$ which is to say that $p \le \frac{1}{3}$. It follows that $e_{10} = 1$ if $p \le \frac{1}{3}$, so that $e_{a0} = 1$ if $p \le \frac{1}{3}$.

When $p > \frac{1}{3}$, we have that $d < 1$, and it is actually the case that $e_{10} = d$, and hence

$$e_{a0} = \left(\frac{-p + \sqrt{p^2+4pq}}{2p}\right)^a \quad \text{if } p > \tfrac{1}{3}.$$

In order to prove this, it suffices to prove that $e_{a0} < 1$ for all large a; this is a minor but necessary chore. Write $T_n = S_n - S_0 = \sum_{i=1}^n X_i$, where X_i is the value of the ith step. Then

$$\begin{aligned}e_{a0} &= \mathbb{P}(T_n \le -a \text{ for some } n \ge 1) = \mathbb{P}(n\mu - T_n \ge n\mu + a \text{ for some } n \ge 1)\\ &\le \sum_{n=1}^{\infty}\mathbb{P}(n\mu - T_n \ge n\mu + a)\end{aligned}$$

where $\mu = \mathbb{E}(X_1) = 2p - q > 0$. As in the theory of large deviations, for $t > 0$,

$$\mathbb{P}(n\mu - T_n \ge n\mu + a) \le e^{-t(n\mu+a)}\{\mathbb{E}(e^{t(\mu-X)})\}^n$$

where X is a typical step. Now $\mathbb{E}(e^{t(\mu-X)}) = 1 + o(t)$ as $t \downarrow 0$, and therefore we may pick $t > 0$ such that $\theta(t) = e^{-t\mu}\mathbb{E}(e^{t(\mu-X)}) < 1$. It follows that $e_{a0} \le \sum_{n=1}^{\infty} e^{-ta}\theta(t)^n$ which is less than 1 for all large a, as required.

(b) Let v_r be the probability of ever visiting $r \ge 1$, having started at 0. By conditioning on the first step,

$$v_n = \begin{cases} qv_2 + pe_{10} & \text{if } n = 1, \\ p + qv_3 & \text{if } n = 2, \\ qv_{n+1} + pv_{n-2} & \text{if } n \ge 3, \end{cases}$$

where e_{10} is given in the first part of the solution. Solving the recursion $v_n = qv_{n+1} + pv_{n-2}$, we obtain the auxiliary cubic $qx^3 - x^2 + p = 0$, with roots $x = 1, r, s$ where

$$r = \frac{p - \sqrt{p^2 + 4pq}}{2q}, \qquad s = \frac{p + \sqrt{p^2 + 4pq}}{2q}.$$

Note that $|r| < 1$ and

$$s \begin{cases} > 1 & \text{for } p > \frac{1}{3}, \\ = 1 & \text{for } p = \frac{1}{3}, \\ \in (0, 1) & \text{for } p < \frac{1}{3}. \end{cases}$$

Suppose $p > \frac{1}{3}$. We have $v_n = A + Br^n$ for $n \ge 1$ (since the $n \ge 3$ recursion includes $v_1, v_2, \dots$, and the v_n are bounded)). The boundary conditions become

$$A + Br = q(A + Br^2) + pe_{10}, \qquad A + Br^2 = p + q(A + Br^3),$$

whence, on eliminating B,

$$A = \frac{re_{10} - 1}{r - 1}.$$

Now,

$$\frac{1}{n}A_n = \frac{1}{n}\sum_{r=1}^{n}(1 - v_r) \to 1 - A,$$

so that

$$a = 1 - A = \frac{r(1 - e_{10})}{r - 1}.$$

When $p = \frac{1}{2}$, $r = -\frac{1}{2}(\sqrt{5} - 1)$ and $e_{10} = \frac{1}{2}(\sqrt{5} - 1)$, so that $A = (5 - \sqrt{5})/(1 + \sqrt{5})$ and $a = \frac{1}{2}(7 - 3\sqrt{5})$.

8. We have that

$$\mathbb{E}(s^X t^Y \mid X + Y = n) = \sum_{k=0}^{n} s^k t^{n-k}\binom{n}{k}p^k q^{n-k} = (ps + qt)^n,$$

where $p + q = 1$. Hence $G_{X,Y}(s, t) = G(ps + qt)$ where G is the probability generating function of $X + Y$. Now X and Y are independent, so that

$$G(ps + qt) = G_X(s)G_Y(t) = G_{X,Y}(s, 1)G_{X,Y}(1, t) = G(ps + q)G(p + qt).$$

Write $f(u) = G(1 + u)$, $x = s - 1$, $y = t - 1$, to obtain $f(px + qy) = f(px)f(qy)$, a functional equation valid at least when $-2 < x, y \le 0$. Now f is continuous within its disc of convergence, and also $f(0) = 1$; the usual argument (see Problem (4.14.5)) implies that $f(x) = e^{\lambda x}$ for some

λ, and therefore $G(s) = f(s-1) = e^{\lambda(s-1)}$. Therefore $X+Y$ has the Poisson distribution with parameter λ. Furthermore, $G_X(s) = G(ps+q) = e^{\lambda p(s-1)}$, whence X has the Poisson distribution with parameter λp. Similarly Y has the Poisson distribution with parameter λq.

9. In the usual notation, $G_{n+1}(s) = G_n(G(s))$. It follows that $G''_{n+1}(1) = G''_n(1)G'(1)^2 + G'_n(1)G''(1)$ so that, after some work, $\text{var}(Z_{n+1}) = \mu^2\,\text{var}(Z_n) + \mu^n\sigma^2$. Iterate to obtain

$$\text{var}(Z_{n+1}) = \sigma^2(\mu^n + \mu^{n+1} + \cdots + \mu^{2n}) = \frac{\sigma^2\mu^n(1-\mu^{n+1})}{1-\mu}, \qquad n \geq 0,$$

for the case $\mu \neq 1$. If $\mu = 1$, then $\text{var}(Z_{n+1}) = \sigma^2(n+1)$.

10. (a) Since the coin is unbiased, we may assume that each player, having won a round, continues to back the same face (heads or tails) until losing. The duration D of the game equals k if and only if k is the first time at which there has been either a run of $r-1$ heads or a run of $r-1$ tails; the probability of this may be evaluated in a routine way. Alternatively, argue as follows. We record S (for 'same') each time a coin shows the same face as its predecessor, and we record C (for 'change') otherwise; start with a C. It is easy to see that each symbol in the resulting sequence is independent of earlier symbols and is equally likely to be S or C. Now $D = k$ if and only if the first run of $r-2$ S's is completed at time k. It is immediate from the result of Problem (5.12.3) that

$$G_D(s) = \frac{(\frac{1}{2}s)^{r-2}(1-\frac{1}{2}s)}{1-s+(\frac{1}{2}s)^{r-1}}.$$

(b) The probability that A_k wins is

$$\pi_k = \sum_{n=1}^{\infty} \mathbb{P}\big(D = n(r-1)+k-1\big).$$

Let ω be a complex $(r-1)$th root of unity, and set

$$W_k(s) = \frac{1}{r-1}\bigg\{G_D(s) + \frac{1}{\omega^{k-1}}G_D(\omega s) + \frac{1}{\omega^{2(k-1)}}G_D(\omega^2 s) + \cdots + \frac{1}{\omega^{(r-2)(k-1)}}G_D(\omega^{r-2}s)\bigg\}.$$

It may be seen (as for Problem (5.12.2)) that the coefficient of s^i in $W_k(s)$ is $\mathbb{P}(D=i)$ if i is of the form $n(r-1)+(k-1)$ for some n, and is 0 otherwise. Therefore $\mathbb{P}(A_k \text{ wins}) = W_k(1)$.
(c) The pool contains £D when it is won. The required mean is therefore

$$\mathbb{E}(D \mid A_k \text{ wins}) = \frac{\mathbb{E}\big(DI_{\{A_k \text{ wins}\}}\big)}{\mathbb{P}(A_k \text{ wins})} = \frac{W'_k(1)}{W_k(1)}.$$

(d) Using the result of Exercise (5.1.2), the generating function of the sequence $\mathbb{P}(D > k)$, $k \geq 0$, is $T(s) = (1 - G_D(s))/(1-s)$. The required probability is the coefficient of s^n in $T(s)$.

11. (a) Let T_n be the total number of people in the first n generations. By considering the size Z_1 of the first generation, we see that

$$T_n = 1 + \sum_{i=1}^{Z_1} T_{n-1}(i)$$

where $T_{n-1}(1), T_{n-1}(2), \ldots$ are independent random variables, each being distributed as T_{n-1}. Using the compounding formula (5.1.25), $H_n(s) = sG(H_{n-1}(s))$.
(b) Let $n \to \infty$ in part (a), and use continuity and monotonicity. (i) Send $s \uparrow 1$ in the functional equation to find that $q = Q(1)$ satisfies $q = G(q)$, whence $q = 1$ (since $\mu = G'(1) < 1$). (ii) Differentiate and let $s \uparrow 1$ to obtain $Q'(1) = 1 + G'(1)Q'(1)$, so that $\mathbb{E}T = 1 + \mu\mathbb{E}T$. (iii) Differentiate a second time.
(c) Solving $qQ^2 - Q + ps =$ for a probability generating function yields

$$Q(s) = \frac{1 - \sqrt{1 - 4pqs}}{2q}.$$

It is easy to see that

$$Q(1) = \begin{cases} p/q & \text{if } p < q, \\ 1 & \text{if } p \geq q. \end{cases}$$

Therefore, Q generates an *improper* distribution if $p < q$, and a proper distribution otherwise.
(d) Using the recursion of part (a), we have

$$H_0(s) = s, \quad H_1(s) = \frac{ps}{1 - qs}, \quad H_2(s) = \frac{(1 - qs)ps}{1 - qs - pqs},$$

and so on. Set $H_n = y_n/x_n$ to obtain

$$\frac{y_n}{x_n} = \frac{ps}{1 - qy_{n-1}/x_{n-1}} = \frac{psx_{n-1}}{x_{n-1} - qy_{n-1}},$$

so that

$$y_n = psx_{n-1}, \quad x_n = x_{n-1} - qy_{n-1}, \qquad n \geq 1.$$

Therefore,

$$x_n = x_{n-1} - pqsx_{n-1}, \quad \text{subject to} \quad x_0 = 1,\ x_1 = 1 - qs.$$

The solution is

$$x_n = A\left(\frac{1+\alpha}{2}\right)^n + B\left(\frac{1-\alpha}{2}\right)^2, \qquad n \geq 0,$$

where

$$\alpha = \sqrt{1 - 4pqs}, \quad A = \frac{\alpha + 1 - 2qs}{2\alpha}, \quad B = \frac{\alpha - 1 + 2qs}{2\alpha}.$$

Finally,

$$Q(s) = \lim_{n\to\infty} \frac{y_n(s)}{x_n(s)} = \frac{2ps}{1 + \sqrt{1 - 4pqs}},$$

in agreement with part (c).

12. We have that

$$\begin{aligned}
\mathbb{P}(Z_n > N \mid Z_m = 0) &= \frac{\mathbb{P}(Z_n > N, Z_m = 0)}{\mathbb{P}(Z_m = 0)} \\
&= \sum_{r=1}^{\infty} \frac{\mathbb{P}(Z_m = 0 \mid Z_n = N + r)\mathbb{P}(Z_n = N + r)}{\mathbb{P}(Z_m = 0)} \\
&= \sum_{r=1}^{\infty} \frac{\mathbb{P}(Z_{m-n} = 0)^{N+r}\mathbb{P}(Z_n = N + r)}{\mathbb{P}(Z_m = 0)} \\
&\leq \frac{\mathbb{P}(Z_m = 0)^{N+1}}{\mathbb{P}(Z_m = 0)} \sum_{r=1}^{\infty} \mathbb{P}(Z_n = N + r) \leq \mathbb{P}(Z_m = 0)^N = G_m(0)^N.
\end{aligned}$$

13. (a) We have that $G_W(s) = G_N(G(s)) = e^{\lambda(G(s)-1)}$. Also, $G_W(s)^{1/n} = e^{\lambda((G(s)-1)/n}$, the same probability generating function as G_W but with λ replaced by λ/n.

(b) We can suppose that $H(0) < 1$, since if $H(0) = 1$ then $H(s) = 1$ for all s, and we may take $\lambda = 0$ and $G(s) = 1$. We may suppose also that $H(0) > 0$. To see this, suppose instead that $H(0) = 0$ so that $H(s) = s^r \sum_{j=0}^{\infty} s^j h_{j+r}$ for some sequence (h_k) and some $r \geq 1$ such that $h_r > 0$. Find a positive integer n such that r/n is non-integral; then $H(s)^{1/n}$ is not a power series, which contradicts the assumption that H is infinitely divisible.

Thus we take $0 < H(0) < 1$, and so $0 < 1 - H(s) < 1$ for $0 \leq s < 1$. Therefore

$$\log H(s) = \log\big(1 - \{1 - H(s)\}\big) = \lambda\big(-1 + A(s)\big)$$

where $\lambda = -\log H(0)$ and $A(s)$ is a power series with $A(0) = 0$, $A(1) = 1$. Writing $A(s) = \sum_{j=1}^{\infty} a_j s^j$, we have that

$$\frac{d^j}{ds^j}\{H(s)e^{\lambda}\}^{1/n}\bigg|_{s=0} = \frac{\lambda}{n} j!\, a_j + \mathrm{o}(n^{-1})$$

as $n \to \infty$. Now $H(s)^{1/n}$ is a probability generating function, so that each such expression is non-negative. Therefore $a_j \geq 0$ for all j, implying that $A(s)$ is a probability generating function, as required.

(c) Suppose $e^{-\lambda(1-G)} = e^{-\mu(1-s)}$ for some $\lambda, \mu > 0$ and some probability generating function G. Then $G(s) = 1 - \rho(1-s)$ where $\rho = \mu/\lambda$. This is the situation when each W_i is 0 with probability $1-\rho$, and 1 otherwise.

14. It is clear from the definition of infinite divisibility that a distribution has this property if and only if, for each n, there exists a characteristic function ψ_n such that $\phi(t) = \psi_n(t)^n$ for all t.

(a) The characteristic functions in question are

$$N(\mu, \sigma^2): \quad \phi(t) = e^{it\mu - \frac{1}{2}\sigma^2 t^2}$$

$$\text{Poisson}(\lambda): \quad \phi(t) = e^{\lambda(e^{it}-1)}$$

$$\Gamma(\lambda, \mu): \quad \phi(t) = \left(\frac{\lambda}{\lambda - it}\right)^{\mu}.$$

In these respective cases, the 'nth root' ψ_n of ϕ is the characteristic function of the $N(\mu/n, \sigma^2/n)$, Poisson (λ/n), and $\Gamma(\lambda, \mu/n)$ distributions.

(b) Suppose that ϕ is the characteristic function of an infinitely divisible distribution, and let ψ_n be a characteristic function such that $\phi(t) = \psi_n(t)^n$. Now $|\phi(t)| \leq 1$ for all t, so that

$$|\psi_n(t)| = |\phi(t)|^{1/n} \to \begin{cases} 1 & \text{if } |\phi(t)| \neq 0, \\ 0 & \text{if } |\phi(t)| = 0. \end{cases}$$

For any value of t such that $\phi(t) \neq 0$, it is the case that $\psi_n(t) \to 1$ as $n \to \infty$. To see this, suppose instead that there exists θ satisfying $0 < \theta < 2\pi$ such that $\psi_n(t) \to e^{i\theta}$ along some subsequence. Then $\psi_n(t)^n$ does not converge along this subsequence, a contradiction. It follows that

$$(*) \qquad \psi(t) = \lim_{n\to\infty} \psi_n(t) = \begin{cases} 1 & \text{if } \phi(t) \neq 0, \\ 0 & \text{if } \phi(t) = 0. \end{cases}$$

Now ϕ is a characteristic function, so that $\phi(t) \neq 0$ on some neighbourhood of the origin. Hence $\psi(t) = 1$ on some neighbourhood of the origin, so that ψ is continuous at the origin. Applying the continuity theorem (5.9.5), we deduce that ψ is itself a characteristic function. In particular, ψ is continuous, and hence $\psi(t) = 1$ for all t, by $(*)$. We deduce that $\phi(t) \neq 0$ for all t.

15. We have that

$$\mathbb{P}(N = n \mid S = N) = \frac{\mathbb{P}(S = n \mid N = n)\mathbb{P}(N = n)}{\sum_k \mathbb{P}(S = k \mid N = k)\mathbb{P}(N = k)} = \frac{p^n \mathbb{P}(N = n)}{\sum_{k=1}^{\infty} p^k \mathbb{P}(N = k)}.$$

Hence $\mathbb{E}(x^N \mid S = N) = G(px)/G(p)$.

If N is Poisson with parameter λ, then

$$\mathbb{E}(x^N \mid S = N) = \frac{e^{\lambda(px-1)}}{e^{\lambda(p-1)}} = e^{\lambda p(x-1)} = G(x)^p.$$

Conversely, suppose that $\mathbb{E}(x^N \mid S = N) = G(x)^p$. Then $G(px) = G(p)G(x)^p$, valid for $|x| \le 1$, $0 < p < 1$. Therefore $f(x) = \log G(x)$ satisfies $f(px) = f(p) + pf(x)$, and in addition f has a power series expansion which is convergent at least for $0 < x \le 1$. Substituting this expansion into the above functional equation for f, and equating coefficients of $p^i x^j$, we obtain that $f(x) = -\lambda(1-x)$ for some $\lambda \ge 0$. It follows that N has a Poisson distribution.

16. Certainly

$$G_X(s) = G_{X,Y}(s,1) = \left(\frac{1-(p_1+p_2)}{1-p_2-p_1 s}\right)^n, \quad G_Y(t) = G_{X,Y}(1,t) = \left(\frac{1-(p_1+p_2)}{1-p_1-p_2 t}\right)^n,$$
$$G_{X+Y}(s) = G_{X,Y}(s,s) = \left(\frac{1-(p_1+p_2)}{1-(p_1+p_2)s}\right)^n,$$

giving that X, Y, and $X+Y$ have distributions similar to the negative binomial distribution. More specifically,

$$\mathbb{P}(X = k) = \binom{n+k-1}{k}\alpha^k(1-\alpha)^n, \quad \mathbb{P}(Y = k) = \binom{n+k-1}{k}\beta^k(1-\beta)^n,$$
$$\mathbb{P}(X+Y = k) = \binom{n+k-1}{k}\gamma^k(1-\gamma)^n,$$

for $k \ge 0$, where $\alpha = p_1/(1-p_2)$, $\beta = p_2/(1-p_1)$, $\gamma = p_1 + p_2$.

Now

$$\mathbb{E}(s^X \mid Y = y) = \frac{\mathbb{E}(s^X I_{\{Y=y\}})}{\mathbb{P}(Y = y)} = \frac{A}{B}$$

where A is the coefficient of t^y in $G_{X,Y}(s,t)$ and B is the coefficient of t^y in $G_Y(t)$. Therefore

$$\mathbb{E}(s^X \mid Y = y) = \left(\frac{1-p_1-p_2}{1-p_1 s}\right)^n \left(\frac{p_2}{1-p_1 s}\right)^y \Big/ \left\{\left(\frac{1-p_1-p_2}{1-p_1}\right)^n \left(\frac{p_2}{1-p_1}\right)^y\right\}$$
$$= \left(\frac{1-p_1}{1-p_1 s}\right)^{n+y}.$$

17. As in the previous solution,

$$G_X(s) = e^{(\alpha+\gamma)(s-1)}, \quad G_Y(s) = e^{(\beta+\gamma)(t-1)}, \quad G_{X+Y}(s) = e^{(\alpha+\beta)(s-1)} e^{\gamma(s^2-1)}.$$

18. (a) Substitute $u = y/a$ to obtain

$$I(a,b) = \int_0^\infty \exp(-y^2 - a^2b^2y^{-2})a^{-1}\,dy = a^{-1}I(1,ab).$$

(b) Differentiating through the integral sign,

$$\begin{aligned}\frac{\partial I}{\partial b} &= \int_0^\infty \left\{-\frac{2b}{u^2}\exp(-a^2u^2 - b^2u^{-2})\right\}du \\ &= -\int_0^\infty 2\exp(-a^2b^2y^{-2} - y^2)\,dy = -2I(1,ab),\end{aligned}$$

by the substitution $u = b/y$.
(c) Hence $\partial I/\partial b = -2aI$, whence $I = c(a)e^{-2ab}$ where

$$c(a) = I(a,0) = \int_0^\infty e^{-a^2u^2}\,du = \frac{\sqrt{\pi}}{2a}.$$

(d) We have that

$$\mathbb{E}(e^{-tX}) = \int_0^\infty e^{-tx}\frac{d}{\sqrt{x}}e^{-c/x-gx}\,dx = 2dI\big(\sqrt{g+t},\sqrt{c}\big)$$

by the substitution $x = y^2$.
(e) Similarly

$$\mathbb{E}(e^{-tX}) = \int_0^\infty e^{-tx}\frac{1}{\sqrt{2\pi x^3}}e^{-1/(2x)}\,dx = \sqrt{\frac{2}{\pi}}I\left(\frac{1}{\sqrt{2}},\sqrt{t}\right)$$

by substituting $x = y^{-2}$.

19. (a) We have that

$$\begin{aligned}\mathbb{E}(e^{itU}) &= \mathbb{E}\big\{\mathbb{E}(e^{itX/Y}\mid Y)\big\} = \mathbb{E}\{\phi_X(t/Y)\} = \mathbb{E}\{e^{-\frac{1}{2}t^2/Y^2}\} \\ &= \int_{-\infty}^\infty e^{-\frac{1}{2}t^2/y^2}\frac{1}{\sqrt{2\pi}}e^{-\frac{1}{2}y^2}\,dy = \sqrt{\frac{2}{\pi}}I\left(\frac{1}{\sqrt{2}},\frac{|t|}{\sqrt{2}}\right) = e^{-|t|}\end{aligned}$$

in the notation of Problem (5.12.18). Hence U has the Cauchy distribution.
(b) Similarly

$$\mathbb{E}(e^{-tV}) = \int_{-\infty}^\infty e^{-tx^{-2}}\frac{1}{\sqrt{2\pi}}e^{-\frac{1}{2}x^2}\,dx = \sqrt{\frac{2}{\pi}}I\left(\frac{1}{\sqrt{2}},\sqrt{t}\right) = e^{-\sqrt{2t}}$$

for $t > 0$. Using the result of Problem (5.12.18e), V has density function

$$f(x) = \frac{1}{\sqrt{2\pi x^3}}e^{-1/(2x)}, \quad x > 0.$$

(c) We have that $W^{-2} = X^{-2} + Y^{-2} + Z^{-2}$. Therefore, using (b),

$$\mathbb{E}(e^{-tW^{-2}}) = e^{-3\sqrt{2t}} = e^{-\sqrt{18t}} = \mathbb{E}(e^{-9Vt})$$

for $t > 0$. It follows that W^{-2} has the same distribution as $9V = 9X^{-2}$, and so W^2 has the same distribution as $\frac{1}{9}X^2$. Therefore, using the fact that both X and W are symmetric random variables, W has the same distribution as $\frac{1}{3}X$, that is $N(0, \frac{1}{9})$.

20. It follows from the inversion theorem that

$$\frac{F(x+h)-F(x)}{h} = \frac{1}{2\pi}\lim_{N\to\infty}\int_{-N}^{N}\frac{1-e^{-ith}}{it}e^{-itx}\phi(t)\,dt.$$

Since $|\phi|$ is integrable, we may use the dominated convergence theorem to take the limit as $h \downarrow 0$ within the integral:

$$f(x) = \frac{1}{2\pi}\lim_{N\to\infty}\int_{-N}^{N} e^{-itx}\phi(t)\,dt.$$

The condition that ϕ be absolutely integrable is stronger than necessary; note that the characteristic function of the exponential distribution fails this condition, in reflection of the fact that its density function has a discontinuity at the origin.

21. Let G_n denote the probability generating function of Z_n. The (conditional) characteristic function of Z_n/μ^n is

$$\mathbb{E}\big(e^{itZ_n/\mu^n} \mid Z_n > 0\big) = \frac{G_n(e^{it/\mu^n}) - G_n(0)}{1 - G_n(0)}.$$

It is a standard exercise (or see Example (5.4.3)) that

$$G_n(s) = \frac{\mu^n - 1 - \mu s(\mu^{n-1} - 1)}{\mu^{n+1} - 1 - \mu s(\mu^n - 1)},$$

whence by an elementary calculation

$$\mathbb{E}\big(e^{itZ_n/\mu^n} \mid Z_n > 0\big) \to \frac{\mu - 1}{\mu - 1 - \mu it} \qquad \text{as } n \to \infty,$$

the characteristic function of the exponential distribution with parameter $1 - \mu^{-1}$.

22. The imaginary part of $\phi_X(t)$ satisfies

$$\tfrac{1}{2}\big\{\phi_X(t) - \overline{\phi_X(t)}\big\} = \tfrac{1}{2}\big\{\phi_X(t) - \phi_X(-t)\big\} = \tfrac{1}{2}\big\{\mathbb{E}(e^{itX}) - \mathbb{E}(e^{-itX})\big\} = 0$$

for all t, if and only if X and $-X$ have the same characteristic function, or equivalently the same distribution.

23. (a) $U = X + Y$ and $V = X - Y$ are independent, so that $\phi_{U+V} = \phi_U\phi_V$, which is to say that $\phi_{2X} = \phi_{X+Y}\phi_{X-Y}$, or

$$\phi(2t) = \big\{\phi(t)^2\big\}\big\{\phi(t)\phi(-t)\big\} = \phi(t)^3\phi(-t).$$

Write $\psi(t) = \phi(t)/\phi(-t)$. Then

$$\psi(2t) = \frac{\phi(2t)}{\phi(-2t)} = \frac{\phi(t)^3\phi(-t)}{\phi(-t)^3\phi(t)} = \psi(t)^2.$$

Therefore

$$\psi(t) = \psi(\tfrac{1}{2}t)^2 = \psi(\tfrac{1}{4}t)^4 = \cdots = \psi(t/2^n)^{2^n} \qquad \text{for } n \ge 0.$$

However, as $h \to 0$,

$$\psi(h) = \frac{\phi(h)}{\phi(-h)} = \frac{1 - \frac{1}{2}h^2 + o(h^2)}{1 - \frac{1}{2}h^2 + o(h^2)} = 1 + o(h^2),$$

so that $\psi(t) = \{1 + o(t^2/2^{2n})\}^{2^n} \to 1$ as $n \to \infty$, whence $\psi(t) = 1$ for all t, giving that $\phi(-t) = \phi(t)$. It follows that

$$\begin{aligned} \phi(t) &= \phi(\tfrac{1}{2}t)^3\phi(-\tfrac{1}{2}t) = \phi(\tfrac{1}{2}t)^4 = \phi(t/2^n)^{2^{2n}} \qquad \text{for } n \ge 1 \\ &= \left\{1 - \frac{1}{2}\cdot\frac{t^2}{2^{2n}} + o(t^2/2^{2n})\right\}^{2^{2n}} \to e^{-\frac{1}{2}t^2} \qquad \text{as } n \to \infty, \end{aligned}$$

so that X and Y are $N(0, 1)$.
(b) With $U = X + Y$ and $V = X - Y$, we have that $\psi(s, t) = \mathbb{E}(e^{isU+itV})$ satisfies

$$(*) \qquad \psi(s, t) = \mathbb{E}(e^{i(s+t)X+i(s-t)Y}) = \phi(s+t)\phi(s-t).$$

Using what is given,

$$\left.\frac{\partial^2\psi}{\partial t^2}\right|_{t=0} = -\mathbb{E}(V^2e^{isU}) = -\mathbb{E}\{e^{isU}\mathbb{E}(V^2 \mid U)\} = -\mathbb{E}(2e^{isU}) = -2\phi(s)^2.$$

However, by $(*)$,

$$\left.\frac{\partial^2\psi}{\partial t^2}\right|_{t=0} = 2\{\phi''(s)\phi(s) - \phi'(s)^2\},$$

yielding the required differential equation, which may be written as

$$\frac{d}{ds}(\phi'/\phi) = -1.$$

Hence $\log\phi(s) = a + bs - \frac{1}{2}s^2$ for constants a, b, whence $\phi(s) = e^{-\frac{1}{2}s^2}$.

24. (a) Using characteristic functions, $\phi_Z(t) = \phi_X(t/n)^n = e^{-|t|}$.
(b) $\mathbb{E}|X_i| = \infty$.

25. (a) See the solution to Problem (5.12.24).
(b) This is much longer. Having established the hint, the rest follows thus:

$$\begin{aligned} f_{X+Y}(y) &= \int_{-\infty}^{\infty} f(x)f(y-x)\,dx \\ &= \frac{1}{\pi(4+y^2)}\int_{-\infty}^{\infty}\{f(x) + f(y-x)\}\,dx + Jg(y) = \frac{2}{\pi(4+y^2)} + Jg(y) \end{aligned}$$

where

$$\begin{aligned} J &= \int_{-\infty}^{\infty}\{xf(x) + (y-x)f(y-x)\}\,dx \\ &= \lim_{M,N\to\infty}\left[\frac{1}{2\pi}\{\log(1+x^2) - \log\bigl(1+(y-x)^2\bigr)\}\right]_{-M}^{N} = 0. \end{aligned}$$

Finally,

$$f_Z(z) = 2f_{X+Y}(2z) = \frac{1}{\pi(1+z^2)}.$$

26. (a) $X_1 + X_2 + \cdots + X_n$.
(b) $X_1 - X_1'$, where X_1 and X_1' are independent and identically distributed.
(c) X_N, where N is a random variable with $\mathbb{P}(N = j) = p_j$ for $1 \le j \le n$, independent of $X_1, X_2, \ldots, X_n$.
(d) $\sum_{j=1}^{M} Z_j$ where $Z_1, Z_2, \ldots$ are independent and distributed as X_1, and M is independent of the Z_j with $\mathbb{P}(M = m) = (\frac{1}{2})^{m+1}$ for $m \ge 0$.
(e) YX_1, where Y is independent of X_1 with the exponential distribution parameter 1.

27. (a) (i) We require

$$\phi(t) = \int_{-\infty}^{\infty} \frac{2e^{itx}}{e^{\pi x} + e^{-\pi x}}\,dx.$$

First method. Consider the contour integral

$$I_K = \int_C \frac{2e^{itz}}{e^{\pi z} + e^{-\pi z}}\,dz$$

where C is a rectangular contour with vertices at $\pm K, \pm K + i$. The integrand has a simple pole at $z = \frac{1}{2}i$, with residue $e^{-\frac{1}{2}t}/(i\pi)$. Hence, by Cauchy's theorem,

$$I_K \to \frac{2e^{-\frac{1}{2}t}}{1+e^{-t}} = \frac{1}{\cosh(\frac{1}{2}t)} \qquad \text{as } K \to \infty.$$

Second method. Expand the denominator to obtain

$$\frac{1}{\cosh(\pi x)} = \sum_{k=0}^{\infty} (-1)^k \exp\{-(2k+1)\pi|x|\}.$$

Multiply by e^{itx} and integrate term by term.
(ii) Define $\phi(t) = 1 - |t|$ for $|t| \le 1$, and $\phi(t) = 0$ otherwise. Then

$$\begin{aligned}\frac{1}{2\pi}\int_{-\infty}^{\infty} e^{-itx}\phi(t)\,dt &= \frac{1}{2\pi}\int_{-1}^{1} e^{-itx}(1-|t|)\,dt \\ &= \frac{1}{\pi}\int_0^1 (1-t)\cos(tx)\,dt = \frac{1}{\pi x^2}(1-\cos x).\end{aligned}$$

Using the inversion theorem, ϕ is the required characteristic function.
(iii) In this case,

$$\int_{-\infty}^{\infty} e^{itx} e^{-x-e^{-x}}\,dx = \int_0^{\infty} y^{-it}e^{-y}\,dy = \Gamma(1-it)$$

where Γ is the gamma function.
(iv) Similarly,

$$\begin{aligned}\int_{-\infty}^{\infty} \tfrac{1}{2}e^{itx}e^{-|x|}\,dx &= \tfrac{1}{2}\left\{\int_0^{\infty} e^{itx}e^{-x}\,dx + \int_0^{\infty} e^{-itx}e^{-x}\,dx\right\} \\ &= \frac{1}{2}\left\{\frac{1}{1-it} + \frac{1}{1+it}\right\} = \frac{1}{1+t^2}.\end{aligned}$$

(v) We have that $\mathbb{E}(X) = -i\phi'(0) = -\Gamma'(1)$. Now, Euler's product for the gamma function states that

$$\Gamma(z) = \lim_{n\to\infty} \frac{n!\,n^z}{z(z+1)\cdots(z+n)}$$

where the convergence is uniform on a neighbourhood of the point $z = 1$. By differentiation,

$$\Gamma'(1) = \lim_{n\to\infty} \left\{ \frac{n}{n+1} \left(\log n - 1 - \frac{1}{2} - \cdots - \frac{1}{n+1} \right) \right\} = -\gamma.$$

(b) By part (a)(ii), this is the characteristic function of Y/π where $f_Y(y) = (1-\cos y)/(\pi y^2)$.

The characteristic function of the given distribution is

$$\phi(t) = \sum_{n=-\infty}^{\infty} e^{int} f(n) = \frac{1}{2} + \frac{4}{\pi^2} \sum_{k=0}^{\infty} \frac{\cos\{(2k+1)t\}}{(2k+1)^2}.$$

Following the usual integrations, this may be seen to be the Fourier cosine series representation of the given function $\psi(t)$. The function ψ is even, and the result follows by the inversion theorem. [We have used standard results of Fourier transform theory, in particular, that a continuous, piecewise continuously differentiable function on $\mathbb{R}$ with period 2π is the sum of its Fourier series.]

28. (a) See Problem (5.12.27b).

(b) Suppose ϕ is the characteristic function of X. Since $\phi'(0) = \phi''(0) = \phi'''(0) = 0$, we have that $\mathbb{E}(X) = \text{var}(X) = 0$, so that $\mathbb{P}(X = 0) = 1$, and hence $\phi(t) = 1$, a contradiction. Hence ϕ is not a characteristic function.

(c) As for (b).

(d) We have that $\cos t = \frac{1}{2}(e^{it} + e^{-it})$, whence ϕ is the characteristic function of a random variable taking values ± 1 each with probability $\frac{1}{2}$.

(e) By the same working as in the solution to Problem (5.12.27b), ϕ is the characteristic function of the density function

$$f(x) = \begin{cases} 1 - |x| & \text{if } |x| < 1, \\ 0 & \text{otherwise.} \end{cases}$$

29. We have that

$$\begin{aligned} |1 - \phi(t)| &\le \mathbb{E}|1 - e^{itX}| = \mathbb{E}\sqrt{(1 - e^{itX})(1 - e^{-itX})} \\ &= \mathbb{E}\sqrt{2\{1 - \cos(tX)\}} \le \mathbb{E}|tX| \end{aligned}$$

since $2(1 - \cos x) \le x^2$ for all x.

30. This is a consequence of Taylor's theorem for functions of two variables:

$$\phi(s,t) = \sum_{\substack{m\le M \\ n\le N}} \frac{s^m t^n}{m!\,n!} \phi_{mn}(0,0) + R_{MN}(s,t)$$

where ϕ_{mn} is the derivative of ϕ in question, and R_{MN} is the remainder. However, subject to appropriate conditions,

$$\phi(s,t) = \sum_{\substack{m\le M \\ n\le N}} \frac{(is)^m (it)^n}{m!\,n!} \mathbb{E}(X^m Y^n) + \mathrm{o}(s^M t^N)$$

whence the claim follows.

31. (a) We have that

$$\frac{x^2}{3} \le \frac{x^2}{2!} - \frac{x^4}{4!} \le 1 - \cos x$$

if $|x| \le 1$, and hence

$$\begin{aligned}\int_{[-t^{-1},t^{-1}]} (tx)^2 \, dF(x) &\le \int_{[-t^{-1},t^{-1}]} 3\{1 - \cos(tx)\} \, dF(x) \\ &\le 3 \int_{-\infty}^{\infty} \{1 - \cos(tx)\} \, dF(x) = 3\{1 - \operatorname{Re} \phi(t)\}.\end{aligned}$$

(b) Using Fubini's theorem,

$$\begin{aligned}\frac{1}{t} \int_0^t \{1 - \operatorname{Re} \phi(v)\} \, dv &= \int_{x=-\infty}^{\infty} \frac{1}{t} \int_{v=0}^{t} \{1 - \cos(vx)\} \, dv \, dF(x) \\ &= \int_{-\infty}^{\infty} \left(1 - \frac{\sin(tx)}{tx}\right) dF(x) \\ &\ge \int_{\substack{x: \\ |tx| \ge 1}} \left(1 - \frac{\sin(tx)}{tx}\right) dF(x)\end{aligned}$$

since $1 - (tx)^{-1} \sin(tx) \ge 0$ if $|tx| < 1$. Also, $\sin(tx) \le (tx) \sin 1$ for $|tx| \ge 1$, whence the last integral is at least

$$\int_{\substack{x: \\ |tx| \ge 1}} (1 - \sin 1) \, dF(x) \ge \tfrac{1}{7} \mathbb{P}(|X| \ge t^{-1}).$$

32. It is easily seen that, if $y > 0$ and n is large,

$$\mathbb{P}\big(n(1 - M_n) > y\big) = \mathbb{P}\left(M_n < 1 - \frac{y}{n}\right) = \prod_{i=1}^{n} \mathbb{P}\left(X_i < 1 - \frac{y}{n}\right) = \left(1 - \frac{y}{n}\right)^n \to e^{-y}.$$

33. (a) The characteristic function of Y_λ is

$$\psi_\lambda(t) = \mathbb{E}\{\exp\big(it(X - \lambda)/\sqrt{\lambda}\big)\} = \exp\{\lambda(e^{it/\sqrt{\lambda}} - 1) - it\sqrt{\lambda}\} = \exp\{-\tfrac{1}{2}t^2 + o(1)\}$$

as $\lambda \to \infty$. Now use the continuity theorem.

(b) In this case,

$$\psi_\lambda(t) = e^{-it\sqrt{\lambda}} \left(1 - \frac{it}{\sqrt{\lambda}}\right)^{-\lambda},$$

so that, as $\lambda \to \infty$,

$$\log \psi_\lambda(t) = -it\sqrt{\lambda} - \lambda \log\left(1 - \frac{it}{\sqrt{\lambda}}\right) = -it\sqrt{\lambda} + \lambda\left(\frac{it}{\sqrt{\lambda}} - \frac{t^2}{2\lambda} + o(\lambda^{-1})\right) \to -\tfrac{1}{2}t^2.$$

(c) Let Z_n be Poisson with parameter n. By part (a),

$$\mathbb{P}\left(\frac{Z_n - n}{\sqrt{n}} \le 0\right) \to \Phi(0) = \tfrac{1}{2}$$

where Φ is the $N(0,1)$ distribution function. The left hand side equals $\mathbb{P}(Z_n \le n) = \sum_{k=0}^{n} e^{-n} n^k / k!$.

34. If you are in possession of $r-1$ different types, the waiting time for the acquisition of the next new type is geometric with probability generating function

$$G_r(s) = \frac{(n-r+1)s}{n-(r-1)s}.$$

Therefore the characteristic function of $U_n = (T_n - n\log n)/n$ is

$$\psi_n(t) = e^{-it\log n} \prod_{r=1}^{n} G_r(e^{it/n}) = n^{-it} \prod_{r=1}^{n} \left\{ \frac{(n-r+1)e^{it/n}}{n-(r-1)e^{it/n}} \right\} = \frac{n^{-it} n!}{\prod_{r=0}^{n-1}(ne^{-it/n} - r)}.$$

The denominator satisfies

$$\prod_{r=0}^{n-1}(ne^{-it/n} - r) = (1+\mathrm{o}(1)) \prod_{r=0}^{n-1}(n - it - r)$$

as $n \to \infty$, by expanding the exponential function, and hence

$$\lim_{n\to\infty} \psi_n(t) = \lim_{n\to\infty} \frac{n^{-it} n!}{\prod_{r=0}^{n-1}(n-it-r)} = \Gamma(1-it),$$

where we have used Euler's product for the gamma function:

$$\frac{n!\, n^z}{\prod_{r=0}^{n}(z+r)} \to \Gamma(z) \quad \text{as } n \to \infty$$

the convergence being uniform on any region of the complex plane containing no singularity of Γ. The claim now follows by the result of part (iii) of Problem (5.12.27a).

35. Let X_n be uniform on $[-n, n]$, with characteristic function

$$\phi_n(t) = \int_{-n}^{n} \frac{1}{2n} e^{itx}\, dx = \begin{cases} \dfrac{\sin(nt)}{nt} & \text{if } t \neq 0, \\ 1 & \text{if } t = 0. \end{cases}$$

It follows that, as $n \to \infty$, $\phi_n(t) \to \delta_{0t}$, the Kronecker delta. The limit function is discontinuous at $t = 0$ and is therefore not itself a characteristic function.

36. (a) Let $G_i(s)$ be the probability generating function of the number shown by the ith die, and suppose that

$$G_1(s)G_2(s) = \sum_{k=2}^{12} \tfrac{1}{11} s^k = \frac{s^2(1-s^{11})}{11(1-s)},$$

so that $1 - s^{11} = 11(1-s)H_1(s)H_2(s)$ where $H_i(s) = s^{-1}G_i(s)$ is a real polynomial of degree 5. However

$$1 - s^{11} = (1-s) \prod_{k=1}^{5} (\omega_k - s)(\overline{\omega}_k - s)$$

where $\omega_1, \overline{\omega}_1, \dots, \omega_5, \overline{\omega}_5$ are the ten complex eleventh roots of unity. The ω_k come in conjugate pairs, and therefore no five of the ten terms in $\prod_{k=1}^{5}(\omega_k - s)(\overline{\omega}_k - s)$ have a product which is a real polynomial. This is a contradiction.

(b) The sum of the scores on two standard dice has generating function

$$\begin{aligned}
G(s) &= \tfrac{1}{36}s^2(1+s+s^2+s^3+s^4+s^5)^2 \\
&= \tfrac{1}{36}s^2(1+s)^2(1-s+s^2)^2(1+s+s^2)^2 \\
&= \left[\tfrac{1}{6}s(1+s)(1+s+s^2)\right]\cdot\left[\tfrac{1}{6}s(1+s)(1-s+s^2)^2(1+s+s^2)\right] \\
&= \left[\tfrac{1}{6}(s+2s^2+2s^3+s^4)\right]\cdot\left[\tfrac{1}{6}(s+s^3+s^4+s^5+s^6+s^8)\right].
\end{aligned}$$

Therefore, two dice labelled 1, 2, 2, 3, 3, 4 and 1, 3, 4, 5, 6, 8, respectively, will generate a random sum with the original distribution.

37. (a) Let H and T be the numbers of heads and tails. The joint probability generating function of H and T is

$$G_{H,T}(s,t) = \mathbb{E}(s^H t^T) = \mathbb{E}(s^H t^{N-H}) = \mathbb{E}\{\mathbb{E}((s/t)^H t^N \mid N)\} = \mathbb{E}\left\{t^N\left(q+\frac{ps}{t}\right)^N\right\}$$

where $p = 1-q$ is the probability of heads on each throw. Hence

$$G_{H,T}(s,t) = G_N(qt+ps) = \exp\{\lambda(qt+ps-1)\}.$$

It follows that

$$G_H(s) = G_{H,T}(s,1) = e^{\lambda p(s-1)}, \qquad G_T(t) = G_{H,T}(1,t) = e^{\lambda q(s-1)},$$

so that $G_{H,T}(s,t) = G_H(s)G_T(t)$, whence H and T are independent.

(b) Suppose conversely that H and T are independent, and write G for the probability generating function of N. From the above calculation, $G_{H,T}(s,t) = G(qt+ps)$, whence $G_H(s) = G(q+ps)$ and $G_T(t) = G(qt+p)$, so that $G(qt+ps) = G(q+ps)G(qt+p)$ for all appropriate s, t. Write $f(x) = G(1-x)$ to obtain $f(x+y) = f(x)f(y)$, valid at least for all $0 \le x, y \le \min\{p,q\}$. The only continuous solutions to this functional equation which satisfy $f(0) = 1$ are of the form $f(x) = e^{\mu x}$ for some μ, whence it is immediate that $G(x) = e^{\lambda(x-1)}$ where $\lambda = -\mu$.

38. The number of such paths π containing exactly n nodes is 2^{n-1}, and each such π satisfies $\mathbb{P}(B(\pi) \ge k) = \mathbb{P}(S_n \ge k)$ where $S_n = Y_1 + Y_2 + \cdots + Y_n$ is the sum of n independent Bernoulli variables having parameter p ($= 1-q$). Therefore $\mathbb{E}\{X_n(k)\} = 2^{n-1}\mathbb{P}(S_n \ge k)$. We set $k = n\beta$, and need to estimate $\mathbb{P}(S_n \ge n\beta)$ as a function of β. By the large deviation theorem (5.11.4), for $p \le \beta < 1$,

$$\mathbb{P}(S_n \ge n\beta)^{1/n} \to \inf_{t>0}\left\{e^{-t\beta}M(t)\right\}$$

where $M(t) = \mathbb{E}(e^{tY_1}) = (q+pe^t)$. With the aid of a little calculus, we find that

$$\mathbb{P}(S_n \ge n\beta)^{1/n} \to \left(\frac{p}{\beta}\right)^\beta\left(\frac{1-p}{1-\beta}\right)^{1-\beta}, \qquad p \le \beta < 1.$$

Hence

$$\mathbb{E}\{X_n(\beta n)\} \to \begin{cases} 0 & \text{if } \gamma(\beta) < 1, \\ \infty & \text{if } \gamma(\beta) > 1, \end{cases}$$

where

$$\gamma(\beta) = 2\left(\frac{p}{\beta}\right)^\beta\left(\frac{1-p}{1-\beta}\right)^{1-\beta}$$

is a decreasing function of β. If $p < \frac{1}{2}$, there is a unique $\beta_c \in [p, 1)$ such that $\gamma(\beta_c) = 1$; if $p \geq \frac{1}{2}$ then $\gamma(\beta) > 1$ for all $\beta \in [p, 1)$ so that we may take $\beta_c = 1$.

Turning to the final part,

$$\mathbb{P}\big(X_n(\beta n) \geq 1\big) \leq \mathbb{E}\{X_n(\beta n)\} \to 0 \qquad \text{if } \beta > \beta_c.$$

It remains to show that $\mathbb{P}(X_n(\beta n) \geq 1) \to 1$ if $\beta < \beta_c$, and here we use some theory of branching processes.

Let $0 < \beta < \beta_c$, so that $\gamma(\beta) > 1$. Since $2\mathbb{P}(S_n \geq n\beta)^{1/n} \to \gamma(\beta) > 1$, we may fix $m \geq 1$ such that $2\mathbb{P}(S_m \geq m\beta)^{1/m} > 1$. From the binary tree T we construct a branching process B as follows. The 0th generation of B contains the root ρ of T. The children of ρ are defined to be those vertices x in the mth generation of T such that there are at least $m\beta$ black vertices on the unique path of T from ρ to x. Thus the mean family-size in B is $\mu := 2^m\mathbb{P}(S_m \geq m\beta) > 1$. Subsequent generations are defined similarly. This branching process is supercritical since $\mu > 1$, and therefore it survives to infinity with some probability $\sigma > 0$. Therefore, $\mathbb{P}(X_{km}(\beta km) \geq 1) \geq \sigma$ for $k \geq 1$.

Some tidying up is now needed. Write $B^y_{\max}(n)$ for the maximum number of black vertices on paths of length n starting at y, and abbreviate $B^\rho_{\max} = B_{\max}$. Note that $X_n(k) \geq 1$ if and only if $B_{\max}(n) \geq k$.

With probability one, there exists $K < \infty$, and a vertex x at depth Km of T, such that the branching process started at x continues forever. Since the path from the root to x has fixed length Km,

$$B_{\max}(km) \geq B^x_{\max}((k-K)m) \qquad \text{for } k \geq K.$$

For $n \geq Km$, choose k such that $km < n \leq (k+1)m$, so that

$$B_{\max}(n) \geq B^x_{\max}((k-K)m).$$

Therefore, for $\beta' < \beta$,

$$\liminf_{n\to\infty}\left(\frac{1}{n}B_{\max}(n)\right) \geq \liminf_{k\to\infty}\left(\frac{1}{(k+1)m}B^x_{\max}((k-K)m)\right) \geq \beta'$$

almost surely. The claim follows.

39. (a) The characteristic function of X_n satisfies

$$\mathbb{E}\big(e^{itX_n}\big) = \left(1 - \frac{\lambda}{n} + \frac{\lambda}{n}e^{it}\right)^n = \left(1 + \frac{\lambda}{n}[e^{it} - 1]\right)^n \to \exp\big(\lambda[e^{it} - 1]\big),$$

the characteristic function of the Poisson distribution.

(b) Similarly,

$$\mathbb{E}(e^{itY_n/n}) = \frac{pe^{it/n}}{1 - (1-p)e^{it/n}} \to \frac{\lambda}{\lambda - it}$$

as $n \to \infty$, the limit being the characteristic function of the exponential distribution.

40. If you cannot follow the hints, take a look at one or more of the following: Moran (1968, p. 389), Breiman (1968, p. 186), Loève (1977, p. 287), Laha and Rohatgi (1979, p. 288).

41. With $Y_k = kX_k$, we have that $\mathbb{E}(Y_k) = 0$, $\mathrm{var}(Y_k) = k^2$, $\mathbb{E}|Y_k^3| = k^3$. Note that $S_n = Y_1 + Y_2 + \cdots + Y_n$ is such that

$$\frac{1}{\{\mathrm{var}(S_n)\}^{3/2}}\sum_{k=1}^n \mathbb{E}|Y_k^3| \sim c\frac{n^4}{n^{9/2}} \to 0$$

as $n \to \infty$, where c is a positive constant. Applying the central limit theorem ((5.10.5) or Problem (5.12.40)), we find that

$$\frac{S_n}{\sqrt{\operatorname{var} S_n}} \xrightarrow{\text{D}} N(0, 1), \qquad \text{as } n \to \infty,$$

where $\operatorname{var} S_n = \sum_{k=1}^{n} k^2 \sim \frac{1}{3}n^3$ as $n \to \infty$.

42. We may suppose that $\mu = 0$ and $\sigma = 1$; if this is not so, then replace X_i by $Y_i = (X_i - \mu)/\sigma$. Let $\mathbf{t} = (t_0, t_1, t_2, \ldots, t_n) \in \mathbb{R}^{n+1}$, and set $\bar{t} = n^{-1}\sum_{j=1}^{n} t_j$. The joint characteristic function of the $n+1$ variables $\overline{X}, Z_1, Z_2, \ldots, Z_n$ is

$$\begin{aligned}\phi(\mathbf{t}) &= \mathbb{E}\left\{\exp\left(it_0\overline{X} + \sum_{j=1}^{n} it_j Z_j\right)\right\} = \mathbb{E}\left\{\prod_{j=1}^{n} \exp\left(i\left[\frac{t_0}{n} + t_j - \bar{t}\right]X_j\right)\right\} \\ &= \prod_{j=1}^{n} \exp\left(-\frac{1}{2}\left[\frac{t_0}{n} + t_j - \bar{t}\right]^2\right)\end{aligned}$$

by independence. Hence

$$\phi(\mathbf{t}) = \exp\left(-\frac{1}{2}\sum_{j=1}^{n}\left[\frac{t_0}{n} + (t_j - \bar{t})\right]^2\right) = \exp\left\{-\frac{t_0^2}{2n} - \frac{1}{2}\sum_{j=1}^{n}(t_j - \bar{t})^2\right\}$$

where we have used the fact that $\sum_{j=1}^{n}(t_j - \bar{t}) = 0$. Therefore

$$\phi(\mathbf{t}) = \mathbb{E}(e^{it_0\overline{X}})\mathbb{E}\left(\exp\left\{i\sum_{1}^{n}(t_j - \bar{t})X_j\right\}\right) = \mathbb{E}(e^{it_0\overline{X}})\mathbb{E}\left(\exp\left\{i\sum_{1}^{n}t_j Z_j\right\}\right),$$

whence $\overline{X}$ is independent of the collection $Z_1, Z_2, \ldots, Z_n$. It follows that $\overline{X}$ is independent of $S^2 = (n-1)^{-1}\sum_{j=1}^{n} Z_j^2$. Compare with Exercise (4.10.5).

43. (i) Clearly, $\mathbb{P}(Y \le y) = \mathbb{P}(X \le \log y) = \Phi(\log y)$ for $y > 0$, where Φ is the $N(0, 1)$ distribution function. The density function of Y follows by differentiating.
(ii) We have that $f_a(x) \ge 0$ if $|a| \le 1$, and

$$\int_0^\infty a\sin(2\pi\log x)\frac{1}{x\sqrt{2\pi}}e^{-\frac{1}{2}(\log x)^2}\,dx = \int_{-\infty}^{\infty}\frac{1}{\sqrt{2\pi}}a\sin(2\pi y)e^{-\frac{1}{2}y^2}\,dy = 0$$

since sine is an odd function. Therefore $\int_{-\infty}^{\infty} f_a(x)\,dx = 1$, so that each such f_a is a density function.

For any positive integer k, the kth moment of f_a is $\int_{-\infty}^{\infty} x^k f(x)\,dx + I_a(k)$ where

$$I_a(k) = \int_{-\infty}^{\infty}\frac{1}{\sqrt{2\pi}}a\sin(2\pi y)e^{ky-\frac{1}{2}y^2}\,dy = 0$$

since the integrand is an odd function of $y - k$. It follows that each f_a has the same moments as f.

44. Here is one way of proving this. Let $X_1, X_2, \ldots$ be the steps of the walk, and let S_n be the position of the walk after the nth step. Suppose $\mu = \mathbb{E}(X_1)$ satisfies $\mu < 0$, and let $e_m = \mathbb{P}(S_n = 0 \text{ for some } n \ge 1 \mid S_0 = -m)$ where $m > 0$. Then $e_m \le \sum_{n=1}^{\infty}\mathbb{P}(T_n > m)$ where $T_n = X_1 + X_2 + \cdots + X_n = S_n - S_0$. Now, for $t > 0$,

$$\mathbb{P}(T_n > m) = \mathbb{P}(T_n - n\mu > m - n\mu) \le e^{-t(m-n\mu)}\mathbb{E}(e^{t(T_n - n\mu)}) = e^{-tm}\left\{e^{t\mu}M(t)\right\}^n$$

where $M(t) = \mathbb{E}(e^{t(X_1-\mu)})$. Now $M(t) = 1+\mathrm{O}(t^2)$ as $t \to 0$, and therefore there exists t (> 0) such that $\theta(t) = e^{t\mu}M(t) < 1$ (remember that $\mu < 0$). With this choice of t, $e_m \le \sum_{n=1}^{\infty} e^{-tm}\theta(t)^n \to 0$ as $m \to \infty$, whence there exists K such that $e_m < \frac{1}{2}$ for $m \ge K$.

Finally, there exist $\delta, \epsilon > 0$ such that $\mathbb{P}(X_1 < -\delta) > \epsilon$, implying that $\mathbb{P}(S_N < -K \mid S_0 = 0) > \epsilon^N$ where $N = \lceil K/\delta \rceil$, and therefore

$$\mathbb{P}(S_n \ne 0 \text{ for all } n \ge 1 \mid S_0 = 0) \ge (1 - e_K)\epsilon^N \ge \tfrac{1}{2}\epsilon^N;$$

therefore the walk is transient. This proof may be shortened by using the Borel–Cantelli lemma.

45. Obviously,

$$L = \begin{cases} a & \text{if } X_1 > a, \\ X_1 + \widetilde{L} & \text{if } X_1 \le a, \end{cases}$$

where $\widetilde{L}$ has the same distribution as L. Therefore,

$$\mathbb{E}(s^L) = s^a\mathbb{P}(X_1 > a) + \sum_{r=1}^{a} s^r\mathbb{E}(s^L)\mathbb{P}(X_1 = r).$$

46. We have that

$$W_n = \begin{cases} W_{n-1} + 1 & \text{with probability } p, \\ W_{n-1} + 1 + \widetilde{W_n} & \text{with probability } q, \end{cases}$$

where $\widetilde{W_n}$ is independent of W_{n-1} and has the same distribution as W_n. Hence $G_n(s) = psG_{n-1}(s) + qsG_{n-1}(s)G_n(s)$. Now $G_0(s) = 1$, and the recurrence relation may be solved by induction. (Alternatively use Problem (5.12.45) with appropriate X_i.)

47. Let W_r be the number of flips until you first see r consecutive heads, so that $\mathbb{P}(L_n < r) = \mathbb{P}(W_r > n)$. Hence,

$$1 + \sum_{n=1}^{\infty} s^n\mathbb{P}(L_n < r) = \sum_{n=0}^{\infty} s^n\mathbb{P}(W > n) = \frac{1 - \mathbb{E}(s^{W_r})}{1 - s},$$

where $\mathbb{E}(s^{W_r}) = G_r(s)$ is given in Problem (5.12.46).

48. We have that

$$X_{n+1} = \begin{cases} \frac{1}{2}X_n & \text{with probability } \frac{1}{2}, \\ \frac{1}{2}X_n + Y_n & \text{with probability } \frac{1}{2}. \end{cases}$$

Hence the characteristic functions satisfy

$$\begin{aligned} \phi_{n+1}(t) = \mathbb{E}(e^{itX_{n+1}}) &= \tfrac{1}{2}\phi_n(\tfrac{1}{2}t) + \tfrac{1}{2}\phi_n(\tfrac{1}{2}t)\frac{\lambda}{\lambda - it} \\ &= \phi_n(\tfrac{1}{2}t)\frac{\lambda - \frac{1}{2}it}{\lambda - it} = \phi_{n-1}(\tfrac{1}{4}t)\frac{\lambda - \frac{1}{4}it}{\lambda - it} = \cdots = \phi_1(t2^{-n})\frac{\lambda - it2^{-n}}{\lambda - it} \to \frac{\lambda}{\lambda - it} \end{aligned}$$

as $n \to \infty$. The limiting distribution is exponential with parameter λ.

49. We have that

$$\int_0^1 G(s)\,ds = \mathbb{E}\left(\int_0^1 s^X\,ds\right) = \mathbb{E}\left(\frac{s^{X+1}}{X+1}\bigg|_0^1\right) = \mathbb{E}\left(\frac{1}{X+1}\right).$$

(a) $(1-e^{-\lambda})/\lambda$, (b) $-(p/q^2)(q+\log p)$, (c) $(1-q^{n+1})/[(n+1)p]$, (d) $-[1+(p/q)\log p]/\log p$.
(e) Not if $\mathbb{P}(X+1>0)=1$, by Jensen's inequality (see Exercise (5.6.1)) and the strict concavity of the function $f(x)=1/x$. If $Y=X+1$ is permitted to be negative, consider the case when $\mathbb{P}(Y=1)=p$, $\mathbb{P}(Y=-2)=\mathbb{P}(Y=-\frac{1}{2})=\frac{1}{2}(1-p)$ for suitable p.

50. By compounding, as in Theorem (5.1.25), the sum has characteristic function

$$G_N(\phi_X(t)) = \frac{p\phi_X(t)}{1-q\phi_X(t)} = \frac{\lambda p}{\lambda p - it},$$

whence the sum is exponentially distributed with parameter λp.

51. Consider the function $G(x)=\{\mathbb{E}(X^2)\}^{-1}\int_{-\infty}^{x} y^2\,dF(y)$. This function is right-continuous and increases from 0 to 1, and is therefore a distribution function. Its characteristic function is

$$\int_{-\infty}^{\infty} \frac{e^{itx}}{\mathbb{E}(X^2)} x^2\,dF(x) = -\frac{1}{\mathbb{E}(X^2)}\frac{d^2}{dt^2}\phi(t).$$

52. By integration, $f_X(x)=f_Y(y)=\frac{1}{2}$, $|x|<1$, $|y|<1$. Since $f(x,y)\neq f_X(x)f_Y(y)$, X and Y are not independent. Now,

$$f_{X+Y}(z) = \int_{-1}^{1} f(x,z-x)\,dx = \begin{cases} \frac{1}{4}(z+2) & \text{if } -2<z<0, \\ \frac{1}{4}(2-z) & \text{if } 0<z<2, \end{cases}$$

the 'triangular' density function on $(-2,2)$. This is the density function of the sum of two independent random variables uniform on $(-1,1)$.

53. By conditioning on X_1,

$$m(x) = 1 + \int_0^1 m(x-u)\,du.$$

Differentiate (with retrospective justification) to obtain

$$m'(x) = \int_0^1 m'(x-u)\,du = m(x) - m(x-1).$$

Now multiply through by e^{-sx} and integrate over $[0,\infty)$, noting that $\lim_{x\downarrow 0} m(x)=1$, to find that

$$-1 + sm^*(s) = m^*(s) - e^{-s}m^*(s),$$

which yields the given m^*.

By swapping the integral and the summation twice, the given summation has Laplace transform

$$\begin{aligned} \int_0^\infty \left(\sum_{r=0}^{\lfloor x\rfloor} \frac{(-1)^r}{r!}(x-r)^r e^{x-r}\right) e^{-sx}\,dx &= \sum_{r=0}^{\infty}\frac{(-1)^r}{r!}\int_0^\infty y^r e^y e^{-s(y+r)}\,dy \\ &= \int_0^\infty e^{y(1-s)}\exp\{-ye^{-s}\}\,dy \\ &= \frac{1}{e^{-s}+s-1}, \qquad s\neq 0, \end{aligned}$$

where we substituted $y=x-r$ in the intermediate step. This equals $m^*(s)$, and the result follows by the Laplace inversion theorem. Note that m is differentiable except on the integers.

54. (a) This holds since, given $X_r \le \frac{1}{2}$ (respectively, $X_r > \frac{1}{2}$), we have that X_r is uniform on $(0, \frac{1}{2})$ (respectively, on $(\frac{1}{2}, 1)$).

(b, c) We have that $\{S_n\} - R_n = k$ if and only if $-\frac{1}{2} + k < S_n - R_n < k + \frac{1}{2}$ (neglecting events of probability zero), which, by (a), has the same probability as $\mathbb{P}(2k - 1 < Y_n < 2k + 1)$, where Y_n is the sum $\sum_{r=1}^{n} Z_r$ of independent random variables with the uniform distribution on $(-1, 1)$. This equals

$$\int_{2k-1}^{2k+1} f_n(z)\,dz = 2f_{n+1}(2k),$$

where we have used the fact that $Y_{n+1} = Y_n + Z_{n+1}$. The characteristic function of f_{n+1} is $(t^{-1}\sin t)^{n+1}$, so by Fourier inversion,

$$\mathbb{P}\big(\{S_n\} - R_n = k\big) = \frac{1}{\pi}\int_{-\infty}^{\infty} \cos(2kt)\left(\frac{\sin t}{t}\right)^{n+1} dt,$$

whence the result of (b) on setting $k = 0$.

55. (a) For a given set of Cartesian coordinates, consider the rotation $U = (V_1 + V_2)/\sqrt{2}$, $W = (V_1 - V_2)/\sqrt{2}$, with V_3 unchanged. By assumption, V_1 and V_2 are independent random variables with equal variances. By Problem (5.12.23), V_1 and V_2 are normally distributed. By a similar rotation in the x/z-plane, V_3 is normal also.

By Problem (4.14.12), $Z = V^2$ has the $\chi^2(3)$ distribution, with density function $f_Z(z) = \sqrt{z}e^{-\frac{1}{2}z}/\sqrt{2\pi}$ for $z > 0$. By a change of variables, $|V|$ has the Maxwell density $f(v) = \sqrt{2/\pi}\,v^2 e^{-\frac{1}{2}v^2}$ for $v > 0$.

(b) Let U be uniformly distributed on $[0, 1]$ and independent of V, and let the first coordinate of $\mathbb{R}^3$ be perpendicular to the planes. The required probability $p(t) = \mathbb{P}\big(U + tV_1 \in [0, 1]\big)$ is the probability that an $N(0, t^2)$ random variable lies in $[-U, 1 - U]$, that is,

$$\begin{aligned} p(t) &= \int_0^1 du \int_{-u}^{1-u} \frac{1}{t\sqrt{2\pi}} e^{-\frac{1}{2}x^2/t^2}\,dx \\ &= \frac{1}{t\sqrt{2\pi}}\left\{\int_{-1}^{1} e^{-\frac{1}{2}x^2/t^2}\,dx - \int_0^1 \left(\int_{-1}^{-u} e^{-\frac{1}{2}x^2/t^2}\,dx + \int_{1-u}^{1} e^{-\frac{1}{2}x^2/t^2}\,dx\right) du\right\}. \end{aligned}$$

This gives the required answer on changing the order of integration in the double integrals.

(c) Finally, $f_T(t) = -dp/dt$. The neater way to answer (b) and (c) is to use Exercise (4.3.12), to obtain that T has the distribution of $W/|V|$, where V has the $N(0, 1)$ distribution, and W is independent and uniformly distributed on $[0, 1]$.

56. Since the joint distribution function $F(x, y)$ of the pair (X, Y) depends only on $r^2 = x^2 + y^2$, the distribution of R is invariant with respect to θ.

(a) By independence and the assumption of rotation invariance, $X\cos\theta$ and $Y\sin\theta$ are independent, whence the characteristic functions satisfy

$$\phi_R(t) = \mathbb{E}(e^{itX\cos\theta})\mathbb{E}(e^{itY\sin\theta}) = \phi_X(t\cos\theta)\phi_Y(t\sin\theta).$$

(b) Set $\theta = 0, \frac{1}{2}\pi, \pi$.

(c) Since X and $-X$ have, by (b), the same distribution,

$$\phi_X(t) = \tfrac{1}{2}\int_{-\infty}^{\infty} (e^{itx} + e^{-itx})\,dF_X(x) = \int_{-\infty}^{\infty} \cos(tx)\,dF_X(x),$$

which is a function of t^2, say $\psi(t^2)$. The function ψ is continuous because ϕ is continuous.
(d) By part (a), $\psi(t^2) = \psi((t\cos\theta)^2)\psi((t\sin\theta)^2)$. Therefore, ψ is a continuous function satisfying $\psi(at^2)\psi(bt^2) = \psi((a+b)t^2)$ for $a, b > 0$. By the result of Problem (4.14.5) or otherwise, $\phi_X(t) = \psi(t^2) = e^{ct^2}$ for some $c \in \mathbb{R}$. It follows that X is $N(0, \sigma^2)$ for some $\sigma^2 \geq 0$.

6

Markov chains

6.1 Solutions. Markov processes

1. The sequence $X_1, X_2, \dots$ of independent random variables satisfies

$$\mathbb{P}(X_{n+1} = j \mid X_1 = i_1, \dots, X_n = i_n) = \mathbb{P}(X_{n+1} = j),$$

whence the sequence is a Markov chain. The chain is homogeneous if the X_i are identically distributed.

2. (a) With Y_n the outcome of the nth throw, $X_{n+1} = \max\{X_n, Y_{n+1}\}$, so that

$$p_{ij} = \begin{cases} 0 & \text{if } j < i \\ \frac{1}{6}i & \text{if } j = i \\ \frac{1}{6} & \text{if } j > i, \end{cases}$$

for $1 \le i, j \le 6$. Similarly,

$$p_{ij}(n) = \begin{cases} 0 & \text{if } j < i \\ (\frac{1}{6}i)^n & \text{if } j = i. \end{cases}$$

If $j > i$, then $p_{ij}(n) = \mathbb{P}(Z_n = j)$, where $Z_n = \max\{Y_1, Y_2, \dots, Y_n\}$, and an elementary calculation yields

$$p_{ij}(n) = \left(\frac{j}{6}\right)^n - \left(\frac{j-1}{6}\right)^n, \qquad i < j \le 6.$$

(b) $N_{n+1} - N_n$ is independent of $N_1, N_2, \dots, N_n$, so that N is Markovian with

$$p_{ij} = \begin{cases} \frac{1}{6} & \text{if } j = i+1, \\ \frac{5}{6} & \text{if } j = i, \\ 0 & \text{otherwise.} \end{cases}$$

(c) The evolution of C is given by

$$C_{r+1} = \begin{cases} 0 & \text{if the die shows 6,} \\ C_r + 1 & \text{otherwise,} \end{cases}$$

whence C is Markovian with

$$p_{ij} = \begin{cases} \frac{1}{6} & \text{if } j = 0, \\ \frac{5}{6} & \text{if } j = i+1, \\ 0 & \text{otherwise.} \end{cases}$$

(d) This time,

$$B_{r+1} = \begin{cases} B_r - 1 & \text{if } B_r > 0, \\ Y_r & \text{if } B_r = 0, \end{cases}$$

where Y_r is a geometrically distributed random variable with parameter $\frac{1}{6}$, independent of the sequence $B_0, B_2, \cdots, B_r$. Hence B is Markovian with

$$p_{ij} = \begin{cases} 1 & \text{if } j = i - 1 \geq 0, \\ (\frac{5}{6})^{j-1}\frac{1}{6} & \text{if } i = 0,\ j \geq 1. \end{cases}$$

3. (i) If $X_n = i$, then $X_{n+1} \in \{i-1, i+1\}$. Now, for $i \geq 1$,

$$(*) \qquad \begin{aligned} \mathbb{P}(X_{n+1} = i+1 \mid X_n = i, B) &= \mathbb{P}(X_{n+1} = i+1 \mid S_n = i, B)\mathbb{P}(S_n = i \mid X_n = i, B) \\ &\quad + \mathbb{P}(X_{n+1} = i+1 \mid S_n = -i, B)\mathbb{P}(S_n = -i \mid X_n = i, B) \end{aligned}$$

where $B = \{X_r = i_r \text{ for } 0 \leq r < n\}$ and $i_0, i_1, \ldots, i_{n-1}$ are integers. Clearly

$$\mathbb{P}(X_{n+1} = i+1 \mid S_n = i, B) = p, \quad \mathbb{P}(X_{n+1} = i+1 \mid S_n = -i, B) = q,$$

where $p\ (= 1 - q)$ is the chance of a rightward step. Let l be the time of the last visit to 0 prior to the time n, $l = \max\{r : i_r = 0\}$. During the time-interval $(l, n]$, the path lies entirely in either the positive integers or the negative integers. If the former, it is required to follow the route prescribed by the event $B \cap \{S_n = i\}$, and if the latter by the event $B \cap \{S_n = -i\}$. The absolute probabilities of these two routes are

$$\pi_1 = p^{\frac{1}{2}(n-l+i)}q^{\frac{1}{2}(n-l-i)}, \quad \pi_2 = p^{\frac{1}{2}(n-l-i)}q^{\frac{1}{2}(n-l+i)},$$

whence

$$\mathbb{P}(S_n = i \mid X_n = i, B) = \frac{\pi_1}{\pi_1 + \pi_2} = \frac{p^i}{p^i + q^i} = 1 - \mathbb{P}(S_n = -i \mid X_n = i, B).$$

Substitute into $(*)$ to obtain

$$\mathbb{P}(X_{n+1} = i+1 \mid X_n = i, B) = \frac{p^{i+1} + q^{i+1}}{p^i + q^i} = 1 - \mathbb{P}(X_{n+1} = i-1 \mid X_n = i, B).$$

Finally $\mathbb{P}(X_{n+1} = 1 \mid X_n = 0, B) = 1$.

(ii) If $Y_n > 0$, then $Y_n - Y_{n+1}$ equals the $(n+1)$th step, a random variable which is independent of the past history of the process. If $Y_n = 0$ then $S_n = M_n$, so that Y_{n+1} takes the values 0 and 1 with respective probabilities p and q, independently of the past history. Therefore Y is a Markov chain with transition probabilities

$$\text{for } i > 0, \quad p_{ij} = \begin{cases} p & \text{if } j = i-1 \\ q & \text{if } j = i+1, \end{cases} \qquad p_{0j} = \begin{cases} p & \text{if } j = 0 \\ q & \text{if } j = 1. \end{cases}$$

The sequence Y is a random walk with a retaining barrier at 0.

4. For any sequence $i_0, i_1, \ldots$ of states,

$$\begin{aligned} \mathbb{P}(Y_{k+1} = i_{k+1} \mid Y_r = i_r \text{ for } 0 \leq r \leq k) &= \frac{\mathbb{P}(X_{n_s} = i_s \text{ for } 0 \leq s \leq k+1)}{\mathbb{P}(X_{n_s} = i_s \text{ for } 0 \leq s \leq k)} \\ &= \frac{\prod_{s=0}^{k} p_{i_s, i_{s+1}}(n_{s+1} - n_s)}{\prod_{s=0}^{k-1} p_{i_s, i_{s+1}}(n_{s+1} - n_s)} \\ &= p_{i_k, i_{k+1}}(n_{k+1} - n_k) = \mathbb{P}(Y_{k+1} = i_{k+1} \mid Y_k = i_k), \end{aligned}$$

where $p_{ij}(n)$ denotes the appropriate n-step transition probability of X.
(a) With the usual notation, the transition matrix of Y is

$$\pi_{ij} = \begin{cases} p^2 & \text{if } j = i+2, \\ 2pq & \text{if } j = i, \\ q^2 & \text{if } j = i-2. \end{cases}$$

(b) With the usual notation, the transition probability π_{ij} is the coefficient of s^j in $G(G(s))^i$.

5. Writing $\mathbf{X} = (X_1, X_2, \dots, X_n)$, we have that

$$\mathbb{P}\big(F \bigm| I(\mathbf{X}) = 1, X_n = i\big) = \frac{\mathbb{P}\big(F, I(\mathbf{X}) = 1, X_n = i\big)}{\mathbb{P}\big(I(\mathbf{X}) = 1, X_n = i\big)}$$

where F is any event defined in terms of $X_n, X_{n+1}, \dots$. Let A be the set of all sequences $\mathbf{x} = (x_1, x_2, \dots, x_{n-1}, i)$ of states such that $I(\mathbf{x}) = 1$. Then

$$\mathbb{P}\big(F, I(\mathbf{X}) = 1, X_n = i\big) = \sum_{\mathbf{x}\in A} \mathbb{P}(F, \mathbf{X} = \mathbf{x}) = \mathbb{P}(F \mid X_n = i) \sum_{\mathbf{x}\in A} \mathbb{P}(\mathbf{X} = \mathbf{x})$$

by the Markov property. Divide through by the final summation to obtain $\mathbb{P}\big(F \bigm| I(\mathbf{X}) = 1, X_n = i\big) = \mathbb{P}(F \mid X_n = i)$.

6. Let $H_n = \{X_k = x_k \text{ for } 0 \le k < n,\ X_n = i\}$. The required probability may be written as

$$\frac{\mathbb{P}(X_{T+m} = j, H_T)}{\mathbb{P}(H_T)} = \frac{\sum_n \mathbb{P}(X_{T+m} = j, H_T, T = n)}{\mathbb{P}(H_T)}.$$

Now $\mathbb{P}(X_{T+m} = j \mid H_T, T = n) = \mathbb{P}(X_{n+m} = j \mid H_n, T = n)$. Let I be the indicator function of the event $H_n \cap \{T = n\}$, an event which depends only upon the values of $X_1, X_2, \dots, X_n$. Using the result of Exercise (6.1.5),

$$\mathbb{P}(X_{n+m} = j \mid H_n, T = n) = \mathbb{P}(X_{n+m} = j \mid X_n = i) = p_{ij}(m).$$

Hence

$$\mathbb{P}(X_{T+m} = j \mid H_T) = \frac{p_{ij}(m) \sum_n \mathbb{P}(H_n, T = n)}{\mathbb{P}(H_T)} = p_{ij}(m).$$

7. Clearly

$$\mathbb{P}(Y_{n+1} = j \mid Y_r = i_r \text{ for } 0 \le r \le n) = \mathbb{P}(X_{n+1} = b \mid X_r = a_r \text{ for } 0 \le r \le n)$$

where $b = h^{-1}(j)$, $a_r = h^{-1}(i_r)$; the claim follows by the Markov property of X.

It is easy to find an example in which h is not one-one, for which X is a Markov chain but Y is not. The first part of Exercise (6.1.3) describes such a case if $S_0 \neq 0$.

8. (a) Not necessarily! Take as example the chains S and Y of Exercise (6.1.3). The sum is $S_n + Y_n = M_n$, which is not a Markov chain.
(b) Still not necessarily! Let X and Y be independent chains (defined on the same probability space) with respective transition matrices

$$\mathbf{P}_X = \begin{pmatrix} a & 1-a \\ 1-b & b \end{pmatrix}, \qquad \mathbf{P}_Y = \begin{pmatrix} p & 1-p \\ 1-q & q \end{pmatrix},$$

and state spaces $\{-1, 1\}$ and $\{0, 2\}$. For general values of a, b, p, q,

$$\mathbb{P}(Z_{n+1} = 1 \mid Z_n = 1,\ Z_{n-1} = 3) \neq \mathbb{P}(Z_{n+1} = 1 \mid Z_n = 1,\ Z_{n-1} = -1).$$

(c) Since Z has independent increments, it is a Markov chain.

9. All of them. (a) Using the Markov property of X,

$$\mathbb{P}(X_{m+r} = k \mid X_m = i_m, \dots, X_{m+r-1} = i_{m+r-1}) = \mathbb{P}(X_{m+r} = k \mid X_{m+r-1} = i_{m+r-1}).$$

(b) Let $\{\text{even}\} = \{X_{2r} = i_{2r} \text{ for } 0 \le r \le m\}$ and $\{\text{odd}\} = \{X_{2r+1} = i_{2r+1} \text{ for } 0 \le r \le m-1\}$. Then,

$$\begin{aligned}\mathbb{P}(X_{2m+2} = k \mid \text{even}) &= \sum{}' \frac{\mathbb{P}(X_{2m+2} = k,\ X_{2m+1} = i_{2m+1},\ \text{even, odd})}{\mathbb{P}(\text{even})} \\ &= \sum{}' \frac{\mathbb{P}(X_{2m+2} = k,\ X_{2m+1} = i_{2m+1} \mid X_{2m} = i_{2m})\mathbb{P}(\text{even, odd})}{\mathbb{P}(\text{even})} \\ &= \mathbb{P}(X_{2m+2} = k \mid X_{2m} = i_{2m}),\end{aligned}$$

where the sum is taken over all possible values of i_s for odd s.
(c) With $Y_n = (X_n, X_{n+1})$,

$$\begin{aligned}\mathbb{P}\big(Y_{n+1} = (k,l) \,\big|\, Y_0 = (i_0, i_1), \dots, Y_n = (i_n, k)\big) &= \mathbb{P}\big(Y_{n+1} = (k,l) \,\big|\, X_{n+1} = k\big) \\ &= \mathbb{P}\big(Y_{n+1} = (k,l) \,\big|\, Y_n = (i_n, k)\big),\end{aligned}$$

by the Markov property of X.

10. We have by Lemma (6.1.8) that, with $\mu_j^{(i)} = \mathbb{P}(X_i = j)$,

$$\text{LHS} = \frac{\mu_{x_1}^{(1)} p_{x_1x_2} \cdots p_{x_{r-1},k} p_{k,x_{r+1}} \cdots p_{x_{n-1}x_n}}{\mu_{x_1}^{(1)} \cdots p_{x_{r-1}x_{r+1}}(2) \cdots p_{x_{n-1}x_n}} = \frac{\mu_{x_{r-1}}^{(r-1)} p_{x_{r-1},k} p_{k,x_{r+1}}}{\mu_{x_{r-1}}^{(r-1)} p_{x_{r-1}x_{r+1}}(2)} = \text{RHS}.$$

11. (a) Since $S_{n+1} = S_n + X_{n+1}$, a sum of independent random variables, S is a Markov chain.
(b) We have that

$$\mathbb{P}(Y_{n+1} = k \mid Y_i = x_i + x_{i-1} \text{ for } 1 \le i \le n) = \mathbb{P}(Y_{n+1} = k \mid X_n = x_n)$$

by the Markov property of X. However, conditioning on X_n is not generally equivalent to conditioning on $Y_n = X_n + X_{n-1}$, so Y does not generally constitute a Markov chain.
(c) $Z_n = nX_1 + (n-1)X_2 + \cdots + X_n$, so Z_{n+1} is the sum of X_{n+1} and a certain linear combination of $Z_1, Z_2, \dots, Z_n$, and so cannot be Markovian.
(d) Since $S_{n+1} = S_n + X_{n+1}$, $Z_{n+1} = Z_n + S_n + X_{n+1}$, and X_{n+1} is independent of $X_1, \dots, X_n$, this is a Markov chain.

12. With $\mathbf{1}$ a row vector of 1's, a matrix $\mathbf{P}$ is stochastic (respectively, doubly stochastic, sub-stochastic) if $\mathbf{P}\mathbf{1}' = \mathbf{1}$ (respectively, $\mathbf{1P} = \mathbf{1}$, $\mathbf{P1}' \le \mathbf{1}$, with inequalities interpreted coordinatewise). By recursion, $\mathbf{P}$ satisfies any of these equations if and only if $\mathbf{P}^n$ satisfies the same equation.

13. Assume X is $\mathcal{C}$-lumpable. For $a, b \in J$, the probability $\mathbb{P}_i(Y_n = C_b)$ does not depend on the choice of $i \in C_a$. Conversely, if the condition holds, then $\mathbb{P}_i(Y_{n+1} = C_{n+1} \mid Y_n = C_n, \dots, Y_1 = C_1)$ is independent of i, so Y is a Markov chain.

6.2 Solutions. Classification of states

1. Let A_k be the event that the last visit to i, prior to n, took place at time k. Suppose that $X_0 = i$, so that $A_0, A_1, \ldots, A_{n-1}$ form a partition of the sample space. It follows, by conditioning on the A_i, that

$$p_{ij}(n) = \sum_{k=0}^{n-1} p_{ii}(k) l_{ij}(n-k)$$

for $i \neq j$. Multiply by s^n and sum over n (≥ 1) to obtain $P_{ij}(s) = P_{ii}(s)L_{ij}(s)$ for $i \neq j$. Now $P_{ij}(s) = F_{ij}(s)P_{jj}(s)$ if $i \neq j$, so that $F_{ij}(s) = L_{ij}(s)$ whenever $P_{ii}(s) = P_{jj}(s)$.

As examples of chains for which $P_{ii}(s)$ does not depend on i, consider a simple random walk on the integers, or a symmetric random walk on a complete graph.

2. Let i ($\neq s$) be a state of the chain, and define $n_i = \min\{n : p_{is}(n) > 0\}$. If $X_0 = i$ and $X_{n_i} = s$ then, with probability one, X makes no visit to i during the intervening period $[1, n_i - 1]$; this follows from the minimality of n_i. Now s is absorbing, and hence

$$\mathbb{P}(\text{no return to } i \mid X_0 = i) \geq \mathbb{P}(X_{n_i} = s \mid X_0 = i) > 0.$$

3. Let I_k be the indicator function of the event $\{X_k = i\}$, so that $N = \sum_{k=0}^{\infty} I_k$ is the number of visits to i. Then

$$\mathbb{E}(N) = \sum_{k=0}^{\infty} \mathbb{E}(I_k) = \sum_{k=0}^{\infty} p_{ii}(k)$$

which diverges if and only if i is recurrent. There is another argument which we shall encounter in some detail when solving Problem (6.15.5).

4. We write $\mathbb{P}_i(\cdot) = \mathbb{P}(\cdot \mid X_0 = i)$. One way is as follows, another is via the calculation of Problem (6.15.5). Note that $\mathbb{P}_i(V_j \geq 1) = \mathbb{P}_i(T_j < \infty)$.

(a) We have that

$$\mathbb{P}_i(V_i \geq 2) = \mathbb{P}_i(V_i \geq 2 \mid V_i \geq 1)\mathbb{P}_i(V_i \geq 1) = \mathbb{P}_i(V_i \geq 1)^2$$

by the strong Markov property (Exercise (6.1.6)) applied at the stopping time T_i. By iteration, $\mathbb{P}_i(V_i \geq n) = \mathbb{P}_i(V_i \geq 1)^n$, and allowing $n \to \infty$ gives the result.

(b) Suppose $i \neq j$. For $m \geq 1$,

$$\mathbb{P}_i(V_j \geq m) = \mathbb{P}_i(V_j \geq m \mid T_j < \infty)\mathbb{P}_i(T_j < \infty) = \mathbb{P}_j(V_j \geq m-1)\mathbb{P}_i(T_j < \infty)$$

by the strong Markov property. Now let $m \to \infty$, and use the result of (a).

5. Let $\theta = \mathbb{P}(T_j < T_i \mid X_0 = i) = \mathbb{P}(T_i < T_j \mid X_0 = j)$, and let N be the number of visits to j before visiting i. Then

$$\mathbb{P}(N \geq 1 \mid X_0 = i) = \mathbb{P}(T_j < T_i \mid X_0 = i) = \theta.$$

Likewise, $\mathbb{P}(N \geq k \mid X_0 = i) = \theta(1-\theta)^{k-1}$ for $k \geq 1$, whence

$$\mathbb{E}(N \mid X_0 = i) = \sum_{k=1}^{\infty} \theta(1-\theta)^{k-1} = 1.$$

6. Let $i \neq j$ and $s \in (0, 1)$. The mean number of visits to j before T is $\sum_n p_{ij}(n)s^n = P_{ij}(s)$. On the other hand, the chain visits j before T if and only if $T > T_j$, where T_j is the time of the first visit to j. The probability of this is $\mathbb{P}(T > T_j) = \sum_n f_{ij}s^n = F_{ij}(s)$. If $T > T_j$, the chain visits j before T a number of times with mean $\sum_n p_{jj}(n)s^n = P_{jj}(s)$. Therefore, $P_{ij}(s) = F_{ij}(s)P_{jj}(s)$, as required. The argument is similar when $i = j$, on remembering to count the initial state.

7. In the obvious notation, $G = G(a, b, \neg c; s)$ satisfies

$$\begin{aligned}\mathbb{E}(s^{T_{ab}}) = F_{ab} &= \mathbb{E}\big(s^{T_{ab}} I(T_{ab} < T_{ac})\big) + \mathbb{E}\big(s^{T_{ab}} I(T_{ab} > T_{ac})\big)\\ &= G + \mathbb{E}\big(s^{T_{ac}} I(T_{ab} > T_{ac})\big) F_{cb},\end{aligned}$$

and similarly,

$$\mathbb{E}(s^{T_{ac}}) = F_{ac} = \mathbb{E}\big(s^{T_{ac}} I(T_{ab} > T_{ac})\big) + G F_{bc}.$$

Now eliminate the brackets and solve for G.

The required probability $G(a, b, \neg c; 1)$ is obtained from the formula by L'Hôpital's rule.

6.3 Solutions. Classification of chains

1. If $r = 1$, then state i is absorbing for $i \geq 1$; also, 0 is transient unless $a_0 = 1$.

Assume $r < 1$ and let $J = \sup\{j : a_j > 0\}$. The states $0, 1, \dots, J$ form an irreducible recurrent class; they are aperiodic if $r > 0$. All other states are transient. For $0 \leq i \leq J$, the recurrence time T_i of i satisfies $\mathbb{P}(T_i = 1) = r$. If $T_i > 1$ then T_i may be expressed as the sum of

$$\begin{aligned} T_i^{(1)} &:= \text{time to reach 0, given that the first step is leftwards,}\\ T_i^{(2)} &:= \text{time spent in excursions from 0 not reaching } i,\\ T_i^{(3)} &:= \text{time taken to reach } i \text{ in final excursion.}\end{aligned}$$

It is easy to see that $\mathbb{E}(T_i^{(1)}) = 1 + (i-1)/(1-r)$ if $i \geq 1$, since the waiting time at each intermediate point has mean $(1-r)^{-1}$. The number N of such 'small' excursions has mass function $\mathbb{P}(N = n) = \alpha_i(1-\alpha_i)^n$, $n \geq 0$, where $\alpha_i = \sum_{j=i}^{\infty} a_j$; hence $\mathbb{E}(N) = (1-\alpha_i)/\alpha_i$. Each such small excursion has mean duration

$$\sum_{j=0}^{i-1}\left(\frac{j}{1-r}+1\right)\frac{a_j}{1-\alpha_i} = 1 + \sum_{j=0}^{i-1}\frac{ja_j}{(1-\alpha_i)(1-r)}$$

and therefore

$$\mathbb{E}(T_i^{(2)}) = \frac{1}{\alpha_i}\left\{(1-\alpha_i) + \sum_{j=0}^{i-1}\frac{ja_j}{1-r}\right\}.$$

By a similar argument,

$$\mathbb{E}(T_i^{(3)}) = \frac{1}{\alpha_i}\sum_{j=i}^{\infty}\left(1 + \frac{j-i}{1-r}\right)a_j.$$

Combining this information, we obtain that

$$\mathbb{E}(T_i) = r + (1-r)\mathbb{E}(T_i^{(1)} + T_i^{(2)} + T_i^{(3)}) = \frac{1}{\alpha_i}\left(1 - r + \sum_{j=0}^{\infty} ja_j\right), \quad i \geq 1,$$

and a similar argument yields $\mathbb{E}(T_0) = 1 + \sum_j ja_j/(1-r)$. The apparent simplicity of these formulae suggests the possibility of an easier derivation; see Exercise (6.4.2). Clearly $\mathbb{E}(T_i) < \infty$ for $i \leq J$ whenever $\sum_j ja_j < \infty$, a condition which certainly holds if $J < \infty$.

2. Assume that $0 < p < 1$. The mean jump-size is $3p - 1$, whence the chain is recurrent if and only if $p = \frac{1}{3}$; see Theorem (5.10.17).

3. (a) All states are absorbing if $p = 0$. Assume henceforth that $p \neq 0$. Diagonalize $\mathbf{P}$ to obtain $\mathbf{P} = \mathbf{B\Lambda B}^{-1}$ where

$$\mathbf{B} = \begin{pmatrix} 1 & 1 & 1 \\ 1 & 0 & -1 \\ 1 & -1 & 1 \end{pmatrix}, \quad \mathbf{B}^{-1} = \begin{pmatrix} \frac{1}{4} & \frac{1}{2} & \frac{1}{4} \\ \frac{1}{2} & 0 & -\frac{1}{2} \\ \frac{1}{4} & -\frac{1}{2} & \frac{1}{4} \end{pmatrix},$$

$$\mathbf{\Lambda} = \begin{pmatrix} 1 & 0 & 0 \\ 0 & 1-2p & 0 \\ 0 & 0 & 1-4p \end{pmatrix}.$$

Therefore

$$\mathbf{P}^n = \mathbf{B\Lambda}^n\mathbf{B}^{-1} = \mathbf{B} \begin{pmatrix} 1 & 0 & 0 \\ 0 & (1-2p)^n & 0 \\ 0 & 0 & (1-4p)^n \end{pmatrix} \mathbf{B}^{-1}$$

whence $p_{ij}(n)$ is easily found.

In particular,

$$p_{11}(n) = \tfrac{1}{4} + \tfrac{1}{2}(1-2p)^n + \tfrac{1}{4}(1-4p)^n, \quad p_{22}(n) = \tfrac{1}{2} + \tfrac{1}{2}(1-4p)^n,$$

and $p_{33}(n) = p_{11}(n)$ by symmetry.

Now $F_{ii}(s) = 1 - P_{ii}(s)^{-1}$, where

$$P_{11}(s) = P_{33}(s) = \frac{1}{4(1-s)} + \frac{1}{2\{1-s(1-2p)\}} + \frac{1}{4\{1-s(1-4p)\}},$$
$$P_{22}(s) = \frac{1}{2(1-s)} + \frac{1}{2\{1-s(1-4p)\}}.$$

After a little work one obtains the mean recurrence times $\mu_i = F'_{ii}(1)$: $\mu_1 = \mu_3 = 4$, $\mu_2 = 2$.

(b) The chain has period 2 (if $p \neq 0$), and all states are positive and recurrent. By symmetry, the mean recurrence times μ_i are equal. One way of calculating their common value (we shall encounter an easier way in Section 6.4) is to observe that the sequence of visits to any given state j is a renewal process (see Example (5.2.15)). Suppose for simplicity that $p \neq 0$. The times between successive visits to j must be even, and therefore we work on a new time-scale in which one new unit equals two old units. Using the renewal theorem (5.2.24), we obtain

$$p_{ij}(2n) \to \frac{2}{\mu_j} \text{ if } |j-i| \text{ is even}, \quad p_{ij}(2n+1) \to \frac{2}{\mu_j} \text{ if } |j-i| \text{ is odd};$$

note that the mean recurrence time of j in the new time-scale is $\frac{1}{2}\mu_j$. Now $\sum_j p_{ij}(m) = 1$ for all m, and so, letting $m = 2n \to \infty$, we find that $4/\mu = 1$ where μ is a typical mean recurrence time.

There is insufficient space here to calculate $p_{ij}(n)$. One way is to diagonalize the transition matrix. Another is to write down a family of difference equations of the form $p_{12}(n) = p \cdot p_{22}(n-1) + (1-p) \cdot p_{42}(n-1)$, and solve them.

4. (a) By symmetry, all states have the same mean-recurrence time. Using the renewal-process argument of the last solution, the common value equals 8, being the number of vertices of the cube. Hence $\mu_v = 8$.

Alternatively, let s be a neighbour of v, and let t be a neighbour of s other than v. In the obvious notation, by symmetry,

$$\mu_v = 1 + \tfrac{3}{4}\mu_{sv}, \qquad \mu_{sv} = 1 + \tfrac{1}{4}\mu_{sv} + \tfrac{1}{2}\mu_{tv},$$
$$\mu_{tv} = 1 + \tfrac{1}{2}\mu_{sv} + \tfrac{1}{4}\mu_{tv} + \tfrac{1}{4}\mu_{wv}, \quad \mu_{wv} = 1 + \tfrac{1}{4}\mu_{wv} + \tfrac{3}{4}\mu_{tv},$$

a system of equations which may be solved to obtain $\mu_v = 8$.
(b) Using the above equations, $\mu_{wv} = \frac{40}{3}$, whence $\mu_{vw} = \frac{40}{3}$ by symmetry.
(c) The required number X satisfies $\mathbb{P}(X = n) = \theta^{n-1}(1-\theta)^2$ for $n \geq 1$, where θ is the probability that the first return of the walk to its starting point precedes its first visit to the diametrically opposed vertex. Therefore

$$\mathbb{E}(X) = \sum_{n=1}^{\infty} n\theta^{n-1}(1-\theta)^2 = 1.$$

5. (a) Let $\mathbb{P}_i(\cdot) = \mathbb{P}(\cdot \mid X_0 = i)$. Since i is recurrent,

$$\begin{aligned} 1 = \mathbb{P}_i(V_i = \infty) &= \mathbb{P}_i(V_j = 0,\ V_i = \infty) + \mathbb{P}_i(V_j > 0,\ V_i = \infty) \\ &\leq \mathbb{P}_i(V_j = 0) + \mathbb{P}_i(T_j < \infty,\ V_i = \infty) \\ &\leq 1 - \mathbb{P}_i(T_j < \infty) + \mathbb{P}_i(T_j < \infty)\mathbb{P}_j(V_i = \infty), \end{aligned}$$

by the strong Markov property. Since $i \to j$, we have that $\mathbb{P}_j(V_i = \infty) \geq 1$, which implies $\eta_{ji} = 1$. Also, $\mathbb{P}_i(T_j < \infty) = 1$, and hence $j \to i$ and j is recurrent. This implies $\eta_{ij} = 1$.
(b) This is an immediate consequence of Exercise (6.2.4b).

6. Let $\mathbb{P}_i(\cdot) = \mathbb{P}(\cdot \mid X_0 = i)$. It is trivial that $\eta_j = 1$ for $j \in A$. For $j \notin A$, condition on the first step and use the Markov property to obtain

$$\eta_j = \sum_{k \in S} p_{jk}\mathbb{P}(T_A < \infty \mid X_1 = k) = \sum_k p_{jk}\eta_k.$$

If $\mathbf{x} = (x_j : j \in S)$ is any non-negative solution of these equations, then $x_j = 1 \geq \eta_j$ for $j \in A$. For $j \notin A$,

$$\begin{aligned} x_j = \sum_{k \in S} p_{jk}x_k &= \sum_{k \in A} p_{jk} + \sum_{k \notin A} p_{jk}x_k = \mathbb{P}_j(T_A = 1) + \sum_{k \notin A} p_{jk}x_k \\ &= \mathbb{P}_j(T_A = 1) + \sum_{k \notin A} p_{jk}\Big\{\sum_{i \in A} p_{ki} + \sum_{i \notin A} p_{ki}x_i\Big\} = \mathbb{P}_j(T_A \leq 2) + \sum_{k \notin A} p_{jk}\sum_{i \notin A} p_{ki}x_i. \end{aligned}$$

We obtain by iteration that, for $j \notin A$,

$$x_j = \mathbb{P}_j(T_A \leq n) + \sum p_{jk_1}p_{k_1k_2}\cdots p_{k_{n-1},k_n}x_{k_n} \geq \mathbb{P}(T_A \leq n),$$

where the sum is over all $k_1, k_2, \dots, k_n \notin A$. We let $n \to \infty$ to find that $x_j \geq \mathbb{P}_j(T_A < \infty) = \eta_j$.

7. The first part follows as in Exercise (6.3.6). Suppose $\mathbf{x} = (x_j : j \in S)$ is a non-negative solution to the equations. As above, for $j \notin A$,

$$\begin{aligned} x_j = 1 + \sum_k p_{jk}x_k &= \mathbb{P}_j(T_A \geq 1) + \sum_{k \notin A} p_{jk}\Big(1 + \sum_{i \notin A} p_{ki}x_i\Big) \\ &= \mathbb{P}_j(T_A \geq 1) + \mathbb{P}_j(T_A \geq 2) + \cdots + \mathbb{P}_j(T_A \geq n) + \sum p_{jk_1}p_{k_1k_2}\cdots p_{k_{k-1}k_n}x_{k_n} \\ &\geq \sum_{m=1}^{n} \mathbb{P}(T_A \geq m), \end{aligned}$$

where the penultimate sum is over all paths of length n that do not visit A. We let $n \to \infty$ to obtain that $x_j \ge \mathbb{E}_j(T_A) = \rho_j$.

8. Yes, because the S_r and T_r are stopping times whenever they are finite. Whether or not the exit times are stopping times depends on their exact definition. The times $U_r = \min\{k > U_{r-1} : X_{U_r} \in A,\ X_{U_r+1} \notin A\}$ are not stopping times, but the times $U_r + 1$ are stopping times.

9. (a) Using the aperiodicity of j, there exist integers $r_1, r_2, \dots, r_s$ having highest common factor 1 and such that $p_{jj}(r_k) > 0$ for $1 \le k \le s$. There exists a positive integer M such that, if $r \ge M$, then $r = \sum_{k=1}^{s} a_k r_k$ for some sequence $a_1, a_2, \dots, a_s$ of non-negative integers. Now, by the Chapman–Kolmogorov equations,

$$p_{jj}(r) \ge \prod_{k=1}^{s} p_{jj}(r_k)^{a_k} > 0,$$

so that $p_{jj}(r) > 0$ for all $r \ge M$.

Finally, find m such that $p_{ij}(m) > 0$. Then

$$p_{ij}(r+m) \ge p_{ij}(m)p_{jj}(r) > 0 \qquad \text{if } r \ge M.$$

(b) Since there are only finitely many pairs i, j, the maximum $R(\mathbf{P}) = \max\{N(i,j) : i, j \in S\}$ is finite. Now $R(\mathbf{P})$ depends only on the positions of the non-negative entries in the transition matrix $\mathbf{P}$. There are only finitely many subsets of entries of $\mathbf{P}$, and so there exists $f(n)$ such that $R(\mathbf{P}) \le f(n)$ for all relevant $n \times n$ transition matrices $\mathbf{P}$.

(c) Consider the two chains with diagrams in the figure beneath. In the case on the left, we have that $p_{11}(5) = 0$, and in the case on the right, we may apply the postage stamp lemma with $a = n$ and $b = n - 1$.

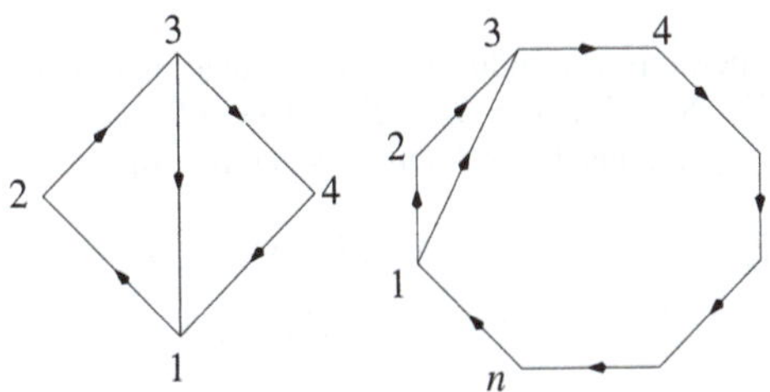

10. Let X_n be the number of green balls after n steps. Let e_j be the probability that X_n is ever zero when $X_0 = j$. By conditioning on the first removal,

$$e_j = \frac{j+2}{2(j+1)} e_{j+1} + \frac{j}{2(j+1)} e_{j-1}, \quad j \ge 1,$$

with $e_0 = 1$. Solving recursively gives

$$(*) \qquad e_j = 1 - (1 - e_1)\left\{1 + \frac{q_1}{p_1} + \cdots + \frac{q_1 q_2 \cdots q_{j-1}}{p_1 p_2 \cdots p_{j-1}}\right\},$$

where

$$p_j = \frac{j+2}{2(j+1)}, \quad q_j = \frac{j}{2(j+1)}.$$

It is easy to see that

$$\sum_{r=0}^{j-1} \frac{q_1 q_2 \cdots q_{j-1}}{p_1 p_2 \cdots p_{j-1}} = 2 - \frac{2}{j+1} \to 2 \quad \text{as } j \to \infty.$$

By the result of Exercise (6.3.6), we seek the minimal non-negative solution (e_j) to $(*)$, which is attained when $2(1-e_1)=1$, that is, $e_1=\frac{1}{2}$. Hence

$$e_j = 1 - \frac{1}{2}\sum_{r=0}^{j-1} \frac{q_1 q_2 \cdots q_{j-1}}{p_1 p_2 \cdots p_{j-1}} = \frac{1}{j+1}.$$

For the second part, let d_j be the expected time until $j-1$ green balls remain, having started with j green balls and $j+2$ red. We condition as above to obtain

$$d_j = 1 + \frac{j}{2(j+1)}\{d_{j+1}+d_j\}.$$

We set $e_j = d_j - (2j+1)$ to find that $(j+2)e_j = je_{j+1}$, whence $e_j = \frac{1}{2}j(j+1)e_1$. The expected time to remove all the green balls is

$$\sum_{j=1}^{n} d_j = \sum_{j=1}^{n}\{e_j + 2(j-1)\} = n(n+2) + e_1\sum_{j=1}^{n}\tfrac{1}{2}j(j+1).$$

The minimal non-negative solution is found by setting $e_1=0$, and the conclusion follows by Exercise (6.3.7).

6.4 Solutions. Stationary distributions and the limit theorem

1. Let Y_n be the number of new errors introduced at the nth stage, and let G be the common probability generating function of the Y_n. Now $X_{n+1} = S_n + Y_{n+1}$ where S_n has the binomial distribution with parameters X_n and q $(=1-p)$. Thus the probability generating function G_n of X_n satisfies

$$\begin{aligned} G_{n+1}(s) &= G(s)\mathbb{E}(s^{S_n}) = G(s)\mathbb{E}\{\mathbb{E}(s^{S_n} \mid X_n)\} = G(s)\mathbb{E}\{(p+qs)^{X_n}\} \\ &= G(s)G_n(p+qs) = G(s)G_n\big(1-q(1-s)\big). \end{aligned}$$

Therefore, for $s<1$,

$$\begin{aligned} G_n(s) &= G(s)G\big(1-q(1-s)\big)G_{n-2}(1-q^2(1-s)) = \cdots \\ &= G_0\big(1-q^n(1-s)\big)\prod_{r=0}^{n-1} G\big(1-q^r(1-s)\big) \to \prod_{r=0}^{\infty} G\big(1-q^r(1-s)\big) \end{aligned}$$

as $n\to\infty$, assuming $q<1$. This infinite product is therefore the probability generating function of the stationary distribution whenever this exists. If $G(s)=e^{\lambda(s-1)}$, then

$$\prod_{r=0}^{\infty} G\big(1-q^r(1-s)\big) = \exp\left\{\lambda(s-1)\sum_{r=0}^{\infty} q^r\right\} = e^{\lambda(s-1)/p},$$

so that the stationary distribution is Poisson with parameter λ/p.

2. (6.3.1): Assume for simplicity that $\sup\{j : a_j > 0\} = \infty$. The chain is irreducible if $r<1$. Look for a stationary distribution $\boldsymbol{\pi}$ with probability generating function G. We have that

$$\pi_0 = a_0\pi_0 + (1-r)\pi_1, \quad \pi_i = a_i\pi_0 + r\pi_i + (1-r)\pi_{i+1} \text{ for } i \ge 1.$$

Hence

$$sG(s) = \pi_0 sA(s) + rs(G(s) - \pi_0) + (1-r)(G(s) - \pi_0)$$

where $A(s) = \sum_{j=0}^{\infty} a_j s^j$, and therefore

$$G(s) = \pi_0 \left(\frac{sA(s) - (1 - r + sr)}{(1-r)(s-1)} \right).$$

Taking the limit as $s \uparrow 1$, we obtain by L'Hôpital's rule that

$$G(1) = \pi_0 \left(\frac{A'(1) + 1 - r}{1 - r} \right).$$

There exists a stationary distribution if and only if $r < 1$ and $A'(1) < \infty$, in which case

$$G(s) = \frac{sA(s) - (1 - r + sr)}{(s-1)(A'(1) + 1 - r)}.$$

Hence the chain is positive recurrent if and only if $r < 1$ and $A'(1) < \infty$. The mean recurrence time μ_i is found by expanding G and setting $\mu_i = 1/\pi_i$.

(6.3.2): Assume that $0 < p < 1$, and suppose first that $p \neq \frac{1}{3}$. Look for a solution $\{y_j : j \neq 0\}$ of the equations

$$(*) \qquad y_i = \sum_{j \neq 0} p_{ij} y_j, \qquad i \neq 0,$$

as in (6.4.10). Away from the origin, this equation is $y_i = q y_{i-1} + p y_{i+2}$ where $p + q = 1$, with auxiliary equation $p\theta^3 - \theta + q = 0$. Now $p\theta^3 - \theta + q = p(\theta - 1)(\theta - \alpha)(\theta - \beta)$ where

$$\alpha = \frac{-p - \sqrt{p^2 + 4pq}}{2p} < -1, \quad \beta = \frac{-p + \sqrt{p^2 + 4pq}}{2p} > 0.$$

Note that $0 < \beta < 1$ if $p > \frac{1}{3}$, while $\beta > 1$ if $p < \frac{1}{3}$.

For $p > \frac{1}{3}$, set

$$y_i = \begin{cases} A + B\beta^i & \text{if } i \geq 1, \\ C + D\alpha^i & \text{if } i \leq -1, \end{cases}$$

the constants A, B, C, D being chosen in such a manner as to 'patch over' the omission of 0 in the equations $(*)$:

$$(**) \qquad y_{-2} = q y_{-3}, \quad y_{-1} = q y_{-2} + p y_1, \quad y_1 = p y_3.$$

The result is a bounded non-zero solution $\{y_j\}$ to $(*)$, and it follows that the chain is transient.

For $p < \frac{1}{3}$, follow the same route with

$$y_i = \begin{cases} 0 & \text{if } i \geq 1, \\ C + D\alpha^i + \mathbb{E}\beta^i & \text{if } i \leq -1, \end{cases}$$

the constants being chosen such that $y_{-2} = q y_{-3}$, $y_{-1} = q y_{-2}$.

Finally suppose that $p = \frac{1}{3}$, so that $\alpha = -2$ and $\beta = 1$. The general solution to $(*)$ is

$$y_i = \begin{cases} A + Bi + C\alpha^i & \text{if } i \geq 1, \\ D + Ei + F\alpha^i & \text{if } i \leq -1, \end{cases}$$

subject to (∗∗). Any bounded solution has $B = E = C = 0$, and (∗∗) implies that $A = D = F = 0$. Therefore the only bounded solution to (∗) is the zero solution, whence the chain is recurrent. The equation $\mathbf{x} = \mathbf{xP}$ is satisfied by the vector $\mathbf{x}$ of 1's; by an appeal to (6.4.6), the walk is null.

(6.3.3): (a) Solve the equation $\boldsymbol{\pi} = \boldsymbol{\pi}\mathbf{P}$ to find a stationary distribution $\boldsymbol{\pi} = (\frac{1}{4}, \frac{1}{2}, \frac{1}{4})$ when $p \neq 0$. Hence the chain is positive and recurrent, with $\mu_1 = \pi_1^{-1} = 4$, and similarly $\mu_2 = 2$, $\mu_3 = 4$.

(b) Similarly, $\boldsymbol{\pi} = (\frac{1}{4}, \frac{1}{4}, \frac{1}{4}, \frac{1}{4})$ is a stationary distribution, and $\mu_i = \pi_i^{-1} = 4$.

(6.3.4): (a) The stationary distribution may be found to be $\pi_i = \frac{1}{8}$ for all i, so that $\mu_v = 8$.

3. The quantities $X_1, X_2, \dots, X_n$ depend only on the initial contents of the reservoir and the rainfalls $Y_0, Y_1, \dots, Y_{n-1}$. The contents on day $n+1$ depend only on the value X_n of the previous contents and the rainfall Y_n. Since Y_n is independent of all earlier rainfalls, the process X is a Markov chain. Its state space is $S = \{0, 1, 2, \dots, K-1\}$ and it has transition matrix

$$\mathbb{P} = \begin{pmatrix} g_0 + g_1 & g_2 & g_3 & \cdots & g_{K-1} & G_K \\ g_0 & g_1 & g_2 & \cdots & g_{K-2} & G_{K-1} \\ 0 & g_0 & g_1 & \cdots & g_{K-3} & G_{K-2} \\ \vdots & \vdots & \vdots & \ddots & \vdots & \vdots \\ 0 & 0 & 0 & \cdots & g_0 & G_1 \end{pmatrix}$$

where $g_i = \mathbb{P}(Y_1 = i)$ and $G_i = \sum_{j=i}^{\infty} g_j$. The equation $\boldsymbol{\pi} = \boldsymbol{\pi}\mathbf{P}$ is as follows:

$$\begin{aligned} \pi_0 &= \pi_0(g_0 + g_1) + \pi_1 g_0, \\ \pi_r &= \pi_0 g_{r+1} + \pi_1 g_r + \cdots + \pi_{r+1} g_0, \qquad 0 < r < K-1, \\ \pi_{K-1} &= \pi_0 G_K + \pi_1 G_{K-1} + \cdots + \pi_{K-1} G_1. \end{aligned}$$

The final equation is a consequence of the previous ones, since $\sum_{i=0}^{K-1} \pi_i = 1$. Suppose then that $\boldsymbol{\nu} = (\nu_1, \nu_2, \dots)$ is an infinite vector satisfying

$$\nu_0 = \nu_0(g_0 + g_1) + \nu_1 g_0, \quad \nu_r = \nu_0 g_{r+1} + \nu_1 g_r + \cdots + \nu_{r+1} g_0 \text{ for } r > 0.$$

Multiply through the equation for ν_r by s^{r+1}, and sum over r to find (after a little work) that

$$N(s) = \sum_{i=0}^{\infty} \nu_i s^i, \qquad G(s) = \sum_{i=0}^{\infty} g_i s^i$$

satisfy $sN(s) = N(s)G(s) + \nu_0 g_0(s-1)$, and hence

$$\frac{1}{\nu_0} N(s) = \frac{g_0(s-1)}{s - G(s)}.$$

The probability generating function of the π_i is therefore a constant multiplied by the coefficients of $s^0, s^1, \dots, s^{K-1}$ in $g_0(s-1)/(s - G(s))$, the constant being chosen in such a way that $\sum_{i=0}^{K-1} \pi_i = 1$.

When $G(s) = p(1 - qs)^{-1}$, then $g_0 = p$ and

$$\frac{g_0(s-1)}{s - G(s)} = \frac{p(1-qs)}{p - qs} = p + \frac{q}{1 - (qs/p)}.$$

The coefficient of s^0 is 1, and of s^i is q^{i+1}/p^i if $i \geq 1$. The stationary distribution is therefore given by $\pi_i = q\pi_0(q/p)^i$ for $i \geq 1$, where

$$\pi_0 = \frac{1}{1 + \sum_1^{K-1} q(q/p)^i} = \frac{p - q}{p - q + q^2(1 - (q/p)^{K-1})}$$

if $p \neq q$, and $\pi_0 = 2/(K+1)$ if $p = q = \frac{1}{2}$.

4. The transition matrices

$$\mathbf{P}_1 = \begin{pmatrix} 1 & 0 \\ 0 & 1 \end{pmatrix}, \qquad \mathbf{P}_2 = \begin{pmatrix} \frac{1}{2} & \frac{1}{2} & 0 & 0 \\ \frac{1}{2} & \frac{1}{2} & 0 & 0 \\ 0 & 0 & \frac{1}{2} & \frac{1}{2} \\ 0 & 0 & \frac{1}{2} & \frac{1}{2} \end{pmatrix}$$

have respective stationary distributions $\boldsymbol{\pi}_1 = (p, 1-p)$ and $\boldsymbol{\pi}_2 = \left(\frac{1}{2}p, \frac{1}{2}p, \frac{1}{2}(1-p), \frac{1}{2}(1-p)\right)$ for any $0 \leq p \leq 1$.

5. (a) Set $i = 1$, and find an increasing sequence $n_1(1), n_1(2), \ldots$ along which $x_1(n)$ converges. Now set $i = 2$, and find a subsequence of $(n_1(j) : j \geq 1)$ along which $x_2(n)$ converges; denote this subsequence by $n_2(1), n_2(2), \ldots$. Continue inductively to obtain, for each i, a sequence $\mathbf{n}_i = (n_i(j) : j \geq 1)$, noting that:

(i) $\mathbf{n}_{i+1}$ is a subsequence of $\mathbf{n}_i$, and

(ii) $\lim_{r\to\infty} x_i(n_i(r))$ exists for all i.

Finally, define $m_k = n_k(k)$. For each $i \geq 1$, the sequence $m_i, m_{i+1}, \ldots$ is a subsequence of $\mathbf{n}_i$, and therefore $\lim_{r\to\infty} x_i(m_r)$ exists.

(b) Let S be the state space of the irreducible Markov chain X. There are countably many pairs i, j of states, and part (a) may be applied to show that there exists a sequence $(n_r : r \geq 1)$ and a family $(\alpha_{ij} : i, j \in S)$, not all zero, such that $p_{ij}(n_r) \to \alpha_{ij}$ as $r \to \infty$.

Now X is recurrent, since otherwise $p_{ij}(n) \to 0$ for all i, j. The coupling argument in the proof of the limit theorem (6.4.20) is valid, so that $p_{aj}(n) - p_{bj}(n) \to 0$ as $n \to \infty$, implying that $\alpha_{aj} = \alpha_{bj}$ for all a, b, j.

6. Just check that $\boldsymbol{\pi}$ satisfies $\boldsymbol{\pi} = \boldsymbol{\pi}\mathbf{P}$ and $\sum_v \pi_v = 1$.

7. Let X_n be the Markov chain which takes the value r if the walk is at any of the 2^r nodes at level r. Then X_n executes a simple random walk with retaining barrier having $p = 1 - q = \frac{2}{3}$, and it is thus transient by Example (6.4.18).

8. Assume that X_n includes particles present just after the entry of the fresh batch Y_n. We may write

$$X_{n+1} = \sum_{i=1}^{X_n} B_{i,n} + Y_n$$

where the $B_{i,n}$ are independent Bernoulli variables with parameter $1 - p$. Therefore X is a Markov chain. It follows also that

$$G_{n+1}(s) = \mathbb{E}(s^{X_{n+1}}) = G_n(p + qs)e^{\lambda(s-1)}.$$

In equilibrium, $G_{n+1} = G_n = G$, where $G(s) = G(p + qs)e^{\lambda(s-1)}$. There is a unique stationary distribution, and it is easy to see that $G(s) = e^{\lambda(s-1)/p}$ must therefore be the solution. The answer is the Poisson distribution with parameter λ/p.

9. The Markov chain X has a uniform transition distribution $p_{jk} = 1/(j+2)$, $0 \leq k \leq j+1$. Therefore,

$$\begin{aligned} \mathbb{E}(X_n) &= \mathbb{E}\big(\mathbb{E}(X_n \mid X_{n-1})\big) = \tfrac{1}{2}\big(1 + \mathbb{E}(X_{n-1})\big) = \cdots \\ &= 1 - (\tfrac{1}{2})^n + (\tfrac{1}{2})^n X_0. \end{aligned}$$

The equilibrium probability generating function satisfies

$$G(s) = \mathbb{E}(s^{X_n}) = \mathbb{E}\big(\mathbb{E}(s^{X_n} \mid X_{n-1})\big) = \mathbb{E}\left\{\frac{1 - s^{X_n+2}}{(1-s)(X_n+2)}\right\},$$

whence

$$\frac{d}{ds}\{(1-s)G(s)\} = -sG(s),$$

subject to $G(1) = 1$. The solution is $G(s) = e^{s-1}$, which is the probability generating function of the Poisson distribution with parameter 1.

10. This is the claim of Theorem (6.4.13). Without loss of generality we may take $s = 0$ and the y_j to be non-negative (since if the y_j solve the equations, then so do $y_j + c$ for any constant c). Let $\mathbf{T}$ be the matrix obtained from $\mathbf{P}$ by deleting the row and column labelled 0, and write $\mathbf{T}^n = (t_{ij}(n) : i, j \neq 0)$. Then $\mathbf{T}^n$ includes all the n-step probabilities of paths that never visit zero.

We claim first that, for all i, j it is the case that $t_{ij}(n) \to 0$ as $n \to \infty$. The quantity $t_{ij}(n)$ may be thought of as the n-step transition probability from i to j in an altered chain in which s has been made absorbing. Since the original chain is assumed irreducible, all states communicate with s, and therefore all states other than s are transient in the altered chain, implying by the summability of $t_{ij}(n)$ (Corollary (6.2.4)) that $t_{ij}(n) \to 0$ as required.

Iterating the inequality $\mathbf{y} \geq \mathbf{Ty}$ yields $\mathbf{y} \geq \mathbf{T}^n\mathbf{y}$, which is to say that

$$y_i \geq \sum_{j=1}^{\infty} t_{ij}(n)y_j \geq \min_{s\geq 1}\{y_{r+s}\} \sum_{j=r+1}^{\infty} t_{ij}(n), \qquad i \geq 1.$$

Let $A_n = \{X_k \neq 0 \text{ for } k \leq n\}$. For $i \geq 1$,

$$\begin{aligned}\mathbb{P}(A_\infty \mid X_0 = i) &= \lim_{n\to\infty} \mathbb{P}(A_n \mid X_0 = i) = \sum_{j=1}^{\infty} t_{ij}(n) \\ &\leq \lim_{n\to\infty}\left\{\sum_{j=1}^{r} t_{ij}(n) + \frac{y_i}{\min_{s\geq 1}\{y_{r+s}\}}\right\}.\end{aligned}$$

Let $\epsilon > 0$, and pick R such that

$$\frac{y_i}{\min_{s\geq 1}\{y_{R+s}\}} < \epsilon.$$

Take $r = R$ and let $n \to \infty$, implying that $\mathbb{P}(A_\infty \mid X_0 = i) = 0$. It follows that 0 is recurrent, and by irreducibility that all states are recurrent.

11. By Exercise (6.4.6), the stationary distribution is $\pi_A = \pi_B = \pi_D = \pi_E = \frac{1}{6}$, $\pi_C = \frac{1}{3}$.

(a) By Theorem (6.4.3), the answer is $\mu_A = 1/\pi_A = 6$.

(b) By the argument around Lemma (6.4.5), the answer is $\rho_D(A) = \pi_D\mu_A = \pi_D/\pi_A = 1$.

(c) Using the same argument, the answer is $\rho_C(A) = \pi_C/\pi_A = 2$.

(d) Let $\mathbb{P}_i(\cdot) = \mathbb{P}(\cdot \mid X_0 = i)$, let T_j be the time of the first passage to state j, and let $\nu_i = \mathbb{P}_i(T_A < T_E)$. By conditioning on the first step,

$$\nu_B = \tfrac{1}{2} + \tfrac{1}{2}\nu_C, \quad \nu_C = \tfrac{1}{4} + \tfrac{1}{4}\nu_B + \tfrac{1}{4}\nu_D, \quad \nu_A = \tfrac{1}{2}\nu_B + \tfrac{1}{2}\nu_C, \quad \nu_D = \tfrac{1}{2}\nu_C,$$

with solution $\nu_A = \frac{5}{8}$, $\nu_B = \frac{3}{4}$, $\nu_C = \frac{1}{2}$, $\nu_D = \frac{1}{4}$.

A typical conditional transition probability $\tau_{ij} = \mathbb{P}_i(X_1 = j \mid T_A < T_E)$ is calculated as follows:

$$\tau_{AB} = \mathbb{P}_A(X_1 = B \mid T_A < T_E) = \frac{\mathbb{P}_A(X_1 = B)\mathbb{P}_B(T_A < T_E)}{\mathbb{P}_A(T_A < T_E)} = \frac{\nu_B}{2\nu_A} = \frac{3}{5},$$

and similarly,

$$\tau_{AC} = \tfrac{2}{5},\ \tau_{BA} = \tfrac{2}{3},\ \tau_{BC} = \tfrac{1}{3},\ \tau_{CA} = \tfrac{1}{2},\ \tau_{CB} = \tfrac{3}{8},\ \tau_{CD} = \tfrac{1}{8},\ \tau_{DC} = 1.$$

We now compute the conditional expectations $\widetilde{\mu}_i = \mathbb{E}_i(T_A \mid T_A < T_E)$ by conditioning on the first step of the conditioned process. This yields equations of the form $\widetilde{\mu}_A = 1 + \frac{3}{5}\widetilde{\mu}_B + \frac{2}{5}\widetilde{\mu}_C$, whose solution gives $\widetilde{\mu}_A = \frac{14}{5}$.

(e) Either use the stationary distribution of the conditional transition matrix $\boldsymbol{\tau}$, or condition on the first step as follows. With N the number of visits to D, and $\eta_i = \mathbb{E}_i(N \mid T_A < T_E)$, we obtain

$$\eta_A = \tfrac{3}{5}\eta_B + \tfrac{2}{5}\eta_C, \quad \eta_B = 0 + \tfrac{1}{3}\eta_C, \quad \eta_C = 0 + \tfrac{3}{8}\eta_B + \tfrac{1}{8}(1 + \eta_D), \quad \eta_D = \eta_C,$$

whence in particular $\eta_A = \frac{1}{10}$.

12. By Exercise (6.4.6), the stationary distribution has $\pi_A = \frac{1}{14}$, $\pi_B = \frac{1}{7}$. Using the argument around Lemma (6.4.5), the answer is $\rho_B(A) = \pi_B\mu_A = \pi_B/\pi_A = 2$.

13. The bottom card B moves upwards one step at a time, at each occasion that the then top card is repositioned beneath the current position of B. When at position $j \ge 2$ (counting from the top), it moves up one step with probability $1 - (j-1)/52$ on any given round, so that the mean time before it arrives at position $j - 1$ is $e_j = 52(52 - j + 1)$. The mean time $\mathbb{E}T$ to reach the top from the bottom is

$$\mathbb{E}T = \sum_{j=2}^{52} e_j = 52\sum_{r=1}^{51}\frac{1}{r} \le 52(1 + \log 52),$$

and the answer is $1 + \mathbb{E}T$.

At any given time, the cards below B are in a uniformly random order. When B is at the top, there are 51! equally likely orders for the other cards. On subsequent reinsertion of B, each of the 52! orders have equal probability.

14. While this may be done as in the previous exercise, it is more interesting to use a coupling argument, as follows. The state space may be taken as the set Π of permutations of $\{1, 2, \dots, 52\}$, and it is easily checked that the Markov chain on Π is irreducible with invariant distribution the uniform distribution $\boldsymbol{\pi}$ on Π.

Consider now two packs of cards, denoted X and Y, such that X is initially in increasing order from top to bottom, and Y has a random order chosen according to $\boldsymbol{\pi}$. We perform the following 'experiment'. At each stage, select $I \in \{1, 2, \dots, 52\}$ uniformly at random (and independently of earlier choices); take card labelled I from each pack and place it on the top of its pack. Each pack evolves thus according to the random-to-top rules, and pack Y retains its distribution $\boldsymbol{\pi}$. Furthermore, once the card labelled k has been selected, it is forevermore in the same positions in the two packs.

There comes a moment at which every card has been selected at least once. After this moment, the two packs are in identical orders, and that of Y is uniformly distributed. Therefore, the ordering of X is also uniformly distributed on Π.

The required time T may be viewed as the collection time in the coupon collector's problem (3.3.2). We have that $\mathbb{E}(T) = n\sum_{r=1}^{n} r^{-1}$.

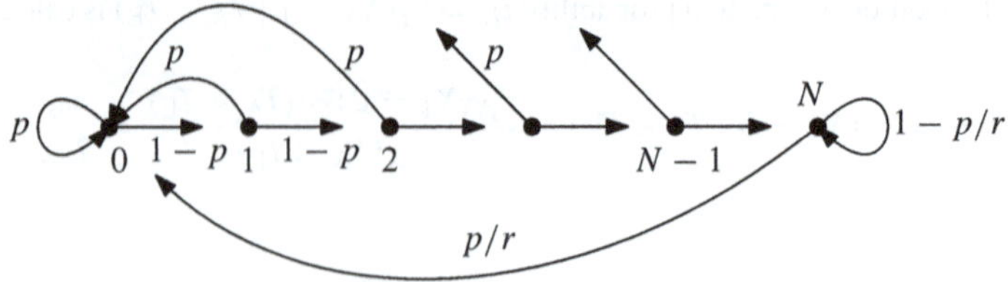

Figure 6.1. A map of the Markov chain of Exercise (6.4.15).

15. (a) The transition probabilities are mapped in Figure 6.1.
(b) Solve for the invariant distribution:

$$\pi_j = (1-p)\pi_{j-1} = (1-p)^j \pi_0, \qquad 1 \le j \le N-1,$$
$$\pi_N = (1-p)\pi_{N-1} + \left(1 - \frac{p}{r}\right)\pi_N.$$

Therefore,

$$\frac{p}{r}\pi_N = (1-p)\pi_{N-1} = (1-p)^N \pi_0,$$

and

$$1 = \sum_{j=0}^{N} \pi_j = \pi_0 \left\{ \sum_{j=0}^{N-1} (1-p)^j + \frac{r}{p}(1-p)^N \right\},$$

whence

$$\pi_0 = \frac{p}{1+(r-1)(1-p)^N}, \qquad \pi_N = \frac{r(1-p)^N}{1+(r-1)(1-p)^N}.$$

The long run probability of inspection is

$$\frac{1}{r}\pi_N + (1-\pi_N) = \frac{1}{1+(r-1)(1-p)^N}.$$

(c) The long run probability of being defective and not inspected is

$$p\pi_N\left(1-\frac{1}{r}\right) = \frac{p(r-1)(1-p)^N}{1+(r-1)(1-p)^N}.$$

16. Let X_n be the position after n steps. Then $X = (X_n)$ is a Markov chain because different jumps are independent. It is evidently irreducible, and is aperiodic since $\mathbb{P}_k(X_1 = k) > 0$. A stationary distribution $\boldsymbol{\pi}$ is a root of the equations

$$\pi_i = \sum_{j=i-1}^{\infty} \frac{\pi_j}{j+2}, \qquad i \ge 0,$$

with the convention that $\pi_{-1} = 0$. One can scratch one's head over these equations, but the solution is simple, namely $\pi_i = A/i!$ for $i \ge 0$ where, by summation, $a = 1/e$.

The mean recurrence time μ_0 of 0 satisfies

$$\mu_0 = \frac{1}{\pi_0} = 1 + \frac{1}{2}\mu,$$

whence $\mu = 2(e-1)$.

17. Let $\mathbf{xP} = \mathbf{x}$ and $x_k = 1 = \rho_k(k)$. Then, for $j \in S$ such that $j \neq k$,

$$\begin{aligned} x_j &= \sum_{i_1 \in S} x_{i_1} p_{i_1,j} = x_k p_{kj} + \sum_{i_1 \neq k} x_{i_1} p_{i_1,j} \\ &= p_{kj} + \sum_{i_1 \neq k} p_{k,i_1} p_{i_1,j} + \sum_{i_1,i_2 \neq k} x_{i_2} p_{i_2,i_1} p_{i_1,j}, \end{aligned}$$

which we iterate to obtain

$$x_j \geq \sum_{n=1}^{m} \mathbb{P}_k(T_j = n,\ T_k > n) \to \rho_j(k) \qquad \text{as } m \to \infty.$$

If the chain is recurrent, $\boldsymbol{\rho}(k)$ is stationary, so that $\mathbf{y} = \mathbf{x} - \boldsymbol{\rho}(k)$ is a non-negative stationary measure with $y_k = 0$. By Lemma (6.4.7c), $\mathbf{y} = \mathbf{0}$ as claimed.

6.5 Solutions. Reversibility

1. Look for a solution to the equations $\pi_i p_{ij} = \pi_j p_{ji}$. The only non-trivial cases of interest are those with $j = i+1$, and therefore $\lambda_i \pi_i = \mu_{i+1}\pi_{i+1}$ for $0 \leq i < b$, with solution

$$\pi_i = \pi_0 \frac{\lambda_0 \lambda_1 \cdots \lambda_{i-1}}{\mu_1 \mu_2 \cdots \mu_i}, \qquad 0 \leq i \leq b,$$

an empty product being interpreted as 1. The constant π_0 is chosen in order that the π_i sum to 1, and the chain is therefore time-reversible.

2. (a) Let $\boldsymbol{\pi}$ be the stationary distribution of X, and suppose X is reversible. We have that $\pi_i p_{ij} = p_{ji}\pi_j$ for all i, j, and furthermore $\pi_i > 0$ for all i. Hence

$$\pi_i p_{ij} p_{jk} p_{ki} = p_{ji}\pi_j p_{jk} p_{ki} = p_{ji} p_{kj} \pi_k p_{ki} = p_{ji} p_{kj} p_{ik} \pi_i$$

as required when $n = 3$. A similar calculation is valid when $n > 3$.

Suppose conversely that the given display holds for all finite sequences of states. Sum over all values of the subsequence $j_2, \dots, j_{n-1}$ to deduce that $p_{ij}(n-1)p_{ji} = p_{ij} p_{ji}(n-1)$, where $i = j_1$, $j = j_n$. Take the limit as $n \to \infty$ to obtain $\pi_j p_{ji} = p_{ij}\pi_i$ as required for time-reversibility.

(b) By (a), the condition on triples is necessary for reversibility. Suppose, conversely, that it holds, and let $c \in S$ be as given. For $A > 0$ and $i \in S$, define $q_i = A p_{ci}/p_{ic}$, noting that $q_c = A$. The constant A is chosen such that $\sum_i q_i = 1$. Now $q_i p_{ic} = q_c p_{ci}$ for $i \in S$. Also, by the triple condition,

$$q_i p_{ij} = A \frac{p_{ci}}{p_{ic}} p_{ij} = A \frac{p_{ci} p_{ij} p_{jc}}{p_{ic} p_{jc}} = A \frac{p_{cj} p_{ji} p_{ic}}{p_{ic} p_{jc}} = q_j p_{ji}.$$

In conclusion, $\mathbf{q} = (q_i)$ is a distribution that satisfies the detailed balance equations. Therefore, X is reversible in equilibrium, and $\boldsymbol{\pi} = \mathbf{q}$.

(c) These are counts of the numbers of cyclic subsets of given size chosen from the state space S. If one of these cycles fails to satisfy the criterion, no more need be tested. One can sometimes do better than this: if the graph of the state space is planar, it suffices to check the condition for all cycles in some base of the cycle vector space of the graph.

(d) Since trees have no cycles, any edge traversed by a closed walk is traversed an equal number of times in each direction. The equation of part (a) follows.

3. With $\boldsymbol{\pi}$ the stationary distribution of X, look for a stationary distribution $\boldsymbol{\nu}$ of Y of the form

$$\nu_i = \begin{cases} c\beta\pi_i & \text{if } i \notin C, \\ c\pi_i & \text{if } i \in C. \end{cases}$$

There are four cases.
(a) $i \in C, j \notin C$: $\nu_i q_{ij} = c\pi_i \beta p_{ij} = c\beta\pi_j p_{ji} = \nu_j q_{ji}$,
(b) $i, j \in C$: $\nu_i q_{ij} = c\pi_i p_{ij} = c\pi_j p_{ji} = \nu_j q_{ji}$,
(c) $i \notin C, j \in C$: $\nu_i q_{ij} = c\beta\pi_i p_{ij} = c\beta\pi_j p_{ji} = \nu_j q_{ji}$,
(d) $i, j \notin C$: $\nu_i q_{ij} = c\beta\pi_i p_{ij} = c\beta\pi_j p_{ji} = \nu_j q_{ji}$.
Hence the modified chain is reversible in equilibrium with stationary distribution $\boldsymbol{\nu}$, when

$$c\left\{\beta \sum_{i\notin C} \pi_i + \sum_{i\in C} \pi_i\right\} = 1.$$

In the limit as $\beta \downarrow 0$, the chain Y never leaves the set C once it has arrived in it.

4. Only if the period is 2, because of the detailed balance equations.

5. With $Y_n = X_n - \frac{1}{2}m$,

$$\begin{aligned} \mathbb{E}(Y_n) &= \mathbb{E}(Y_{n-1}) + \mathbb{E}(X_n - X_{n-1}) \\ &= \mathbb{E}(Y_{n-1}) + \mathbb{E}\left\{\left(1 - \frac{X_{n-1}}{m}\right) - \frac{X_{n-1}}{m}\right\} = \mathbb{E}(Y_{n-1}) - \frac{2}{m}\mathbb{E}(Y_{n-1}). \end{aligned}$$

Now iterate.

6. (a) The distribution $\pi_1 = \beta/(\alpha+\beta)$, $\pi_2 = \alpha/(\alpha+\beta)$ satisfies the detailed balance equations, so this chain is reversible.
(b) By symmetry, the stationary distribution is $\boldsymbol{\pi} = (\frac{1}{3}, \frac{1}{3}, \frac{1}{3})$, which satisfies the detailed balance equations if and only if $p = \frac{1}{2}$.
(c) This chain is reversible if and only if $p = \frac{1}{2}$.

7. A simple random walk which moves rightwards with probability p has a stationary measure $\pi_n = A(p/q)^n$, in the sense that $\boldsymbol{\pi}$ is a vector satisfying $\boldsymbol{\pi} = \boldsymbol{\pi}\mathbf{P}$. It is not necessarily the case that this $\boldsymbol{\pi}$ has finite sum. It may then be checked that the recipe given in the solution to Exercise (6.5.3) yields $\pi(i,j) = \rho_1^i \rho_2^j / \sum_{(r,s)\in C} \rho_1^r \rho_2^s$ as stationary distribution for the given process, where C is the relevant region of the plane, and $\rho_i = p_i/q_i$ and p_i $(= 1 - q_i)$ is the chance that the ith walk moves rightwards on any given step.

8. Since the chain is irreducible with a finite state space, we have that $\pi_i > 0$ for all i. Assume the chain is reversible. The balance equations $\pi_i p_{ij} = \pi_j p_{ji}$ give $p_{ij} = \pi_j p_{ji}/\pi_i$. Let $\mathbf{D}$ be the matrix with entries $1/\pi_i$ on the diagonal, and $\mathbf{S}$ the matrix $(\pi_j p_{ji})$, and check that $\mathbf{P} = \mathbf{DS}$.

Conversely, if $\mathbf{P} = \mathbf{DS}$, then $d_i^{-1} p_{ij} = d_j^{-1} p_{ji}$, whence $\pi_i = d_i^{-1}/\sum_k d_k^{-1}$ satisfies the detailed balance equations.

Note that

$$p_{ij} = \frac{1}{\sqrt{\pi_i}} \sqrt{\frac{\pi_i}{\pi_j}}\, p_{ij} \sqrt{\pi_j}.$$

If the chain is reversible in equilibrium, the matrix $\mathbf{M} = \left(\sqrt{\pi_i/\pi_j}\, p_{ij}\right)$ is symmetric, and therefore $\mathbf{M}$, and, by the above, $\mathbf{P}$, has real eigenvalues. An example of the failure of the converse is the transition matrix

$$\mathbf{P} = \begin{pmatrix} 0 & 1 & 0 \\ \frac{3}{4} & 0 & \frac{1}{4} \\ 1 & 0 & 0 \end{pmatrix},$$

which has real eigenvalues 1, and $-\frac{1}{2}$ (twice), and stationary distribution $\boldsymbol{\pi} = (\frac{4}{9}, \frac{4}{9}, \frac{1}{9})$. However, $\pi_1 p_{13} = 0 \neq \frac{1}{9} = \pi_3 p_{31}$, so that such a chain is not reversible.

9. Simply check the detailed balance equations $\pi_i p_{ij} = \pi_j p_{ji}$.

10. Try solving the detailed balance equations $\pi_i p_{i,i+1} = \pi_{i+1} p_{i+1,i}$. You will find that $\pi_i = c/[i(i+2)]$ where c is determined by

$$1 = \sum_{i=1}^{\infty} \pi_i = \frac{c}{2} \sum_{i=1}^{\infty} \left(\frac{1}{i} - \frac{1}{i+2} \right) = \frac{c}{2} \left(1 + \frac{1}{2} \right),$$

so that $c = \frac{4}{3}$. The chain is positive recurrent with $\mu_i = \pi_i^{-1} = \frac{3}{4} i(i+2)$.

11. We have $p_{ic} > 0$ for $i \neq c$. It is enough to check Kelly's reversibility condition of Exercise (6.5.2b) for the four cycles cne, ces, csw, and cwn, and hence the chain is reversible in equilibrium.

Using the detailed balance equations $\pi_i p_{ic} = \pi_c p_{ci}$, the stationary distribution $\boldsymbol{\pi}$ satisfies $\pi_i = \pi_c p_{ci} / p_{ic}$ for $i \neq c$. By summing the π_i, we find $\pi_c = \frac{2}{11}$. The result follows by the Theorem (6.4.3).

12. (a) First check that $\mathbf{L} = (l_{ij})$ is a stochastic matrix: it has non-negative entries with row-sums 1. It is irreducible since $l_{ij}(n) \geq a^n p_{ij}(n)$, and aperiodic since $l_{ii} \geq 1 - a > 0$ for $i \in S$.
(b) We have that $\pi_i p_{ij} = \pi_j p_{ji}$ for $i, j \in S$. Therefore,

$$\pi_i l_{ij} = a\pi_i p_{ij} + (1-a)\pi_i \delta_{ij} = a\pi_j p_{ji} + (1-a)\pi_j \delta_{ij} = \pi_j l_{ji},$$

where δ_{ij} is the Kronecker delta. The chain is called *lazy* because it hangs around a lot.

6.6 Solutions. Chains with finitely many states

1. Let $\mathbf{P} = (p_{ij} : 1 \leq i, j \leq n)$ be a stochastic matrix and let C be the subset of $\mathbb{R}^n$ containing all vectors $\mathbf{x} = (x_1, x_2, \ldots, x_n)$ satisfying $x_i \geq 0$ for all i and $\sum_{i=1}^{n} x_i = 1$; for $\mathbf{x} \in C$, let $\|\mathbf{x}\| = \max_j \{x_j\}$. Define the linear mapping $T : C \to \mathbb{R}^n$ by $T(\mathbf{x}) = \mathbf{xP}$. Let us check that T is a continuous function from C into C. First,

$$\|T(\mathbf{x})\| = \max_j \left\{ \sum_i x_i p_{ij} \right\} \leq \alpha \|\mathbf{x}\|$$

where

$$\alpha = \max_j \left\{ \sum_i p_{ij} \right\};$$

hence $\|T(\mathbf{x}) - T(\mathbf{y})\| \leq \alpha \|\mathbf{x} - \mathbf{y}\|$. Secondly, $T(\mathbf{x})_j \geq 0$ for all j, and

$$\sum_j T(\mathbf{x})_j = \sum_j \sum_i x_i p_{ij} = \sum_i x_i \sum_j p_{ij} = 1.$$

Applying the given theorem, there exists a point $\boldsymbol{\pi}$ in C such that $T(\boldsymbol{\pi}) = \boldsymbol{\pi}$, which is to say that $\boldsymbol{\pi} = \boldsymbol{\pi}\mathbf{P}$.

2. Let $\mathbf{P}$ be a stochastic $m \times m$ matrix and let $\mathbf{T}$ be the $m \times (m+1)$ matrix with (i, j)th entry

$$t_{ij} = \begin{cases} p_{ij} - \delta_{ij} & \text{if } j \leq m, \\ 1 & \text{if } j = m+1, \end{cases}$$

where δ_{ij} is the Kronecker delta. Let $\mathbf{v} = (0, 0, \ldots, 0, 1) \in \mathbb{R}^{m+1}$. If statement (ii) of the question is valid, there exists $\mathbf{y} = (y_1, y_2, \ldots, y_{m+1})$ such that

$$y_{m+1} < 0, \qquad \sum_{j=1}^{m}(p_{ij} - \delta_{ij})y_j + y_{m+1} \geq 0 \text{ for } 1 \leq i \leq m;$$

this implies that

$$\sum_{j=1}^{m} p_{ij} y_j \geq y_i - y_{m+1} > y_i \qquad \text{for all } i,$$

and hence the impossibility that $\sum_{j=1}^{m} p_{ij} y_j > \max_i\{y_i\}$. It follows that statement (i) holds, which is to say that there exists a non-negative vector $\mathbf{x} = (x_1, x_2, \ldots, x_m)$ such that $\mathbf{x}(\mathbf{P} - \mathbf{I}) = \mathbf{0}$ and $\sum_{i=1}^{m} x_i = 1$; such an $\mathbf{x}$ is the required eigenvector.

3. Thinking of x_{n+1} as the amount you may be sure of winning, you seek a betting scheme $\mathbf{x}$ such that x_{n+1} is maximized subject to the inequalities

$$x_{n+1} \leq \sum_{i=1}^{n} x_i t_{ij} \quad \text{for } 1 \leq j \leq m.$$

Writing $a_{ij} = -t_{ij}$ for $1 \leq i \leq n$ and $a_{n+1,j} = 1$, we obtain the linear program:

$$\text{maximize} \quad x_{n+1} \quad \text{subject to} \quad \sum_{i=1}^{n+1} x_i a_{ij} \leq 0 \quad \text{for } 1 \leq j \leq m.$$

The dual linear program is:

$$\text{minimize} \quad 0 \quad \text{subject to} \quad \sum_{j=1}^{m} a_{ij} y_j = 0 \quad \text{for } 1 \leq i \leq n,$$
$$\sum_{j=1}^{m} a_{n+1,j} y_j = 1, \quad y_j \geq 0 \quad \text{for } 1 \leq j \leq m.$$

Re-expressing the a_{ij} in terms of the t_{ij} as above, the dual program takes the form:

$$\text{minimize} \quad 0 \quad \text{subject to} \quad \sum_{j=1}^{m} t_{ij} p_j = 0 \quad \text{for } 1 \leq i \leq n,$$
$$\sum_{j=1}^{m} p_j = 1, \quad p_j \geq 0 \quad \text{for } 1 \leq j \leq m.$$

The vector $\mathbf{x} = \mathbf{0}$ is a feasible solution of the primal program. The dual program has a feasible solution if and only if statement (a) holds. Therefore, if (a) holds, the dual program has minimal value 0, whence by the duality theorem of linear programming, the maximal value of the primal program is 0, in contradiction of statement (b). On the other hand, if (a) does not hold, the dual has no feasible solution, and therefore the primal program has no optimal solution. That is, the objective function of the primal is unbounded, and therefore (b) holds. [This was proved by De Finetti in 1937.]

4. Use induction, the claim being evidently true when $n = 1$. Suppose it is true for $n = m$. Certainly $\mathbf{P}^{m+1}$ is of the correct form, and the equation $\mathbf{P}^{m+1}\mathbf{x}' = \mathbf{P}(\mathbf{P}^m\mathbf{x}')$ with $\mathbf{x} = (1, \omega, \omega^2)$ yields in its first row

$$a_{1,m+1} + a_{2,m+1}\omega + a_{3,m+1}\omega^2 = (1 - p + p\omega)^m (\mathbf{P}\mathbf{x}')_1 = (1 - p + p\omega)^{m+1}$$

as required.

5. The first part follows from the fact that $\boldsymbol{\pi}\mathbf{1}' = 1$ if and only if $\boldsymbol{\pi}\mathbf{U} = \mathbf{1}$. The second part follows from the fact that $\pi_i > 0$ for all i if $\mathbf{P}$ is finite and irreducible, since this implies the invertibility of $\mathbf{I} - \mathbf{P} + \mathbf{U}$.

6. The chessboard corresponds to a graph with $8 \times 8 = 64$ vertices, pairs of which are connected by edges when the corresponding move is legitimate for the piece in question. By Exercises (6.4.6), (6.5.9), we need only check that the graph is connected, and to calculate the degree of a corner vertex.

(a) For the king there are 4 vertices of degree 3, 24 of degree 5, 36 of degree 8. Hence, the number of edges is 210 and the degree of a corner is 3. Therefore $\mu(\text{king}) = 420/3 = 140$.

(b) $\mu(\text{queen}) = (28 \times 21 + 20 \times 23 + 12 \times 25 + 4 \times 27)/21 = 208/3$.

(c) We restrict ourselves to the set of 32 vertices accessible from a given corner. Then $\mu(\text{bishop}) = (14 \times 7 + 10 \times 9 + 6 \times 11 + 2 \times 13)/7 = 40$.

(d) $\mu(\text{knight}) = (4 \times 2 + 8 \times 3 + 20 \times 4 + 16 \times 6 + 16 \times 8)/2 = 168$.

(e) $\mu(\text{rook}) = 64 \times 14/14 = 64$.

7. They are walking on a product space of 8×16 vertices. Of these, 6×16 have degree 6×3 and 16×2 have degree 6×5. Hence

$$\mu(C) = (6 \times 16 \times 6 \times 3 + 16 \times 2 \times 6 \times 5)/18 = 448/3.$$

8. $|\mathbf{P} - \lambda I| = (\lambda - 1)(\lambda + \frac{1}{2})(\lambda + \frac{1}{6})$. Tedious computation yields the eigenvectors, and thus

$$\mathbf{P}^n = \frac{1}{3}\begin{pmatrix} 1 & 1 & 1 \\ 1 & 1 & 1 \\ 1 & 1 & 1 \end{pmatrix} + \tfrac{1}{12}(-\tfrac{1}{2})^n \begin{pmatrix} 8 & -4 & -4 \\ 2 & -1 & -1 \\ -10 & 5 & 5 \end{pmatrix} + \tfrac{1}{4}(-\tfrac{1}{6})^n \begin{pmatrix} 0 & 0 & 0 \\ -2 & 3 & -1 \\ 2 & -3 & 1 \end{pmatrix}.$$

9. (a) By construction,

$$p_{ij} = \begin{cases} \dfrac{b l_{ij}}{d_i} + \dfrac{1-b}{n} & \text{if } d_i \neq 0, \\[2ex] \dfrac{1}{n} = \dfrac{b}{n} + \dfrac{1-b}{n} & \text{if } d_i = 0. \end{cases}$$

(b) Since the state space is finite and irreducible, the chain possesses a unique stationary distribution $\boldsymbol{\pi}$, which satisfies

$$(*) \qquad \boldsymbol{\pi} = \boldsymbol{\pi}\mathbf{P} = \boldsymbol{\pi} b\mathbf{Q} + \boldsymbol{\pi}\mathbf{v}'\mathbf{e}.$$

Here,

$$\boldsymbol{\pi}\mathbf{v}' = \frac{1-b}{n}\sum_i \pi_i = \frac{1-b}{n},$$

whence $\boldsymbol{\pi} = \{(1-b)/n\}\mathbf{e}(\mathbf{I} - b\mathbf{Q})^{-1}$, where the inverse exists for $b < 1$.

(c) This is because π_i measures the long run relative frequency with which the web-surfer in question visits site i.

10. Since $\boldsymbol{\pi}$ is an eigenvector, it is not the vector of zeros. Let $A = \{i : \pi_i \geq 0\}$ and $B = \{i : \pi_i < 0\}$. Since $\boldsymbol{\pi}\mathbf{P} = \boldsymbol{\pi}$,

$$\begin{aligned} 0 &= \sum_{j \in B}\Big(\sum_{i \in A} \pi_i p_{i,j} + \sum_{i \in B} \pi_i p_{i,j} - \pi_j\Big) \\ &= \sum_{i \in A} \pi_i \sum_{j \in B} p_{i,j} + \sum_{i \in B} \pi_i \Big(\sum_{j \in B} p_{i,j} - 1\Big) \\ &= \sum_{i \in A} \pi_i \sum_{j \in B} p_{i,j} - \sum_{i \in B} \pi_i \sum_{j \in A} p_{i,j}. \end{aligned}$$

Since both double sums are non-negative, they are both zero. Therefore, $p_{i,j} = 0$ for $i \in B$, $j \in A$, and by a similar argument $p_{i,j} = 0$ when $\pi_i > 0$ and $\pi_j \leq 0$. This contradicts irreducibility unless either all elements of $\boldsymbol{\pi}$ are strictly positive or all are strictly negative. In either case, there must exist a stationary distribution $\boldsymbol{\pi}$ with strictly positive entries.

If there exist two distinct stationary distributions $\boldsymbol{\pi}$, $\boldsymbol{\nu}$, then $\boldsymbol{\pi} - \boldsymbol{\nu}$ is a left eigenvector of $\mathbf{P}$ with both positive and negative components, which is impossible by the above. Therefore, $\boldsymbol{\pi} = \boldsymbol{\nu}$.

6.7 Solutions. Branching processes revisited

1. We have (using Example (5.4.3), or the fact that $G_{n+1}(s) = G(G_n(s))$) that the probability generating function of Z_n is

$$G_n(s) = \frac{n - (n-1)s}{n + 1 - ns},$$

so that

$$\mathbb{P}(Z_n = k) = \left(\frac{n}{n+1}\right)^{k+1} - \left(\frac{n-1}{n+1}\right)\left(\frac{n}{n+1}\right)^{k-1} = \frac{n^{k-1}}{(n+1)^{k+1}}$$

for $k \geq 1$. Therefore, for $y > 0$, as $n \to \infty$,

$$\mathbb{P}(Z_n \leq 2yn \mid Z_n > 0) = \frac{1}{1 - G_n(0)} \sum_{k=1}^{\lfloor 2yn \rfloor} \frac{n^{k-1}}{(n+1)^{k+1}} = 1 - \left(1 + \frac{1}{n}\right)^{-\lfloor 2yn \rfloor} \to 1 - e^{-2y}.$$

2. Using the independence of different lines of descent,

$$\mathbb{E}(s^{Z_n} \mid \text{extinction}) = \sum_{k=0}^{\infty} \frac{s^k \mathbb{P}(Z_n = k, \text{extinction})}{\mathbb{P}(\text{extinction})} = \sum_{k=0}^{\infty} \frac{s^k \mathbb{P}(Z_n = k)\eta^k}{\eta} = \frac{1}{\eta} G_n(s\eta),$$

where G_n is the probability generating function of Z_n.

3. We have that $\eta = G(\eta)$. In this case $G(s) = q(1 - ps)^{-1}$, and therefore $\eta = q/p$. Hence

$$\begin{aligned}\frac{1}{\eta} G_n(s\eta) &= \frac{p}{q} \cdot \frac{q\{p^n - q^n - p(sq/p)(p^{n-1} - q^{n-1})\}}{p^{n+1} - q^{n+1} - p(sq/p)(p^n - q^n)} \\ &= \frac{p\{q^n - p^n - qs(q^{n-1} - p^{n-1})\}}{q^{n+1} - p^{n+1} - qs(q^n - p^n)},\end{aligned}$$

which is $G_n(s)$ with p and q interchanged.

4. (a) Using the fact that $\text{var}(X \mid X > 0) \geq 0$,

$$\mathbb{E}(X^2) = \mathbb{E}(X^2 \mid X > 0)\mathbb{P}(X > 0) \geq \mathbb{E}(X \mid X > 0)^2 \mathbb{P}(X > 0) = \mathbb{E}(X)\mathbb{E}(X \mid X > 0).$$

(b) Hence

$$\mathbb{E}(Z_n/\mu^n \mid Z_n > 0) \leq \frac{\mathbb{E}(Z_n^2)}{\mu^n \mathbb{E}(Z_n)} = \mathbb{E}(W_n^2)$$

where $W_n = Z_n/\mathbb{E}(Z_n)$. By an easy calculation (see Lemma (5.4.2)),

$$\mathbb{E}(W_n^2) = \frac{\sigma^2(1 - \mu^{-n})}{\mu^2 - \mu} + 1 \leq \frac{\sigma^2}{\mu^2 - \mu} + 1 = \frac{2p}{p - q}$$

where $\sigma^2 = \text{var}(Z_1) = p/q^2$.
(c) Doing the calculation exactly,

$$\mathbb{E}(Z_n/\mu^n \mid Z_n > 0) = \frac{\mathbb{E}(Z_n/\mu^n)}{\mathbb{P}(Z_n > 0)} = \frac{1}{1 - G_n(0)} \to \frac{1}{1-\eta}$$

where $\eta = \mathbb{P}(\text{ultimate extinction}) = q/p$.

6.8 Solutions. Birth processes and the Poisson process

1. Let F and W be the incoming Poisson processes, and let $N(t) = F(t)+W(t)$. Certainly $N(0) = 0$ and N is non-decreasing. Arrivals of flies during $[0, s]$ are independent of arrivals during $(s, t]$, if $s < t$; similarly for wasps. Therefore the aggregated arrival process during $[0, s]$ is independent of the aggregated process during $(s, t]$. Now

$$\mathbb{P}\big(N(t+h) = n+1 \,\big|\, N(t) = n\big) = \mathbb{P}(A \triangle B)$$

where

$$A = \big\{\text{one fly arrives during } (t, t+h]\big\}, \quad B = \big\{\text{one wasp arrives during } (t, t+h]\big\}.$$

We have that

$$\begin{aligned}\mathbb{P}(A \triangle B) &= \mathbb{P}(A) + \mathbb{P}(B) - \mathbb{P}(A \cap B) \\ &= \lambda h + \mu h - (\lambda h)(\mu h) + \mathrm{o}(h) = (\lambda + \mu)h + \mathrm{o}(h).\end{aligned}$$

Finally

$$\mathbb{P}\big(N(t+h) > n+1 \,\big|\, N(t) = n\big) \le \mathbb{P}(A \cap B) + \mathbb{P}(C \cup D),$$

where $C = \{\text{two or more flies arrive in } (t, t+h]\}$ and $D = \{\text{two or more wasps arrive in } (t, t+h]\}$. This probability is no greater than $(\lambda h)(\mu h) + \mathrm{o}(h) = \mathrm{o}(h)$.

2. Let I be the incoming Poisson process, and let G be the process of arrivals of green insects. Matters of independence are dealt with as above. Finally,

$$\begin{aligned}&\mathbb{P}\big(G(t+h) = n+1 \,\big|\, G(t) = n\big) = p\mathbb{P}\big(I(t+h) = n+1 \,\big|\, I(t) = n\big) + \mathrm{o}(h) = p\lambda h + \mathrm{o}(h), \\ &\mathbb{P}\big(G(t+h) > n+1 \,\big|\, G(t) = n\big) \le \mathbb{P}\big(I(t+h) > n+1 \,\big|\, I(t) = n\big) = \mathrm{o}(h).\end{aligned}$$

3. Conditioning on T_1 and using the time-homogeneity of the process,

$$\mathbb{P}\big(E(t) > x \,\big|\, T_1 = u\big) = \begin{cases} \mathbb{P}\big(E(t-u) > x\big) & \text{if } u \le t, \\ 0 & \text{if } t < u \le t+x, \\ 1 & \text{if } u > t+x, \end{cases}$$

(draw a diagram to help you see this). Therefore

$$\begin{aligned}\mathbb{P}\big(E(t) > x\big) &= \int_0^\infty \mathbb{P}\big(E(t) > x \mid T_1 = u\big)\lambda e^{-\lambda u}\,du \\ &= \int_0^t \mathbb{P}\big(E(t-u) > x\big)\lambda e^{-\lambda u}\,du + \int_{t+x}^\infty \lambda e^{-\lambda u}\,du.\end{aligned}$$

You may solve the integral equation using Laplace transforms. Alternately you may guess the answer and then check that it works. The answer is $\mathbb{P}(E(t) \le x) = 1 - e^{-\lambda x}$, the exponential distribution. Actually this answer is obvious since $E(t) > x$ if and only if there is no arrival in $[t, t+x]$, an event having probability $e^{-\lambda x}$.

4. The forward equation is

$$p'_{ij}(t) = \lambda(j-1)p_{i,j-1}(t) - \lambda j p_{ij}(t), \qquad i \le j,$$

with boundary conditions $p_{ij}(0) = \delta_{ij}$, the Kronecker delta. We write $G_i(s,t) = \sum_j s^j p_{ij}(t)$, the probability generating function of $B(t)$ conditional on $B(0) = i$. Multiply through the differential equation by s^j and sum over j:

$$\frac{\partial G_i}{\partial t} = \lambda s^2 \frac{\partial G_i}{\partial s} - \lambda s \frac{\partial G_i}{\partial s},$$

a partial differential equation with boundary condition $G_i(s,0) = s^i$. This may be solved in the usual way to obtain $G_i(s,t) = g(e^{\lambda t}(1-s^{-1}))$ for some function g. Using the boundary condition, we find that $g(1-s^{-1}) = s^i$ and so $g(u) = (1-u)^{-i}$, yielding

$$G_i(s,t) = \frac{1}{\{1 - e^{\lambda t}(1-s^{-1})\}^i} = \frac{(se^{-\lambda t})^i}{\{1 - s(1-e^{-\lambda t})\}^i}.$$

The coefficient of s^j is, by the binomial series,

$$(*) \qquad p_{ij}(t) = e^{-i\lambda t}\binom{j-1}{i-1}(1-e^{-\lambda t})^{j-i}, \qquad j \ge i,$$

as required.

Alternatively use induction. Set $j = i$ to obtain $p'_{ii}(t) = -\lambda i p_{ii}(t)$ (remember $p_{i,i-1}(t) = 0$), and therefore $p_{ii}(t) = e^{-\lambda i t}$. Rewrite the differential equation as

$$\frac{d}{dt}\left(p_{ij}(t)e^{\lambda j t}\right) = \lambda(j-1)p_{i,j-1}(t)e^{\lambda j t}.$$

Set $j = i+1$ and solve to obtain $p_{i,i+1}(t) = ie^{-\lambda i t}\left(1 - e^{-\lambda t}\right)$. Hence $(*)$ holds, by induction.

The mean is

$$\mathbb{E}(B(t)) = \left.\frac{\partial}{\partial s} G_I(s,t)\right|_{s=1} = Ie^{\lambda t},$$

by an easy calculation. Similarly $\text{var}(B(t)) = A + \mathbb{E}(B(t)) - \mathbb{E}(B(t))^2$ where

$$A = \left.\frac{\partial^2}{\partial s^2} G_I(s,t)\right|_{s=1}.$$

Alternatively, note that $B(t)$ has the negative binomial distribution with parameters $e^{-\lambda t}$ and I.

5. The forward equations are

$$p'_n(t) = \lambda_{n-1}p_{n-1}(t) - \lambda_n p_n(t), \qquad n \ge 0,$$

where $\lambda_i = i\lambda + \nu$. The process is honest, and therefore $m(t) = \sum_n np_n(t)$ satisfies

$$\begin{aligned} m'(t) &= \sum_{n=1}^{\infty} n\big[(n-1)\lambda + \nu\big] p_{n-1}(t) - \sum_{n=0}^{\infty} n(n\lambda + \nu)p_n(t) \\ &= \sum_{n=0}^{\infty} \big\{\lambda[(n+1)n - n^2] + \nu[(n+1) - n]\big\} p_n(t) \\ &= \sum_{n=0}^{\infty} (\lambda n + \nu)p_n(t) = \lambda m(t) + \nu. \end{aligned}$$

Solve subject to $m(0) = 0$ to obtain $m(t) = \nu(e^{\lambda t} - 1)/\lambda$.

6. Using the fact that the time to the nth arrival is the sum of exponential interarrival times (or using equation (6.8.15)), we have that

$$\widehat{p}_n(\theta) = \int_0^\infty e^{-\theta t} p_n(t)\, dt$$

is given by

$$\widehat{p}_n(\theta) = \frac{1}{\lambda_n} \prod_{i=0}^{n} \frac{\lambda_i}{\lambda_i + \theta}$$

which may be expressed, using partial fractions, as

$$\widehat{p}_n(\theta) = \frac{1}{\lambda_n} \sum_{i=0}^{n} \frac{a_i \lambda_i}{\lambda_i + \theta}$$

where

$$a_i = \prod_{\substack{j=0 \\ j \neq i}}^{n} \frac{\lambda_j}{\lambda_j - \lambda_i}$$

so long as $\lambda_i \neq \lambda_j$ whenever $i \neq j$. The Laplace transform $\widehat{p}_n$ may now be inverted as

$$p_n(t) = \frac{1}{\lambda_n} \sum_{i=0}^{n} a_i \lambda_i e^{-\lambda_i t}.$$

See also Exercise (4.8.4).

7. Let T_n be the time of the nth arrival, and let $T = \lim_{n\to\infty} T_n = \sup\{t : N(t) < \infty\}$. Now, as in Exercise (6.8.6),

$$\lambda_n \widehat{p}_n(\theta) = \prod_{i=0}^{n} \frac{\lambda_i}{\lambda_i + \theta} = \mathbb{E}(e^{-\theta T_n})$$

since $T_n = X_0 + X_1 + \cdots + X_n$ where X_k is the $(k+1)$th interarrival time, a random variable which is exponentially distributed with parameter λ_k. Using the continuity theorem, $\mathbb{E}(e^{-\theta T_n}) \to \mathbb{E}(e^{-\theta T})$ as $n \to \infty$, whence $\lambda_n \widehat{p}_n(\theta) \to \mathbb{E}(e^{-\theta T})$ as $n \to \infty$, which may be inverted to obtain $\lambda_n p_n(t) \to f(t)$ as $n \to \infty$ where f is the density function of T. Now

$$\mathbb{E}\big(N(t) \,\big|\, N(t) < \infty\big) = \frac{\sum_{n=0}^{\infty} np_n(t)}{\sum_{n=0}^{\infty} p_n(t)}$$

which converges or diverges according to whether or not $\sum_n np_n(t)$ converges. However $p_n(t) \sim \lambda_n^{-1} f(t)$ as $n \to \infty$, so that $\sum_n np_n(t) < \infty$ if and only if $\sum_n n\lambda_n^{-1} < \infty$.

When $\lambda_n = (n + \frac{1}{2})^2$, we have that

$$\mathbb{E}(e^{-\theta T}) = \prod_{n=0}^{\infty} \left\{ 1 + \frac{\theta}{(n + \frac{1}{2})^2} \right\}^{-1} = \operatorname{sech}\left(\pi\sqrt{\theta}\right).$$

Inverting the Laplace transform (or consulting a table of such transforms) we find that

$$f(t) = -\frac{1}{\pi^2} \frac{\partial}{\partial \nu} \theta_1\left(\tfrac{1}{2}\nu \,\middle|\, t/\pi^2\right)\Big|_{\nu=0}$$

where θ_1 is the first Jacobi theta function.

8. Let N be the number of events in the intervals $(0, x)$. The light is green at time x if and only if N is even, which occurs with probability $p_e(x) = \frac{1}{2}(1 + e^{-2\lambda x})$. If N is odd, then by the lack-of-memory property $W(x)$ is exponentially distributed with parameter λ. Therefore,

$$\mathbb{P}(W(x) \le w) = \begin{cases} p_e(x) & \text{if } w = 0, \\ (1 - p_e(x))(1 - e^{-\lambda w}) & \text{if } w > 0. \end{cases}$$

9. Let the birth times be $T_1, T_2, \dots$ and the interbirth times $W_1 = T_1$, $W_2 = T_2 - T_1, \dots$. The W_r are independent, and W_r has the exponential distribution with parameter $r\lambda$. The joint density function $g = g(t_1, t_2, \dots, t_b)$ of $T_1, T_2, \dots, T_b$ conditional on $T_b < t < T_{b+1}$ satisfies

$$\begin{aligned} \log g &= C_1 - \sum_{r=1}^{b} \lambda r(t_r - t_{r-1}) - \lambda(b+1)(t - t_b) \\ &= C_2 + \sum_{r=1}^{b} \lambda t_r = C_3 + \sum_{r=1}^{b} \log f(t_r). \end{aligned}$$

for $0 = t_0 < t_1 < t_2 < \cdots < t_b < t$ and some $C_i = C_i(b, t)$. The claim follows by consideration of the order-statistic density function.

10. Let $\mathbb{E}_A$ (respectively, $\mathbb{E}_F$) denote expectation conditional on an ascent (respectively, fall) at time 0. Let X be the time of the first subsequent occurrence (either fall or ascent), and I_F (respectively, I_A) the indicator function that it is a fall (respectively, ascent). By conditioning on the next event, the time T to the first coincidence satisfies

$$\begin{aligned} \mathbb{E}_A(T) &= \left[\mathbb{E}_A(XI_F I(X < c)) + \mathbb{E}_A(TI_F I(X > c))\right] + \mathbb{E}_A(TI_A) \\ &= \mathbb{E}(XI_F) + \mathbb{E}_F(T)\mathbb{E}(I_F I(X > c)) + \mathbb{E}_A(TI_A) \\ &= \frac{1}{\nu} + \mathbb{E}_F(T) \int_c^\infty \lambda e^{-\nu x}\, dx + \frac{\mu}{\nu}\mathbb{E}_A(T) \\ &= \frac{1}{\nu} + \mathbb{E}_F(T)\frac{\lambda}{\nu} e^{-\nu c} + \frac{\mu}{\nu}\mathbb{E}_A(T), \end{aligned}$$

where $\nu = \lambda + \mu$ is the rate of the aggregate Poisson process of events. Likewise,

$$\mathbb{E}_F(T) = \frac{1}{\nu} + \mathbb{E}_A(T)\frac{\mu}{\nu} e^{-\nu c} + \frac{\lambda}{\nu}\mathbb{E}_F(T).$$

We solve to obtain

$$\frac{\mu}{\nu}\mathbb{E}_A(T) = \frac{\mu + \lambda e^{-\nu c}}{\lambda\nu(1 - e^{-2\nu c})}, \qquad \frac{\lambda}{\nu}\mathbb{E}_F(T) = \frac{\lambda + \mu e^{-\nu c}}{\mu\nu(1 - e^{-2\nu c})},$$

which we insert into

$$\mathbb{E}(T) = \mathbb{E}(X) + \frac{\mu}{\nu}\mathbb{E}_A(T) + \frac{\lambda}{\nu}\mathbb{E}_F(T)$$

to obtain the result. By expanding the answer in powers of c, the last line follows.

The idea of the direct proof is to declare a coincidence if some fall occurs in $(t, t + dt)$ with an ascent in the interval $(t - c, t + c)$. The probability of this is $(\lambda dt + \mathrm{O}(dt^2))(1 - e^{-2\mu c}) = 2\lambda\mu c dt + \mathrm{O}(c^2) + \mathrm{O}(dt^2)$.

11. By independence and stationarity,

$$\begin{aligned}\mathbb{P}(s < S_1 \le t < S_2) &= \mathbb{P}\big(N(t) - N(s) = 1,\ N(s) = 0\big)\\ &= \mathbb{P}(N(s) = 0)\mathbb{P}(N(t - s) = 1,\\ &= e^{-\lambda s}\lambda(t - s)e^{-\lambda(t-s)},\end{aligned}$$

which we differentiate with respect to s and t to obtain the joint density function

$$f(s, t) = \lambda^2 e^{-\lambda t}, \qquad 0 < s < t < \infty.$$

12. By conditioning on the instant U of the first commission, the probability $r(S)$ of ultimate bankruptcy satisfies

$$r(S) = e^{-\lambda S} + \int_0^S r(R + S - u)\lambda e^{-\lambda u}\,du.$$

Substitute $y = u - S$ and differentiate with respect to S to obtain $r'(S) + \lambda r(S) = \lambda r(R + S)$. We have $r(0) = 1$, and $r(S) \le 1$ for all $S > 0$. Look for a solution of the form $r(S) = e^{-\lambda\theta S}$ with $\theta \ge 0$. The equation and the boundary conditions are satisfied when $e^{-\lambda\theta R} = 1 - \theta$, which has one root at $\theta = 0$ when $\lambda R \le 1$, and an additional root at some $\psi \in (0, 1)$ when $\lambda R > 1$. The general solution subject to $r(0) = 1$ is

$$r(S) = \begin{cases} 1 & \text{if } \lambda R \le 1,\\ 1 - A + Ae^{-\lambda\psi S} & \text{if } \lambda R > 1.\end{cases}$$

It is the case that $A = 1$ when $\lambda R > 1$, and here is an explanation of that. The process is composed of cycles comprising a period of time during which costs accrue linearly, followed by a sale. The mean profit per cycle is $\mu := R - (1/\lambda)$. If $\mu > 0$ then, by the law of large numbers, the mean aggregate profit increases linearly with time. As the initial capital S becomes larger, the chance of bankruptcy diminishes to 0. Therefore, $A = 1$ in this case. The details are omitted.

In either case, we have $r(S) = e^{-\lambda\psi S} = (1-\psi)^{S/R}$ where ψ is the largest root of $e^{-\lambda\theta R} = 1-\theta$, and the given solution follows with $a = 1 - \psi$. Another approach to this problem may be found in Problem (12.9.27).

6.9 Solutions. Continuous-time Markov chains

1. (a) We have that

$$p_{11}' = -\mu p_{11} + \lambda p_{12}, \quad p_{22}' = -\lambda p_{22} + \mu p_{21},$$

where $p_{12} = 1 - p_{11}$, $p_{21} = 1 - p_{22}$. Solve these subject to $p_{ij}(t) = \delta_{ij}$, the Kronecker delta, to obtain that the matrix $\mathbf{P}_t = (p_{ij}(t))$ is given by

$$\mathbf{P}_t = \frac{1}{\lambda+\mu}\begin{pmatrix} \lambda + \mu e^{-(\lambda+\mu)t} & \mu - \mu e^{-(\lambda+\mu)t} \\ \lambda - \lambda e^{-(\lambda+\mu)t} & \mu + \lambda e^{-(\lambda+\mu)t} \end{pmatrix}.$$

(b) There are many ways of calculating $\mathbf{G}^n$; let us use generating functions. Note first that $\mathbf{G}^0 = \mathbf{I}$, the identity matrix. Write

$$\mathbf{G}^n = \begin{pmatrix} a_n & b_n \\ c_n & d_n \end{pmatrix}, \qquad n \geq 0,$$

and use the equation $\mathbf{G}^{n+1} = \mathbf{G} \cdot \mathbf{G}^n$ to find that

$$a_{n+1} = -\mu a_n + \mu c_n, \quad c_{n+1} = \lambda a_n - \lambda c_n.$$

Hence $a_{n+1} = -(\mu/\lambda)c_{n+1}$ for $n \geq 0$, and the first difference equation becomes $a_{n+1} = -(\lambda+\mu)a_n$, $n \geq 1$, which, subject to $a_1 = -\mu$, has solution $a_n = (-1)^n \mu(\lambda + \mu)^{n-1}$, $n \geq 1$. Therefore $c_n = (-1)^{n+1}\lambda(\lambda + \mu)^{n-1}$ for $n \geq 1$, and one may see similarly that $b_n = -a_n$, $d_n = -c_n$ for $n \geq 1$. Using the facts that $a_0 = d_0 = 1$ and $b_0 = c_0 = 0$, we deduce that $\sum_{n=0}^{\infty}(t^n/n!)\mathbf{G}^n = \mathbf{P}_t$ where $\mathbf{P}_t$ is given in part (a).

(c) With $\boldsymbol{\pi} = (\pi_1, \pi_2)$, we have that $-\mu\pi_1 + \lambda\pi_2 = 0$ and $\mu\pi_1 - \lambda\pi_2 = 0$, whence $\pi_1 = (\lambda/\mu)\pi_2$. In addition, $\pi_1 + \pi_2 = 1$ if $\pi_1 = \lambda/(\lambda + \mu) = 1 - \pi_2$.

2. (a) The required probability is

$$\frac{\mathbb{P}\big(X(t) = 2, X(3t) = 1 \mid X(0) = 1\big)}{\mathbb{P}\big(X(3t) = 1 \mid X(0) = 1\big)} = \frac{p_{12}(t)p_{21}(2t)}{p_{11}(3t)}$$

using the Markov property and the homogeneity of the process.

(b) Likewise, the required probability is

$$\frac{p_{12}(t)p_{21}(2t)p_{11}(t)}{p_{11}(3t)p_{11}(t)},$$

the same as in part (a).

3. The interarrival times and runtimes are independent and exponentially distributed. It is the lack-of-memory property which guarantees that X has the Markov property.

The state space is $S = \{0, 1, 2, \dots\}$ and the generator is

$$\mathbf{G} = \begin{pmatrix} -\lambda & \lambda & 0 & 0 & \cdots \\ \mu & -(\lambda+\mu) & \lambda & 0 & \cdots \\ 0 & \mu & -(\lambda+\mu) & \lambda & \cdots \\ \vdots & \vdots & \vdots & \vdots & \ddots \end{pmatrix}.$$

Solutions of the equation $\boldsymbol{\pi}\mathbf{G} = \mathbf{0}$ satisfy

$$-\lambda\pi_0 + \mu\pi_1 = 0, \quad \lambda\pi_{j-1} - (\lambda+\mu)\pi_j + \mu\pi_{j+1} = 0 \text{ for } j \geq 1,$$

with solution $\pi_i = \pi_0(\lambda/\mu)^i$. We have in addition that $\sum_i \pi_i = 1$ if $\lambda < \mu$ and $\pi_0 = 1 - (\lambda/\mu)$.

4. One may use the strong Markov property. Alternatively, by the Markov property,

$$\mathbb{P}(Y_{n+1} = j \mid Y_n = i, T_n = t, B) = \mathbb{P}(Y_{n+1} = j \mid Y_n = i, T_n = t)$$

for any event B defined in terms of $\{X(s) : s \le T_n\}$. Hence

$$\mathbb{P}(Y_{n+1} = j \mid Y_n = i, B) = \int_0^\infty \mathbb{P}(Y_{n+1} = j \mid Y_n = i, T_n = t) f_{T_n}(t)\, dt$$
$$= \mathbb{P}(Y_{n+1} = j \mid Y_n = i),$$

so that Y is a Markov chain. Now $q_{ij} = \mathbb{P}(Y_{n+1} = j \mid Y_n = i)$ is given by

$$q_{ij} = \int_0^\infty p_{ij}(t)\lambda e^{-\lambda t}\, dt,$$

by conditioning on the $(n+1)$th interarrival time of N; here, as usual, $p_{ij}(t)$ is a transition probability of X. Now

$$\sum_i \pi_i q_{ij} = \int_0^\infty \Big(\sum_i \pi_i p_{ij}(t)\Big)\lambda e^{-\lambda t}\, dt = \int_0^\infty \pi_j \lambda e^{-\lambda t}\, dt = \pi_j.$$

5. The jump chain $Z = \{Z_n : n \ge 0\}$ has transition probabilities $h_{ij} = g_{ij}/g_i$, $i \ne j$. The chance that Z ever reaches A from j is also η_j, and $\eta_j = \sum_k h_{jk}\eta_k$ for $j \notin A$, by Exercise (6.3.6). Hence $g_j\eta_j = \sum_{k\ne j} g_{jk}\eta_k$, as required.

6. Let $T_1 = \inf\{t : X(t) \ne X(0)\}$, and more generally let T_m be the time of the mth change in value of X. For $j \notin A$,

$$\mu_j = \mathbb{E}_j(T_1) + \sum_{k\ne j} h_{jk}\mu_k,$$

where $\mathbb{E}_j$ denotes expectation conditional on $X_0 = j$. Now $\mathbb{E}_j(T_1) = g_j^{-1}$, and the given equations follow. Suppose next that $(a_k : k \in S)$ is another non-negative solution of these equations. With $U_i = T_{i+1} - T_i$ and $R = \min\{n \ge 1 : Z_n \in A\}$, we have for $j \notin A$ that

$$a_j = \frac{1}{g_j} + \sum_{k\notin A} h_{jk}a_k = \frac{1}{g_j} + \sum_{k\notin A} h_{jk}\left\{\frac{1}{g_k} + \sum_{m\notin A} h_{km}a_m\right\}$$
$$= \mathbb{E}_j(U_0) + \mathbb{E}_j(U_1 I_{\{R>1\}}) + \mathbb{E}_j(U_2 I_{\{R>2\}}) + \cdots + \mathbb{E}_j(U_n I_{\{R>n\}}) + \Sigma,$$

where Σ is a sum of non-negative terms. It follows that

$$a_j \ge \mathbb{E}_j(U_0) + \mathbb{E}_j(U_1 I_{\{R>1\}}) + \cdots + \mathbb{E}_j(U_n I_{\{R>n\}})$$
$$= \mathbb{E}_j\left(\sum_{r=0}^n U_r I_{\{R>r\}}\right) = \mathbb{E}_j\big(\min\{T_n, H_A\}\big) \to \mathbb{E}_j(H_A)$$

as $n \to \infty$, by monotone convergence. Therefore, $\boldsymbol{\mu}$ is the minimal non-negative solution.

7. First note that i is recurrent if and only if it is also a recurrent state in the jump chain Z. The integrand being positive, we can write

$$\int_0^\infty p_{ii}(t)\, dt = \mathbb{E}\left[\int_0^\infty I_{\{X(t)=i\}}\, dt \,\middle|\, X(0) = i\right] = \mathbb{E}\left[\sum_{n=0}^\infty (T_{n+1} - T_n) I_{\{Y_n = i\}} \,\middle|\, Y_0 = i\right]$$

where $\{T_n : n \ge 1\}$ are the times of the jumps of X. The right side equals

$$\sum_{n=0}^\infty \mathbb{E}(T_1 \mid X(0) = i) h_{ii}(n) = \frac{1}{g_i}\sum_{n=0}^\infty h_{ii}(n)$$

where $\mathbf{H} = (h_{ij})$ is the transition matrix of Z. The sum diverges if and only if i is recurrent for Z.

8. Since the imbedded jump walk is recurrent, so is X. The probability of visiting m during an excursion is $\alpha = (2m)^{-1}$, since such a visit requires an initial step to the right, followed by a visit to m before 0, cf. Example (3.9.6). Having arrived at m, the chance of returning to m before visiting 0 is $1 - \alpha$, by the same argument with 0 and m interchanged. In this way one sees that the number N of visits to m during an excursion from 0 has distribution given by $\mathbb{P}(N \ge k) = \alpha(1-\alpha)^{k-1}$, $k \ge 1$. The 'total jump rate' from any state is λ, whence T may be expressed as $\sum_{i=0}^{N} V_i$ where the V_i are exponential with parameter λ. Therefore,

$$\mathbb{E}(e^{\theta T}) = G_N\left(\frac{\lambda}{\lambda - \theta}\right) = (1-\alpha) + \alpha\frac{\alpha\lambda}{\alpha\lambda - \theta}.$$

The distribution of T is a mixture of an atom at 0 and the exponential distribution with parameter $\alpha\lambda$.

9. The number N of sojourns in i has a geometric distribution $\mathbb{P}(N = k) = f^{k-1}(1-f)$, $k \ge 1$, for some $f < 1$. The length of each of these sojourns has the exponential distribution with some parameter g_i. By the independence of these lengths, the total time T in state i has moment generating function

$$\mathbb{E}(e^{\theta T}) = \sum_{k=1}^{\infty} f^{k-1}(1-f)\left(\frac{g_i}{g_i - \theta}\right)^k = \frac{g_i(1-f)}{g_i(1-f) - \theta}.$$

The distribution of T is exponential with parameter $g_i(1-f)$.

10. The jump chain is the simple random walk with probabilities $\lambda/(\lambda+\mu)$ and $\mu/(\lambda+\mu)$, and with $p_{01} = 1$. By Corollary (5.3.6), the chance of ever hitting 0 having started at 1 is μ/λ, whence the probability of returning to 0 having started there is $f = \mu/\lambda$. By the result of Exercise (6.9.9),

$$\mathbb{E}(e^{\theta V_0}) = \frac{\lambda - \mu}{\lambda - \mu - \theta},$$

as required. Having started at 0, the walk visits the state $r \ge 1$ with probability 1. The probability of returning to r having started there is

$$f_r = \frac{\mu}{\lambda+\mu} + \frac{\lambda}{\lambda+\mu}\cdot\frac{\mu}{\lambda} = \frac{2\mu}{\lambda+\mu},$$

and each sojourn is exponentially distributed with parameter $g_r = \lambda + \mu$. Now $g_r(1 - f_r) = \lambda - \mu$, whence, as above,

$$\mathbb{E}(e^{\theta V_r}) = \frac{\lambda - \mu}{\lambda - \mu - \theta}.$$

The probability of ever reaching 0 from $X(0)$ is $(\mu/\lambda)^{X(0)}$, and the time spent there subsequently is exponential with parameter $\lambda - \mu$. Therefore, the mean total time spent at 0 is

$$\mathbb{E}\left(\frac{(\mu/\lambda)^{X(0)}}{\lambda - \mu}\right) = \frac{G(\mu/\lambda)}{\lambda - \mu}.$$

11. Using the usual notation,

$$p_{ii}(h) \ge \mathbb{P}_i(U_0 > h) = e^{-g_i h} = 1 - g_i h + o(h),$$

and, for $i \ne j$,

$$\begin{aligned} p_{ij}(h) &\ge \mathbb{P}_i\big(U_0 < h < U_0 + U_1,\ X(U_0) = j\big) \\ &\ge (1 - e^{-g_i h})y_{ij}e^{-g_j h} = g_i h y_{ij}(1 - g_j h) + o(h) \\ &= g_{ij}h + o(h). \end{aligned}$$

Adding the above equations and recalling $g_{ii} = -g_i$, we obtain

$$1 = \sum_{j \in S} p_{ij}(h) = 1 + h \sum_{j \in S} g_{ij} + o(h) = 1 + o(h).$$

Since the last holds with equality, each of the previous equations holds with equality.

12. The jump chain is a symmetric random walk on the integers, and every state is recurrent. By Theorem (6.9.17c), the original chain does not explode.

13. The jump chain has transition probabilities $p_{01} = 1$ and

$$p_{n,n+1} = \frac{n+1}{n+2}, \quad p_{n,0} = \frac{1}{n+2}, \qquad n \geq 1.$$

The probability that the jump chain, started from 0, first returns to 0 at time m is

$$f_{00}(m) = 1 \cdot \frac{2}{3} \cdots \frac{m-1}{m} \cdot \frac{1}{m+1} = \frac{2}{m(m+1)}, \qquad m \geq 2.$$

Therefore, the time T of the first return to 0 by the jump chain satisfies

$$\mathbb{P}_0(T < \infty) = \sum_{m=1}^{\infty} f_{00}(m) = 1, \qquad \mathbb{E}_0(T) = \sum_{m=1}^{\infty} m f_{00}(m) = \infty,$$

so that the jump chain is null recurrent. Hence the original chain is recurrent. Since $g_n \leq 1$ for all n, the mean return time m_0 of 0 in the original chain is at least $\mathbb{E}_0(T) - 1 = \infty$.

14. Let $X(0) = Z_0 = i$ and assume to avoid triviality that $g_i > 0$. If i is recurrent for Z, then a.s. Z visits i infinitely often, whence i is recurrent for X. Conversely, if i is recurrent for X, there exists a.s. an unbounded, increasing sequence (T_m) of times at which X hits i, such that $T_{m+1} - T_m > h$. For given m, let $kh < T_m \leq (k+1)h$. Conditional on T_m, the probability that $Z_{(k+1)h} = i$ is at least $\mathbb{P}_i(U_0 > h) = e^{-g_i h} > 0$. Hence, Z visits i infinitely often a.s., whence i is recurrent for Z.

Evidently X is irreducible if Z is irreducible. Assume X is irreducible, and to avoid triviality that there exist two or more states. For states i, j, find $t > 0$ such that $p_{ij}(t) > 0$, and let $kh < t \leq (k+1)h$. As above, $\mathbb{P}_i(Z_{(k+1)h} = j) \geq p_{ij}(t) e^{-g_j h} > 0$.

6.10 Solutions. Kolmogorov equations and the limit theorem

1. (a) These inequalities hold since, for any column vector $\mathbf{x}$,

$$|(Q_1 + Q_2)\mathbf{x}| \leq |Q_1 \mathbf{x}| + |Q_2 \mathbf{x}| \leq \big(|Q_1| + |Q_2|\big)|\mathbf{x}|,$$
$$|Q_1 Q_2 \mathbf{x}| \leq |Q_1| \cdot |Q_2 \mathbf{x}| \leq |Q_1| \cdot |Q_2| \cdot |\mathbf{x}|.$$

(b) Note first that $|Q| \leq N \max_{i,j} |q_{ij}| < \infty$. For each pair i, j of indices,

$$\big|(E_n - E_m)_{ij}\big| \leq \left|\sum_{m+1}^{n} \frac{Q^k}{k!}\right| \leq \sum_{m+1}^{n} \frac{|Q|^k}{k!} \to 0 \qquad \text{as } m, n \to \infty.$$

Therefore, each given entry of E_n forms a Cauchy sequence, and hence converges. The limit matrix $E(Q) := e^Q$ is then well defined. Furthermore,

$$|E(Q) - E_n| \leq \sum_{k=n+1}^{\infty} \frac{|Q|^k}{k!} \to \infty \qquad \text{as } n \to \infty.$$

(c) Let $Q_1, Q_2 \in \mathcal{Q}$ be a commuting pair. Then

$$e^{Q_1+Q_2} = \sum_{k=0}^{\infty} \frac{(Q_1+Q_2)^k}{k!} = \sum_{k=0}^{\infty} \frac{1}{k!} \sum_{r=0}^{k} \binom{k}{r} Q_1^r Q_2^{k-r}$$
$$= \sum_{r=0}^{\infty} \frac{Q_1^r}{r!} \sum_{k=r}^{\infty} \frac{Q_2^{k-r}}{(k-r)!} = e^{Q_1} e^{Q_2}.$$

2. Define the generator $\mathbf{G}$ by $g_{ii} = -\nu_i$, $g_{ij} = \nu_i y_{ij}$, so that the imbedded chain has transition matrix $\mathbf{Y}$. A root of the equation $\boldsymbol{\pi}\mathbf{G} = \mathbf{0}$ satisfies

$$0 = \sum_i \pi_i g_{ij} = -\pi_j \nu_j + \sum_{i:i \neq j} (\pi_i \nu_i) h_{ij}$$

whence the vector $\boldsymbol{\zeta} = (\pi_j \nu_j : j \in S)$ satisfies $\boldsymbol{\zeta} = \boldsymbol{\zeta}\mathbf{Y}$. Therefore $\boldsymbol{\zeta} = \alpha \boldsymbol{\nu}$, which is to say that $\pi_j \nu_j = \alpha \nu_j$, for some constant α. Now $\nu_j > 0$ for all j, so that $\pi_j = \alpha$, which implies that $\sum_j \pi_j \neq 1$. Therefore the continuous-time chain X with generator $\mathbf{G}$ has no stationary distribution.

3. The jump chain Y is a simple symmetric random walk on $\mathbb{Z}$. Since transience and recurrence are equivalent for the jump and the continuous-time chain, X is recurrent, and hence non-explosive. The (non-probability) measure $\mu_i \equiv 1$ is invariant for the jump chain, whence $\lambda_i := \mu_i / g_i$ is invariant for X. Now $\lambda_i = (i^2+1)^{-1}$ is summable over $i \in \mathbb{Z}$. Therefore, X has an invariant distribution and is non-explosive, and hence positive recurrent.

4. The jump chain Y is a biased random walk on $\mathbb{Z}$ with $p_{i,i+1} = \frac{2}{3} = 1 - p_{i,i-1}$. Since Y is transient, so is X. The easiest way to find an invariant distribution is to use the theory of reversibility for continuous-time chains; see Problem (6.15.16). By the results of that problem, since X is a birth–death chain on $\mathbb{Z}$, it has an invariant distribution if and only if it is reversible, or equivalently if there exists a distribution satisfying the detailed balance equations $\pi_i g_{ij} = \pi_j g_{ji}$. Therefore,

$$\pi_i = \begin{cases} \left(\frac{2}{3}\right)^i \pi_0 & \text{if } i \geq 0, \\ \left(\frac{1}{6}\right)^{-i} \pi_0 & \text{if } i < 0, \end{cases}$$

and we choose π_0 such that $\boldsymbol{\pi}$ is a probability distribution. Despite transience, the chain has an invariant distribution, whence the chain must explode. See also Example (6.10.21).

5. Since $\mathbf{G}$ is a generator, it has row sums 0, so that

$$\mathbf{G}f(i) = \sum_{j \in S} g_{ij} f(j) = \sum_{j \in S} g_{ij} \big(f(j) - f(i)\big).$$

The interchanges of limits and summations are justified in the following since S is assumed finite. Let $\mathbf{P} = (P_t : t \geq 0)$ be the transition semigroup associated with $\mathbf{G}$. Then

$$\mathbb{E}_i f(X_t) - f(i) = \sum_{j \in S} p_{ij}(t)\big(f(j) - f(i)\big).$$

On letting $t \to 0$, by Exercise (6.9.11),

$$\lim_{t \to 0} \frac{1}{t}\big[\mathbb{E}_i f(X_t) - f(i)\big] = \sum_{j \in S} g_{ij}\big(f(j) - f(i)\big) = \mathbf{G}f(i).$$

For the final part, let $\psi(s) = \mathbb{E}_i f(X_s)$. It suffices to show that $\psi'(s) = \mathbb{E}(\mathbf{G}f(X_s))$, and the claim follows from this by integrating over the interval $[0, t]$. We have that

$$\begin{aligned}\psi'(s) &= \lim_{u\to 0}\left(\frac{1}{u}\big[\mathbb{E}_i f(X_{s+u}) - \mathbb{E}_i f(X_s)\big]\right)\\ &= \lim_{u\to 0}\left(\sum_{j\in S} p_{ij}(s)\frac{1}{u}\big[\mathbb{E}_i\big(f(X_{s+u})\,\big|\, X_s = j\big) - f(j)\big]\right)\\ &= \sum_{j\in S} p_{ij}(s)\lim_{u\to 0}\left(\frac{1}{u}\big[\mathbb{E}_j f(X_u) - f(j)\big]\right) = \sum_{j\in S} p_{ij}(s)\mathbf{G}f(j) = \mathbb{E}_i(\mathbf{G}f(X_s)).\end{aligned}$$

6.11 Solutions. Birth–death processes and imbedding

1. The jump chain is a walk $\{Z_n\}$ on the set $S = \{0, 1, 2, \dots\}$ satisfying, for $i \ge 1$,

$$\mathbb{P}(Z_{n+1} = j \mid Z_n = i) = \begin{cases} p_i & \text{if } j = i+1,\\ 1 - p_i & \text{if } j = i-1,\end{cases}$$

where $p_i = \lambda_i/(\lambda_i + \mu_i)$. Also $\mathbb{P}(Z_{n+1} = 1 \mid Z_n = 0) = 1$.

2. The transition matrix $\mathbf{H} = (h_{ij})$ of Z is given by

$$h_{ij} = \begin{cases} \dfrac{i\mu}{\lambda + i\mu} & \text{if } j = i-1,\\[2mm] \dfrac{\lambda}{\lambda + i\mu} & \text{if } j = i+1.\end{cases}$$

To find the stationary distribution of Y, either solve the equation $\boldsymbol{\pi} = \boldsymbol{\pi}\mathbf{Q}$, or look for a solution of the detailed balance equations $\pi_i h_{i,i+1} = \pi_{i+1} h_{i+1,i}$. Following the latter route, we have that

$$\pi_i = \pi_0 \frac{h_{01}h_{12}\cdots h_{i-1,i}}{h_{i,i-1}\cdots h_{21}h_{10}}, \qquad i \ge 1,$$

whence $\pi_i = \pi_0 \rho^i (1 + i/\rho)/i!$ for $i \ge 1$. Choosing π_0 accordingly, we obtain the result.

It is a standard calculation that X has stationary distribution $\boldsymbol{\nu}$ given by $\nu_i = \rho^i e^{-\rho}/i!$ for $i \ge 0$. The difference between $\boldsymbol{\pi}$ and $\boldsymbol{\nu}$ arises from the fact that the holding-times of X have distributions which depend on the current state.

3. We have, by conditioning on $X(h)$, that

$$\begin{aligned}\eta(t+h) &= \mathbb{E}\{\mathbb{P}\big(X(t+h) = 0 \,\big|\, X(h)\big)\}\\ &= \mu h \cdot 1 + (1 - \lambda h - \mu h)\eta(t) + \lambda h \xi(t) + o(h)\end{aligned}$$

where $\xi(t) = \mathbb{P}(X(t) = 0 \mid X(0) = 2)$. The process X may be thought of as a collection of particles each of which dies at rate μ and divides at rate λ, different particles enjoying a certain independence; this is a consequence of the linearity of λ_n and μ_n. Hence $\xi(t) = \eta(t)^2$, since each of the initial pair is required to have no descendants at time t. Therefore

$$\eta'(t) = \mu - (\lambda + \mu)\eta(t) + \lambda\eta(t)^2$$

subject to $\eta(0) = 0$.

Rewrite the equation as

$$\frac{\eta'}{(1-\eta)(\mu-\lambda\eta)} = 1$$

and solve using partial fractions to obtain

$$\eta(t) = \begin{cases} \dfrac{\lambda t}{\lambda t + 1} & \text{if } \lambda = \mu, \\ \dfrac{\mu(1 - e^{t(\mu-\lambda)})}{\lambda - \mu e^{t(\mu-\lambda)}} & \text{if } \lambda \neq \mu. \end{cases}$$

Finally, if $0 < t < u$,

$$\mathbb{P}\big(X(t) = 0 \,\big|\, X(u) = 0\big) = \mathbb{P}\big(X(u) = 0 \,\big|\, X(t) = 0\big)\frac{\mathbb{P}(X(t) = 0)}{\mathbb{P}(X(u) = 0)} = \frac{\eta(t)}{\eta(u)}.$$

4. The random variable $X(t)$ has generating function

$$G(s,t) = \frac{\mu(1-s) - (\mu - \lambda s)e^{-t(\lambda-\mu)}}{\lambda(1-s) - (\mu - \lambda s)e^{-t(\lambda-\mu)}}$$

as usual. The generating function of $X(t)$, conditional on $\{X(t) > 0\}$, is therefore

$$\sum_{n=1}^{\infty} s^n \frac{\mathbb{P}(X(t) = n)}{\mathbb{P}(X(t) > 0)} = \frac{G(s,t) - G(0,t)}{1 - G(0,t)}.$$

Substitute for G and take the limit as $t \to \infty$ to obtain as limit

$$H(s) = \frac{(\mu - \lambda)s}{\mu - \lambda s} = \sum_{n=1}^{\infty} s^n p_n$$

where, with $\rho = \lambda/\mu$, we have that $p_n = \rho^{n-1}(1-\rho)$ for $n \geq 1$.

5. Extinction is certain if $\lambda < \mu$, and in this case, by Theorem (6.11.16),

$$\begin{aligned} \mathbb{E}(T) &= \int_0^\infty \mathbb{P}(T > t)\,dt = \int_0^\infty \{1 - \mathbb{E}(s^{X(t)})|_{s=0}\}\,dt \\ &= \int_0^\infty \frac{(\mu-\lambda)e^{(\lambda-\mu)t}}{\mu - \lambda e^{(\lambda-\mu)t}}\,dt = \frac{1}{\lambda}\log\left(\frac{\mu}{\mu-\lambda}\right). \end{aligned}$$

If $\lambda > \mu$ then $\mathbb{P}(T < \infty) = \mu/\lambda$, so

$$\mathbb{E}(T \mid T < \infty) = \int_0^\infty \left\{1 - \frac{\lambda}{\mu}\mathbb{E}(s^{X(t)})|_{s=0}\right\} dt = \int_0^\infty \frac{(\lambda-\mu)e^{(\mu-\lambda)t}}{\lambda - \mu e^{(\mu-\lambda)t}}\,dt = \frac{1}{\mu}\log\left(\frac{\lambda}{\lambda-\mu}\right).$$

In the case $\lambda = \mu$, $\mathbb{P}(T < \infty) = 1$ and $\mathbb{E}(T) = \infty$.

6. By considering the imbedded random walk, we find that the probability of ever returning to 1 is $\max\{\lambda, \mu\}/(\lambda + \mu)$, so that the number of visits is geometric with parameter $\min\{\lambda, \mu\}/(\lambda + \mu)$. Each visit has an exponentially distributed duration with parameter $\lambda + \mu$, and a short calculation using moment generating functions shows that $V_1(\infty)$ is exponential with parameter $\min\{\lambda, \mu\}$.

Next, by a change of variables, Theorem (6.11.16), and some calculation,

$$\begin{aligned}\sum_r s^r \mathbb{E}(V_r(t)) &= \mathbb{E}\left(\sum_r \int_0^t s^r I_{\{X(u)=r\}}\,du\right) = \mathbb{E}\left(\int_0^t s^{X(u)}\,du\right)\\ &= \int_0^t \mathbb{E}(s^{X(u)})\,du = \frac{\mu t}{\lambda} - \frac{1}{\lambda}\log\left\{\frac{\lambda(1-s)-(\mu-\lambda s)e^{-(\lambda-\mu)t}}{\lambda-\mu}\right\}\\ &= -\frac{1}{\lambda}\log\left\{1-\frac{\lambda s(e^{\rho t}-1)}{\mu e^{\rho t}-\lambda}\right\} + \text{terms not involving } s,\end{aligned}$$

where $\rho = \mu - \lambda$. We take the limit as $t \to \infty$ and we pick out the coefficient of s^r.

7. If $\lambda = \mu$ then, by Theorem (6.11.16),

$$\mathbb{E}(s^{X(t)}) = \frac{\lambda t(1-s)+s}{\lambda t(1-s)+1} = 1 - \frac{1-s}{\lambda t(1-s)+1},$$

and

$$\begin{aligned}\int_0^t \mathbb{E}(s^{X(u)})\,du &= t - \frac{1}{\lambda}\log\{\lambda t(1-s)+1\}\\ &= -\frac{1}{\lambda}\log\left\{1-\frac{\lambda t s}{1+\lambda t}\right\} + \text{terms not involving } s.\end{aligned}$$

Letting $t \to \infty$ and picking out the coefficient of s^r gives $\mathbb{E}(V_r(\infty)) = (r\lambda)^{-1}$. An alternative method utilizes the imbedded simple random walk and the exponentiality of the sojourn times.

8. (a) By conditioning on the holding time at n,

$$d_n = \frac{\mu_n}{\lambda_n+\mu_n}\cdot\frac{1}{\lambda_n+\mu_n} + \frac{\lambda_n}{\lambda_n+\mu_n}\left\{\frac{1}{\lambda_n+\mu_n} + d_{n+1} + d_n\right\}.$$

(b) Similarly,

$$M_n(\theta) = \frac{\mu_n}{\lambda_n+\mu_n}M(\theta) + \frac{\lambda_n}{\lambda_n+\mu_n}M(\theta)M_{n+1}(\theta)M_n(\theta),$$

where

$$M(\theta) = \frac{\lambda_n+\mu_n}{\lambda_n+\mu_n-\theta}, \qquad \theta < \lambda,$$

is the moment generating function of the holding time at n.

9. When $X(0) = i$, denote by d_i the mean time until X takes the value $i-1$. By Exercise (6.11.8),

$$d_{i+1} = \frac{\mu}{\lambda(n-i)}d_i - \frac{1}{\lambda i(n-i)}, \qquad i \geq 1.$$

By iteration, subject to the boundary condition $d_n = 1/(n\mu)$, we find that

$$d_n = \frac{1}{n\mu} = \left(\frac{\mu}{\lambda}\right)^{n-1}\frac{d_1}{(n-1)!} - \frac{\mu^{n-2}}{\lambda^{n-1}(n-1)!} - \frac{\mu^{n-3}}{\lambda^{n-2}2\,(n-2)!} - \cdots - \frac{\mu}{\lambda^2(n-2)\,2!} - \frac{1}{\lambda(n-1)}.$$

Thus, for example, $d_1 \geq (\lambda/\mu)^n(n-1)!/(n\lambda)$, implying that extinction takes a very long time on average if the number of niches is large. This is essentially an epidemic without immunity.

Another way to study the process is to allow transitions from 0 to strictly positive states. The ensuing process is irreducible, and the mean passage times may be calculated from knowledge of its invariant distribution.

6.12 Solutions. Special processes

1. The jump chain is simple random walk with step probabilities $\lambda/(\lambda+\mu)$ and $\mu/(\lambda+\mu)$. The expected time μ_{10} to pass from 1 to 0 satisfies

$$\mu_{10} = 1 + \frac{\lambda}{\lambda+\mu}(\mu_{21}+\mu_{10}) = 1 + \frac{2\lambda}{\lambda+\mu}\mu_{10},$$

whence $\mu_{10} = (\mu+\lambda)/(\mu-\lambda)$. Since each sojourn is exponentially distributed with parameter $\mu+\lambda$, the result follows by an easy calculation. See also Theorem (11.3.17).

2. We apply the method of Theorem (6.12.7) with

$$G_N(s,u) = \frac{se^{-\lambda u}}{1-s+se^{-\lambda u}},$$

the probability generating function of the population size at time u in a simple birth generating function of the ensuing population size at time v is

$$H(s,v) = \exp\left(\nu\int_0^v [G_N(s,u)-1]\,du\right) = \{s+(1-s)e^{\lambda v}\}^{-\nu/\lambda}.$$

The individuals alive at time t arose subsequent to the most recent disaster at time $t-D$, where D has density function $\delta e^{-\delta x}$, $x>0$. Therefore,

$$\mathbb{E}(s^{X(t)}) = \mathbb{E}(H(s,D)) = \int_0^t \frac{\delta e^{-\delta x}e^{-\nu t}\,dx}{(1-s+se^{-\lambda t})^{\nu/\delta}} + e^{-\delta t}\frac{e^{-\nu t}}{(1-s+se^{-\lambda t})^{\nu/\delta}}.$$

3. The mean number of descendants after time t of a single progenitor at time 0 is $e^{(\lambda-\mu)t}$. The expected number due to the arrival of a single individual at a uniformly distributed time in the interval on $[0,x]$ is therefore

$$\frac{1}{x}\int_0^x e^{(\lambda-\mu)u}\,du = \frac{e^{(\lambda-\mu)x}-1}{(\lambda-\mu)x}.$$

The aggregate effect at time x of N earlier arrivals is the same, by Theorem (6.8.11), as that of N arrivals at independent times which are uniformly distributed on $[0,x]$. Since $\mathbb{E}(N)=\nu x$, the mean population size at time x is $\nu[e^{(\lambda-\mu)x}-1]/(\lambda-\mu)$. The most recent disaster occurred at time $t-D$, where D has density function $\delta e^{-\delta x}$, $x>0$, and it follows that

$$\mathbb{E}(X(t)) = \int_0^t \delta e^{-\delta x}\frac{\nu}{\lambda-\mu}[e^{(\lambda-\mu)x}-1]\,dx + \frac{\nu}{\lambda-\mu}e^{-\delta x}[e^{(\lambda-\mu)t}-1].$$

This is bounded as $t\to\infty$ if and only if $\delta > \lambda-\mu$.

4. Let N be the number of clients who arrive during the interval $[0,t]$. Conditional on the event $\{N=n\}$, the arrival times have, by Theorem (6.8.11), the same joint distribution as n independent variables chosen uniformly from $[0,t]$. The probability that an arrival at a uniform time in $[0,t]$ is still in service at time t is $\beta = \int_0^t [1-G(t-x)]t^{-1}\,dx$, whence, conditional on $\{N=n\}$, the total number M still in service is $\text{bin}(n,\beta)$. Therefore,

$$\mathbb{E}(e^{\theta M}) = \mathbb{E}\big(\mathbb{E}(e^{\theta M}\mid N)\big) = \mathbb{E}\big((\beta e^{\theta}+1-\beta)^N\big) = G_N(\beta e^{\theta}+1-\beta) = e^{\lambda\beta t(e^{\theta}-1)},$$

whence M has the Poisson distribution with parameter $\lambda\beta t = \lambda\int_0^t[1-G(x)]\,dx$. The last part holds since $\int_0^t[1-G(x)]\,dx \to \mathbb{E}(S)$ as $t\to\infty$.

6.13 Solutions. Spatial Poisson processes

1. It is easy to check from the axioms that the combined process $N(t) = B(t) + G(t)$ is a Poisson process with intensity $\beta + \gamma$.

(a) The time S (respectively, T) until the arrival of the first brown (respectively, grizzly) bear is exponentially distributed with parameter β (respectively, γ), and these times are independent. Now,

$$\mathbb{P}(S < T) = \int_0^\infty \beta e^{-\beta s} e^{-\gamma s}\, ds = \frac{\beta}{\beta + \gamma}.$$

(b) Using (a), and the lack-of-memory of the process, the required probability is

$$\left(\frac{\gamma}{\beta+\gamma}\right)^r \frac{\beta}{\beta+\gamma}.$$

(c) Using Theorem (6.12.7),

$$\mathbb{E}\big(\min\{S, T\} \,\big|\, B(1) = 1\big) = \mathbb{E}\left\{\frac{1}{G(1)+2}\right\} = \frac{\gamma - 1 + e^{-\gamma}}{\gamma^2}.$$

2. (a) Let B_r be the ball with centre $\mathbf{0}$ and radius r, and let $N_r = |\Pi \cap B_r|$. We have by Theorem (6.13.11) that $S_r = \sum_{\mathbf{x}\in\Pi\cap B_r} g(\mathbf{x})$ satisfies

$$\mathbb{E}(S_r \mid N_r) = N_r \int_{B_r} g(\mathbf{x}) \frac{\lambda(\mathbf{x})}{\Lambda(B_r)}\, d\mathbf{x},$$

where $\Lambda(B) = \int_{\mathbf{y}\in B} \lambda(\mathbf{y})\, d\mathbf{y}$. Therefore, $\mathbb{E}(S_r) = \int_{B_r} g(\mathbf{x})\lambda(\mathbf{x})\, d\mathbf{x}$, implying by monotone convergence that $\mathbb{E}(S) = \int_{\mathbb{R}^d} g(\mathbf{x})\lambda(\mathbf{x})\, d\mathbf{x}$. Similarly,

$$\begin{aligned}
\mathbb{E}(S_r^2 \mid N_r) &= \mathbb{E}\left(\left\{\sum_{\mathbf{x}\in\Pi\cap B_r} g(\mathbf{x})\right\}^2\right)\\
&= \mathbb{E}\left(\sum_{\mathbf{x}\in\Pi\cap B_r} g(\mathbf{x})^2\right) + \mathbb{E}\left(\sum_{\substack{\mathbf{x}\neq\mathbf{y}\\ \mathbf{x},\mathbf{y}\in\Pi\cap B_r}} g(\mathbf{x})g(\mathbf{y})\right)\\
&= N_r \int_{B_r} g(\mathbf{x})^2 \frac{\lambda(\mathbf{x})}{\Lambda(B_r)}\, d\mathbf{x} + N_r(N_r - 1) \iint_{\mathbf{x},\mathbf{y}\in B_r} g(\mathbf{x})g(\mathbf{y}) \frac{\lambda(\mathbf{x})\lambda(\mathbf{y})}{\Lambda(B_r)^2}\, d\mathbf{x}\, d\mathbf{y},
\end{aligned}$$

whence

$$\mathbb{E}(S_r^2) = \int_{B_r} g(\mathbf{x})^2\lambda(\mathbf{x})\, d\mathbf{x} + \left(\int_{B_r} g(\mathbf{x})\lambda(\mathbf{x})\, d\mathbf{x}\right)^2.$$

By monotone convergence,

$$\mathbb{E}(S^2) = \int_{\mathbb{R}^d} g(\mathbf{x})^2\lambda(\mathbf{x})\, d\mathbf{x} + \mathbb{E}(S)^2,$$

and the formula for the variance follows.

(b) Let $t > 0$. Given that $N_r = n$, by the conditional property, Theorem (6.13.11),

$$\mathbb{E}(e^{-tS_r}) = \mathbb{E}\big(\mathbb{E}(e^{-tS_r} \mid N_r)\big) = \mathbb{E}\left(\left\{\int_{B_r} e^{-tg(\mathbf{x})} \frac{\lambda(\mathbf{x})}{\Lambda(B_r)}\, d\mathbf{x}\right\}^{N_r}\right)$$
$$= \exp\left\{-\int_{B_r} (1 - e^{-tg(\mathbf{x})})\lambda(\mathbf{x})\, d\mathbf{x}\right\},$$

if $\Lambda(B_r) > 0$. Since $S_r \to S$ as $r \to \infty$,

$$(*) \qquad \mathbb{E}(e^{-tS}) = \lim_{r\to\infty} \mathbb{E}(e^{-tS_r}) = \exp\left\{-\int_{\mathbb{R}^d} (1 - e^{-tg(\mathbf{x})})\lambda(\mathbf{x})\, d\mathbf{x}\right\}, \qquad t > 0.$$

Now $1 - e^{-x} \le \min\{1, x\}$ for $x \in \mathbb{R}$. By monotone convergence, $\mathbb{E}(e^{-tS}) \to 1$ as $t \downarrow 0$ if the integral condition holds. The claim follows.

(c) Subject to the integral condition, we have $\mathbb{P}(0 \le S < \infty) = 1$. The two sides of (*) are analytic in $t \in \mathbb{C}$ on the region $\mathrm{Re}(t) \ge 0$, and part (c) follows with $\mathrm{Re}(t) \in \mathbb{R}$.

3. If $B_1, B_2, \dots, B_n$ are disjoint regions of the disc, then the numbers of projected points therein are Poisson-distributed and independent, since they originate from disjoint regions of the sphere. By elementary coordinate geometry, the intensity function in plane polar coordinates is $2\lambda/\sqrt{1-r^2}$, $0 \le r \le 1$, $0 \le \theta < 2\pi$.

4. The same argument is valid with resulting intensity function $2\lambda\sqrt{1-r^2}$.

5. The Mercator projection represents the spherical coordinates (θ, ϕ) as Cartesian coordinates in the range $0 \le \phi < 2\pi$, $0 \le \theta \le \pi$. (Recall that θ is the angle made with the axis through the north pole.) Therefore a uniform intensity on the globe corresponds to an intensity function $\lambda \sin\theta$ on the map. Likewise, a uniform intensity on the map corresponds to an intensity $\lambda/\sin\theta$ on the globe.

6. Let the X_r have characteristic function ϕ. Conditional on the value of $N(t)$, the corresponding arrival times have the same distribution as $N(t)$ independent variables with the uniform distribution, whence

$$\mathbb{E}(e^{i\theta S(t)}) = \mathbb{E}\{\mathbb{E}(e^{i\theta S(t)} \mid N(t))\} = \mathbb{E}\{\mathbb{E}(e^{i\theta Xe^{-\alpha U}})^{N(t)}\}$$
$$= \exp\{\lambda t(\mathbb{E}(e^{i\theta Xe^{-\alpha U}}) - 1)\} = \exp\left\{\lambda \int_0^t \{\phi(\theta e^{-\alpha u}) - 1\}\, du\right\},$$

where U is uniformly distributed on $[0, t]$. By differentiation,

$$\mathbb{E}(S(t)) = -i\phi'_{S(t)}(0) = \frac{\lambda}{\alpha}\mathbb{E}(X)(1 - e^{-\alpha t}),$$
$$\mathbb{E}(S(t)^2) = -\phi''_{S(t)}(0) = \mathbb{E}(S(t))^2 + \frac{\lambda \mathbb{E}(X^2)}{2\alpha}(1 - e^{-2\alpha t}).$$

Now, for $s < t$, $S(t) = S(s)e^{-\alpha(t-s)} + \widehat{S}(t-s)$ where $\widehat{S}(t-s)$ is independent of $S(s)$ with the same distribution as $S(t-s)$. Hence, for $s < t$,

$$\mathrm{cov}\big(S(s), S(t)\big) = \mathrm{var}(S(s))e^{-\alpha(t-s)} = \frac{\lambda\mathbb{E}(X^2)}{2\alpha}(1 - e^{-2\alpha s})e^{-\alpha(t-s)} \to \frac{\lambda\mathbb{E}(X^2)}{2\alpha}e^{-\alpha v}$$

as $s \to \infty$ with $v = t - s$ fixed. Therefore, $\rho(S(s), S(s+v)) \to e^{-\alpha v}$ as $s \to \infty$.

7. The first two arrival times T_1, T_2 satisfy

$$\mathbb{P}(T_1 \le x, T_2 - T_1 > y) = \int_0^x \lambda(u)e^{-\Lambda(u)}e^{-(\Lambda(u+y)-\Lambda(u))}\, du = \int_0^x \lambda(u)e^{-\Lambda(u+y)}\, du.$$

Differentiate with respect to x and y to obtain the joint density function $\lambda(x)\lambda(x+y)e^{-\Lambda(x+y)}$, $x, y \ge 0$. Since this does not generally factorize as the product of a function of x and a function of y, T_1 and T_2 are dependent in general.

8. Let X_i be the time of the first arrival in the process N_i. Then

$$\begin{aligned}\mathbb{P}(I = 1,\ T \ge t) &= \mathbb{P}\big(t \le X_1 < \inf\{X_2, X_3, \dots\}\big)\\ &= \int_t^\infty \mathbb{P}\big(\inf\{X_2, X_3, \dots\} > x\big)\lambda_1 e^{-\lambda_1 x}\,dx = \frac{\lambda_1}{\lambda}e^{-\lambda t}.\end{aligned}$$

9. (a) For given (p, θ), the line L passing through $P = (p\cos\theta, p\sin\theta)$ perpendicular to OP has equation $x\cos\theta + y\sin\theta = p$, whence

$$p = \frac{1}{\sqrt{u^2+v^2}}, \qquad \tan\theta = \frac{v}{u}.$$

We call (p, θ) the *coordinates* of the line L.

(b) A 'uniform' line process is one that corresponds to a uniform Poisson process on $S^\perp$. Since the Jacobian of the mapping $(p, \theta) \mapsto (u, v)$ has modulus

$$|J| = \frac{1}{(u^2+v^2)^{3/2}},$$

the claim follows.

(c) Consider first a translation of $\mathbb{L}$ that shifts the origin to the point (a, b). This maps the coordinates (p, θ) of the line L to $(p - a\cos\theta - b\sin\theta, \theta)$. The Jacobian of this mapping is 1, so it is area-preserving, and the image Poisson process is uniform also. Consider next a rotation T_α through an angle α around the origin. This maps (p, θ) to $((-1)^k p, \theta + \alpha - k\pi)$ where k is such that $0 \le \theta + \alpha - k\pi < \pi$. For given α, the integer k takes no more than 2 values, so that any set can be partitioned as two sets each of which is rotated by T_α. Thus T_α preserves area, and the claim holds as before.

10. A lorry of mass M parked at position x exerts a force $M\,\mathrm{sign}(x)/x^2$ on the pedestrian. Note that $g(x) = 1/x^2$ is integrable over $\mathbb{R}$. By the Campbell–Hardy theorem (see Exercise (6.13.2c) and Theorem (6.13.23)) or by direct computation,

$$\begin{aligned}\phi_G(t) &= \exp\left\{-\int_{-\infty}^0 \left[1 - \mathbb{E}\left(\exp\left\{\frac{-itM}{x^2}\right\}\right)\right] dx - \int_0^\infty \left[1 - \mathbb{E}\left(\exp\left\{\frac{itM}{x^2}\right\}\right)\right] dx\right\}\\ &= \exp\left\{-2\int_0^\infty \left[1 - \mathbb{E}\left(\cos\left(\frac{|t|M}{x^2}\right)\right)\right] dx\right\},\end{aligned}$$

since cosine is an even function, and sine is odd. Take the expectation outside the integral, and make the substitution $y = |t|M/x^2$ to find that $\phi_G(t) = \exp(-c|t|^{1/2})$ where

$$c = \mathbb{E}\sqrt{M}\int_0^\infty \frac{1}{y^{3/2}}(1 - \cos y)\,dy = \mathbb{E}\sqrt{2\pi M}.$$

In calculating the integral, it may be useful to know that $\int_0^\infty \sin^2\psi\,d\psi = \frac{1}{2}\sqrt{\pi/2}$. Thus $c = \infty$ if and only if $\mathbb{E}\sqrt{M} = \infty$. [See also Problems (6.15.56), (6.15.57), and (8.10.9).]

6.14 Solutions. Markov chain Monte Carlo

1. If $\mathbf{P}$ is reversible then

$$\text{RHS} = \sum_i \Big(\sum_j p_{ij} x_j\Big) y_i \pi_i = \sum_{i,j} \pi_i p_{ij} x_j y_i = \sum_{i,j} \pi_j p_{ji} y_i x_j = \sum_j \pi_j x_j \Big(\sum_i p_{ji} y_i\Big) = \text{LHS}.$$

Suppose conversely that $\langle \mathbf{x}, \mathbf{Py} \rangle = \langle \mathbf{Px}, \mathbf{y} \rangle$ for all $\mathbf{x}, \mathbf{y} \in l^2(\pi)$. Choose $\mathbf{x}, \mathbf{y}$ to be unit vectors with 1 in the ith and jth place respectively, to obtain the detailed balance equations $\pi_i p_{ij} = \pi_j p_{ji}$.

2. Just check that $0 \le b_{ij} \le 1$ and that the $p_{ij} = g_{ij} b_{ij}$ satisfy the detailed balance equations (6.14.3).

3. It is immediate that $p_{jk} = |A_{jk}|$, the Lebesgue measure of A_{jk}. This is a method for simulating a Markov chain with a given transition matrix.

4. (a) Note first from equation (4.12.7) that $d(\mathbf{U}) = \frac{1}{2}\sup_{i \ne j} d_{\text{TV}}(u_{i\cdot}, u_{j\cdot})$, where $u_{i\cdot}$ is the mass function u_{it}, $t \in T$. The required inequality may be hacked out, but instead we will use the maximal coupling of Exercises (4.12.4, 5); see also Problem (7.11.16). Thus requires a little notation. For $i, j \in S$, $i \ne j$, we find a pair (X_i, X_j) of random variables taking values in T according to the marginal mass functions $u_{i\cdot}$, $u_{j\cdot}$, and such that $\mathbb{P}(X_i \ne X_j) = \frac{1}{2} d_{\text{TV}}(u_{i\cdot}, u_{j\cdot})$. The existence of such a pair was proved in Exercise (4.12.5). Note that the value of X_i depends on j, but this fact has been suppressed from the notation for ease of reading. Having found (X_i, X_j), we find a pair $(Y(X_i), Y(X_j))$ taking values in U according to the marginal mass functions $v_{X_i\cdot}$, $v_{X_j\cdot}$, and such that $\mathbb{P}(Y(X_i) \ne Y(X_j) \mid X_i, X_j) = \frac{1}{2} d_{\text{TV}}(v_{X_i\cdot}, v_{X_j\cdot})$. Now, taking a further liberty with the notation,

$$\begin{aligned}
\mathbb{P}\big(Y(X_i) \ne Y(X_j)\big) &= \sum_{\substack{r,s \in S \\ r \ne s}} \mathbb{P}(X_i = r,\ X_j = s)\mathbb{P}\big(Y(r) \ne Y(s)\big) \\
&= \sum_{\substack{r,s \in S \\ r \ne s}} \mathbb{P}(X_i = r,\ X_j = s)\tfrac{1}{2} d_{\text{TV}}(v_{r\cdot}, v_{s\cdot}) \\
&\le \big\{\tfrac{1}{2}\sup_{r \ne s} d_{\text{TV}}(v_{r\cdot}, v_{s\cdot})\big\}\mathbb{P}(X_i \ne X_j),
\end{aligned}$$

whence

$$d(\mathbf{UV}) = \sup_{i \ne j} \mathbb{P}\big(Y(X_i) \ne Y(X_j)\big) \le \big\{\tfrac{1}{2}\sup_{r \ne s} d_{\text{TV}}(v_{r\cdot}, v_{s\cdot})\big\}\big\{\sup_{i,j} \mathbb{P}(X_i \ne X_j)\big\}$$

and the claim follows.

(b) Write $S = \{1, 2, \dots, m\}$, and take

$$\mathbf{U} = \begin{pmatrix} \mathbb{P}(X_0 = 1) & \mathbb{P}(X_0 = 2) & \cdots & \mathbb{P}(X_0 = m) \\ \mathbb{P}(Y_0 = 1) & \mathbb{P}(Y_0 = 2) & \cdots & \mathbb{P}(Y_0 = m) \end{pmatrix}.$$

The claim follows by repeated application of the result of part (a).

It may be shown with the aid of a little matrix theory that the second largest eigenvalue of a finite stochastic matrix $\mathbf{P}$ is no larger in modulus that $d(\mathbf{P})$; cf. the equation prior to Theorem (6.14.9).

5. (a) The Markov chains beginning at distinct $i, j \in \Theta$ evolve independently until they meet, and after meeting they stick together. Since the state space is finite, and the chain is aperiodic and irreducible, they meet a.s. in finite time. This holds for all pairs of starting states, and the claim follows.

(b) In the first example, let W be a random permutation of Θ. This requires $\mathbf{P}$ to be doubly stochastic, and so the stationary distribution is the constant vector. In the second example, take $\Theta = \{1, 2, 3\}$ and

$$\mathbf{P} = \begin{pmatrix} 0 & \frac{1}{2} & \frac{1}{2} \\ 1 & 0 & 0 \\ 1 & 0 & 0 \end{pmatrix}.$$

It is immediate that the chains starting at 2 and 3 coalesce in one step, but the chain starting at 1 never meets either. The chain is periodic with period 2, and this weakness is fixed by considering the minor 'lazy' variant of Exercise (6.5.12) with transition matrix $\frac{1}{2}\mathbf{P} + \frac{1}{2}\mathbf{I}$ where $\mathbf{I}$ is the identity matrix.

6. In the notation of Example (6.14.2), let

$$\alpha(\theta) = \sum_{v \sim w} \theta_v \theta_w, \qquad \theta \in \Theta,$$

and let us prove that

$$(*) \qquad \alpha(\theta \vee \psi) - \alpha(\theta) \geq \alpha(\psi) - \alpha(\theta \wedge \psi), \qquad \theta, \psi \in \Theta,$$

from which the FKG lattice condition (6.14.20) follows by exponentiation. Let $\theta, \psi \in \Theta$, and $D = \{v \in V : \theta_v \neq \psi_v\}$. Evidently (*) holds when $D = \varnothing$, and we show first that (*) holds when $|D| = 1$. Suppose $D = \{v\}$, and assume without loss of generality that $\theta_v = -1$ and $\psi_v = +1$.

Let d be the degree of v (that is, the number of edges touching v) in the graph G. Let $N_+(\theta)$ (respectively, $N_-(\theta)$) be the number of neighbours w of v with $\theta_w = +1$ (respectively, $\theta_w = -1$). Since θ and ψ disagree only at v, and $\theta \wedge \psi \leq \psi$, it must be the case that $(\theta \wedge \psi)_v = -1$. Since $\psi_v = +1$, we have that

$$(**) \quad \alpha(\psi) - \alpha(\theta \wedge \psi) = \big(N_+(\psi) - N_-(\psi)\big) - \big(N_-(\psi) - N_+(\psi)\big) = 4N_+(\psi) - 2d,$$

and (*) may be written as $N_+(\theta \vee \psi) \geq N_+(\psi)$, which holds trivially since $\theta \vee \psi \geq \psi$.

One may either deduce the case of general D from the above by an induction, or argue directly as follows. Let $N_+^v(\theta)$ be given as above with the role of $v \in V$ highlighted. As in (**),

$$\alpha(\psi) - \alpha(\theta \wedge \psi) = 4 \sum_{v \in D} \big(N_+^v(\psi) - N_+^v(\theta \wedge \psi)\big).$$

Inequality (*) follows on noting that

$$N_+^v(\theta \vee \psi) - N_+^v(\theta) \geq N_+^v(\psi) - N_+^v(\theta \wedge \psi), \qquad v \in D.$$

This is best seen by considering each neighbour w of v in turn.

6.15 Solutions to problems

1. (a) The state 4 is absorbing. The state 3 communicates with 4, and is therefore transient. The set $\{1, 2\}$ is finite, closed, and aperiodic, and hence ergodic. We have that $f_{34}(n) = (\frac{1}{4})^{n-1}\frac{1}{2}$, so that $f_{34} = \sum_n f_{34}(n) = \frac{2}{3}$.

(b) The chain is irreducible with period 2. All states are positive recurrent. Solve the equation $\boldsymbol{\pi} = \boldsymbol{\pi}\mathbf{P}$ to find the stationary distribution $\boldsymbol{\pi} = \big(\frac{3}{8}, \frac{3}{16}, \frac{5}{16}, \frac{1}{8}\big)$ whence the mean recurrence times are $\frac{8}{3}, \frac{16}{3}, \frac{16}{5}, 8$, in order.

2. (a) Let $\mathbf{P}$ be the transition matrix, assumed to be doubly stochastic. Then

$$\sum_i p_{ij}(n) = \sum_i \sum_k p_{ik}(n-1)p_{kj} = \sum_k \Big(\sum_i p_{ik}(n-1)\Big) p_{kj}$$

whence, by induction, the n-step transition matrix $\mathbf{P}^n$ is doubly stochastic for all $n \geq 1$.

If j is not positive recurrent, then $p_{ij}(n) \to 0$ as $n \to \infty$, for all i, implying that $\sum_i p_{ij}(n) \to 0$, a contradiction. Therefore all states are positive recurrent.

If in addition the chain is irreducible and aperiodic then $p_{ij}(n) \to \pi_j$, where $\boldsymbol{\pi}$ is the unique stationary distribution. However, it is easy to check that $\boldsymbol{\pi} = (N^{-1}, N^{-1}, \dots, N^{-1})$ is a stationary distribution if $\mathbf{P}$ is doubly stochastic.

(b) Suppose the chain is recurrent. In this case there exists a positive root of the equation $\mathbf{x} = \mathbf{xP}$, this root being unique up to a multiplicative constant (see Theorem (6.4.6) and the forthcoming Problem (6.15.7)). Since the transition matrix is doubly stochastic, we may take $\mathbf{x} = \mathbf{1}$, the vector of 1's. By the above uniqueness of $\mathbf{x}$, there can exist no stationary distribution, and therefore the chain is null. We deduce that the chain cannot be positive recurrent.

3. By the Chapman–Kolmogorov equations,

$$p_{ii}(m+r+n) \geq p_{ij}(m)p_{jj}(r)p_{ji}(n), \qquad m, r, n \geq 0.$$

Choose two states i and j, and pick m and n such that $\alpha = p_{ij}(m)p_{ji}(n) > 0$. Then

$$p_{ii}(m+r+n) \geq \alpha p_{jj}(r).$$

Set $r = 0$ to find that $p_{ii}(m+n) > 0$, and so $d(i) \mid (m+n)$. If $d(i) \nmid r$ then $p_{ii}(m+r+n) = 0$, so that $p_{jj}(r) = 0$; therefore $d(i) \mid d(j)$. Similarly $d(j) \mid d(i)$, giving that $d(i) = d(j)$.

4. (a) See the solution to Exercise (6.3.9a).

(b) Let $i, j, r, s \in S$, and choose $N(i,r)$ and $N(j,s)$ according to part (a). Then

$$\mathbb{P}\big(Z_n = (r,s) \,\big|\, Z_0 = (i,j)\big) = p_{ir}(n)p_{js}(n) > 0$$

if $n \geq \max\{N(i,r), N(j,s)\}$, so that the chain is irreducible and aperiodic.

(c) Suppose $S = \{1, 2\}$ and

$$\mathbf{P} = \begin{pmatrix} 0 & 1 \\ 1 & 0 \end{pmatrix}.$$

In this case $\big\{\{1,1\},\{2,2\}\big\}$ and $\big\{\{1,2\},\{2,1\}\big\}$ are closed sets of states for the bivariate chain.

5. Clearly $\mathbb{P}(N = 0) = 1 - f_{ij}$, while, by conditioning on the time of the nth visit to j, we have that $\mathbb{P}(N \geq n+1 \mid N \geq n) = f_{jj}$ for $n \geq 1$, whence the answer is immediate. Now $\mathbb{P}(N = \infty) = 1 - \sum_{n=0}^{\infty} \mathbb{P}(N = n)$ which equals 1 if and only if $f_{ij} = f_{jj} = 1$.

6. Fix $i \neq j$ and let $m = \min\{n : p_{ij}(n) > 0\}$. If $X_0 = i$ and $X_m = j$ then there can be no intermediate visit to i (with probability one), since such a visit would contradict the minimality of m.

Suppose $X_0 = i$, and note that $(1 - f_{ji})p_{ij}(m) \leq 1 - f_{ii}$, since if the chain visits j at time m and subsequently does not return to i, then no return to i takes place at all. However $f_{ii} = 1$ if i is recurrent, so that $f_{ji} = 1$.

7. (a) We may take $S = \{0, 1, 2, \dots\}$. Note that $q_{ij}(n) \geq 0$, and

$$\sum_j q_{ij}(n) = 1, \qquad q_{ij}(n+1) = \sum_{l=0}^{\infty} q_{il}(1)q_{lj}(n),$$

whence $\mathbf{Q} = (q_{ij}(1))$ is the transition matrix of a Markov chain, and $\mathbf{Q}^n = (q_{ij}(n))$. This chain is recurrent since

$$\sum_n q_{ii}(n) = \sum_n p_{ii}(n) = \infty \qquad \text{for all } i,$$

and irreducible since i communicates with j in the new chain whenever j communicates with i in the original chain.

That

$$(*) \qquad g_{ij}(n) = \frac{x_j}{x_i} l_{ji}(n), \qquad i \neq j,\ n \geq 1,$$

is evident when $n = 1$ since both sides are $q_{ij}(1)$. Suppose it is true for $n = m$ where $m \geq 1$. Now

$$l_{ji}(m+1) = \sum_{k:k\neq j} l_{jk}(m) p_{ki}, \qquad i \neq j,$$

so that

$$\frac{x_j}{x_i} l_{ji}(m+1) = \sum_{k:k\neq j} g_{kj}(m) q_{ik}(1), \qquad i \neq j,$$

which equals $g_{ij}(m+1)$ as required.

(b) Sum $(*)$ over n to obtain that

$$(**) \qquad 1 = \frac{x_j}{x_i} \rho_i(j), \qquad i \neq j,$$

where $\rho_i(j)$ is the mean number of visits to i between two visits to j; we have used the fact that $\sum_n g_{ij}(n) = 1$, since the chain is recurrent (see Problem (6.15.6)). It follows that $x_i = x_0 \rho_i(0)$ for all i, and therefore $\mathbf{x}$ is unique up to scalar multiplication.

(c) The claim is trivial when $i = j$, and we assume therefore that $i \neq j$. Let $N_i(j)$ be the number of visits to i before reaching j for the first time, and write $\mathbb{P}_k$ and $\mathbb{E}_k$ for probability and expectation conditional on $X_0 = k$. Clearly, $\mathbb{P}_j(N_i(j) \geq r) = h_{ji}(1 - h_{ij})^{r-1}$ for $r \geq 1$, whence

$$\rho_i(j) = \mathbb{E}_j(N_i(j)) = \sum_{r=1}^{\infty} \mathbb{P}_j(N_i(j) \geq r) = \frac{h_{ji}}{h_{ij}}.$$

The claim follows by $(**)$.

8. (a) If such a Markov chain exists, then

$$u_n = \sum_{i=1}^{n} f_i u_{n-i}, \qquad n \geq 1,$$

where f_i is the probability that the first return of X to its recurrent starting point s takes place at time i. Certainly $u_0 = 1$.

Conversely, suppose u is a renewal sequence with respect to the collection $(f_m : m \geq 1)$. Let X be a Markov chain on the state space $S = \{0, 1, 2, \dots\}$ with transition matrix

$$p_{ij} = \begin{cases} \mathbb{P}(T \geq i+2 \mid T \geq i+1) & \text{if } j = i+1, \\ 1 - \mathbb{P}(T \geq i+2 \mid T \geq i+1) & \text{if } j = 0, \end{cases}$$

where T is a random variable having mass function $f_m = \mathbb{P}(T = m)$. With $X_0 = 0$, the chance that the first return to 0 takes place at time n is

$$\mathbb{P}\left(X_n = 0, \prod_1^{n-1} X_i \neq 0 \,\middle|\, X_0 = 0\right) = p_{01}p_{12}\cdots p_{n-2,n-1}p_{n-1,0}$$

$$= \left(1 - \frac{G(n+1)}{G(n)}\right)\prod_{i=1}^{n-1}\frac{G(i+1)}{G(i)}$$

$$= G(n) - G(n+1) = f_n$$

where $G(m) = \mathbb{P}(T \geq m) = \sum_{n=m}^{\infty} f_n$. Now $v_n = \mathbb{P}(X_n = 0 \mid X_0 = 0)$ satisfies

$$v_0 = 1, \qquad v_n = \sum_{i=1}^{n} f_i v_{n-i} \qquad \text{for } n \geq 1,$$

whence $v_n = u_n$ for all n.

(b) Let X and Y be the two Markov chains which are associated (respectively) with u and v in the above sense. We shall assume that X and Y are independent. The product $(u_n v_n : n \geq 1)$ is now the renewal sequence associated with the bivariate Markov chain (X_n, Y_n).

9. Of the first $2n$ steps, let there be i rightwards, j upwards, and k inwards. Now $\mathbf{X}_{2n} = \mathbf{0}$ if and only if there are also i leftwards, j downwards, and k outwards. The number of such possible combinations is $(2n)!/\{(i!\,j!\,k!)^2\}$, and each such combination has probability $(\frac{1}{6})^{2(i+j+k)} = (\frac{1}{6})^{2n}$. The first equality follows, and the second is immediate.

Now

$$(*) \qquad \mathbb{P}(\mathbf{X}_{2n} = \mathbf{0}) \leq \left(\frac{1}{2}\right)^{2n}\binom{2n}{n} M \sum_{i+j+k=n}\frac{n!}{3^n i!\,j!\,k!}$$

where

$$M = \max\left\{\frac{n!}{3^n i!\,j!\,k!} : i, j, k \geq 0,\ i + j + k = n\right\}.$$

It is not difficult to see that the maximum M is attained when i, j, and k are all closest to $\frac{1}{3}n$, so that

$$M \leq \frac{n!}{3^n(\lfloor\frac{1}{3}n\rfloor!)^3}.$$

Furthermore the summation in $(*)$ equals 1, since the summand is the probability that, in allocating n balls randomly to three urns, the urns contain respectively i, j, and k balls. It follows that

$$\mathbb{P}(\mathbf{X}_{2n} = \mathbf{0}) \leq \frac{(2n)!}{12^n n!\,(\lfloor\frac{1}{3}n\rfloor!)^3}$$

which, by an application of Stirling's formula, is no bigger than $Cn^{-\frac{3}{2}}$ for some constant C. Hence $\sum_n \mathbb{P}(\mathbf{X}_{2n} = \mathbf{0}) < \infty$, so that the origin is transient.

10. No. The line of ancestors of any cell-state is a random walk in three dimensions. The difference between two such lines of ancestors is also a type of random walk, which in three dimensions is transient.

11. There are one or two absorbing states according as whether one or both of α and β equal zero. If $\alpha\beta \neq 0$, the chain is irreducible and recurrent. It is periodic if and only if $\alpha = \beta = 1$, in which case it has period 2.

If $0 < \alpha\beta < 1$ then

$$\boldsymbol{\pi} = \left(\frac{\beta}{\alpha+\beta}, \frac{\alpha}{\alpha+\beta}\right)$$

is the stationary distribution. There are various ways of calculating $\mathbf{P}^n$; see Exercise (6.3.3) for example. In this case the answer is given by

$$(\alpha+\beta)\mathbf{P}^n = (1-\alpha-\beta)^n \begin{pmatrix} \alpha & -\alpha \\ -\beta & \beta \end{pmatrix} + \begin{pmatrix} \beta & \alpha \\ \beta & \alpha \end{pmatrix};$$

proof by induction. Hence

$$(\alpha+\beta)\mathbf{P}^n \to \begin{pmatrix} \beta & \alpha \\ \beta & \alpha \end{pmatrix} \qquad \text{as } n \to \infty.$$

The chain is reversible in equilibrium if and only if $\pi_1 p_{12} = \pi_2 p_{21}$, which is to say that $\alpha\beta = \beta\alpha$!

12. The transition matrix is given by

$$p_{ij} = \begin{cases} \left(\dfrac{N-i}{N}\right)^2 & \text{if } j = i+1, \\ 1 - \left(\dfrac{i}{N}\right)^2 - \left(\dfrac{N-i}{N}\right)^2 & \text{if } j = i, \\ \left(\dfrac{i}{N}\right)^2 & \text{if } j = i-1, \end{cases}$$

for $0 \le i \le N$. This process is a birth–death process in discrete time, and by Exercise (6.5.1) is reversible in equilibrium. Its stationary distribution satisfies the detailed balance equation $\pi_i p_{i,i+1} = \pi_{i+1} p_{i+1,i}$ for $0 \le i < N$, whence $\pi_i = \pi_0 \binom{N}{i}^2$ for $0 \le i \le N$, where

$$\frac{1}{\pi_0} = \sum_{i=0}^{N} \binom{N}{i}^2 = \binom{2N}{N}.$$

13. (a) The chain X is irreducible; all states are therefore of the same type. The state 0 is aperiodic, and so therefore is every other state. Suppose that $X_0 = 0$, and let T be the time of the first return to 0. Then $\mathbb{P}(T > n) = a_0 a_1 \cdots a_{n-1} = b_n$ for $n \ge 1$, so that 0 is recurrent if and only if $b_n \to 0$ as $n \to \infty$.

(b) The mean of T is

$$\mathbb{E}(T) = \sum_{n=0}^{\infty} \mathbb{P}(T > n) = \sum_{n=0}^{\infty} b_n.$$

The stationary distribution $\boldsymbol{\pi}$ satisfies

$$\pi_0 = \sum_{k=0}^{\infty} \pi_k (1 - a_k), \qquad \pi_n = \pi_{n-1} a_{n-1} \text{ for } n \ge 1.$$

Hence $\pi_n = \pi_0 b_n$ and $\pi_0^{-1} = \sum_{n=0}^{\infty} b_n$ if this sum is finite.

(c) Suppose a_i has the stated form for $i \geq I$. Then

$$b_n = b_I \prod_{i=I}^{n-1} (1 - Ai^{-\beta}), \qquad n \geq I.$$

Hence $b_n \to 0$ if and only if $\sum_i Ai^{-\beta} = \infty$, which is to say that $\beta \leq 1$. The chain is therefore recurrent if and only if $\beta \leq 1$.

(d) We have that $1 - x \leq e^{-x}$ for $x \geq 0$, and therefore

$$\sum_{n=I}^{\infty} b_n \leq b_I \sum_{n=I}^{\infty} \exp\left\{-A \sum_{i=I}^{n-1} i^{-\beta}\right\} \leq b_I \sum_{n=I}^{\infty} \exp\left\{-An \cdot n^{-\beta}\right\} < \infty \quad \text{if } \beta < 1.$$

(e) If $\beta = 1$ and $A > 1$, there is a constant c_I such that

$$\sum_{n=I}^{\infty} b_n \leq b_I \sum_{n=I}^{\infty} \exp\left\{-A \sum_{i=I}^{n-1} \frac{1}{i}\right\} \leq c_I \sum_{n=I}^{\infty} \exp\{-A \log n\} = c_I \sum_{n=I}^{\infty} n^{-A} < \infty,$$

giving that the chain is positive.

(f) If $\beta = 1$ and $A \leq 1$,

$$b_n = b_I \prod_{i=I}^{n-1} \left(1 - \frac{A}{i}\right) \geq b_I \prod_{i=I}^{n-1} \left(\frac{i-1}{i}\right) = b_I \left(\frac{I-1}{n-1}\right).$$

Therefore $\sum_n b_n = \infty$, and the chain is null.

14. By Lemma (6.10.23), the transition probabilities $p_{ij}(t)$ are uniformly continuous in t. Now $\log x$ is continuous for $0 < x \leq 1$, and therefore g is continuous. Certainly $g(0) = 0$. In addition $p_{ii}(s+t) \geq p_{ii}(s)p_{ii}(t)$ for $s, t \geq 0$, whence $g(s+t) \leq g(s) + g(t)$, $s, t \geq 0$.

For the last part

$$\frac{1}{t}(p_{ii}(t) - 1) = \frac{g(t)}{t} \cdot \frac{p_{ii}(t) - 1}{-\log\{1 - (1 - p_{ii}(t))\}} \to -\lambda$$

as $t \downarrow 0$, since $x/\log(1-x) \to -1$ as $x \downarrow 0$.

15. We may assume that the generator is such that $g_i > 0$ for all i, since otherwise the claim is trivial. By Theorem (6.9.20), the continuous-time chain X is irreducible if and only if its jump chain Y is irreducible. Now Y is irreducible if and only if, for distinct $i, j \in S$, there exists a sequence $i, k_1, k_2, \dots, k_n, j$ of distinct states such that the transition matrix $\mathbf{Y} = (y_{uv} : u, v \in S)$ satisfies $y_{i,k_1} y_{k_1,k_2} \cdots y_{k_n,j} > 0$. By (6.9.7), $y_{uv} = g_{uv}/g_u$, and the claim follows.

16. Note that $\pi_i > 0$ for $i \in S$, by Lemma (6.10.15).

(a) (i) The given $\widehat{\mathbf{G}}$ is a generator, since it has non-negative off-diagonal entries, and row sums

$$\sum_i \widehat{g}_{ji} = \sum_i \frac{\pi_i}{\pi_j} g_{ij} = \frac{1}{\pi_j} \sum_i \pi_i g_{ij} = 0,$$

since $\boldsymbol{\pi}\mathbf{G} = \mathbf{0}$. Let $0 = t_0 < t_1 < t_2 < \cdots < t_n = T$ and $i_0, i_1, \dots, i_n \in S$. Then,

$$\begin{aligned}
\mathbb{P}\big(Y(t_0) = i_0, Y(t_1) = i_1, \dots, Y(t_n) = i_n\big) \\
&= \mathbb{P}\big(X(T) = i_0, X(T - t_1) = i_1, \dots, X(0) = i_n\big) \\
&= \pi_{i_n} p_{i_n, i_{n-1}}(s_n) p_{i_{n-1}, i_{n-2}}(s_{n-1}) \cdots p_{i_1, i_0}(s_1) \\
&= \pi_{i_0} \widehat{p}_{i_0, i_1}(s_1) \widehat{p}_{i_1, i_2}(s_2) \cdots \widehat{p}_{i_{n-1}, i_n}(s_n),
\end{aligned}$$

where $s_i = t_i - t_{i-1}$ and

$$\widehat{p}_{uv}(s) = \frac{\pi_v}{\pi_u} p_{vu}(s).$$

Since X is right-continuous, Y is left-continuous, and this needs fixing for Y to be a Markov chain.

The family $(\widehat{\mathbf{P}}_t)$ is the stochastic semigroup of the chain Y, which inherits its irreducibility and non-explosivity from X. It therefore satisfies the Kolmogorov backward equation with generator $\widehat{\mathbf{P}}_0'$, which is easily checked to equal $\widehat{\mathbf{G}}$.

(ii) Let X be reversible, so that X and Y have the same distributions. Then $\widehat{\mathbf{G}} = \mathbf{G}$, whence the detailed balance equations hold.

Conversely, if $\widehat{\mathbf{G}} = \mathbf{G}$, then X and Y have the same generator and the same initial distribution. Therefore, they have the same joint distributions, so that X is reversible.

(iii) If $\boldsymbol{\nu}$ satisfies the detailed balance equations, then $\sum_i \nu_i g_{ij} = \sum_i \nu_j g_{ji} = \nu_j \sum_i g_{ji} = 0$.

(b) (i) Let X be non-explosive with stationary distribution $\boldsymbol{\pi}$, and let $X(0)$ have distribution $\boldsymbol{\pi}$. The process X is reversible if and only if the detailed balance equations hold, which implies by iteration that

$$\pi_{k_1} g_{k_1,k_2} g_{k_2,k_3} \cdots g_{k_n,k_1} = g_{k_2,k_1} g_{k_3,k_2} \cdots g_{k_1,k_n} \pi_{k_1},$$

for any $k_1, k_2, \dots, k_n \in S$. The required equation follows on dividing by π_{k_1}.

Conversely, suppose the given equation holds. Fix a reference state i_0. For $i \in S$, by irreducibility and Problem (6.15.15), there exists a sequence $i, i_n, i_{n-1}, \dots, i_1, i_0$ of states (not necessarily distinct) such that $g_{i,i_n} g_{i_n,i_{n-1}} \cdots g_{i_1,i_0} > 0$, and we call such a sequence a *walk* from i to i_0. Define

$$\nu_i = \frac{g_{i_0,i_1} g_{i_1,i_2} \cdots g_{i_n,i}}{g_{i,i_n} g_{i_n,i_{n-1}} \cdots g_{i_1,i_0}}.$$

Note that ν_j does not depend on the choice of walk from i to i_0, since if $i, i'_m, i'_{m-1} \dots, i'_1, i_0$ is another such walk, then

$$\frac{g_{i_0,i_1} g_{i_1,i_2} \cdots g_{i_n,i}}{g_{i,i_n} g_{i_n,i_{n-1}} \cdots g_{i_1,i_0}} = \frac{g_{i_0,i'_1} g_{i'_1,i'_2} \cdots g_{i'_m,i}}{g_{i,i'_m} g_{i'_m,i'_{m-1}} \cdots g_{i'_1,i_0}}$$

by the given condition. We show next that the ν_i satisfy the detailed balance equations

$$(*) \qquad \nu_j g_{ji} = \nu_i g_{ij} \qquad \text{for } i \neq j.$$

Certainly (*) holds if $g_{ij} = g_{ji} = 0$. Assume $g_{ji} > 0$. We can express ν_j as

$$\nu_j = \frac{g_{i_0,i_1} g_{i_1,i_2} \cdots g_{i_n,i} g_{ij}}{g_{ji} g_{i,i_n} g_{i_n,i_{n-1}} \cdots g_{i_1,i_0}} = (g_{ij}/g_{ji})\nu_i,$$

as required.

Since $\boldsymbol{\nu}$ satisfies the detailed balance equations, it satisfies $\boldsymbol{\nu}\mathbf{G} = \mathbf{0}$. By Theorem (6.10.15), $\boldsymbol{\pi}$ is the unique measure satisfying this equation. Therefore, $\boldsymbol{\nu}$ and $\boldsymbol{\pi}$ differ by a scalar multiple, implying that $\boldsymbol{\pi}$ satisfies the detailed balance equations as required.

(ii) It is left to the reader to adapt the solution of Exercise (6.5.2b).

(c) Let $S = \{1, 2\}$ and

$$\mathbf{G} = \begin{pmatrix} -\alpha & \alpha \\ \beta & -\beta \end{pmatrix}$$

where $\alpha\beta > 0$. The chain is non-explosive with stationary distribution

$$\boldsymbol{\pi} = \left(\frac{\beta}{\alpha + \beta}, \frac{\alpha}{\alpha + \beta} \right),$$

and therefore $\pi_1 g_{12} = \pi_2 g_{21}$.
(d) Let X be a non-explosive birth–death process with birth rates λ_i and death rates μ_i. The stationary distribution $\boldsymbol{\pi}$ satisfies

$$\pi_1\mu_1 - \pi_0\lambda_0 = 0, \quad \pi_{k+1}\mu_{k+1} - \pi_k\lambda_k = \pi_k\mu_k - \pi_{k-1}\lambda_{k-1} \text{ for } k \geq 1.$$

Therefore $\pi_{k+1}\mu_{k+1} = \pi_k\lambda_k$ for $k \geq 0$, the detailed balance conditions.

17. Consider the continuous-time chain with generator

$$\mathbf{G} = \begin{pmatrix} -\beta & \beta \\ \gamma & -\gamma \end{pmatrix}.$$

It is a standard calculation (Exercise (6.9.1)) that the associated semigroup satisfies

$$(\beta+\gamma)\mathbf{P}_t = \begin{pmatrix} \gamma + \beta h(t) & \beta(1-h(t)) \\ \gamma(1-h(t)) & \beta + \gamma h(t) \end{pmatrix}$$

where $h(t) = e^{-t(\beta+\gamma)}$. Now $\mathbf{P}_1 = \mathbf{P}$ if and only if $\gamma + \beta h(1) = \beta + \gamma h(1) = \alpha(\beta+\gamma)$, which is to say that $\beta = \gamma = -\frac{1}{2}\log(2\alpha - 1)$, a solution which requires that $\alpha > \frac{1}{2}$.

18. The forward equations for $p_n(t) = \mathbb{P}(X(t) = n)$ are

$$p_0'(t) = \mu p_1 - \lambda p_0,$$
$$p_n'(t) = \lambda p_{n-1} - (\lambda + n\mu)p_n + \mu(n+1)p_{n+1}, \qquad n \geq 1.$$

In the usual way,

$$\frac{\partial G}{\partial t} = (s-1)\left(\lambda G - \mu\frac{\partial G}{\partial s}\right)$$

with boundary condition $G(s,0) = s^I$. The characteristics are given by

$$dt = \frac{ds}{\mu(s-1)} = \frac{dG}{\lambda(s-1)G},$$

and therefore $G = e^{\rho(s-1)} f\big((s-1)e^{-\mu t}\big)$, for some function f, determined by the boundary condition to satisfy $e^{\rho(s-1)} f(s-1) = s^I$. The claim follows.

As $t \to \infty$, $G(s,t) \to e^{\rho(s-1)}$, the generating function of the Poisson distribution, parameter ρ.

19. (a) The forward equations are

$$\frac{\partial}{\partial t} p_{ii}(s,t) = -\lambda(t)p_{ii}(s,t),$$
$$\frac{\partial}{\partial t} p_{ij}(s,t) = -\lambda(t)p_{ij}(s,t) + \lambda(t)p_{i,j-1}(t), \qquad i < j.$$

Assume $N(s) = i$ and $s < t$. In the usual way,

$$G(s,t;x) = \sum_{j=i}^{\infty} x^j \mathbb{P}\big(N(t) = j \mid N(s) = i\big)$$

satisfies

$$\frac{\partial G}{\partial t} = \lambda(t)(x-1)G.$$

We integrate subject to the boundary condition to obtain

$$G(s, t; x) = x^i \exp\left\{(x-1)\int_s^t \lambda(u)\,du\right\},$$

whence $p_{ij}(t)$ is found to be the probability that $A = j - i$ where A has the Poisson distribution with parameter $\int_s^t \lambda(u)\,du$.

The backward equations are

$$\frac{\partial}{\partial s} p_{ij}(s,t) = \lambda(s)p_{i+1,j}(s,t) - \lambda(s)p_{ij}(s,t);$$

using the fact that $p_{i+1,j}(t) = p_{i,j-1}(t)$, we are led to

$$-\frac{\partial G}{\partial s} = \lambda(s)(x-1)G.$$

The solution is the same as above.
(b) We have that

$$\mathbb{P}(T > t) = p_{00}(t) = \exp\left\{-\int_0^t \lambda(u)\,du\right\},$$

so that

$$f_T(t) = \lambda(t)\exp\left\{-\int_0^t \lambda(u)\,du\right\}, \qquad t \geq 0.$$

In the case $\lambda(t) = c/(1+t)$,

$$\mathbb{E}(T) = \int_0^\infty \mathbb{P}(T > t)\,dt = \int_0^\infty \frac{du}{(1+u)^c}$$

which is finite if and only if $c > 1$.

20. Let $s > 0$. Each offer has probability $1 - F(s)$ of exceeding s, and therefore the first offer exceeding s is the Mth offer overall, where $\mathbb{P}(M = m) = F(s)^{m-1}[1 - F(s)]$, $m \geq 1$. Conditional on $\{M = m\}$, the value of X_M is independent of the values of $X_1, X_2, \dots, X_{M-1}$, with

$$\mathbb{P}(X_M > u \mid M = m) = \frac{1 - F(u)}{1 - F(s)}, \qquad 0 < s \leq u,$$

and $X_1, X_2, \dots, X_{M-1}$ have shared (conditional) distribution function

$$G(u \mid s) = \frac{F(u)}{F(s)}, \qquad 0 \leq u \leq s.$$

For any event B defined in terms of $X_1, X_2, \dots, X_{M-1}$, we have that

$$\begin{aligned}
\mathbb{P}(X_M > u, B) &= \sum_{m=1}^\infty \mathbb{P}(X_M > u, B \mid M = m)\mathbb{P}(M = m) \\
&= \sum_{m=1}^\infty \mathbb{P}(X_M > u \mid M = m)\mathbb{P}(B \mid M = m)\mathbb{P}(M = m) \\
&= \mathbb{P}(X_M > u)\sum_{m=1}^\infty \mathbb{P}(B \mid M = m)\mathbb{P}(M = m) \\
&= \mathbb{P}(X_M > u)\mathbb{P}(B), \qquad 0 < s \leq u,
\end{aligned}$$

where we have used the fact that $\mathbb{P}(X_M > u \mid M = m)$ is independent of m. It follows that the first record value exceeding s is independent of all record values not exceeding s. By a similar argument (or an iteration of the above) all record values exceeding s are independent of all record values not exceeding s.

The chance of a record value in $(s, s+h]$ is

$$\mathbb{P}(s < X_M \le s+h) = \frac{F(s+h) - F(s)}{1 - F(s)} = \frac{f(s)h}{1-F(s)} + \mathrm{o}(h).$$

A very similar argument works for the runners-up. Let $X_{M_1}, X_{M_2}, \dots$ be the values, in order, of offers exceeding s. It may be seen that this sequence is independent of the sequence of offers not exceeding s, whence it follows that the sequence of runners-up is a non-homogeneous Poisson process. There is a runner-up in $(s, s+h]$ if (neglecting terms of order $\mathrm{o}(h)$) the first offer exceeding s is larger than $s+h$, and the second is in $(s, s+h]$. The probability of this is

$$\left(\frac{1-F(s+h)}{1-F(s)}\right)\left(\frac{F(s+h)-F(s)}{1-F(s)}\right) + \mathrm{o}(h) = \frac{f(s)h}{1-F(s)} + \mathrm{o}(h).$$

21. Let $F_t(x) = \mathbb{P}(N^*(t) \le x)$, and let A be the event that N has a arrival during $(t, t+h)$. Then

$$F_{t+h}(x) = \lambda h \mathbb{P}\big(N^*(t+h) \le x \bigm| A\big) + (1-\lambda h)F_t(x) + \mathrm{o}(h)$$

where

$$\mathbb{P}\big(N^*(t+h) \le x \bigm| A\big) = \int_{-\infty}^{\infty} F_t(x-y) f(y)\, dy.$$

Hence

$$\frac{\partial}{\partial t} F_t(x) = -\lambda F_t(x) + \lambda \int_{-\infty}^{\infty} F_t(x-y) f(y)\, dy.$$

Take Fourier transforms to find that $\phi_t(\theta) = \mathbb{E}(e^{i\theta N^*(t)})$ satisfies

$$\frac{\partial \phi_t}{\partial t} = -\lambda\phi_t + \lambda\phi_t\phi,$$

an equation which may be solved subject to $\phi_0(\theta) = 1$ to obtain $\phi_t(\theta) = e^{\lambda t(\phi(\theta)-1)}$.

Alternatively, using conditional expectation,

$$\phi_t(\theta) = \mathbb{E}\{\mathbb{E}(e^{i\theta N^*(t)} \mid N(t))\} = \mathbb{E}\{\phi(\theta)^{N(t)}\}$$

where $N(t)$ is Poisson with parameter λt.

22. (a) We have that

$$\mathbb{E}(s^{N(t)}) = \mathbb{E}\{\mathbb{E}(s^{N(t)} \mid \Lambda)\} = \tfrac{1}{2}\{e^{\lambda_1 t(s-1)} + e^{\lambda_2 t(s-1)}\},$$

whence $\mathbb{E}(N(t)) = \frac{1}{2}(\lambda_1+\lambda_2)t$ and $\mathrm{var}(N(t)) = \frac{1}{2}(\lambda_1+\lambda_2)t + \frac{1}{4}(\lambda_1-\lambda_2)^2t^2$.
(b) In the notation of (a), $G_{N(t)}(s) = \mathbb{E}(e^{\Lambda t(s-1)}) = M_\Lambda(t(s-1))$. Therefore,

$$\mathrm{var}(N(t)) = t^2 M_\Lambda''(0) + tM_\Lambda'(0) - t^2 M_\Lambda'(0)^2 = t^2\,\mathrm{var}(\Lambda) + \mathbb{E}(N(t)).$$

(c) Since the interarrival times of M^* are not exponentially distributed, M^* is not a Poisson process. Let I_t be the indicator function that $M(t)$ is odd. Then $M^* = \frac{1}{2}(M - I)$, so that $\mathbb{E}(M^*(t)) \sim \frac{1}{2}t$ and $\mathrm{var}(M^*(t)) \sim \frac{1}{4}t$ as $t \to \infty$. This violates the inequality of part (b) for large t.

23. Conditional on $\{X(t) = i\}$, the next arrival in the birth process takes place at rate λi.

24. The forward equations for $p_n(t) = \mathbb{P}(X(t) = n)$ are

$$p_n'(t) = \frac{1+\mu(n-1)}{1+\mu t} p_{n-1}(t) - \frac{1+\mu n}{1+\mu t} p_n(t), \qquad n \ge 0,$$

with the convention that $p_{-1}(t) = 0$. Multiply by s^n and sum to deduce that

$$(1+\mu t)\frac{\partial G}{\partial t} = sG + \mu s^2 \frac{\partial G}{\partial s} - G - \mu s \frac{\partial G}{\partial s}$$

as required.

Differentiate with respect to s and take the limit as $s \uparrow 1$. If $\mathbb{E}(X(t)^2) < \infty$, then

$$m(t) = \mathbb{E}(X(t)) = \left.\frac{\partial G}{\partial s}\right|_{s=1}$$

satisfies $(1+\mu t)m'(t) = 1 + \mu m(t)$ subject to $m(0) = I$. Solving this in the usual way, we obtain $m(t) = I + (1+\mu I)t$.

Differentiate again to find that

$$n(t) = \mathbb{E}\big(X(t)(X(t)-1)\big) = \left.\frac{\partial^2 G}{\partial s^2}\right|_{s=1}$$

satisfies $(1+\mu t)n'(t) = 2\big(m(t) + \mu m(t) + \mu n(t)\big)$ subject to $n(0) = I(I-1)$. The solution is

$$n(t) = I(I-1) + 2I(1+\mu I)t + (1+\mu I)(1+\mu+\mu I)t^2.$$

The variance of $X(t)$ is $n(t) + m(t) - m(t)^2$.

25. (a) Condition on the value of the first step:

$$\eta_j = \frac{\lambda_j}{\lambda_j+\mu_j} \cdot \eta_{j+1} + \frac{\mu_j}{\lambda_j+\mu_j} \cdot \eta_{j-1}, \qquad j \ge 1,$$

as required. Set $x_i = \eta_{i+1} - \eta_i$ to obtain $\lambda_j x_j = \mu_j x_{j-1}$ for $j \ge 1$, so that

$$x_j = x_0 \prod_{i=1}^{j} \frac{\mu_i}{\lambda_i}, \qquad j \ge 1.$$

It follows that

$$\eta_{j+1} = \eta_0 + \sum_{k=0}^{j} x_k = 1 + (\eta_1 - 1)\sum_{k=0}^{j} e_k.$$

The η_j are probabilities, and lie in $[0, 1]$. If $\sum_1^\infty e_k = \infty$ then we must have $\eta_1 = 1$, which implies that $\eta_j = 1$ for all j.

(b) By conditioning on the first step, the probability η_j, of visiting 0 having started from j, satisfies

$$\eta_j = \frac{(j+1)^2\eta_{j+1} + j^2\eta_{j-1}}{j^2 + (j+1)^2}.$$

Hence, $(j+1)^2(\eta_{j+1}-\eta_j) = j^2(\eta_j - \eta_{j-1})$, giving $(j+1)^2(\eta_{j+1}-\eta_j) = \eta_1 - \eta_0$. Therefore,

$$1 - \eta_{j+1} = (1-\eta_1)\sum_{k=0}^{j} \frac{1}{(k+1)^2} \to (1-\eta_1)\tfrac{1}{6}\pi^2 \qquad \text{as } j \to \infty.$$

By Exercise (6.3.6), we seek the minimal non-negative solution, which is achieved when $\eta_1 = 1 - (6/\pi^2)$.

26. We may suppose that $X(0) = 0$. Let $T_n = \inf\{t : X(t) = n\}$. Suppose $T_n = T$, and let $Y = T_{n+1} - T$. Condition on all possible occurrences during the interval $(T, T+h)$ to find that

$$\mathbb{E}(Y) = (\lambda_n h)h + \mu_n h(h + \mathbb{E}(Y')) + (1 - \lambda_n h - \mu_n h)(h + \mathbb{E}(Y)) + o(h),$$

where Y' is the mean time which elapses before reaching $n+1$ from $n-1$. Set $m_n = \mathbb{E}(T_{n+1} - T_n)$ to obtain that

$$m_n = \mu_n h(m_{n-1} + m_n) + m_n + h\{1 - (\lambda_n + \mu_n)m_n\} + o(h).$$

Divide by h and take the limit as $h \downarrow 0$ to find that $\lambda_n m_n = 1 + \mu_n m_{n-1}$, $n \geq 1$. Therefore

$$m_n = \frac{1}{\lambda_n} + \frac{\mu_n}{\lambda_n} m_{n-1} = \cdots = \frac{1}{\lambda_n} + \frac{\mu_n}{\lambda_n \lambda_{n-1}} + \cdots + \frac{\mu_n \mu_{n-1} \cdots \mu_1}{\lambda_n \lambda_{n-1} \cdots \lambda_0},$$

since $m_0 = \lambda_0^{-1}$. The process is dishonest if $\sum_{n=0}^{\infty} m_n < \infty$, since in this case $T_\infty = \lim T_n$ has finite mean, so that $\mathbb{P}(T_\infty < \infty) = 1$.

On the other hand, the process grows no faster than a birth process with birth rates λ_i, which is honest if $\sum_{n=0}^{\infty} 1/\lambda_n = \infty$. Can you find a better condition?

27. We know that, conditional on $X(0) = I$, $X(t)$ has generating function

$$G(s,t) = \left(\frac{\lambda t(1-s) + s}{\lambda t(1-s) + 1}\right)^I,$$

so that

$$\mathbb{P}(T \leq x \mid X(0) = I) = \mathbb{P}(X(x) = 0 \mid X(0) = I) = G(0,x) = \left(\frac{\lambda x}{\lambda x + 1}\right)^I.$$

It follows that, in the limit as $x \to \infty$,

$$\mathbb{P}(T \leq x) = \sum_{I=0}^{\infty} \left(\frac{\lambda x}{\lambda x + 1}\right)^I \mathbb{P}(X(0) = I) = G_{X(0)}\left(\frac{\lambda x}{\lambda x + 1}\right) \to 1.$$

For the final part, the required probability is $\{xI/(xI+1)\}^I = \{1 + (xI)^{-1}\}^{-I}$, which tends to $e^{-1/x}$ as $I \to \infty$.

28. Let Y be an immigration–death process *without* disasters, with $Y(0) = 0$. We have from Problem (6.15.18) that $Y(t)$ has generating function $G(s,t) = \exp\{\rho(s-1)(1-e^{-\mu t})\}$ where $\rho = \lambda/\mu$. As seen earlier, and as easily verified by taking the limit as $t \to \infty$, Y has a stationary distribution.

From the process Y we may generate the process X in the following way. At the epoch of each disaster, we paint every member of the population grey. At any given time, the unpainted individuals constitute X, and the aggregate population constitutes Y. When constructed in this way, it is the case that $Y(t) \leq X(t)$, so that Y is a Markov chain which is dominated by a chain having a stationary distribution. It follows that X has a stationary distribution $\boldsymbol{\pi}$ (the state 0 is recurrent for X, and therefore recurrent for Y also).

Suppose X is in equilibrium. The times of disasters form a Poisson process with intensity δ. At any given time t, the elapsed time T since the last disaster is exponentially distributed with parameter δ. At the time of this disaster, the value of $X(t)$ is reduced to 0 whatever its previous value.

It follows by averaging over the value of T that the generating function $H(s) = \sum_{n=0}^{\infty} s^n \pi_n$ of $X(t)$ is given by

$$H(s) = \int_0^\infty \delta e^{-\delta u} G(s,u)\,du = \frac{\delta}{\mu} e^{\rho(s-1)} \int_0^1 x^{(\delta/\mu)-1} e^{-\rho(s-1)x}\,dx$$

by the substitution $x = e^{-\mu u}$. The mean of $X(t)$ is

$$H'(1) = \int_0^\infty \delta e^{-\delta u} \mathbb{E}(Y(u))\,du = \int_0^\infty \delta e^{-\delta u} \rho(1-e^{-\mu u})\,du = \frac{\rho\mu}{\delta+\mu} = \frac{\lambda}{\delta+\mu}.$$

29. Let $G(|B|, s)$ be the generating function of $X(B)$. If $B \cap C = \varnothing$, then $X(B \cup C) = X(B) + X(C)$, so that $G(\alpha+\beta, s) = G(\alpha, s)G(\beta, s)$ for $|s| \le 1, \alpha, \beta \ge 0$. The only solutions to this equation which are monotone in α are of the form $G(\alpha, s) = e^{\alpha\lambda(s)}$ for $|s| \le 1$, and for some function $\lambda(s)$. Now any interval may be divided into n equal sub-intervals, and therefore $G(\alpha, s)$ is the generating function of an infinitely divisible distribution. Using the result of Problem (5.12.13b), $\lambda(s)$ may be written in the form $\lambda(s) = (A(s) - 1)\lambda$ for some λ and some probability generating function $A(s) = \sum_0^\infty a_i s^i$. We now use (iii): if $|B| = \alpha$,

$$\frac{\mathbb{P}(X(B) \ge 1)}{\mathbb{P}(X(B) = 1)} = \frac{1 - e^{\alpha\lambda(a_0-1)}}{\alpha\lambda a_1 e^{\alpha\lambda(a_0-1)}} \to 1$$

as $\alpha \downarrow 0$. Therefore $a_0 + a_1 = 1$, and hence $A(s) = a_0 + (1-a_0)s$, and $X(B)$ has a Poisson distribution with parameter proportional to $|B|$.

30. (a) Let $M(r, s)$ be the number of points of the resulting process on $\mathbb{R}_+$ lying in the interval $(r, s]$. Since disjoint intervals correspond to disjoint annuli of the plane, the process M has independent increments in the sense that $M(r_1, s_1), M(r_2, s_2), \dots, M(r_n, s_n)$ are independent whenever $r_1 < s_1 < r_2 < \cdots < r_n < s_n$. Furthermore, for $r < s$ and $k \ge 0$,

$$\mathbb{P}\big(M(r,s) = k\big) = \mathbb{P}\big(N \text{ has } k \text{ points in the corresponding annulus}\big) = \frac{\{\lambda\pi(s-r)\}^k e^{-\lambda\pi(s-r)}}{k!}.$$

(b) We have similarly that

$$\mathbb{P}(R_{(k)} \le x) = \mathbb{P}(N \text{ has least } k \text{ points in circle of radius } x) = \sum_{r=k}^{\infty} \frac{(\lambda\pi x^2)^r e^{-\lambda\pi x^2}}{r!},$$

and the claim follows by differentiating, and utilizing the successive cancellation.

31. The number $X(S)$ of points within the sphere with volume S and centre at the origin has the Poisson distribution with parameter λS. Hence $\mathbb{P}(X(S) = 0) = e^{-\lambda S}$, implying that the volume V of the largest such empty ball has the exponential distribution with parameter λ.

It follows that $\mathbb{P}(R > r) = \mathbb{P}(V > cr^n) = e^{-\lambda c r^n}$ for $r \ge 0$, where c is the volume of the unit ball in n dimensions. Therefore

$$f_R(r) = \lambda n c r^{n-1} e^{-\lambda c r^n}, \qquad r \ge 0.$$

Finally, $\mathbb{E}(R) = \int_0^\infty e^{-\lambda c r^n}\,dr$, and we set $v = \lambda c r^n$.

32. The time between the kth and $(k+1)$th infection has mean λ_k^{-1}, whence

$$\mathbb{E}(T) = \sum_{k=1}^{N} \frac{1}{\lambda_k}.$$

Now

$$\sum_{k=1}^{N} \frac{1}{k(N+1-k)} = \frac{1}{N+1}\left\{\sum_{k=1}^{N}\frac{1}{k} + \sum_{k=1}^{N}\frac{1}{N+1-k}\right\}$$
$$= \frac{2}{N+1}\sum_{k=1}^{N}\frac{1}{k} = \frac{2}{N+1}\{\log N + \gamma + \mathrm{O}(N^{-1})\}.$$

It may be shown with more work (as in the solution to Problem (5.12.34)) that the moment generating function of $\lambda(N+1)T - 2\log N$ converges as $N \to \infty$, the limit being $\{\Gamma(1-\theta)\}^2$.

33. (a) The forward equations for $p_n(t) = \mathbb{P}(V(t) = n + \frac{1}{2})$ are

$$p_n'(t) = (n+1)p_{n+1}(t) - (2n+1)p_n(t) + np_{n-1}(t), \qquad n \ge 0,$$

with the convention that $p_{-1}(t) = 0$. It follows as usual that

$$\frac{\partial G}{\partial t} = \frac{\partial G}{\partial s} - \left(2s\frac{\partial G}{\partial s} + G\right) + \left(s^2\frac{\partial G}{\partial s} + sG\right)$$

as required. The general solution is

$$G(s,t) = \frac{1}{1-s} f\left(t + \frac{1}{1-s}\right)$$

for some function f. The boundary condition is $G(s,0) = 1$, and the solution is as given.

(b) Clearly

$$m_n(T) = \mathbb{E}\left(\int_0^T I_{nt}\,dt\right) = \int_0^T \mathbb{E}(I_{nt})\,dt$$

by Fubini's theorem, where I_{nt} is the indicator function of the event that $V(t) = n + \frac{1}{2}$.

As for the second part,

$$\sum_{n=0}^{\infty} s^n m_n(T) = \int_0^T G(s,t)\,dt = \frac{\log[1+(1-s)T]}{1-s},$$

so that, in the limit as $T \to \infty$,

$$\sum_{n=0}^{\infty} s^n\big(m_n(T) - \log T\big) = \frac{1}{1-s}\log\left(\frac{1+(1-s)T}{T}\right) \to \frac{\log(1-s)}{1-s} = -\sum_{n=1}^{\infty} s^n a_n$$

if $|s| < 1$, where $a_n = \sum_{i=1}^{n} i^{-1}$, as required.

(c) The mean velocity at time t is

$$\frac{1}{2} + \left.\frac{\partial G}{\partial s}\right|_{s=1} = t + \frac{1}{2}.$$

34. (a) No, since

$$\mathbb{P}\big(X_{n+1} = 1 \,\big|\, X_n = 1, X_{n-1} = a\big) = \begin{cases} 1 & \text{if } a = 0. \\ 0 & \text{if } a = 1. \end{cases}$$

(b) Make a map of the state space, with transitions and probabilities as indicated below. Started at $Y_1 = (0, 1)$, the sequence X_n visits 0 before 3 if and only if Y makes some number m, say, of traversals of the triangle labelled $*$, before heading for the pair $(1, 0)$. Each such traversal has probability $\frac{1}{8}$, so that

$$\mathbb{P}(3 \text{ before } 0) = 1 - \tfrac{1}{2} \sum_{m=0}^{\infty} \left(\tfrac{1}{8}\right)^m = \tfrac{3}{7}.$$

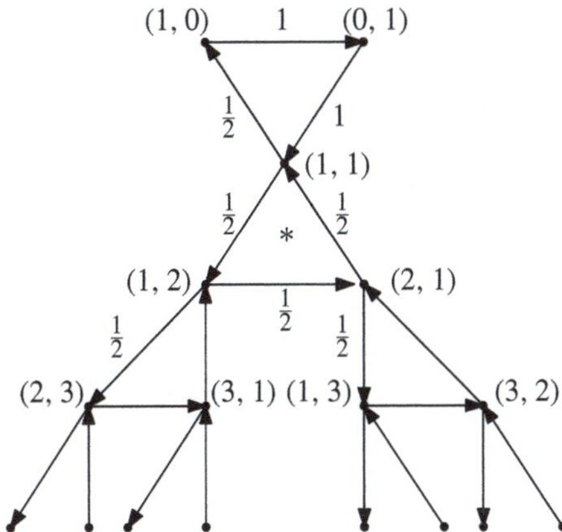

(c) Let the hitting probability of $(1, 1)$ be λ from $(1, 2)$ and μ from $(2, 1)$. The diagram below $(1, 1)$ has a self-similar structure, so that, by conditioning on the first step,

$$\lambda = \tfrac{1}{2}\mu + \tfrac{1}{2}\lambda^2, \qquad \mu = \tfrac{1}{2} + \tfrac{1}{2}\lambda\mu.$$

Solving for λ, we find $\lambda^3 - 4\lambda^2 + 4\lambda - 1 = 0$, with roots $\lambda = 1, \frac{1}{2}(3 \pm \sqrt{5})$. The only roots that can be probabilities are $\lambda = 1, \frac{1}{2}(3 - \sqrt{5})$, and the value $\lambda = 1$ is eliminated as follows. Consider the map from the state $(1, 1)$ downwards. Every step downwards corresponds to an addition, and every step rightwards or upwards to a subtraction. Let D be the set of states reached from $(1, 1)$ by sequences of moves ending in an addition. The imbedded process on D is a transient random walk. Therefore, Y is transient. On the other hand, if $\lambda = 1$, then $\mu = 1$, and the chain is recurrent, a contradiction.

35. We write A, 1, 2, 3, 4, 5 for the vertices of the hexagon in clockwise order. Let $T_i = \min\{n \geq 1 : X_n = i\}$ and $\mathbb{P}_i(\cdot) = \mathbb{P}(\cdot \mid X_0 = i)$.

(a) By symmetry, the probabilities $p_i = \mathbb{P}_i(T_\text{A} < T_\text{C})$ satisfy

$$p_\text{A} = \tfrac{2}{3}p_1, \quad p_1 = \tfrac{1}{3} + \tfrac{1}{3}p_2, \quad p_2 = \tfrac{1}{3}p_1 + \tfrac{1}{3}p_3, \quad p_3 = \tfrac{2}{3}p_2,$$

whence $p_\text{A} = \frac{7}{27}$.

(b) By Exercise (6.4.6), the stationary distribution is $\pi_\text{C} = \frac{1}{4}$, $\pi_i = \frac{1}{8}$ for $i \neq \text{C}$, whence $\mu_\text{A} = \pi_\text{A}^{-1} = 8$.

(c) By the argument leading to Lemma (6.4.7), this equals $\mu_\text{A}\pi_\text{C} = 2$.

(d) We condition on the event $E = \{T_\text{A} < T_\text{C}\}$. The probabilities $b_i = \mathbb{P}_i(E)$ satisfy

$$b_1 = \tfrac{1}{3} + \tfrac{1}{3}b_2, \quad b_2 = \tfrac{1}{3}b_1 + \tfrac{1}{3}b_3, \quad b_3 = \tfrac{2}{3}b_2,$$

yielding $b_1 = \frac{7}{18}$, $b_2 = \frac{1}{6}$, $b_3 = \frac{1}{9}$. The transition probabilities conditional on E are now found by equations of the form

$$\tau_{12} = \frac{\mathbb{P}_2(E)p_{12}}{\mathbb{P}_1(E)} = \frac{b_2}{3b_1} = \frac{1}{7},$$

and similarly $\tau_{21} = \frac{7}{9}$, $\tau_{23} = \frac{2}{9}$, $\tau_{32} = \frac{1}{2}$, $\tau_{1A} = \frac{6}{7}$. Hence, with the obvious notation,

$$\mu_{2A} = 1 + \tfrac{7}{9}\mu_{1A} + \tfrac{2}{9}\mu_{3A}, \quad \mu_{3A} = 1 + \mu_{2A}, \quad \mu_{1A} = 1 + \tfrac{1}{7}\mu_{2A},$$

giving $\mu_{1A} = \frac{10}{7}$, and the required answer is $1 + \mu_{1A} = 1 + \frac{10}{7} = \frac{17}{7}$.

36. (a) We have that

$$p_{i,i+1} = \frac{\beta(m-i)^2}{m^2}, \quad p_{i+1,i} = \frac{\alpha(i+1)^2}{m^2}.$$

Look for a solution to the detailed balance equations

$$\pi_i \frac{\beta(m-i)^2}{m^2} = \pi_{i+1}\frac{\alpha(i+1)^2}{m^2}$$

to find the stationary distribution

$$\pi_i = \binom{m}{i}^2 (\beta/\alpha)^i \pi_0, \quad \text{where} \quad \pi_0 = \left\{\sum_{i=0}^{m} \binom{m}{i}^2 (\beta/\alpha)^i\right\}^{-1}.$$

(b) In this case,

$$p_{i,i+1} = \frac{\beta(m-i)}{m}, \quad p_{i+1,i} = \frac{\alpha(i+1)}{m}.$$

Look for a solution to the detailed balance equations

$$\pi_i \frac{\beta(m-i)}{m} = \pi_{i+1}\frac{\alpha(i+1)}{m},$$

yielding the stationary distribution

$$\pi_i = (\beta/\alpha)^i \binom{m}{i} \pi_0, \quad \text{where} \quad \pi_0 = \left\{\sum_{i=0}^{m} \binom{m}{i} (\beta/\alpha)^i\right\}^{-1} = \left(\frac{\alpha}{\alpha+\beta}\right)^m.$$

37. We have that

$$\begin{aligned}
d(s+t) &= \sum_k \pi_k c\left(\frac{a_k(s+t)}{\pi_k}\right) \\
&= \sum_k \pi_k c\left(\sum_j \frac{a_j(s)p_{jk}(t)}{\pi_j}\frac{\pi_j}{\pi_k}\right) && \text{by the Chapman–Kolmogorov equations} \\
&\ge \sum_k \pi_k \sum_j \frac{\pi_j p_{jk}(t)}{\pi_k} c\left(\frac{a_j(s)}{\pi_j}\right) && \text{by the concavity of } c \\
&= \sum_j \pi_j c\left(\frac{a_j(s)}{\pi_j}\right) = d(s),
\end{aligned}$$

where we have used the fact that $\sum_j \pi_j p_{jk}(t) = \pi_k$. Now $a_j(s) \to \pi_j$ as $s \to \infty$, and therefore $d(t) \to c(1)$.

We apply the first part with $c(x) = -x \log x$ for $x \in (0, 1]$ to find that

$$-D(a(t); \boldsymbol{\pi}) = -\sum_j \pi_j \frac{a_j(t)}{\pi_j} \log \frac{a_j(t)}{\pi_j} \uparrow 0 \qquad \text{as } t \to \infty.$$

38. By the Chapman–Kolmogorov equations and the reversibility,

$$\begin{aligned} u_0(2t) &= \sum_j \mathbb{P}\big(X(2t) = 0 \,\big|\, X(t) = j\big)\mathbb{P}\big(X(t) = j \,\big|\, X(0) = 0\big) \\ &= \sum_j \frac{\pi_0}{\pi_j}\mathbb{P}\big(X(2t) = j \,\big|\, X(t) = 0\big)u_j(t) = \pi_0 \sum_j \pi_j \left(\frac{u_j(t)}{\pi_j}\right)^2. \end{aligned}$$

The function $c(x) = -x^2$ is concave, and the claim follows by the result of the previous problem.

39. This may be done in a variety of ways, by breaking up the distribution of a typical displacement and using the superposition theorem (6.13.5), by the colouring theorem (6.13.14), or by Rényi's theorem (6.13.17) as follows. Let B be a closed bounded region of $\mathbb{R}^d$. We colour a point of Π at $\mathbf{x} \in \mathbb{R}^d$ *black* with probability $\mathbb{P}(\mathbf{x} + X \in B)$, where X is a typical displacement. By the colouring theorem, the number of black points has a Poisson distribution with parameter

$$\begin{aligned} \int_{\mathbb{R}^d} \lambda \mathbb{P}(\mathbf{x} + X \in B)\, d\mathbf{x} &= \lambda \int_{\mathbf{y} \in B} d\mathbf{y} \int_{\mathbf{x} \in \mathbb{R}^d} \mathbb{P}(X \in d\mathbf{y} - \mathbf{x}) \\ &= \lambda \int_{\mathbf{y} \in B} d\mathbf{y} \int_{\mathbf{v} \in \mathbb{R}^d} \mathbb{P}(X \in d\mathbf{v}) = \lambda |B|, \end{aligned}$$

by the change of variables $\mathbf{v} = \mathbf{y} - \mathbf{x}$. Therefore the probability that no displaced point lies in B is $e^{-\lambda|B|}$, and the claim follows by Rényi's theorem.

40. Conditional on the number $N(s)$ of points originally in the interval $(0, s)$, the positions of these points are jointly distributed as uniform random variables, so the mean number of these points which lie in $(-\infty, a)$ after the perturbation satisfies

$$\lambda s \int_0^s \frac{1}{s}\mathbb{P}(X + u \le a)\, du \to \lambda \int_0^\infty F_X(a - u)\, du = \mathbb{E}(\mathrm{R_L}) \qquad \text{as } s \to \infty,$$

where X is a typical displacement. Likewise, $\mathbb{E}(\mathrm{L_R}) = \lambda \int_0^\infty [1 - F_X(a + u)]\, du$. Equality is valid if and only if

$$\int_a^\infty [1 - F_X(v)]\, dv = \int_{-\infty}^a F_X(v)\, dv,$$

which is equivalent to $a = \mathbb{E}(X)$, by Exercise (4.3.5).

The last part follows immediately on setting $X_r = V_r t$, where V_r is the velocity of the rth car.

41. Conditional on the number $N(t)$ of arrivals by time t, the arrival times of these ants are distributed as independent random variables with the uniform distribution. Let U be a typical arrival time, so that U is uniformly distributed on $(0, t)$. The arriving ant is in the pantry at time t with probability $\pi = \mathbb{P}(U + X > t)$, or in the sink with probability $\rho = \mathbb{P}(U + X < t < U + X + Y)$, or departed with probability $1 - \rho - \pi$. Thus,

$$\begin{aligned} \mathbb{E}(x^{A(t)} y^{B(t)}) &= \mathbb{E}\big\{\mathbb{E}(x^{A(t)} y^{B(t)} \,\big|\, N(t))\big\} \\ &= \mathbb{E}\big\{(\pi x + \rho y + 1 - \pi - \rho)^{N(t)}\big\} = e^{\lambda \pi t(x-1)} e^{\lambda \rho t(y-1)}. \end{aligned}$$

Thus $A(t)$ and $B(t)$ are independent Poisson-distributed random variables.

Set $x = y$ to see that the number of ants in the kitchen at time t has the Poisson distribution with parameter $t(\pi + \rho) = t\mathbb{P}(X + Y + U > t)$. Now,

$$\begin{aligned} t\mathbb{P}(X+Y+U > t) &= \int_0^t [1 - F_Z(t-u)]\,du \qquad \text{where } Z = X+Y, \\ &= \int_0^t [1 - F_Z(v)]\,dv \\ &\to \mathbb{E}Z = \mathbb{E}(X+Y) \qquad \text{as } t \to \infty. \end{aligned}$$

This limit is finite if and only if the number of ants has a limiting distribution.

If the ants arrive in pairs and then separate,

$$\mathbb{E}\big(x^{A(t)}y^{B(t)} \mid N(t)\big) = \big\{\pi^2x^2 + 2\pi\rho xy + \rho^2y^2 + 2\gamma\pi x + 2\gamma\rho y + \gamma^2\big\}^{N(t)}$$

where $\gamma = 1 - \pi - \rho$. Hence,

$$\mathbb{E}(x^{A(t)}y^{B(t)}) = \exp\big\{\lambda\{(\pi x + \rho y + \gamma)^2 - 1\}\big\},$$

whence $A(t)$ and $B(t)$ are *not* independent in this case.

42. The sequence $\{X_r\}$ generates a Poisson process $N(t) = \max\{n : S_n \le t\}$. The statement that $S_n = t$ is equivalent to saying that there are $n - 1$ arrivals in $(0, t)$, and in addition an arrival at t. By Theorem (6.12.7) or Theorem (6.13.11), the first $n - 1$ arrival times have the required distribution.

Part (b) follows similarly, on noting that $f_{\mathbf{U}}(\mathbf{u})$ depends on $\mathbf{u} = (u_1, u_2, \dots, u_n)$ only through the constraints on the u_r.

43. Let Y be a Markov chain independent of X, having the same transition matrix and such that Y_0 has the stationary distribution $\boldsymbol{\pi}$. Let $T = \min\{n \ge 1 : X_n = Y_n\}$ and suppose $X_0 = i$. As in the proof of Theorem (6.4.17),

$$|p_{ij}(n) - \pi_j| = \left|\sum_k \pi_k\big(p_{ij}(n) - p_{kj}(n)\big)\right| \le \sum_k \pi_k\mathbb{P}(T > n) = \mathbb{P}(T > n).$$

Now,

$$\mathbb{P}(T > r+1 \mid T > r) \le 1 - \epsilon^2 \qquad \text{for } r \ge 0,$$

where $\epsilon = \min_{ij}\{p_{ij}\} > 0$. The claim follows with $\lambda = 1 - \epsilon^2$.

44. Let $I_k(n)$ be the indicator function of a visit to k at time n, so that $\mathbb{E}(I_k(n)) = \mathbb{P}(X_n = k) = a_k(n)$, say. By Problem (6.15.43), $|a_k(n) - \pi_k| \le \lambda^n$. Now,

$$\begin{aligned} \mathbb{E}\left(\left|\frac{1}{n}V_i(n) - \pi_i\right|^2\right) &= \frac{1}{n^2}\mathbb{E}\left(\left[\sum_{r=0}^{n-1}\{I_i(r) - \pi_i\}\right]^2\right) \\ &= \frac{1}{n^2}\sum_r\sum_m \mathbb{E}\{(I_i(r) - \pi_i)(I_i(m) - \pi_i)\}. \end{aligned}$$

Let $s = \min\{m, r\}$ and $t = |m - r|$. The last summation equals

$$\frac{1}{n^2}\sum_r\sum_m\{a_i(s)p_{ii}(t) - a_i(r)\pi_i - a_i(m)\pi_i + \pi_i^2\}$$

$$= \frac{1}{n^2}\sum_r\sum_m\Big\{(a_i(s)-\pi_i)(p_{ii}(t)-\pi_i) + \pi_i(p_{ii}(t)-\pi_i)$$
$$+\pi_i(a_i(s)-\pi_i) - \pi_i(a_i(r)-\pi_i) - \pi_i(a_i(m)-\pi_i)\Big\}$$
$$\le \frac{1}{n^2}\sum_r\sum_m(\lambda^{s+t} + \lambda^t + \lambda^s + \lambda^r + \lambda^m)$$
$$\le \frac{An}{n^2} \to 0 \qquad \text{as } n \to \infty,$$

where $0 < A < \infty$. For the last part, use the fact that $\sum_{r=0}^{n-1} f(X_r) = \sum_{i\in S} f(i)V_i(n)$. The result is obtained by Minkowski's inequality (Problem (4.14.27b)) and the first part.

45. We have by the Markov property that $f(X_{n+1} \mid X_n, X_{n-1}, \dots, X_0) = f(X_{n+1} \mid X_n)$, whence

$$\mathbb{E}\big(\log f(X_{n+1} \mid X_n, X_{n-1}, \dots, X_0) \,\big|\, X_n, \dots, X_0\big) = \mathbb{E}\big(\log f(X_{n+1} \mid X_n) \,\big|\, X_n\big).$$

Taking the expectation of each side gives the result. Furthermore,

$$H(X_{n+1} \mid X_n) = -\sum_{i,j}(p_{ij}\log p_{ij})\mathbb{P}(X_n = i).$$

Now X has a unique stationary distribution $\boldsymbol{\pi}$, so that $\mathbb{P}(X_n = i) \to \pi_i$ as $n \to \infty$. The state space is finite, and the claim follows.

46. Let $T = \inf\{t : X_t = Y_t\}$. Since X and Y are recurrent, and since each process moves by distance 1 at continuously distributed times, it is the case that $\mathbb{P}(T < \infty) = 1$. We define

$$Z_t = \begin{cases} X_t & \text{if } t < T, \\ Y_t & \text{if } t \ge T, \end{cases}$$

noting that the processes X and Z have the same distributions.

(a) By the above remarks,

$$|\mathbb{P}(X_t = k) - \mathbb{P}(Y_t = k)| = |\mathbb{P}(Z_t = k) - \mathbb{P}(Y_t = k)|$$
$$\le \big|\mathbb{P}(Z_t = k,\ T \le t) + \mathbb{P}(Z_t = k,\ T > t) - \mathbb{P}(Y_t = k,\ T \le t) - \mathbb{P}(Y_t = k,\ T > t)\big|$$
$$\le \mathbb{P}(X_t = k,\ T > t) + \mathbb{P}(Y_t = k,\ T > t).$$

We sum over $k \in A$, and let $t \to \infty$.

(b) We have in this case that $Z_t \le Y_t$ for all t. The claim follows from the fact that X and Z are processes with the same distributions.

47. We reformulate the problem in the following way. Suppose there are two containers, W and N, containing n particles in all. During the time interval $(t, t + dt)$, any particle in W moves to N with probability $\mu dt + o(dt)$, and any particle in N moves to W with probability $\lambda dt + o(dt)$. The particles move independently of one another. The number $Z(t)$ of particles in W has the same rules of evolution as the process X in the original problem. Now, $Z(t)$ may be expressed as the sum of two independent random variables U and V, where U is $\text{bin}(r, \theta_t)$, V is $\text{bin}(n - r, \psi_t)$, and θ_t is the probability that a particle starting in W is in W at time t, ψ_t is the probability that a particle starting in N at 0 is in W at t. By considering the two-state Markov chain of Exercise (6.9.1),

$$\theta_t = \frac{\lambda + \mu e^{-(\lambda+\mu)t}}{\lambda+\mu}, \qquad \psi_t = \frac{\lambda - \lambda e^{-(\lambda+\mu)t}}{\lambda+\mu},$$

and therefore

$$\mathbb{E}(s^{X(t)}) = \mathbb{E}(s^U)\mathbb{E}(s^V) = (s\theta_t + 1 - s)^r(s\psi_t + 1 - s)^{n-r}.$$

Also, $\mathbb{E}(X(t)) = r\theta_t + (n-r)\psi_t$ and $\text{var}(X(t)) = r\theta_t(1-\theta_t) + (n-r)\psi_t(1-\psi_t)$. In the limit as $n \to \infty$, the distribution of $X(t)$ approaches the $\text{bin}(n, \lambda/(\lambda+\mu))$ distribution.

48. Solving the equations

$$\pi_0 = q_1\pi_1 + p_2\pi_2, \quad \pi_1 = q_2\pi_2 + p_0\pi_0, \quad \sum_i \pi_i = 1,$$

gives the first claim. We have that $\gamma = \sum_i (p_i - q_i)\pi_i$, and the formula for γ follows.

Considering the three walks in order, we have that:

A. $\pi_i = \frac{1}{3}$ for each i, and $\gamma_A = -2a < 0$.

B. Substitution in the formula for γ_B gives the numerator as $3\{-\frac{49}{40}a + o(a)\}$, which is negative for small a whereas the denominator is positive.

C. The transition probabilities are the averages of those for A and B, namely, $p_0 = \frac{1}{2}(\frac{1}{10} - a) + \frac{1}{2}(\frac{1}{2} - a) = \frac{3}{10} - a$, and so on. The numerator in the formula for γ_C equals $\frac{9}{160} + o(1)$, which is positive for small a.

49. Call a car *green* if it satisfies the given condition. The chance that a green car arrives on the scene during the time interval $(u, u+h)$ is $\lambda h\mathbb{P}(V < x/(t-u))$ for $u < t$. Therefore, the arrival process of green cars is an non-homogeneous Poisson process with rate function

$$\lambda(u) = \begin{cases} \lambda\mathbb{P}(V < x/(t-u)) & \text{if } u < t, \\ 0 & \text{if } u \ge t. \end{cases}$$

Hence the required number has the Poisson distribution with mean

$$\begin{aligned} \lambda\int_0^t \mathbb{P}\left(V < \frac{x}{t-u}\right) du &= \lambda\int_0^t \mathbb{P}\left(V < \frac{x}{u}\right) du \\ &= \lambda\int_0^t \mathbb{E}(I_{\{Vu<x\}})\,du = \lambda\mathbb{E}\big(V^{-1}\min\{x, Vt\}\big). \end{aligned}$$

50. The answer is the probability of exactly one arrival in the interval (s, t), which equals $g(s) = \lambda(t-s)e^{-\lambda(t-s)}$. By differentiation, g has its maximum at $\bar{s} = \max\{0, t - \lambda^{-1}\}$, and $g(\bar{s}) = e^{-1}$ when $t \ge \lambda^{-1}$.

51. We measure money in millions and time in hours. The number of available houses has the Poisson distribution with parameter 30λ, whence the number A of affordable houses has the Poisson distribution with parameter $\frac{1}{6}\cdot 30\lambda = 5\lambda$ (cf. Exercise (3.5.2)). Since each viewing time T has moment generating function $\mathbb{E}(e^{\theta T}) = (e^{2\theta} - e^{\theta})/\theta$, the answer is

$$G_A\big(\mathbb{E}(e^{\theta T})\big) = \exp\{5\lambda(e^{2\theta} - e^{\theta} - \theta)/\theta\}.$$

52. Let $\mathbf{P}$ be the transition matrix of X, and let $\mu_{ij} = \mathbb{E}_i(\min\{n \geq 1 : X_n = j\})$, the mean first passage time from i to j. By the Markov chain limit theorem, $\mu_{ii} = 1/\pi_i$. By conditioning on the first step, the mean first passage time from i to Z is $\mathbb{E}(\mu_{iZ}) = 1 + \sum_j p_{ij} K_j$. Alternatively,

$$\mathbb{E}(\mu_{iZ}) = \mu_{ii}\pi_i + \sum_{j \neq i} h_{ij}\pi_j = 1 + K_i,$$

whence $K_i = \sum_j p_{ij} K_j$. That is, $\mathbf{K} = (K_i : i \in S)$ is a right eigenvector of $\mathbf{P}$ corresponding to the eigenvalue 1. Therefore, $\mathbf{K}$ has constant entries, as required. [For an ergodic Markov chain on an infinite state space, the same equation holds, and the same result follows by the maximum principle of potential theory.]

53. (a) The process is a Markov chain because the occasions of rain are independent with constant density. The transition matrix is obtained case by case as

$$p_{ij} = \begin{cases} 1 & \text{if } i = 0,\ j = r, \\ 1 - p & \text{if } i + j = r,\ i > 0, \\ p & \text{if } i + j = r + 1, \\ 0 & \text{otherwise.} \end{cases}$$

(b) Look for a solution $\boldsymbol{\pi}$ to the detailed balance equations, to find that $\pi_i = \pi_j$ for $1 \leq i, j \leq r$ and $\pi_0 = (1-p)\pi_r$, Therefore, $r\pi_r + (1-p)\pi_r = 1$, whence $\boldsymbol{\pi}$ is as given. The long run probability of getting wet is $p\pi_0$.

(c) Let e_i be the mean number of walks until the first wet one, starting with $i \in \{0, 1\}$ umbrellas. Then

$$\begin{aligned} e_0 &= p \cdot 1 + (1-p)[1 + e_1] = 1 + (1-p)e_1, \\ e_1 &= p[1 + e_1] + (1-p)[1 + e_0] = 1 + pe_1 + (1-p)e_0. \end{aligned}$$

Therefore, $e_1 = (2-p)/[p(1-p)]$.

54. Let X be the time of the first flush. Then

$$T = \begin{cases} h & \text{if } X \geq h, \\ X + T' & \text{if } X < h, \end{cases} \qquad N = \begin{cases} 0 & \text{if } X \geq h, \\ 1 + N' & \text{if } X < h, \end{cases}$$

where T' and N' have the same distribution as T and N, respectively, and are independent of X. By conditional expectation, $G = G(\theta, s) = \mathbb{E}(e^{-\theta T} s^N)$ satisfies

$$\begin{aligned} G &= \mathbb{E}\big(\mathbb{E}(e^{-\theta T} s^N \mid X)\big) \\ &= e^{-\theta h} e^{-\lambda h} + sG \int_0^h e^{-\theta x} \lambda e^{-\lambda x}\, dx \\ &= e^{-(\lambda+\theta)h} + \frac{\lambda s G}{\lambda + \theta}\big(1 - e^{-(\lambda+\theta)h}\big), \end{aligned}$$

as required. Denoting the coefficient of s^n in G by $c_n(\theta)$,

$$H(\theta) := \mathbb{E}(e^{-\theta T} \mid N = n) = \frac{c_n(\theta)}{c_n(0)} = e^{-\theta h} \left\{ \frac{\lambda(1 - e^{-(\lambda+\theta)h})}{(\lambda+\theta)(1 - e^{-\lambda h})} \right\}^n,$$

and $\mathbb{E}(T \mid N = n) = -H'(0)$.

55. Since $\boldsymbol{\pi} = \boldsymbol{\pi}\mathbf{P}$ and $\mathbf{P}$ is a stochastic matrix,

$$\pi_i \sum_{j\in A} p_{ij} + \pi_i \sum_{j\notin A} p_{ij} = \pi_i = \sum_{j\in A} \pi_j p_{ji} + \sum_{j\notin A} \pi_j p_{ji}.$$

Sum over $i \in A$ and note that the first term on the left cancels with that on the right. The remaining terms are as required.

56. This is the two-dimensional version of Exercise (6.13.10), but with a twist. A boulder at position $\mathbf{x}$ is weighted by its gravitational attraction $g(\mathbf{x})$, but this function behaves like $1/r^2$ and is not integrable over $\mathbb{R}^2$. It follows as in (6.13.2b) that the total gravitational attraction on the hiker is not absolutely convergent. By restricting the question to the limit of regions with increasing radii, the divergences cancel.

We shall adapt the Campbell–Hardy method to polar coordinates in two dimensions. Note first that the characteristic function $\phi_{G_R}(t)$ is symmetric, and that

$$(*) \qquad e^{ic} + e^{-ic} = 2\cos c = 2\cos|c|, \qquad c \in \mathbb{R}.$$

As in the solution to (6.13.10), and by the preceding remarks,

$$\begin{aligned}\phi_{G_R}(t) &= \exp\left\{-\int_{r=0}^{R}\int_{\theta=0}^{2\pi}\left[1-\mathbb{E}\left(\exp\left\{\frac{itM\cos\theta}{r^2}\right\}\right)\right] r\,dr\,d\theta\right\}\\ &= \exp\left\{-\mathbb{E}\int_{r=0}^{R}\int_{\theta=0}^{2\pi}\left[1-\cos\left(\frac{|t\cos\theta|M}{r^2}\right)\right] r\,dr\,d\theta\right\}.\end{aligned}$$

Make the substitution $y = |t\cos\theta|M/r^2$ to find that

$$\begin{aligned}\phi_{G_R}(t) &= \exp\left\{-|t|\mathbb{E}(M)\left(\int_{\theta=0}^{2\pi}\int_{y=|t\cos\theta|M/R^2}^{\infty}\left(\frac{1}{2y^2}(1-\cos y)\right)|\cos\theta|\,dy\,d\theta\right)\right\}\\ &\to \exp\{-c|t|\} \qquad \text{as } R\to\infty,\end{aligned}$$

where

$$c = 2\mathbb{E}(M)\int_0^\infty \frac{1}{y^2}(1-\cos y)\,dy = \pi\mathbb{E}(M).$$

It may help to know that $\int_0^\infty \psi^{-1}\sin\psi\,d\psi = \frac{1}{2}\pi$.

57. Follow the same strategy as in Problem (6.15.56) to find, using spherical polar coordinates, that

$$\phi_{G_R}(t) = \exp\left\{-\mathbb{E}\iiint\left[1-\cos\left(\frac{M|t\sin\theta\cos\phi|}{r^2}\right)\right] r^2\sin\theta\,dr\,d\theta\,d\phi\right\},$$

where the triple integral is over the ball with radius R and centre at the origin. Substitute $y = M|t\sin\theta\cos\phi|/r^2$ and pass to the limit as $R\to\infty$, to obtain that $\phi_{G_R}(t)\to\exp\{-c|t|^{3/2}\}$ where

$$\begin{aligned}c &= \frac{1}{2}\mathbb{E}(M^{3/2})\left(\int_0^\infty \frac{1}{y^{5/2}}(1-\cos y)\,dy\right)\left(\int_0^\pi \sin^{5/2}\theta\,d\theta\right)\left(\int_0^{2\pi}|\cos\phi|^{3/2}\,d\phi\right)\\ &= \frac{4}{15}\mathbb{E}\left[(2\pi M)^{3/2}\right].\end{aligned}$$

The reader may care to check that the above three integrals are equal to $(2/3)\sqrt{2\pi}$, $2^{3/2}B(\frac{7}{4},\frac{7}{4})$, and $[2/(3\sqrt{2\pi})]\Gamma(\frac{1}{4})^2$, respectively, where $B(a,b)$ is the beta function of paragraph (4.4.8). An alternative approach to this and similar problems may be found in Problem (8.10.9).

7

Convergence of random variables

7.1 Solutions. Introduction

1. (a) $\mathbb{E}|(cX)^r| = |c|^r \cdot \{\|X\|_r\}^r$.

(b) This is Minkowski's inequality.

(c) Let $\epsilon > 0$. Certainly $|X| \geq I_\epsilon$ where I_ϵ is the indicator function of the event $\{|X| > \epsilon\}$. Hence $\mathbb{E}|X^r| \geq \mathbb{E}|I_\epsilon^r| = \mathbb{P}(|X| > \epsilon)$, implying that $\mathbb{P}(|X| > \epsilon) = 0$ for all $\epsilon > 0$. The converse is trivial.

2. (a) $\mathbb{E}(\{aX + bY\}Z) = a\mathbb{E}(XZ) + b\mathbb{E}(YZ)$.

(b) $\mathbb{E}(\{X + Y\}^2) + \mathbb{E}(\{X - Y\}^2) = 2\mathbb{E}(X^2) + 2\mathbb{E}(Y^2)$.

(c) Clearly

$$\mathbb{E}\left(\left\{\sum_{i=1}^{n} X_i\right\}^2\right) = \sum_{i=1}^{n} \mathbb{E}(X_i^2) + 2\sum_{i<j} \mathbb{E}(X_i X_j).$$

3. Let $f(u) = \frac{2}{3}\epsilon$, $g(u) = 0$, $h(u) = -\frac{2}{3}\epsilon$, for all u. Then $d_\epsilon(f, g) + d_\epsilon(g, h) = 0$ whereas $d_\epsilon(f, h) = 1$.

4. Either argue directly, or as follows. With any distribution function F, we may associate a graph $\widetilde{F}$ obtained by adding to the graph of F vertical line segments connecting the two endpoints at each discontinuity of F. By drawing a picture, you may see that $\sqrt{2}\, d(F, G)$ equals the maximum distance between $\widetilde{F}$ and $\widetilde{G}$ measured along lines of slope -1. It is now clear that $d(F, G) = 0$ if and only if $F = G$, and that $d(F, G) = d(G, F)$. Finally, by the triangle inequality for real numbers, we have that $d(F, H) \leq d(F, G) + d(G, H)$.

5. Take X to be any random variable satisfying $\mathbb{E}(X^2) = \infty$, and define $X_n = X$ for all n.

7.2 Solutions. Modes of convergence

1. (a) By Minkowski's inequality,

$$\left\{\mathbb{E}|X^r|\right\}^{1/r} \leq \left\{\mathbb{E}(|X_n - X|^r)\right\}^{1/r} + \left\{\mathbb{E}|X_n^r|\right\}^{1/r};$$

let $n \to \infty$ to obtain $\liminf_{n\to\infty} \mathbb{E}|X_n^r| \geq \mathbb{E}|X^r|$. By another application of Minkowski's inequality,

$$\left\{\mathbb{E}|X_n^r|\right\}^{1/r} \leq \left\{\mathbb{E}(|X_n - X|^r)\right\}^{1/r} + \left\{\mathbb{E}|X^r|\right\}^{1/r},$$

whence $\limsup_{n\to\infty} \mathbb{E}|X_n^r| \leq \mathbb{E}|X^r|$.

(b) We have that

$$|\mathbb{E}(X_n) - \mathbb{E}(X)| = |\mathbb{E}(X_n - X)| \le \mathbb{E}|X_n - X| \to 0$$

as $n \to \infty$. The converse is clearly false. If each X_n takes the values ± 1, each with probability $\frac{1}{2}$, then $\mathbb{E}(X_n) = 0$, but $\mathbb{E}|X_n - 0| = 1$.

(c) By part (a), $\mathbb{E}(X_n^2) \to \mathbb{E}(X^2)$. Now $X_n \xrightarrow{1} X$ by Theorem (7.2.3), and therefore $\mathbb{E}(X_n) \to \mathbb{E}(X)$ by part (b). Therefore $\text{var}(X_n) = \mathbb{E}(X_n^2) - \mathbb{E}(X_n)^2 \to \text{var}(X)$.

2. Assume that $X_n \xrightarrow{P} X$. Since $|X_n| \le Z$ for all n, it is the case that $|X| \le Z$ a.s. Therefore $Z_n = |X_n - X|$ satisfies $Z_n \le 2Z$ a.s. In addition, if $\epsilon > 0$,

$$\mathbb{E}|Z_n| = \mathbb{E}\left(Z_n I_{\{Z_n \le \epsilon\}}\right) + \mathbb{E}\left(Z_n I_{\{Z_n > \epsilon\}}\right) \le \epsilon + 2\mathbb{E}\left(Z I_{\{Z_n > \epsilon\}}\right).$$

As $n \to \infty$, $\mathbb{P}(|Z_n| > \epsilon) \to 0$, and therefore the last term tends to 0; to see this, use the fact that $\mathbb{E}(Z) < \infty$, together with the result of Exercise (5.6.5). Now let $\epsilon \downarrow 0$ to obtain that $\mathbb{E}|Z_n| \to 0$ as $n \to \infty$.

3. (a) We have that $X - n^{-1} \le X_n \le X$, so that $\mathbb{E}(X_n) \to \mathbb{E}(X)$, and similarly $\mathbb{E}(Y_n) \to \mathbb{E}(Y)$. By the independence of X_n and Y_n,

$$\mathbb{E}(X_n Y_n) = \mathbb{E}(X_n)\mathbb{E}(Y_n) \to \mathbb{E}(X)\mathbb{E}(Y).$$

Finally, $(X - n^{-1})(Y - n^{-1}) \le X_n Y_n \le XY$, and

$$\mathbb{E}\left\{\left(X - \frac{1}{n}\right)\left(Y - \frac{1}{n}\right)\right\} = \mathbb{E}(XY) - \frac{\mathbb{E}(X) + \mathbb{E}(Y)}{n} + \frac{1}{n^2} \to \mathbb{E}(XY)$$

as $n \to \infty$, so that $\mathbb{E}(X_n Y_n) \to \mathbb{E}(XY)$.

(b) Take $X = Y = 1/\sqrt{U}$ where U is uniformly distributed on $(0, 1)$.

4. Let $F_1, F_2, \dots$ be distribution functions. As in Section 5.9, we write $F_n \to F$ if $F_n(x) \to F(x)$ for all x at which F is continuous. We are required to prove that $F_n \to F$ if and only if $d(F_n, F) \to 0$.

Suppose that $d(F_n, F) \to 0$. Then, for $\epsilon > 0$, there exists N such that

$$F(x - \epsilon) - \epsilon \le F_n(x) \le F(x + \epsilon) + \epsilon \qquad \text{for all } x.$$

Take the limits as $n \to \infty$ and $\epsilon \to 0$ in that order, to find that $F_n(x) \to F(x)$ whenever F is continuous at x.

Suppose that $F_n \to F$. Let $\epsilon > 0$, and find real numbers $a = x_1 < x_2 < \cdots < x_n = b$, each being points of continuity of F, such that

(i) $F_i(a) < \epsilon$ for all i, $F(b) > 1 - \epsilon$,

(ii) $|x_{i+1} - x_i| < \epsilon$ for $1 \le i < n$.

In order to pick a such that $F_i(a) < \epsilon$ for all i, first choose a' such that $F(a') < \frac{1}{2}\epsilon$ and F is continuous at a', then find M such that $|F_m(a') - F(a')| < \frac{1}{2}\epsilon$ for $m \ge M$, and lastly find a continuity point a of F such that $a \le a'$ and $F_m(a) < \epsilon$ for $1 \le m < M$.

There are finitely many points x_i, and therefore there exists N such that $|F_m(x_i) - F(x_i)| < \epsilon$ for all i and $m \ge N$. Now, if $m \ge N$ and $x_i \le x < x_{i+1}$,

$$F_m(x) \le F_m(x_{i+1}) < F(x_{i+1}) + \epsilon \le F(x + \epsilon) + \epsilon,$$

and similarly

$$F_m(x) \ge F_m(x_i) > F(x_i) - \epsilon \ge F(x - \epsilon) - \epsilon.$$

Similar inequalities hold if $x \le a$ or $x \ge b$, and it follows that $d(F_m, F) < \epsilon$ if $m \ge N$. Therefore $d(F_m, F) \to 0$ as $m \to \infty$.

5. (a) Suppose $c > 0$ and pick δ such that $0 < \delta < c$. Find N such that $\mathbb{P}(|Y_n - c| > \delta) < \delta$ for $n \ge N$. Now, for $x \ge 0$,

$$\mathbb{P}(X_n Y_n \le x) \le \mathbb{P}\big(X_n Y_n \le x, |Y_n - c| \le \delta\big) + \mathbb{P}\big(|Y_n - c| > \delta\big) \le \mathbb{P}\left(X_n \le \frac{x}{c-\delta}\right) + \delta,$$

and similarly

$$\mathbb{P}(X_n Y_n > x) \le \mathbb{P}\big(X_n Y_n > x, |Y_n - c| \le \delta\big) + \delta \le \mathbb{P}\left(X_n > \frac{x}{c+\delta}\right) + \delta.$$

Taking the limits as $n \to \infty$ and $\delta \downarrow 0$, we find that $\mathbb{P}(X_n Y_n \le x) \to \mathbb{P}(X \le x/c)$ if x/c is a point of continuity of the distribution function of X. A similar argument holds if $x < 0$, and we conclude that $X_n Y_n \xrightarrow{\text{D}} cX$ if $c > 0$. No extra difficulty arises if $c < 0$, and the case $c = 0$ is similar.

For the second part, it suffices to prove that $Y_n^{-1} \xrightarrow{\text{P}} c^{-1}$ if $Y_n \xrightarrow{\text{P}} c\ (\ne 0)$. This is immediate from the fact that $|Y_n^{-1} - c^{-1}| < \epsilon/\{|c|(|c| - \epsilon)\}$ if $|Y_n - c| < \epsilon\ (< |c|)$.

(b) By a standard argument of analysis, g is uniformly continuous on any closed bounded subset of $\mathbb{R}^2$. Let $\epsilon > 0$ and $N < \infty$. By uniform continuity, there exists $\delta > 0$ such that

$$|g(X_n, Y_n) - g(0, Y)| < \epsilon$$

if $|X_n| \le \delta$, $|Y_n - Y| \le \delta$, and $|Y| \le N$. It follows that

$$\begin{aligned} \mathbb{P}\big(|g(X_n, Y_n) - g(0, Y)| \ge \epsilon\big) &\le \mathbb{P}(|X_n| > \delta) + \mathbb{P}\big(|Y_n - Y| > \delta\big) + \mathbb{P}(|Y| > N) \\ &\to \mathbb{P}(|Y| \ge N) \qquad \text{as } n \to \infty \\ &\to 0 \qquad\qquad\quad\ \ \text{as } N \to \infty. \end{aligned}$$

Therefore, $g(X_n, Y_n) \xrightarrow{\text{P}} g(0, Y)$ as $n \to \infty$.

6. The subset A of the sample space Ω may be expressed thus:

$$A = \bigcap_{k=1}^{\infty} \bigcup_{n=1}^{\infty} \bigcap_{m=1}^{\infty} \{|X_{n+m} - X_n| < k^{-1}\},$$

a countable sequence of intersections and unions of events.

For the last part, define

$$X(\omega) = \begin{cases} \lim_{n\to\infty} X_n(\omega) & \text{if } \omega \in A \\ 0 & \text{if } \omega \notin A. \end{cases}$$

The function X is $\mathcal{F}$-measurable since $A \in \mathcal{F}$.

7. (a) If $X_n(\omega) \to X(\omega)$ then $c_n X_n(\omega) \to cX(\omega)$.

(b) We have by Minkowski's inequality that, as $n \to \infty$,

$$\mathbb{E}\big(|c_n X_n - cX|^r\big) \le |c_n|^r \mathbb{E}\big(|X_n - X|^r\big) + |c_n - c|^r \mathbb{E}|X^r| \to 0.$$

(c) If $c = 0$, the claim is nearly obvious. Otherwise $c \neq 0$, and we may assume that $c > 0$. For $0 < \epsilon < c$, there exists N such that $|c_n - c| < \epsilon$ whenever $n \geq N$. By the triangle inequality, $|c_n X_n - cX| \leq |c_n(X_n - X)| + |(c_n - c)X|$, so that, for $n \geq N$,

$$\begin{aligned}\mathbb{P}\big(|c_n X_n - cX| > \epsilon\big) &\leq \mathbb{P}\big(c_n|X_n - X| > \tfrac{1}{2}\epsilon\big) + \mathbb{P}\big(|c_n - c| \cdot |X| > \tfrac{1}{2}\epsilon\big)\\ &\leq \mathbb{P}\left(|X_n - X| > \frac{\epsilon}{2(c+\epsilon)}\right) + \mathbb{P}\left(|X| > \frac{\epsilon}{2|c_n - c|}\right)\\ &\to 0 \qquad \text{as } n \to \infty.\end{aligned}$$

(d) A neat way is to use the Skorokhod representation (7.2.14). If $X_n \xrightarrow{\text{D}} X$, find random variables Y_n, Y with the same distributions such that $Y_n \xrightarrow{\text{a.s.}} Y$. Then $c_n Y_n \xrightarrow{\text{a.s.}} cY$, so that $c_n Y_n \xrightarrow{\text{D}} cY$, implying the same conclusion for the X's.

8. If X is not a.s. constant, there exist real numbers c and ϵ such that $0 < \epsilon < \frac{1}{2}$ and $\mathbb{P}(X < c) > 2\epsilon$, $\mathbb{P}(X > c + \epsilon) > 2\epsilon$. Since $X_n \xrightarrow{\text{P}} X$, there exists N such that

$$\mathbb{P}(X_n < c) > \epsilon, \quad \mathbb{P}(X_n > c + \epsilon) > \epsilon, \quad \text{if } n \geq N.$$

Also, by the triangle inequality, $|X_r - X_s| \leq |X_r - X| + |X_s - X|$; therefore there exists M such that $\mathbb{P}\big(|X_r - X_s| > \epsilon\big) < \epsilon^3$ for $r, s \geq M$. Assume now that the X_n are independent. Then, for $r, s \geq \max\{M, N\}$, $r \neq s$,

$$\epsilon^3 > \mathbb{P}(|X_r - X_s| > \epsilon) \geq \mathbb{P}(X_r < c, X_s > c + \epsilon) = \mathbb{P}(X_r < c)\mathbb{P}(X_s > c + \epsilon) > \epsilon^2,$$

a contradiction.

9. Either use the fact (Exercise (4.12.3)) that convergence in total variation implies convergence in distribution, together with Theorem (7.2.19), or argue directly thus. Since $|u(\cdot)| \leq K < \infty$,

$$\big|\mathbb{E}(u(X_n)) - \mathbb{E}(u(X))\big| = \left|\sum_k u(k)\{f_n(k) - f(k)\}\right| \leq K \sum_k |f_n(k) - f(k)| \to 0.$$

10. The partial sum $S_n = \sum_{r=1}^n X_r$ is Poisson-distributed with parameter $\sigma_n = \sum_{r=1}^n \lambda_r$. For fixed x, the event $\{S_n \leq x\}$ is decreasing in n, whence by Theorem (1.3.5), if $\sigma_n \to \sigma < \infty$ and x is a non-negative integer,

$$\mathbb{P}\left(\sum_{r=1}^\infty X_r \leq x\right) = \lim_{n\to\infty} \mathbb{P}(S_n \leq x) = \sum_{j=0}^x \frac{e^{-\sigma}\sigma^j}{j!}.$$

Hence if $\sigma < \infty$, $\sum_{r=1}^\infty X_r$ converges to a Poisson random variable. On the other hand, if $\sigma_n \to \infty$, then $e^{-\sigma_n}\sum_{j=0}^x \sigma_n^j/j! \to 0$, giving that $\mathbb{P}\big(\sum_{r=1}^\infty X_r > x\big) = 1$ for all x, and therefore the sum diverges with probability 1, as required.

11. We have that

$$\mathbb{P}(I_m > x\sqrt{m}) = \prod_{r=2}^{\lfloor x\sqrt{m}\rfloor}\left(1 - \frac{r-1}{m}\right), \qquad 0 < x < \sqrt{m}.$$

Now, $\log(1-y) = -y + \mathrm{O}(y^2)$ as $y \downarrow 0$, whence

$$\begin{aligned}\log \mathbb{P}(I_m > x\sqrt{m}) &= \mathrm{O}(m^{-1/2}) - \sum_{r=1}^{\lfloor x\sqrt{m}\rfloor - 1} \frac{r}{m}\\ &= \mathrm{O}(m^{-1/2}) - \frac{x\sqrt{m}(x\sqrt{m}-1)}{2m} \to -\tfrac{1}{2}x^2 \qquad \text{as } m \to \infty.\end{aligned}$$

Exponentiating and differentiating yields the Rayleigh density function.

12. (a) For $x > 0$ and a non-negative integer n, by Markov's inequality,

$$\mathbb{P}(X \geq 2x) \leq \frac{\mathbb{E}(X^n)}{(2x)^n}.$$

With $x = 1$,

$$\sum_{n=0}^{\infty} \frac{1}{\mathbb{E}(X^n)} \leq \sum_{n=0}^{\infty} \frac{1}{2^n \mathbb{P}(X \geq 2)} < \infty.$$

(b) We have

$$\begin{aligned}\mathbb{E}(x^M) &= \sum_{m=0}^{\infty} x^m \mathbb{P}(M = m) = \sum_{m=0}^{\infty} \frac{x^m}{c\mathbb{E}(X^m)} \\ &\leq \sum_{m=0}^{\infty} \frac{1}{c2^m \mathbb{P}(X \geq 2x)} < \infty, \qquad x > 0.\end{aligned}$$

On the other hand,

$$\mathbb{E}(X^M) = \sum_{m=0}^{\infty} \mathbb{E}(X^m)\mathbb{P}(M = m) = \sum_{m=0}^{\infty} \frac{1}{c} = \infty.$$

7.3 Solutions. Some ancillary results

1. (a) If $|X_n - X_m| > \epsilon$ then either $|X_n - X| > \frac{1}{2}\epsilon$ or $|X_m - X| > \frac{1}{2}\epsilon$, so that

$$\mathbb{P}\big(|X_n - X_m| > \epsilon\big) \leq \mathbb{P}\big(|X_n - X| > \tfrac{1}{2}\epsilon\big) + \mathbb{P}\big(|X_m - X| > \tfrac{1}{2}\epsilon\big) \to 0$$

as $n, m \to \infty$, for $\epsilon > 0$.

Conversely, suppose that $\{X_n\}$ is Cauchy convergent in probability. For each positive integer k, there exists n_k such that

$$\mathbb{P}\big(|X_n - X_m| \geq 2^{-k}\big) < 2^{-k} \qquad \text{for } n, m \geq n_k.$$

The sequence (n_k) may not be increasing, and we work instead with the sequence defined by $N_1 = n_1$, $N_{k+1} = \max\{N_k + 1, n_{k+1}\}$. We have that

$$\sum_k \mathbb{P}\big(|X_{N_{k+1}} - X_{N_k}| \geq 2^{-k}\big) < \infty,$$

whence, by the first Borel–Cantelli lemma, a.s. only finitely many of the events $\{|X_{N_{k+1}} - X_{N_k}| \geq 2^{-k}\}$ occur. Therefore, the expression

$$X = X_{N_1} + \sum_{k=1}^{\infty} (X_{N_{k+1}} - X_{N_k})$$

converges absolutely on an event C having probability one. Define $X(\omega)$ accordingly for $\omega \in C$, and $X(\omega) = 0$ for $\omega \notin C$. We have, by the definition of X, that $X_{N_k} \xrightarrow{\text{a.s.}} X$ as $k \to \infty$. Finally, we 'fill in the gaps'. As before, for $\epsilon > 0$,

$$\mathbb{P}\big(|X_n - X| > \epsilon\big) \leq \mathbb{P}\big(|X_n - X_{N_k}| > \tfrac{1}{2}\epsilon\big) + \mathbb{P}\big(|X_{N_k} - X| > \tfrac{1}{2}\epsilon\big) \to 0$$

as $n, k \to \infty$, where we are using the assumption that $\{X_n\}$ is Cauchy convergent in probability.
(b) Since $X_n \xrightarrow{\text{P}} X$, the sequence $\{X_n\}$ is Cauchy convergent in probability. Hence

$$\mathbb{P}(|Y_n - Y_m| > \epsilon) = \mathbb{P}(|X_n - X_m| > \epsilon) \to 0 \quad \text{as } n, m \to \infty,$$

for $\epsilon > 0$. Therefore $\{Y_n\}$ is Cauchy convergent also, and the sequence converges in probability to some limit Y. Finally, $X_n \xrightarrow{\text{D}} X$ and $Y_n \xrightarrow{\text{D}} X$, so that X and Y have the same distribution.

2. Since $A_n \subseteq \bigcup_{m=n}^{\infty} A_m$, we have that

$$\limsup_{n\to\infty} \mathbb{P}(A_n) \le \lim_{n\to\infty} \mathbb{P}\left(\bigcup_{m=n}^{\infty} A_m\right) = \mathbb{P}\left(\lim_{n\to\infty} \bigcup_{m=n}^{\infty} A_m\right) = \mathbb{P}(A_n \text{ i.o.}),$$

where we have used the continuity of $\mathbb{P}$. Alternatively, apply Fatou's lemma to the sequence $I_{A_n^c}$ of indicator functions.

3. (a) Suppose $X_{2n} = 1$, $X_{2n+1} = -1$, for $n \ge 1$. Then $\{S_n = 0 \text{ i.o.}\}$ occurs if $X_1 = -1$, and not if $X_1 = 1$. The event is therefore not in the tail σ-field of the X's.
(b) Here is a way. As usual, $\mathbb{P}(S_{2n} = 0) = \binom{2n}{n}\{p(1-p)\}^n$, so that

$$\sum_n \mathbb{P}(S_{2n} = 0) < \infty \quad \text{if } p \ne \tfrac{1}{2},$$

implying by the first Borel–Cantelli lemma that $\mathbb{P}(S_n = 0 \text{ i.o.}) = 0$.
(c) Changing the values of any finite collection of the steps has no effect on $I = \liminf T_n$ and $J = \limsup T_n$, since such changes are extinguished in the limit by the denominator '$\sqrt{n}$'. Hence I and J are tail functions, and are measurable with respect to the tail σ-field. In particular, $\{I \le -x\} \cap \{J \ge x\}$ lies in the σ-field.

Take $x = 1$, say. Then, $\mathbb{P}(I \le -1) = \mathbb{P}(J \ge 1)$ by symmetry; using Exercise (7.3.2) and the central limit theorem,

$$\mathbb{P}(J \ge 1) \ge \mathbb{P}(S_n \ge \sqrt{n} \text{ i.o.}) \ge \limsup_{n\to\infty} \mathbb{P}(S_n \ge \sqrt{n}) = 1 - \Phi(1) > 0,$$

where Φ is the $N(0, 1)$ distribution function. Since $\{J \ge 1\}$ is a tail event of an independent sequence, it has probability either 0 or 1, and therefore $\mathbb{P}(I \le -1) = \mathbb{P}(J \ge 1) = 1$, and also $\mathbb{P}(I \le -1, J \ge 1) = 1$. That is, on an event having probability one, each visit of the walk to the left of $-\sqrt{n}$ is followed by a visit of the walk to the right of $\sqrt{n}$, and vice versa. It follows that the walk visits 0 infinitely often, with probability one.

4. Let A be exchangeable. Since A is defined in terms of the X_i, it follows by a standard result of measure theory that, for each n, there exists an event $A_n \in \sigma(X_1, X_2, \dots, X_n)$, such that $\mathbb{P}(A \triangle A_n) \to 0$ as $n \to \infty$. We may express A_n and A in the form

$$A_n = \{\mathbf{X}_n \in B_n\}, \quad A = \{\mathbf{X} \in B\},$$

where $\mathbf{X}_n = (X_1, X_2, \dots, X_n)$, and B_n and B are appropriate subsets of $\mathbb{R}^n$ and $\mathbb{R}^\infty$. Let

$$A'_n = \{\mathbf{X}'_n \in B_n\}, \quad A' = \{\mathbf{X}' \in B\},$$

where $\mathbf{X}'_n = (X_{n+1}, X_{n+2}, \dots, X_{2n})$ and $\mathbf{X}' = (X_{n+1}, X_{n+2}, \dots, X_{2n}, X_1, X_2, \dots, X_n, X_{2n+1}, X_{2n+2}, \dots)$.

Now $\mathbb{P}(A_n \cap A'_n) = \mathbb{P}(A_n)\mathbb{P}(A'_n)$, by independence. Also, $\mathbb{P}(A_n) = \mathbb{P}(A'_n)$, and therefore

$$(*) \qquad \mathbb{P}(A_n \cap A'_n) = \mathbb{P}(A_n)^2 \to \mathbb{P}(A)^2 \quad \text{as } n \to \infty.$$

By the exchangeability of A, we have that $\mathbb{P}(A \,\triangle\, A'_n) = \mathbb{P}(A' \,\triangle\, A'_n)$, which in turn equals $\mathbb{P}(A \,\triangle\, A_n)$, using the fact that the X_i are independent and identically distributed. Therefore,

$$|\mathbb{P}(A_n \cap A'_n) - \mathbb{P}(A)| \le \mathbb{P}(A \,\triangle\, A_n) + \mathbb{P}(A \,\triangle\, A'_n) \to 0 \quad \text{as } n \to \infty.$$

Combining this with $(*)$, we obtain that $\mathbb{P}(A) = \mathbb{P}(A)^2$, and hence $\mathbb{P}(A)$ equals 0 or 1.

5. The value of S_n does not depend on the order of the first n steps, but only on their sum. If $S_n = 0$ i.o., then $S'_n = 0$ i.o. for all walks $\{S'_n\}$ obtained from $\{S_n\}$ by permutations of finitely many steps.

6. Since f is continuous on a closed interval, it is bounded: $|f(y)| \le c$ for all $y \in [0, 1]$ for some c. Furthermore f is uniformly continuous on $[0, 1]$, which is to say that, if $\epsilon > 0$, there exists δ (> 0), such that $|f(y) - f(z)| < \epsilon$ if $|y - z| \le \delta$. With this choice of ϵ, δ, we have that $|\mathbb{E}(ZI_{A^c})| < \epsilon$, and

$$|\mathbb{E}(ZI_A)| \le 2c\mathbb{P}(A) \le 2c \cdot \frac{x(1-x)}{n\delta^2}$$

by Chebyshov's inequality. Therefore

$$|\mathbb{E}(Z)| < \epsilon + \frac{2c}{n\delta^2},$$

which is less than 2ϵ for values of n exceeding $2c/(\epsilon\delta^2)$.

7. If $\{X_n\}$ converges completely to X then, by the first Borel–Cantelli lemma, $|X_n - X| > \epsilon$ only finitely often with probability one, for all $\epsilon > 0$. This implies that $X_n \xrightarrow{\text{a.s.}} X$; see Theorem (7.2.4c).

Suppose conversely that $\{X_n\}$ is a sequence of independent variables which converges almost surely to X. By Exercise (7.2.8), X is almost surely constant, and we may therefore suppose that $X_n \xrightarrow{\text{a.s.}} c$ where $c \in \mathbb{R}$. It follows that, for $\epsilon > 0$, only finitely many of the (independent) events $\{|X_n - c| > \epsilon\}$ occur, with probability one. Using the second Borel–Cantelli lemma,

$$\sum_n \mathbb{P}\big(|X_n - c| > \epsilon\big) < \infty.$$

8. Of the various ways of doing this, here is one. We have that

$$\binom{n}{2}^{-1} \sum_{1 \le i < j \le n} X_i X_j = \frac{n}{n-1}\left(\frac{1}{n}\sum_{i=1}^n X_i\right)^2 - \frac{1}{n(n-1)}\sum_{i=1}^n X_i^2.$$

Now $n^{-1}\sum_1^n X_i \xrightarrow{\text{D}} \mu$, by the law of large numbers (5.10.2); hence $n^{-1}\sum_1^n X_i \xrightarrow{\text{P}} \mu$ (use Theorem (7.2.4a)). It follows that $(n^{-1}\sum_1^n X_i)^2 \xrightarrow{\text{P}} \mu^2$; to see this, either argue directly or use Problem (7.11.3). Now use Exercise (7.2.7) to find that

$$\frac{n}{n-1}\left(\frac{1}{n}\sum_{i=1}^n X_i\right)^2 \xrightarrow{\text{P}} \mu^2.$$

Arguing similarly,

$$\frac{1}{n(n-1)}\sum_{i=1}^n X_i^2 \xrightarrow{\text{P}} 0,$$

and the result follows by the fact (Theorem (7.3.9)) that the sum of these two expressions converges in probability to the sum of their limits.

9. Evidently,

$$\mathbb{P}\left(\frac{X_n}{\log n} \geq 1+\epsilon\right) = \frac{1}{n^{1+\epsilon}}, \qquad \text{for } |\epsilon| < 1.$$

By the Borel–Cantelli lemmas, the events $A_n = \{X_n/\log n \geq 1+\epsilon\}$ occur a.s. infinitely often for $-1 < \epsilon \leq 0$, and a.s. only finitely often for $\epsilon > 0$.

10. (a) Mills's ratio (Exercise (4.4.8) or Problem (4.14.1c)) informs us that $1 - \Phi(x) \sim x^{-1}\phi(x)$ as $x \to \infty$. Therefore,

$$\mathbb{P}\big(|X_n| \geq \sqrt{2\log n}(1+\epsilon)\big) \sim \frac{1}{\sqrt{2\pi \log n}(1+\epsilon)n^{(1+\epsilon)^2}}.$$

The sum over n of these terms converges if and only if $\epsilon > 0$, and the Borel–Cantelli lemmas imply the claim.

(b) This is an easy implication of the Borel–Cantelli lemmas.

11. Let X be uniformly distributed on the interval $[-1, 1]$, and define $X_n = I(X \leq (-1)^n/n)$. The distribution of X_n approaches the Bernoulli distribution which takes the values ± 1 with equal probabilities $\frac{1}{2}$. The median of X_n is 1 if n is even and 0 if n is odd.

12. (i) We have that

$$\sum_{r=1}^{\infty} \mathbb{P}\left(\frac{X_r}{r} \geq x\right) = \sum_{r=1}^{\infty} \mathbb{P}\left(\frac{X_r}{x} \geq r\right) = \frac{\mathbb{E}(X_r)}{x} = \infty$$

for $x > 0$. The result follows by the second Borel–Cantelli lemma.

(ii) (a) The stationary distribution $\boldsymbol{\pi}$ is found in the usual way to satisfy

$$\pi_k = \frac{k-1}{k+1}\pi_{k-1} = \cdots = \frac{2}{k(k+1)}\pi_1, \qquad k \geq 2.$$

Hence $\pi_k = \{k(k+1)\}^{-1}$ for $k \geq 1$, a distribution with mean $\sum_{k=1}^{\infty}(k+1)^{-1} = \infty$.

(b) By construction, $\mathbb{P}(X_n \leq X_0 + n) = 1$ for all n, whence

$$\mathbb{P}\left(\limsup_{n\to\infty} \frac{X_n}{n} \leq 1\right) = 1.$$

It may in fact be shown that $\mathbb{P}\big(\limsup_{n\to\infty} X_n/n = 0\big) = 1$.

13. We divide the numerator and denominator by $\sqrt{n}\sigma$. By the central limit theorem, the former converges in distribution to the $N(0, 1)$ distribution. We expand the new denominator, squared, as

$$\frac{1}{n\sigma^2}\sum_{r=1}^{n}(X_r - \mu)^2 - \frac{2}{n\sigma^2}(\overline{X} - \mu)\sum_{r=1}^{n}(X_r - \mu) + \frac{1}{\sigma^2}(\overline{X} - \mu)^2.$$

By the weak law of large numbers (Theorem (5.10.2), combined with Theorem (7.2.3)), the first term converges in probability to 1, and the other terms to 0. Their sum converges to 1, by Theorem (7.3.9), and the result follows by Slutsky's theorem, Exercise (7.2.5).

14. For $t > 0$,

$$\mathbb{P}(|X| > t) = \mathbb{P}(X^2 + a > t^2 + a) \leq \frac{\mathbb{E}(X^4) + 2a\sigma^2 + a^2}{(t^2 + a)^2}, \qquad a \in \mathbb{R},$$

by Markov's inequality. We choose a to minimize this, namely,

$$a = \frac{\mathbb{E}(X^4) - \sigma^2 t^2}{t^2 - \sigma^2}, \qquad t \neq \sigma.$$

On substituting into the above,

$$\mathbb{P}(|X| > t) \le \frac{(\mathbb{E}(X^4) - \sigma^4)(\mathbb{E}(X^4) - 2\sigma^2 t^2 + t^4)}{(\mathbb{E}(X^4) - 2\sigma^2 t^2 + t^4)^2}, \qquad t \neq \sigma,$$

which gives the inequality when $t \neq \sigma$, on noting that $\mathbb{E}(X^4) - 2\sigma^2 t^2 + t^4 > (\sigma^2 - t^2)^2 \neq 0$. The inequality with $t = \sigma$ holds by the continuity in t of the right side.

7.4 Solutions. Laws of large numbers

1. Let $S_n = X_1 + X_2 + \cdots + X_n$. Then

$$\mathbb{E}(S_n^2) = \sum_{i=2}^{n} \frac{i}{\log i} \le \frac{n^2}{\log n}$$

and therefore $S_n/n \xrightarrow{\text{m.s.}} 0$. On the other hand, $\sum_i \mathbb{P}(|X_i| \ge i) = 1$, so that $|X_i| \ge i$ i.o., with probability one, by the second Borel–Cantelli lemma. For such a value of i, we have that $|S_i - S_{i-1}| \ge i$, implying that S_n/n does not converge, with probability one.

2. Let the X_n satisfy

$$\mathbb{P}(X_n = -n) = 1 - \frac{1}{n^2}, \qquad \mathbb{P}(X_n = n^3 - n) = \frac{1}{n^2},$$

whence they have zero mean. However,

$$\sum_n \mathbb{P}\left(\frac{X_n}{n} \neq -1\right) = \sum_n \frac{1}{n^2} < \infty,$$

implying by the first Borel–Cantelli lemma that $\mathbb{P}(X_n/n \to -1) = 1$. It is an elementary result of real analysis that $n^{-1}\sum_{r=1}^{n} x_n \to -1$ if $x_n \to -1$, and the claim follows.

3. The random variable $N(S)$ has mean and variance $\lambda|S| = cr^d$, where c is a constant depending only on d. By Chebyshov's inequality,

$$\mathbb{P}\left(\left|\frac{N(S)}{|S|} - \lambda\right| \ge \epsilon\right) \le \frac{\lambda}{\epsilon^2 |S|} = \left(\frac{\lambda}{\epsilon}\right)^2 \frac{1}{cr^d}.$$

By the first Borel–Cantelli lemma, $\big||S_k|^{-1}N(S_k) - \lambda\big| \ge \epsilon$ for only finitely many integers k, a.s., where S_k is the sphere of radius k. It follows that $N(S_k)/|S_k| \xrightarrow{\text{a.s.}} \lambda$ as $k \to \infty$. The same conclusion holds as $k \to \infty$ through the reals, since $N(S)$ isd non-decreasing in the radius of S.

4. Let X_n be a sequence of independent random variables each taking the values 1.3 and 0.75 with equal probabilities $\frac{1}{2}$. The gambler's fortune W_n after n bets satisfies

$$\mathbb{E}(W_n) = \mathbb{E}\left(W_0 \prod_{r=1}^{n} X_r\right) = \mathbb{E}(W_0)\{\tfrac{1}{2}(1.3 + 0.75)\}^n \to \infty \qquad \text{as } n \to \infty.$$

On the other hand,

$$\log W_n = \log W_0 + \sum_{r=1}^{n} \log X_r,$$

so that

$$\frac{1}{n}\log W_n \xrightarrow{\text{a.s.}} \mathbb{E}(\log X_1) \simeq -0.01 \qquad \text{as } n \to \infty.$$

Therefore, $W_n \xrightarrow{\text{a.s.}} 0$.

5. (a) This holds since

$$\mathbb{P}(B) = \mathbb{P}(B \cap A^{\text{c}}) + \mathbb{P}(B \cap A) \le \mathbb{P}(B) + \mathbb{P}(B \mid A).$$

(b) The event A^{c} occurs if and only if, for some $j = 1, 2, \dots, n$, we have $|X_j| > \delta n$. Now use the union bound.
(c) On the event A, S_n equals the sum of the Y_j. By Chebyshov's inequality,

$$\mathbb{P}(B \mid A) \le \frac{n \operatorname{var}(Y_1)}{(n\epsilon)^2} \le \frac{\mathbb{E}(Y_1^2)}{n\epsilon^2}.$$

Suppose $\eta := \mathbb{E}|X_1| < \infty$, and note that

(*) $$\mathbb{P}(B \mid A) \le \frac{\delta n \mathbb{E}(Y_1)}{n\epsilon^2} = \frac{\delta\eta}{\epsilon^2}.$$

Since $\eta < \infty$, we have that $\mu_n := \mathbb{E}(Y_1)$ satisfies $\mu_n \to \mu$ as $n \to \infty$. Pick N such that $|\mu_n - \mu| \le \epsilon$ for $n \ge N$. With $T_n = \sum_{j=1}^{n} Y_j$ and $n \ge N$,

$$\mathbb{P}\big(|S_n/n - \mu| \ge 2\epsilon\big) \le \mathbb{P}(B) \qquad \text{for } n \ge N.$$

By (a)–(c) and (*),

$$\mathbb{P}\big(|S_n/n - \mu| \ge 2\epsilon\big) \le n\mathbb{P}(|X_1| > \delta n) + \frac{\delta\eta}{\epsilon^2} \to \frac{\delta\eta}{\epsilon^2} \qquad \text{as } n \to \infty,$$

by Exercise (5.6.4). Since $\delta > 0$ is arbitrary, we may take the limit as $\delta \downarrow 0$.

7.5 Solutions. The strong law

1. Let I_{ij} be the indicator function of the event that X_j lies in the ith interval. Then

$$\log R_m = \sum_{i=1}^{n} Z_m(i) \log p_i = \sum_{i=1}^{n}\sum_{j=1}^{m} I_{ij} \log p_i = \sum_{j=1}^{m} Y_j$$

where, for $1 \le j \le m$, $Y_j = \sum_{i=1}^{n} I_{ij} \log p_i$ is the sum of independent identically distributed variables with mean

$$\mathbb{E}(Y_j) = \sum_{i=1}^{n} p_i \log p_i = -h.$$

By the strong law, $m^{-1} \log R_m \xrightarrow{\text{a.s.}} -h$.

2. The following two observations are clear:
(a) $N(t) < n$ if and only if $T_n > t$,
(b) $T_{N(t)} \le t < T_{N(t)+1}$.
If $\mathbb{E}(X_1) < \infty$, then $\mathbb{E}(T_n) < \infty$, so that $\mathbb{P}(T_n > t) \to 0$ as $t \to \infty$. Therefore, by (a),

$$\mathbb{P}(N(t) < n) = \mathbb{P}(T_n > t) \to 0 \quad \text{as } t \to \infty,$$

implying that $N(t) \xrightarrow{\text{a.s.}} \infty$ as $t \to \infty$.

Secondly, by (b),

$$\frac{T_{N(t)}}{N(t)} \le \frac{t}{N(t)} < \frac{T_{N(t)+1}}{N(t)+1} \cdot (1 + N(t)^{-1}).$$

Take the limit as $t \to \infty$, using the fact that $T_n/n \xrightarrow{\text{a.s.}} \mathbb{E}(X_1)$ by the strong law, to deduce that $t/N(t) \xrightarrow{\text{a.s.}} \mathbb{E}(X_1)$.

3. By the strong law, $S_n/n \xrightarrow{\text{a.s.}} \mathbb{E}(X_1) \ne 0$. In particular, with probability 1, $S_n = 0$ only finitely often.

7.6 Solution. The law of the iterated logarithm

1. The sum S_n is approximately $N(0, n)$, so that

$$\mathbb{P}\big(S_n > \sqrt{\alpha n \log n}\big) = 1 - \Phi\big(\sqrt{\alpha \log n}\big) < \frac{n^{-\frac{1}{2}\alpha}}{\sqrt{\alpha \log n}}$$

for all large n, by the tail estimate of Exercise (4.4.8) or Problem (4.14.1c) for the normal distribution. This is summable if $\alpha > 2$, and the claim follows by an application of the first Borel–Cantelli lemma.

7.7 Solutions. Martingales

1. Suppose $i < j$. Then

$$\begin{aligned}\mathbb{E}(X_j X_i) &= \mathbb{E}\Big\{\mathbb{E}\big[(S_j - S_{j-1})X_i \mid S_0, S_1, \dots, S_{j-1}\big]\Big\} \\ &= \mathbb{E}\Big\{X_i\big[\mathbb{E}(S_j \mid S_0, S_1, \dots, S_{j-1}) - S_{j-1}\big]\Big\} = 0\end{aligned}$$

by the martingale property.

2. Clearly $\mathbb{E}|S_n| < \infty$ for all n. Also, for $n \ge 0$,

$$\begin{aligned}\mathbb{E}(S_{n+1} \mid Z_0, Z_1, \dots, Z_n) &= \frac{1}{\mu^{n+1}}\left\{\mathbb{E}(Z_{n+1} \mid Z_0, \dots, Z_n) - m\left(\frac{1-\mu^{n+1}}{1-\mu}\right)\right\} \\ &= \frac{1}{\mu^{n+1}}\left\{m + \mu Z_n - m\left(\frac{1-\mu^{n+1}}{1-\mu}\right)\right\} = S_n.\end{aligned}$$

3. Certainly $\mathbb{E}|S_n| < \infty$ for all n. Secondly, for $n \ge 1$,

$$\begin{aligned}\mathbb{E}(S_{n+1} \mid X_0, X_1, \dots, X_n) &= \alpha\mathbb{E}(X_{n+1} \mid X_0, \dots, X_n) + X_n \\ &= (\alpha a + 1)X_n + \alpha b X_{n-1},\end{aligned}$$

which equals S_n if $\alpha = (1-a)^{-1}$.

4. The gambler stakes $Z_i = f_{i-1}(X_1, \dots, X_{i-1})$ on the ith play, at a return of X_i per unit. Therefore $S_i = S_{i-1} + X_i Z_i$ for $i \ge 2$, with $S_1 = X_1 Y$. Secondly,

$$\mathbb{E}(S_{n+1} - S_n \mid X_1, \dots, X_n) = Z_{n+1}\mathbb{E}(X_{n+1} \mid X_1, \dots, X_n) = 0,$$

where we have used the fact that Z_{n+1} depends only on $X_1, X_2, \dots, X_n$.

5. Insert n uniformly at random into $(\pi_1, \pi_2, \dots, \pi_{n-1})$ to obtain a random permutation of the sequence $(1, 2, \dots, n)$. In R_{n-1} of the n possible placements, n falls to the right of an existing run, and then $R_n = R_{n-1}$. Otherwise, it splits an existing run to yield $R_n = R_{n-1} + 1$. Hence,

$$\begin{aligned}\mathbb{E}(M_n \mid M_1, \dots, M_{n-1}) &= nR_{n-1} \cdot \frac{R_{n-1}}{n} + n(R_{n-1}+1)\left(1 - \frac{R_{n-1}}{n}\right) - \tfrac{1}{2}n(n+1) \\ &= (n-1)R_{n-1} - \tfrac{1}{2}n(n-1) = M_{n-1},\end{aligned}$$

as required. We have that $\mathbb{E}(M_n) = \mathbb{E}(M_1) = 0$, so that $\mathbb{E}(R_n) = \frac{1}{2}(n+1)$. Finally,

$$\begin{aligned}\mathbb{E}(R_n^2 \mid M_1, \dots, M_{n-1}) &= R_{n-1}^2 \cdot \frac{R_{n-1}}{n} + (R_{n-1}+1)^2\left(1 - \frac{R_{n-1}}{n}\right) \\ &= R_{n-1}^2\left(1 - \frac{2}{n}\right) + 1 + R_{n-1}\left(2 - \frac{1}{n}\right).\end{aligned}$$

Take expectations to find that $r_n = \mathbb{E}(R_n^2)$ satisfies

$$r_n = r_{n-1}\left(1 - \frac{2}{n}\right) + n + \tfrac{1}{2}, \qquad n \ge 2.$$

In particular, $r_2 = \frac{5}{2}$ (the term $r_1 = 1$ does not feature in these difference equations). The solution is $r_n = \frac{1}{12}(3n+4)(n+1)$ for $n \ge 2$.

7.8 Solutions. Martingale convergence theorem

1. It is easily checked that S_n defines a martingale with respect to itself, and the claim follows from the Doob–Kolmogorov inequality, using the fact that

$$\mathbb{E}(S_n^2) = \sum_{j=1}^{n} \operatorname{var}(X_j).$$

2. It would be easy but somewhat perverse to use the martingale convergence theorem, and so we give a direct proof based on Kolmogorov's inequality of Exercise (7.8.1). Applying this inequality to the sequence $Z_m, Z_{m+1}, \dots$ where $Z_i = (X_i - \mathbb{E}X_i)/i$, we obtain that $S_n = Z_1 + Z_2 + \cdots + Z_n$ satisfies, for $\epsilon > 0$,

$$\mathbb{P}\left(\max_{m \le n \le r} |S_n - S_m| > \epsilon\right) \le \frac{1}{\epsilon^2}\sum_{n=m+1}^{r} \operatorname{var}(Z_n).$$

We take the limit as $r \to \infty$, using the continuity of $\mathbb{P}$, to obtain

$$\mathbb{P}\left(\sup_{n \ge m} |S_n - S_m| > \epsilon\right) \le \frac{1}{\epsilon^2}\sum_{n=m+1}^{\infty} \frac{1}{n^2}\operatorname{var}(X_n).$$

Now let $m \to \infty$ to obtain (after a small step)

$$\mathbb{P}\left(\lim_{m\to\infty} \sup_{n\geq m} |S_n - S_m| \leq \epsilon\right) = 1 \qquad \text{for all } \epsilon > 0.$$

Any real sequence (x_n) satisfying

$$\lim_{m\to\infty} \sup_{n\geq m} |x_n - x_m| \leq \epsilon \qquad \text{for all } \epsilon > 0,$$

is Cauchy convergent, and hence convergent. It follows that S_n converges a.s. to some limit Y.

The last part is an immediate consequence, using Kronecker's lemma.

3. By the martingale convergence theorem, $S = \lim_{n\to\infty} S_n$ exists a.s., and $S_n \xrightarrow{\text{m.s.}} S$. Using Exercise (7.2.1c), $\text{var}(S_n) \to \text{var}(S)$, and therefore $\text{var}(S) = 0$.

7.9 Solutions. Prediction and conditional expectation

1. (a) Clearly the best predictors are $\mathbb{E}(X \mid Y) = Y^2$, $\mathbb{E}(Y \mid X) = 0$.

(b) We have, after expansion, that

$$\mathbb{E}\{(X - aY - b)^2\} = \text{var}(Y^2) + a^2\mathbb{E}(Y^2) + \{b - \mathbb{E}(Y^2)\}^2,$$

since $\mathbb{E}(Y) = \mathbb{E}(Y^3) = 0$. This is a minimum when $b = \mathbb{E}(Y^2) = \frac{1}{3}$, and $a = 0$. The best linear predictor of X given Y is therefore $\frac{1}{3}$.

Note that $\mathbb{E}(Y \mid X) = 0$ *is* a linear function of X; it is therefore the best linear predictor of Y given X.

2. (a) By the result of Problem (4.14.13), $\mathbb{E}(Y \mid X) = \mu_2 + \rho\sigma_2(X - \mu_1)/\sigma_1$, in the natural notation.

(b) We have $\mu_1 = \mu_2 = 0$ and

$$\rho = \frac{\sum_i a_i b_i}{\sigma_1\sigma_2}, \quad \sigma_1^2 = \sum_i a_i^2, \quad \sigma_2^2 = \sum_i b_i^2.$$

3. Write

$$g(\mathbf{a}) = \sum_{i=1}^n a_i X_i = \mathbf{a}\mathbf{X}',$$

and

$$v(\mathbf{a}) = \mathbb{E}\{(Y - g(\mathbf{a}))^2\} = \mathbb{E}(Y^2) - 2\mathbf{a}\mathbb{E}(Y\mathbf{X}') + \mathbf{a}\mathbf{V}\mathbf{a}'.$$

Let $\widehat{\mathbf{a}}$ be a vector satisfying $\mathbf{V}\widehat{\mathbf{a}}' = \mathbb{E}(Y\mathbf{X}')$. Then

$$\begin{aligned} v(\mathbf{a}) - v(\widehat{\mathbf{a}}) &= \mathbf{a}\mathbf{V}\mathbf{a}' - 2\mathbf{a}\mathbb{E}(Y\mathbf{X}') + 2\widehat{\mathbf{a}}\mathbb{E}(Y\mathbf{X}') - \widehat{\mathbf{a}}\mathbf{V}\widehat{\mathbf{a}}' \\ &= \mathbf{a}\mathbf{V}\mathbf{a}' - 2\mathbf{a}\mathbf{V}\widehat{\mathbf{a}}' + \widehat{\mathbf{a}}\mathbf{V}\widehat{\mathbf{a}}' = (\mathbf{a} - \widehat{\mathbf{a}})\mathbf{V}(\mathbf{a} - \widehat{\mathbf{a}})' \geq 0, \end{aligned}$$

since $\mathbf{V}$ is non-negative definite. Hence $v(\mathbf{a})$ is a minimum when $\mathbf{a} = \widehat{\mathbf{a}}$, and the answer is $g(\widehat{\mathbf{a}})$. If $\mathbf{V}$ is non-singular, $\widehat{\mathbf{a}} = \mathbb{E}(Y\mathbf{X})\mathbf{V}^{-1}$.

4. Recall that $Z = \mathbb{E}(Y \mid \mathcal{G})$ is the ('almost') unique $\mathcal{G}$-measurable random variable with finite mean and satisfying $\mathbb{E}\{(Y - Z)I_G\} = 0$ for all $G \in \mathcal{G}$.

(a) $\Omega \in \mathcal{G}$, and hence $\mathbb{E}\{\mathbb{E}(Y \mid \mathcal{G})I_\Omega\} = \mathbb{E}(ZI_\Omega) = \mathbb{E}(YI_\Omega)$.

(b) $U = \alpha\mathbb{E}(Y \mid \mathcal{G}) + \beta\mathbb{E}(Z \mid \mathcal{G})$ satisfies

$$\begin{aligned}\mathbb{E}(UI_G) &= \alpha\mathbb{E}\{\mathbb{E}(Y \mid \mathcal{G})I_G\} + \beta\mathbb{E}\{\mathbb{E}(Z \mid \mathcal{G})I_G\} \\ &= \alpha\mathbb{E}(YI_G) + \beta\mathbb{E}(ZI_G) = \mathbb{E}\{(\alpha Y + \beta Z)I_G\}, \qquad G \in \mathcal{G}.\end{aligned}$$

Also, U is $\mathcal{G}$-measurable.

(c) Suppose there exists m (> 0) such that $G = \{\mathbb{E}(Y \mid \mathcal{G}) < -m\}$ has strictly positive probability. Then $G \in \mathcal{G}$, and so $\mathbb{E}(YI_G) = \mathbb{E}\{\mathbb{E}(Y \mid \mathcal{G})I_G\}$. However $YI_G \geq 0$, whereas $\mathbb{E}(Y \mid \mathcal{G})I_G < -m$. We obtain a contradiction on taking expectations.

(d) Just check the definition of conditional expectation.

(e) If Y is independent of $\mathcal{G}$, then $\mathbb{E}(YI_G) = \mathbb{E}(Y)\mathbb{P}(G)$ for $G \in \mathcal{G}$. Hence $\mathbb{E}\{(Y - \mathbb{E}(Y))I_G\} = 0$ for $G \in \mathcal{G}$, as required.

(f) If g is convex then, for all $a \in \mathbb{R}$, there exists $\lambda(a)$ such that

$$g(y) \geq g(a) + (y - a)\lambda(a);$$

furthermore λ may be chosen to be a measurable function of a. Set $a = \mathbb{E}(Y \mid \mathcal{G})$ and $y = Y$, to obtain

$$g(Y) \geq g\{\mathbb{E}(Y \mid \mathcal{G})\} + \{Y - \mathbb{E}(Y \mid \mathcal{G})\}\lambda\{\mathbb{E}(Y \mid \mathcal{G})\}.$$

Take expectations conditional on $\mathcal{G}$, and use the fact that $\mathbb{E}(Y \mid \mathcal{G})$ is $\mathcal{G}$-measurable.

(g) We have that

$$\big|\mathbb{E}(Y_n \mid \mathcal{G}) - \mathbb{E}(Y \mid \mathcal{G})\big| \leq \mathbb{E}\{|Y_n - Y| \,\big|\, \mathcal{G}\} \leq V_n$$

where $V_n = \mathbb{E}\{\sup_{m \geq n} |Y_m - Y| \,\big|\, \mathcal{G}\}$. Now $\{V_n : n \geq 1\}$ is non-increasing and bounded below. Hence $V = \lim_{n\to\infty} V_n$ exists and satisfies $V \geq 0$. Also

$$\mathbb{E}(V) \leq \mathbb{E}(V_n) = \mathbb{E}\left\{\sup_{m \geq n} |Y_m - Y|\right\},$$

which tends to 0 as $m \to \infty$, by the dominated convergence theorem. Therefore $\mathbb{E}(V) = 0$, and hence $\mathbb{P}(V = 0) = 1$. The claim follows.

5. $\mathbb{E}(Y \mid X) = X$.

6. (a) Let $\{X_n : n \geq 1\}$ be a sequence of members of H which is Cauchy convergent in mean square, that is, $\mathbb{E}\{|X_n - X_m|^2\} \to 0$ as $m, n \to \infty$. By Chebyshov's inequality, $\{X_n : n \geq 1\}$ is Cauchy convergent in probability, and therefore converges in probability to some limit X (see Exercise (7.3.1)). It follows that there exists a subsequence $\{X_{n_k} : k \geq 1\}$ which converges to X almost surely. Since each X_{n_k} is $\mathcal{G}$-measurable, we may assume that X is $\mathcal{G}$-measurable. Now, as $n \to \infty$,

$$\mathbb{E}\{|X_n - X|^2\} = \mathbb{E}\left\{\liminf_{k\to\infty} |X_n - X_{n_k}|^2\right\} \leq \liminf_{k\to\infty} \mathbb{E}\{|X_n - X_{n_k}|^2\} \to 0,$$

where we have used Fatou's lemma and Cauchy convergence in mean square. Therefore $X_n \xrightarrow{\text{m.s.}} X$. That $\mathbb{E}(X^2) < \infty$ is a consequence of Exercise (7.2.1a).

(b) That (i) implies (ii) is obvious, since $I_G \in H$. Suppose that (ii) holds. Any Z ($\in H$) may be written as the limit, as $n \to \infty$, of random variables of the form

$$Z_n = \sum_{i=1}^{m(n)} a_i(n) I_{G_i(n)}$$

for reals $a_i(n)$ and events $G_i(n)$ in $\mathcal{G}$; furthermore we may assume that $|Z_n| \le |Z|$. It is easy to see that $\mathbb{E}\{(Y-M)Z_n\} = 0$ for all n. By dominated convergence, $\mathbb{E}\{(Y-M)Z_n\} \to \mathbb{E}\{(Y-M)Z\}$, and the claim follows.

7. (a) Suppose $m(X,Y) = 0$. Let $f(x) = I(x \le u)$, $g(y) = I(y \le v)$, and apply the bound $m(X,Y) = 0$ to the pairs (f, g) and $(1-f, g)$, to deduce that $\mathbb{P}(X \le u, Y \le v) = \mathbb{P}(X \le u)\mathbb{P}(Y \le v)$, whence. X and Y are independent. The converse is elementary.

(b, c) Without loss of generality, we consider functions $f(X)$ and $g(Y)$ with mean 0 and variance 1. By the Cauchy–Schwarz inequality, with $Z = \mathbb{E}(g(Y) \mid X)$,

$$(*) \qquad \mathbb{E}(f(X)g(Y)) = \mathbb{E}(f(X)Z) \le \sqrt{\mathbb{E}(f(X)^2)\mathbb{E}(Z^2)} = \sqrt{\mathbb{E}(Z^2)},$$

with equality if and only if there exists $c \in \mathbb{R}$ such that $\mathbb{P}(f(X) = cZ) = 1$. Let $\widehat{f}, \widehat{g}$ be the extremal functions f, g in the definition of m. By (*), $m(X,Y)^2 = \sup_g \operatorname{var}(Z)$.

Now, with $\widehat{Z} = \mathbb{E}(\widehat{g}(Y) \mid X)$,

$$m(X,Y) = \mathbb{E}(\widehat{f}(X)\widehat{g}(Y)) = \mathbb{E}(\widehat{f}(X)\widehat{Z}) = \frac{1}{c}\mathbb{E}(\widehat{f}(X)^2) = \frac{1}{c},$$

whence $c = 1/m(X,Y)$, so that $m\widehat{f}(X) = \widehat{Z}$ a.s.

(d) By the above with f and g interchanged, $m\widehat{g}(Y) = \mathbb{E}(\widehat{f}(X) \mid Y)$ a.s. Take conditional expectation given X to obtain

$$m^2\widehat{f}(X) = m\mathbb{E}(\widehat{g}(Y) \mid X) = \mathbb{E}\big(\mathbb{E}(\widehat{f}(X) \mid Y) \bigm| X\big),$$

with a similar expression with f and g interchanged.

(e) Suppose X and Z are conditionally independent given Y. For arbitrary $g(X)$ and $h(Z)$, assumed to have means 0 and variances 1 and σ^2, respectively,

$$\begin{aligned}\rho(g(X), h(Z)) &= \frac{1}{\sigma}\mathbb{E}\big(\mathbb{E}(g(X)h(Z) \mid Y)\big)\\ &= \frac{1}{\sigma}\mathbb{E}\big(\mathbb{E}(g(X) \mid Y)\mathbb{E}(h(Z) \mid Y)\big)\\ &\le \sqrt{\mathbb{E}\big(\mathbb{E}(g(X) \mid Y)^2\big)\mathbb{E}\big(\mathbb{E}(h(Z) \mid Y)^2/\sigma^2\big)}\\ &\le m(X,Y)m(Y,Z),\end{aligned}$$

by the conditional independence, the Cauchy–Schwarz inequality, and part (b). If (X,Y) and (Z,Y) have the same distribution,

$$\begin{aligned}m(X,Z) &\ge \operatorname{cov}(g(X), g(Z)) = \operatorname{cov}\big(\mathbb{E}(g(X) \mid Y), \mathbb{E}(g(Z) \mid Y)\big)\\ &= \operatorname{var}\big(\mathbb{E}(g(X) \mid Y)\big) \qquad \text{by symmetry.}\end{aligned}$$

It follows by (b) that

$$m(X,Z) \ge m(X,Y)^2 = m(X,Y)m(Y,Z),$$

which combines with the above to imply equality.

(f) Write $m(\rho) = m(U,V)$, and let $\rho_1, \rho_2 \in (-1,1)$. Let A, B, C be independent $N(0,1)$ random variables, and define

$$X = A, \quad Y = X\rho_1 + B\sqrt{1-\rho_1^2}, \quad Z = Y\rho_2 + C\sqrt{1-\rho_2^2}.$$

Note that, given Y, the random variables X and Z are independent, so that X, Y, Z is a Markov chain. By part (e), $m(\rho_1\rho_2) \le m(\rho_1)m(\rho_2)$, and hence $m(\rho) \le m(\rho')$ if $|\rho| \le |\rho'|$.

8. Suppose $\rho_{\text{mon}}(X, Y) = 0$, and let $f(x) = I(x \le u)$, $g(y) = I(y \le v)$, where $u, v \in \mathbb{R}$. If either $\rho(f(X), g(Y)) > 0$ or $\rho(f(X), -g(X)) > 0$, then we cannot have $\rho_{\text{mon}}(X, Y) = 0$. Therefore, $\rho(f(X), g(Y)) = 0$, and hence $\mathbb{P}(X \le u,\ Y \le v) = \mathbb{P}(X \le u)\mathbb{P}(Y \le v)$ for $u, v \in \mathbb{R}$. It follows that X and Y are independent. The converse is elementary, as are the final inequalities.

9. (a) The MMSE predictor is $\mathbb{E}(X \mid Y = y)$. The conditional density function of X given $Y = y$ is

$$f_{X|Y}(x \mid y) = \frac{f(x, y)}{f_Y(y)} = \frac{2e^{-(x+y)}}{2e^{-y}(1 - e^{-y})} = \frac{e^{-x}}{1 - e^{-y}}, \qquad 0 < x < y,$$

which has mean

$$\mathbb{E}(X \mid Y = y) = \frac{1 - e^{-y} - ye^{-y}}{1 - e^{-y}}.$$

(b) The best linear predictor of X given $Y = y$ is

$$L(y) = \mathbb{E}(X) + \frac{\text{cov}(X, Y)}{\text{var}(Y)}(y - \mathbb{E}(Y)) = \frac{1}{5}(1 + y).$$

You may find the solution to Problem (4.14.7) to be useful in your calculation.
(c) The linear predictor is not bad for small y, but is very poor for large y.

7.10 Solutions. Uniform integrability

1. It is easily checked by considering whether $|x| \le a$ or $|y| \le a$ that, for $a > 0$,

$$|x + y|I_{\{|x+y| \ge 2a\}} \le 2\left(|x|I_{\{|x| \ge a\}} + |y|I_{\{|y| \ge a\}}\right).$$

Now substitute $x = X_n$ and $y = Y_n$, and take expectations.

2. (a) Let $\epsilon > 0$. There exists N such that $\mathbb{E}(|X_n - X|^r) < \epsilon$ if $n > N$. Now $\mathbb{E}|X^r| < \infty$, by Exercise (7.2.1a), and therefore there exists δ (> 0) such that

$$\mathbb{E}(|X|^r I_A) < \epsilon, \quad \mathbb{E}(|X_n|^r I_A) < \epsilon \quad \text{for } 1 \le n \le N,$$

for all events A such that $\mathbb{P}(A) < \delta$. By Minkowski's inequality,

$$\left\{\mathbb{E}(|X_n|^r I_A)\right\}^{1/r} \le \left\{\mathbb{E}(|X_n - X|^r I_A)\right\}^{1/r} + \left\{\mathbb{E}(|X|^r I_A)\right\}^{1/r} \le 2\epsilon^{1/r} \qquad \text{if } n > N$$

if $\mathbb{P}(A) < \delta$. Therefore $\{|X_n|^r : n \ge 1\}$ is uniformly integrable.

If r is an integer then $\{X_n^r : n \ge 1\}$ is uniformly integrable also. Also $X_n^r \xrightarrow{\text{P}} X^r$ since $X_n \xrightarrow{\text{P}} X$ (use the result of Problem (7.11.3)). Therefore $\mathbb{E}(X_n^r) \to \mathbb{E}(X^r)$ as required.

(b) Suppose now that the collection $\{|X_n|^r : n \ge 1\}$ is uniformly integrable and $X_n \xrightarrow{\text{P}} X$. We show first that $\mathbb{E}|X^r| < \infty$, as follows. There exists a subsequence $\{X_{n_k} : k \ge 1\}$ which converges to X almost surely. By Fatou's lemma,

$$\mathbb{E}|X^r| = \mathbb{E}\left(\liminf_{k\to\infty} |X_{n_k}|^r\right) \le \liminf_{k\to\infty} \mathbb{E}|X_{n_k}^r| \le \sup_n \mathbb{E}|X_n^r| < \infty.$$

If $\epsilon > 0$, there exists δ (> 0) such that

$$\mathbb{E}(|X^r|I_A) < \epsilon, \quad \mathbb{E}(|X_n^r|I_A) < \epsilon \qquad \text{for all } n,$$

whenever A is such that $\mathbb{P}(A) < \delta$. There exists N such that $B_n(\epsilon) = \{|X_n - X| > \epsilon\}$ satisfies $\mathbb{P}(B_n(\epsilon)) < \delta$ for $n > N$. Consequently

$$\mathbb{E}(|X_n - X|^r) \le \epsilon^r + \mathbb{E}\left(|X_n - X|^r I_{B_n(\epsilon)}\right), \qquad n > N,$$

of which the final term satisfies

$$\left\{\mathbb{E}\left(|X_n - X|^r I_{B_n(\epsilon)}\right)\right\}^{1/r} \le \left\{\mathbb{E}\left(|X_n^r| I_{B_n(\epsilon)}\right)\right\}^{1/r} + \left\{\mathbb{E}\left(|X^r| I_{B_n(\epsilon)}\right)\right\}^{1/r} \le 2\epsilon^{1/r}.$$

Therefore, $X_n \xrightarrow{r} X$.

3. Fix $\epsilon > 0$, and find a real number a such that $g(x) > x/\epsilon$ if $x > a$. If $b \ge a$,

$$\mathbb{E}\left(|X_n| I_{\{|X_n|>b\}}\right) < \epsilon \mathbb{E}\{g(|X_n|)\} \le \epsilon \sup_n \mathbb{E}\{g(|X_n|)\},$$

whence the left side approaches 0, uniformly in n, as $b \to \infty$.

4. Here is a quick way. Extinction is (almost) certain for such a branching process, so that $Z_n \xrightarrow{\text{a.s.}} 0$, and hence $Z_n \xrightarrow{\text{P}} 0$. If $\{Z_n : n \ge 0\}$ were uniformly integrable, it would follow that $\mathbb{E}(Z_n) \to 0$ as $n \to \infty$; however $\mathbb{E}(Z_n) = 1$ for all n.

5. We may suppose that X_n, Y_n, and Z_n have finite means, for all n. We have that $0 \le Y_n - X_n \le Z_n - X_n$ where, by Theorem (7.3.9c), $Z_n - X_n \xrightarrow{\text{P}} Z - X$. Also

$$\mathbb{E}|Z_n - X_n| = \mathbb{E}(Z_n - X_n) \to \mathbb{E}(Z - X) = \mathbb{E}|Z - X|,$$

so that $\{Z_n - X_n : n \ge 1\}$ is uniformly integrable, by Theorem (7.10.3). It follows that $\{Y_n - X_n : n \ge 1\}$ is uniformly integrable. Also $Y_n - X_n \xrightarrow{\text{P}} Y - X$, and therefore by Theorem (7.10.3), $\mathbb{E}|Y_n - X_n| \to \mathbb{E}|Y - X|$, which is to say that $\mathbb{E}(Y_n) - \mathbb{E}(X_n) \to \mathbb{E}(Y) - \mathbb{E}(X)$; hence $\mathbb{E}(Y_n) \to \mathbb{E}(Y)$.

It is not necessary to use uniform integrability; try doing it using the 'more primitive' Fatou's lemma.

6. For any event A, $\mathbb{E}(|X_n| I_A) \le \mathbb{E}(Z I_A)$ where $Z = \sup_n |X_n|$. The uniform integrability follows by the assumption that $\mathbb{E}(Z) < \infty$.

7. Let $\{A_n : n \ge 1\}$ and $\{B_n : n \ge 1\}$ be independent sequences of independent random variables taking non-negative values such that: (i) $\mathbb{E}(A_n) \to 0$ as $n \to \infty$, but A_n does not converge to 0 a.s., and (ii) $\mathbb{E}(B_n) = 1$ and, for each elementary event ω there exists $N = N(\omega)$ such that $B_n(\omega) = 0$ for $n \ge N$. Set $X_n = A_n B_n$ and let $\mathcal{G}$ be the σ-field generated by the A_m. Then $\mathbb{E}(X_n \mid \mathcal{G}) = A_n \mathbb{E}(B_n) = A_n$, whereas $X_n = 0$ for $n \ge N$.

7.11 Solutions to problems

1. $\mathbb{E}|X_n^r| = \infty$ for $r \ge 1$, so there is no convergence in any mean. However, if $\epsilon > 0$,

$$\mathbb{P}(|X_n| > \epsilon) = 1 - \frac{2}{\pi} \tan^{-1}(n\epsilon) \to 0 \qquad \text{as } n \to \infty,$$

so that $X_n \xrightarrow{\text{P}} 0$.

You have insufficient information to decide whether or not X_n converges almost surely:

(a) Let X be Cauchy, and let $X_n = X/n$. Then X_n has the required density function, and $X_n \xrightarrow{\text{a.s.}} 0$.

(b) Let the X_n be independent with the specified density functions. For $\epsilon > 0$,

$$\mathbb{P}(|X_n| > \epsilon) = \frac{2}{\pi}\sin^{-1}\left(\frac{1}{\sqrt{1+n^2\epsilon^2}}\right) \sim \frac{2}{\pi n\epsilon},$$

so that $\sum_n \mathbb{P}(|X_n| > \epsilon) = \infty$. By the second Borel–Cantelli lemma, $|X_n| > \epsilon$ i.o. with probability one, implying that X_n does not converge a.s. to 0.

2. Assume all the random variables are defined on the same probability space; otherwise it is meaningless to add or multiply them.

(i) (a) Clearly $X_n(\omega) + Y_n(\omega) \to X(\omega) + Y(\omega)$ whenever $X_n(\omega) \to X(\omega)$ and $Y_n(\omega) \to Y(\omega)$. Therefore

$$\{X_n + Y_n \nrightarrow X + Y\} \subseteq \{X_n \nrightarrow X\} \cup \{Y_n \nrightarrow Y\},$$

a union of events having zero probability.

(b) Use Minkowski's inequality to obtain that

$$\{\mathbb{E}(|X_n + Y_n - X - Y|^r)\}^{1/r} \le \{\mathbb{E}(|X_n - X|^r)\}^{1/r} + \{\mathbb{E}(|Y_n - Y|^r)\}^{1/r}.$$

(c) If $\epsilon > 0$, we have that

$$\{|X_n + Y_n - X - Y| > \epsilon\} \subseteq \{|X_n - X| > \tfrac{1}{2}\epsilon\} \cup \{|Y_n - Y| > \tfrac{1}{2}\epsilon\},$$

and the probability of the right side tends to 0 as $n \to \infty$.

(d) If $X_n \xrightarrow{D} X$ and the X_n are symmetric, then $-X_n \xrightarrow{D} X$. However $X_n + (-X_n) \xrightarrow{D} 0$, which generally differs from $2X$ in distribution.

(ii) (e) Almost-sure convergence follows as in (a) above.

(f) The corresponding statement for convergence in rth mean is false in general. Find a random variable Z such that $\mathbb{E}|Z^r| < \infty$ but $\mathbb{E}|Z^{2r}| = \infty$, and define $X_n = Y_n = Z$ for all n.

(g) Suppose $X_n \xrightarrow{P} X$ and $Y_n \xrightarrow{P} Y$. Let $\epsilon > 0$. Then

$$\begin{aligned}\mathbb{P}(|X_nY_n - XY| > \epsilon) &= \mathbb{P}\Big(\big|(X_n - X)(Y_n - Y) + (X_n - X)Y + X(Y_n - Y)\big| > \epsilon\Big)\\ &\le \mathbb{P}(|X_n - X|\cdot|Y_n - Y| > \tfrac{1}{3}\epsilon) + \mathbb{P}(|X_n - X|\cdot|Y| > \tfrac{1}{3}\epsilon)\\ &\quad + \mathbb{P}(|X|\cdot|Y_n - Y| > \tfrac{1}{3}\epsilon).\end{aligned}$$

Now, for $\delta > 0$,

$$\mathbb{P}(|X_n - X|\cdot|Y| > \tfrac{1}{3}\epsilon) \le \mathbb{P}(|X_n - X| > \epsilon/(3\delta)) + \mathbb{P}(|Y| > \delta),$$

which tends to 0 in the limit as $n \to \infty$ and $\delta \to \infty$ in that order. Together with two similar facts, we obtain that $X_nY_n \xrightarrow{P} XY$.

(h) The example of (d) above indicates that the corresponding statement is false for convergence in distribution.

3. Let $\epsilon > 0$. We may pick M such that $\mathbb{P}(|X| \ge M) \le \epsilon$. The continuous function g is uniformly continuous on the bounded interval $[-M, M]$. There exists $\delta > 0$ such that

$$|g(x) - g(y)| \le \epsilon \quad \text{if } |x - y| \le \delta \text{ and } |x| \le M.$$

If $|g(X_n) - g(X)| > \epsilon$, then either $|X_n - X| > \delta$ or $|X| \ge M$. Therefore

$$\mathbb{P}(|g(X_n) - g(X)| > \epsilon) \le \mathbb{P}(|X_n - X| > \delta) + \mathbb{P}(|X| \ge M) \to \mathbb{P}(|X| \ge M) \le \epsilon,$$

in the limit as $n \to \infty$. It follows that $g(X_n) \xrightarrow{\text{P}} g(X)$.

4. Clearly

$$\mathbb{E}(e^{itX_n}) = \prod_{j=1}^{n} \mathbb{E}(e^{itY_j/10^j}) = \prod_{j=1}^{n} \left\{ \frac{1}{10} \cdot \frac{1 - e^{it/10^{j-1}}}{1 - e^{it/10^j}} \right\}$$
$$= \frac{1 - e^{it}}{10^n(1 - e^{it/10^n})} \to \frac{e^{it} - 1}{it}$$

as $n \to \infty$. The limit is the characteristic function of the uniform distribution on $[0, 1]$.

Now $X_n \le X_{n+1} \le 1$ for all n, so that $Y(\omega) = \lim_{n\to\infty} X_n(\omega)$ exists for all ω. Therefore $X_n \xrightarrow{\text{a.s.}} Y$; hence $X_n \xrightarrow{\text{D}} Y$, whence Y has the uniform distribution.

5. (a) Suppose $s < t$. Then

$$\mathbb{E}\big(N(s)N(t)\big) = \mathbb{E}(N(s)^2) + \mathbb{E}\{N(s)(N(t) - N(s))\} = \mathbb{E}(N(s)^2) + \mathbb{E}(N(s))\mathbb{E}\big(N(t) - N(s)\big),$$

since N has independent increments. Therefore

$$\operatorname{cov}(N(s), N(t)) = \mathbb{E}\big(N(s)N(t)\big) - \mathbb{E}(N(s))\mathbb{E}(N(t))$$
$$= (\lambda s)^2 + \lambda s + \lambda s\{\lambda(t - s)\} - (\lambda s)(\lambda t) = \lambda s.$$

In general, $\operatorname{cov}(N(s), N(t)) = \lambda \min\{s, t\}$.

(b) $N(t + h) - N(t)$ has the same distribution as $N(h)$, if $h > 0$. Hence

$$\mathbb{E}\left(\{N(t + h) - N(t)\}^2\right) = \mathbb{E}(N(h)^2) = (\lambda h)^2 + \lambda h$$

which tends to 0 as $h \to 0$.

(c) By Markov's inequality,

$$\mathbb{P}(|N(t + h) - N(t)| > \epsilon) \le \frac{1}{\epsilon^2}\mathbb{E}\left(\{N(t + h) - N(t)\}^2\right),$$

which tends to 0 as $h \to 0$, if $\epsilon > 0$.

(d) Let $\epsilon > 0$. For $0 < h < \epsilon^{-1}$,

$$\mathbb{P}\left(\left|\frac{N(t + h) - N(t)}{h}\right| > \epsilon\right) = \mathbb{P}\big(N(t + h) - N(t) \ge 1\big) = \lambda h + \mathrm{o}(h),$$

which tends to 0 as $h \to 0$.

On the other hand,

$$\mathbb{E}\left(\left\{\frac{N(t + h) - N(t)}{h}\right\}^2\right) = \frac{1}{h^2}\{(\lambda h)^2 + \lambda h\}$$

which tends to ∞ as $h \downarrow 0$.

6. By Markov's inequality, $S_n = \sum_{i=1}^{n} X_i$ satisfies

$$\mathbb{P}(|S_n| > n\epsilon) \le \frac{\mathbb{E}(S_n^4)}{(n\epsilon)^4}$$

for $\epsilon > 0$. Using the properties of the X's,

$$\begin{aligned}\mathbb{E}(S_n^4) &= n\mathbb{E}(X_1^4) + 4\binom{n}{2}\mathbb{E}(X_1^3 X_2) + \binom{4}{2}\binom{n}{2}\mathbb{E}(X_1^2 X_2^2) \\ &\quad + 3\binom{4}{2}\binom{n}{3}\mathbb{E}(X_1^2 X_2 X_3) + 4!\binom{n}{4}\mathbb{E}(X_1 X_2 X_3 X_4) \\ &= n\mathbb{E}(X_1^4) + \binom{4}{2}\binom{n}{2}\mathbb{E}(X_1^2 X_2^2),\end{aligned}$$

since $\mathbb{E}(X_i) = 0$ for all i. Therefore there exists a constant C such that

$$\sum_n \mathbb{P}(|n^{-1}S_n| > \epsilon) \le \sum_n \frac{C}{n^2} < \infty,$$

implying (via the first Borel–Cantelli lemma) that $n^{-1}S_n \xrightarrow{\text{a.s.}} 0$.

7. We have by Markov's inequality that

$$\sum_n \mathbb{P}(|X_n - X| > \epsilon) \le \sum_n \frac{\mathbb{E}\{|X_n - X|^r\}}{\epsilon^r} < \infty$$

for $\epsilon > 0$, so that $X_n \xrightarrow{\text{a.s.}} X$ (via the first Borel–Cantelli lemma).

8. Either use the Skorokhod representation or characteristic functions. Following the latter route, the characteristic function of $aX_n + b$ is

$$\mathbb{E}(e^{it(aX_n+b)}) = e^{itb}\phi_n(at) \to e^{itb}\phi_X(at) = \mathbb{E}(e^{it(aX+b)})$$

where ϕ_n is the characteristic function of X_n. The result follows by the continuity theorem.

9. (a) For any positive reals c, t,

$$\mathbb{P}(X \ge t) = \mathbb{P}(X + c \ge t + c) \le \frac{\mathbb{E}\{(X+c)^2\}}{(t+c)^2}.$$

Set $c = \sigma^2/t$ to obtain the required inequality. See also Exercise (3.6.11).

(b) Set $t = \sigma + \epsilon > \sigma$ to find $\mathbb{P}(X - \mu \ge \sigma + \epsilon) < \frac{1}{2}$. Likewise, $\mathbb{P}(X - \mu \le -\sigma - \epsilon) < \frac{1}{2}$, and the result follows.

(c) We have

$$\begin{aligned}|\mu - m| = |\mathbb{E}(X - m)| \le \mathbb{E}|X - m| &\quad \text{by Jensen's inequality} \\ \le \mathbb{E}|X - \mu| &\quad \text{since } a = \mu \text{ minimizes } \mathbb{E}|X - a| \\ \le \sqrt{\mathbb{E}[(X-\mu)^2]} = \sigma &\quad \text{by the Cauchy–Schwarz inequality.}\end{aligned}$$

10. Note that $g(u) = u/(1+u)$ is an increasing function on $[0, \infty)$. Therefore, for $\epsilon > 0$,

$$\mathbb{P}(|X_n| > \epsilon) = \mathbb{P}\left(\frac{|X_n|}{1+|X_n|} > \frac{\epsilon}{1+\epsilon}\right) \le \frac{1+\epsilon}{\epsilon}\cdot\mathbb{E}\left(\frac{|X_n|}{1+|X_n|}\right)$$

by Markov's inequality. If this expectation tends to 0 then $X_n \xrightarrow{\text{P}} 0$.

Suppose conversely that $X_n \xrightarrow{\text{P}} 0$. Then

$$\mathbb{E}\left(\frac{|X_n|}{1+|X_n|}\right) \le \frac{\epsilon}{1+\epsilon} \cdot \mathbb{P}(|X_n| \le \epsilon) + 1 \cdot \mathbb{P}(|X_n| > \epsilon) \to \frac{\epsilon}{1+\epsilon}$$

as $n \to \infty$, for $\epsilon > 0$. However ϵ is arbitrary, and hence the expectation has limit 0.

11. (i) The argument of the solution to Exercise (7.9.6a) shows that $\{X_n\}$ converges in mean square if it is mean-square Cauchy convergent. Conversely, suppose that $X_n \xrightarrow{\text{m.s.}} X$. By Minkowski's inequality,

$$\left\{\mathbb{E}\big((X_m - X_n)^2\big)\right\}^{1/2} \le \left\{\mathbb{E}\big((X_m - X)^2\big)\right\}^{1/2} + \left\{\mathbb{E}\big((X_n - X)^2\big)\right\}^{1/2} \to 0$$

as $m, n \to \infty$, so that $\{X_n\}$ is mean-square Cauchy convergent.

(ii) The corresponding result is valid for convergence almost surely, in rth mean, and in probability. For a.s. convergence, it is self-evident by the properties of Cauchy-convergent sequences of real numbers. For convergence in probability, see Exercise (7.3.1). For convergence in rth mean ($r \ge 1$), just adapt the argument of (i) above.

12. If $\operatorname{var}(X_i) \le M$ for all i, the variance of $n^{-1}\sum_{i=1}^n X_i$ is

$$\frac{1}{n^2}\sum_{i=1}^n \operatorname{var}(X_i) \le \frac{M}{n} \to 0 \quad \text{as } n \to \infty.$$

13. (a) We have that

$$\mathbb{P}(M_n \le a_n x) = F(a_n x)^n \to H(x) \quad \text{as } n \to \infty.$$

If $x \le 0$ then $F(a_n x)^n \to 0$, so that $H(x) = 0$. Suppose that $x > 0$. Then

$$-\log H(x) = -\lim_{n\to\infty}\left\{n \log\big[1 - (1 - F(a_n x))\big]\right\} = \lim_{n\to\infty}\{n(1 - F(a_n x))\}$$

since $-y^{-1}\log(1-y) \to 1$ as $y \downarrow 0$. Setting $x = 1$, we obtain $n(1 - F(a_n)) \to -\log H(1)$, and the second limit follows.

(b) This is immediate from the fact that it is valid for all sequences $\{a_n\}$.

(c) We have that

$$\frac{1 - F(te^{x+y})}{1 - F(t)} = \frac{1 - F(te^{x+y})}{1 - F(te^x)} \cdot \frac{1 - F(te^x)}{1 - F(t)} \to \frac{\log H(e^y)}{\log H(1)} \cdot \frac{\log H(e^x)}{\log H(1)}$$

as $t \to \infty$. Therefore $g(x + y) = g(x)g(y)$. Now g is non-increasing with $g(0) = 1$. Therefore $g(x) = e^{-\beta x}$ for some β, and hence $H(u) = \exp(-\alpha u^{-\beta})$ for $u > 0$, where $\alpha = -\log H(1)$.

14. Either use the result of Problem (7.11.13) or do the calculations directly thus. We have that

$$\mathbb{P}\big(M_n \le xn/\pi\big) = \left\{\frac{1}{2} + \frac{1}{\pi}\tan^{-1}\left(\frac{xn}{\pi}\right)\right\}^n = \left\{1 - \frac{1}{\pi}\tan^{-1}\left(\frac{\pi}{xn}\right)\right\}^n$$

if $x > 0$, by elementary trigonometry. Now $\tan^{-1} y = y + \mathrm{o}(y)$ as $y \to 0$, and therefore

$$\mathbb{P}\big(M_n \le xn/\pi\big) = \left(1 - \frac{1}{xn} + \mathrm{o}(n^{-1})\right)^n \to e^{-1/x} \qquad \text{as } n \to \infty.$$

15. The characteristic function of the average satisfies

$$\phi(t/n)^n = \left(1 + \frac{i\mu t}{n} + \mathrm{o}(n^{-1})\right)^n \to e^{i\mu t} \qquad \text{as } n \to \infty.$$

By the continuity theorem, the average converges in distribution to the *constant* μ, and hence in probability also.

16. (a) With $u_n = u(x_n)$, we have that

$$\left|\mathbb{E}(u(X)) - \mathbb{E}(u(Y))\right| \le \sum_n |u_n| \cdot |f_n - g_n| \le \sum_n |f_n - g_n|$$

if $\|u\|_\infty \le 1$. There is equality if u_n equals the sign of $f_n - g_n$. The second equality holds as in Problem (2.7.13) and Exercise (4.12.3).

(b) Similarly, if $\|u\|_\infty \le 1$,

$$\left|\mathbb{E}(u(X)) - \mathbb{E}(u(Y))\right| \le \int_{-\infty}^{\infty} |u(x)| \cdot |f(x) - g(x)|\,dx \le \int_{-\infty}^{\infty} |f(x) - g(x)|\,dx$$

with equality if $u(x)$ is the sign of $f(x) - g(x)$. Secondly, we have that

$$\left|\mathbb{P}(X \in A) - \mathbb{P}(Y \in A)\right| = \tfrac{1}{2}\left|\mathbb{E}(u(X)) - \mathbb{E}(u(Y))\right| \le \tfrac{1}{2} d_{\mathrm{TV}}(X, Y),$$

where

$$u(x) = \begin{cases} 1 & \text{if } x \in A, \\ -1 & \text{if } x \notin A. \end{cases}$$

Equality holds when $A = \{x \in \mathbb{R} : f(x) \ge g(x)\}$.

(c) Suppose $d_{\mathrm{TV}}(X_n, X) \to 0$. Fix $a \in \mathbb{R}$, and let u be the indicator function of the interval $(-\infty, a]$. Then $|\mathbb{E}(u(X_n)) - \mathbb{E}(u(X))| = |\mathbb{P}(X_n \le a) - \mathbb{P}(X \le a)|$, and the claim follows.

On the other hand, if $X_n = n^{-1}$ with probability one, then $X_n \xrightarrow{\mathrm{D}} 0$. However, by part (a), $d_{\mathrm{TV}}(X_n, 0) = 2$ for all n.

(d) This is tricky without a knowledge of Radon–Nikodým derivatives, and we therefore restrict ourselves to the case when X and Y are discrete. (The continuous case is analogous.) As in the solution to Exercise (4.12.4), $\mathbb{P}(X \ne Y) \ge \frac{1}{2} d_{\mathrm{TV}}(X, Y)$. That equality is possible was proved for Exercise (4.12.5), and we rephrase that solution here. Let $\mu_n = \min\{f_n, g_n\}$ and $\mu = \sum_n \mu_n$, and note that

$$d_{\mathrm{TV}}(X, Y) = \sum_n |f_n - g_n| = \sum_n \{f_n + g_n - 2\mu_n\} = 2(1 - \mu).$$

It is easy to see that

$$\tfrac{1}{2} d_{\mathrm{TV}}(X, Y) = \mathbb{P}(X \ne Y) = \begin{cases} 1 & \text{if } \mu = 0, \\ 0 & \text{if } \mu = 1, \end{cases}$$

and therefore we may assume that $0 < \mu < 1$. Let U, V, W be random variables with mass functions

$$\mathbb{P}(U = x_n) = \frac{\mu_n}{\mu}, \quad \mathbb{P}(V = x_n) = \frac{\max\{f_n - g_n, 0\}}{1 - \mu}, \quad \mathbb{P}(W = x_n) = \frac{-\min\{f_n - g_n, 0\}}{1 - \mu},$$

and let Z be a Bernoulli variable with parameter μ, independent of (U, V, W). We now choose the pair X', Y' by

$$(X', Y') = \begin{cases} (U, U) & \text{if } Z = 1, \\ (V, W) & \text{if } Z = 0. \end{cases}$$

It may be checked that X' and Y' have the same distributions as X and Y, and furthermore that $\mathbb{P}(X' \neq Y') = \mathbb{P}(Z = 0) = 1 - \mu = \frac{1}{2} d_{\text{TV}}(X, Y)$.

(e) By part (d), we may find independent pairs (X_i', Y_i'), $1 \leq i \leq n$, having the same marginals as (X_i, Y_i), respectively, and such that $\mathbb{P}(X_i' \neq Y_i') = \frac{1}{2} d_{\text{TV}}(X_i, Y_i)$. Now,

$$\begin{aligned} d_{\text{TV}}\left(\sum_{i=1}^n X_i, \sum_{i=1}^n Y_i\right) &= d_{\text{TV}}\left(\sum_{i=1}^n X_i', \sum_{i=1}^n Y_i'\right) \\ &\leq 2\mathbb{P}\left(\sum_{i=1}^n X_i' \neq \sum_{i=1}^n Y_i'\right) \leq 2\sum_{i=1}^n \mathbb{P}(X_i' \neq Y_i') = \sum_{i=1}^n d_{\text{TV}}(X_i, Y_i). \end{aligned}$$

17. If $X_1, X_2, \ldots$ are independent variables having the Poisson distribution with parameter λ, then $S_n = X_1 + X_2 + \cdots + X_n$ has the Poisson distribution with parameter λn. Now $n^{-1} S_n \xrightarrow{\text{D}} \lambda$, so that $\mathbb{E}(g(n^{-1} S_n)) \to g(\lambda)$ for all bounded continuous g. The result follows.

18. The characteristic function ψ_{mn} of

$$U_{mn} = \frac{(X_n - n) - (Y_m - m)}{\sqrt{m+n}}$$

satisfies

$$\log \psi_{mn}(t) = n\left(e^{it/\sqrt{m+n}} - 1\right) + m\left(e^{-it/\sqrt{m+n}} - 1\right) + \frac{(m-n)it}{\sqrt{m+n}} \to -\tfrac{1}{2}t^2$$

as $m, n \to \infty$, implying by the continuity theorem that $U_{mn} \xrightarrow{\text{D}} N(0, 1)$. Now $X_n + Y_m$ is Poisson-distributed with parameter $m + n$, and therefore

$$V_{mn} = \sqrt{\frac{X_n + Y_m}{m+n}} \xrightarrow{\text{P}} 1 \qquad \text{as } m, n \to \infty$$

by the law of large numbers and Problem (7.11.3). It follows by Slutsky's theorem (7.2.5a) that $U_{mn}/V_{mn} \xrightarrow{\text{D}} N(0, 1)$ as required.

19. (a) The characteristic function of X_n is $\phi_n(t) = \exp\{i\mu_n t - \frac{1}{2}\sigma_n^2 t^2\}$ where μ_n and σ_n^2 are the mean and variance of X_n. Now, $\lim_{n\to\infty} \phi_n(1)$ exists. However $\phi_n(1)$ has modulus $e^{-\frac{1}{2}\sigma_n^2}$, and therefore $\sigma^2 = \lim_{n\to\infty} \sigma_n^2$ exists. The remaining component $e^{i\mu_n t}$ of $\phi_n(t)$ converges as $n \to \infty$, say $e^{i\mu_n t} \to \theta(t)$ as $n \to \infty$ where $\theta(t)$ lies on the unit circle of the complex plane. Now $\phi_n(t) \to \theta(t)e^{-\frac{1}{2}\sigma^2 t^2}$, which is required to be a characteristic function; therefore θ is a continuous function of t. Of the various ways of showing that $\theta(t) = e^{i\mu t}$ for some μ, here is one. The sequence $\psi_n(t) = e^{i\mu_n t}$ is a sequence of characteristic functions whose limit $\theta(t)$ is continuous at $t = 0$. Therefore θ is a characteristic function. However ψ_n is the characteristic function of the constant μ_n, which must converge in distribution as $n \to \infty$; it follows that the real sequence $\{\mu_n\}$ converges to some limit μ, and $\theta(t) = e^{i\mu t}$ as required.

This proves that $\phi_n(t) \to \exp\{i\mu t - \frac{1}{2}\sigma^2 t^2\}$, and therefore the limit X is $N(\mu, \sigma^2)$.

(b) Each linear combination $sX_n + tY_n$ converges in probability, and hence in distribution, to $sX + tY$. Now $sX_n + tY_n$ has a normal distribution, implying by part (a) that $sX + tY$ is normal. Therefore the joint characteristic function of X and Y satisfies

$$\begin{aligned} \phi_{X,Y}(s,t) = \phi_{sX+tY}(1) &= \exp\left\{i\mathbb{E}(sX + tY) - \tfrac{1}{2}\operatorname{var}(sX + tY)\right\} \\ &= \exp\left\{i(s\mu_X + t\mu_Y) - \tfrac{1}{2}(s^2\sigma_X^2 + 2st\rho_{XY}\sigma_X\sigma_Y + t^2\sigma_Y^2)\right\} \end{aligned}$$

in the natural notation. Viewed as a function of s and t, this is the joint characteristic function of a bivariate normal distribution.

When working in such a context, the technique of using linear combinations of X_n and Y_n is sometimes called the 'Cramér–Wold device'.

20. (i) Write $Y_i = X_i - \mathbb{E}(X_i)$ and $T_n = \sum_{i=1}^n Y_i$. It suffices to show that $n^{-1}T_n \xrightarrow{\text{m.s.}} 0$. Now, as $n \to \infty$,

$$\mathbb{E}(T_n^2/n^2) = \frac{1}{n^2}\sum_{i=1}^n \operatorname{var}(X_i) + \frac{2}{n^2}\sum_{1\le i<j\le n} \operatorname{cov}(X_i, X_j) \le \frac{nc}{n^2} \to 0.$$

(ii) Let $\epsilon > 0$. There exists I such that $|\rho(X_i, X_j)| \le \epsilon$ if $|i-j| \ge I$. Now

$$\sum_{i,j=1}^n \operatorname{cov}(X_i, X_j) \le \sum_{\substack{|i-j|\le I\\ 1\le i,j\le n}} \operatorname{cov}(X_i, X_j) + \sum_{\substack{|i-j|>I\\ 1\le i,j\le n}} \operatorname{cov}(X_i, X_j) \le 2nIc + n^2\epsilon c,$$

since $\operatorname{cov}(X_i, X_j) \le |\rho(X_i, X_j)|\sqrt{\operatorname{var}(X_i)\cdot\operatorname{var}(X_j)}$. Therefore,

$$\mathbb{E}(T_n^2/n^2) \le \frac{2Ic}{n} + \epsilon c \to \epsilon c \qquad \text{as } n \to \infty.$$

This is valid for all positive ϵ, and the result follows.

21. The integral

$$\int_2^\infty \frac{c}{x\log|x|}\,dx$$

diverges, and therefore $\mathbb{E}(X_1)$ does not exist.

The characteristic function ϕ of X_1 may be expressed as

$$\phi(t) = 2c\int_2^\infty \frac{\cos(tx)}{x^2\log x}\,dx$$

whence

$$\frac{\phi(t)-\phi(0)}{2c} = -\int_2^\infty \frac{1-\cos(tx)}{x^2\log x}\,dx.$$

Now $0 \le 1-\cos\theta \le \min\{2, \theta^2\}$, and therefore

$$\left|\frac{\phi(t)-\phi(0)}{2c}\right| \le \int_2^{1/t} \frac{t^2}{\log x}\,dx + \int_{1/t}^\infty \frac{2}{x^2\log x}\,dx, \qquad \text{if } t > 0.$$

Now

$$\frac{1}{u}\int_2^u \frac{dx}{\log x} \to 0 \qquad \text{as } u \to \infty,$$

and

$$\int_u^\infty \frac{2}{x^2\log x}\,dx \le \frac{1}{\log u}\int_u^\infty \frac{2}{x^2}\,dx = \frac{2}{u\log u}, \qquad u > 1.$$

Therefore

$$\left|\frac{\phi(t)-\phi(0)}{2c}\right| = o(t) \qquad \text{as } t \downarrow 0.$$

Now ϕ is an even function, and hence $\phi'(0)$ exists and equals 0. Use the result of Problem (7.11.15) to deduce that $n^{-1}\sum_1^n X_i$ converges in distribution to 0, and therefore in probability also, since 0 is constant. The X_i do not obey the strong law since they have no mean.

22. If the two points are **U** and **V** then

$$\mathbb{E}\{(U_1 - V_1)^2\} = \int_0^1 \int_0^1 (u - v)^2 \, du \, dv = \tfrac{1}{6},$$

and therefore

$$\frac{1}{n} X_n^2 = \frac{1}{n} \sum_{i=1}^{n} (U_i - V_i)^2 \xrightarrow{\text{P}} \frac{1}{6} \qquad \text{as } n \to \infty,$$

by the independence of the components. It follows that $X_n/\sqrt{n} \xrightarrow{\text{P}} 1/\sqrt{6}$ either by the result of Problem (7.11.3) or by the fact that

$$\left| \frac{X_n^2}{n} - \frac{1}{6} \right| = \left| \frac{X_n}{\sqrt{n}} - \frac{1}{\sqrt{6}} \right| \cdot \left| \frac{X_n}{\sqrt{n}} + \frac{1}{\sqrt{6}} \right| \geq \frac{1}{\sqrt{6}} \left| \frac{X_n}{\sqrt{n}} - \frac{1}{\sqrt{6}} \right|.$$

23. The characteristic function of $Y_j = X_j^{-1}$ is

$$\phi(t) = \tfrac{1}{2} \int_0^1 (e^{it/x} + e^{-it/x}) \, dx = \int_0^1 \cos(t/x) \, dx = |t| \int_{|t|}^{\infty} \frac{\cos y}{y^2} \, dy$$

by the substitution $x = |t|/y$. Therefore

$$\phi(t) = 1 - |t| \int_{|t|}^{\infty} \frac{1 - \cos y}{y^2} \, dy = 1 - I|t| + \text{o}(|t|) \qquad \text{as } t \to 0,$$

where, integrating by parts,

$$I = \int_0^{\infty} \frac{1 - \cos y}{y^2} \, dy = \int_0^{\infty} \frac{\sin u}{u} \, du = \frac{\pi}{2}.$$

It follows that $T_n = n^{-1} \sum_{j=1}^{n} X_j^{-1}$ has characteristic function

$$\phi(t/n)^n = \left(1 - \frac{\pi |t|}{2n} + \text{o}(n^{-1})\right)^n \to e^{-\frac{1}{2}\pi |t|}$$

as $t \to \infty$, whence $2T_n/\pi$ is asymptotically Cauchy-distributed. In particular,

$$\mathbb{P}\big(|2T_n/\pi| > 1\big) \to \frac{2}{\pi} \int_1^{\infty} \frac{du}{1 + u^2} = \frac{1}{2} \qquad \text{as } t \to \infty.$$

24. Let m_n be a non-decreasing sequence of integers satisfying $1 \leq m_n < n$, $m_n \to \infty$, and define

$$Y_{nk} = \begin{cases} X_k & \text{if } |X_k| \leq m_n \\ \text{sign}(X_k) & \text{if } |X_k| > m_n, \end{cases}$$

noting that Y_{nk} takes the value ± 1 each with probability $\frac{1}{2}$ whenever $m_n < k \leq n$. Let $Z_n = \sum_{k=1}^{n} Y_{nk}$. Then

$$\mathbb{P}(U_n \neq Z_n) \leq \sum_{k=1}^{n} \mathbb{P}(|X_k| \geq m_n) \leq \sum_{k=m_n}^{n} \frac{1}{k^2} \to 0 \qquad \text{as } n \to \infty,$$

from which it follows that $U_n/\sqrt{n} \xrightarrow{D} N(0,1)$ if and only if $Z_n/\sqrt{n} \xrightarrow{D} N(0,1)$. Now

$$Z_n = \sum_{k=1}^{m_n} Y_{nk} + B_{n-m_n}$$

where B_{n-m_n} is the sum of $n - m_n$ independent summands each of which takes the values ± 1, each possibility having probability $\frac{1}{2}$. Furthermore

$$\left| \frac{1}{\sqrt{n}} \sum_{k=1}^{m_n} Y_{nk} \right| \le \frac{m_n^2}{\sqrt{n}}$$

which tends to 0 if m_n is chosen to be $m_n = \lfloor n^{1/5} \rfloor$; with this choice for m_n, we have that $n^{-1} B_{n-m_n} \xrightarrow{D} N(0,1)$, and the result follows.

Finally,

$$\operatorname{var}(U_n) = \sum_{k=1}^{n} \left(2 - \frac{1}{k^2}\right)$$

so that

$$\operatorname{var}\big(U_n/\sqrt{n}\big) = 2 - \frac{1}{n} \sum_{k=1}^{n} \frac{1}{k^2} \to 2.$$

25. (i) Let ϕ_n and ϕ be the characteristic functions of X_n and X. The characteristic function ψ_k of X_{N_k} is

$$\psi_k(t) = \sum_{j=1}^{\infty} \phi_j(t) \mathbb{P}(N_k = j)$$

whence

$$|\psi_k(t) - \phi(t)| \le \sum_{j=1}^{\infty} |\phi_j(t) - \phi(t)| \mathbb{P}(N_k = j).$$

Let $\epsilon > 0$. We have that $\phi_j(t) \to \phi(t)$ as $j \to \infty$, and hence for any $T > 0$, there exists $J(T)$ such that $|\phi_j(t) - \phi(t)| < \epsilon$ if $j \ge J(T)$ and $|t| \le T$. Finally, there exists $K(T)$ such that $\mathbb{P}(N_k \le J(T)) \le \epsilon$ if $k \ge K(T)$. It follows that

$$|\psi_k(t) - \phi(t)| \le 2\mathbb{P}\big(N_k \le J(T)\big) + \epsilon \mathbb{P}\big(N_k > J(T)\big) \le 3\epsilon$$

if $|t| \le T$ and $k \ge K(T)$; therefore $\psi_k(t) \to \phi(t)$ as $k \to \infty$.

(ii) Let $Y_n = \sup_{m \ge n} |X_m - X|$. For $\epsilon > 0$, $n \ge 1$,

$$\begin{aligned} \mathbb{P}\big(|X_{N_k} - X| > \epsilon\big) &\le \mathbb{P}(N_k \le n) + \mathbb{P}\big(|X_{N_k} - X| > \epsilon, N_k > n\big) \\ &\le \mathbb{P}(N_k \le n) + \mathbb{P}(Y_n > \epsilon) \to \mathbb{P}(Y_n > \epsilon) \qquad \text{as } k \to \infty. \end{aligned}$$

Now take the limit as $n \to \infty$ and use the fact that $Y_n \xrightarrow{\text{a.s.}} 0$.

26. (a) We have that

$$\frac{a(n-k,n)}{a(n+1,n)} = \prod_{i=0}^{k} \left(1 - \frac{i}{n}\right) \le \exp\left(-\sum_{i=0}^{k} \frac{i}{n}\right).$$

(b) The expectation is

$$E_n = \sum_j g\left(\frac{j-n}{\sqrt{n}}\right)\frac{n^j e^{-n}}{j!}$$

where the sum is over all j satisfying $n - M\sqrt{n} \le j \le n$. For such a value of j,

$$g\left(\frac{j-n}{\sqrt{n}}\right)\frac{n^j e^{-n}}{j!} = \frac{e^{-n}}{\sqrt{n}}\left(\frac{n^{j+1}}{j!} - \frac{n^j}{(j-1)!}\right),$$

whence E_n has the form given.

(c) Now g is continuous on the interval $[-M, 0]$, and it follows by the central limit theorem that

$$E_n \to \int_{-M}^{0} g(x)\frac{1}{\sqrt{2\pi}}e^{-\frac{1}{2}x^2}\,dx = \int_0^M \frac{x}{\sqrt{2\pi}}e^{-\frac{1}{2}x^2}\,dx = \frac{1-e^{-\frac{1}{2}M^2}}{\sqrt{2\pi}}.$$

Also,

$$E_n \le \frac{e^{-n}}{\sqrt{n}}a(n+1,n) \le E_n + \frac{e^{-n}}{\sqrt{n}}a(n-k,n) \le E_n + \frac{e^{-n-k^2/(2n)}}{\sqrt{n}}a(n+1,n)$$

where $k = \lfloor M\sqrt{n}\rfloor$. Take the limits as $n \to \infty$ and $M \to \infty$ in that order to obtain

$$\frac{1}{\sqrt{2\pi}} \le \lim_{n\to\infty}\left\{\frac{n^{n+\frac{1}{2}}e^{-n}}{n!}\right\} \le \frac{1}{\sqrt{2\pi}}.$$

27. Clearly

$$\mathbb{E}(R_{n+1} \mid R_0, R_1, \dots, R_n) = R_n + \frac{R_n}{n+2}$$

since a red ball is added with probability $R_n/(n+2)$. Hence

$$\mathbb{E}(S_{n+1} \mid R_0, R_1, \dots, R_n) = S_n,$$

and also $0 \le S_n \le 1$. Using the martingale convergence theorem, $S = \lim_{n\to\infty} S_n$ exists almost surely and in mean square.

28. Let $0 < \epsilon < \frac{1}{3}$, and let

$$k(t) = \lfloor \theta t\rfloor,\ m(t) = \lceil (1-\epsilon^3)k(t)\rceil,\ n(t) = \lfloor (1+\epsilon^3)k(t)\rfloor$$

and let $I_{mn}(t)$ be the indicator function of the event $\{m(t) \le M(t) < n(t)\}$. Since $M(t)/t \xrightarrow{\text{P}} \theta$, we may find T such that $\mathbb{E}(I_{mn}(t)) > 1-\epsilon$ for $t \ge T$.

We may approximate $S_{M(t)}$ by the random variable $S_{k(t)}$ as follows. With $A_j = \{|S_j - S_{k(t)}| > \epsilon\sqrt{k(t)}\}$,

$$\begin{aligned}
\mathbb{P}(A_{M(t)}) &\le \mathbb{P}(A_{M(t)}, I_{mn}(t) = 1) + \mathbb{P}(A_{M(t)}, I_{mn}(t) = 0)\\
&\le \mathbb{P}\Big(\bigcup_{j=m(t)}^{k(t)-1} A_j\Big) + \mathbb{P}\Big(\bigcup_{j=k(t)}^{n(t)-1} A_j\Big) + \mathbb{P}(I_{mn}(t) = 0)\\
&\le \frac{\{k(t)-m(t)\}\sigma^2}{\epsilon^2 k(t)} + \frac{\{n(t)-k(t)\}\sigma^2}{\epsilon^2 k(t)} + \epsilon\\
&\le \epsilon(1+2\sigma^2), \qquad \text{if } t \ge T,
\end{aligned}$$

by Kolmogorov's inequality (Exercise (7.8.1) and Problem (7.11.29)). Send $t \to \infty$ to find that

$$D_t = \frac{S_{M(t)} - S_{k(t)}}{\sqrt{k(t)}} \xrightarrow{\text{P}} 0 \qquad \text{as } t \to \infty.$$

Now $S_{k(t)}/\sqrt{k(t)} \xrightarrow{\text{D}} N(0, \sigma^2)$ as $t \to \infty$, by the usual central limit theorem. Therefore

$$\frac{S_{M(t)}}{\sqrt{k(t)}} = D_t - \frac{S_{k(t)}}{\sqrt{k(t)}} \xrightarrow{\text{D}} N(0, \sigma^2),$$

which implies the first claim, since $k(t)/(\theta t) \to 1$ (see Exercise (7.2.7)). The second part follows by Slutsky's theorem (7.2.5a).

29. We have that $S_n = S_k + (S_n - S_k)$, and so, for $n \geq k$,

$$\mathbb{E}(S_n^2 I_{A_k}) = \mathbb{E}(S_k^2 I_{A_k}) + 2\mathbb{E}\{S_k(S_n - S_k)I_{A_k}\} + \mathbb{E}\{(S_n - S_k)^2 I_{A_k}\}.$$

Now $S_k^2 I_{A_k} \geq c^2 I_{A_k}$; the second term on the right side is 0, by the independence of the X's, and the third term is non-negative. The first inequality of the question follows. Summing over k, we obtain $\mathbb{E}(S_n^2) \geq c^2 \mathbb{P}(M_n > c)$ as required.

30. (i) With $S_n = \sum_{i=1}^n X_i$, we have by Kolmogorov's inequality that

$$\mathbb{P}\left(\max_{1 \leq k \leq n} |S_{m+k} - S_m| > \epsilon\right) \leq \frac{1}{\epsilon^2} \sum_{k=m}^{m+n} \mathbb{E}(X_k^2)$$

for $\epsilon > 0$. Take the limit as $m, n \to \infty$ to obtain in the usual way that $\{S_r : r \geq 0\}$ is a.s. Cauchy convergent, and therefore a.s. convergent, if $\sum_1^\infty \mathbb{E}(X_k^2) < \infty$. It is shorter to use the martingale convergence theorem, noting that S_n is a martingale with uniformly bounded second moments.

(ii) Apply part (i) to the sequence $Y_k = X_k/b_k$ to deduce that $\sum_{k=1}^\infty X_k/b_k$ converges a.s. The claim now follows by Kronecker's lemma (see Exercise (7.8.2)).

31. (a) This is immediate by the observation that

$$e^{\lambda(\mathbf{P})} = f_{X_0} \prod_{i,j} p_{ij}^{N_{ij}}.$$

(b) Clearly $\sum_j p_{ij} = 1$ for each i, and we introduce Lagrange multipliers $\{\mu_i : i \in S\}$ and write $V = \lambda(\mathbf{P}) + \sum_i \mu_i \sum_j p_{ij}$. Differentiating V with respect to each p_{ij} yields a stationary (maximum) value when $(N_{ij}/p_{ij}) + \mu_i = 0$. Hence $\sum_k N_{ik} = -\mu_k$, and

$$\widehat{p}_{ij} = -\frac{N_{ij}}{\mu_i} = \frac{N_{ij}}{\sum_k N_{ik}}.$$

(c) We have that $N_{ij} = \sum_{r=1}^{\sum_k N_{ik}} I_r$ where I_r is the indicator function of the event that the rth transition out of i is to j. By the Markov property, the I_r are independent with constant mean p_{ij}. Using the strong law of large numbers and the fact that $\sum_k N_{ik} \xrightarrow{\text{a.s.}} \infty$ as $n \to \infty$, $\widehat{p}_{ij} \xrightarrow{\text{a.s.}} \mathbb{E}(I_1) = p_{ij}$.

32. (a) If X is transient then $V_i(n) < \infty$ a.s., and $\mu_i = \infty$, whence $V_i(n)/n \xrightarrow{\text{a.s.}} 0 = \mu_i^{-1}$. If X is recurrent, then without loss of generality we may assume $X_0 = i$. Let $T(r)$ be the duration of the rth excursion from i. By the strong Markov property, the $T(r)$ are independent and identically distributed with mean μ_i. Furthermore,

$$\frac{1}{V_i(n)} \sum_{r=1}^{V_i(n)-1} T(r) \le \frac{n}{V_i(n)} \le \frac{1}{V_i(n)} \sum_{r=1}^{V_i(n)} T(r).$$

By the strong law of large numbers and the fact that $V_i(n) \xrightarrow{\text{a.s.}} \infty$ as $n \to \infty$, the two outer terms sandwich the central term, and the result follows.

(b) Note that $\sum_{r=0}^{n-1} f(X_r) = \sum_{i \in S} f(i) V_i(n)$. With Q a finite subset of S, and $\pi_i = \mu_i^{-1}$, the unique stationary distribution,

$$\left| \sum_{r=0}^{n-1} \frac{f(X_r)}{n} - \sum_i \frac{f(i)}{\mu_i} \right| = \left| \sum_i \left(\frac{V_i(n)}{n} - \frac{1}{\mu_i} \right) f(i) \right|$$
$$\le \left\{ \sum_{i \in Q} \left| \frac{V_i(n)}{n} - \frac{1}{\mu_i} \right| + \sum_{i \notin Q} \left(\frac{V_i(n)}{n} + \frac{1}{\mu_i} \right) \right\} \|f\|_\infty,$$

where $\|f\|_\infty = \sup\{|f(i)| : i \in S\}$. The sum over $i \in Q$ converges a.s. to 0 as $n \to \infty$, by part (a). The other sum satisfies

$$\sum_{i \notin Q} \left(\frac{V_i(n)}{n} + \frac{1}{\mu_i} \right) = 2 - \sum_{i \in Q} \left(\frac{V_i(n)}{n} + \pi_i \right)$$

which approaches 0 a.s., in the limits as $n \to \infty$ and $Q \uparrow S$.

33. (a) Since the chain is recurrent, we may assume without loss of generality that $X_0 = j$. Define the times $R_1, R_2, \dots$ of return to j, the sojourn lengths $S_1, S_2, \dots$ in j, and the times $V_1, V_2, \dots$ between visits to j. By the Markov property and the strong law of large numbers,

$$\frac{1}{n} \sum_{r=1}^{n} S_r \xrightarrow{\text{a.s.}} \frac{1}{g_j}, \qquad \frac{1}{n} R_n = \frac{1}{n} \sum_{r=1}^{n} V_r \xrightarrow{\text{a.s.}} m_j.$$

Also, $R_n / R_{n+1} \xrightarrow{\text{a.s.}} 1$, since $m_j = \mathbb{E}(R_1) < \infty$. If $R_n < t < R_{n+1}$, then

$$\frac{R_n}{R_{n+1}} \cdot \frac{\sum_{r=1}^{n} S_r}{\sum_{r=1}^{n} V_r} \le \frac{1}{t} \int_0^t I_{\{X(s)=j\}} \, ds \le \frac{R_{n+1}}{R_n} \cdot \frac{\sum_{r=1}^{n+1} S_r}{\sum_{r=1}^{n+1} V_r}.$$

Let $n \to \infty$ to obtain the result.

(b) Note by Theorem (6.10.22) that $p_{ij}(t) \to \pi_j$ as $t \to \infty$. We take expectations of the integral in part (a), and the claim follows as in Corollary (6.4.25).

(c) Use the fact that

$$\int_0^t f(X(s)) \, ds = \sum_{j \in S} \int_0^t I_{\{X(s)=j\}} \, ds$$

together with the method of solution of Problem (7.11.32b).

34. (a) By the first Borel–Cantelli lemma, $X_n = Y_n$ for all but finitely many values of n, almost surely. Off an event of probability zero, the sequences are identical for all large n.

(b) This follows immediately from part (a), since $X_n - Y_n = 0$ for all large n, almost surely.

(c) By the above, $a_n^{-1} \sum_{r=1}^{\infty} (X_r - Y_r) \xrightarrow{\text{a.s.}} 0$, which implies the claim.

35. Let $Y_n = X_n I_{\{|X_n| \le a\}}$. Then,

$$\sum_n \mathbb{P}(X_n \neq Y_n) = \sum_n \mathbb{P}(|X_n| > a) < \infty$$

by assumption (a), whence $\{X_n\}$ and $\{Y_n\}$ are tail-equivalent (see Problem (7.11.34)). By assumption (b) and the martingale convergence theorem (7.8.1) applied to the partial sums $\sum_{n=1}^{N}(Y_n - \mathbb{E}(Y_n))$, the infinite sum $\sum_{n=1}^{\infty}(Y_n - \mathbb{E}(Y_n))$ converges almost surely. Finally, $\sum_{n=1}^{\infty}\mathbb{E}(Y_n)$ converges by assumption (c), and therefore $\sum_{n=1}^{\infty} Y_n$, and hence $\sum_{n=1}^{\infty} X_n$, converges a.s.

36. (a) Let $n_1 < n_2 < \cdots < n_r = n$. Since the I_k take only two values, it suffices to show that

$$\mathbb{P}(I_{n_s} = 1 \text{ for } 1 \le s \le r) = \prod_{s=1}^{r} \mathbb{P}(I_{n_s} = 1).$$

Since F is continuous, the X_i take distinct values with probability 1, and furthermore the ranking of $X_1, X_2, \dots, X_n$ is equally likely to be any of the $n!$ available. Let $x_1, x_2, \dots, x_n$ be distinct reals, and write $A = \{X_i = x_i \text{ for } 1 \le i \le n\}$. Now,

$$\begin{aligned}\mathbb{P}(I_{n_s} &= 1 \text{ for } 1 \le s \le r \mid A)\\ &= \frac{1}{n!}\left\{\binom{n-1}{n_{s-1}}(n-1-n_{s-1})!\right\}\left\{\binom{n_{s-1}-1}{n_{s-2}}(n_{s-1}-1-n_{s-2})!\right\}\cdots(n_1-1)!\\ &= \frac{1}{n_s}\cdot\frac{1}{n_{s-1}}\cdots\frac{1}{n_1},\end{aligned}$$

and the claim follows on averaging over the x_i.

(b) We have that $\mathbb{E}(I_k) = \mathbb{P}(I_k = 1) = k^{-1}$ and $\text{var}(I_k) = k^{-1}(1-k^{-1})$, whence $\sum_k \text{var}(I_k/\log k) < \infty$. By the independence of the I_k and the martingale convergence theorem (7.8.1), $\sum_{k=1}^{\infty}(I_k - k^{-1})/\log k$ converges a.s. Therefore, by Kronecker's lemma (see Exercise (7.8.2)),

$$\frac{1}{\log n}\sum_{j=1}^{n}\left(I_j - \frac{1}{j}\right) \xrightarrow{\text{a.s.}} 0 \qquad \text{as } n \to \infty.$$

The result follows on recalling that $\sum_{j=1}^{n} j^{-1} \sim \log n$ as $n \to \infty$.

37. By an application of the three series theorem of Problem (7.11.35), the series converges almost surely.

38. We have $f_X(x) = x^{s-1}e^{-x}/\Gamma(s)$ for $x > 0$, so that, by a change of variable,

$$f_Y(y) = \frac{1}{\Gamma(s)} f_X\big(y\sqrt{s}+s\big)\sqrt{s}, \qquad y > -\sqrt{s}.$$

Since f_Y integrates to 1,

$$\frac{\Gamma(s)e^s\sqrt{s}}{s^s} = \int_{-\sqrt{s}}^{\infty}\left(\frac{y\sqrt{s}+s}{s}\right)^{s-1} e^{-y\sqrt{s}}\,dy = \int_{-\sqrt{s}}^{\infty} e^{-u(y)}\,dy,$$

where $u(y) = y\sqrt{s} - (s-1)\log\big(1 + y/\sqrt{s}\big)$. Now $\log(1+z) = z - \frac{1}{2}z^2 + O(z^3)$ as $z \to 0$, so that $e^{-u(y)} \to e^{-\frac{1}{2}y^2}$ uniformly on any bounded interval $[-c, c]$, as $s \to \infty$. We shall use the hint to control the other parts of the integral from $-\sqrt{s}$ to ∞.

Now,

$$u'(y) = \sqrt{s} - \frac{s-1}{y+\sqrt{s}} = \frac{y\sqrt{s}+1}{y+\sqrt{s}}$$

is positive and increasing in $y \in (c, \infty)$, and tends to y as $s \to \infty$. By the hint,

$$\int_c^{\infty} e^{-u(y)}\,dy \le \frac{1}{u'(c)}\big[-e^{-u(y)}\big]_c^{\infty} = \frac{1}{u'(c)} e^{-u(c)} \to \frac{1}{c} e^{-\frac{1}{2}c^2} \qquad \text{as } s \to \infty.$$

The integral $\int_{-\sqrt{s}}^{-c} e^{-u(y)}\,dy$ is treated likewise, so that, as $s \to \infty$,

$$\frac{\Gamma(s)e^s\sqrt{s}}{s^s} \to \int_{-\infty}^{\infty} e^{-\frac{1}{2}y^2}\,dy = \sqrt{2\pi},$$

which is Stirling's formula for the gamma function. [This proof is due to P. Diaconis and D. Freedman. The presence of probability theory in the argument is illusory.]

39. (a) The characteristic function of S_n/D_n is

$$\phi_n(t) = \prod_{r=1}^{n} \tfrac{1}{2}\big(e^{itc_r/D_n} + e^{-itc_r/D_n}\big) = \prod_{r=1}^{n} \cos(tc_r/D_n).$$

By the hint,

$$-\frac{2}{3}t^4\frac{B_n}{D_n^4} \le \frac{1}{2}t^2 + \log\phi_n(t) \le -\frac{1}{12}t^4\frac{B_n}{D_n^4}.$$

If $B_n/D_n^4 \to 0$, then $\phi_n(t) \to e^{-\frac{1}{2}t^2}$ as $n \to \infty$, and the claim follows by the continuity theorem. Conversely, if $\phi_n(t) \to e^{-\frac{1}{2}t^2}$, then, by the above inequalities, we must have $B_n/D_n^4 \to 0$.

(b) Use binary expansions to see that the answer is uniform on $(0, 1)$.

40. The inequality is elementary by Markov's inequality when $\rho = 0$, so we assume $\rho \neq 0$. Let $R = \{(x, y) : |x| \vee |y| > \epsilon\}$, and $|t| \le 1$. By the properties of g, the required probability satisfies

$$\iint_R f_{X,Y}(x,y)\,dx\,dy \le \iint_R g(x,y) f_{X,Y}(x,y)\,dx\,dy \le \mathbb{E}(g(X,Y)) = \frac{2(1-t\rho)}{\epsilon^2(1-t^2)}.$$

The last is minimized by choosing $t = \big(1-\sqrt{1-\rho^2}\big)/\rho \le 1$, and this choice of t yields the given inequality.

41. (a) By Markov's inequality, for $\theta \ge 0$,

$$\mathbb{P}(X \ge t) = \mathbb{P}(e^{\theta X} \ge e^{\theta t}) \le \frac{\mathbb{E}(e^{\theta X})}{e^{\theta t}} = \exp\{-1 + e^{\theta} - \theta t\},$$

which is minimized by setting $\theta = \log t$.

(b) We use the fact that $\mathbb{P}(M_n < t) = \big(1 - \mathbb{P}(X \ge t)\big)^n$. Set $t = (1+a)\log n/\log\log n$ with $a > 0$ to obtain, by part (a),

$$\mathbb{P}(M_n < t) \ge \left(1 - \frac{e^{t-1}}{t^t}\right)^n = \left(1 - n^{-(1+a)(1+o(1))}\right)^n \to 1 \qquad \text{as } n \to \infty.$$

With t given similarly (and assumed for simplicity to be an integer) and $-1 < a < 0$,

$$\begin{aligned}\mathbb{P}(M_n < t) \le \big(1 - \mathbb{P}(X = t)\big)^n &= \left(1 - \frac{1}{e\,t!}\right)^n = \left(1 - n^{-(1+a)(1+o(1))}\right)^n \\ &\le \exp\big(-n^{-a(1+o(1))}\big) \to 0 \qquad \text{as } n \to \infty.\end{aligned}$$

We have used the fact that $1 - x \le e^{-x}$ for $x > 0$, at the last stage.

8

Random processes

8.2 Solutions. Stationary processes

1. With $a_i(n) = \mathbb{P}(X_n = i)$, we have that

$$\begin{aligned}\operatorname{cov}(X_m, X_{m+n}) &= \mathbb{P}(X_{m+n} = 1 \mid X_m = 1)\mathbb{P}(X_m = 1) - \mathbb{P}(X_{m+n} = 1)\mathbb{P}(X_m = 1)\\ &= a_1(m)p_{11}(n) - a_1(m)a_1(m+n),\end{aligned}$$

and therefore,

$$\rho(X_m, X_{m+n}) = \frac{a_1(m)p_{11}(n) - a_1(m)a_1(m+n)}{\sqrt{a_1(m)(1-a_1(m))a_1(m+n)(1-a_1(m+n))}}.$$

Now, $a_1(m) \to \alpha/(\alpha+\beta)$ as $m \to \infty$, and

$$p_{11}(n) = \frac{\alpha}{\alpha+\beta} + \frac{\beta}{\alpha+\beta}(1-\alpha-\beta)^n,$$

whence $\rho(X_m, X_{m+n}) \to (1-\alpha-\beta)^n$ as $m \to \infty$. Finally,

$$\lim_{n\to\infty} \frac{1}{n}\sum_{r=1}^{n} \mathbb{P}(X_r = 1) = \frac{\alpha}{\alpha+\beta}.$$

The process is strictly stationary if and only if X_0 has the stationary distribution.

2. We have that $\mathbb{E}(T(t)) = 0$ and $\operatorname{var}(T(t)) = \operatorname{var}(T_0) = 1$. Hence:
(a) $\rho(T(s), T(s+t)) = \mathbb{E}(T(s)T(s+t)) = \mathbb{E}\big[(-1)^{N(t+s)-N(s)}\big] = e^{-2\lambda t}$.
(b) Evidently, $\mathbb{E}(X(t)) = 0$, and

$$\begin{aligned}\mathbb{E}[X(t)^2] &= \mathbb{E}\left(\int_0^t\int_0^t T(u)T(v)\,du\,dv\right)\\ &= 2\int_{0<u<v<t} \mathbb{E}\big(T(u)T(v)\big)\,du\,dv = 2\int_{v=0}^{t}\int_{u=0}^{v} e^{-2\lambda(v-u)}\,du\,dv\\ &= \frac{1}{\lambda}\left(t - \frac{1}{2\lambda} + \frac{1}{2\lambda}e^{-2\lambda t}\right) \sim \frac{t}{\lambda} \quad \text{as } t \to \infty.\end{aligned}$$

3. We show first the existence of the limit $\lambda = \lim_{t\downarrow 0} g(t)/t$, where $g(t) = \mathbb{P}(N(t) > 0)$. Clearly,

$$\begin{aligned}g(x+y) &= \mathbb{P}\big(N(x+y) > 0\big)\\ &= \mathbb{P}\big(N(x) > 0\big) + \mathbb{P}\big(N(x) = 0 \text{ and } N(x+y) - N(x) > 0\big)\\ &\le g(x) + g(y) \qquad \text{for } x, y \ge 0.\end{aligned}$$

Such a function g is called subadditive, and the existence of λ follows by the *subadditive limit theorem* discussed in Problem (6.15.14). Note that $\lambda = \infty$ is a possibility.

Next, we partition the interval $(0, 1]$ into n equal sub-intervals, and let $I_n(r)$ be the indicator function of the event that at least one arrival lies in $\big((r-1)/n, r/n\big]$, $1 \le r \le n$. Then $\sum_{r=1}^n I_n(r) \uparrow N(1)$ as $n \to \infty$, with probability 1. By stationarity and monotone convergence,

$$\mathbb{E}(N(1)) = \mathbb{E}\left(\lim_{n\to\infty} \sum_{r=1}^{n} I_n(r)\right) = \lim_{n\to\infty} \mathbb{E}\left(\sum_{r=1}^{n} I_n(r)\right) = \lim_{n\to\infty} ng(n^{-1}) = \lambda.$$

4. We use the fact that, when X_0 has the stationary distribution $\boldsymbol{\pi}$, the vectors $(X_0, X_1, \dots, X_{n-1})$ and $(X_1, X_2, \dots, X_n)$ have the same distributions. For $n \ge 1$, we have that

$$\begin{aligned}
\mathbb{P}(T_A = n) &= \mathbb{P}(X_1 \notin A, \dots, X_{n-1} \notin A, X_n \in A) \\
&= \mathbb{P}(X_1 \notin A, \dots, X_{n-1} \notin A) - \mathbb{P}(X_1 \notin A, \dots, X_{n-1} \notin A, X_n \notin A) \\
&= \mathbb{P}(X_1 \notin A, \dots, X_{n-1} \notin A) - \mathbb{P}(X_0 \notin A, \dots, X_{n-2} \notin A, X_{n-1} \notin A) \\
&= \mathbb{P}(X_0 \in A, X_1 \notin A, \dots, X_{n-1} \notin A) \\
&= \pi(A)\mathbb{P}(T_A \ge n \mid X_0 \in A).
\end{aligned}$$

Now sum over n. A version of this formula holds more generally for suitable stationary processes that need not be Markovian.

8.3 Solutions. Renewal processes

1. See Problem (6.15.8).

2. With X a certain inter-event time, independent of the chain so far,

$$B_{n+1} = \begin{cases} X - 1 & \text{if } B_n = 0, \\ B_n - 1 & \text{if } B_n > 0. \end{cases}$$

Therefore, B is a Markov chain with transition probabilities $p_{i,i-1} = 1$ for $i > 0$, and $p_{0j} = f_{j+1}$ for $j \ge 0$, where $f_n = \mathbb{P}(X = n)$. The stationary distribution satisfies $\pi_j = \pi_{j+1} + \pi_0 f_{j+1}$, $j \ge 0$, with solution $\pi_j = \mathbb{P}(X > j)/\mathbb{E}(X)$, provided $\mathbb{E}(X)$ is finite.

The transition probabilities of B when reversed in equilibrium are

$$\widetilde{p}_{i,i+1} = \frac{\pi_{i+1}}{\pi_i} = \frac{\mathbb{P}(X > i+1)}{\mathbb{P}(X > i)}, \quad \widetilde{p}_{i0} = \frac{f_{i+1}}{\mathbb{P}(X > i)}, \qquad \text{for } i \ge 0.$$

These are the transition probabilities of the chain U of Exercise (8.3.1) with the f_j as given.

3. We have that $\rho^n u_n = \sum_{r=1}^n \rho^{n-k} u_{n-k} \rho^k f_k$, whence $v_n = \rho^n u_n$ defines a renewal sequence provided $\rho > 0$ and $\sum_n \rho^n f_n = 1$. By Exercise (8.3.1), there exists a Markov chain U and a state s such that $v_n = \mathbb{P}(U_n = s) \to \pi_s$, as $n \to \infty$, as required.

4. Noting that $N(0) = 0$,

$$\begin{aligned}
\sum_{r=0}^{\infty} \mathbb{E}(N(r))s^r &= \sum_{r=1}^{\infty}\sum_{k=1}^{r} u_k s^r = \sum_{k=1}^{\infty} u_k \sum_{r=k}^{\infty} s^r \\
&= \sum_{k=1}^{\infty} \frac{u_k s^k}{1-s} = \frac{U(s)-1}{1-s} = \frac{F(s)U(s)}{1-s}.
\end{aligned}$$

Let $S_m = \sum_{k=1}^{m} X_k$ and $S_0 = 0$. Then $\mathbb{P}(N(r) = n) = \mathbb{P}(S_n \le r) - \mathbb{P}(S_{n+1} \le r)$, and

$$\begin{aligned}\sum_{t=0}^{\infty} s^t \mathbb{E}\left[\binom{N(t)+k}{k}\right] &= \sum_{t=0}^{\infty} s^t \sum_{n=0}^{\infty} \binom{n+k}{k} \big(\mathbb{P}(S_n \le t) - \mathbb{P}(S_{n+1} \le t)\big) \\ &= \sum_{t=0}^{\infty} s^t \left[1 + \sum_{n=1}^{\infty} \binom{n+k-1}{k-1} \mathbb{P}(S_n \le t)\right].\end{aligned}$$

Now,

$$\sum_{t=0}^{\infty} s^t \mathbb{P}(S_n \le t) = \sum_{t=0}^{\infty} s^t \sum_{i=0}^{t} \mathbb{P}(S_n = i) = \sum_{i=0}^{\infty} \mathbb{P}(S_n = i) \sum_{t=i}^{\infty} s^t = \frac{F(s)^n}{1-s},$$

whence, by the negative binomial theorem,

$$\sum_{t=0}^{\infty} s^t \mathbb{E}\left[\binom{N(t)+k}{k}\right] = \frac{1}{(1-s)(1-F(s))^k} = \frac{U(s)^k}{1-s}.$$

5. This is an immediate consequence of the fact that the interarrival times of a Poisson process are exponentially distributed, since this specifies the distribution of the process.

8.4 Solutions. Queues

1. We use the lack-of-memory property repeatedly, together with the fact that, if X and Y are independent exponential variables with respective parameters λ and μ, then $\mathbb{P}(X < Y) = \lambda/(\lambda+\mu)$.
(a) In this case,

$$p = \frac{1}{2}\left\{\frac{\lambda}{\lambda+\mu}\cdot\frac{\mu}{\lambda+\mu} + \frac{\mu}{\lambda+\mu}\right\} + \frac{1}{2}\left\{\frac{\mu}{\lambda+\mu}\cdot\frac{\lambda}{\lambda+\mu} + \frac{\lambda}{\lambda+\mu}\right\} = \frac{1}{2} + \frac{2\lambda\mu}{(\lambda+\mu)^2}.$$

(b) If $\lambda < \mu$, and you pick the quicker server, $p = 1 - \left(\dfrac{\mu}{\lambda+\mu}\right)^2$.

(c) And finally, $p = \dfrac{2\lambda\mu}{(\lambda+\mu)^2}$.

2. The given event occurs if the time X to the next arrival is less than t, and also less than the time Y of service of the customer present. Now,

$$\mathbb{P}(X \le t,\ X \le Y) = \int_0^t \lambda e^{-\lambda x} e^{-\mu x}\,dx = \frac{\lambda}{\lambda+\mu}\big(1 - e^{-(\lambda+\mu)t}\big).$$

3. By conditioning on the time of passage of the first vehicle,

$$\mathbb{E}(T) = \int_0^a \big(x + \mathbb{E}(T)\big)\lambda e^{-\lambda x}\,dx + ae^{-\lambda a},$$

and the result follows. If it takes a time b to cross the other lane, and so $a+b$ to cross both, then, with an obvious notation,

(a) $$\mathbb{E}(T_a) + \mathbb{E}(T_b) = \frac{e^{a\lambda}-1}{\lambda} + \frac{e^{b\mu}-1}{\mu},$$

(b) $$\mathbb{E}(T_{a+b}) = \frac{e^{(a+b)(\lambda+\mu)}-1}{\lambda+\mu}.$$

The latter must be the greater, by a consideration of the problem, or by turgid calculation.

4. Look for a solution of the detailed balance equations

$$\mu\pi_{n+1} = \frac{\lambda(n+1)}{n+2}\pi_n, \qquad n \geq 0.$$

to find that $\pi_n = \rho^n \pi_0/(n+1)$ is a stationary distribution if $\rho < 1$, in which case $\pi_0 = -\rho/\log(1-\rho)$. Hence $\sum_n n\pi_n = \lambda\pi_0/(\mu - \lambda)$, and by the lack-of-memory property the mean time spent waiting for service is $\rho\pi_0/(\mu - \lambda)$. An arriving customer joins the queue with probability

$$\sum_{n=0}^{\infty} \frac{n+1}{n+2}\pi_n = \frac{\rho + \log(1-\rho)}{\rho\log(1-\rho)}.$$

5. By considering possible transitions during the interval $(t, t+h)$, the probability $p_i(t)$ that exactly i demonstrators are busy at time t satisfies:

$$\begin{aligned} p_2(t+h) &= p_1(t)2h + p_2(t)(1-2h) + \mathrm{o}(h), \\ p_1(t+h) &= p_0(t)2h + p_1(t)(1-h)(1-2h) + p_2(t)2h + \mathrm{o}(h), \\ p_0(t+h) &= p_0(t)(1-2h) + p_1(t)h + \mathrm{o}(h). \end{aligned}$$

Hence,

$$p_2'(t) = 2p_1(t) - 2p_2(t), \quad p_1'(t) = 2p_0(t) - 3p_1(t) + 2p_2(t), \quad p_0'(t) = -2p_0(t) + p_1(t),$$

and therefore $p_2(t) = a + be^{-2t} + ce^{-5t}$ for some constants a, b, c. By considering the values of p_2 and its derivatives at $t = 0$, the boundary conditions are found to be $a + b + c = 0$, $-2b - 5c = 0$, $4b + 25c = 4$, and the claim follows.

8.5 Solutions. The Wiener process

1. We might as well assume that W is standard, in that $\sigma^2 = 1$. Because the joint distribution is multivariate normal, we may use Exercise (4.7.5) for the first part, and Exercise (4.9.8) for the second, giving the answer

$$\frac{1}{8} + \frac{1}{4\pi}\left\{\sin^{-1}\sqrt{\frac{s}{t}} + \sin^{-1}\sqrt{\frac{s}{u}} + \sin^{-1}\sqrt{\frac{t}{u}}\right\}.$$

2. Writing $W(s) = \sqrt{s}X$, $W(t) = \sqrt{t}Z$, and $W(u) = \sqrt{u}Y$, we obtain random variables X, Y, Z with the standard trivariate normal distribution, with correlations $\rho_1 = \sqrt{s/u}$, $\rho_2 = \sqrt{t/u}$, $\rho_3 = \sqrt{s/t}$. By the solution to Exercise (4.9.9),

$$\mathrm{var}(Z \mid X, Y) = \frac{(u-t)(t-s)}{t(u-s)},$$

yielding $\mathrm{var}(W(t) \mid W(s), W(u))$ as required. Also,

$$\mathbb{E}\big(W(t)W(u) \mid W(s), W(v)\big) = \mathbb{E}\left\{\left[\frac{(u-t)W(s) + (t-s)W(u)}{u-s}\right] W(u) \,\middle|\, W(s), W(v)\right\},$$

which yields the conditional correlation after some algebra.

3. Whenever $a^2 + b^2 = 1$.

4. We may take $\sigma = 1$, without loss of generality. Let $\Delta_j(n) = W((j+1)t/n) - W(jt/n)$. By the independence of these increments,

$$\begin{aligned}
\mathbb{E}\left(\sum_{j=0}^{n-1} \Delta_j(n)^2 - t\right)^2 &= \sum_{j=0}^{n-1} \mathbb{E}\left(\Delta_j(n)^2 - \frac{t}{n}\right)^2 && \text{because } \mathbb{E}(\Delta_j(n)^2) = \frac{t}{n} \\
&= \sum_{j=0}^{n-1}\left(\frac{3t^2}{n^2} - \frac{2t^2}{n^2} + \frac{t^2}{n^2}\right) && \text{because } \mathbb{E}(\Delta_j(n)^4) = \frac{3t^2}{n^2} \\
&= \frac{2t^2}{n} \to 0 && \text{as } n \to \infty.
\end{aligned}$$

The same proof may be used to show the more general fact that the above holds with $\Delta_j(n) = W(t_{j+1}) - W(t_j)$, where $0 = t_0 < t_1 < \cdots < t_n = t$ is a partition of $[0, t]$ with mesh size $\epsilon = \max_j |t_{j+1} - t_j|$ satisfying $\epsilon \to 0$. Further discussion of quadratic variation may be found in, for example, Mörters and Peres 2010.

5. They all have mean zero and variance t, but only (a) has independent normally distributed increments.

6. A linear combination of normal variables is normal, and with mean 0 if the summands have mean 0. Furthermore, by considering characteristic functions, the property of being normal is preserved under distributional limits. Therefore, by passing to limits, $\int_\epsilon^t [W(u)/u]\,du$ is normally distributed with mean 0. It therefore has a normally distributed limit as $\epsilon \downarrow 0$, and it remains to compute the variance of the limit. Now,

$$\begin{aligned}
\mathbb{E}\left(\int_\epsilon^t \frac{W(u)}{u}\,du \int_\epsilon^t \frac{W(v)}{v}\,dv\right) &= \int_\epsilon^t \int_\epsilon^t \frac{u \wedge v}{uv}\,dv\,du && \text{by Lemma (8.5.1)} \\
&= 2\int_\epsilon^t du \int_\epsilon^u \frac{v}{uv}\,dv \to 2t && \text{as } \epsilon \downarrow 0.
\end{aligned}$$

The answer is $N(0, 2t)$.

7. Clearly, $AW(0) = 0$, and AW inherits the property of independent increments from W. One may complete the argument using the rotation invariance of the n-dimensional normal distribution, but instead we will use characteristic functions. We have that $AW(t) - AW(s)$ has characteristic function

$$\begin{aligned}
\mathbb{E}\Big(\exp\big\{i\boldsymbol{\theta} A\big(W(t) - W(s)\big)\big\}\Big) &= \mathbb{E}\Big(\exp\big\{i(\boldsymbol{\theta} A)\big(W(t) - W(s)\big)\big\}\Big) \\
&= \exp\big\{-\tfrac{1}{2}(t-s)|\boldsymbol{\theta} A|^2\big\} = \exp\big\{-\tfrac{1}{2}(t-s)|\boldsymbol{\theta}|^2\big\} \\
&= \mathbb{E}\Big(\exp\big\{i\boldsymbol{\theta}\big(W(t) - W(s)\big)\big\}\Big),
\end{aligned}$$

where $\boldsymbol{\theta}$ is a row vector, and since W is a Wiener process and A is orthonormal.

8. Since $W(t) - W(s)$ is distributed as $N(0, |t-s|)$,

$$\mathbb{E}\big(|W(t) - W(s)|^p\big) = \frac{1}{\sqrt{2\pi|t-s|}} \int_{\mathbb{R}} |x|^p \exp\left\{-\frac{|x|^2}{2|t-s|}\right\} dx, \qquad s \neq t.$$

Make the change of variables $y = x/\sqrt{|t-s|}$ to obtain

$$\mathbb{E}\big(|W(t) - W(s)|^p\big) = \frac{|t-s|^{p/2}}{\sqrt{2\pi}} \int_{\mathbb{R}} |y|^p \exp\left\{-\frac{|y|^2}{2}\right\} dy = c_p|t-s|^{p/2}.$$

8.6 Solutions. Lévy processes and subordinators

1. For $0 \le s < t < \infty$,

$$|\phi(t,\theta) - \phi(s,\theta)| = \left|\mathbb{E}\{e^{i\theta X(s)}(e^{i\theta D} - 1)\}\right| \le |\mathbb{E}e^{i\theta X(s)}| \cdot \mathbb{E}|e^{i\theta D} - 1|,$$

where $D = X(t) - X(s)$ has the same distribution as $X(t-s)$, and we have used the independence of increments. Let $\delta > 0$. The last term satisfies

$$\begin{aligned}\mathbb{E}|e^{i\theta D} - 1| &\le \mathbb{E}\{|e^{i\theta D} - 1| \cdot I(|D| > \delta)\} + \mathbb{E}\{|e^{i\theta D} - 1| \cdot I(|D| \le \delta)\} \\ &\le 2\mathbb{P}(|D| > \delta) + \sup_{|y| \le \delta} |e^{i\theta y} - 1|.\end{aligned}$$

Let $\epsilon > 0$ and choose $\delta > 0$ such that $|e^{i\theta y} - 1| \le \frac{1}{3}\epsilon$ for $|y| \le \delta$, and with this choice of δ we choose τ such that $\mathbb{P}(|D| > \delta) \le \frac{1}{3}\epsilon$ for $0 < t - s < \tau$. (We have used the fact that $X(u) \xrightarrow{\mathrm{P}} 0$ as $u \to 0$ here.) In conclusion,

$$\mathbb{E}|e^{i\theta D} - 1| \le \epsilon \qquad \text{for } 0 \le t - s \le \tau,$$

and the claim follows.

2. By the result of Problem (6.15.21), the characteristic function of the compound Poisson process $X(t)$ is $\phi(t,\theta) = \exp\{\lambda t(\psi(\theta) - 1)\}$ where ψ is the characteristic function of the summands. The Lévy symbol is thus $\lambda(\psi(\theta) - 1)$.

3. We have that

$$\begin{aligned}\log M_u(\theta) = -u\log(1+\theta) &= -u\int_0^\theta \frac{dx}{1+x} = -u\int_0^\theta dx \int_0^\infty e^{-y(1+x)}\,dy \\ &= -u\int_{y=0}^\infty e^{-y}\int_{x=0}^\theta e^{-yx}\,dx\,dy = -u\int_0^\infty (1 - e^{-\theta y})\frac{1}{y}e^{-y}\,dy.\end{aligned}$$

4. Let $Y(t) = X(T(t))$ where X is a Lévy process and T is an independent subordinator. Since X and T are continuous in probability, so is Y. Let $0 \le s_1 < t_1 \le s_2 < t_2 \le \cdots \le s_n < t_n$, and consider the joint characteristic function of the increments $I_r = Y(t_r) - Y(s_r)$:

$$\begin{aligned}&\mathbb{E}\left(\exp\left\{i\sum_{r=1}^n \theta_r I_r\right\}\right) \\ &\quad = \mathbb{E}\left[\mathbb{E}\left(\exp\left\{i\sum_{r=1}^n \theta_r\big(X(T(t_r)) - X(T(s_r))\big)\,\middle|\, T(s_1), T(t_1), \dots, T(t_n)\right\}\right)\right] \\ &\quad = \prod_{r=1}^n \mathbb{E}\Big(\exp\{i\theta_r X(T(t_r) - T(s_r))\}\Big) = \prod_{r=1}^n \mathbb{E}\Big(\exp\{i\theta_r Y(t_r - s_r)\}\Big),\end{aligned}$$

where we have used the fact that X and T have stationary independent increments. The final factorization indicates that Y also has stationary independent increments.

5. Let $Y(t) = N(T(t))$ be a subordinated Poisson process with time-change $T(t)$ having the given gamma density. In the notation of the question, for $s < 2$,

$$\mathbb{E}(s^Y) = \sum_{n=0}^\infty s^n \int_0^\infty \frac{e^{-x}x^n}{n!} \cdot \frac{e^{-x}x^{t-1}}{\Gamma(t)}\,dx = \int_0^\infty \frac{1}{\Gamma(t)}x^{t-1}e^{-(2-s)x}\,dx = \frac{1}{(2-s)^t},$$

since we recognise the integrand, when multiplied by the correct factor, as the density function of the $\Gamma(2-s, t)$ distribution.

6. (a) With $\mathcal{F}_t = \sigma(X(s) : s \le t)$, for $s < t$,

$$\mathbb{E}(X(t) \mid \mathcal{F}_s) = \mathbb{E}(X(t) - X(s) \mid \mathcal{F}_s) + \mathbb{E}(X(s) \mid \mathcal{F}_s) = \mathbb{E}(X(t) - X(s)) + X(s),$$

where we have used the independence of $X(t) - X(s)$ and $\mathcal{F}_s$.
(b) Similarly, by the properties of increments,

$$\begin{aligned}\mathbb{E}(Z(t)^2 \mid \mathcal{F}_s) &= Z(s)^2 + \mathbb{E}\big((Z(t) - Z(s))^2 \,\big|\, \mathcal{F}_s\big) + 2\mathbb{E}\big((Z(t) - Z(s))Z(s) \,\big|\, \mathcal{F}_s\big) \\ &= Z(s)^2 + f(s, t),\end{aligned}$$

for some deterministic function f. On taking expectations, we find that $f(s,t) = \mathbb{E}(Z(t)^2) - \mathbb{E}(Z(s)^2)$, and the proof is complete.
(c) Fix θ and write $M(t) = e^{i\theta X(t)}/\phi(t,\theta)$. For $s < t$, by the properties of increments,

$$\begin{aligned}\mathbb{E}(M(t) \mid \mathcal{F}_s) &= \frac{1}{\phi(t,\theta)} e^{i\theta X(s)} \mathbb{E}\big(e^{i\theta(X(t)-X(s))} \,\big|\, \mathcal{F}_s\big) \\ &= \frac{1}{\phi(t,\theta)} e^{i\theta X(s)} \mathbb{E}\big(e^{i\theta(X(t)-X(s))}\big) \\ &= \frac{\phi(t-s,\theta)}{\phi(t,\theta)} e^{i\theta X(s)} = \frac{1}{\phi(s,\theta)} e^{i\theta X(s)} = M(s),\end{aligned}$$

since $\phi(t,\theta) = \phi(s,\theta)\phi(t-s,\theta)$.

7. Let $\mathcal{F}_t = \sigma(Y(s) : s \le t)$. By conditioning, for $s < t$,

$$\begin{aligned}\mathbb{E}(Y(t) \mid \mathcal{F}_s) &= \mathbb{E}\Big(\mathbb{E}\big(X(T(t)) \,\big|\, \{T(u) : u \le s\},\ \mathcal{F}_s\big) \,\Big|\, \mathcal{F}_s\Big) \\ &= \mathbb{E}\big(X(T(s)) \,\big|\, \mathcal{F}_s\big) = Y(s).\end{aligned}$$

Since by assumption $\mathbb{E}|Y(t)| < \infty$, Y is a martingale with respect to the filtration $(\mathcal{F}_t)$. If X is positive, then $\mathbb{E}|Y(t)| = \mathbb{E}(X(T(t))) = \mathbb{E}(X(0)) < \infty$ since X is a martingale.

8.7 Solutions. Self-similarity and stability

1. We prove this by induction on n. It is trivially true when $n = 1$. Assume it holds for $n = N \ge 1$. By the induction hypothesis,

$$\sum_{r=1}^{N+1} X_r = B_N X + X_{N+1} + A_N,$$

where X is independent of X_{N+1}. The claim follows by the definition of stability.

2. By self-similarity, $X(t) \overset{D}{=} t^H X(1)$. Therefore, $\operatorname{var}(X(t)) = t^{2H} \operatorname{var}(X(1))$. Also, for $s < t$,

$$\begin{aligned}\mathbb{E}(X(s)X(t)) &= \tfrac{1}{2}\Big[\mathbb{E}(X(t)^2) + \mathbb{E}(X(s)^2) - \mathbb{E}\big((X(t) - X(s))^2\big)\Big] \\ &= \tfrac{1}{2}\Big[t^{2H}\mathbb{E}(X(1)^2) + s^{2H}\mathbb{E}(X(1)^2) - \mathbb{E}(X(t-s)^2)\Big] \\ &= \tfrac{1}{2}\big(t^{2H} + s^{2H} - (t-s)^{2H}\big)\mathbb{E}(X(1)^2),\end{aligned}$$

by the stationarity of increments and self-similarity.

3. Let F be the distribution function of a stable law, and let $n \geq 2$. By the definition of stability, or by Exercise (8.7.1), for independent random variables $X_1, X_2, \dots, X_n$ with distribution function F, there exist $B_n > 0$ and C_n such that $\sum_{r=1}^{n} Y_r$ has distribution function F, where $Y_r = (X_r - C_n)/B_n$. Therefore, F is infinitely divisible (see Problem (5.12.14)).

4. Let X be a self-similar Lévy process with $\text{var}(X(1)) < \infty$. By self-similarity, $\text{var}(X(2)) = 2^{2H}\,\text{var}(X(1))$, and by the property of independent stationary increments, $\text{var}(X(2)) = 2\,\text{var}(X(1)$. Therefore, $H = \frac{1}{2}$, and X is the Wiener process.

5. Since ϕ is real-valued, the distribution is symmetric (use either the inversion theorem or Problem (5.12.22)). For strict α-stability, we need that $X_1 + X_2 + \cdots + X_m \stackrel{\text{D}}{=} m^{1/\alpha} X_1$ for $m \geq 1$. The characteristic function of the sum satisfies

$$\phi(\theta)^m = e^{-m|\theta|^\alpha} = \phi(m^{1/\alpha}\theta),$$

as required.

6. By conditioning on Y,

$$\phi_Z(t) = \mathbb{E}\big(\mathbb{E}(e^{itXY^{1/\alpha}} \mid Y)\big) = \mathbb{E}\big(\phi_X(tY^{1/a})\big) = \mathbb{E}\big(e^{-Y|t|^\alpha}\big) = \exp\big(-k|t|^{\alpha\beta}\big),$$

which is the characteristic function of a symmetric, stable random variable with exponent $\alpha\beta$.

We have that U is symmetric and stable with exponent $\alpha = 2$, and, by the result of Problem (5.12.19b), $Y = 1/V^2$ is positive and stable with exponent $\beta = \frac{1}{2}$. Therefore, $Z = U/|V| = UY^{1/2}$ is symmetric and stable with exponent $\alpha\beta = 1$, and hence Z has the Cauchy distribution. By an argument using symmetry, U/V has the Cauchy distribution also.

Note further that, if X is normally distributed, and Y is positive and $\frac{1}{2}\alpha$-stable, then $X\sqrt{Y}$ is symmetric and α-stable. This may be viewed as a representation of a symmetric, stable distribution in terms of the $N(0, 1)$ distribution.

8.8 Solutions. Time changes

1. For $t_0 < t_1 < \cdots < t_n < t$ and $s > 0$,

$$\begin{aligned}
&\mathbb{P}\big(X(t+s) = k \,\big|\, X(t_0), X(t_1), \dots, X(t_n), X(t)\big) \\
&\quad = \mathbb{E}\Big(\mathbb{P}\big(X(t+s) = k \,\big|\, Z(T(t_0)), \dots, Z(T(t_n)), Z(T(t))\big) \,\Big|\, T(t)\Big) \\
&\quad = \mathbb{E}\Big(\mathbb{P}\big(X(t+s) = k \,\big|\, Z(T(t))\big) \,\Big|\, T(t)\Big) \qquad \text{by the Markov property for } X \\
&\quad = \mathbb{P}\big(X(t+s) = k \,\big|\, X(t)\big).
\end{aligned}$$

2. Since N is an independent subordinator, X is a Markov chain as in Exercise (8.8.1). The transition probability $p_{ij}(t)$ is an elementary computation by conditioning on the value of $N(t)$.

3. Since W and T are Lévy processes, so is Y. By conditioning on $T(t)$,

$$\begin{aligned}
\mathbb{E}(e^{i\theta Y(t)}) &= \mathbb{E}\big\{\mathbb{E}(e^{i\theta W(T(t))} \mid T(t))\big\} = \mathbb{E}(e^{-\frac{1}{2}\theta^2 T(t)}) \\
&= \exp\big\{-t(\tfrac{1}{2}\theta^2)^{a/2}\big\} = \exp\{-t2^{-a/2}|\theta|^a\},
\end{aligned}$$

which is the characteristic function of a symmetric Lévy process.

4. This is an important technicality. Let $\mathcal{F} = (\mathcal{F}_t)$ be the natural filtration, and recall that $\mathcal{F}$ is right-continuous.

(a) Let $s \geq 0$. Given that $X(s) > 0$, we have $X(s+t) = X(s) + (t-s)$; given that $X(s) = 0$, $X(t)$ has the same distribution as the unconditional distribution of $X(t-s)$. (The last holds since V has the lack-of-memory property.) Therefore, X is a Markov process.

(b) The random variable V is a stopping time since $\{V > t\} \in \bigcap_{s>t} \mathcal{F}_s = \mathcal{F}_t$.

(c) It is clear, by definition of X and V, that $X_1(0) = X_2(0) = 0$. On the other hand,

$$\mathbb{P}(X_1(1) = 0) = \mathbb{P}(V \geq 1) > 0, \qquad \mathbb{P}(X_2(1) = 0) = 0.$$

This is an example of a process X that has the arguably undesirable feature of satisfying the Markov property but not the strong Markov property. It is usual to introduce a further property, called the *Feller property*, that implies the strong Markov property. The Feller property requires that the distribution of $X(t)$ varies continuously with the starting state $X(0)$.

8.10 Solutions to problems

1. $\mathbb{E}(Y_n) = 0$, and $\text{cov}(Y_m, Y_{m+n}) = \sum_{i=0}^{r} \alpha_i \alpha_{n+i}$ for $m, n \geq 0$, with the convention that $\alpha_k = 0$ for $k > r$. The covariance does not depend on m, and therefore the sequence is stationary.

2. We have, by iteration, that $Y_n = S_n(m) + \alpha^{m+1} Y_{n-m-1}$ where $S_n(m) = \sum_{j=0}^{m} \alpha^j Z_{n-j}$. There are various ways of showing that the sequence $\{S_n(m) : m \geq 1\}$ converges in mean square and almost surely, and the shortest is as follows. We have that $\alpha^{m+1} Y_{n-m-1} \to 0$ in m.s. and a.s. as $m \to \infty$; to see this, use the facts that $\text{var}(\alpha^{m+1} Y_{n-m-1}) = \alpha^{2(m+1)} \text{var}(Y_0)$, and

$$\sum_m \mathbb{P}(\alpha^{m+1} Y_{n-m-1} > \epsilon) \leq \sum_m \frac{\alpha^{2(m+1)} \mathbb{E}(Y_0^2)}{\epsilon^2} < \infty, \qquad \epsilon > 0.$$

It follows that $S_n(m) = Y_n - \alpha^{m+1} Y_{n-m-1}$ converges in m.s. and a.s. as $m \to \infty$. A longer route to the same conclusion is as follows. For $r < s$,

$$\mathbb{E}\big(|S_n(s) - S_n(r)|^2\big) = \mathbb{E}\left\{\left(\sum_{j=r+1}^{s} \alpha^j Z_{n-j}\right)^2\right\} = \sum_{j=r+1}^{s} \alpha^{2j} \leq \frac{\alpha^{2r}}{1-\alpha^2},$$

whence $\{S_n(m) : m \geq 1\}$ is Cauchy convergent in mean square, and therefore converges in mean square. In order to show the almost sure convergence of $S_n(m)$, one may argue as follows. Certainly

$$\mathbb{E}\left(\sum_{j=0}^{m} |\alpha^j Z_{n-j}|\right) = \sum_{j=0}^{m} \mathbb{E}|\alpha^j Z_{n-j}| \to \sum_{j=0}^{\infty} |\alpha|^j \mathbb{E}|Z_{n-j}| \leq \sum_{j=0}^{\infty} |\alpha|^j < \infty,$$

whence $\sum_{j=0}^{\infty} \alpha^j Z_{n-j}$ is a.s. absolutely convergent, and therefore a.s. convergent also. We may express $\lim_{m\to\infty} S_n(m)$ as $\sum_{j=0}^{\infty} \alpha^j Z_{n-j}$. Also, $\alpha^{m+1} Y_{n-m-1} \to 0$ in mean square and a.s. as $m \to \infty$, and we may therefore express Y_n as

$$Y_n = \sum_{j=0}^{\infty} \alpha^j Z_{n-j} \qquad \text{a.s.}$$

It follows that $\mathbb{E}(Y_n) = \lim_{m\to\infty} \mathbb{E}(S_n(m)) = 0$. Finally, for $r > 0$, the autocovariance function is given by

$$c(r) = \text{cov}(Y_n, Y_{n-r}) = \mathbb{E}\big\{(\alpha Y_{n-1} + Z_n) Y_{n-r}\big\} = \alpha c(r-1),$$

whence

$$c(r) = \alpha^{|r|}c(0) = \frac{\alpha^{|r|}}{1-\alpha^2}, \qquad r = \dots, -1, 0, 1, \dots,$$

since $c(0) = \text{var}(Y_n)$.

3. If t is a non-negative integer, $N(t)$ is the number of 0's and 1's preceding the $(t+1)$th 1. Therefore $N(t) + 1$ has the negative binomial distribution with mass function

$$f(k) = \binom{k-1}{t} p^{t+1}(1-p)^{k-1-t}, \qquad k \ge t+1.$$

If t is not an integer, then $N(t) = N(\lfloor t \rfloor)$.

4. We have that

$$\mathbb{P}\big(Q(t+h) = j \mid Q(t) = i\big) = \begin{cases} \lambda h + o(h) & \text{if } j = i+1, \\ \mu i h + o(h) & \text{if } j = i-1, \\ 1 - (\lambda + \mu i)h + o(h) & \text{if } j = i, \end{cases}$$

an immigration–death process with constant birth rate λ and death rates $\mu_i = i\mu$.

Either calculate the stationary distribution in the usual way, or use the fact that birth–death processes are reversible in equilibrium. Hence $\lambda\pi_i = \mu(i+1)\pi_{i+1}$ for $i \ge 0$, whence

$$\pi_i = \frac{1}{i!}\left(\frac{\lambda}{\mu}\right)^i e^{-\lambda/\mu}, \qquad i \ge 0.$$

5. We have that $\widetilde{X}(t) = R\cos(\Psi)\cos(\theta t) - R\sin(\Psi)\sin(\theta t)$. Consider the transformation $u = r\cos\psi$, $v = -r\sin\psi$, which maps $[0,\infty) \times [0, 2\pi)$ to $\mathbb{R}^2$. The Jacobian is

$$\begin{vmatrix} \dfrac{\partial u}{\partial r} & \dfrac{\partial u}{\partial \psi} \\ \dfrac{\partial v}{\partial r} & \dfrac{\partial v}{\partial \psi} \end{vmatrix} = -r,$$

whence $U = R\cos\Psi$, $V = -R\sin\Psi$ have joint density function satisfying

$$r f_{U,V}(r\cos\psi, -r\sin\psi) = f_{R,\Psi}(r,\psi).$$

Substitute $f_{U,V}(u,v) = e^{-\frac{1}{2}(u^2+v^2)}/(2\pi)$, to obtain

$$f_{R,\Psi}(r,\psi) = \frac{1}{2\pi} r e^{-\frac{1}{2}r^2}, \qquad r > 0,\ 0 \le \psi < 2\pi.$$

Thus R and Ψ are independent, the latter being uniform on $[0, 2\pi)$.

6. A customer arriving at time u is designated *green* if he is in state A at time t, an event having probability $p(u, t-u)$. By the colouring theorem (6.13.14), the arrival times of green customers form a non-homogeneous Poisson process with intensity function $\lambda(u)p(u, t-u)$, and the claim follows.

7. By conditioning on the Poissonian events in the interval $(0, h)$,

$$\begin{aligned} r(y) &= \lambda h \mathbb{E}\big(r(y+h-X_1)\big) + (1-\lambda h)r(y+h) + \mathrm{o}(h) \\ &= \lambda h\left\{\int_0^{y+h} r(y+h-x)\,dF(x) + \mathbb{P}(X_1 > y+h)\right\} + (1-\lambda h)r(y+h) + \mathrm{o}(h), \end{aligned}$$

whence, on subtracting $r(y)$, dividing by h and letting $h \downarrow 0$.

$$r'(y) = \lambda r(y) - \lambda\mathbb{P}(X_1 > y) - \lambda \int_0^y r(y+h)\,dF(x).$$

This may be solved using Lebesgue-Stieltjes transforms as in Exercise (10.1.7), or using martingale theory as in Problem (12.9.27).

8. (a) The normality holds since fBM is Gaussian. By the result of Exercise (8.7.2),

$$\mathbb{E}\big(|X(t) - X(s)|^2\big) = t^{2H} + s^{2H} - (t^{2H} + s^{2H} - |t-s|^{2H}) = |t-s|^{2H}.$$

(b) By part (a), for $s \neq t$,

$$\begin{aligned}\mathbb{E}\big(|X(t) - X(s)|^r\big) &= \frac{1}{|t-s|^H\sqrt{2\pi}} \int_{-\infty}^{\infty} |x|^r \exp\left\{-\frac{x^2}{2|t-s|^{2H}}\right\} dx \\ &= |t-s|^{rH} \cdot \frac{1}{\sqrt{2\pi}} \int_{-\infty}^{\infty} |y|^r e^{-\frac{1}{2}y^2}\,dy,\end{aligned}$$

by the substitution $x = y|t-s|^H$. It is easily seen that $C = (2k)!/(k!\,2^k)$ when $r = 2k$ is even.

(c) The distribution of a zero-mean Gaussian process is specified by its covariance function. For the standard Wiener process, the covariance function is $c(s,t) = \min\{s,t\}$, which agrees with part (a) if and only if $H = \frac{1}{2}$.

9. (a) By the superposition theorem (6.13.5), the union of two independent Poisson processes is a Poisson process with the sum of the intensities. Their combined gravitational force is the sum of the two separate forces.

(b) Changing the intensity of a Poisson process from 1 to λ amounts to changing the length scale from 1 to $1/\lambda^{1/3}$. This changes the scale of the force from 1 to $\lambda^{2/3}$.

(c) It is tempting to argue as follows. For $a, b > 0$, we have by the above that $aG_1' + bG_1'' \stackrel{\text{D}}{=} cG_1$ where $c = (a^{3/2} + b^{3/2})^{2/3}$. Therefore, G_1 is strictly α-stable with $\alpha = \frac{2}{3}$. However, as in the solution to Problems (6.15.56)–(6.15.57), the aggregate gravitational attraction at the origin is not absolutely convergent. We may instead follow Holtsmark in restricting space to the R-ball centred at the origin, and passing to the limit $R \to \infty$, thereby allowing the divergences to cancel. The weak limit is then α-stable.

[See Problem (6.15.57) for an alternative approach. The above argument of Feller (1974, p. 174) is applicable in other dimensions with adjusted exponents.]

9

Stationary processes

9.1 Solutions. Introduction

1. We examine sequences W_n of the form

$$(*) \qquad W_n = \sum_{k=0}^{\infty} a_k Z_{n-k}$$

for the real sequence $\{a_k : k \geq 0\}$. Substitute, to obtain $a_0 = 1, a_1 = \alpha, a_r = \alpha a_{r-1} + \beta a_{r-2}, r \geq 2$, with solution

$$a_r = \begin{cases} (1+r)\lambda_1^r & \text{if } \alpha^2 + 4\beta = 0, \\ \dfrac{\lambda_1^{r+1} - \lambda_2^{r+1}}{\lambda_1 - \lambda_2} & \text{otherwise,} \end{cases}$$

where λ_1 and λ_2 are the (possibly complex) roots of the quadratic $x^2 - \alpha x - \beta = 0$ (these roots are distinct if and only if $\alpha^2 + 4\beta \neq 0$).

Using the method in the solution to Problem (8.10.2), the sum in $(*)$ converges in mean square and almost surely if $|\lambda_1| < 1$ and $|\lambda_2| < 1$. Assuming this holds, we have from $(*)$ that $\mathbb{E}(W_n) = 0$ and the autocovariance function is

$$c(m) = \mathbb{E}(W_n W_{n-m}) = \alpha c(m-1) + \beta c(m-2), \qquad m \geq 1,$$

by the independence of the Z_n. Therefore W is weakly stationary, and the autocovariance function may be expressed in terms of α and β.

2. We adopt the convention that, if the binary expansion of U is non-unique, then we take the (unique) non-terminating such expansion. It is clear that X_i takes values in $\{0, 1\}$, and

$$\mathbb{P}\big(X_{n+1} = 1 \,\big|\, X_i = x_i \text{ for } 1 \leq i \leq n\big) = \tfrac{1}{2}$$

for all $x_1, x_2, \dots, x_n$; therefore the X's are independent Bernoulli random variables. For any sequence $k_1 < k_2 < \cdots < k_r$, the joint distribution of $V_{k_1}, V_{k_2}, \dots, V_{k_r}$ depends only on that of $X_{k_1+1}, X_{k_1+2}, \dots$. Since this distribution is the same as the distribution of $X_1, X_2, \dots$, we have that $(V_{k_1}, V_{k_2}, \dots, V_{k_r})$ has the same distribution as $(V_0, V_{k_2-k_1}, \dots, V_{k_r-k_1})$. Therefore V is strongly stationary.

Clearly $\mathbb{E}(V_n) = \mathbb{E}(V_0) = \frac{1}{2}$, and, by the independence of the X_i,

$$\text{cov}(V_0, V_n) = \sum_{i=1}^{\infty} 2^{-2i-n} \,\text{var}(X_i) = \tfrac{1}{12}(\tfrac{1}{2})^n.$$

3. (i) For mean-square convergence, we show that $S_k = \sum_{n=0}^{k} a_n X_n$ is mean-square Cauchy convergent as $k \to \infty$. We have that, for $r < s$,

$$\mathbb{E}\{(S_s - S_r)^2\} = \sum_{i,j=r+1}^{s} a_i a_j c(i-j) \le c(0)\left\{\sum_{i=r+1}^{s} |a_i|\right\}^2$$

since $|c(m)| \le c(0)$ for all m, by the Cauchy–Schwarz inequality. The last sum tends to 0 as $r, s \to \infty$ if $\sum_i |a_i| < \infty$. Hence S_k converges in mean square as $k \to \infty$.

Secondly,

$$\mathbb{E}\left(\sum_{k=1}^{n} |a_k X_k|\right) \le \sum_{k=1}^{n} |a_k| \cdot \mathbb{E}|X_k| \le \sqrt{\mathbb{E}(X_0^2)} \sum_{k=1}^{n} |a_k|$$

which converges as $n \to \infty$ if the $|a_k|$ are summable. It follows that $\sum_{k=1}^{n} |a_k X_k|$ converges absolutely (almost surely), and hence $\sum_{k=1}^{n} a_k X_k$ converges a.s.

(ii) Each sum converges a.s. and in mean square, by part (i). Now

$$c_Y(m) = \sum_{j,k=0}^{\infty} a_j a_k c(m+k-j)$$

whence

$$\sum_m |c_Y(m)| \le c(0)\left\{\sum_{j=0}^{\infty} |a_j|\right\}^2 < \infty.$$

4. Clearly X_n has distribution $\boldsymbol{\pi}$ for all n, so that $\{f(X_n) : n \ge m\}$ has fdds which do not depend on the value of m. Therefore the sequence is strongly stationary.

5. The covariance is

$$\text{cov}(U, V) = \mathbb{E}(U\overline{V}) = \mathbb{E}\big((WY + XZ) - i(XY - WZ)\big) = 0.$$

If U, V are independent, then the vectors (W, X), (Y, Z) are independent. However, $\text{cov}(W, Z) = -1$, a contradiction.

Example (4.5.9) is concerned with real-valued random variables, and U, V are complex-valued. They have neither univariate distributions nor univariate density functions.

6. Since cosine has period 2π and U is uniformly distributed on $(-\pi, \pi)$, we have that $\mathbb{E}(X_n) = 0$. The autocovariance function is

$$\begin{aligned} c(m, m+n) = \mathbb{E}(X_m X_{m+n}) &= \mathbb{E}\big(\cos(mS + U)\cos((m+n)S + U)\big) \\ &= \tfrac{1}{2}\mathbb{E}\big(\cos((2m+n)S + 2U) + \cos(nS)\big) \\ &= 0 + \tfrac{1}{2}\mathbb{E}\big(\cos(nS)\big) = \tfrac{1}{2}\int_{-\pi}^{\pi} g(s)\cos(ns)\, ds, \end{aligned}$$

as before. Thus $c(m, m+n)$ depends on n only, and X is weakly stationary. The autocorrelation function ρ_X is given by

$$(*) \qquad \rho_X(m, m+n) = \frac{c(m, m+n)}{c(m, m)} = \int_{-\pi}^{\pi} g(s)\cos(ns)\, ds.$$

The autocorrelation function of Y is easily seen to be

$$\rho_Y(m, m+n) = \begin{cases} 1 & \text{if } n = 0, \\ \dfrac{a}{1+a^2} & \text{if } n = 1, \\ 0 & \text{if } n \ge 2. \end{cases}$$

It is natural to try setting g to be a linear combination of a constant and the cosine function, say $g(s) = \alpha + \beta \cos s$. Since g is a density function on $(-\pi, \pi)$, we must have $\alpha = 1/(2\pi)$. By (*) with this choice for g,

$$\rho_X(m, m+n) = \begin{cases} 1 & \text{if } n = 0, \\ \pi\beta & \text{if } n = 1, \\ 0 & \text{if } n \geq 2, \end{cases}$$

so that $\rho_X = \rho_Y$ if

$$g(s) = \frac{1}{2\pi} + \frac{a \cos s}{\pi(1+a^2)}, \qquad s \in (-\pi, \pi).$$

9.2 Solutions. Linear prediction

1. (i) We have that

$$(*) \qquad \mathbb{E}\{(X_{n+1} - \alpha X_n)^2\} = (1+\alpha^2)c(0) - 2\alpha c(1),$$

which is minimized by setting $\alpha = c(1)/c(0)$. Hence $\widehat{X}_{n+1} = c(1)X_n/c(0)$.
(ii) Similarly

$$(**) \qquad \mathbb{E}\{(X_{n+1} - \beta X_n - \gamma X_{n-1})^2\} = (1+\beta^2+\gamma^2)c(0) + 2\beta(\gamma-1)c(1) - 2\gamma c(2),$$

an expression which is minimized by the choice

$$\beta = \frac{c(1)\big(c(0) - c(2)\big)}{c(0)^2 - c(1)^2}, \qquad \gamma = \frac{c(0)c(2) - c(1)^2}{c(0)^2 - c(1)^2};$$

$\widetilde{X}_{n+1}$ is given accordingly.
(iii) Substitute α, β, γ into $(*)$ and $(**)$, and subtract to obtain, after some manipulation,

$$D = \frac{\{c(1)^2 - c(0)c(2)\}^2}{c(0)\{c(0)^2 - c(1)^2\}}.$$

(a) In this case $c(0) = \frac{1}{2}$, and $c(1) = c(2) = 0$. Therefore $\widehat{X}_{n+1} = \widetilde{X}_{n+1} = 0$, and $D = 0$.
(b) In this case $D = 0$ also.

In both (a) and (b), little of substance is gained by using $\widetilde{X}_{n+1}$ in place of $\widehat{X}_{n+1}$.

2. Let $\{Z_n : n = \dots, -1, 0, 1, \dots\}$ be independent random variables with zero means and unit variances, and define the moving-average process

$$(*) \qquad X_n = \frac{Z_n + aZ_{n-1}}{\sqrt{1+a^2}}.$$

It is easily checked that X has the required autocovariance function.

By the projection theorem, $X_n - \widehat{X}_n$ is orthogonal to the collection $\{X_{n-r} : r \geq 1\}$, so that $\mathbb{E}\{(X_n - \widehat{X}_n)X_{n-r}\} = 0$, $r \geq 1$. Set $\widehat{X}_n = \sum_{s=1}^{\infty} b_s X_{n-s}$ to obtain that

$$\alpha = b_1 + b_2\alpha, \quad 0 = b_{s-1}\alpha + b_s + b_{s+1}\alpha \qquad \text{for } s \geq 2,$$

where $\alpha = a/(1+a^2)$. The unique bounded solution to the above difference equation is $b_s = (-1)^{s+1}a^s$, and therefore

$$\widehat{X}_n = \sum_{s=1}^{\infty}(-1)^{s+1}a^s X_{n-s}.$$

The mean squared error of prediction is

$$\mathbb{E}\{(X_n - \widehat{X}_n)^2\} = \mathbb{E}\left\{\left(\sum_{s=0}^{\infty}(-a)^s X_{n-s}\right)^2\right\} = \frac{1}{1+a^2}\mathbb{E}(Z_n^2) = \frac{1}{1+a^2}.$$

Clearly $\mathbb{E}(\widehat{X}_n) = 0$ and

$$\operatorname{cov}(\widehat{X}_n, \widehat{X}_{n-m}) = \sum_{r,s=1}^{\infty} b_r b_s c(m+r-s), \qquad m \ge 0,$$

so that $\widehat{X}$ is weakly stationary.

9.3 Solutions. Autocovariances and spectra

1. It is clear that $\mathbb{E}(X_n) = 0$ and $\operatorname{var}(X_n) = 1$. Also

$$\operatorname{cov}(X_m, X_{m+n}) = \cos(m\lambda)\cos\{(m+n)\lambda\} + \sin(m\lambda)\sin\{(m+n)\lambda\} = \cos(n\lambda),$$

so that X is stationary, and the spectrum of X is the singleton $\{\lambda\}$.

2. Certainly $\phi_U(t) = (e^{it\pi} - e^{-it\pi})/(2\pi i t)$, so that $\mathbb{E}(X_n) = \phi_U(1)\phi_V(n) = 0$. Also

$$\operatorname{cov}(X_m, X_{m+n}) = \mathbb{E}(X_m \overline{X}_{m+n}) = \mathbb{E}\big(e^{i\{U - Vm - U + V(m+n)\}}\big) = \phi_V(n),$$

whence X is stationary. Finally, the autocovariance function is

$$c(n) = \phi_V(n) = \int e^{in\lambda}\, dF(\lambda),$$

whence F is the spectral distribution function.

3. The characteristic functions of these distributions are

(i) $$\rho(t) = e^{-\frac{1}{2}t^2},$$

(ii) $$\rho(t) = \frac{1}{2}\left(\frac{1}{1-it} + \frac{1}{1+it}\right) = \frac{1}{1+t^2}.$$

4. (i) We have that

$$\operatorname{var}\left(\frac{1}{n}\sum_{j=1}^{n} X_j\right) = \frac{1}{n^2}\sum_{j,k=1}^{n}\operatorname{cov}(X_j, X_k) = \frac{c(0)}{n^2}\int_{(-\pi,\pi]}\left(\sum_{j,k=1}^{n} e^{i(j-k)\lambda}\right) dF(\lambda).$$

The integrand is

$$\left|\sum_{j=1}^{n} e^{ij\lambda}\right|^2 = \left(\frac{e^{in\lambda}-1}{e^{i\lambda}-1}\right)\left(\frac{e^{-in\lambda}-1}{e^{-i\lambda}-1}\right) = \frac{1-\cos(n\lambda)}{1-\cos\lambda},$$

whence

$$\operatorname{var}\left(\frac{1}{n}\sum_{j=1}^{n} X_j\right) = c(0)\int_{(-\pi,\pi]}\left(\frac{\sin(n\lambda/2)}{n\sin(\lambda/2)}\right)^2 dF(\lambda).$$

It is easily seen that $|\sin\theta| \le |\theta|$, and therefore the integrand is no larger than

$$\left(\frac{\lambda/2}{\sin(\lambda/2)}\right)^2 \le (\tfrac{1}{2}\pi)^2.$$

As $n \to \infty$, the integrand converges to the function which is zero everywhere except at the origin, where (by continuity) we may assign it the value 1. It may be seen, using the dominated convergence theorem, that the integral converges to $F(0) - F(0-)$, the size of the discontinuity of F at the origin, and therefore the variance tends to 0 if and only if $F(0) - F(0-) = 0$.

Using a similar argument,

$$\frac{1}{n}\sum_{j=0}^{n-1} c(j) = \frac{c(0)}{n}\int_{(-\pi,\pi]}\left(\sum_{j=0}^{n-1} e^{ij\lambda}\right) dF(\lambda) = c(0)\int_{(-\pi,\pi]} g_n(\lambda)\, dF(\lambda)$$

where

$$g_n(\lambda) = \begin{cases} 1 & \text{if } \lambda = 0, \\ \dfrac{e^{in\lambda} - 1}{n(e^{i\lambda} - 1)} & \text{if } \lambda \neq 0, \end{cases}$$

is a bounded sequence of functions which converges as before to the Kronecker delta function $\delta_{\lambda 0}$. Therefore

$$\frac{1}{n}\sum_{j=0}^{n-1} c(j) \to c(0)\big(F(0) - F(0-)\big) \qquad \text{as } n \to \infty.$$

5. (a) By iteration, for $k \ge 0$,

$$X_n = Z_n + \sum_{r=1}^{k} \phi^{r-1}(\theta + \phi) Z_{n-r} + \phi^k \theta Z_{n-k-1} + \phi^{k+1} X_{n-k-1}.$$

We require $|\phi| < 1$ for convergence as $k \to \infty$. Assume $|\phi| < 1$, so that

$$(*) \qquad X_n = Z_n + (\theta + \phi)\sum_{r=1}^{\infty} \phi^{r-1} Z_{n-r},$$

implying by its form that X is stationary with zero mean. The value of θ is immaterial.
(b) We have $Z_n = -\theta Z_{n-1} - X_n - \phi X_{n-1}$. Subject to the change of signs, this is as before with θ and ϕ interchanged, We therefore require that $|\theta| < 1$.
(c) Square (*) and apply $\mathbb{E}$, to find that

$$\mathbb{E}(X_n^2) = 1 + \sum_{r=1}^{\infty} \phi^{2(r-1)}(\theta + \phi)^2 = \frac{1 + \theta^2 + 2\theta\phi}{1 - \phi^2}.$$

Similarly,

$$c(n, n+1) = \mathbb{E}(X_n X_{n+1}) = \frac{(\theta + \phi)(1 + \theta\phi)}{1 - \phi^2},$$

and, more generally,

$$c(m, m+n) = \mathbb{E}(X_m X_{m+n}) = \phi c(m, m+n-1), \qquad n \geq 2.$$

The autocorrelation function of X is symmetric and given by

$$\rho(n) = \frac{c(m, m+n)}{c(0,0)} = \begin{cases} 1 & \text{if } n = 0, \\ \dfrac{(\theta+\phi)(1+\theta\phi)}{1+\theta^2+2\theta\phi} & \text{if } n = 1, \\ \phi^{n-1}\rho(1) & \text{if } n \geq 2. \end{cases}$$

The spectral density function is

$$\begin{aligned} f(\lambda) &= \frac{1}{2\pi} + \frac{1}{\pi}\rho(1)\cos\lambda + \frac{1}{\pi}\sum_{n=2}^{\infty} \rho(1)\phi^{n-1}(e^{in\lambda} + e^{-in\lambda}) \\ &= \frac{1}{2\pi} + \frac{1}{\pi}\rho(1)\cos\lambda + \frac{\rho(1)}{\pi}\left\{\frac{\phi e^{2i\lambda}}{1-\phi e^{i\lambda}} + \frac{\phi e^{-2i\lambda}}{1-\phi e^{-i\lambda}}\right\}, \qquad \lambda \in (-\pi, \pi]. \end{aligned}$$

6. Substitute the sum into the recurrence and equate coefficients of the Z_r to obtain

$$d_0 = 1, \qquad d_1 = \phi d_0, \qquad d_k = \phi d_{k-1} + \theta d_{k-2} \quad \text{for } k \geq 2.$$

Hence $d_k = a_1 r_2^k + a_2 r_2^k$ where the r_i are the roots of the quadratic $x^2 - \phi x - \theta = 0$. Therefore,

$$d_0 = 1 = a_1 + a_2, \qquad d_1 = \phi = a_1 r_1 + a_2 r_2.$$

Suppose $|r_1|, |r_2| < 1$. Since the Z_r are uncorrelated,

$$\mathbb{E}(X_n^2) = \sum_{k=0}^{\infty} d_k^2 = \frac{a_1^2}{1-r_1^2} + \frac{2a_1 a_2}{1-r_1 r_2} + \frac{a_2^2}{1-r_1^2},$$

on summing the geometric series. The given answer follows after a calculation using the properties of the r_i.

7. As in Example (9.2.5), we have $\operatorname{var}(X_{n+1} - \widehat{X}_{n+1}) = 1$. By Example (9.3.23), the given exponential equals

$$\exp\left\{\frac{1}{2\pi}\int_{-\pi}^{\pi} -\log\left(1 - \alpha e^{i\lambda} - \alpha e^{-i\lambda} + \alpha^2\right) d\lambda\right\} = 1.$$

To check this, write the logarithm as

$$\begin{aligned} \log\left[(1-\alpha e^{i\lambda})(1-\alpha e^{-i\lambda})\right] &= \log(1-\alpha e^{i\lambda}) + \log(1-\alpha e^{-i\lambda}) \\ &= \sum_{r=1}^{\infty} \frac{\alpha^r e^{ir\lambda}}{r} + \sum_{r=1}^{\infty} \frac{\alpha^r e^{-ir\lambda}}{r}, \end{aligned}$$

and integrate term-by-term over $(-\pi, \pi)$ to obtain the answer 0.

9.4 Solutions. Stochastic integration and the spectral representation

1. Let H_X be the space of all linear combinations of the X_i, and let $\overline{H}_X$ be the closure of this space, that is, H_X together with the limits of all mean-square Cauchy-convergent sequences in H_X. All members of H_X have zero mean, and therefore all members of $\overline{H}_X$ also. Now $S(\lambda) \in \overline{H}_X$ for all λ, whence $\mathbb{E}(S(\lambda) - S(\mu)) = 0$ for all λ and μ.

2. First, each Y_m lies in the space $\overline{H}_X$ containing all linear combinations of the X_n and all limits of mean-square Cauchy-convergent sequences of the same form. As in the solution to Exercise (9.4.1), all members of $\overline{H}_X$ have zero mean, and therefore $\mathbb{E}(Y_m) = 0$ for all m. Secondly,

$$\mathbb{E}(Y_m \overline{Y}_n) = \int_{(-\pi,\pi]} \frac{e^{im\lambda}e^{-in\lambda}}{2\pi f(\lambda)} f(\lambda)\, d\lambda = \delta_{mn}.$$

As for the last part,

$$\sum_{j=-\infty}^{\infty} a_j Y_{n-j} = \int_{(-\pi,\pi]} \left(\sum_j a_j e^{-ij\lambda}\right) \frac{e^{in\lambda}}{\sqrt{2\pi f(\lambda)}}\, dS(\lambda) = \int_{(-\pi,\pi]} e^{in\lambda}\, dS(\lambda) = X_n.$$

This proves that such a sequence X_n may be expressed as a moving average of an orthonormal sequence.

3. Let $\overline{H}_X$ be the space of all linear combinations of the X_n, together with all limits of (mean-square) Cauchy-convergent sequences of such combinations. Using the result of Problem (7.11.19), all elements in $\overline{H}_X$ are normally distributed. In particular, all increments of the spectral process are normal. Similarly, all pairs in $\overline{H}_X$ are jointly normally distributed, and therefore two members of $\overline{H}_X$ are independent if and only if they are uncorrelated. Increments of the spectral process have zero means (by Exercise (9.4.1)) and are orthogonal. Therefore they are uncorrelated, and hence independent.

9.5 Solutions. The ergodic theorem

1. With the usual shift operator τ, it is obvious that $\tau^{-1}\varnothing = \varnothing$, so that $\varnothing \in \mathcal{I}$. Secondly, if $A \in \mathcal{I}$, then $\tau^{-1}(A^{c}) = (\tau^{-1}A)^{c} = A^{c}$, whence $A^{c} \in \mathcal{I}$. Thirdly, suppose $A_1, A_2, \ldots \in \mathcal{I}$. Then

$$\tau^{-1}\left(\bigcup_{i=1}^{\infty} A_i\right) = \bigcup_{i=1}^{\infty} \tau^{-1}A_i = \bigcup_{i=1}^{\infty} A_i,$$

so that $\bigcup_1^\infty A_i \in \mathcal{I}$.

2. The left-hand side is the sum of covariances, $c(0)$ appearing n times, and $c(i)$ appearing $2(n-i)$ times for $0 < i < n$, in agreement with the right-hand side.

Let $\epsilon > 0$. If $\overline{c}(j) = j^{-1}\sum_{i=0}^{j-1} c(i) \to \sigma^2$ as $j \to \infty$, there exists J such that $|\overline{c}(j) - \sigma^2| < \epsilon$ when $j \geq J$. Now

$$\frac{2}{n^2}\sum_{j=1}^{n} j\overline{c}(j) \leq \frac{2}{n^2}\left\{\sum_{j=1}^{J} j\overline{c}(j) + \sum_{j=J+1}^{n} j(\sigma^2+\epsilon)\right\} \to \sigma^2 + \epsilon$$

as $n \to \infty$. A related lower bound is proved similarly, and the claim follows since ϵ (> 0) is arbitrary.

3. It is easily seen that $S_m = \sum_{i=0}^{m} \alpha_i X_{n+i}$ constitutes a martingale with respect to the X's, and

$$\mathbb{E}(S_m^2) = \sum_{i=0}^{m} \alpha_i^2 \mathbb{E}(X_{n+i}^2) \le \sum_{i=0}^{\infty} \alpha_i^2,$$

whence S_m converges a.s. and in mean square as $m \to \infty$.

Since the X_n are independent and identically distributed, the sequence Y_n is strongly stationary; also $\mathbb{E}(Y_n) = 0$, and so $n^{-1} \sum_{i=1}^{n} Y_i \to Z$ a.s. and in mean, for some random variable Z with mean zero. For any fixed m (≥ 1), the contribution of $X_1, X_2, \dots, X_m$ towards $\sum_{i=1}^{n} Y_i$ is, for large n, no larger than

$$C_m = \left| \sum_{j=1}^{m} (\alpha_0 + \alpha_1 + \cdots + \alpha_{j-1}) X_j \right|.$$

Now $n^{-1} C_m \to 0$ as $n \to \infty$, so that Z is defined in terms of the subsequence $X_{m+1}, X_{m+2}, \dots$ for all m, which is to say that Z is a tail function of a sequence of independent random variables. Therefore Z is a.s. constant, and so $Z = 0$ a.s.

9.6 Solutions. Gaussian processes

1. The quick way is to observe that c is the autocovariance function of a Poisson process with intensity 1. Alternatively, argue as follows. The sum is unchanged by taking complex conjugates, and hence is real. Therefore it equals

$$\begin{aligned}\sum_{j=1}^{n} t_j \left(|z_j|^2 + z_j \sum_{k=j+1}^{n} \bar{z}_k + \bar{z}_j \sum_{k=j+1}^{n} z_k \right) &= \sum_{j=1}^{n} t_j \left(\left| \sum_{k=j}^{n} z_k \right|^2 - \left| \sum_{k=j+1}^{n} z_k \right|^2 \right) \\ &= \sum_{j=1}^{n} (t_j - t_{j-1}) \left| \sum_{k=j}^{n} z_k \right|^2\end{aligned}$$

where $t_0 = 0$.

2. For $s, t \ge 0$, $X(s)$ and $X(s+t)$ have a bivariate normal distribution with zero means, unit variances, and covariance $c(t)$. It is standard (see Problem (4.14.13)) that $\mathbb{E}(X(s+t) \mid X(s)) = c(t)X(s)$. Now

$$\begin{aligned}c(s+t) = \mathbb{E}\big(X(0)X(s+t)\big) &= \mathbb{E}\Big\{\mathbb{E}\big(X(0)X(s+t) \mid X(0), X(s)\big)\Big\} \\ &= \mathbb{E}\big(X(0)c(t)X(s)\big) = c(s)c(t)\end{aligned}$$

by the Markov property. Therefore c satisfies $c(s+t) = c(s)c(t)$, $c(0) = 1$, whence $c(s) = c(1)^{|s|} = \rho^{|s|}$. Using the inversion formula, the spectral density function is

$$f(\lambda) = \frac{1}{2\pi} \sum_{s=-\infty}^{\infty} c(s) e^{-is\lambda} = \frac{1 - \rho^2}{2\pi |1 - \rho e^{i\lambda}|^2}, \qquad |\lambda| \le \pi.$$

Note that X has the same autocovariance function as a certain autoregressive process. Indeed, stationary Gaussian Markov processes have such a representation.

3. If X is Gaussian and strongly stationary, then it is weakly stationary since it has a finite variance. Conversely suppose X is Gaussian and weakly stationary. Then $c(s,t) = \operatorname{cov}(X(s), X(t))$ depends on $t - s$ only. The joint distribution of $X(t_1), X(t_2), \dots, X(t_n)$ depends only on the common mean and the covariances $c(t_i, t_j)$. Now $c(t_i, t_j)$ depends on $t_j - t_i$ only, whence $X(t_1), X(t_2), \dots, X(t_n)$ have the same joint distribution as $X(s + t_1), X(s + t_2), \dots, X(s + t_n)$. Therefore X is strongly stationary.

4. (a) If $s, t > 0$, we have from Problem (4.14.13) that

$$\mathbb{E}\big(X(s+t)^2 \mid X(s)\big) = X(s)^2 c(t)^2 + 1 - c(t)^2,$$

whence

$$\begin{aligned}\operatorname{cov}\big(X(s)^2, X(s+t)^2\big) &= \mathbb{E}\big(X(s)^2 X(s+t)^2\big) - 1\\ &= \mathbb{E}\Big\{X(s)^2 \mathbb{E}\big(X(s+t)^2 \mid X(s)\big)\Big\} - 1\\ &= c(t)^2 \mathbb{E}(X(s)^4) + (1 - c(t)^2)\mathbb{E}(X(s)^2) - 1 = 2c(t)^2\end{aligned}$$

by an elementary calculation.

(b) Likewise $\operatorname{cov}(X(s)^3, X(s+t)^3) = 3(3 + 2c(t)^2)c(t)$.

5. (a) Certainly, $X(0) = W(0) = 0$, and X is Gaussian because it has independent normally-distributed increments. Furthermore,

$$\phi_{X(t)}(\theta) = \exp\{-\tfrac{1}{2}\theta^2 T(t)\},$$

and

$$c(s, t) = \mathbb{E}\big(X(s)X(t)\big) = \min\{T(s), T(t)\}.$$

(b) As in Exercise (8.6.4), X is a Lévy process. By Theorem (8.7.18), it suffices for stability that

$$\mathbb{E}(e^{i\theta X(t)}) = \mathbb{E}\big(\mathbb{E}(e^{i\theta X(t)} \mid T(t))\big) = \mathbb{E}\big(e^{\frac{1}{2}\theta^2 2T(t)}\big) = e^{-t|\theta|^{2\alpha}}.$$

6. Express $Y(R)^2$ in terms of the $X(\cdot, \cdot)$ and use the covariance function of the Wiener process to obtain

$$\mathbb{E}(Y(R)^2) = (u - s)(v - t) = |R|,$$

the area of R. Likewise, for $R \cap R' = \varnothing$, we calculate term-by-term to find that $\mathbb{E}(Y(R)Y(R')) = 0$. Since $Y(R)$ and $Y(R')$ are bivariate normally distributed with zero covariance, they are independent.

9.7 Solutions to problems

1. It is easily seen that $Y_n = X_n + (\alpha - \beta)X_{n-1} + \beta Y_{n-1}$, whence the autocovariance function c of Y is given by

$$c(k) = \begin{cases} \dfrac{1 + \alpha^2 - \beta^2}{1 - \beta^2} & \text{if } k = 0,\\[2ex] \beta^{|k|-1}\left\{\dfrac{\alpha(1 + \alpha\beta - \beta^2)}{1 - \beta^2}\right\} & \text{if } k \neq 0.\end{cases}$$

Set $\widehat{Y}_{n+1} = \sum_{i=0}^{\infty} a_i Y_{n-i}$ and find the a_i for which it is the case that $\mathbb{E}\{(Y_{n+1} - \widehat{Y}_{n+1})Y_{n-k}\} = 0$ for $k \geq 0$. These equations yield

$$c(k+1) = \sum_{i=0}^{\infty} a_i c(k - i), \qquad k \geq 0,$$

which have solution $a_i = \alpha(\beta - \alpha)^i$ for $i \ge 0$.

2. The autocorrelation functions of X and Y satisfy

$$\sigma_X^2 \rho_X(n) = \sigma_Y^2 \sum_{j,k=0}^{r} a_j a_k \rho_Y(n+k-j).$$

Therefore

$$\begin{aligned} \sigma_X^2 f_X(\lambda) &= \frac{\sigma_Y^2}{2\pi} \sum_{n=-\infty}^{\infty} e^{-in\lambda} \sum_{j,k=0}^{r} a_j a_k \rho_Y(n+k-j) \\ &= \frac{\sigma_Y^2}{2\pi} \sum_{j,k=0}^{r} a_j a_k e^{i(k-j)\lambda} \sum_{n=-\infty}^{\infty} e^{-i(n+k-j)\lambda} \rho_Y(n+k-j) \\ &= \sigma_Y^2 |G_a(e^{i\lambda})|^2 f_Y(\lambda). \end{aligned}$$

In the case of exponential smoothing, $G_a(e^{i\lambda}) = (1-\mu)/(1-\mu e^{i\lambda})$, so that

$$f_X(\lambda) = \frac{c(1-\mu)^2 f_Y(\lambda)}{1 - 2\mu \cos\lambda + \mu^2}, \qquad |\lambda| < \pi,$$

where $c = \sigma_Y^2/\sigma_X^2$ is a constant chosen to make this a density function.

3. Consider the sequence $\{X_n\}$ defined by

$$X_n = Y_n - \widehat{Y}_n = Y_n - \alpha Y_{n-1} - \beta Y_{n-2}.$$

Now X_n is orthogonal to $\{Y_{n-k} : k \ge 1\}$, so that the X_n are uncorrelated random variables with spectral density function $f_X(\lambda) = (2\pi)^{-1}$, $\lambda \in (-\pi, \pi)$. By the result of Problem (9.7.2),

$$\sigma_X^2 f_X(\lambda) = \sigma_Y^2 |1 - \alpha e^{i\lambda} - \beta e^{2i\lambda}|^2 f_Y(\lambda),$$

whence

$$f_Y(\lambda) = \frac{\sigma_X^2/\sigma_Y^2}{2\pi |1 - \alpha e^{i\lambda} - \beta e^{2i\lambda}|^2}, \qquad -\pi < \lambda < \pi.$$

4. Let $\{X_n' : n \ge 1\}$ be the interarrival times of such a process counted from a time at which a meteorite falls. Then $X_1', X_2', \dots$ are independent and distributed as X_2. Let Y_n' be the indicator function of the event $\{X_m' = n \text{ for some } m\}$. Then

$$\begin{aligned} \mathbb{E}(Y_m Y_{m+n}) &= \mathbb{P}(Y_m = 1, Y_{m+n} = 1) \\ &= \mathbb{P}(Y_{m+n} = 1 \mid Y_m = 1)\mathbb{P}(Y_m = 1) = \mathbb{P}(Y_n' = 1)\alpha \end{aligned}$$

where $\alpha = \mathbb{P}(Y_m = 1)$. The autocovariance function of Y is therefore $c(n) = \alpha\{\mathbb{P}(Y_n' = 1) - \alpha\}$, $n \ge 0$, and Y is stationary.

The spectral density function of Y satisfies

$$f_Y(\lambda) = \frac{1}{2\pi} \sum_{n=-\infty}^{\infty} e^{-in\lambda} \frac{c(n)}{\alpha(1-\alpha)} = \operatorname{Re}\left\{ \frac{1}{\pi\alpha(1-\alpha)} \sum_{n=0}^{\infty} e^{in\lambda} c(n) - \frac{1}{2\pi} \right\}.$$

Now

$$\sum_{n=0}^{\infty} e^{in\lambda} Y_n' = \sum_{n=0}^{\infty} e^{i\lambda T_n'}$$

where $T_n' = X_1' + X_2' + \cdots + X_n'$; just check the non-zero terms. Therefore

$$\sum_{n=0}^{\infty} e^{in\lambda} c(n) = \alpha \mathbb{E}\left\{ \sum_{n=0}^{\infty} e^{i\lambda T_n'} \right\} - \frac{\alpha^2}{1 - e^{i\lambda}} = \frac{\alpha}{1 - \phi(\lambda)} - \frac{\alpha^2}{1 - e^{i\lambda}}$$

when $e^{i\lambda} \neq 1$, where ϕ is the characteristic function of X_2. It follows that

$$f_Y(\lambda) = \frac{1}{\pi(1-\alpha)} \operatorname{Re}\left\{ \frac{1}{1-\phi(\lambda)} - \frac{\alpha}{1 - e^{i\lambda}} \right\} - \frac{1}{2\pi}, \qquad |\lambda| < \pi.$$

5. We have that

$$\mathbb{E}(\cos(nU)) = \int_{-\pi}^{\pi} \frac{1}{2\pi} \cos(nu)\, du = 0, \quad \mathbb{E}(\cos^2(nU)) = \int_{-\pi}^{\pi} \frac{1}{2\pi} \cos^2(nu)\, du = \tfrac{1}{2},$$

for $n \geq 1$. Also

$$\mathbb{E}(\cos(mU)\cos(nU)) = \mathbb{E}\left\{ \tfrac{1}{2}\left(\cos[(m+n)U] + \cos[(m-n)U]\right) \right\} = 0$$

if $m \neq n$. Hence X is stationary with autocorrelation function $\rho(k) = \delta_{k0}$, and spectral density function $f(\lambda) = (2\pi)^{-1}$ for $|\lambda| < \pi$. Finally

$$\begin{aligned}
\mathbb{E}\{\cos(mU)\cos(nU)\cos(rU)\} &= \tfrac{1}{2}\mathbb{E}\left\{ \left(\cos[(m+n)U] + \cos[(m-n)U]\right)\cos(rU) \right\} \\
&= \tfrac{1}{4}\{\rho(m+n-r) + \rho(m-n-r)\}
\end{aligned}$$

which takes different values in the two cases $(m, n, r) = (1, 2, 3),\ (2, 3, 4)$.

6. (a) The increments of N during any collection of intervals $\{(u_i, v_i) : 1 \leq i \leq n\}$ have the same fdds if all the intervals are shifted by the same constant. Therefore X is strongly stationary. Certainly $\mathbb{E}(X(t)) = \lambda\alpha$ for all t, and the autocovariance function is

$$c(t) = \operatorname{cov}(X(0), X(t)) = \begin{cases} 0 & \text{if } t > \alpha, \\ \lambda(\alpha - t) & \text{if } 0 \leq t \leq \alpha. \end{cases}$$

Therefore the autocorrelation function is

$$\rho(t) = \begin{cases} 0 & \text{if } |t| > \alpha, \\ 1 - |t/\alpha| & \text{if } |t| \leq \alpha, \end{cases}$$

which we recognize as the characteristic function of the spectral density $f(\lambda) = \{1 - \cos(\alpha\lambda)\}/(\alpha\pi\lambda^2)$; see Problems (5.12.27b, 28a).

(b) We have that $\mathbb{E}(X(t)) = 0$; furthermore, for $s \leq t$, the correlation of $X(s)$ and $X(t)$ is

$$\begin{aligned}
\frac{1}{\sigma^2}\operatorname{cov}(X(s), X(t)) &= \frac{1}{\sigma^2}\operatorname{cov}(W(s) - W(s-1), W(t) - W(t-1)) \\
&= s - \min\{s, t-1\} - (s-1) + (s-1) \\
&= \begin{cases} 1 & \text{if } s \leq t-1, \\ s - t + 1 & \text{if } t - 1 < s \leq t. \end{cases}
\end{aligned}$$

This depends on $t - s$ only, and therefore X is stationary; X is Gaussian and therefore strongly stationary also.

The autocorrelation function is

$$\rho(h) = \begin{cases} 0 & \text{if } |h| \geq 1, \\ 1 - |h| & \text{if } |h| < 1, \end{cases}$$

which we recognize as the characteristic function of the density function $f(\lambda) = (1 - \cos\lambda)/(\pi\lambda^2)$.

7. We have from Problem (8.10.1) that the general moving-average process of part (b) is stationary with autocovariance function $c(k) = \sum_{j=0}^{r} \alpha_j \alpha_{k+j}$, $k \geq 0$, with the convention that $\alpha_s = 0$ if $s < 0$ or $s > r$.

(a) In this case, the autocorrelation function is

$$\rho(k) = \begin{cases} 1 & \text{if } k = 0, \\ \dfrac{\alpha}{1+\alpha^2} & \text{if } |k| = 1, \\ 0 & \text{if } |k| > 1, \end{cases}$$

whence the spectral density function is

$$f(\lambda) = \frac{1}{2\pi}\Big(\rho(0) + e^{i\lambda}\rho(1) + e^{-i\lambda}\rho(-1)\Big) = \frac{1}{2\pi}\left(1 + \frac{2\alpha\cos\lambda}{1+\alpha^2}\right), \quad |\lambda| < \pi.$$

(b) We have that

$$f(\lambda) = \frac{1}{2\pi}\sum_{k=-\infty}^{\infty} e^{-ik\lambda}\rho(k) = \frac{1}{2\pi c(0)}\sum_j \alpha_j e^{ij\lambda}\sum_k \alpha_{k+j}e^{-i(k+j)\lambda} = \frac{|A(e^{i\lambda})|^2}{2\pi c(0)}$$

where $c(0) = \sum_j \alpha_j^2$ and $A(z) = \sum_j \alpha_j z^j$. See Problem (9.7.2) also.

8. The spectral density function f is given by the inversion theorem (5.9.1) as

$$f(x) = \frac{1}{2\pi}\int_{-\infty}^{\infty} e^{-itx}\rho(t)\,dt$$

under the condition $\int_0^\infty |\rho(t)|\,dt < \infty$; see Problem (5.12.20). Now

$$|f(x)| \leq \frac{1}{2\pi}\int_{-\infty}^{\infty}|\rho(t)|\,dt$$

and

$$|f(x+h) - f(x)| \leq \frac{1}{2\pi}\int_{-\infty}^{\infty}|e^{ith} - 1|\cdot|\rho(t)|\,dt.$$

The integrand is dominated by the integrable function $2|\rho(t)|$. Using the dominated convergence theorem, we deduce that $|f(x+h) - f(x)| \to 0$ as $h \to 0$, uniformly in x.

9. By Exercise (9.5.2), $\text{var}\big(n^{-1}\sum_{j=1}^n X_j\big) \to \sigma^2$ if $C_n = n^{-1}\sum_{j=1}^n \text{cov}(X_1, X_j) \to \sigma^2$. If $\text{cov}(X_1, X_n) \to 0$ then $C_n \to 0$, and the result follows.

10. Let $X_1, X_2, \dots$ be independent identically distributed random variables with mean μ. The sequence X is stationary, and it is a consequence of the ergodic theorem that $n^{-1}\sum_{j=1}^n X_j \to Z$ a.s.

and in mean, where Z is a tail function of $X_1, X_2, \dots$ with mean μ. Using the zero–one law, Z is a.s. constant, and therefore $\mathbb{P}(Z = \mu) = 1$.

11. We have from the ergodic theorem that $n^{-1}\sum_{i=1}^{n} Y_i \to \mathbb{E}(Y \mid \mathcal{I})$ a.s. and in mean, where $\mathcal{I}$ is the σ-field of invariant events. The condition of the question is therefore

$$(*) \qquad \mathbb{E}(Y \mid \mathcal{I}) = \mathbb{E}(Y) \qquad \text{a.s., for all appropriate } Y.$$

Suppose $(*)$ holds. Pick $A \in \mathcal{I}$, and set $Y = I_A$ to obtain $I_A = \mathbb{Q}(A)$ a.s. Now I_A takes the values 0 and 1, so that $\mathbb{Q}(A)$ equals 0 or 1, implying that $\mathbb{Q}$ is ergodic. Conversely, suppose $\mathbb{Q}$ is ergodic. Then $\mathbb{E}(Y \mid \mathcal{I})$ is measurable on a trivial σ-field, and therefore equals $\mathbb{E}(Y)$ a.s.

12. Suppose $\mathbb{Q}$ is strongly mixing. If A is an invariant event then $A = \tau^{-n}A$. Therefore $\mathbb{Q}(A) = \mathbb{Q}(A \cap \tau^{-n}A) \to \mathbb{Q}(A)^2$ as $n \to \infty$, implying that $\mathbb{Q}(A)$ equals 0 or 1, and therefore $\mathbb{Q}$ is ergodic.

13. The vector $\mathbf{X} = (X_1, X_2, \dots)$ induces a probability measure $\mathbb{Q}$ on $(\mathbb{R}^T, \mathbb{B}^T)$. Since T is measure-preserving, $\mathbb{Q}$ is stationary. Let $Y : \mathbb{R}^T \to \mathbb{R}$ be given by $Y(\mathbf{x}) = x_1$ for $\mathbf{x} = (x_1, x_2, \dots)$, and define $Y_i(\mathbf{x}) = Y(\tau^{i-1}(\mathbf{x}))$ where τ is the usual shift operator on $\mathbb{R}^T$. The vector $\mathbf{Y} = (Y_1, Y_2, \dots)$ has the same distributions as the vector $\mathbf{X}$. By the ergodic theorem for $\mathbf{Y}$, $n^{-1}\sum_{i=1}^{n} Y_i \to \mathbb{E}(Y \mid \mathcal{J})$ a.s. and in mean, where $\mathcal{J}$ is the invariant σ-field of τ. It follows that the limit

$$(*) \qquad Z = \lim_{n\to\infty} \frac{1}{n}\sum_{i=1}^{n} X_i$$

exists a.s. and in mean. Now $U = \limsup_{n\to\infty}\left(n^{-1}\sum_1^n X_i\right)$ is invariant, since

$$\frac{1}{n}\left\{\sum_{i=1}^{n}\big(X_i(\omega) - X_i(T\omega)\big)\right\} = \frac{1}{n}\{X(\omega) - X(T^n\omega)\} \to 0 \qquad \text{a.s.,}$$

implying that $U(\omega) = U(T\omega)$ a.s. It follows that U is $\mathcal{I}$-measurable, and it is the case that $Z = U$ a.s. Take conditional expectations of $(*)$, given $\mathcal{I}$, to obtain $U = \mathbb{E}(X \mid \mathcal{I})$ a.s.

If T is ergodic, then $\mathcal{I}$ is trivial, so that $\mathbb{E}(X \mid \mathcal{I})$ is a.s. constant; therefore $\mathbb{E}(X \mid \mathcal{I}) = \mathbb{E}(X)$ a.s.

14. (a) For $(a, b) \subseteq [0, 1)$, we have $T^{-1}(a, b) = (\frac{1}{2}a, \frac{1}{2}b) \cup (\frac{1}{2} + \frac{1}{2}a, \frac{1}{2} + \frac{1}{2}b)$, and therefore T is measurable. Secondly,

$$\mathbb{P}\big(T^{-1}(a, b)\big) = 2(\tfrac{1}{2}b - \tfrac{1}{2}a) = b - a = \mathbb{P}\big((a, b)\big),$$

so that T^{-1} preserves the measure of intervals. The intervals generate $\mathcal{B}$, and it is then standard that T^{-1} preserves the measures of all events.

(b) Let A be invariant, in that $A = T^{-1}A$. Let $0 \le \omega < \frac{1}{2}$; it is easily seen that $T(\omega) = T(\omega + \frac{1}{2})$. Therefore $\omega \in A$ if and only if $\omega + \frac{1}{2} \in A$, implying that $A \cap [\frac{1}{2}, 1) = \frac{1}{2} + \{A \cap [0, \frac{1}{2})\}$; hence

$$\mathbb{P}(A \cap E) = \tfrac{1}{2}\mathbb{P}(A) = \mathbb{P}(A)\mathbb{P}(E) \qquad \text{for } E = [0, \tfrac{1}{2}),\ [\tfrac{1}{2}, 1).$$

This proves that A is independent of both $[0, \frac{1}{2})$ and $[\frac{1}{2}, 1)$. A similar proof gives that A is independent of any set E which is, for some n, the union of intervals of the form $[k2^{-n}, (k+1)2^{-n})$ for $0 \le k < 2^n$. It is a fundamental result of measure theory that there exists a sequence $E_1, E_2, \dots$ of events such that

(i) E_n is of the above form, for each n,

(ii) $\mathbb{P}(A \mathbin{\triangle} E_n) \to 0$ as $n \to \infty$.

Choosing the E_n accordingly, it follows that

$$\mathbb{P}(A \cap E_n) = \mathbb{P}(A)\mathbb{P}(E_n) \to \mathbb{P}(A)^2 \qquad \text{by independence,}$$
$$|\mathbb{P}(A \cap E_n) - \mathbb{P}(A)| \le \mathbb{P}(A \,\triangle\, E_n) \to 0.$$

Therefore $\mathbb{P}(A) = \mathbb{P}(A)^2$ so that $\mathbb{P}(A)$ equals 0 or 1.

For $\omega \in \Omega$, expand ω in base 2, $\omega = 0.\omega_1\omega_2\cdots$, and define $Y(\omega) = \omega_1$. It is easily seen that $Y(T^{n-1}\omega) = \omega_n$, whence the ergodic theorem (Problem (9.7.13)) yields that $n^{-1}\sum_{i=1}^{n}\omega_i \to \frac{1}{2}$ as $n \to \infty$ for all ω in some event of probability 1.

(c) By part (b), the set N of numbers in (0, 1) with the given property (the set of 'normal numbers') has Lebesgue measure 1. The random variable $Y' = Y \cdot I(Y \in N)$ has the same distribution as Y, and a.s. takes values in N.

(d) One may check from the definition that Z is a random variable. It cannot be continuous since, by the law of large numbers, it does not satisfy (c) above. It cannot be discrete, since there is no countable set in which it takes values with probability 1. There is a general result due to Lebesgue that every distribution function can be expressed as a convex combination of three distribution functions, one continuous, one discrete, and one 'singular'. By the above, Z has a distribution with a non-trivial singular component.

The random variable Z is in fact purely singular, in that $\mathbb{P}(Z = z) = 0$ for $z \in [0, 1]$, and there exists an uncountable set N_p with Lebesgue measure 0 such that $\mathbb{P}(Z \in N_p) = 1$. The first claim is elementary. For the second, the argument is simple. For $\pi \in [0, 1]$, let S_π be the set of binary sequences such that the average of the first n terms converges to π as $n \to \infty$, and let N_π be the corresponding subset of [0, 1]. By part (b), $N_{1/2}$ has Lebesgue measure 1. By the strong law of large numbers, we have $\mathbb{P}(Z \in N_p) = 1$. However, $N_p \cap N_{1/2} = \varnothing$, whence N_p has Lebesgue measure 0.

15. We may as well assume that $0 < \alpha < 1$. Let $T : [0, 1) \to [0, 1)$ be given by $T(x) = x + \alpha$ (mod 1). It is easily seen that T is invertible and measure-preserving. Furthermore $T(X)$ is uniform on [0, 1], and it follows that the sequence $Z_1, Z_2, \dots$ has the same fdds as $Z_2, Z_3, \dots$, which is to say that Z is stationary. It therefore suffices to prove that T is an ergodic shift, since this will imply by the ergodic theorem that

$$\frac{1}{n}\sum_{j=1}^{n} Z_j \to \mathbb{E}(Z_1) = \int_0^1 g(u)\,du.$$

We use Fourier analysis. Let A be an invariant subset of [0, 1). The indicator function of A has a Fourier series:

$$(*) \qquad I_A(x) \sim \sum_{n=-\infty}^{\infty} a_n e_n(x)$$

where $e_n(x) = e^{2\pi i n x}$ and

$$a_n = \frac{1}{2\pi}\int_0^1 I_A(x)e_{-n}(x)\,dx = \frac{1}{2\pi}\int_A e_{-n}(x)\,dx.$$

Similarly the indicator function of $T^{-1}A$ has a Fourier series,

$$I_{T^{-1}A}(x) \sim \sum_n b_n e_n(x)$$

where, using the substitution $y = T(x)$,

$$b_n = \frac{1}{2\pi}\int_0^1 I_{T^{-1}A}(x)e_{-n}(x)\,dx = \frac{1}{2\pi}\int_0^1 I_A(y)e_{-n}(T^{-1}(y))\,dy = a_n e^{-2\pi i n\alpha},$$

since $e_m(y-\alpha) = e^{-2\pi i m\alpha} e_m(y)$. Therefore $I_{T^{-1}A}$ has Fourier series

$$I_{T^{-1}A}(x) \sim \sum_n e^{-2\pi i n\alpha} a_n e_n(x).$$

Now $I_A = I_{T^{-1}A}$ since A is invariant. We compare the previous formula with that of $(*)$, and deduce that $a_n = e^{-2\pi i n\alpha} a_n$ for all n. Since α is irrational, it follows that $a_n = 0$ if $n \neq 0$, and therefore I_A has Fourier series a_0, a constant. Therefore I_A is a.s. constant, which is to say that either $\mathbb{P}(A) = 0$ or $\mathbb{P}(A) = 1$.

16. Let $G_t(z) = \mathbb{E}(z^{X(t)})$, the probability generating function of $X(t)$. Since X has stationary independent increments, for any n (≥ 1), $X(t)$ may be expressed as the sum

$$X(t) = \sum_{i=1}^{n} \{X(it/n) - X((i-1)t/n)\}$$

of independent identically distributed variables. Hence $X(t)$ is infinitely divisible. By Problem (5.12.13), we may write

$$(*) \qquad G_t(z) = e^{-\lambda(t)(1-A(z))}$$

for some probability generating function A, and some $\lambda(t)$.

Similarly, $X(s+t) = X(s) + \{X(s+t) - X(s)\}$, whence $G_{s+t}(z) = G_s(z)G_t(z)$, implying that $G_t(z) = e^{\mu(z)t}$ for some $\mu(z)$; we have used a little monotonicity here. Combining this with $(*)$, we obtain that $G_t(z) = e^{-\lambda t(1-A(z))}$ for some λ.

Finally, $X(t)$ has jumps of unit magnitude only, whence the probability generating function A is given by $A(z) = z$.

17. (a) We have that

$$(*) \qquad X(t) - X(0) = \{X(s) - X(0)\} + \{X(t) - X(s)\}, \qquad 0 \leq s \leq t,$$

whence, by stationarity,

$$\{m(t) - m(0)\} = \{m(s) - m(0)\} + \{m(t-s) - m(0)\}.$$

Now m is continuous, so that $m(t) - m(0) = \beta t$, $t \geq 0$, for some β; see Problem (4.14.5).

(b) Take variances of $(*)$ to obtain $v(t) = v(s) + v(t-s)$, $0 \leq s \leq t$, whence $v(t) = \sigma^2 t$ for some σ^2.

18. In the context of this chapter, a process Z is a standard Wiener process if it is Gaussian with $Z(0) = 0$, with zero means, and autocovariance function $c(s,t) = \min\{s,t\}$.

(a) $Z(t) = \alpha W(t/\alpha^2)$ satisfies $Z(0) = 0$, $\mathbb{E}(Z(t)) = 0$, and

$$\operatorname{cov}\big(Z(s), Z(t)\big) = \alpha^2 \min\{s/\alpha^2, t/\alpha^2\} = \min\{s,t\}.$$

(b) The only calculation of any interest here is

$$\begin{aligned}\operatorname{cov}\big(W(s+\alpha) - &W(\alpha), W(t+\alpha) - W(\alpha)\big) \\ &= c(s+\alpha, t+\alpha) - c(\alpha, t+\alpha) - c(s+\alpha, \alpha) + c(\alpha,\alpha) \\ &= (s+\alpha) - \alpha - \alpha + \alpha = s, \qquad s \leq t.\end{aligned}$$

(c) $V(0) = 0$, and $\mathbb{E}(V(t)) = 0$. Finally, if $s, t > 0$,

$$\operatorname{cov}\big(V(s), V(t)\big) = st\operatorname{cov}\big(W(1/s), W(1/t)\big) = st\min\{1/s, 1/t\} = \min\{t, s\}.$$

(d) $Z(t) = W(1) - W(1-t)$ satisfies $Z(0) = 0$, $\mathbb{E}(Z(t)) = 0$. Also Z is Gaussian, and

$$\begin{aligned}\operatorname{cov}\big(Z(s), Z(t)\big) &= 1 - (1-s) - (1-t) + \min\{1-s, 1-t\}\\ &= \min\{s, t\}, \qquad 0 \le s, t \le 1.\end{aligned}$$

19. The process W has stationary independent increments, and $G(t) = \mathbb{E}(|W(t)|^2)$ satisfies $G(t) = t \to 0$ as $t \to 0$; hence $\int_0^\infty \phi(u)\,dW(u)$ is well defined for any ϕ satisfying

$$\int_0^\infty |\phi(u)|^2\,dG(u) = \int_0^\infty \phi(u)^2\,du < \infty.$$

It is obvious that $\phi(u) = I_{[0,t]}(u)$ and $\phi(u) = e^{-(t-u)}I_{[0,t]}(u)$ are such functions.

Now $X(t)$ is the limit (in mean-square) of the sequence

$$S_n(t) = \sum_{j=0}^{n-1}\big\{W((j+1)t/n) - W(jt/n)\big\}, \qquad n \ge 1.$$

However $S_n(t) = W(t)$ for all n, and therefore $S_n(t) \xrightarrow{\text{m.s.}} W(t)$ as $n \to \infty$.

Finally, $Y(s)$ is the limit (in mean-square) of a sequence of normal random variables with mean 0, and therefore is Gaussian with mean 0. If $s < t$,

$$\begin{aligned}\operatorname{cov}\big(Y(s), Y(t)\big) &= \int_0^\infty \big(e^{-(s-u)}I_{[0,s]}(u)\big)\big(e^{-(t-u)}I_{[0,t]}(u)\big)\,dG(u)\\ &= \int_0^s e^{2u-s-t}\,du = \tfrac{1}{2}(e^{s-t} - e^{-s-t}).\end{aligned}$$

Y is an Ornstein–Uhlenbeck process.

20. (a) $W(t)$ is $N(0, t)$, so that

$$\begin{aligned}\mathbb{E}|W(t)| &= \int_{-\infty}^{\infty} \frac{|u|}{\sqrt{2\pi t}} e^{-\frac{1}{2}(u^2/t)}\,du = \sqrt{2t/\pi},\\ \operatorname{var}(|W(t)|) &= \mathbb{E}(W(t)^2) - \frac{2t}{\pi} = t\left(1 - \frac{2}{\pi}\right).\end{aligned}$$

The process X is never negative, and therefore it is not Gaussian. It is Markov since, if $s < t$ and B is an event defined in terms of $\{X(u) : u \le s\}$, then the conditional distribution function of $X(t)$ satisfies

$$\begin{aligned}\mathbb{P}\big(X(t) \le y \mid X(s) = x, B\big) &= \mathbb{P}\big(X(t) \le y \mid W(s) = x, B\big)\mathbb{P}\big(W(s) = x \mid X(s) = x, B\big)\\ &\quad + \mathbb{P}\big(X(t) \le y \mid W(s) = -x, B\big)\mathbb{P}\big(W(s) = -x \mid X(s) = x, B\big)\\ &= \tfrac{1}{2}\Big\{\mathbb{P}\big(X(t) \le y \mid W(s) = x\big) + \mathbb{P}\big(X(t) \le y \mid W(s) = -x\big)\Big\},\end{aligned}$$

which does not depend on B.

(b) Certainly,

$$\mathbb{E}(Y(t)) = \int_{-\infty}^{\infty} \frac{e^u}{\sqrt{2\pi t}} e^{-\frac{1}{2}(u^2/t)}\, du = e^{\frac{1}{2}t}.$$

Secondly, $W(s) + W(t) = 2W(s) + \{W(t) - W(s)\}$ is $N(0, 3s + t)$ if $s < t$, implying that

$$\mathbb{E}\big(Y(s)Y(t)\big) = \mathbb{E}\left(e^{W(s)+W(t)}\right) = e^{\frac{1}{2}(3s+t)},$$

and therefore

$$\operatorname{cov}\big(Y(s), Y(t)\big) = e^{\frac{1}{2}(3s+t)} - e^{\frac{1}{2}(s+t)}, \qquad s < t.$$

$W(1)$ is $N(0, 1)$, and therefore $Y(1)$ has the log-normal distribution. Therefore Y is not Gaussian. It is Markov since W is Markov, and $Y(t)$ is a one–one function of $W(t)$.

(c) We shall assume that the random function W is a.s. continuous, a point to which we return in Chapter 13. Certainly,

$$\mathbb{E}(Z(t)) = \int_0^t \mathbb{E}(W(u))\, du = 0,$$

$$\begin{aligned}\mathbb{E}\big(Z(s)Z(t)\big) &= \iint_{\substack{0\le u\le s\\ 0\le v\le t}} \mathbb{E}\big(W(u)W(v)\big)\, du\, dv \\ &= \int_{u=0}^{s}\left\{\int_{v=0}^{u} v\, dv + \int_{v=u}^{t} u\, dv\right\} du = \tfrac{1}{6}s^2(3t - s), \qquad s < t,\end{aligned}$$

since $\mathbb{E}(W(u)W(v)) = \min\{u, v\}$.

(d) The process Z is Gaussian, as the following argument indicates. The single random variable $Z(t)$ may be expressed as a limit of the form

$$(*) \qquad \lim_{n\to\infty} \sum_{i=1}^{n} \left(\frac{t}{n}\right) W(it/n),$$

each such summation being normal. The limit of normal variables is normal (see Problem (7.11.19)), and therefore $Z(t)$ is normal. The limit in $(*)$ exists a.s., and hence in probability. By an appeal to (7.11.19b), pairs $(Z(s), Z(t))$ are bivariate normal, and a similar argument is valid for all n-tuples of the $Z(u)$. See the related Exercise (8.6.6).

On the other hand, Z is not Markov. An increment $Z(t) - Z(s)$ depends very much on $W(s)$, and the collection $\{Z(u) : u \le s\}$ contains much information about $W(s)$ in excess of the information contained in the single value $Z(s)$.

(e) Since $Z(t)$ is normal with zero mean, its odd moments are zero. For $n = 2$,

$$\begin{aligned}\mathbb{E}(Z(t)^2) &= \mathbb{E}\left(\int_0^t W(x)\, dx \int_0^t W(y)\, dy\right) = \iint_{(0,t)^2} \mathbb{E}(W(x)W(y))\, dx\, dy \\ &= \iint_{(0,t)^2} \min\{x, y\}\, dx\, dy = 2\int_0^t dx \int_0^x y\, dy = \tfrac{1}{3}t^3.\end{aligned}$$

Hence $\mathbb{E}(e^{\theta Z(t)}) = \exp\left(\tfrac{1}{6}\theta^2 t^3\right)$, and the coefficient of θ^{2r} is the $(2r)$th moment, namely

$$\mathbb{E}(Z(t)^{2r}) = (2r-1)(2r-3)\cdots 1\cdot(\tfrac{1}{3}t^3)^r = \frac{(2r)!}{2^r\, r!}\cdot(\tfrac{1}{3}t^3)^r.$$

21. Let $U_i = X(t_i)$. The random variables $A = U_1$, $B = U_2 - U_1$, $C = U_3 - U_2$, $D = U_4 - U_3$ are independent and normal with zero means and respective variances $t_1, t_2 - t_1, t_3 - t_2, t_4 - t_3$. The Jacobian of the transformation is 1, and it follows that U_1, U_2, U_3, U_4 have joint density function

$$f_{\mathbf{U}}(\mathbf{u}) = \frac{e^{-\frac{1}{2}Q}}{(2\pi)^2\sqrt{t_1(t_2-t_1)(t_3-t_2)(t_4-t_3)}}$$

where

$$Q = \frac{u_1^2}{t_1} + \frac{(u_2-u_1)^2}{t_2-t_1} + \frac{(u_3-u_2)^2}{t_3-t_2} + \frac{(u_4-u_3)^2}{t_4-t_3}.$$

Likewise U_1 and U_4 have joint density function

$$\frac{e^{-\frac{1}{2}R}}{2\pi\sqrt{t_1(t_4-t_1)}} \quad \text{where} \quad R = \frac{u_1^2}{t_1} + \frac{(u_4-u_1)^2}{t_4-t_1}.$$

Hence the joint density function of U_2 and U_3, given $U_1 = U_4 = 0$, is

$$g(u_2, u_3) = \frac{e^{-\frac{1}{2}S}}{2\pi}\sqrt{\frac{t_4-t_1}{(t_2-t_1)(t_3-t_2)(t_4-t_3)}}$$

where

$$S = \frac{u_2^2}{t_2-t_1} + \frac{(u_3-u_2)^2}{t_3-t_2} + \frac{u_3^2}{t_4-t_3}.$$

Now g is the density function of a bivariate normal distribution with zero means, marginal variances

$$\sigma_1^2 = \frac{(t_2-t_1)(t_3-t_2)}{t_3-t_1}, \qquad \sigma_2^2 = \frac{(t_4-t_3)(t_3-t_2)}{t_4-t_2}$$

and correlation

$$\rho = \frac{\sigma_1\sigma_2}{t_3-t_2} = \sqrt{\frac{(t_4-t_3)(t_2-t_1)}{(t_4-t_2)(t_3-t_1)}}.$$

See also Exercise (8.5.2).

22. (a) The random variables $\{I_j(x) : 1 \le j \le n\}$ are independent, so that

$$\mathbb{E}(F_n(x)) = x, \qquad \text{var}(F_n(x)) = \frac{1}{n}\,\text{var}(I_1(x)) = \frac{x(1-x)}{n}.$$

By the central limit theorem, $\sqrt{n}\{F_n(x) - x\} \xrightarrow{\text{D}} Y(x)$, where $Y(x)$ is $N(0, x(1-x))$.

(b) The limit distribution is multivariate normal. There are general methods for showing this, and here is a sketch. If $0 \le x_1 < x_2 \le 1$, then the number M_2 $(= nF(x_2))$ of the I_j not greater than x_2 is approximately $N(nx_2, nx_2(1-x_2))$. Conditional on $\{M_2 = m\}$, the number $M_1 = nF(x_1)$ is approximately $N(mu, mu(1-u))$ where $u = x_1/x_2$. It is now a small exercise to see that the pair (M_1, M_2) is approximately bivariate normal with means nx_1, nx_2, with variances $nx_1(1-x_1)$, $nx_2(1-x_2)$, and such that

$$\mathbb{E}(M_1M_2) = \mathbb{E}\{M_2\mathbb{E}(M_1 \mid M_2)\} = \mathbb{E}(M_2^2x_1/x_2) \sim nx_1(1-x_2) + n^2x_1x_2,$$

whence $\text{cov}(M_1, M_2) \sim nx_1(1-x_2)$. It follows similarly that the limit of the general collection is multivariate normal with mean 0, variances $x_i(1-x_i)$, and covariances $c_{ij} = x_i(1-x_j)$.

(c) The autocovariance function of the limit distribution is $c(s,t) = \min\{s,t\} - st$, whereas, for $0 \le s \le t \le 1$, we have that $\text{cov}(Z(s), Z(t)) = s - ts - st + st = \min\{s,t\} - st$. It may be shown that the limit of the process $\{\sqrt{n}(F_n(x) - x) : n \ge 1\}$ exists as $n \to \infty$, in a certain sense, the limit being a Brownian bridge; such a limit theorem for processes is called a 'functional limit theorem'.

23. (a) The claim holds evidently when $n = 1$ and $r = 0, 1$. Suppose it holds for $n = N - 1$ and $r = 0, 1, 2, \dots, N - 1$. Then, with $A_r = \{X_r = x_r\}$,

$$\begin{aligned}\mathbb{P}(A_1 \cap \cdots \cap A_N) &= \mathbb{P}(A_N \mid A_1 \cap \cdots \cap A_{N-1})\mathbb{P}(A_1 \cap \cdots \cap A_{N-1}) \\ &= p_{N,s}(x_N) \cdot \frac{s!\,(N-1-s)!}{N!},\end{aligned}$$

where $s = \sum_{k=1}^{N-1} X_k$, and

$$p_{N,s}(x) = \begin{cases} \dfrac{N-s}{N+1} & \text{if } x = 0, \\ \dfrac{s+1}{N+1} & \text{if } x = 1, \end{cases}$$

as required.

(b) Since the answer in (a) depends only on r and N, the sequence $X_1, X_2, \dots, X_N$ is exchangeable . (See Exercise (7.3.4) for a definition of an exchangeable event; a sequence is said to be *exchangeable* if its joint distribution is invariant under permutations of the indices.) It is then automatic that the sequence X is stationary, and the convergence follows by the ergodic theorem, on noting that the X_i are bounded.

To see that exchangeability implies stationarity, one argues as follows. Stationarity is equivalent to: for $n \ge 1$, the distribution of $(X_1, X_2, \dots, X_n)$ is the same as that of $(X_2, X_3, \dots, X_{n+1})$. This is a consequence of exchangeability by an evident permutation of the indices $1, 2, \dots, n+1$.

(c) By the result of (a), $S_n = \sum_{r=1}^{n} X_r$ satisfies

$$\mathbb{P}(S_n = m) = \binom{n}{m} \cdot \frac{m!\,(n-m)!}{(n+1)!} = \frac{1}{n+1},$$

so that $R = \lim_{n\to\infty} S_n/n$ is uniformly distributed on $(0, 1)$.

10

Renewals

10.1 Solutions. The renewal equation

1. Since $\mathbb{E}(X_1) > 0$, there exists ϵ (> 0) such that $\mathbb{P}(X_1 \geq \epsilon) > \epsilon$. Let $X_k' = \epsilon I_{\{X_k \geq \epsilon\}}$, and denote by N' the related renewal process. Now $N(t) \leq N'(t)$, so that $\mathbb{E}(e^{\theta N(t)}) \leq \mathbb{E}(e^{\theta N'(t)})$, for $\theta > 0$. Let Z_m be the number of renewals (in N') between the times at which N' reaches the values $(m-1)\epsilon$ and $m\epsilon$. The Z's are independent with

$$\mathbb{E}(e^{\theta Z_m}) = \frac{\epsilon e^{\theta}}{1-(1-\epsilon)e^{\theta}}, \qquad \text{if } (1-\epsilon)e^{\theta} < 1,$$

whence $\mathbb{E}(e^{\theta N'(t)}) \leq \left(\epsilon e^{\theta}\left\{1-(1-\epsilon)e^{\theta}\right\}^{-1}\right)^{t/\epsilon}$ for sufficiently small positive θ.

2. Let X_1 be the time of the first arrival. If $X_1 > s$, then $W = s$. On the other hand if $X_1 < s$, then the process starts off afresh at the new starting time X_1. Therefore, by conditioning on the value of X_1,

$$\begin{aligned} F_W(x) &= \int_0^{\infty} \mathbb{P}(W \leq x \mid X_1 = u)\, dF(u) = \int_0^{s} \mathbb{P}(W \leq x-u)\, dF(u) + \int_s^{\infty} 1 \cdot dF(u) \\ &= \int_0^{s} \mathbb{P}(W \leq x-u)\, dF(u) + \{1-F(s)\} \end{aligned}$$

if $x \geq s$. It is clear that $F_W(x) = 0$ if $x < s$. This integral equation for F_W may be written in the standard form

$$F_W(x) = H(x) + \int_0^{x} F_W(x-u)\, d\widehat{F}(u)$$

where H and $\widehat{F}$ are given by

$$H(x) = \begin{cases} 0 & \text{if } x < s, \\ 1-F(s) & \text{if } x \geq s, \end{cases} \qquad \widehat{F}(x) = \begin{cases} F(x) & \text{if } x < s, \\ F(s) & \text{if } x \geq s. \end{cases}$$

This renewal-type equation may be solved in the usual way by the method of Laplace–Stieltjes transforms. We have that $F_W^* = H^* + F_W^* \widehat{F}^*$, whence $F_W^* = H^*/(1-\widehat{F}^*)$. If N is a Poisson process then $F(x) = 1 - e^{-\lambda x}$. In this case

$$H^*(\theta) = \int_0^{\infty} e^{-\theta x}\, dH(x) = e^{-(\lambda+\theta)s},$$

since H is constant apart from a jump at $x = s$. Similarly

$$\widehat{F}^*(\theta) = \int_0^{s} e^{-\theta x}\, dF(x) = \frac{\lambda}{\lambda+\theta}\left(1 - e^{-(\lambda+\theta)s}\right),$$

so that

$$F_W^*(\theta) = \frac{(\lambda + \theta)e^{-(\lambda+\theta)s}}{\theta + \lambda e^{-(\lambda+\theta)s}}.$$

Finally, replace θ with $-\theta$, and differentiate to find the mean.

3. We have as usual that $\mathbb{P}(N(t) = n) = \mathbb{P}(S_n \le t) - \mathbb{P}(S_{n+1} \le t)$. In the respective cases,

(a) $$\mathbb{P}(N(t) = n) = \sum_{r=0}^{\lfloor t \rfloor} \frac{1}{r!}\{e^{-\lambda n}(\lambda n)^r - e^{-\lambda(n+1)}[\lambda(n+1)]^r\},$$

(b) $$\mathbb{P}(N(t) = n) = \int_0^t \left\{ \frac{\lambda^{nb}x^{nb-1}}{\Gamma(nb)} - \frac{\lambda^{(n+1)b}x^{(n+1)b-1}}{\Gamma((n+1)b)} \right\} e^{-\lambda x}\, dx.$$

4. By conditioning on X_1, $m(t) = \mathbb{E}(N(t))$ satisfies

$$m(t) = \int_0^t \big(1 + m(t-x)\big)\, dx = t + \int_0^t m(x)\, dx, \qquad 0 \le t \le 1.$$

Hence $m' = 1 + m$, with solution $m(t) = e^t - 1$, for $0 \le t \le 1$. (For larger values of t, $m(t) = 1 + \int_0^1 m(t-x)\, dx$, and a tiresome iteration is in principle possible.)

With $v(t) = \mathbb{E}(N(t)^2)$,

$$v(t) = \int_0^t \big[v(t-x) + 2m(t-x) + 1\big]\, dx = t + 2(e^t - t - 1) + \int_0^t v(x)\, dx, \quad 0 \le t \le 1.$$

Hence $v' = v + 2e^t - 1$, with solution $v(t) = 1 - e^t + 2te^t$ for $0 \le t \le 1$.

5. By conditioning on the last arrival time,

$$\mathbb{P}(X_{N+1} > x \mid S_N = t - y) = \begin{cases} 1 & \text{if } y \ge x, \\ \mathbb{P}(X_1 > x \mid X_1 > y) = \dfrac{1 - F(x)}{1 - F(y)} & \text{if } y \le x, \end{cases}$$

where $N = N(t)$. Thus,

$$\mathbb{P}(X_{N+1} > x) \ge 1 - F(x) = \mathbb{P}(X_1 > x), \qquad x \ge 0.$$

6. We have $m(t) = \sum_{k=1}^{\infty} F_k(t)$, where F_k is the distribution function of the kth arrival time S_k. Now,

$$\widehat{F}_k(\theta) = \int_0^\infty e^{-\theta t} F_k(t)\, dt = \frac{1}{\theta}\int_0^\infty e^{-\theta t} f_{S_k}(t)\, dt = \frac{1}{\theta}\widehat{f}(\theta)^k,$$

so that

$$\widehat{m}(\theta) = \sum_{k=0}^{\infty} \widehat{F}_k(\theta) = \frac{1}{\theta}\sum_{k=1}^{\infty} \widehat{f}(\theta)^k,$$

with the given sum for $\theta > 0$.

7. This follows by taking Laplace–Stieltjes transforms of the equation for r, and recalling the basic properties of such transforms (see Appendix I).

10.2 Solutions. Limit theorems

1. Let Z_i be the number of passengers in the ith plane, and assume that the Z_i are independent of each other and of the arrival process. The number of passengers who have arrived by time t is $S(t) = \sum_{i=1}^{N(t)} Z_i$. Now

$$\frac{1}{t}S(t) = \frac{N(t)}{t} \cdot \frac{S(t)}{N(t)} \to \frac{\mathbb{E}(Z_1)}{\mu} \qquad \text{a.s.}$$

by the law of the large numbers, since $N(t)/t \to 1/\mu$ a.s., and $N(t) \to \infty$ a.s.

2. We have that

$$\mathbb{E}(T_M^2) = \mathbb{E}\left\{\left(\sum_{i=1}^{\infty} Z_i I_{\{M\ge i\}}\right)^2\right\} = \sum_{i=1}^{\infty} \mathbb{E}\big(Z_i^2 I_{\{M\ge i\}}\big) + 2\sum_{1\le i<j<\infty} \mathbb{E}\big(Z_i Z_j I_{\{M\ge j\}}\big)$$

since $I_{\{M\ge i\}} I_{\{M\ge j\}} = I_{\{M\ge i\vee j\}}$, where $i \vee j = \max\{i, j\}$. Now

$$\mathbb{E}\big(Z_i^2 I_{\{M\ge i\}}\big) = \mathbb{E}\big(Z_i^2 I_{\{M\le i-1\}^{\mathrm{c}}}\big) = \mathbb{E}(Z_i^2)\mathbb{P}(M \ge i),$$

since $\{M \le i-1\}$ is defined in terms of $Z_1, Z_2, \dots, Z_{i-1}$, and is therefore independent of Z_i. Similarly $\mathbb{E}(Z_i Z_j I_{\{M\ge j\}}) = \mathbb{E}(Z_j)\mathbb{E}(Z_i I_{\{M\ge j\}}) = 0$ if $i < j$. It follows that

$$\mathbb{E}(T_M^2) = \sum_{i=1}^{\infty} \mathbb{E}(Z_i^2)\mathbb{P}(M \ge i) = \sigma^2 \sum_{i=1}^{\infty} \mathbb{P}(M \ge i) = \sigma^2\mathbb{E}(M).$$

3. (i) The shortest way is to observe that $N(t) + k$ is a stopping time if $k \ge 1$. Alternatively, we have by Wald's equation that $\mathbb{E}(T_{N(t)+1}) = \mu(m(t) + 1)$. Also

$$\mathbb{E}(X_{N(t)+k}) = \mathbb{E}\{\mathbb{E}(X_{N(t)+k} \mid N(t))\} = \mu, \qquad k \ge 2,$$

and therefore, for $k \ge 1$,

$$\mathbb{E}(T_{N(t)+k}) = \mathbb{E}(T_{N(t)+1}) + \sum_{j=2}^{k} \mathbb{E}(X_{N(t)+j}) = \mu\big(m(t) + k\big).$$

(ii) Suppose $p \ne 1$ and

$$\mathbb{P}(X_1 = a) = \begin{cases} p & \text{if } a = 1, \\ 1-p & \text{if } a = 2. \end{cases}$$

Then $\mu = 2 - p \ne 1$. Also

$$\mathbb{E}(T_{N(1)}) = (1-p)\mathbb{E}(T_0 \mid N(1) = 0) + p\mathbb{E}(T_1 \mid N(1) = 1) = p,$$

whereas $m(1) = p$. Therefore $\mathbb{E}(T_{N(1)}) \ne \mu m(1)$.

4. Let $V(t) = N(t) + 1$, and let $W_1, W_2, \dots$ be defined inductively as follows. $W_1 = V(1)$, W_2 is obtained similarly to W_1 but relative to the renewal process starting at the $V(1)$th renewal, i.e. at time $T_{N(1)+1}$, and W_n is obtained similarly:

$$W_n = N(T_{X_{n-1}} + 1) - N(T_{X_{n-1}}) + 1, \qquad n \ge 2,$$

where $X_m = W_1 + W_2 + \cdots + W_m$. For each n, W_n is independent of the sequence $W_1, W_2, \dots, W_{n-1}$, and therefore the W_n are independent copies of $V(1)$. It is easily seen, by measuring the time-intervals covered, that $V(t) \le \sum_{i=1}^{\lceil t \rceil} W_i$, and hence

$$\frac{1}{t} V(t) \le \frac{1}{t} \sum_{i=1}^{\lceil t \rceil} W_i \to \mathbb{E}(V(1)) \qquad \text{a.s. and in mean, as } t \to \infty.$$

It follows that the family $\{m^{-1} \sum_{i=1}^{m} W_i : m \ge 1\}$ is uniformly integrable (see Theorem (7.10.3)). Now $N(t) \le V(t)$, and so $\{N(t)/t : t \ge 0\}$ is uniformly integrable also.

Since $N(t)/t \xrightarrow{\text{a.s.}} \mu^{-1}$, it follows by uniform integrability that there is also convergence in mean.

5. (a) Using the fact that $\mathbb{P}(N(t) = k) = \mathbb{P}(S_k \le t) - \mathbb{P}(S_{k+1} \le t)$, we find that

$$\begin{aligned} \mathbb{E}(s^{N(T)}) &= \int_0^\infty \left(\sum_{k=0}^{\infty} s^k \mathbb{P}(N(t) = k) \right) \nu e^{-\nu t} \, dt \\ &= \sum_{k=0}^{\infty} s^k \left\{ \int_0^\infty \left[\mathbb{P}(S_k \le t) - \mathbb{P}(S_{k+1} \le t) \right] \nu e^{-\nu t} \, dt \right\}. \end{aligned}$$

By integration by parts, $\int_0^\infty \mathbb{P}(S_k \le t) \nu e^{-\nu t} \, dt = M(-\nu)^k$ for $k \ge 0$. Therefore,

$$\mathbb{E}(s^{N(T)}) = \sum_{k=0}^{\infty} s^k \{ M(-\nu)^k - M(-\nu)^{k+1} \} = \frac{1 - M(-\nu)}{1 - sM(-\nu)}.$$

(b) In this case, $\mathbb{E}(s^{N(T)}) = \mathbb{E}(e^{\lambda T(s-1)}) = M_T(\lambda(s-1))$. When T has the given gamma distribution, $M_T(\theta) = \{\nu/(\nu - \theta)\}^b$, and

$$\mathbb{E}(s^{N(T)}) = \left(\frac{\nu}{\nu + \lambda} \right)^b \left(1 - \frac{\lambda s}{\nu + \lambda} \right)^b.$$

The coefficient of s^k may be found by use of the binomial theorem.

10.3 Solutions. Excess life

1. Let $g(y) = \mathbb{P}(E(t) > y)$, assumed not to depend on t. By the integral equation for the distribution of $E(t)$,

$$g(y) = 1 - F(t + y) + g(y) \int_0^t dF(x).$$

Write $h(x) = 1 - F(x)$ to obtain $g(y)h(t) = h(t + y)$, for $y, t \ge 0$. With $t = 0$, we have that $g(y)h(0) = h(y)$, whence $g(y) = h(y)/h(0)$ satisfies $g(t + y) = g(t)g(y)$, for $y, t \ge 0$. Now g is left-continuous, and we deduce as usual that $g(t) = e^{-\lambda t}$ for some λ. Hence $F(t) = 1 - e^{-\lambda t}$, and the renewal process is a Poisson process.

2. (a) Examine a sample path of E. If $E(t) = x$, then the sample path decreases (with slope -1) until it reaches the value 0, at which point it jumps to a height X, where X is the next interarrival time. Since X is independent of all previous interarrival times, the process is Markovian.

(b) In contrast, C has sample paths which increase (with slope 1) until a renewal occurs, at which they drop to 0. If $C(s) = x$ and, in addition, we know the entire history of the process up to time s, the time of the next renewal depends only on the length of the spent period (i.e. x) of the interarrival time in process. Hence C is Markovian.

3. (a) We have that

$$(*) \qquad \mathbb{P}\big(E(t) \le y\big) = F(t+y) - \int_0^t G(t+y-x)\,dm(x)$$

where $G(u) = 1 - F(u)$. Check the conditions of the key renewal theorem (10.2.7): $g(t) = G(t+y)$ satisfies:

(i) $g(t) \ge 0$,

(ii) $\int_0^\infty g(t)\,dt \le \int_0^\infty [1 - F(u)]\,du = \mathbb{E}(X_1) < \infty$,

(iii) g is non-increasing.

We conclude, by that theorem, that

$$\lim_{t\to\infty} \mathbb{P}\big(E(t) \le y\big) = 1 - \frac{1}{\mu}\int_0^\infty g(x)\,dx = \int_0^y \frac{1}{\mu}[1 - F(x)]\,dx.$$

(b) Integrating by parts,

$$\int_0^\infty \frac{x^r}{\mu}[1 - F(x)]\,dx = \frac{1}{\mu}\int_0^\infty \frac{x^{r+1}}{r+1}\,dF(x) = \frac{\mathbb{E}(X_1^{r+1})}{\mu(r+1)}.$$

See Exercise (4.3.3).

(c) As in Exercise (4.3.3), we have that $\mathbb{E}(E(t)^r) = \int_0^\infty ry^{r-1}\mathbb{P}(E(t) > y)\,dy$, implying by $(*)$ that

$$\mathbb{E}(E(t)^r) = \mathbb{E}\big(\{(X_1 - t)^+\}^r\big) + \int_{y=0}^\infty \int_{x=0}^t ry^{r-1}\mathbb{P}(X_1 > t+y-x)\,dm(x)\,dy,$$

whence the given integral equation is valid with

$$h(u) = \int_0^\infty ry^{r-1}\mathbb{P}(X_1 > u+y)\,dy = \mathbb{E}\big(\{(X_1 - u)^+\}^r\big).$$

Now h satisfies the conditions of the key renewal theorem, whence

$$\begin{aligned}\lim_{t\to\infty} \mathbb{E}(E(t)^r) &= \frac{1}{\mu}\int_0^\infty h(u)\,du = \frac{1}{\mu}\iint_{0<u,y<\infty} y^r\,dF(u+y)\,du \\ &= \frac{1}{\mu}\int_0^\infty y^r\mathbb{P}(X_1 > y)\,dy = \frac{\mathbb{E}(X_1^{r+1})}{\mu(r+1)}.\end{aligned}$$

4. We have that

$$\mathbb{P}\big(E(t) > y \,\big|\, C(t) = x\big) = \mathbb{P}(X_1 > y + x \mid X_1 > x) = \frac{1 - F(y+x)}{1 - F(x)},$$

whence

$$\mathbb{E}\big(E(t) \,\big|\, C(t) = x\big) = \int_0^\infty \frac{1 - F(y+x)}{1 - F(x)}\,dy = \frac{\mathbb{E}\{(X_1 - x)^+\}}{1 - F(x)}.$$

5. (a) Apply Exercise (10.2.2) to the sequence $X_i - \mu$, $1 \le i < \infty$, to obtain $\text{var}(T_{M(t)} - \mu M(t)) = \sigma^2\mathbb{E}(M(t))$.

(b) Clearly $T_{M(t)} = t + E(t)$, where E is excess lifetime, and hence $\mu M(t) = (t + E(t)) - (T_{M(t)} - \mu M(t))$, implying in turn that

$$(*) \qquad \mu^2 \operatorname{var}(M(t)) = \operatorname{var}(E(t)) + \operatorname{var}(S_{M(t)}) - 2\operatorname{cov}\big(E(t), S_{M(t)}\big),$$

where $S_{M(t)} = T_{M(t)} - \mu M(t)$. Now

$$\operatorname{var}(E(t)) \le \mathbb{E}(E(t)^2) \to \frac{\mathbb{E}(X_1^3)}{3\mu} \qquad \text{as } t \to \infty$$

if $\mathbb{E}(X_1^3) < \infty$ (see Exercise (10.3.3c)), implying that

$$(**) \qquad \frac{1}{t}\operatorname{var}(E(t)) \to 0 \qquad \text{as } t \to \infty.$$

This is valid under the weaker assumption that $\mathbb{E}(X_1^2) < \infty$, as the following argument shows. By Exercise (10.3.3c),

$$\mathbb{E}(E(t)^2) = \alpha(t) + \int_0^t \alpha(t-u)\, dm(u),$$

where $\alpha(u) = \mathbb{E}(\{(X_1 - u)^+\}^2)$. Now use the key renewal theorem together with the fact that $\alpha(t) \le \mathbb{E}(X_1^2 I_{\{X_1 > t\}}) \to 0$ as $t \to \infty$.

Using the Cauchy–Schwarz inequality,

$$\frac{1}{t}\big|\operatorname{cov}\big(E(t), S_{M(t)}\big)\big| \le \frac{1}{t}\sqrt{\operatorname{var}(E(t))\operatorname{var}(S_{M(t)})} \to 0$$

as $t \to \infty$, by part (a) and $(**)$. Returning to $(*)$, we have that

$$\frac{\mu^2}{t}\operatorname{var}(M(t)) \to \lim_{t\to\infty}\left\{\frac{\sigma^2}{t}\big(m(t)+1\big)\right\} = \frac{\sigma^2}{\mu}.$$

10.4 Solutions. Applications

1. Visualize a renewal as arriving after two stages, type 1 stages being exponential parameter λ and type 2 stages being exponential parameter μ. The 'stage' process is the flip–flop two-state Markov process of Exercise (6.9.1). With an obvious notation,

$$p_{11}(t) = \frac{\lambda}{\lambda+\mu}e^{-(\lambda+\mu)t} + \frac{\mu}{\lambda+\mu}.$$

Hence the excess lifetime distribution is a mixture of the exponential distribution with parameter μ, and the distribution of the sum of two exponential random variables, thus,

$$f_{E(t)}(x) = p_{11}(t)g(x) + (1 - p_{11}(t))\mu e^{-\mu x},$$

where $g(x)$ is the density function of a typical interarrival time. By Wald's equation,

$$\mathbb{E}(t + E(t)) = \mathbb{E}(S_{N(t)+1}) = \mathbb{E}(X_1)\mathbb{E}(N(t)+1) = \left(\frac{1}{\lambda} + \frac{1}{\mu}\right)(m(t)+1).$$

We substitute

$$\mathbb{E}(E(t)) = p_{11}(t)\left(\frac{1}{\lambda} + \frac{1}{\mu}\right) + (1 - p_{11}(t))\frac{1}{\mu} = \frac{1}{\mu} + \frac{p_{11}(t)}{\lambda}$$

to obtain the required expression.

2. (a) Note that the process is stationary, and $E(t) \le x$ if and only if $C(t+x) < x$. By Theorem (10.4.17), E has density function h, so that C and E have the same densities.

(b) The total life D and current life C satisfy

$$\mathbb{P}(D \ge y \mid C = x) = \mathbb{P}(X \ge y \mid X \ge x) = \frac{1 - F(y)}{1 - F(x)}, \qquad y \ge x.$$

The conditional density function of D follows by differentiation. By (a), the unconditional density function is

$$f_D(y) = \int_0^y \frac{f(y)}{1 - F(x)} \cdot \frac{1 - F(x)}{\mu}\, dx = \frac{y}{\mu} f(y), \qquad y > 0.$$

(c) We have that

$$\begin{aligned}\mathbb{P}(UD > x) &= \int_0^1 \mathbb{P}(D > x/u)\, du = \int_0^1 \left(\int_{x/u}^\infty \frac{w}{\mu} f(w)\, dw\right) du \\ &= \frac{1}{\mu}\int_x^\infty \left(\int_{x/w}^1 du\right) w f(w)\, dw = \frac{1}{\mu}\int_x^\infty (w - x) f(w)\, dw,\end{aligned}$$

and differentiation gives the density as $(1 - F(x))/\mu$ as required.

We expect an interarrival interval to include the time t with probability proportional to its length, so that $f_D(y) \propto y f(y)$, and we expect t to be uniformly distributed over the interval in which it lies.

3. (a) Let T be the time of the earliest renewal in the interval $(a, b]$, with $T = \infty$ if there is no arrival in the interval. Since there is no arrival during (a, T), we have that

$$N(b) - N(a) \le \begin{cases} 1 + N(b+T) - N(a+T) & \text{if } T < \infty, \\ 1 & \text{if } T = \infty. \end{cases}$$

Now take expectations.

(b) This holds since $m(t) < \infty$, and $\lim_{t\to\infty} m(t)/t$ is finite.

(c) We couple two renewal processes N and N' with renewal functions m and m'. The first process has interarrival times $X_1, X_2, \dots$ and the second is a stationary renewal process with interarrival times $Y, X_1, X_2, \dots$ where Y has density function $g(y) = (1 - F(y))/\mu$ for $y > 0$. Thus, N' is obtained from N by introducing a renewal at time Y, and then shifting the renewals of N by a time Y. By Theorem (10.4.17), N' is stationary, whence $m'(t) = t/\mu$.

By considering the above coupling of N and N', and taking account the two possibilities $Y \le t$ and $Y > t$, we find that

$$N'(t) = N(t) - N\big(t - (Y \wedge t), t\big] + I(Y \le t),$$

where $N(a, b]$ is the number of renewals of N in the interval $(a, b]$. Take expectations to obtain

$$(*) \qquad m(t) - \frac{t}{\mu} = \mathbb{E}\big(N(t - Y \wedge t, t]\big) - \mathbb{P}(Y \le t).$$

By Blackwell's renewal theorem (10.2.5),

$$\mathbb{E}\big(N(t - Y \wedge t, t] \,\big|\, Y\big) \xrightarrow{\text{a.s.}} \frac{Y}{\mu} \qquad \text{as } t \to \infty.$$

By part (a),

$$\mathbb{E}\big(N(t - Y \wedge t, t] \,\big|\, Y\big) \le 1 + Z,$$

where $Z = \mathbb{E}(N(Y) \mid Y)$ has mean satisfying $\mathbb{E}(Z) \le A(1 + \mathbb{E}(Y)) < \infty$ by part (b). We may therefore apply the dominated convergence theorem to obtain

$$\mathbb{E}\big(N(t - Y \wedge t, t]\big) \to \frac{\mathbb{E}(Y)}{\mu} \qquad \text{as } t \to \infty,$$

and the claim follows from (*) on noting that

$$\mathbb{E}(Y) = \int_0^\infty \frac{1}{\mu} y(1 - F(y))\,dy = \frac{\mathbb{E}(X_1^2)}{2\mu} = \frac{\sigma^2 + \mu^2}{2\mu}.$$

We note that, in certain other works, the limit in part (c) is stated with $\sigma^2 + \mu^2$ (rather than $\sigma^2 - \mu^2$) in the numerator. The difference arises because of different conventions over whether $t = 0$ is counted as a renewal of the process. The reader is reminded that $t = 0$ is not considered a renewal in the current work.

10.5 Solutions. Renewal–reward processes

1. Suppose, at time s, you are paid a reward at rate $u(X(s))$. By Theorem (10.5.10), equation (10.5.7), and Theorem (6.10.15b),

$$(*) \qquad \frac{1}{t}\int_0^t I_{\{X(s)=j\}}\,ds \xrightarrow{\text{a.s.}} \frac{1}{\mu_j g_j} = \pi_j.$$

Suppose $|u(i)| \le K < \infty$ for all $i \in S$, and let F be a finite subset of the state space. Then

$$\begin{aligned}\left|\frac{1}{t}\int_0^t u(X(s))\,ds - \sum_i \pi_i u(i)\right| &= \left|\sum_i u(i)\left(\frac{1}{t}\int_0^t I_{\{X(s)=i\}}\,ds - \pi_i\right)\right| \\ &\le K\sum_{i\in F}\left|\frac{1}{t}\int_0^t I_{\{X(s)=i\}}\,ds - \pi_i\right| + K\left(\frac{t - T_t(F)}{t}\right) + K\sum_{i\notin F}\pi_i,\end{aligned}$$

where $T_t(F)$ is the total time spent in F up to time t. Take the limit as $t \to \infty$ using (*), and then as $F \uparrow S$, to obtain the required result.

2. Suppose you are paid a reward at unit rate during every interarrival time of type X, i.e. at all times t at which $M(t)$ is even. By the renewal–reward theorem (10.5.1),

$$\frac{1}{t}\int_0^t I_{\{M(s)\text{ is even}\}}\,ds \xrightarrow{\text{a.s.}} \frac{\mathbb{E}(\text{reward during interarrival time})}{\mathbb{E}(\text{length of interarrival time})} = \frac{\mathbb{E}X_1}{\mathbb{E}X_1 + \mathbb{E}Y_1}.$$

The answer to the question is yes, by the renewal–reward theorem.

3. Suppose, at time t, you are paid a reward at rate $C(t)$. The expected reward during an interval (cycle) of length X is $\int_0^X s\,ds = \frac{1}{2}X^2$, since the age C is the same at the time s into the interval. The result follows by the renewal–reward theorem (10.5.1) and equation (10.5.7). The same conclusion is valid for the excess lifetime $E(s)$, the integral in this case being $\int_0^X (X - s)\,ds = \frac{1}{2}X^2$.

4. Suppose $X_0 = j$. Let $V_1 = \min\{n \ge 1 : X_n = j,\ X_m = k \text{ for some } 1 \le m < n\}$, the first visit to j *subsequent to* a visit to k, and let $V_{r+1} = \min\{n \ge V_r : X_n = j,\ X_m = k \text{ for some } V_r + 1 \le$

$m < n\}$. The V_r are the times of a renewal process. Suppose a reward of one ecu is paid at every visit to k. By the renewal–reward theorem and equation (10.5.7),

$$(*) \qquad \pi_k = \frac{1}{\mathbb{E}(V_1 \mid X_0 = j)} \mathbb{E}\left(\sum_{m=1}^{V_1 - 1} I_{\{X_m = k\}} \right).$$

By considering the time of the first visit to k,

$$\mathbb{E}(V_1 \mid X_0 = j) = \mathbb{E}(T_k \mid X_0 = j) + \mathbb{E}(T_j \mid X_0 = k).$$

The latter expectation in $(*)$ is the mean of a random variable N having the geometric distribution $\mathbb{P}(N = n) = p(1 - p)^{n-1}$ for $n \geq 1$, where $p = \mathbb{P}(T_j < T_k \mid X_0 = k)$. Since $\mathbb{E}(N) = p^{-1}$, we deduce as required that

$$\pi_k = \frac{1/\mathbb{P}(T_j < T_k \mid X_0 = k)}{\mathbb{E}(T_k \mid X_0 = j) + \mathbb{E}(T_j \mid X_0 = k)}.$$

5. Let the process accumulate a reward 1 each time the current life is no bigger than y, and 0 otherwise. The expected reward per cycle is $\int_0^y xf(x)\,dx$, and the expected cycle-length is μ. Therefore,

$$F_D(y) = \lim_{n \to \infty} \frac{1}{n} \mathbb{E}\left(\sum_{i=1}^{n} I(X_i \leq y) \right) = \frac{1}{\mu} \int_0^y xf(x)\,dx.$$

See the related Exercise (10.4.2b).

6. (a) The required mean is

$$\int_0^\infty \mathbb{P}(X \wedge d > x)\,dx = \int_0^d e^{-\lambda x}\,dx = \frac{1}{\lambda}(1 - e^{-\lambda d}).$$

(b) Let c be the normal cost of a tyre replacement. Under the first option, and assuming four wheels, and that tyres fail independently of one another, by the Poisson superposition theorem the process of failures is Poisson with rate $\frac{1}{2}$. The mean long-run cost per unit time is $(4c)/2 = 2c$.

Under the second option, a new cycle starts at time $T = \min\{X, 2\}$, where X is the time of the first failure. By part (a), $\mathbb{E}(T) = 2(1 - e^{-1})$. The cost is $(4c)/20$ at time 2 with probability e^{-1}, or $4c(1 + (1/20))$ at $X < 2$ with probability $1 - e^{-1}$. The mean long-run cost per unit time is, by the renewal–reward theorem,

$$\left(\frac{(1 - e^{-1})4 + (1/5)}{2(1 - e^{-1})} \right) c > 2c,$$

so choose option 1.

7. By the lack-of-memory property of the exponential distribution, we may take the cycles for the renewal–reward theorem to be $(0, m), (m, 2m), \dots$. The proportion of a cycle spent working is $(X \wedge m)/m$, where X is exponentially distributed. The uptime-ratio is therefore

$$\mathbb{E}\left(\frac{X \wedge m}{m} \right) = \frac{1}{m} \int_0^m e^{-\lambda x}\,dx.$$

One may alternatively take the repair times as regeneration points for cycles. The inter-repair time has mean $m/(1 - e^{-\lambda m})$, and the mean uptime per cycle is $1/\lambda$, so that the uptime ratio is $(1 - e^{-\lambda m})/(\lambda m)$.

10.6 Solutions to problems

1. (a) For any n, $\mathbb{P}(N(t) < n) \le \mathbb{P}(T_n > t) \to 0$ as $t \to \infty$.
(b) Either use Exercise (10.1.1), or argue as follows. Since $\mu > 0$, there exists ϵ (> 0) such that $\mathbb{P}(X_1 > \epsilon) > 0$. For all n,

$$\mathbb{P}(T_n \le n\epsilon) = 1 - \mathbb{P}(T_n > n\epsilon) \le 1 - \mathbb{P}(X_1 > \epsilon)^n < 1,$$

so that, if $t > 0$, there exists $n = n(t)$ such that $\mathbb{P}(T_n \le t) < 1$.

Fix t and let n be chosen accordingly. Any positive integer k may be expressed in the form $k = \alpha n + \beta$ where $0 \le \beta < n$. Now $\mathbb{P}(T_k \le t) \le \mathbb{P}(T_n \le t)^\alpha$ for $\alpha n \le k < (\alpha + 1)n$, and hence

$$m(t) = \sum_{k=1}^{\infty} \mathbb{P}(T_k \le t) \le \sum_{\alpha=0}^{\infty} n\mathbb{P}(T_n \le t)^\alpha < \infty.$$

(c) It is easiest to use Exercise (10.1.1), which implies the stronger conclusion that the moment generating function of $N(t)$ is finite in a neighbourhood of the origin.

2. (i) Condition on X_1 to obtain

$$v(t) = \int_0^t \mathbb{E}\{(N(t-u)+1)^2\}\,dF(u) = \int_0^t \{v(t-u) + 2m(t-u) + 1\}\,dF(u).$$

Take Laplace–Stieltjes transforms to find that $v^* = (v^* + 2m^* + 1)F^*$, where $m^* = F^* + m^*F^*$ as usual. Therefore $v^* = m^*(1 + 2m^*)$, which may be inverted to obtain the required integral equation.
(ii) If N is a Poisson process with intensity λ, then $m(t) = \lambda t$, and therefore $v(t) = (\lambda t)^2 + \lambda t$.

3. Fix $x \in \mathbb{R}$. Then

$$\mathbb{P}\left(\frac{N(t) - (t/\mu)}{\sqrt{t\sigma^2/\mu^3}} \ge x\right) = \mathbb{P}\left(N(t) \ge (t/\mu) + x\sqrt{t\sigma^2/\mu^3}\right) = \mathbb{P}(T_{a(t)} \le t)$$

where $a(t) = \lfloor (t/\mu) + x\sqrt{t\sigma^2/\mu^3} \rfloor$. Now,

$$\mathbb{P}(T_{a(t)} \le t) = \mathbb{P}\left(\frac{T_{a(t)} - \mu a(t)}{\sigma\sqrt{a(t)}} \le \frac{t - \mu a(t)}{\sigma\sqrt{a(t)}}\right).$$

However $a(t) \sim t/\mu$ as $t \to \infty$, and therefore

$$\frac{t - \mu a(t)}{\sigma\sqrt{a(t)}} \to -x \qquad \text{as } t \to \infty,$$

implying by the usual central limit theorem that

$$\mathbb{P}\left(\frac{N(t) - (t/\mu)}{\sqrt{t\sigma^2/\mu^3}} \ge x\right) \to \Phi(-x) \qquad \text{as } t \to \infty$$

where Φ is the $N(0, 1)$ distribution function.

An alternative proof makes use of Anscombe's theorem (7.11.28).

4. We have that, for $y \le t$,

$$\mathbb{P}\big(C(t) \ge y\big) = \mathbb{P}\big(E(t-y) > y\big) \to \lim_{u\to\infty} \mathbb{P}\big(E(u) > y\big) \qquad \text{as } t \to \infty$$
$$= \int_y^\infty \frac{1}{\mu}[1 - F(x)]\,dx$$

by Exercise (10.3.3a). The current and excess lifetimes have the same asymptotic distributions.

5. Using the lack-of-memory property of the Poisson process, the current lifetime $C(t)$ is independent of the excess lifetime $E(t)$, the latter being exponentially distributed with parameter λ. To derive the density function of $C(t)$ either solve (without difficulty in this case) the relevant integral equation, or argue as follows. Looking *backwards* in time from t, the arrival process looks like a Poisson process up to distance t (at the origin) where it stops. Therefore $C(t)$ may be expressed as $\min\{Z, t\}$ where Z is exponential with parameter λ; hence

$$f_{C(t)}(s) = \begin{cases} \lambda e^{-\lambda s} & \text{if } s \le t, \\ 0 & \text{if } s > t, \end{cases}$$

and $\mathbb{P}(C(t) = t) = e^{-\lambda t}$. Now $D(t) = C(t) + E(t)$, whose distribution is easily found (by the convolution formula) to be as given.

6. The ith interarrival time may be expressed in the form $T + Z_i$ where Z_i is exponential with parameter λ. In addition, $Z_1, Z_2, \dots$ are independent, by the lack-of-memory property. Now

$$1 - \widetilde{F}(x) = \mathbb{P}(T + Z_1 > x) = \mathbb{P}(Z_1 > x - T) = e^{-\lambda(x-T)}, \qquad x \ge T.$$

Taking into account the (conventional) dead period beginning at time 0, we have that

$$\mathbb{P}(\widetilde{N}(t) \ge k) = \mathbb{P}\Big(kT + \sum_{i=1}^{k} Z_i \le t\Big) = \mathbb{P}\big(N(t - kT) \ge k\big), \qquad t \ge kT,$$

where N is a Poisson process.

7. We have that $\widetilde{X}_1 = L + E(L)$ where L is the length of the dead period beginning at 0, and $E(L)$ is the excess lifetime at L. Therefore, conditioning on L,

$$\mathbb{P}(\widetilde{X}_1 \le x) = \int_0^x \mathbb{P}\big(E(l) \le x - l\big)\, dF_L(l).$$

We have that

$$\mathbb{P}\big(E(t) \le y\big) = F(t+y) - \int_0^t \big\{1 - F(t+y-x)\big\}\, dm(x).$$

By the renewal equation,

$$m(t+y) = F(t+y) + \int_0^{t+y} F(t+y-x)\, dm(x),$$

whence, by subtraction,

$$\mathbb{P}\big(E(t) \le y\big) = \int_t^{t+y} \big\{1 - F(t+y-x)\big\}\, dm(x).$$

It follows that

$$\begin{aligned}\mathbb{P}\big(\widetilde{X}_1 \le x\big) &= \int_{l=0}^{x} \int_{y=l}^{x} \big\{1 - F(x-y)\big\}\, dm(y)\, dF_L(l) \\ &= \left[F_L(l) \int_l^x \big\{1 - F(x-y)\big\}\, dm(y)\right]_{l=0}^{x} + \int_0^x F_L(l)\big\{1 - F(x-l)\big\}\, dm(l)\end{aligned}$$

using integration by parts. The term in square brackets equals 0.

8. (a) Each interarrival time has the same distribution as the sum of two independent random variables with the exponential distribution. Therefore $N(t)$ has the same distribution as $\lfloor \frac{1}{2} M(t) \rfloor$ where M is a Poisson process with intensity λ. Therefore $m(t) = \frac{1}{2}\mathbb{E}(M(t)) - \frac{1}{2}\mathbb{P}(M(t) \text{ is odd})$. Now $\mathbb{E}(M(t)) = \lambda t$, and

$$\mathbb{P}\big(M(t) \text{ is odd}\big) = \sum_{n=0}^{\infty} \frac{(\lambda t)^{2n+1} e^{-\lambda t}}{(2n+1)!} = \tfrac{1}{2} e^{-\lambda t}(e^{\lambda t} - e^{-\lambda t}).$$

With more work, one may establish the probability generating function of $N(t)$.

(b) Doing part (a) as above, one may see that $\widetilde{m}(t) = m(t)$.

9. Clearly $C(t)$ and $E(t)$ are independent if the process N is a Poisson process. Conversely, suppose that $C(t)$ and $E(t)$ are independent, for each fixed choice of t. The event $\{C(t) \geq y\} \cap \{E(t) \geq x\}$ occurs if and only if $E(t-y) \geq x + y$. Therefore

$$\mathbb{P}\big(C(t) \geq y\big)\mathbb{P}\big(E(t) \geq x\big) = \mathbb{P}\big(E(t-y) \geq x + y\big).$$

Take the limit as $t \to \infty$, remembering Exercise (10.3.3) and Problem (10.6.4), to obtain that $G(y)G(x) = G(x+y)$ if $x, y \geq 0$, where

$$G(u) = \int_u^\infty \frac{1}{\mu}[1 - F(v)]\, dv.$$

Now $1 - G$ is a distribution function, and hence has the lack-of-memory property (Problem (4.14.5)), implying that $G(u) = e^{-\lambda u}$ for some λ. This implies in turn that $[1 - F(u)]/\mu = -G'(u) = \lambda e^{-\lambda u}$, whence $\mu = 1/\lambda$ and $F(u) = 1 - e^{-\lambda u}$.

10. Clearly N is a renewal process if N_2 is Poisson. Suppose that N is a renewal process, and write λ for the intensity of N_1, and F_2 for the interarrival time distribution of N_2. By considering the time X_1 to the first arrival of N,

$$(*) \qquad 1 - F(x) = \mathbb{P}\big(N_1(x) = N_2(x) = 0\big) = e^{-\lambda x}\big(1 - F_2(x)\big).$$

Writing E, E_i for the excess lifetimes of N, N_i, we have that

$$\mathbb{P}\big(E(t) > x\big) = \mathbb{P}\big(E_1(t) > x,\, E_2(t) > x\big) = e^{-\lambda x}\mathbb{P}\big(E_2(t) > x\big).$$

Take the limit as $t \to \infty$, using Exercise (10.3.3), to find that

$$\int_x^\infty \frac{1}{\mu}[1 - F(u)]\, du = e^{-\lambda x} \int_x^\infty \frac{1}{\mu_2}[1 - F_2(u)]\, du,$$

where μ_2 is the mean of F_2. Differentiate, and use $(*)$, to obtain

$$\frac{1}{\mu} e^{-\lambda x}[1 - F_2(x)] = \lambda e^{-\lambda x} \int_x^\infty \frac{1}{\mu_2}[1 - F_2(u)]\, du + \frac{e^{-\lambda x}}{\mu_2}[1 - F_2(x)],$$

which simplifies to give $1 - F_2(x) = c\int_x^\infty [1 - F_2(u)]\, du$ where $c = \lambda\mu/(\mu_2 - \mu)$; this integral equation has solution $F_2(x) = 1 - e^{-cx}$.

11. (i) Taking transforms of the renewal equation in the usual way, we find that

$$m^*(\theta) = \frac{F^*(\theta)}{1 - F^*(\theta)} = \frac{1}{1 - F^*(\theta)} - 1$$

where

$$F^*(\theta) = \mathbb{E}(e^{-\theta X_1}) = 1 - \theta\mu + \tfrac{1}{2}\theta^2(\mu^2 + \sigma^2) + \mathrm{o}(\theta^2)$$

as $\theta \to 0$. Substitute this into the above expression to obtain

$$m^*(\theta) = \frac{1}{\theta\mu\{1 - \frac{1}{2}\theta(\mu + \sigma^2/\mu) + \mathrm{o}(\theta)\}} - 1$$

and expand to obtain the given expression. A formal inversion yields the expression for m.
(ii) The transform of the right-hand side of the integral equation is

$$(*) \qquad \frac{1}{\mu\theta} - F_E^*(\theta) + m^*(\theta) - F_E^*(\theta)m^*(\theta).$$

By Exercise (10.3.3), $F_E^*(\theta) = [1 - F^*(\theta)]/(\mu\theta)$, and $(*)$ simplifies to $m^*(\theta) - (m^* - m^*F^* - F^*)/(\mu\theta)$, which equals $m^*(\theta)$ since the quotient is 0 (by the renewal equation).

Using the key renewal theorem, as $t \to \infty$,

$$\int_0^t [1 - F_E(t-x)]\,dm(x) \to \frac{1}{\mu}\int_0^\infty [1 - F_E(u)]\,du = \frac{\mathbb{E}(X_1^2)}{2\mu^2} = \frac{\sigma^2 + \mu^2}{2\mu^2}$$

by Exercise (10.3.3b). Therefore,

$$m(t) - \frac{t}{\mu} \to -1 + \frac{\sigma^2 + \mu^2}{2\mu^2} = \frac{\sigma^2 - \mu^2}{2\mu^2}.$$

12. (i) Conditioning on X_1, we obtain

$$m^{\mathrm{d}}(t) = F^{\mathrm{d}}(t) + \int_0^t m(t-x)\,dF^{\mathrm{d}}(x).$$

Therefore $m^{\mathrm{d}*} = F^{\mathrm{d}*} + m^* F^{\mathrm{d}*}$. Also $m^* = F^* + m^*F^*$, so that

$$(*) \qquad m^{\mathrm{d}*} = F^{\mathrm{d}*}\left(1 + \frac{F^*}{1 - F^*}\right),$$

whence $m^{\mathrm{d}*} = F^{\mathrm{d}*} + m^{\mathrm{d}*}F^*$, the transform of the given integral equation.
(ii) Arguing as in Problem (10.6.2), $v^{\mathrm{d}*} = F^{\mathrm{d}*} + 2m^*F^{\mathrm{d}*} + v^*F^{\mathrm{d}*}$ where $v^* = F^*(1+2m^*)/(1-F^*)$ is the corresponding object in the ordinary renewal process. We eliminate v^* to find that

$$v^{\mathrm{d}*} = \frac{F^{\mathrm{d}*}(1 + 2m^*)}{1 - F^*} = m^{\mathrm{d}*}(1 + 2m^*)$$

by $(*)$. Now invert.

13. Taking into account the structure of the process, it suffices to deal with the case $I = 1$. Refer to Example (10.4.22) for the basic notation and analysis. It is easily seen that $\beta = (\nu - 1)\lambda$. Now $\widetilde{F}(t) = 1 - e^{-\nu\lambda t}$. Solve the renewal equation (10.4.24) to obtain

$$g(t) = h(t) + \int_0^t h(t-x)\,d\widetilde{m}(x)$$

where $\widetilde{m}(x) = \nu\lambda x$ is the renewal function associated with the interarrival time distribution $\widetilde{F}$. Therefore $g(t) = 1$, and $m(t) = e^{\beta t}$.

14. We have from Lemma (10.4.5) that $p^* = 1 - F_Z^* + p^* F^*$, where $F^* = F_Y^* F_Z^*$. Solve to obtain

$$p^* = \frac{1 - F_Z^*}{1 - F_Y^* F_Z^*}.$$

15. The first locked period begins at the time of arrival of the first particle. Since all future events may be timed relative to this arrival time, we may take this time to be 0. We shall therefore assume that a particle arrives at 0; call this the 0th particle, with locking time Y_0.

We shall condition on the time X_1 of the arrival of the next particle. Now

$$\mathbb{P}(L > t \mid X_1 = u) = \begin{cases} \mathbb{P}(Y_0 > t) & \text{if } u > t, \\ \mathbb{P}(Y_0 > u)\mathbb{P}(L' > t - u) & \text{if } u \le t, \end{cases}$$

where L' has the same distribution as L; the second part is a consequence of the fact that the process 'restarts' at each arrival. Therefore

$$\mathbb{P}(L > t) = \big(1 - G(t)\big)\mathbb{P}(X_1 > t) + \int_0^t \mathbb{P}(L > t - u)\big(1 - G(u)\big) f_{X_1}(u)\, du,$$

the required integral equation.

If $G(x) = 1 - e^{-\mu x}$, the solution is $\mathbb{P}(L > t) = e^{-\mu t}$, so that L has the same distribution as the locking times of individual particles. This striking fact may be attributed to the lack-of-memory property of the exponential distribution.

16. (a) It is clear that $M(tp)$ is a renewal process whose interarrival times are distributed as $X_1 + X_2 + \cdots + X_R$ where $\mathbb{P}(R = r) = pq^{r-1}$ for $r \ge 1$. It follows that $M(t)$ is a renewal process whose first interarrival time

$$X(p) = \inf\{t : M(t) = 1\} = p \inf\{t : M(tp) = 1\}$$

has distribution function

$$\mathbb{P}\big(X(p) \le x\big) = \sum_{r=1}^{\infty} \mathbb{P}(R = r) F_r(x/p).$$

(b) The characteristic function ϕ_p of F_p is given by

$$\phi_p(t) = \sum_{r=1}^{\infty} pq^{r-1} \int_{-\infty}^{\infty} e^{ixt}\, dF_r(t/p) = \sum_{r=1}^{\infty} pq^{r-1} \phi(pt)^r = \frac{p\phi(pt)}{1 - q\phi(pt)}$$

where ϕ is the characteristic function of F. Now $\phi(pt) = 1 + i\mu pt + o(p)$ as $p \downarrow 0$, so that

$$\phi_p(t) = \frac{1 + i\mu pt + o(p)}{1 - i\mu t + o(1)} = \frac{1 + o(1)}{1 - i\mu t}$$

as $p \downarrow 0$. The limit is the characteristic function of the exponential distribution with mean μ, and the continuity theorem tells us that the process M converges in distribution as $p \downarrow 0$ to a Poisson process with intensity $1/\mu$ (in the sense that the interarrival time distribution converges to the appropriate limit).

(c) If M and N have the same fdds, then $\phi_p(t) = \phi(t)$, which implies that $\phi(pt) = \phi(t)/(p + q\phi(t))$. Hence $\psi(t) = \phi(t)^{-1}$ satisfies $\psi(pt) = q + p\psi(t)$ for $t \in \mathbb{R}$. Now ψ is continuous, and it follows as in the solution to Problem (5.12.15) that ψ has the form $\psi(t) = 1+\beta t$, implying that $\phi(t) = (1+\beta t)^{-1}$ for some $\beta \in \mathbb{C}$. The only characteristic function of this form is that of an exponential distribution, and the claim follows.

17. (a) Let $N(t)$ be the number of times the sequence has been typed up to the tth keystroke. Then N is a renewal process whose interarrival times have the required mean μ; we have that $\mathbb{E}(N(t))/t \to \mu^{-1}$ as $t \to \infty$. Now each epoch of time marks the completion of such a sequence with probability $(\frac{1}{100})^{14}$, so that

$$\frac{1}{t}\mathbb{E}(N(t)) = \frac{1}{t}\sum_{n=14}^{t}\left(\frac{1}{100}\right)^{14} \to \left(\frac{1}{100}\right)^{14} \qquad \text{as } t \to \infty,$$

implying that $\mu = 10^{28}$.

The problem with 'omo' is 'omomo' (i.e. appearances may overlap). Let us call an epoch of time a 'renewal point' if it marks the completion of the word 'omo', disjoint from the words completed at previous renewal points. In each appearance of 'omo', either the first 'o' or the second 'o' (but not both) is a renewal point. Therefore the probability u_n, that n is a renewal point, satisfies $(\frac{1}{100})^3 = u_n + u_{n-2}(\frac{1}{100})^2$. Average this over n to obtain

$$\left(\frac{1}{100}\right)^3 = \lim_{n\to\infty}\frac{1}{t}\sum_{n=1}^{t}\left\{u_n + u_{n-2}\left(\frac{1}{100}\right)^2\right\} = \frac{1}{\mu} + \frac{1}{\mu}\left(\frac{1}{100}\right)^2,$$

and therefore $\mu = 10^6 + 10^2$.

(b) (i) Arguing as for 'omo', we obtain $p^3 = u_n + pu_{n-1} + p^2u_{n-2}$, whence $p^3 = (1+p+p^2)/\mu$.
(ii) Similarly, $p^2q = u_n + pqu_{n-2}$, so that $\mu = (1+pq)/(p^2q)$.

18. The fdds of $\{N(u) - N(t) : u \ge t\}$ depend on the distributions of $E(t)$ and of the interarrival times. In a stationary renewal process, the distribution of $E(t)$ does not depend on the value of t, whence $\{N(u) - N(t) : u \ge t\}$ has the same fdds as $\{N(u) : u \ge 0\}$, implying that X is strongly stationary.

19. We use the renewal–reward theorem. The mean time between expeditions is $B\mu$, and this is the mean length of a cycle of the process. The mean cost of keeping the bears during a cycle is $\frac{1}{2}B(B-1)c\mu$, whence the long-run average cost is $\{d + B(B-1)c\mu/2\}/(B\mu)$.

20. Consider a renewal process with interevent times distributed as T. By the renewal–reward theorem, the time W_n spent in state j up to time n satisfies

$$\lim_{n\to\infty}\frac{1}{n}\mathbb{E}_k(W_n) = \frac{\mathbb{E}(V_j(k))}{\mathbb{E}_k(T)}.$$

By the limit theorem (6.4.20) for Markov chains (and comments (b), (d) following), the left side equals π_j. See the related equation (6.4.5).

21. Since cars are independent and their properties are identically distributed, the times of purchases may be taken as regeneration points of cycles in a renewal process. Let C_i be the total cost of owning the ith car, and $C(t)$ the total cost of car ownership up to time t. By the renewal–reward theorem,

$$(*) \qquad \lim_{t\to\infty}\frac{\mathbb{E}(C(t))}{t} = \frac{\mathbb{E}(C_1)}{\mathbb{E}(Y)}.$$

Now,

$$\mathbb{E}(C_1) = \mathbb{E}\left(c - c\lambda^Y + \sum_{k=1}^{Y} r\mu^{k-1}\right) = c - cG(\lambda) + r\mathbb{E}\left(\frac{1-\mu^Y}{1-\mu}\right),$$

Insert this into the right side of (*) and choose m to minimize it.

11

Queues

11.2 Solutions. M/M/1

1. The stationary distribution satisfies $\boldsymbol{\pi} = \boldsymbol{\pi}\mathbf{P}$ when it exists, where $\mathbf{P}$ is the transition matrix. The equations

$$\pi_0 = \frac{\pi_1}{1+\rho}, \qquad \pi_1 = \pi_0 + \frac{\pi_2}{1+\rho}, \qquad \pi_n = \frac{\rho\pi_{n-1}}{1+\rho} + \frac{\pi_{n+1}}{1+\rho} \quad \text{for } n \geq 2,$$

with $\sum_{i=0}^{\infty} \pi_i = 1$, have the given solution. If $\rho \geq 1$, no such solution exists. It is slightly shorter to use the fact that such a walk is reversible in equilibrium, from which it follows that $\boldsymbol{\pi}$ satisfies

$$(*) \qquad \pi_0 = \frac{\pi_1}{1+\rho}, \qquad \frac{\rho\pi_n}{1+\rho} = \frac{\pi_{n+1}}{1+\rho} \quad \text{for } n \geq 1.$$

2. (i) This continuous-time walk is a Markov chain with generator given by $g_{01} = \theta_0$, $g_{n,n+1} = \theta_n\rho/(1+\rho)$ and $g_{n,n-1} = \theta_n/(1+\rho)$ for $n \geq 1$, other off-diagonal terms being 0. Such a process is reversible in equilibrium (see Problem (6.15.16)), and its stationary distribution $\boldsymbol{\nu}$ must satisfy $\nu_n g_{n,n+1} = \nu_{n+1} g_{n+1,n}$. These equations may be written as

$$\nu_0\theta_0 = \frac{\nu_1\theta_1}{1+\rho}, \qquad \frac{\nu_n\theta_n\rho}{1+\rho} = \frac{\nu_{n+1}\theta_{n+1}}{1+\rho} \quad \text{for } n \geq 1.$$

These are identical to the equations labelled $(*)$ in the previous solution, with π_n replaced by $\nu_n\theta_n$. It follows that $\nu_n = C\pi_n/\theta_n$ for some positive constant C.
(ii) If $\theta_0 = \lambda$, $\theta_n = \lambda + \mu$ for $n \geq 1$, we have that

$$1 = \sum_n \nu_n = C\left\{\frac{\pi_0}{\lambda} + \frac{1-\pi_0}{\mu+\lambda}\right\} = \frac{C}{2\lambda},$$

whence $C = 2\lambda$ and the result follows.

3. Let Q be the number of people ahead of the arriving customer at the time of his arrival. Using the lack-of-memory property of the exponential distribution, the customer in service has residual service-time with the exponential distribution, parameter μ, whence W may be expressed as $S_1+S_2+\cdots+S_Q$, the sum of independent exponential variables, parameter μ. The characteristic function of W is

$$\begin{aligned}\phi_W(t) &= \mathbb{E}\{\mathbb{E}(e^{itW} \mid Q)\} = \mathbb{E}\left\{\left(\frac{\mu}{\mu - it}\right)^Q\right\} \\ &= \frac{1-\rho}{1-\rho\mu/(\mu-it)} = (1-\rho) + \rho\left(\frac{\mu-\lambda}{\mu-\lambda-it}\right).\end{aligned}$$

This is the characteristic function of the given distribution. The atom at 0 corresponds to the possibility that $Q = 0$.

4. We prove this by induction on the value of $i + j$. If $i + j = 0$ then $i = j = 0$, and it is easy to check that $\pi(0; 0, 0) = 1$ and $A(0; 0, 0) = 1$, $A(n; 0, 0) = 0$ for $n \geq 1$. Suppose then that $K \geq 1$, and that the claim is valid for all pairs (i, j) satisfying $i + j = K$. Let i and j satisfy $i + j = K + 1$. The last ball picked has probability $i/(i + j)$ of being red; conditioning on the colour of the last ball, we have that

$$\pi(n; i, j) = \frac{i}{i+j}\pi(n-1; i-1, j) + \frac{j}{i+j}\pi(n+1; i, j-1).$$

Now $(i - 1) + j = K = i + (j - 1)$. Applying the induction hypothesis, we find that

$$\pi(n; i, j) = \frac{i}{i+j}\Big\{A(n-1; i-1, j) - A(n; i-1, j)\Big\} + \frac{j}{i+j}\Big\{A(n+1; i, j-1) - A(n+2; i, j-1)\Big\}.$$

Substitute to obtain the required answer, after a little cancellation and collection of terms. Can you see a more natural way?

5. Let A and B be independent Poisson process with intensities λ and μ respectively. These processes generate a queue-process as follows. At each arrival time of A, a customer arrives in the shop. At each arrival-time of B, the customer being served completes his service and leaves; if the queue is empty at this moment, then nothing happens. It is not difficult to see that this queue-process is $M(\lambda)/M(\mu)/1$. Suppose that $A(t) = i$ and $B(t) = j$. During the time-interval $[0, t]$, the order of arrivals and departures follows the schedule of Exercise (11.2.4), arrivals being marked as red balls and departures as lemon balls. The imbedded chain has the same distributions as the random walk of that exercise, and it follows that $\mathbb{P}\big(Q(t) = n \mid A(t) = i, B(t) = j\big) = \pi(n; i, j)$. Therefore $p_n(t) = \sum_{i,j} \pi(n; i, j)\mathbb{P}(A(t) = i)\mathbb{P}(B(t) = j)$.

6. With $\rho = \lambda/\mu$, the stationary distribution of the imbedded chain is, as in Exercise (11.2.1),

$$\widehat{\pi}_n = \begin{cases} \frac{1}{2}(1-\rho) & \text{if } n = 0, \\ \frac{1}{2}(1-\rho^2)\rho^{n-1} & \text{if } n \geq 1. \end{cases}$$

In the usual notation of continuous-time Markov chains, $g_0 = \lambda$ and $g_n = \lambda + \mu$ for $n \geq 1$, whence, by Exercise (6.10.11), there exists a constant c such that

$$\pi_0 = \frac{c}{2\lambda}(1-\rho), \quad \pi_n = \frac{c}{2(\lambda+\mu)}(1-\rho^2)\rho^{n-1} \quad \text{for } n \geq 1.$$

Now $\sum_i \pi_i = 1$, and therefore $c = 2\lambda$ and $\pi_n = (1 - \rho)\rho^n$ as required. The working is reversible.

7. (a) Let $Q_i(t)$ be the number of people in the ith queue at time t, including any currently in service. The process Q_1 is reversible in equilibrium, and departures in the original process correspond to arrivals in the reversed process. It follows that the departure process of the first queue is a Poisson process with intensity λ, and that the departure process of Q_1 is independent of the current value of Q_1.

(b) We have from part (a) that, for any given t, the random variables $Q_1(t)$, $Q_2(t)$ are independent. Consider an arriving customer when the queues are in equilibrium, and let W_i be his waiting time (before service) in the ith queue. With T the time of arrival, and recalling Exercise (11.2.3),

$$\begin{aligned} \mathbb{P}(W_1 = 0, W_2 = 0) &> \mathbb{P}(Q_i(T) = 0 \text{ for } i = 1, 2) = \mathbb{P}(Q_1(T) = 0)\mathbb{P}(Q_2(T) = 0) \\ &= (1-\rho_1)(1-\rho_2) = \mathbb{P}(W_1 = 0)\mathbb{P}(W_2 = 0). \end{aligned}$$

Therefore W_1 and W_2 are not independent. There is a slight complication arising from the fact that T is a random variable. However, T is independent of everybody who has gone before, and in particular of the earlier values of the queue processes Q_i.

11.3 Solutions. M/G/1

1. In equilibrium, the queue-length Q_n just after the nth departure satisfies

$$(*) \qquad Q_{n+1} = A_n + Q_n - h(Q_n)$$

where A_n is the number of arrivals during the $(n+1)$th service period, and $h(m) = 1 - \delta_{m0}$. Now Q_n and Q_{n+1} have the same distribution. Take expectations to obtain

$$(**) \qquad 0 = \mathbb{E}(A_n) - \mathbb{P}(Q_n > 0),$$

where $\mathbb{E}(A_n) = \lambda d$, the mean number of arrivals in an interval of length d. Next, square $(*)$ and take expectations:

$$0 = \mathbb{E}(A_n^2) + \mathbb{E}\big(h(Q_n)^2\big) + 2\Big\{\mathbb{E}(A_n Q_n) - \mathbb{E}\big(A_n h(Q_n)\big) - \mathbb{E}\big(Q_n h(Q_n)\big)\Big\}.$$

Use the facts that A_n is independent of Q_n, and that $Q_n h(Q_n) = Q_n$, to find that

$$0 = \{(\lambda d)^2 + \lambda d\} + \mathbb{P}(Q_n > 0) + 2\big\{(\lambda d - 1)\mathbb{E}(Q_n) - \lambda d\,\mathbb{P}(Q_n > 0)\big\}$$

and therefore, by $(**)$,

$$\mathbb{E}(Q_n) = \frac{2\rho - \rho^2}{2(1-\rho)}.$$

2. From the standard theory, M_B satisfies $M_B(s) = M_S(s - \lambda + \lambda M_B(s))$, where $M_S(\theta) = \mu/(\mu - \theta)$. Substitute to find that $x = M_B(s)$ is a root of the quadratic $\lambda x^2 - x(\lambda + \mu - s) + \mu = 0$. For some small positive s, $M_B(s)$ is smooth and non-decreasing. Therefore $M_B(s)$ is the root given.

3. (a) Let T_n be the instant of time at which the server is freed for the nth time. By the lack-of-memory property of the exponential distribution, the time of the first arrival after T_n is independent of all arrivals prior to T_n, whence T_n is a 'regeneration point' of the queue (so to say). It follows that the times which elapse between such regeneration points are independent, and it is easily seen that they have the same distribution.

(b) The total time T is the sum of the waiting time W and the service time S. Since these are independent, T has moment generating function $M_S(s)M_W(s)$ where M_W is given by Theorem (11.3.16).

4. Since all frustrated customers leave immediately, the queue is empty at the end of every service period. Consider, in the sense of the renewal–reward theorem, time partitioned into cycles where a cycle comprises a consecutive idle and service period of the server. During a typical cycle, the mean number of arrivals is $1 + \lambda\mathbb{E}(S)$, and the mean number lost is $\lambda\mathbb{E}(S)$. Therefore, the proportion of arrivals lost is $\lambda\mathbb{E}(S)/(1 + \lambda\mathbb{E}(S))$.

11.4 Solutions. G/M/1

1. The transition matrix of the imbedded chain obtained by observing queue-lengths just before arrivals is

$$\mathbf{P}_A = \begin{pmatrix} 1-\alpha_0 & \alpha_0 & 0 & 0 & \cdots \\ 1-\alpha_0-\alpha_1 & \alpha_1 & \alpha_0 & 0 & \cdots \\ 1-\alpha_0-\alpha_1-\alpha_2 & \alpha_2 & \alpha_1 & \alpha_0 & \cdots \\ \vdots & \vdots & \vdots & \vdots & \ddots \end{pmatrix}.$$

The equation $\boldsymbol{\pi} = \boldsymbol{\pi}\mathbf{P}_A$ may be written as

$$\pi_0 = \sum_{i=0}^{\infty} \pi_i \left(1 - \sum_{j=0}^{i} \alpha_j\right), \qquad \pi_n = \sum_{i=0}^{\infty} \alpha_i \pi_{n+i-1} \quad \text{for } n \ge 1.$$

It is easily seen, by adding, that the first equation is a consequence of the remaining equations, taken in conjunction with $\sum_0^\infty \pi_i = 1$. Therefore $\boldsymbol{\pi}$ is specified by the equation for π_n, $n \ge 1$.

The indicated substitution gives

$$\theta^n = \theta^{n-1} \sum_{i=0}^{\infty} \alpha_i \theta^i$$

which is satisfied whenever θ satisfies

$$\theta = \sum_{i=0}^{\infty} \alpha_i \theta^i = \sum_{i=0}^{\infty} \mathbb{E}\left\{ \frac{(\mu X \theta)^i e^{-\mu X}}{i!} \right\} = \mathbb{E}(e^{-\mu X} e^{\mu X \theta}) = M_X\big(\mu(\theta - 1)\big).$$

It is easily seen that $A(\theta) = M_X(\mu(\theta - 1))$ is convex and non-decreasing on $[0, 1]$, and satisfies $A(0) > 0$, $A(1) = 1$. Now $A'(1) = \mu \mathbb{E}(X) = \rho^{-1} > 1$, implying that there is a unique $\eta \in (0, 1)$ such that $A(\eta) = \eta$. With this value of η, the vector $\boldsymbol{\pi}$ given by $\pi_j = (1 - \eta)\eta^j$, $j \ge 0$, is a stationary distribution of the imbedded chain. This $\boldsymbol{\pi}$ is the unique such distribution because the chain is irreducible.

2. (i) The equilibrium distribution is $\pi_n = (1-\eta)\eta^n$ for $n \ge 0$, with mean $\sum_{n=0}^\infty n\pi_n = \eta/(1-\eta)$.
(ii) Using the lack-of-memory property of the service time in progress at the time of the arrival, we see that the waiting time may be expressed as $W = S_1 + S_2 + \cdots + S_Q$ where Q has distribution $\boldsymbol{\pi}$, given above, and the S_n are service times independent of Q. Therefore

$$\mathbb{E}(W) = \mathbb{E}(S_1)\mathbb{E}(Q) = \frac{\eta/\mu}{(1-\eta)}.$$

3. We have that $Q(n+) = 1 + Q(n-)$ a.s. for each integer n, whence $\lim_{t\to\infty} \mathbb{P}(Q(t) = m)$ cannot exist.

Since the traffic intensity is less than 1, the imbedded chain is ergodic with stationary distribution as in Exercise (11.4.1).

11.5 Solutions. G/G/1

1. Let T_n be the starting time of the nth busy period. Then T_n is an arrival time, and also the beginning of a service period. Conditional on the value of T_n, the future evolution of the queue is independent of the past, whence the random variables $\{T_{n+1} - T_n : n \ge 1\}$ are independent. It is easily seen that they are identically distributed.

2. If the server is freed at time T, the time I until the next arrival has the exponential distribution with parameter μ (since arrivals form a Poisson process).

By the duality theory of queues, the waiting time in question has moment generating function $M_W(s) = (1-\zeta)/(1-\zeta M_I(s))$ where $M_I(s) = \mu/(\mu - s)$ and $\zeta = \mathbb{P}(W > 0)$. Therefore,

$$M_W(s) = \frac{\zeta\mu(1-\zeta)}{\mu(1-\zeta) - s} + (1-\zeta),$$

the moment generating function of a mixture of an atom at 0 and an exponential distribution with parameter $\mu(1-\zeta)$.

If G is the probability generating function of the equilibrium queue-length, then, using the lack-of-memory property of the exponential distribution, we have that $M_W(s) = G(\mu/(\mu-s))$, since W is the sum of the (residual) service times of the customers already present. Set $u = \mu/(\mu-s)$ to find that $G(u) = (1-\zeta)/(1-\zeta u)$, the generating function of the mass function $f(k) = (1-\zeta)\zeta^k$ for $k \ge 0$. It may of course be shown that ζ is the smallest positive root of the equation $x = M_X(\mu(x-1))$, where X is a typical interarrival time.

3. We have that

$$1 - G(y) = \mathbb{P}(S - X > y) = \int_0^\infty \mathbb{P}(S > u + y)\, dF_X(u), \qquad y \in \mathbb{R},$$

where S and X are typical (independent) service and interarrival times. Hence, formally,

$$dG(y) = -\int_0^\infty d\mathbb{P}(S > u + y)\, dF_X(u) = dy \int_{-y}^\infty \mu e^{-\mu(u+y)}\, dF_X(u),$$

since $f_S(u+y) = e^{-\mu(u+y)}$ if $u > -y$, and is 0 otherwise.

With F as given,

$$\int_{-\infty}^x F(x-y)\, dG(y) = \iint_{\substack{-\infty<y\le x\\ -y<u<\infty}} \{1 - \eta e^{-\mu(1-\eta)(x-y)}\}\mu e^{-\mu(u+y)}\, dF_X(u)\, dy.$$

First integrate over y, then over u (noting that $F_X(u) = 0$ for $u < 0$), and the double integral collapses to $F(x)$, when $x \ge 0$.

11.6 Solution. Heavy traffic

1. Q_ρ has characteristic function

$$\phi_\rho(t) = \sum_{n=0}^\infty e^{itn}\rho^n(1-\rho) = \frac{1-\rho}{1-\rho e^{it}}.$$

Therefore the characteristic function of $(1-\rho)Q_\rho$ satisfies

$$\phi_\rho\big((1-\rho)t\big) = \frac{1-\rho}{1-\rho e^{i(1-\rho)t}} \to \frac{1}{1-it} \qquad \text{as } \rho \uparrow 1.$$

The limit characteristic function is that of the exponential distribution, and the result follows by the continuity theorem.

11.7 Solutions. Networks of queues

1. The first observation follows as in Example (11.7.4). The equilibrium distribution is given as in Theorem (11.7.14) by

$$\pi(\mathbf{n}) = \prod_{i=1}^c \frac{\alpha_i^{n_i} e^{-\alpha_i}}{n_i!}, \qquad \text{for } \mathbf{n} = (n_1, n_2, \dots, n_c) \in \mathbb{Z}^c,$$

the product of Poisson distributions. This is related to Bartlett's theorem (see Problem (8.7.6)) by defining the state A as 'being in station i at some given time'.

2. The number of customers in the queue is a birth–death process, and is therefore reversible in equilibrium. The claims follow in the same manner as was argued in the solution to Exercise (11.2.7).

3. (a) We may take as state space the set $\{0, 1', 1'', 2, 3, \dots\}$, where $i \in \{0, 2, 3, \dots\}$ is the state of having i people in the system including any currently in service, and $1'$ (respectively $1''$) is the state of having exactly one person in the system, this person being served by the first (respectively second) server. It is straightforward to check that this process is reversible in equilibrium, whence the departure process is as stated, by the argument used in Exercise (11.2.7).

(b) This time, we take as state space the set $\{0', 0'', 1', 1'', 2, 3, \dots\}$ having the same states as in part (a) with the difference that $0'$ (respectively $0''$) is the state in which there are no customers present and the first (respectively second) server has been free for the shorter time. It is easily seen that transition from $0'$ to $1''$ has strictly positive probability whereas transition from $1''$ to $0'$ has zero probability, implying that the process is not reversible. By drawing a diagram of the state space, or otherwise, it may be seen that the time-reversal of the process has the same structure as the original, with the unique change that states $0'$ are $0''$ are interchanged. Since departures in the original process correspond to arrivals in the time-reversal, the required properties follow in the same manner as in Exercise (11.2.7).

4. The total time spent by a given customer in service may be expressed as the sum of geometrically distributed number of exponential random variables, and this is easily shown to be exponential with parameter $\delta\mu$. The queue is therefore in effect a M(λ)/M($\delta\mu$)/1 system, and the stationary distribution is the geometric distribution with parameter $\rho = \lambda/(\delta\mu)$, provided $\rho < 1$. As in Exercise (11.2.7), the process of departures is Poisson.

Assume that rejoining customers go to the end of the queue, and note that the number of customers present constitutes a Markov chain. However, the composite process of arrivals is not Poisson, since increments are no longer independent. This may be seen as follows. In equilibrium, the probability of an arrival of either kind during the time interval $(t, t+h)$ is $\lambda h + \rho\mu(1-\delta)h + o(h) = (\lambda/\delta)h + o(h)$. If there were an arrival of either kind during $(t-h, t)$, then (with conditional probability $1 - O(h)$) the queue is non-empty at time t, whence the conditional probability of an arrival of either kind during $(t, t+h)$ is $\lambda h + \mu(1-\delta)h + o(h)$; this is of a larger order of magnitude than the earlier probability $(\lambda/\delta)h + o(h)$.

5. For stations r, s, we write $r \to s$ if an individual at r visits s at a later time with a strictly positive probability. Let C comprise the station j together with all stations i such that $i \to j$. The process restricted to C is an open migration process in equilibrium. By Theorem (11.7.20), the restricted process is reversible, whence the process of departures from C via j is a Poisson process with some intensity ζ. Individuals departing C via j proceed directly to k with probability

$$\frac{\lambda_{jk}\phi_j(n_j)}{\mu_j\phi_j(n_j) + \sum_{r\notin C}\lambda_{jr}\phi_j(n_j)} = \frac{\lambda_{jk}}{\mu_j + \sum_{r\notin C}\lambda_{jr}},$$

independently of the number n_j of individuals currently at j. Such a thinned Poisson process is a Poisson process also (cf. Exercise (6.8.2)), and the claim follows.

6. The number of individuals in the system at time t is no larger than the total number of arrivals, which grows in the manner of a birth process with parameter $V = \sum_j \nu_j < \infty$. The result follows by Theorem (6.8.17).

7. Let $\boldsymbol{\pi}$ satisfy $\sum_{i\in T}\pi_i g_{ij} = 0$ for $j \neq \infty$. Since $\mathbf{G}$ has zero row-sums,

$$\sum_{i\in T}\pi_i g_{i\infty} = \sum_{i\in T}\pi_i\Big(-\sum_{j\neq\infty} g_{ij}\Big) = -\sum_{j\neq\infty}\Big(\sum_{i\in T}\pi_i g_{ij}\Big) = 0.$$

11.8 Solutions to problems

1. Although the two cases may be done together, we choose to do them separately. When $k = 1$, the equilibrium distribution $\boldsymbol{\pi}$ satisfies:

$$\begin{aligned} \mu\pi_1 - \lambda\pi_0 &= 0, \\ \mu\pi_{n+1} - (\lambda + \mu)\pi_n + \lambda\pi_{n-1} &= 0, \qquad 1 \le n < N, \\ -\mu\pi_N + \lambda\pi_{N-1} &= 0, \end{aligned}$$

a system of equations with solution $\pi_n = \pi_0(\lambda/\mu)^n$ for $0 \le n \le N$, where (if $\lambda \neq \mu$)

$$\pi_0^{-1} = \sum_{n=0}^{N} (\lambda/\mu)^n = \frac{1 - (\lambda/\mu)^{N+1}}{1 - (\lambda/\mu)}.$$

Now let $k = 2$. The queue is a birth–death process with rates

$$\lambda_i = \begin{cases} \lambda & \text{if } i < N, \\ 0 & \text{if } i \ge N, \end{cases} \qquad \mu_i = \begin{cases} \mu & \text{if } i = 1, \\ 2\mu & \text{if } i \ge 2. \end{cases}$$

It is reversible in equilibrium, and its stationary distribution satisfies $\lambda_i\pi_i = \mu_{i+1}\pi_{i+1}$. We deduce that $\pi_i = 2\rho^i\pi_0$ for $1 \le i \le N$, where $\rho = \lambda/(2\mu)$ and

$$\pi_0^{-1} = 1 + \sum_{i=1}^{N} 2\rho^i.$$

2. The answer is obtainable in either case by following the usual method. It is shorter to use the fact that such processes are reversible in equilibrium.

(a) The stationary distribution $\boldsymbol{\pi}$ satisfies $\pi_n\lambda p(n) = \pi_{n+1}\mu$ for $n \ge 0$, whence $\pi_n = \pi_0\rho^n/n!$ where $\rho = \lambda/\mu$. Therefore $\pi_n = \rho^n e^{-\rho}/n!$.

(b) Similarly,

$$\pi_n = \pi_0\rho^n \prod_{m=0}^{n-1} p(m) = \pi_0\rho^n 2^{-\frac{1}{2}n(n-1)}, \qquad n \ge 0,$$

where

$$\pi_0^{-1} = \sum_{n=0}^{\infty} \rho^n \left(\tfrac{1}{2}\right)^{\frac{1}{2}n(n-1)}.$$

At the instant of arrival of a potential customer, the probability q that she joins the queue is obtained by conditioning on its length:

$$q = \sum_{n=0}^{\infty} p(n)\pi_n = \pi_0 \sum_{n=0}^{\infty} \rho^n 2^{-n-\frac{1}{2}n(n-1)} = \pi_0 \sum_{n=0}^{\infty} \rho^n 2^{-\frac{1}{2}n(n+1)} = \pi_0 \frac{1}{\rho}\{\pi_0^{-1} - 1\}.$$

3. *First method.* Let (Q_1, Q_2) be the queue-lengths, and suppose they are in equilibrium. Since Q_1 is a birth–death process, it is reversible, and we write $\widehat{Q}_1(t) = Q_1(-t)$. The sample paths of Q_1 have increasing jumps of size 1 at times of a Poisson process with intensity λ; these jumps mark arrivals at the cash desk. By reversibility, $\widehat{Q}_1$ has the same property; such increasing jumps for $\widehat{Q}_1$ are decreasing jumps for Q_1, and therefore the times of departures from the cash desk form a Poisson process with intensity λ. Using the same argument, the quantity $Q_1(t)$ together with the departures prior to t have the same joint distribution as the quantity $\widehat{Q}_1(-t)$ together with all arrivals after $-t$. However $\widehat{Q}_1(-t)$ is independent of its subsequent arrivals, and therefore $Q_1(t)$ is independent of its earlier departures.

It follows that arrivals at the second desk are in the manner of a Poisson process with intensity λ, and that $Q_2(t)$ is independent of $Q_1(t)$. Departures from the second desk form a Poisson process also.

Hence, in equilibrium, Q_1 is M(λ)/M(μ_1)/1 and Q_2 is M(λ)/M(μ_2)/1, and they are independent at any given time. Therefore their joint stationary distribution is

$$\pi_{mn} = \mathbb{P}\big(Q_1(t) = m,\, Q_2(t) = n\big) = (1-\rho_1)(1-\rho_2)\rho_1^m \rho_2^n$$

where $\rho_i = \lambda/\mu_i$.

Second method. The pair $(Q_1(t), Q_2(t))$ is a bivariate Markov chain. A stationary distribution $(\pi_{mn} : m, n \geq 0)$ satisfies

$$(\lambda + \mu_1 + \mu_2)\pi_{mn} = \lambda\pi_{m-1,n} + \mu_1\pi_{m+1,n-1} + \mu_2\pi_{m,n+1}, \qquad m, n \geq 1,$$

together with other equations when $m = 0$ or $n = 0$. It is easily checked that these equations have the solution given above, when $\rho_i < 1$ for $i = 1, 2$.

4. Let D_n be the time of the nth departure, and let $Q_n = Q(D_n+)$ be the number of waiting customers immediately after D_n. We have in the usual way that $Q_{n+1} = A_n + Q_n - h(Q_n)$, where A_n is the number of arrivals during the $(n+1)$th service time, and $h(x) = \min\{x, m\}$. Let $G(s) = \sum_{i=0}^{\infty} \pi_i s^i$ be the equilibrium probability generating function of the Q_n. Then, since Q_n is independent of A_n,

$$G(s) = \mathbb{E}(s^{A_n})\mathbb{E}(s^{Q_n - h(Q_n)})$$

where

$$\mathbb{E}(s^{A_n}) = \int_0^\infty e^{\lambda u(s-1)} f_S(u)\, du = M_S\big(\lambda(s-1)\big),$$

M_S being the moment generating function of a service time, and

$$\mathbb{E}(s^{Q_n - h(Q_n)}) = \sum_{i=0}^{m} \pi_i + \sum_{i=m+1}^{\infty} s^{i-m}\pi_i = s^{-m}\left\{G(s) + \sum_{i=0}^{m}(s^m - s^i)\pi_i\right\}.$$

Combining these relations, we obtain that G satisfies

$$s^m G(s) = M_S\big(\lambda(s-1)\big)\left\{G(s) + \sum_{i=0}^{m}(s^m - s^i)\pi_i\right\},$$

whenever it exists.

Finally suppose that $m = 2$ and $M_S(\theta) = \mu/(\mu - \theta)$. In this case,

$$G(s) = \frac{\mu\{\pi_0(s+1) + \pi_1 s\}}{\mu(s+1) - \lambda s^2}.$$

Now $G(1) = 1$, whence $\mu(2\pi_0 + \pi_1) = 2\mu - \lambda$; this implies in particular that $2\mu - \lambda > 0$. Also $G(s)$ converges for $|s| \leq 1$. Therefore any zero of the denominator in the interval $[-1, 1]$ is also a zero of the numerator. There exists exactly one such zero, since the denominator is a quadratic which takes the value $-\lambda$ at $s = -1$ and the value $2\mu - \lambda$ at $s = 1$. The zero in question is at

$$s_0 = \frac{\mu - \sqrt{\mu^2 + 4\lambda\mu}}{2\lambda},$$

and it follows that $\pi_0 + (\pi_0 + \pi_1)s_0 = 0$. Solving for π_0 and π_1, we obtain

$$G(s) = \frac{1-\alpha}{1-\alpha s},$$

where $\alpha = 2\lambda/\{\mu + \sqrt{\mu^2 + 4\lambda\mu}\}$.

5. Recalling standard M/G/1 theory, the moment generating function M_B satisfies

$$(*) \qquad M_B(s) = M_S\big(s - \lambda + \lambda M_B(s)\big) = \frac{\mu}{\mu - \{s - \lambda + \lambda M_B(s)\}}$$

whence $M_B(s)$ is one of

$$\frac{(\lambda + \mu - s) \pm \sqrt{(\lambda + \mu - s)^2 - 4\lambda\mu}}{2\lambda}.$$

Now $M_B(s)$ is non-decreasing in s, and therefore it is the value with the minus sign. The density function of B may be found by inverting the moment generating function; see Feller (1971, p. 482), who has also an alternative derivation of M_B.

As for the mean and variance, either differentiate M_B, or differentiate $(*)$. Following the latter route, we obtain the following relations involving M $(= M_B)$:

$$2\lambda M M' + M + (s - \lambda - \mu)M' = 0,$$
$$2\lambda M M'' + 2\lambda(M')^2 + 2M' + (s - \lambda - \mu)M'' = 0.$$

Set $s = 0$ to obtain $M'(0) = (\mu - \lambda)^{-1}$ and $M''(0) = 2\mu(\mu - \lambda)^{-3}$, whence the claims are immediate.

6. (i) This question is closely related to Exercise (11.3.1). With the same notation as in that solution, we have that

$$(*) \qquad Q_{n+1} = A_n + Q_n - h(Q_n)$$

where $h(x) = \min\{1, x\}$. Taking expectations, we obtain $\mathbb{P}(Q_n > 0) = \mathbb{E}(A_n)$ where

$$\mathbb{E}(A_n) = \int_0^\infty \mathbb{E}(A_n \mid S = s)\, dF_S(s) = \lambda\mathbb{E}(S) = \rho,$$

and S is a typical service time. Square $(*)$ and take expectations to obtain

$$\mathbb{E}(Q_n) = \frac{\rho(1 - 2\rho) + \mathbb{E}(A_{n+1}^2)}{2(1 - \rho)},$$

where $\mathbb{E}(A_n^2)$ is found (as above) to equal $\rho + \lambda^2\mathbb{E}(S^2)$.

(ii) If a customer waits for time W and is served for time S, he leaves behind him a queue-length which is Poisson with parameter $\lambda(W + S)$. In equilibrium, its mean satisfies $\lambda\mathbb{E}(W + S) = \mathbb{E}(Q_n)$, whence $\mathbb{E}(W)$ is given as claimed.

(iii) $\mathbb{E}(W)$ is a minimum when $\mathbb{E}(S^2)$ is minimized, which occurs when S is concentrated at its mean. Deterministic service times minimize mean waiting time.

7. Condition on arrivals in $(t, t+h)$. If there are no arrivals, then $W_{t+h} \le x$ if and only if $W_t \le x+h$. If there is an arrival, and his service time is S, then $W_{t+h} \le x$ if and only if $W_t \le x+h-S$. Therefore

$$F(x; t+h) = (1 - \lambda h)F(x + h; t) + \lambda h \int_0^{x+h} F(x + h - s; t)\, dF_S(s) + o(h).$$

Subtract $F(x;t)$, divide by h, and take the limit as $h \downarrow 0$, to obtain the differential equation.

We take Laplace–Stieltjes transforms. Integrating by parts, for $\theta \le 0$,

$$\int_{(0,\infty)} e^{\theta x}\,dh(x) = -h(0) - \theta\{M_U(\theta) - H(0)\},$$

$$\int_{(0,\infty)} e^{\theta x}\,dH(x) = M_U(\theta) - H(0),$$

$$\int_{(0,\infty)} e^{\theta x}\,d\mathbb{P}(U+S \le x) = M_U(\theta)M_S(\theta),$$

and therefore

$$0 = -h(0) - \theta\{M_U(\theta) - H(0)\} + \lambda H(0) + \lambda M_U(\theta)\{M_S(\theta) - 1\}.$$

Set $\theta = 0$ to obtain that $h(0) = \lambda H(0)$, and therefore

$$H(0) = -\frac{1}{\theta} M_U(\theta)\big\{\lambda\big(M_S(\theta) - 1\big) - \theta\big\}.$$

Take the limit as $\theta \to 0$, using L'Hôpital's rule, to obtain $H(0) = 1 - \lambda\mathbb{E}(S) = 1 - \rho$. The moment generating function of U is given accordingly. Note that M_U is the same as the moment generating function of the equilibrium distribution of *actual* waiting time. That is to say, *virtual* and *actual* waiting times have the same equilibrium distributions in this case.

8. In this case U takes the values 1 and -2 each with probability $\frac{1}{2}$ (as usual, $U = S - X$ where S and X are typical (independent) service and interarrival times). The integral equation for the limiting waiting time distribution function F becomes

$$F(0) = \tfrac{1}{2}F(2), \qquad F(x) = \tfrac{1}{2}\big\{F(x-1) + F(x+2)\big\} \qquad \text{for } x = 1, 2, \dots.$$

The auxiliary equation is $\theta^3 - 2\theta + 1 = 0$, with roots 1 and $-\frac{1}{2}(1 \pm \sqrt{5})$. Only roots lying in $[-1, 1]$ can contribute, whence

$$F(x) = A + B\left(\frac{-1+\sqrt{5}}{2}\right)^x$$

for some constants A and B. Now $F(x) \to 1$ as $x \to \infty$, since the queue is stable, and therefore $A = 1$. Using the equation for $F(0)$, we find that $B = \frac{1}{2}(1 - \sqrt{5})$.

9. Q is a M(λ)/M(μ)/∞ queue, otherwise known as an immigration–death process (see Exercise (6.11.2) and Problem (6.15.18)). As found in (6.15.18), $Q(t)$ has probability generating function

$$G(s,t) = \big\{1 + (s-1)e^{-\mu t}\big\}^I \exp\big\{\rho(s-1)(1 - e^{-\mu t})\big\}$$

where $\rho = \lambda/\mu$. Hence

$$\mathbb{E}(Q(t)) = Ie^{-\mu t} + \rho(1 - e^{-\mu t}),$$

$$\mathbb{P}(Q(t) = 0) = (1 - e^{-\mu t})^I \exp\big\{-\rho(1 - e^{-\mu t})\big\},$$

$$\mathbb{P}(Q(t) = n) \to \frac{1}{n!}\rho^n e^{-\rho} \qquad \text{as } t \to \infty.$$

If $\mathbb{E}(I)$ and $\mathbb{E}(B)$ denote the mean lengths of an idle period and a busy period in equilibrium, we have that the proportion of time spent idle is $\mathbb{E}(I)/\{\mathbb{E}(I) + \mathbb{E}(B)\}$. This in turn equals

$\lim_{t\to\infty} \mathbb{P}(Q(t) = 0) = e^{-\rho}$. Now $\mathbb{E}(I) = \lambda^{-1}$, by the lack-of-memory property of the arrival process, so that $\mathbb{E}(B) = (e^{\rho} - 1)/\lambda$.

10. We have in the usual way that

$$(*) \qquad Q(t+1) = A_t + Q(t) - \min\{1, Q(t)\}$$

where A_t has the Poisson distribution with parameter λ. When the queue is in equilibrium, $\mathbb{E}(Q(t)) = \mathbb{E}(Q(t+1))$, and hence

$$\mathbb{P}(Q(t) > 0) = \mathbb{E}\big(\min\{1, Q(t)\}\big) = \mathbb{E}(A_t) = \lambda.$$

We have from $(*)$ that the probability generating function $G(s)$ of the equilibrium distribution of $Q(t)$ $(\equiv Q)$ is

$$G(s) = \mathbb{E}(s^{A_t})\mathbb{E}(s^{Q-\min\{1,Q\}}) = e^{\lambda(s-1)}\big\{\mathbb{E}(s^{Q-1}I_{\{Q\geq 1\}}) + \mathbb{P}(Q=0)\big\}.$$

Also,

$$G(s) = \mathbb{E}(s^{Q}I_{\{Q\geq 1\}}) + \mathbb{P}(Q=0),$$

and hence

$$G(s) = e^{\lambda(s-1)}\left\{\frac{1}{s}G(s) + \left(1 - \frac{1}{s}\right)(1-\lambda)\right\}$$

whence

$$G(s) = \frac{(1-s)(1-\lambda)}{1 - se^{-\lambda(s-1)}}.$$

The mean queue length is $G'(1) = \frac{1}{2}\lambda(2-\lambda)/(1-\lambda)$. Since service times are of unit length, and arrivals form a Poisson process, the mean residual service time of the customer in service at an arrival time is $\frac{1}{2}$, so long as the queue is non-empty. Hence

$$\mathbb{E}(W) = \mathbb{E}(Q) - \tfrac{1}{2}\mathbb{P}(Q > 0) = \frac{\lambda}{2(1-\lambda)}.$$

11. The length B of a typical busy period has moment generating function satisfying $M_B(s) = \exp\{s - \lambda + \lambda M_B(s)\}$; this fact may be deduced from the standard theory of M/G/1, or alternatively by a random-walk approach. Now T may be expressed as $T = I + B$ where I is the length of the first idle period, a random variable with the exponential distribution, parameter λ. It follows that $M_T(s) = \lambda M_B(s)/(\lambda - s)$. Therefore, as required,

$$(*) \qquad (\lambda - s)M_T(s) = \lambda \exp\big\{s - \lambda + (\lambda - s)M_T(s)\big\}.$$

If $\lambda \geq 1$, the queue-length at moments of departure is either null recurrent or transient, and it follows that $\mathbb{E}(T) = \infty$. If $\lambda < 1$, we differentiate $(*)$ and set $s = 0$ to obtain $\lambda\mathbb{E}(T) - 1 = \lambda^2\mathbb{E}(T)$, whence $\mathbb{E}(T) = \{\lambda(1-\lambda)\}^{-1}$.

12. (a) Q is a birth–death process with parameters $\lambda_i = \lambda$, $\mu_i = \mu$, and is therefore reversible in equilibrium; see Problems (6.15.16) and (11.8.3).

(b) The equilibrium distribution satisfies $\lambda\pi_i = \mu\pi_{i+1}$ for $i \geq 0$, whence $\pi_i = (1-\rho)\rho^i$ where $\rho = \lambda/\mu$. A typical waiting time W is the sum of Q independent service times, so that

$$M_W(s) = G_Q\big(M_S(s)\big) = \frac{1-\rho}{1 - \rho\mu/(\mu - s)} = \frac{(1-\rho)(\mu - s)}{\mu(1-\rho) - s}.$$

(c) See the solution to Problem (11.8.3).

(d) Follow the solution to Problem (11.8.3) (either method) to find that, at any time t in equilibrium, the queue lengths are independent, the jth having the equilibrium distribution of M(λ)/M(μ_j)/1. The joint mass function is therefore

$$f(x_1, x_2, \dots, x_K) = \prod_{j=1}^{K} (1-\rho_j)\rho_j^{x_j}$$

where $\rho_j = \lambda/\mu_j$.

13. The size of the queue is a birth–death process with rates $\lambda_i = \lambda$, $\mu_i = \mu \min\{i, k\}$. Either solve the equilibrium equations in order to find a stationary distribution $\boldsymbol{\pi}$, or argue as follows. The process is reversible in equilibrium (see Problem (6.15.16)), and therefore $\lambda_i \pi_i = \mu_{i+1}\pi_{i+1}$ for all i. These 'balance equations' become

$$\lambda \pi_i = \begin{cases} \mu(i+1)\pi_{i+1} & \text{if } 0 \le i < k, \\ \mu k \pi_{i+1} & \text{if } i \ge k. \end{cases}$$

These are easily solved iteratively to obtain

$$\pi_i = \begin{cases} \pi_0 \alpha^i / i! & \text{if } 0 \le i \le k, \\ \pi_0 (\alpha/k)^i k^k / k! & \text{if } i \ge k \end{cases}$$

where $\alpha = \lambda/\mu$. Therefore there exists a stationary distribution if and only if $\lambda < k\mu$, and it is given accordingly, with

$$\pi_0^{-1} = \sum_{i=0}^{k-1} \frac{\alpha^i}{i!} + \frac{k^k}{k!} \sum_{i=k}^{\infty} (\alpha/k)^i.$$

The cost of having k servers is

$$C_k = Ak + B\pi_0 \sum_{i=k}^{\infty} (i-k+1) \frac{(\alpha/k)^i k^k}{k!}$$

where $\pi_0 = \pi_0(k)$. One finds, after a little computation, that

$$C_1 = A + \frac{B\alpha}{1-\alpha}, \qquad C_2 = 2A + \frac{2B\alpha^2}{4-\alpha^2}.$$

Therefore

$$C_2 - C_1 = \frac{\alpha^3(A-B) + \alpha^2(2B-A) - 4\alpha(A+B) + 4A}{(1-\alpha)(4-\alpha^2)}.$$

Viewed as a function of α, the numerator is a cubic taking the value $4A$ at $\alpha = 0$ and the value $-3B$ at $\alpha = 1$. This cubic has a unique zero at some $\alpha^* \in (0, 1)$, and $C_1 < C_2$ if and only if $0 < \alpha < \alpha^*$.

14. The state of the system is the number $Q(t)$ of customers within it at time t. The state 1 may be divided into two sub-states, being σ_1 and σ_2, where σ_i is the state in which server i is occupied but the other server is not. The state space is therefore $S = \{0, \sigma_1, \sigma_2, 2, 3, \dots\}$.

The usual way of finding the stationary distribution, when it exists, is to solve the equilibrium equations. An alternative is to argue as follows. If there exists a stationary distribution, then the process is reversible in equilibrium if and only if

$$(*) \qquad g_{i_1,i_2} g_{i_2,i_3} \cdots g_{i_k,i_1} = g_{i_1,i_k} g_{i_k,i_{k-1}} \cdots g_{i_2,i_1}$$

for all sequences $i_1, i_2, \ldots, i_k$ of states, where $\mathbf{G} = (g_{uv})_{u,v \in S}$ is the generator of the process (this may be shown in very much the same way as was the corresponding claim for discrete-time chains in Exercise (6.5.2a); see also Problem (6.15.16)). It is clear that $(*)$ is satisfied by this process for all sequences of states which do not include both σ_1 and σ_2; this holds since the terms g_{uv} are exactly those of a birth–death process in such a case. In order to see that $(*)$ holds for a sequence containing both σ_1 and σ_2, it suffices to perform the following calculation:

$$g_{0,\sigma_1} g_{\sigma_1,2} g_{2,\sigma_2} g_{\sigma_2,0} = (\tfrac{1}{2}\lambda)\lambda\mu_2\mu_1 = g_{0,\sigma_2} g_{\sigma_2,2} g_{2,\sigma_1} g_{\sigma_1,0}.$$

Since the process is reversible in equilibrium, the stationary distribution $\boldsymbol{\pi}$ satisfies $\pi_u g_{uv} = \pi_v g_{vu}$ for all $u, v \in S$, $u \neq v$. Therefore

$$\pi_u \lambda = \pi_{u+1}(\mu_1 + \mu_2), \qquad u \geq 2,$$
$$\tfrac{1}{2}\pi_0\lambda = \pi_{\sigma_1}\mu_1 = \pi_{\sigma_2}\mu_2, \quad \pi_{\sigma_1}\lambda = \pi_2\mu_2, \quad \pi_{\sigma_2}\lambda = \pi_2\mu_1,$$

and hence

$$\pi_{\sigma_1} = \frac{\lambda}{2\mu_1}\pi_0, \quad \pi_{\sigma_2} = \frac{\lambda}{2\mu_2}\pi_0, \quad \pi_u = \frac{\lambda^2}{2\mu_1\mu_2}\left(\frac{\lambda}{\mu_1+\mu_2}\right)^{u-2}\pi_0 \quad \text{for } u \geq 2.$$

This gives a stationary distribution if and only if $\lambda < \mu_1 + \mu_2$, under which assumption π_0 is easily calculated.

A similar analysis is valid if there are s servers and an arriving customer is equally likely to go to any free server, otherwise waiting in turn. This process also is reversible in equilibrium, and the stationary distribution is similar to that given above.

15. We have from the standard theory that Q_μ has as mass function $\pi_j = (1-\eta)\eta^j$, $j \geq 0$, where η is the smallest positive root of the equation $x = e^{\mu(x-1)}$. The moment generating function of $(1-\mu^{-1})Q_\mu$ is

$$M_\mu(\theta) = \mathbb{E}\big(\exp\{\theta(1-\mu^{-1})Q_\mu\}\big) = \frac{1-\eta}{1-\eta e^{\theta(1-\mu^{-1})}}.$$

Writing $\mu = 1+\epsilon$, we have by expanding $e^{\mu(\eta-1)}$ as a Taylor series that $\eta = \eta(\epsilon) = 1 - 2\epsilon + o(\epsilon)$ as $\epsilon \downarrow 0$. This gives

$$M_\mu(\theta) = \frac{2\epsilon + o(\epsilon)}{1-(1-2\epsilon)(1+\theta\epsilon)+o(\epsilon)} = \frac{2\epsilon+o(\epsilon)}{(2-\theta)\epsilon + o(\epsilon)} \to \frac{2}{2-\theta}$$

as $\epsilon \downarrow 0$, implying the result, by the continuity theorem.

16. The numbers P (of passengers) and T (of taxis) up to time t have the Poisson distribution with respective parameters πt and τt. The required probabilities $p_n = \mathbb{P}(P = T+n)$ have generating function

$$\begin{aligned}\sum_{n=-\infty}^{\infty} p_n z^n &= \sum_{n=-\infty}^{\infty}\sum_{m=0}^{\infty} \mathbb{P}(P = m+n)\mathbb{P}(T=m)z^n \\ &= \sum_{m=0}^{\infty} \mathbb{P}(T=m)z^{-m}G_P(z) \\ &= G_T(z^{-1})G_P(z) = e^{-(\pi+\tau)t}e^{(\pi z + \tau z^{-1})t},\end{aligned}$$

in which the coefficient of z^n is easily found to be that given.

17. Let $N(t)$ be the number of machines which have arrived by time t. Given that $N(t) = n$, the times $T_1, T_2, \dots, T_n$ of their arrivals may be thought of as the order statistics of a family of independent uniform variables on $[0, t]$, say $U_1, U_2, \dots, U_n$; see Theorem (6.8.11). The machine which arrived at time U_i is, at time t,

$$\left.\begin{array}{r}\text{in the } X\text{-stage}\\ \text{in the } Y\text{-stage}\\ \text{repaired}\end{array}\right\} \text{ with probability } \left\{\begin{array}{l}\alpha(t)\\ \beta(t)\\ 1-\alpha(t)-\beta(t)\end{array}\right.$$

where $\alpha(t) = \mathbb{P}(U + X > t)$ and $\beta(t) = \mathbb{P}(U + X \le t < U + X + Y)$, where U is uniform on $[0, t]$, and (X, Y) is a typical repair pair, independent of U. Therefore

$$\mathbb{P}\big(U(t) = j, V(t) = k \,\big|\, N(t) = n\big) = \frac{n!\,\alpha(t)^j \beta(t)^k (1-\alpha(t)-\beta(t))^{n-k-j}}{j!\,k!\,(n-j-k)!},$$

implying that

$$\begin{aligned}\mathbb{P}\big(U(t) = j, V(t) = k\big) &= \sum_{n=0}^{\infty} \frac{e^{-\lambda t}(\lambda t)^n}{n!} \mathbb{P}\big(U(t) = j, V(t) = k \,\big|\, N(t) = n\big)\\ &= \frac{\{\lambda t\alpha(t)\}^j e^{-\lambda t\alpha(t)}}{j!} \cdot \frac{\{\lambda t\beta(t)\}^k e^{-\lambda t\beta(t)}}{k!}.\end{aligned}$$

18. The maximum deficit M_n seen up to and including the time of the nth claim satisfies

$$M_n = \max\left\{M_{n-1}, \sum_{j=1}^{n}(K_j - X_j)\right\} = \max\{0, U_1, U_1 + U_2, \dots, U_1 + U_2 + \cdots + U_n\},$$

where the X_j are the inter-claim times, and $U_j = K_j - X_j$. We have as in the analysis of G/G/1 that M_n has the same distribution as $V_n = \max\{0, U_n, U_n + U_{n-1}, \dots, U_n + U_{n-1} + \cdots + U_1\}$, whence M_n has the same distribution as the $(n+1)$th waiting time in a M(λ)/G/1 queue with service times K_j and interarrival times X_j. The result follows by Theorem (11.3.16).

19. (a) Look for a solution to the detailed balance equations $\lambda\pi_i = (i+1)\mu\pi_{i+1}$, $0 \le i < s$, to find that the stationary distribution is given by $\pi_i = (\rho^i/i!)\pi_0$.
(b) Let p_c be the required fraction. We have by Little's theorem (10.5.18) that

$$p_c = \frac{\lambda(\pi_{c-1} - \pi_c)}{\mu} = \rho(\pi_{c-1} - \pi_c), \quad c \ge 2,$$

and $p_1 = \pi_1$, where π_s is the probability that channels $1, 2, \dots, s$ are busy in a queue M/M/s having the property that further calls are lost when all s servers are occupied.

20. The equilibrium distribution is $\pi_n = \rho^n(1-\rho)$, where $\rho = \lambda/\mu$. Let T be the time until the queue is first empty. In particular, $T = 0$ with probability π_0. While the queue is non-empty, its length evolves in the manner of a random walk on the non-negative integers that steps one step rightwards with probability $p = \lambda/(\lambda+\mu) = \rho/(1+\rho) < \frac{1}{2}$ and one step leftwards otherwise. Each step occupies a holding time with mean $1/(\lambda+\mu)$.

Let K be the number of jumps to pass from position 1 to position 0. By conditioning on the first step, we have that $\mathbb{E}(K) = 1 + 2p\mathbb{E}(K)$, whence $\mathbb{E}(K) = 1/(1-2p)$. Therefore, the mean number of steps to pass from position $n \ge 0$ to position 0 is $n\mathbb{E}(K) = n/(1-2p)$. Hence,

$$\mathbb{E}(T) = \frac{1}{\lambda+\mu}\sum_{n=1}^{\infty} \frac{n\pi_n}{1-2p} = \frac{\lambda}{(\mu-\lambda)^2}.$$

21. Since busy periods are equidistributed for this queue as for the queue in equilibrium, we may assume the queue is in equilibrium. By Exercise (6.12.4), the number of customers being served at any fixed time has the Poisson distribution with mean ρ. In particular, the probability that the queue is empty is $e^{-\rho}$.

Let a cycle commence at the end of every busy period, so that a cycle comprises an idle period I plus a busy period B, and the reward of a cycle is taken as the length of the idle period. Note that $\mathbb{E}(I) = 1/\lambda$. By the lack-of-memory property of the exponential distribution, we may apply the renewal–reward theorem to find that the density of idle periods is $\lambda^{-1}/(\mathbb{E}(B) + \lambda^{-1})$. Using the first observation above,

$$e^{-\rho} = \frac{\lambda^{-1}}{\mathbb{E}(B) + \lambda^{-1}},$$

and the given expression for $\mathbb{E}(B)$ follows.

12

Martingales

12.1 Solutions. Introduction

1. (i) We have that $\mathbb{E}(Y_m) = \mathbb{E}\{\mathbb{E}(Y_{m+1} \mid \mathcal{F}_m)\} = \mathbb{E}(Y_{m+1})$, and the result follows by induction.
(ii) For a submartingale, $\mathbb{E}(Y_m) \le \mathbb{E}\{\mathbb{E}(Y_{m+1} \mid \mathcal{F}_m)\} = \mathbb{E}(Y_{m+1})$, and the result for supermartingales follows similarly.

2. We have that

$$\mathbb{E}(Y_{n+m} \mid \mathcal{F}_n) = \mathbb{E}\big\{\mathbb{E}(Y_{n+m} \mid \mathcal{F}_{n+m-1}) \,\big|\, \mathcal{F}_n\big\} = \mathbb{E}(Y_{n+m-1} \mid \mathcal{F}_n)$$

if $m \ge 1$, since $\mathcal{F}_n \subseteq \mathcal{F}_{n+m-1}$. Iterate to obtain $\mathbb{E}(Y_{n+m} \mid \mathcal{F}_n) = \mathbb{E}(Y_n \mid \mathcal{F}_n) = Y_n$.

3. (i) $Z_n\mu^{-n}$ has mean 1, and

$$\mathbb{E}\big(Z_{n+1}\mu^{-(n+1)} \,\big|\, \mathcal{F}_n\big) = \mu^{-(n+1)}\mathbb{E}(Z_{n+1} \mid \mathcal{F}_n) = \mu^{-n}Z_n,$$

where $\mathcal{F}_n = \sigma(Z_1, Z_2, \dots, Z_n)$.
(ii) Certainly $\eta^{Z_n} \le 1$, and therefore it has finite mean. Also,

$$\mathbb{E}\big(\eta^{Z_{n+1}} \,\big|\, \mathcal{F}_n\big) = \mathbb{E}\left(\eta^{\sum_1^{Z_n} X_i} \,\middle|\, \mathcal{F}_n\right) = G(\eta)^{Z_n}$$

where the X_i are independent family sizes with probability generating function G. Now $G(\eta) = \eta$, and the claim follows.

4. (i) With X_n denoting the size of the nth jump,

$$\mathbb{E}(S_{n+1} \mid \mathcal{F}_n) = S_n + \mathbb{E}(X_{n+1} \mid \mathcal{F}_n) = S_n$$

where $\mathcal{F}_n = \sigma(X_1, X_2, \dots, X_n)$. Also $\mathbb{E}|S_n| \le n$, so that $\{S_n\}$ is a martingale.
(ii) Similarly $\mathbb{E}(S_n^2) = \text{var}(S_n) = n$, and

$$\mathbb{E}\big(S_{n+1}^2 - (n+1) \,\big|\, \mathcal{F}_n\big) = S_n^2 + \mathbb{E}(X_{n+1}^2) + 2S_n\mathbb{E}(X_{n+1}) - (n+1) = S_n^2 - n.$$

(iii) Suppose the walk starts at k, and there are absorbing barriers at 0 and N ($\ge k$). Let T be the time at which the walk is absorbed, and make the assumptions that $\mathbb{E}(S_T) = S_0$, $\mathbb{E}(S_T^2 - T) = S_0^2$. Then the probability p_k of ultimate ruin satisfies

$$0 \cdot p_k + N \cdot (1 - p_k) = k, \qquad 0 \cdot p_k + N^2 \cdot (1 - p_k) - \mathbb{E}(T) = k^2,$$

and therefore $p_k = 1 - (k/N)$ and $\mathbb{E}(T) = k(N-k)$.

5. (i) By Exercise (12.1.2), for $r \geq i$,

$$\mathbb{E}(Y_r Y_i) = \mathbb{E}\{\mathbb{E}(Y_r Y_i \mid \mathcal{F}_i)\} = \mathbb{E}\{Y_i \mathbb{E}(Y_r \mid \mathcal{F}_i)\} = \mathbb{E}(Y_i^2),$$

an answer which is independent of r. Therefore

$$\mathbb{E}\{(Y_k - Y_j)Y_i\} = \mathbb{E}(Y_k Y_i) - \mathbb{E}(Y_j Y_i) = 0 \qquad \text{if } i \leq j \leq k.$$

(ii) We have that

$$\mathbb{E}\{(Y_k - Y_j)^2 \mid \mathcal{F}_i\} = \mathbb{E}(Y_k^2 \mid \mathcal{F}_i) - 2\mathbb{E}(Y_k Y_j \mid \mathcal{F}_i) + \mathbb{E}(Y_j^2 \mid \mathcal{F}_i).$$

Now $\mathbb{E}(Y_k Y_j \mid \mathcal{F}_i) = \mathbb{E}\{\mathbb{E}(Y_k Y_j \mid \mathcal{F}_j) \mid \mathcal{F}_i\} = \mathbb{E}(Y_j^2 \mid \mathcal{F}_i)$, and the claim follows.
(iii) Taking expectations of the last conclusion,

$$(*) \qquad 0 \leq \mathbb{E}\{(Y_k - Y_j)^2\} = \mathbb{E}(Y_k^2) - \mathbb{E}(Y_j^2), \qquad j \leq k.$$

Now $\{\mathbb{E}(Y_n^2) : n \geq 1\}$ is non-decreasing and bounded, and therefore converges. Therefore, by $(*)$, $\{Y_n : n \geq 1\}$ is Cauchy convergent in mean square, and therefore convergent in mean square, by Problem (7.11.11).

6. (i) Using Jensen's inequality (Exercise (7.9.4)),

$$\mathbb{E}\big(u(Y_{n+1}) \mid \mathcal{F}_n\big) \geq u\big(\mathbb{E}(Y_{n+1} \mid \mathcal{F}_n)\big) = u(Y_n).$$

(ii) It suffices to note that $|x|$, x^2, and x^+ are convex functions of x; draw pictures if you are in doubt about these functions.

7. (i) This follows just as in Exercise (12.1.6), using the fact that $u\{\mathbb{E}(Y_{n+1} \mid \mathcal{F}_n)\} \geq u(Y_n)$ in this case.
(ii) The function x^+ is convex and non-decreasing. Finally, let $\{S_n : n \geq 0\}$ be a simple random walk whose steps are $+1$ with probability $p\ (= 1 - q > \frac{1}{2})$ and -1 otherwise. If $S_n < 0$, then

$$\mathbb{E}\left(|S_{n+1}| \mid \mathcal{F}_n\right) = p(|S_n| - 1) + q(|S_n| + 1) = |S_n| - (p - q) < |S_n|;$$

note that $\mathbb{P}(S_n < 0) > 0$ if $n \geq 1$. The same example suffices in the remaining case.

8. Clearly $\mathbb{E}|\lambda^{-n}\psi(X_n)| \leq \lambda^{-n}\sup\{|\psi(j)| : j \in S\}$. Also,

$$\mathbb{E}\big(\psi(X_{n+1}) \mid \mathcal{F}_n\big) = \sum_{j \in S} p_{X_n, j}\psi(j) \leq \lambda\psi(X_n)$$

where $\mathcal{F}_n = \sigma(X_1, X_2, \ldots, X_n)$. Divide by λ^{n+1} to obtain that the given sequence is a supermartingale.

9. Since $\text{var}(Z_1) > 0$, the function G, and hence also G_n, is a strictly increasing function on $[0, 1]$. Since $s = G_{n+1}(H_{n+1}(s)) = G_n(G(H_{n+1}(s)))$ and $G_n(H_n(s)) = s$, we have that $G(H_{n+1}(s)) = H_n(s)$. With $\mathcal{F}_m = \sigma(Z_k : 0 \leq k \leq m)$,

$$\mathbb{E}\big(H_{n+1}(s)^{Z_{n+1}} \mid \mathcal{F}_n\big) = G(H_{n+1}(s))^{Z_n} = H_n(s)^{Z_n}.$$

12.2 Solutions. Martingale differences and Hoeffding's inequality

1. Let $\mathcal{F}_i = \sigma(\{V_j, W_j : 1 \le j \le i\})$ and $Y_i = \mathbb{E}(Z \mid \mathcal{F}_i)$. With $Z(j)$ the maximal worth attainable without using the jth object, we have that

$$\mathbb{E}\big(Z(j) \,\big|\, \mathcal{F}_j\big) = \mathbb{E}\big(Z(j) \,\big|\, \mathcal{F}_{j-1}\big), \qquad Z(j) \le Z \le Z(j) + M.$$

Take conditional expectations of the second inequality, given $\mathcal{F}_j$ and given $\mathcal{F}_{j-1}$, and deduce that $|Y_j - Y_{j-1}| \le M$. Therefore Y is a martingale with bounded differences, and Hoeffding's inequality yields the result.

2. Let $\mathcal{F}_i$ be the σ-field generated by the (random) edges joining pairs (v_a, v_b) with $1 \le a, b \le i$, and let $\chi_i = \mathbb{E}(\chi \mid \mathcal{F}_i)$. We write $\chi(j)$ for the minimal number of colours required in order to colour each vertex in the graph obtained by deleting v_j. The argument now follows that of the last exercise, using the fact that $\chi(j) \le \chi \le \chi(j) + 1$.

3. (a) We shall use the inequalities $e^{-x} \le 1 - x + \frac{1}{2}x^2$ for $x \ge 0$, and $1 + x \le e^x$ for $x \in \mathbb{R}$. For $\theta \ge 0$,

$$\begin{aligned}
\mathbb{E}(e^{\theta X} \mid Y) &= e^{b\theta}\mathbb{E}(e^{-\theta(b-X)} \mid Y) \\
&\le e^{b\theta}\left[1 - \theta\mathbb{E}(b - X \mid Y) + \tfrac{1}{2}\theta^2\mathbb{E}((b-X)^2 \mid Y)\right] \\
&\le e^{b\theta}\exp\{-b\theta + \theta\mathbb{E}(X \mid Y) + \tfrac{1}{2}\theta^2\mathbb{E}((b-X)^2 \mid Y)\} \\
&\le \exp\{\tfrac{1}{2}\theta^2(b^2 + \sigma^2)\}.
\end{aligned}$$

(b) One may extend the proof of Hoeffding's inequality, Theorem (12.2.3). Alternatively, one may use induction on n. For $t, \theta \ge 0$, by part (a),

$$(*) \qquad \mathbb{P}(M_1 \ge t) \le e^{-\theta t}\mathbb{E}(e^{\theta D_1}) \le e^{-\theta t}\exp\{\tfrac{1}{2}\theta^2(b_1^2 + \sigma_1^2)\},$$

which is minimized by the choice $\theta = t/(b_1^2 + \sigma_1^2)$.

Let $\theta \ge 0$. We shall prove by induction that, for $n \ge 1$,

$$\mathrm{H}_n: \qquad \mathbb{E}(e^{\theta M_n}) \le \exp\left\{\tfrac{1}{2}\theta^2\sum_{r=1}^{n}(b_r^2 + \sigma_r^2)\right\}.$$

Trivially, H_0 holds. Assume that H_n holds for some $n \ge 0$. Then

$$\mathbb{E}(e^{\theta M_{n+1}} \mid M_n) = e^{\theta M_n}\mathbb{E}(e^{\theta D_{n+1}} \mid M_n),$$

so that

$$\mathbb{E}(e^{\theta M_{n+1}}) \le \exp\left\{\tfrac{1}{2}\theta^2\sum_{r=1}^{n}(b_r^2 + \sigma_r^2)\right\}\exp\{\tfrac{1}{2}\theta^2(b_{n+1}^2 + \sigma_{n+1}^2)\},$$

by H_n and part (a). H_{n+1} follows. The argument is completed as in (*).

4. Note that, for $r < s$,

$$\mathbb{E}(D_r D_s) = \mathbb{E}\big(\mathbb{E}(D_r D_s \mid \mathcal{F}_{s-1})\big) = \mathbb{E}\big(D_r\mathbb{E}(D_s \mid \mathcal{F}_{s-1})\big) = 0,$$

by the pull-through property and the definition of a martingale. Therefore,

$$X_n = M_n^2 - Q_n = 2\sum_{1 \le r < s \le n} D_r D_s$$

has zero mean and satisfies $\mathbb{E}(X_n \mid \mathcal{F}_{n-1}) = X_{n-1}$, as required. The martingale property follows likewise for (Y_n) on noting that

$$\mathbb{E}\big(\mathbb{E}(D_r^2 \mid \mathcal{F}_{r-1}) \,\big|\, \mathcal{F}_{n-1}\big) = \mathbb{E}(D_r^2 \mid \mathcal{F}_{r-1}) \qquad \text{if } r < n.$$

12.3 Solutions. Crossings and convergence

1. Let $T_1 = \min\{n : Y_n \geq b\}$, $T_2 = \min\{n > T_1 : Y_n \leq a\}$, and define T_k inductively by

$$T_{2k-1} = \min\{n > T_{2k-2} : Y_n \geq b\}, \quad T_{2k} = \min\{n > T_{2k-1} : Y_n \leq a\}.$$

The number of downcrossings by time n is $D_n(a, b; Y) = \max\{k : T_{2k} \leq n\}$.
(a) Between each pair of upcrossings of $[a, b]$, there must be a downcrossing, and *vice versa*. Hence $|D_n(a, b; Y) - U_n(a, b; Y)| \leq 1$.
(b) Let I_i be the indicator function of the event that $i \in (T_{2k-1}, T_{2k}]$ for some k, and let

$$Z_n = \sum_{i=1}^{n} I_i(Y_i - Y_{i-1}), \qquad n \geq 0.$$

It is easily seen that

$$Z_n \leq -(b - a)D_n(a, b; Y) + (Y_n - b)^+,$$

whence

$$(*) \qquad (b - a)\mathbb{E}D_n(a, b; Y) \leq \mathbb{E}\{(Y_n - b)^+\} - \mathbb{E}(Z_n).$$

Now I_i is $\mathcal{F}_{i-1}$-measurable, since

$$\{I_i = 1\} = \bigcup_k \big(\{T_{2k-1} \leq i - 1\} \setminus \{T_{2k} \leq i - 1\}\big).$$

Therefore,

$$\mathbb{E}(Z_n - Z_{n-1}) = \mathbb{E}\{\mathbb{E}(I_n(Y_n - Y_{n-1}) \mid \mathcal{F}_{n-1})\} = \mathbb{E}\{I_n(\mathbb{E}(Y_n \mid \mathcal{F}_{n-1}) - Y_{n-1})\} \geq 0$$

since $I_n \geq 0$ and Y is a submartingale. It follows that $\mathbb{E}(Z_n) \geq \mathbb{E}(Z_{n-1}) \geq \cdots \geq \mathbb{E}(Z_0) = 0$, and the final inequality follows from $(*)$.

2. If Y is a supermartingale, then $-Y$ is a submartingale. Upcrossings of $[a, b]$ by Y correspond to downcrossings of $[-b, -a]$ by $-Y$, so that

$$\mathbb{E}U_n(a, b; Y) = \mathbb{E}D_n(-b, -a; -Y) \leq \frac{\mathbb{E}\{(-Y_n + a)^+\}}{b - a} = \frac{\mathbb{E}\{(Y_n - a)^-\}}{b - a},$$

by Exercise (12.3.1). If $a, Y_n \geq 0$ then $(Y_n - a)^- \leq a$.

3. The random sequence $\{\psi(X_n) : n \geq 1\}$ is a bounded supermartingale, which converges a.s. to some limit Y. The chain is irreducible and recurrent, so that each state is visited infinitely often a.s.; it follows that $\lim_{n\to\infty} \psi(X_n)$ cannot exist (a.s.) unless ψ is a constant function.

4. The sequence Y is a martingale since Y_n is the sum of independent variables with zero means. Also $\sum_1^\infty \mathbb{P}(Z_n \neq 0) = \sum_1^\infty n^{-2} < \infty$, implying by the Borel–Cantelli lemma that $Z_n = 0$ except for finitely many values of n (a.s.); therefore the partial sum Y_n converges a.s. as $n \to \infty$ to some finite limit.

It is easily seen that $a_n = 5a_{n-1}$ and therefore $a_n = 8 \cdot 5^{n-2}$, if $n \geq 3$. It follows that $|Y_n| \geq \frac{1}{2}a_n$ if and only if $|Z_n| = a_n$. Therefore

$$\mathbb{E}|Y_n| \geq \tfrac{1}{2}a_n\mathbb{P}(|Y_n| \geq \tfrac{1}{2}a_n) = \tfrac{1}{2}a_n\mathbb{P}(|Z_n| = a_n) = \frac{a_n}{2n^2}$$

which tends to infinity as $n \to \infty$.

5. Take $\mathcal{F}_n = \sigma(X_1, X_2, \dots, X_n)$. For $n \ge r$,

$$\begin{aligned}\mathbb{E}(M_{n+1} \mid \mathcal{F}_n) &= \frac{1}{(n+1)(n+2)}\mathbb{E}(S_n + X_{n+1} \mid \mathcal{F}_n) \\ &= \frac{1}{(n+1)(n+2)}\left[S_n + \frac{2}{n}(X_1 + \cdots + X_n)\right] = \frac{S_n}{n(n+1)} = M_n,\end{aligned}$$

and

$$\mathbb{E}(M_n) = \mathbb{E}(M_r) = \frac{1}{r(r+1)}(x_1 + x_2 + \cdots + x_r).$$

Further details of random adding may be found in Problem (3.11.42), and in Clifford and Stirzaker 2019.

6. Clearly, $\mathbb{E}|M_n| < \infty$. Furthermore, with $\mathcal{F}_n = \sigma(B_m, R_m : m \le n)$,

$$\begin{aligned}\mathbb{E}(M_{n+1} \mid \mathcal{F}_n) &= \frac{B_n}{B_n + R_n}(B_n - R_n - 1)(B_n + R_n) + \frac{R_n}{B_n + R_n}(B_n - R_n + 1)(B_n + R_n) \\ &= M_n.\end{aligned}$$

Now $B_n + R_n - 1 = n \to \infty$, and $B_n - R_n \ne 0$ infinitely often by construction, so almost-sure convergence is impossible.

12.4 Solutions. Stopping times

1. We have that

$$\begin{aligned}\{T_1 + T_2 = n\} &= \bigcup_{k=0}^{n} \big(\{T_1 = k\} \cap \{T_2 = n - k\}\big), \\ \big\{\max\{T_1, T_2\} \le n\big\} &= \{T_1 \le n\} \cap \{T_2 \le n\}, \\ \big\{\min\{T_1, T_2\} \le n\big\} &= \{T_1 \le n\} \cup \{T_2 \le n\}.\end{aligned}$$

Each event on the right-hand side lies in $\mathcal{F}_n$.

2. Let $\mathcal{F}_n = \sigma(X_1, X_2, \dots, X_n)$ and $S_n = X_1 + X_2 + \cdots + X_n$. Now

$$\{N(t) + 1 = n\} = \{S_{n-1} \le t\} \cap \{S_n > t\} \in \mathcal{F}_n.$$

3. $(Y^+, \boldsymbol{\mathcal{F}})$ is a submartingale, and $T = \min\{k : Y_k \ge x\}$ is a stopping time. Now $0 \le T \wedge n \le n$, so that $\mathbb{E}(Y_0^+) \le \mathbb{E}(Y_{T\wedge n}^+) \le \mathbb{E}(Y_n^+)$, whence

$$\mathbb{E}(Y_n^+) \ge \mathbb{E}\big(Y_{T\wedge n}^+ I_{\{T \le n\}}\big) \ge x\mathbb{P}(T \le n).$$

4. We may suppose that $\mathbb{E}(Y_0) < \infty$. With the notation of the previous solution, we have that

$$\mathbb{E}(Y_0) \ge \mathbb{E}\big(Y_{T\wedge n}\big) \ge \mathbb{E}\big(Y_{T\wedge n} I_{\{T \le n\}}\big) \ge x\mathbb{P}(T \le n).$$

5. It suffices to prove that $\mathbb{E}Y_S \le \mathbb{E}Y_T$, since the other inequalities are of the same form but with different choices of pairs of stopping times. Let I_m be the indicator function of the event $\{S < m \le T\}$, and define

$$Z_n = \sum_{m=1}^{n} I_m(Y_m - Y_{m-1}), \qquad 0 \le n \le N.$$

Note that I_m is $\mathcal{F}_{m-1}$-measurable, so that

$$\mathbb{E}(Z_n - Z_{n-1}) = \mathbb{E}\{I_n \mathbb{E}(Y_n - Y_{n-1} \mid \mathcal{F}_{n-1})\} \ge 0,$$

since Y is a submartingale. Therefore $\mathbb{E}(Z_N) \ge \mathbb{E}(Z_{N-1}) \ge \cdots \ge \mathbb{E}(Z_0) = 0$. On the other hand, $Z_N = Y_T - Y_S$, and therefore $\mathbb{E}(Y_T) \ge \mathbb{E}(Y_S)$.

6. De Moivre's martingale is $Y_n = (q/p)^{S_n}$, where $q = 1 - p$. Now $Y_n \ge 0$, and $\mathbb{E}(Y_0) = 1$, and the maximal inequality gives that

$$\mathbb{P}\left(\max_{0 \le m \le n} S_m \ge x\right) = \mathbb{P}\left(\max_{0 \le m \le n} Y_m \ge (q/p)^x\right) \le (p/q)^x.$$

Take the limit as $n \to \infty$ to find that $S_\infty = \sup_m S_m$ satisfies

$$(*) \qquad \mathbb{E}(S_\infty) = \sum_{x=1}^{\infty} \mathbb{P}(S_\infty \ge x) \le \frac{p}{q-p}.$$

We can calculate $\mathbb{E}(S_\infty)$ exactly as follows. It is the case that $S_\infty \ge x$ if and only if the walk ever visits the point x, an event with probability f^x for $x \ge 0$, where $f = p/q$ (see Exercise (5.3.1)). The inequality of $(*)$ may be replaced by equality.

7. (a) First, $\varnothing \cap \{T \le n\} = \varnothing \in \mathcal{F}_n$. Secondly, if $A \cap \{T \le n\} \in \mathcal{F}_n$ then

$$A^{c} \cap \{T \le n\} = \{T \le n\} \setminus \big(A \cap \{T \le n\}\big) \in \mathcal{F}_n.$$

Thirdly, if $A_1, A_2, \ldots$ satisfy $A_i \cap \{T \le n\} \in \mathcal{F}_n$ for each i, then

$$\left(\bigcup_i A_i\right) \cap \{T \le n\} = \bigcup_i \big(A_i \cap \{T \le n\}\big) \in \mathcal{F}_n.$$

Therefore $\mathcal{F}_T$ is a σ-field.

For each integer m, it is the case that

$$\{T \le m\} \cap \{T \le n\} = \begin{cases} \{T \le n\} & \text{if } m > n, \\ \{T \le m\} & \text{if } m \le n, \end{cases}$$

an event lying in $\mathcal{F}_n$. Therefore $\{T \le m\} \in \mathcal{F}_T$ for all m.

(b) Let $A \in \mathcal{F}_S$. Then, for any n,

$$\big(A \cap \{S \le T\}\big) \cap \{T \le n\} = \bigcup_{m=0}^{n} \big(A \cap \{S \le m\}\big) \cap \{T = m\},$$

the union of events in $\mathcal{F}_n$, which therefore lies in $\mathcal{F}_n$. Hence $A \cap \{S \le T\} \in \mathcal{F}_T$.

(c) We have $\{S \le T\} = \Omega$, and (b) implies that $A \in \mathcal{F}_T$ whenever $A \in \mathcal{F}_S$.

8. (a) For any vector $\mathbf{x} \in \mathbb{R}^r$,

$$\begin{aligned} \mathbb{P}\big((X_{T+1}, \ldots, X_{T+r}) = \mathbf{x}\big) &= \sum_{t=1}^{n-r} \mathbb{P}\big(\{T = t\} \cap \{(X_{T+1}, \ldots, X_{T+r}) = \mathbf{x}\}\big) \\ &= \sum_{t=1}^{n-r} \mathbb{P}\big(\{T = t\} \cap \{(X_{n-r+1}, \ldots, X_n) = \mathbf{x}\}\big) \\ &= \mathbb{P}\big((X_{n-r+1}, \ldots, X_n) = \mathbf{x}\big) \\ &= \mathbb{P}\big((X_1, \ldots, X_r) = \mathbf{x}\big) \end{aligned}$$

where we used exchangeability twice and the fact $\mathbb{P}(T \le n - r) = 1$.

(b) The sequence $I_1, I_2, \dots, I_{v+w}$ of indicators of whiteness of the drawn ball is exchangeable, and T is a stopping time. Therefore, by part (a),

$$\mathbb{P}(I_{T+1} = 1) = \mathbb{P}(I_1 = 1) = \frac{w}{v+w}.$$

12.5 Solutions. Optional stopping

1. Under the conditions of (a) or (b), the family $\{Y_{T\wedge n} : n \ge 0\}$ is uniformly integrable. Now $T \wedge n \to T$ as $n \to \infty$, so that $Y_{T\wedge n} \to Y_T$ a.s. Using uniform integrability, $\mathbb{E}(Y_{T\wedge n}) \to \mathbb{E}(Y_T)$, and the claim follows by the fact that $\mathbb{E}(Y_{T\wedge n}) = \mathbb{E}(Y_0)$.

2. It suffices to prove that $\{Y_{T\wedge n} : n \ge 0\}$ is uniformly integrable. Recall that $\{X_n : n \ge 0\}$ is uniformly integrable if

$$\lim_{a\to\infty}\left\{\sup_n \mathbb{E}\left(|X_n| I_{\{|X_n|\ge a\}}\right)\right\} \to 0 \quad \text{as } a \to \infty.$$

(a) Now,

$$\begin{aligned}\mathbb{E}\left(|Y_{T\wedge n}| I_{\{|Y_{T\wedge n}|\ge a\}}\right) &= \mathbb{E}\left(|Y_T| I_{\{T\le n, |Y_T|\ge a\}}\right) + \mathbb{E}\left(|Y_n| I_{\{T>n, |Y_n|\ge a\}}\right)\\ &\le \mathbb{E}\left(|Y_T| I_{\{|Y_T|\ge a\}}\right) + \mathbb{E}\left(|Y_n| I_{\{T>n\}}\right) = g(a) + h(n),\end{aligned}$$

say. We have that $g(a) \to 0$ as $a \to \infty$, since $\mathbb{E}|Y_T| < \infty$. Also $h(n) \to 0$ as $n \to \infty$, so that $\sup_{n\ge N} h(n)$ may be made arbitrarily small by suitable choice of N. On the other hand, $\mathbb{E}\left(|Y_n| I_{\{|Y_n|\ge a\}}\right) \to 0$ as $a \to \infty$ *uniformly* in $n \in \{0, 1, \dots, N\}$, and the claim follows.

(b) Since Y_n^+ defines a submartingale, we have that $\sup_n \mathbb{E}(Y_{T\wedge n}^+) \le \sup_n \mathbb{E}(Y_n^+) < \infty$, the second inequality following by the uniform integrability of $\{Y_n\}$. Using the martingale convergence theorem, $Y_{T\wedge n} \to Y_T$ a.s. where $\mathbb{E}|Y_T| < \infty$. Now

$$\mathbb{E}|Y_{T\wedge n} - Y_T| = \mathbb{E}\big(|Y_n - Y_T| I_{\{T>n\}}\big) \le \mathbb{E}\big(|Y_n| I_{\{T>n\}}\big) + \mathbb{E}\big(|Y_T| I_{\{T>n\}}\big).$$

Also $\mathbb{P}(T > n) \to 0$ as $n \to \infty$, so that the final two terms tend to 0 (by the uniform integrability of the Y_i and the finiteness of $\mathbb{E}|Y_T|$ respectively). Therefore $Y_{T\wedge n} \xrightarrow{1} Y_T$, and the claim follows by the standard theorem (7.10.3).

3. By uniform integrability, $Y_\infty = \lim_{n\to\infty} Y_n$ exists a.s. and in mean, and $Y_n = \mathbb{E}(Y_\infty \mid \mathcal{F}_n)$.

(a) On the event $\{T = n\}$ it is the case that $Y_T = Y_n$ and $\mathbb{E}(Y_\infty \mid \mathcal{F}_T) = \mathbb{E}(Y_\infty \mid \mathcal{F}_n)$; for the latter statement, use the definition of conditional expectation. It follows that $Y_T = \mathbb{E}(Y_\infty \mid \mathcal{F}_T)$, irrespective of the value of T.

(b) We have from Exercise (12.4.7) that $\mathcal{F}_S \subseteq \mathcal{F}_T$. Now $Y_S = \mathbb{E}(Y_\infty \mid \mathcal{F}_S) = \mathbb{E}\{\mathbb{E}(Y_\infty \mid \mathcal{F}_T) \mid \mathcal{F}_S) = \mathbb{E}(Y_T \mid \mathcal{F}_S)$.

4. Let T be the time until absorption, and note that $\{S_n\}$ is a bounded, and therefore uniformly integrable, martingale. Also $\mathbb{P}(T < \infty) = 1$ since T is no larger than the waiting time for N consecutive steps in the same direction. It follows that $\mathbb{E}(S_0) = \mathbb{E}(S_T) = N\mathbb{P}(S_T = N)$, so that $\mathbb{P}(S_T = N) = \mathbb{E}(S_0)/N$. Secondly, $\{S_n^2 - n : n \ge 0\}$ is a martingale (see Exercise (12.1.4)), and the optional stopping theorem (if it may be applied) gives that

$$\mathbb{E}(S_0^2) = \mathbb{E}(S_T^2 - T) = N^2\mathbb{P}(S_T = N) - \mathbb{E}(T),$$

and hence $\mathbb{E}(T) = N\mathbb{E}(S_0) - \mathbb{E}(S_0^2)$ as required.

It remains to check the conditions of the optional stopping theorem. Certainly $\mathbb{P}(T < \infty) = 1$, and in addition $\mathbb{E}(T^2) < \infty$ by the argument above. We have that $\mathbb{E}|S_T^2 - T| \le N^2 + \mathbb{E}(T) < \infty$. Finally,

$$\mathbb{E}\{(S_n^2 - n)I_{\{T>n\}}\} \le (N^2 + n)\mathbb{P}(T > n) \to 0$$

as $n \to \infty$, since $\mathbb{E}(T^2) < \infty$.

5. Let $\mathcal{F}_n = \sigma(S_1, S_2, \dots, S_n)$. It is immediate from the identity $\cos(A + \lambda) + \cos(A - \lambda) = 2\cos A \cos \lambda$ that

$$\mathbb{E}(Y_{n+1} \mid \mathcal{F}_n) = \frac{\cos[\lambda(S_n + 1 - \frac{1}{2}(b-a))] + \cos[\lambda(S_n - 1 - \frac{1}{2}(b-a))]}{2(\cos\lambda)^{n+1}} = Y_n,$$

and therefore Y is a martingale (it is easy to see that $\mathbb{E}|Y_n| < \infty$ for all n).

Suppose that $0 < \lambda < \pi/(a+b)$, and note that $0 \le |\lambda\{S_n - \frac{1}{2}(b-a)\}| < \frac{1}{2}\lambda(a+b) < \frac{1}{2}\pi$ for $n \le T$. Now $Y_{T\wedge n}$ constitutes a martingale which satisfies

$$(*) \qquad \frac{\cos\{\frac{1}{2}\lambda(a+b)\}}{(\cos\lambda)^{T\wedge n}} \le Y_{T\wedge n} \le \frac{1}{(\cos\lambda)^T}.$$

If we can prove that $\mathbb{E}\{(\cos\lambda)^{-T}\} < \infty$, it will follow that $\{Y_{T\wedge n}\}$ is uniformly integrable. This will imply in turn that $\mathbb{E}(Y_T) = \lim_{n\to\infty} \mathbb{E}(Y_{T\wedge n}) = \mathbb{E}(Y_0)$, and therefore

$$\cos\{\tfrac{1}{2}\lambda(a+b)\}\mathbb{E}\{(\cos\lambda)^{-T}\} = \cos\{\tfrac{1}{2}\lambda(b-a)\}$$

as required. We have from $(*)$ that

$$\mathbb{E}(Y_0) = \mathbb{E}(Y_{T\wedge n}) \ge \cos\{\tfrac{1}{2}\lambda(a+b)\}\mathbb{E}\{(\cos\lambda)^{-T\wedge n}\}.$$

Now $T \wedge n \to T$ as $n \to \infty$, implying by Fatou's lemma that

$$\mathbb{E}\{(\cos\lambda)^{-T}\} \le \frac{\mathbb{E}(Y_0)}{\cos\{\frac{1}{2}\lambda(a+b)\}} = \frac{\cos\{\frac{1}{2}\lambda(a-b)\}}{\cos\{\frac{1}{2}\lambda(a+b)\}}.$$

6. (a) The occurrence of the event $\{U = n\}$ depends on $S_1, S_2, \dots, S_n$ only, and therefore U is a stopping time. Think of U as the time until the first sequence of five consecutive heads in a sequence of coins tosses. Using the renewal-theory argument of Problem (10.6.17), we find that $\mathbb{E}(U) = 62$.
(b) Knowledge of $S_1, S_2, \dots, S_n$ is insufficient to determine whether or not $V = n$, and therefore V is not a stopping time. Now $\mathbb{E}(V) = \mathbb{E}(U) - 5 = 57$.
(c) W is a stopping time, since it is a first-passage time. Also $\mathbb{E}(W) = \infty$ since the walk is *null* recurrent.

7. With the usual notation,

$$\begin{aligned}\mathbb{E}(M_{m+n} \mid \mathcal{F}_m) &= \mathbb{E}\left(\sum_{r=0}^{m} S_r + \sum_{r=m+1}^{m+n} S_r - \tfrac{1}{3}(S_{m+n} - S_m + S_m)^3 \,\middle|\, \mathcal{F}_m\right) \\ &= M_m + nS_m - S_m\mathbb{E}\{(S_{m+n} - S_m)^2\} \\ &= M_m + nS_m - nS_m\mathbb{E}(X_1^2) = M_m.\end{aligned}$$

Thus $\{M_n : n \geq 0\}$ is a martingale, and evidently T is a stopping time. The conditions of the optional stopping theorem (12.5.1) hold, and therefore, by a result of Example (3.9.6),

$$a - \tfrac{1}{3}a^3 = M_0 = \mathbb{E}(M_T) = \mathbb{E}\left(\sum_{r=0}^{T} S_r\right) - \tfrac{1}{3}K^3 \cdot \frac{a}{K}.$$

8. We partition the sequence into consecutive batches of $a+b$ flips. If any such batch contains only 1's, then the game is over. Hence $\mathbb{P}(T > n(a+b)) \leq \{1 - (\frac{1}{2})^{a+b}\}^n \to 0$ as $n \to \infty$. Therefore,

$$\mathbb{E}|S_T^2 - T| \leq \mathbb{E}(S_T^2) + \mathbb{E}(T) \leq (a+b)^2 + \mathbb{E}(T) < \infty,$$

and

$$\mathbb{E}\big[(S_T^2 - T)I_{\{T>n\}}\big] \leq (a+b)^2\mathbb{P}(T > n) + \mathbb{E}(T I_{\{T>n\}}) \to 0 \qquad \text{as } n \to \infty.$$

9. The sequence $G_n - nq$ is a martingale, and real-world constraints entail $\mathbb{E}(T) < \infty$. By the optional stopping theorem, $\mathbb{E}(G_T) = q\mathbb{E}(T)$, and likewise $\mathbb{E}(B_T) = p\mathbb{E}(T)$.

We cannot know $\mathbb{E}(G_T/B_T)$ without further knowledge of the definition of T. Contrast "stop at the first boy", for which $\mathbb{E}(G_T/B_T) = q/p$, with "stop at the first girl", for which $\mathbb{E}(G_T/B_T) = \infty$.

10. By the martingale property, $p_{00} = 1$, since otherwise $\mathbb{E}(X_{n+1} \mid X_n = 0) > X_n$. Likewise, $p_{bb} = 1$. Since X_n is uniformly bounded, it converges a.s. to some X, so that $X_n = X$ for all but finitely many values of n. By the given communication property, $X \in \{0, b\}$. By the optional stopping theorem with T the time of absorption, $X_0 = \mathbb{E}(X_T) = b\mathbb{P}(X_T = b) + 0$.

12.6 Solution. The maximal inequality

1. (a) For $r < s$,

$$\mathbb{E}(D_r D_s) = \mathbb{E}\big(\mathbb{E}(D_r D_s \mid \mathcal{F}_{s-1})\big) = \mathbb{E}\big(D_r \mathbb{E}(D_s \mid \mathcal{F}_{s-1})\big) = 0.$$

(b) By the Doob–Kolmogorov inequality (7.8.2), for $\epsilon > 0$,

$$\mathbb{P}\left(\max_{1\leq r\leq n} |M_r| > n\epsilon\right) \leq \frac{1}{n^2\epsilon^2}\mathbb{E}(M_n^2) = \frac{1}{n^2\epsilon^2}\sum_{r=1}^{n}\sigma_r^2 \to 0 \qquad \text{as } n \to \infty,$$

by the assumption and part (a).

(c) This is the martingale version of Exercise (7.8.2). The sequence $Z_n = \sum_{r=1}^n D_r/r$ is a zero-mean martingale with variance satisfying

$$\mathbb{E}(Z_n^2) = \sum_{r=1}^{n}\frac{1}{r^2}\sigma_r^2 \to \sum_{r=1}^{\infty}\frac{1}{r^2}\sigma_r^2 < \infty \qquad \text{as } n \to \infty,$$

by the assumption and part (a). By the martingale convergence theorem (12.3.1) (see also Theorem (7.8.1)), the limit $Z_n \xrightarrow{\text{a.s.}} Z$ exists as $n \to \infty$, and it follows by Kronecker's lemma that $n^{-1}\sum_{r=1}^n D_r = n^{-1}M_n \xrightarrow{\text{a.s.}} 0$.

12.7 Solutions. Backward martingales and continuous-time martingales

1. Let $s \le t$. We have that $\mathbb{E}(\eta(X(t)) \mid \mathcal{F}_s, X_s = i) = \sum_j p_{ij}(t-s)\eta(j)$. Hence

$$\frac{d}{dt}\mathbb{E}\big(\eta(X(t)) \,\big|\, \mathcal{F}_s, X_s = i\big) = \big(\mathbf{P}_{t-s}\mathbf{G}\boldsymbol{\eta}'\big)_i = 0,$$

so that $\mathbb{E}(\eta(X(t)) \mid \mathcal{F}_s, X_s = i) = \eta(i)$, which is to say that $\mathbb{E}(\eta(X(t)) \mid \mathcal{F}_s) = \eta(X(s))$.

2. Let $W(t) = \exp\{-\theta N(t) + \lambda t(1-e^{-\theta})\}$ where $\theta \ge 0$. It may be seen that $W(t \wedge T_a)$, $t \ge 0$, constitutes a martingale. Furthermore

$$|W(t \wedge T_a)| \le \exp\big\{\lambda(t \wedge T_a)(1-e^{-\theta})\big\} \uparrow \exp\big\{\lambda T_a(1-e^{-\theta})\big\} \quad \text{as } t \to \infty,$$

where, by assumption, the limit has finite expectation for sufficiently small positive θ (this fact may be checked easily). In this case, $\{W(t \wedge T_a) : t \ge 0\}$ is uniformly integrable. Now $W(t \wedge T_a) \to W(T_a)$ a.s. as $t \to \infty$, and it follows by the optional stopping theorem that

$$1 = \mathbb{E}(W(0)) = \mathbb{E}\big(W(t \wedge T_a)\big) \to \mathbb{E}(W(T_a)) = e^{-\theta a}\mathbb{E}\{e^{\lambda T_a(1-e^{-\theta})}\}.$$

Write $s = e^{-\theta}$ to obtain $s^{-a} = \mathbb{E}\{e^{\lambda T_a(1-s)}\}$. Differentiate at $s = 1$ to find that $a = \lambda\mathbb{E}(T_a)$ and $a(a+1) = \lambda^2\mathbb{E}(T_a^2)$, whence the claim is immediate.

The last part is elementary since T_a is the sum of a independent random variables with the exponential distribution with parameter λ. Another way is to make the change of variables $e^{\theta} = \lambda/(\lambda - is)$ in the above martingale. Then check the conditions of the optional stopping theorem, and apply that theorem to obtain the answer.

3. Let $\mathcal{G}_m$ be the σ-field generated by the two sequences of random variables $S_m, S_{m+1} \dots, S_n$ and $U_{m+1}, U_{m+2}, \dots, U_n$. It is a straightforward exercise in conditional density functions to see that

$$\mathbb{E}(S_m \mid \mathcal{G}_{m+1}) = \frac{m}{m+1}S_{m+1}, \quad \mathbb{E}(U_{m+1}^{-1} \mid \mathcal{G}_{m+1}) = \int_0^{U_{m+2}} \frac{(m+1)x^{m-1}}{(U_{m+2})^{m+1}}\,dx = \frac{m+1}{mU_{m+2}},$$

whence $\mathbb{E}(R_m \mid \mathcal{G}_{m+1}) = R_{m+1}$ as required. [The integrability condition is elementary.]

Let $T = \max\{m : R_m \ge 1\}$ with the convention that $T = 1$ if $R_m < 1$ for all m. As in the closely related Example (12.7.6), T is a stopping time. We apply the optional stopping theorem (12.7.5) to the backward martingale R to obtain that $\mathbb{E}(R_T \mid \mathcal{G}_n) = R_n = S_n/t$. Now, $R_T \ge 1$ on the event $\{R_m \ge 1 \text{ for some } m \le n\}$, whence

$$\frac{y}{t} = \mathbb{E}(R_T \mid S_n = y) \ge \mathbb{P}\big(R_m \ge 1 \text{ for some } m \le n \,\big|\, S_n = y\big).$$

[Equality may be shown to hold. See Karlin and Taylor (1981, pp. 110–113), and Example (12.7.6).]

4. (a) Use the facts that:

$$\begin{aligned} W(t) &= \big[W(t+s) - W(s)\big] + W(s), \\ W(t)^2 - t &= \big[2W(s)(W(t) - W(s)) + (W(t) - W(s))^2 - (t-s)\big] + W(s)^2 - s, \end{aligned}$$

and likewise for cases (iii) and (iv). Now take conditional expectation given $\mathcal{F}_s$ to obtain the martingale property. It is elementary that each has finite expectation.

(b) Find s such that $\mathbb{P}(|W(s)| \ge a) \ge \frac{1}{2}$. Then $\mathbb{P}(T > ks) \le (\frac{1}{2})^k$, so that $\mathbb{E}(T^n) < \infty$. By the optional stopping theorem applied to the martingale $W(t)^2 - t$ with bounded stopping time $T \wedge t$, we have that

$$0 = \mathbb{E}\big(W(T \wedge t)^2\big) - \mathbb{E}(T \wedge t) \to a^2 - \mathbb{E}(T) \qquad \text{as } t \to \infty,$$

where we have applied the bounded convergence theorem to the first term, and the monotone convergence theorem to the second. Hence, $\mathbb{E}(T) = a^2$. By the same argument applied to the martingale of case (iv), we obtain $a^4 - 6a^2\mathbb{E}(T) = -3\mathbb{E}(T^2)$. The formula $\mathbb{E}(T^2) = \frac{5}{3}a^4$ follows.

5. It is elementary that $\mathbb{E}|M(t)| < \infty$. By independence, for $s < t$,

$$\mathbb{E}\big(Y(t)X(t)^2 \mid \mathcal{F}_s\big) = \mathbb{E}(Y(t) \mid \mathcal{F}_s)\mathbb{E}(X(t)^2 \mid \mathcal{F}_s) = Y(s)\big[X(s)^2 + t - s\big],$$

since $Y(t)$ and $X(t)^2 - t$ are martingales. (See Exercise (12.7.4a).) Also, for $s < t$,

$$\mathbb{E}\left(\int_0^t Y(u)\,du \,\middle|\, \mathcal{F}_s\right) = \int_0^s Y(u)\,du + \int_s^t Y(s)\,du = \int_0^s Y(u)\,du + (t-s)Y(s),$$

and the result follows.

6. Let

$$M(t) = |X(t) + iY(t)|^2 - 2t = X(t)^2 + Y(t)^2 - 2t.$$

As in Example (12.7.4a), M is a martingale with respect to the natural filtration $\mathcal{F}_t = \sigma(X(s), Y(s) : s \le t)$. The random variable T is a stopping time, and is a.s. finite since

$$\mathbb{P}(T > t) \le \mathbb{P}(|X(t)| < 1) \to 0 \qquad \text{as } t \to \infty.$$

For fixed $t > 0$, M is uniformly bounded on the interval $[0, T \wedge t]$, so that

$$\begin{aligned}\mathbb{E}(M(T\wedge t)) &= \mathbb{E}(X(T\wedge t)^2) + \mathbb{E}(Y(T\wedge t)^2) - 2\mathbb{E}(T\wedge t)) \\ &= \mathbb{E}(M(0)) = |z|^2.\end{aligned}$$

Take the limit as $t \to \infty$, and use the dominated and monotone convergence theorems where appropriate, to find that

$$1 = \mathbb{E}\big(X(T)^2 + Y(T)^2\big) = |z|^2 + 2\mathbb{E}(T).$$

7. (a) It is easy to check that Q is a martingale. Let $T = \inf\{t > 0 : Q(t) \in \{-m, n\}\}$, so that Q is uniformly bounded on the time interval $[0, T]$. By the optional stopping theorem, with the obvious notation, $p_{-m} + p_n = 1$ and

$$-mp_{-m} + np_n = 0, \qquad \text{whence} \qquad p_n = \frac{m}{m+n} = 1 - p_{-m}.$$

(b) One may check that the process $Q(t)^2 - 2\lambda t$ is also a martingale. By optional stopping again,

$$2\lambda\mathbb{E}(T) = \mathbb{E}(Q(T)^2) = \frac{m^2 n}{m+n} + \frac{n^2 m}{m+n} = mn,$$

whence $\mathbb{E}(T) = mn/(2\lambda)$. These results may also be obtained using the imbedded random walk.

8. (a) The increment $N((r+1)t/n) - N(rt/n)$ is independent of $\mathcal{F}_{rt/n}$ with the Poisson distribution with second moment $(\lambda t/n) + (\lambda t/n)^2$.

(b) One may define the optional quadratic variation of a continuous-time martingale very much as in the discrete case of Exercise (12.2.4). With probability 1, $N(t) - \lambda t$ is finite, and the path of the process over the interval $[0, t]$ is continuous except for $N(t)$ distinct unit jumps upwards. Therefore, for a partition $0 = s_0 < s_1 < \cdots < s_n = t$ of $[0, t]$ with $\epsilon = \max\{s_{i+1} - s_i : i = 0, 1, \dots, n-1\}$ and $\epsilon > 0$ sufficiently small,

$$\sum_{i=0}^{n-1}\big[N(s_{i+1}) - N(s_i) - \lambda(s_{i+1} - s_i)\big]^2 = N(t) + o(\epsilon).$$

(c) By Example (12.7.9), $(N(t) - \lambda t)^2 - \lambda t$ and $N(t) - \lambda t$ are martingales with respect to the same natural filtration. Therefore, their difference $M(t)$ is also a martingale.

12.9 Solutions to problems

1. Clearly $\mathbb{E}(Z_n) \le (\mu + m)^n$, and hence $\mathbb{E}|Y_n| < \infty$. Secondly, Z_{n+1} may be expressed as $\sum_{i=1}^{Z_n} X_i + A$, where $X_1, X_2, \dots$ are the family sizes of the members of the nth generation, and A is the number of immigrants to the $(n+1)$th generation. Therefore $\mathbb{E}(Z_{n+1} \mid Z_n) = \mu Z_n + m$, whence

$$\mathbb{E}(Y_{n+1} \mid Z_n) = \frac{1}{\mu^{n+1}} \left\{ \mu Z_n + m\left(1 - \frac{1-\mu^{n+1}}{1-\mu}\right)\right\} = Y_n.$$

2. Each birth in the $(n+1)$th generation is to an individual, say the sth, in the nth generation. Hence, for each r, $B_{(n+1),r}$ may be expressed in the form $B_{(n+1),r} = B_{n,s} + B_j'(s)$, where $B_j'(s)$ is the age of the parent when its jth child is born. Therefore

$$\mathbb{E}\left\{\sum_r e^{-\theta B_{(n+1),r}} \,\middle|\, \mathcal{F}_n\right\} = \mathbb{E}\left\{\sum_{s,j} e^{-\theta(B_{n,s}+B_j'(s))} \,\middle|\, \mathcal{F}_n\right\} = \sum_s e^{-\theta B_{n,s}} M_1(\theta),$$

which gives that $\mathbb{E}(Y_{n+1} \mid \mathcal{F}_n) = Y_n$. Finally, $\mathbb{E}(Y_1(\theta)) = 1$, and hence $\mathbb{E}(Y_n(\theta)) = 1$.

3. If $x, c > 0$, then

$$(*) \qquad \mathbb{P}\left(\max_{1\le k\le n} Y_k > x\right) \le \mathbb{P}\left(\max_{1\le k\le n} (Y_k + c)^2 > (x+c)^2\right).$$

Now $(Y_k + c)^2$ is a convex function of Y_k, and therefore defines a submartingale (Exercise (12.1.7)). Applying the maximal inequality to this submartingale, we obtain an upper bound of $\mathbb{E}\{(Y_n + c)^2\}/(x+c)^2$ for the right-hand side of $(*)$. We set $c = \mathbb{E}(Y_n^2)/x$ to obtain the result.

4. (a) Note that $Z_n = Z_{n-1} + c_n\{X_n - \mathbb{E}(X_n \mid \mathcal{F}_{n-1})\}$, so that $(Z, \mathcal{F})$ is a martingale. Let T be the stopping time $T = \min\{k : c_k Y_k \ge x\}$. Then $\mathbb{E}(Z_{T\wedge n}) = \mathbb{E}(Z_0) = 0$, so that

$$0 \ge \mathbb{E}\left\{c_{T\wedge n} Y_{T\wedge n} - \sum_{k=1}^{T\wedge n} c_k \mathbb{E}(X_k \mid \mathcal{F}_{k-1})\right\}$$

since the final term in the definition of Z_n is non-negative. Therefore

$$x\mathbb{P}(T \le n) \le \mathbb{E}\{c_{T\wedge n} Y_{T\wedge n}\} \le \sum_{k=1}^{n} c_k \mathbb{E}\{\mathbb{E}(X_k \mid \mathcal{F}_{k-1})\},$$

where we have used the facts that $Y_n \ge 0$ and $\mathbb{E}(X_k \mid \mathcal{F}_{k-1}) \ge 0$. The claim follows.

(b) Let $X_1, X_2, \dots$ be independent random variables, with zero means and finite variances, and let $Y_j = \sum_{i=1}^{j} X_i$. Then Y_j^2 defines a non-negative submartingale, whence

$$\mathbb{P}\left(\max_{1\le k\le n} |Y_k| \ge x\right) = \mathbb{P}\left(\max_{1\le k\le n} Y_k^2 \ge x^2\right) \le \frac{1}{x^2}\sum_{k=1}^{n} \mathbb{E}(Y_k^2 - Y_{k-1}^2) = \frac{1}{x^2}\sum_{k=1}^{n} \mathbb{E}(X_k^2).$$

5. The function $h(u) = |u|^r$ is convex, and therefore $Y_i(m) = |S_i - S_m|^r$, $i \ge m$, defines a submartingale with respect to the filtration $\mathcal{F}_i = \sigma(\{X_j : 1 \le j \le i\})$. Apply the HRC inequality of Problem (12.9.4), with $c_k = 1$, to obtain the required inequality.

If $r = 1$, we have that

$$(*) \qquad \mathbb{E}\big(|S_{m+n} - S_m|\big) \le \sum_{k=m+1}^{m+n} \mathbb{E}|Z_k|$$

by the triangle inequality. Let $m, n \to \infty$ to find, in the usual way, that the sequence $\{S_n\}$ converges a.s.; Kronecker's lemma (see Exercise (7.8.2)) then yields the final claim.

Suppose $1 < r \le 2$, in which case a little more work is required. The function h is differentiable, and therefore

$$h(v) - h(u) = (v-u)h'(u) + \int_0^{v-u} \{h'(u+x) - h'(u)\}\,dx.$$

Now $h'(y) = r|y|^{r-1}\text{sign}(y)$ has a derivative decreasing in $|y|$. It follows (draw a picture) that $h'(u+x) - h'(u) \le 2h'(\frac{1}{2}x)$ if $x \ge 0$, and therefore the above integral is no larger than $2h(\frac{1}{2}(v-u))$. Apply this with $v = S_{m+k+1} - S_m$ and $u = S_{m+k} - S_m$, to obtain

$$\mathbb{E}\big(|S_{m+k+1} - S_m|^r\big) - \mathbb{E}\big(|S_{m+k} - S_m|^r\big) \le \mathbb{E}\big(Z_{m+k+1}h'(S_{m+k} - S_m)\big) + 2\mathbb{E}\big(|\tfrac{1}{2}Z_{m+k+1}|^r\big).$$

Sum over k and use the fact that

$$\mathbb{E}\big(Z_{m+k+1}h'(S_{m+k} - S_m)\big) = \mathbb{E}\big\{h'(S_{m+k} - S_m)\mathbb{E}(Z_{m+k+1} \mid \mathcal{F}_{m+k})\big\} = 0,$$

to deduce that

$$\mathbb{E}\big(|S_{m+n} - S_m|^r\big) \le 2^{2-r} \sum_{k=m+1}^{m+n} \mathbb{E}\big(|Z_k|^r\big).$$

The argument is completed as after $(*)$.

6. With $I_k = I_{\{Y_k=0\}}$, we have that

$$\begin{aligned}\mathbb{E}(Y_n \mid \mathcal{F}_{n-1}) &= \mathbb{E}\Big(X_n I_{n-1} + nY_{n-1}|X_n|(1 - I_{n-1}) \,\Big|\, \mathcal{F}_{n-1}\Big)\\ &= I_{n-1}\mathbb{E}(X_n) + nY_{n-1}(1 - I_{n-1})\mathbb{E}|X_n| = Y_{n-1}\end{aligned}$$

since $\mathbb{E}(X_n) = 0$, $\mathbb{E}|X_n| = n^{-1}$. Also $\mathbb{E}|Y_n| \le \mathbb{E}\big\{|X_n|(1 + n|Y_{n-1}|)\big\}$ and $\mathbb{E}|Y_1| < \infty$, whence $\mathbb{E}|Y_n| < \infty$. Therefore $(Y, \mathcal{F})$ is a martingale.

Now $Y_n = 0$ if and only if $X_n = 0$. Therefore $\mathbb{P}(Y_n = 0) = \mathbb{P}(X_n = 0) = 1 - n^{-1} \to 1$ as $n \to \infty$, implying that $Y_n \xrightarrow{\text{P}} 0$. On the other hand, $\sum_n \mathbb{P}(X_n \ne 0) = \infty$, and therefore $\mathbb{P}(Y_n \ne 0 \text{ i.o.}) = 1$ by the second Borel–Cantelli lemma. However, Y_n takes only integer values, and therefore Y_n does not converge to 0 a.s. The martingale convergence theorem is inapplicable since $\sup_n \mathbb{E}|Y_n| = \infty$.

7. Assume that $t > 0$ and $M(t) = 1$. Then $Y_n = e^{tS_n}$ defines a positive martingale (with mean 1) with respect to $\mathcal{F}_n = \sigma(X_1, X_2, \dots, X_n)$. By the maximal inequality,

$$\mathbb{P}\left(\max_{1\le k\le n} S_k \ge x\right) = \mathbb{P}\left(\max_{1\le k\le n} Y_k \ge e^{tx}\right) \le e^{-tx}\mathbb{E}(Y_n),$$

and the result follows by taking the limit as $n \to \infty$.

8. The sequence $Y_n = \xi^{Z_n}$ defines a martingale; this may be seen easily, as in Example (7.7.5). Now $\{Y_n\}$ is uniformly bounded, and therefore $Y_\infty = \lim_{n\to\infty} Y_n$ exists a.s. and satisfies $\mathbb{E}(Y_\infty) = \mathbb{E}(Y_0) = \xi$.

Suppose $0 < \xi < 1$. In this case Z_1 is not a.s. zero, so that Z_n cannot converge a.s. to a constant c unless $c \in \{0, \infty\}$. Therefore the a.s. convergence of Y_n entails the a.s. convergence of Z_n to a limit random variable taking values 0 and ∞. In this case, $\mathbb{E}(Y_\infty) = 1 \cdot \mathbb{P}(Z_n \to 0) + 0 \cdot \mathbb{P}(Z_n \to \infty)$, implying that $\mathbb{P}(Z_n \to 0) = \xi$, and therefore $\mathbb{P}(Z_n \to \infty) = 1 - \xi$.

9. It is a consequence of the maximal inequality that $\mathbb{P}(Y_n^* \ge x) \le x^{-1}\mathbb{E}(Y_n I_{\{Y_n^* \ge x\}})$ for $x > 0$. Therefore

$$\begin{aligned}
\mathbb{E}(Y_n^*) &= \int_0^\infty \mathbb{P}(Y_n^* \ge x)\,dx \le 1 + \int_1^\infty \mathbb{P}(Y_n^* \ge x)\,dx \\
&\le 1 + \mathbb{E}\left\{ Y_n \int_1^\infty x^{-1} I_{(1,Y_n^*]}(x)\,dx \right\} \\
&= 1 + \mathbb{E}(Y_n \log^+ Y_n^*) \le 1 + \mathbb{E}(Y_n \log^+ Y_n) + \mathbb{E}(Y_n^*)/e.
\end{aligned}$$

10. (a) We have, as in Exercise (12.7.1), that

$$(*) \qquad \mathbb{E}\big(h(X(t)) \,\big|\, B, X(s) = i\big) = \sum_j p_{ij}(t) h(j) \quad \text{for } s < t,$$

for any event B defined in terms of $\{X(u) : u \le s\}$. The derivative of this expression, with respect to t, is $(\mathbf{P}_t \mathbf{G} \mathbf{h}')_i$, where $\mathbf{P}_t$ is the transition semigroup, $\mathbf{G}$ is the generator, and $\mathbf{h} = (h(j) : j \ge 0)$. In this case,

$$(\mathbf{G}\mathbf{h}')_j = \sum_k g_{jk} h(k) = \lambda_j \{h(j+1) - h(j)\} - \mu_j \{h(j) - h(j-1)\} = 0$$

for all j. Therefore the left side of $(*)$ is constant for $t \ge s$, and is equal to its value at time s, i.e. $X(s)$. Hence $h(X(t))$ defines a martingale.

(b) We apply the optional stopping theorem with $T = \min\{t : X(t) \in \{0, n\}\}$ to obtain $\mathbb{E}(h(X(T))) = \mathbb{E}(h(X(0)))$, and therefore $(1 - \pi(m))h(n) = h(m)$ as required. It is necessary but not difficult to check the conditions of the optional stopping theorem.

11. (a) Since Y is a submartingale, so is Y^+ (see Exercise (12.1.6)). Now

$$\mathbb{E}(Y_{n+m+1}^+ \mid \mathcal{F}_n) = \mathbb{E}\{\mathbb{E}(Y_{n+m+1}^+ \mid \mathcal{F}_{n+1}) \,|\, \mathcal{F}_n\} \ge \mathbb{E}(Y_{n+m}^+ \mid \mathcal{F}_n).$$

Therefore $\{\mathbb{E}(Y_{n+m}^+ \mid \mathcal{F}_n) : m \ge 0\}$ is (a.s.) non-decreasing, and therefore converges (a.s.) to a limit M_n. Also, by monotone convergence of conditional expectation,

$$\mathbb{E}(M_{n+1} \mid \mathcal{F}_n) = \lim_{m\to\infty} \mathbb{E}\{\mathbb{E}(Y_{n+m+1}^+ \mid \mathcal{F}_{n+1}) \,|\, \mathcal{F}_n\} = \lim_{m\to\infty} \mathbb{E}(Y_{n+m+1}^+ \mid \mathcal{F}_n) = M_n,$$

and furthermore $\mathbb{E}(M_n) = \lim_{m\to\infty} \mathbb{E}(Y_{m+n}^+) \le M$. It is the case that M_n is $\mathcal{F}_n$-measurable, and therefore it is a martingale.

(b) We have that $Z_n = M_n - Y_n$ is the difference of a martingale and a submartingale, and is therefore a supermartingale. Also $M_n \ge Y_n^+ \ge 0$, and the decomposition for Y_n follows.

(c) In this case Z_n is a martingale, being the difference of two martingales. Also $M_n \ge \mathbb{E}(Y_n^+ \mid \mathcal{F}_n) = Y_n^+ \ge Y_n$ a.s., and the claim follows.

12. We may as well assume that $\mu < P$ since the inequality is trivial otherwise. The moment generating function of $P - C_1$ is $M(t) = e^{t(P-\mu)+\frac{1}{2}\sigma^2 t^2}$, and we choose t such that $M(t) = 1$, i.e. $t = -2(P - \mu)/\sigma^2$. Now define $Z_n = \min\{e^{tY_n}, 1\}$ and $\mathcal{F}_n = \sigma(C_1, C_2, \ldots, C_n)$. Certainly $\mathbb{E}|Z_n| < \infty$; also

$$\mathbb{E}(Z_{n+1} \mid \mathcal{F}_n) \le \mathbb{E}(e^{tY_{n+1}} \mid \mathcal{F}_n) = e^{tY_n} M(t) = e^{tY_n}$$

and $\mathbb{E}(Z_{n+1} \mid \mathcal{F}_n) \le 1$, implying that $\mathbb{E}(Z_{n+1} \mid \mathcal{F}_n) \le Z_n$. Therefore $(Z_n, \mathcal{F}_n)$ is a positive supermartingale. Let $T = \inf\{n : Y_n \le 0\} = \inf\{n : Z_n = 1\}$. Then $T \wedge m$ is a bounded stopping time, whence $\mathbb{E}(Z_0) \ge \mathbb{E}(Z_{T\wedge m}) \ge \mathbb{P}(T \le m)$. Let $m \to \infty$ to obtain the result.

13. Let $\mathcal{F}_n = \sigma(R_1, R_2, \dots, R_n)$.

(a) $0 \le Y_n \le 1$, and Y_n is $\mathcal{F}_n$-measurable. Also

$$\mathbb{E}(R_{n+1} \mid R_n) = R_n + \frac{R_n}{n+r+b},$$

whence Y_n satisfies $\mathbb{E}(Y_{n+1} \mid \mathcal{F}_n) = Y_n$. Therefore $\{Y_n : n \ge 0\}$ is a uniformly integrable martingale, and therefore converges a.s. and in mean.

(b) In order to apply the optional stopping theorem, it suffices that $\mathbb{P}(T < \infty) = 1$ (since Y is uniformly integrable). However $\mathbb{P}(T > n) = \frac{1}{2} \cdot \frac{2}{3} \cdots \frac{n}{n+1} = (n+1)^{-1} \to 0$. Using that theorem, $\mathbb{E}(Y_T) = \mathbb{E}(Y_0)$, which is to say that $\mathbb{E}\{T/(T+2)\} = \frac{1}{2}$, and the result follows.

(c) Apply the maximal inequality.

14. As in the previous solution, with $\mathcal{G}_n$ the σ-field generated by $A_1, A_2, \dots$ and $\mathcal{F}_n$,

$$\begin{aligned}\mathbb{E}(Y_{n+1} \mid \mathcal{G}_n) &= \left(\frac{R_n + A_n}{R_n + B_n + A_n}\right)\left(\frac{R_n}{R_n + B_n}\right) + \left(\frac{R_n}{R_n + B_n + A_n}\right)\left(\frac{B_n}{R_n + B_n}\right) \\ &= \frac{R_n}{R_n + B_n} = Y_n,\end{aligned}$$

so that $\mathbb{E}(Y_{n+1} \mid \mathcal{F}_n) = \mathbb{E}\{\mathbb{E}(Y_{n+1} \mid \mathcal{G}_n) \mid \mathcal{F}_n\} = Y_n$. Also $|Y_n| \le 1$, and therefore Y_n is a martingale.

We need to show that $\mathbb{P}(T < \infty) = 1$. Let I_n be the indicator function of the event $\{T > n\}$. We have by conditioning on the A_n that

$$\mathbb{E}(I_n \mid \mathbf{A}) = \prod_{j=0}^{n-1}\left(1 - \frac{1}{2 + S_j}\right) \to \prod_{j=0}^{\infty}\left(1 - \frac{1}{2 + S_j}\right)$$

as $n \to \infty$, where $S_j = \sum_{i=1}^{j} A_i$. The infinite product equals 0 a.s. if and only if $\sum_j (2+S_j)^{-1} = \infty$ a.s. By monotone convergence, $\mathbb{P}(T < \infty) = 1$ under this condition. If this holds, we may apply the optional stopping theorem to obtain that $\mathbb{E}(Y_T) = \mathbb{E}(Y_0)$, which is to say that

$$\mathbb{E}\left(1 - \frac{1 + A_T}{2 + S_T}\right) = \frac{1}{2}.$$

15. At each stage k, let L_k be the length of the sequence 'in play', and let Y_k be the sum of its entries, so that $L_0 = n$, $Y_0 = \sum_{i=1}^{n} x_i$. If you lose the $(k+1)$th gamble, then $L_{k+1} = L_k + 1$ and $Y_{k+1} = Y_k + Z_k$ where Z_k is the stake on that play, whereas if you win, then $L_{k+1} = L_k - 2$ and $Y_{k+1} = Y_k - Z_k$; we have assumed that $L_k \ge 2$, similar relations being valid if $L_k = 1$. Note that L_k is a random walk with mean step size -1, implying that the first-passage time T to 0 is a.s. finite, and has all moments finite. Your profits at time k amount to $Y_0 - Y_k$, whence your profit at time T is Y_0, since $Y_T = 0$.

Since the games are fair, Y_k constitutes a martingale. Therefore $\mathbb{E}(Y_{T\wedge m}) = \mathbb{E}(Y_0) \ne 0$ for all m. However $T \wedge m \to T$ a.s. as $m \to \infty$, so that $Y_{T\wedge m} \to Y_T$ a.s. Now $\mathbb{E}(Y_T) = 0 \ne \lim_{m\to\infty} \mathbb{E}(Y_{T\wedge m})$, and it follows that $\{Y_{T\wedge m} : m \ge 1\}$ is not uniformly integrable. Therefore $\mathbb{E}(\sup_m Y_{T\wedge m}) = \infty$; see Exercise (7.10.6).

Further results for the Labouchere system may be found in Han and Wang 2019.

16. Since the game is fair, $\mathbb{E}(S_{n+1} \mid S_n) = S_n$. Also $|S_n| \leq 1 + 2 + \cdots + n < \infty$. Therefore S_n is a martingale. The occurrence of the event $\{N = n\}$ depends only on the outcomes of the coin-tosses up to and including the nth; therefore N is a stopping time.

A tail appeared at time $N - 3$, followed by three heads. Therefore the gamblers $G_1, G_2, \dots, G_{N-3}$ have forfeited their initial capital by time N, while G_{N-i} has had $i + 1$ successful rounds for $0 \leq i \leq 2$. Therefore $S_N = N - (p^{-1} + p^{-2} + p^{-3})$, after a little calculation. It is easy to check that N satisfies the conditions of the optional stopping theorem, and it follows that $\mathbb{E}(S_N) = \mathbb{E}(S_0) = 0$, which is to say that $\mathbb{E}(N) = p^{-1} + p^{-2} + p^{-3}$.

In order to deal with HTH, the gamblers are re-programmed to act as follows. If they win on their first bet, they bet their current fortune on *tails*, returning to heads thereafter. In this case, $S_N = N - (p^{-1} + p^{-2}q^{-1})$ where $q = 1 - p$ (remember that the game is fair), and therefore $\mathbb{E}(N) = p^{-1} + p^{-2}q^{-1}$.

17. Let $\mathcal{F}_n = \sigma(\{X_i, Y_i : 1 \leq i \leq n\})$, and note that T is a stopping time with respect to this filtration. Furthermore $\mathbb{P}(T < \infty) = 1$ since T is no larger than the first-passage time to 0 of either of the two single-coordinate random walks, each of which has mean 0 and is therefore recurrent.

Let $\sigma_1^2 = \operatorname{var}(X_1)$ and $\sigma_2^2 = \operatorname{var}(Y_1)$. We have that $U_n - U_0$ and $V_n - V_0$ are sums of independent summands with means 0 and variances σ_1^2 and σ_2^2 respectively. It follows by considering the martingales $(U_n - U_0)^2 - n\sigma_1^2$ and $(V_n - V_0)^2 - n\sigma_2^2$ (see equation (12.5.14) and Exercise (10.2.2)) that

$$\mathbb{E}\{(U_T - U_0)^2\} = \sigma_1^2 \mathbb{E}(T), \quad \mathbb{E}\{(V_T - V_0)^2\} = \sigma_2^2 \mathbb{E}(T).$$

Applying the same argument to $(U_n + V_n) - (U_0 + V_0)$, we obtain

$$\mathbb{E}\big\{(U_T + V_T - U_0 - V_0)^2\big\} = \mathbb{E}(T)\mathbb{E}\{(X_1 + Y_1)^2\} = \mathbb{E}(T)(\sigma_1^2 + 2c + \sigma_2^2).$$

Subtract the two earlier equations to obtain

$$(*) \qquad \mathbb{E}\big\{(U_T - U_0)(V_T - V_0)\big\} = c\mathbb{E}(T)$$

if $\mathbb{E}(T) < \infty$. Now $U_T V_T = 0$, and in addition $\mathbb{E}(U_T) = U_0$, $\mathbb{E}(V_T) = V_0$, by Wald's equation and the fact that $\mathbb{E}(X_1) = \mathbb{E}(Y_1) = 0$. It follows that $-\mathbb{E}(U_0 V_0) = c\mathbb{E}(T)$ if $\mathbb{E}(T) < \infty$, in which case $c < 0$.

Suppose conversely that $c < 0$. Then $(*)$ is valid with T replaced throughout by the bounded stopping time $T \wedge m$, and hence

$$0 \leq \mathbb{E}(U_{T\wedge m} V_{T\wedge m}) = \mathbb{E}(U_0 V_0) + c\mathbb{E}(T \wedge m).$$

Therefore $\mathbb{E}(T \wedge m) \leq \mathbb{E}(U_0 V_0)/(2|c|)$ for all m, implying that $\mathbb{E}(T) = \lim_{m\to\infty} \mathbb{E}(T \wedge m) < \infty$, and so $\mathbb{E}(T) = -\mathbb{E}(U_0 V_0)/c$ as before.

18. Certainly $0 \leq X_n \leq 1$, and in addition X_n is measurable with respect to the σ-field $\mathcal{F}_n = \sigma(R_1, R_2, \dots, R_n)$. Also $\mathbb{E}(R_{n+1} \mid R_n) = R_n - R_n/(52 - n)$, whence $\mathbb{E}(X_{n+1} \mid \mathcal{F}_n) = X_n$. Therefore X_n is a martingale.

A strategy corresponds to a stopping time. If the player decides to call at the stopping time T, he wins with (conditional) probability X_T, and therefore $\mathbb{P}(\text{wins}) = \mathbb{E}(X_T)$, which equals $\mathbb{E}(X_0)$ $(= \frac{1}{2})$ by the optional stopping theorem.

Here is a trivial solution to the problem. It may be seen that the chance of winning is the same for a player who, after calling "Red Now", picks the card placed at the bottom of the pack rather than that at the top. The bottom card is red with probability $\frac{1}{2}$, irrespective of the strategy of the player.

19. (a) A sum s of money in week t is equivalent to a sum $s/(1+\alpha)^t$ in week 0, since the latter sum may be invested now to yield s in week t. If he sells in week t, his discounted costs are $\sum_{n=1}^{t} c/(1+\alpha)^n$

and his discounted profit is $X_t/(1+\alpha)^t$. He wishes to find a stopping time for which his mean discounted gain is a maximum.

Now

$$-\sum_{n=1}^{T}(1+\alpha)^{-n}c = \frac{c}{\alpha}\{(1+\alpha)^{-T}-1\},$$

so that $\mu(T) = \mathbb{E}\{(1+\alpha)^{-T}Z_T\} - (c/\alpha)$.

(b) The function $h(\gamma) = \alpha\gamma - \int_\gamma^\infty \mathbb{P}(Z_n > y)\,dy$ is continuous and strictly increasing on $[0,\infty)$, with $h(0) = -\mathbb{E}(Z_n) < 0$ and $h(\gamma) \to \infty$ as $\gamma \to \infty$. Therefore there exists a unique γ (> 0) such that $h(\gamma) = 0$, and we choose γ accordingly.

(c) Let $\mathcal{F}_n = \sigma(Z_1, Z_2, \dots, Z_n)$. We have that

$$\mathbb{E}\big(\max\{Z_n, \gamma\}\big) = \gamma + \int_\gamma^\infty [1 - G(y)]\,dy = (1+\alpha)\gamma$$

where $G(y) = \mathbb{P}(Z_n \le y)$. Therefore $\mathbb{E}(V_{n+1} \mid \mathcal{F}_n) = (1+\alpha)^{-n}\gamma \le V_n$, so that $(V_n, \mathcal{F}_n)$ is a non-negative supermartingale.

Let $\mu(\tau)$ be the mean gain of following the strategy 'accept the first offer exceeding $\tau - (c/\alpha)$'. The corresponding stopping time T satisfies $\mathbb{P}(T = n) = G(\tau)^n(1 - G(\tau))$, and therefore

$$\begin{aligned}\mu(\tau) + (c/\alpha) &= \sum_{n=0}^{\infty}\mathbb{E}\{(1+\alpha)^{-T}Z_T I_{\{T=n\}}\}\\ &= \sum_{n=0}^{\infty}(1+\alpha)^{-n}G(\tau)^n(1-G(\tau))\mathbb{E}(Z_1 \mid Z_1 > \tau)\\ &= \frac{1+\alpha}{1+\alpha-G(\tau)}\left\{\tau(1-G(\tau)) + \int_\tau^\infty (1-G(y))\,dy\right\}.\end{aligned}$$

Differentiate with care to find that the only value of τ lying in the support of Z_1 such that $\mu'(\tau) = 0$ is the value $\tau = \gamma$. Furthermore this value gives a maximum for $\mu(\tau)$. Therefore, amongst strategies of the above sort, the best is that with $\tau = \gamma$. Note that $\mu(\gamma) = \gamma(1+\alpha) - (c/\alpha)$.

Consider now a general strategy with corresponding stopping time T, where $\mathbb{P}(T < \infty) = 1$. For any positive integer m, $T \wedge m$ is a bounded stopping time, whence $\mathbb{E}(V_{T\wedge m}) \le \mathbb{E}(V_0) = \gamma(1+\alpha)$. Now $|V_{T\wedge m}| \le \sum_{i=0}^\infty |V_i|$, and $\sum_{i=0}^\infty \mathbb{E}|V_i| < \infty$. Therefore $\{V_{T\wedge m} : m \ge 0\}$ is uniformly integrable. Also $V_{T\wedge m} \to V_T$ a.s. as $m \to \infty$, and it follows that $\mathbb{E}(V_{T\wedge m}) \to \mathbb{E}(V_T)$. We conclude that $\mu(T) = \mathbb{E}(V_T) - (c/\alpha) \le \gamma(1+\alpha) - (c/\alpha) = \mu(\gamma)$. Therefore the strategy given above is optimal.

(d) In the special case, $\mathbb{P}(Z_1 > y) = (y-1)^{-2}$ for $y \ge 2$, whence $\gamma = 10$. The target price is therefore 9, and the mean number of weeks before selling is $G(\gamma)/(1 - G(\gamma)) = 80$.

20. Since G is convex on $[0,\infty)$ wherever it is finite, and since $G(1) = 1$ and $G'(1) < 1$, there exists a unique value of η (> 1) such that $G(\eta) = \eta$. Furthermore, $Y_n = \eta^{Z_n}$ defines a martingale with mean $\mathbb{E}(Y_0) = \eta$. Using the maximal inequality (12.6.6),

$$\mathbb{P}\big(\sup_n Z_n \ge k\big) = \mathbb{P}\big(\sup_n Y_n \ge \eta^k\big) \le \frac{1}{\eta^{k-1}}$$

for positive integers k. Therefore

$$\mathbb{E}\big(\sup_n Z_n\big) \le \sum_{k=1}^{\infty}\frac{1}{\eta - 1}.$$

21. Let M_n be the number present after the nth round, so $M_0 = K$, and $M_{n+1} = M_n - X_{n+1}, n \geq 1$, where X_n is the number of matches in the nth round. By the result of Problem (3.11.17), $\mathbb{E}X_n = 1$ for all n, whence

$$\mathbb{E}(M_{n+1} + n + 1 \mid \mathcal{F}_n) = M_n + n,$$

where $\mathcal{F}_n$ is the σ-field generated by $M_0, M_1, \ldots, M_n$. Thus the sequence $\{M_n + n\}$ is a martingale. Now, N is clearly a stopping time, and therefore $K = M_0 + 0 = \mathbb{E}(M_N + N) = \mathbb{E}N$.

We have that

$$\begin{aligned}
&\mathbb{E}\big\{(M_{n+1} + n + 1)^2 + M_{n+1} \,\big|\, \mathcal{F}_n\big\} \\
&\quad = (M_n + n)^2 - 2(M_n + n)\mathbb{E}(X_{n+1} - 1) + M_n + \mathbb{E}\big\{(X_{n+1} - 1)^2 - X_{n+1} \,\big|\, \mathcal{F}_n\big\} \\
&\quad \leq (M_n + n)^2 + M_n,
\end{aligned}$$

where we have used the fact that

$$\operatorname{var}(X_{n+1} \mid \mathcal{F}_n) = \begin{cases} 1 & \text{if } M_n > 1, \\ 0 & \text{if } M_n = 1. \end{cases}$$

Hence the sequence $\{(M_n + n)^2 + M_n\}$ is a supermartingale. By an optional stopping theorem for supermartingales,

$$K^2 + K = M_0^2 + M_0 \geq \mathbb{E}\big\{(M_N + N)^2 + M_N\big\} = \mathbb{E}(N^2),$$

and therefore $\operatorname{var}(N) \leq K$.

22. In the usual notation,

$$\begin{aligned}
\mathbb{E}\big(M(s+t) \,\big|\, \mathcal{F}_s\big) &= \mathbb{E}\left(\int_0^s W(u)\,du + \int_s^{s+t} W(u)\,du - \tfrac{1}{3}\big\{W(s+t) - W(s) + W(s)\big\}^3 \,\middle|\, \mathcal{F}_s\right) \\
&= M(s) + tW(s) - W(s)\mathbb{E}\big([W(s+t) - W(s)]^2 \,\big|\, \mathcal{F}_s\big) = M(s)
\end{aligned}$$

as required. We apply the optional stopping theorem (12.7.12) with the stopping time $T = \inf\{u : W(u) \in \{a, b\}\}$. The hypotheses of the theorem follow easily from the boundedness of the process for $t \in [0, T]$, and it follows that

$$\mathbb{E}\left(\int_0^T W(u)\,du - \tfrac{1}{3}W(T)^3\right) = 0.$$

Hence the required area A has mean

$$\mathbb{E}(A) = \mathbb{E}\left(\int_0^T W(u)\,du\right) = \frac{1}{3}\mathbb{E}(W(T)^3) = \frac{1}{3}a^3\left(\frac{-b}{a-b}\right) + \frac{1}{3}b^3\left(\frac{a}{a-b}\right).$$

[We have used the optional stopping theorem twice actually, in that $\mathbb{E}(W(T)) = 0$ and therefore $\mathbb{P}(W(T) = a) = -b/(a-b)$.]

23. With $\mathcal{F}_s = \sigma(W(u) : 0 \leq u \leq s)$, we have for $s < t$ that

$$\mathbb{E}(R(t)^2 \mid \mathcal{F}_s) = \mathbb{E}\big(|W(s)|^2 + |W(t) - W(s)|^2 + 2W(s)\cdot(W(t) - W(s)) \,\big|\, \mathcal{F}_s\big) = R(s)^2 + (t - s),$$

and the first claim follows. We apply the optional stopping theorem (12.7.12) with $T = \inf\{u : |W(u)| = a\}$, as in Problem (12.9.22), to find that $0 = \mathbb{E}(R(T)^2 - T) = a^2 - \mathbb{E}(T)$.

24. We apply the optional stopping theorem to the martingale $W(t)$ with the stopping time T to find that $\mathbb{E}(W(T)) = -a(1-p_b) + bp_b = 0$, where $p_b = \mathbb{P}(W(T) = b)$. By Example (12.7.10), $W(t)^2 - t$ is a martingale, and therefore, by the optional stopping theorem again,

$$\mathbb{E}\big((W(T)^2 - T\big) = a^2(1-p_b) + b^2 p_b - \mathbb{E}(T) = 0,$$

whence $\mathbb{E}(T) = ab$. For the final part, we take $a = b$ and apply the optional stopping theorem to the martingale $\exp[\theta W(t) - \frac{1}{2}\theta^2 t]$ to obtain

$$\mathbb{E}\big(\exp[\theta W(T) - \tfrac{1}{2}\theta^2 T]\big) = \big\{e^{-b\theta}(1-p_b) + e^{b\theta}p_b\big\}\mathbb{E}(e^{-\frac{1}{2}\theta^2 T}) = 1,$$

on noting that the conditional distribution of T given $W(T) = b$ is the same as that given $W(T) = -b$. Therefore, $\mathbb{E}(e^{-\frac{1}{2}\theta^2 T}) = 1/\cosh(b\theta)$, and the answer follows by substituting $s = \frac{1}{2}\theta^2$.

25. (a) Note that $0 < X_n < 1$ for all n. Now,

$$\mathbb{P}(X_n > U_{n+1} \mid \mathcal{F}_n) = X_n = 1 - \mathbb{P}(X_n < U_{n+1} \mid \mathcal{F}_n),$$

so that

$$\mathbb{E}(X_{n+1} \mid \mathcal{F}_n) = [(1-a_n)X_n + a_n]X_n + (1-a_n)X_n(1-X_n) = X_n.$$

Thus X is a uniformly bounded martingale, which therefore converges a.s. and in mean square to some X_∞. By mean square convergence, we have $\mathbb{E}(X_n^2) \to \mathbb{E}(X_\infty^2)$.

(b) By the Doob decomposition (12.1.10), the bounded submartingale X_n^2 may be expressed in the form $X_n^2 = M_n + A_n$, where $(M, \mathcal{F})$ is a bounded martingale and A_n is an increasing bounded sequence of $\mathcal{F}_n$-measurable random variables. Furthermore,

$$A_{n+1} - A_n = \mathbb{E}(X_{n+1}^2 \mid \mathcal{F}_n) - X_n^2 = \mathbb{E}\big((X_{n+1} - X_n)^2 \,\big|\, \mathcal{F}_n\big),$$

whence

$$A := \lim_{n\to\infty} A_n \quad \text{satisfies} \quad A = X_0^2 + \sum_{n=0}^{\infty} \mathbb{E}\big((X_{n+1} - X_n)^2 \,\big|\, \mathcal{F}_n\big).$$

Finally,

$$\mathbb{E}(A) = \lim_{n\to\infty}\big(\mathbb{E}(X_n^2) - \mathbb{E}(M_n)\big) = \mathbb{E}(X_\infty^2) - \mathbb{E}(M_0)$$

where $M_0 = X_0^2 = \rho^2$ by the Doob decomposition.

(c) By a calculation as in part (a),

$$\mathbb{E}\big((X_{n+1} - X_n)^2 \,\big|\, \mathcal{F}_n\big) = a_n^2 X_n(1-X_n), \qquad n \ge 1.$$

By part (b), $\mathbb{P}(S = A < \infty) = 1$.

(d) On taking expectations,

$$\mathbb{E}(A) = \rho^2 + \sum_{n=0}^{\infty} a_n^2 \mathbb{E}\big(X_n(1-X_n)\big),$$

and, in addition, $\mathbb{E}(X_n(1-X_n)) \to \mathbb{E}(X_\infty(1-X_\infty))$ as $n \to \infty$. If $\sum_n a_n^2 = \infty$, it must be the case that $\mathbb{E}(X_\infty(1-X_\infty)) = 0$, whence X_∞ takes values 0, 1 only. Now, $\mathbb{E}(X_\infty) = \mathbb{E}(X_0) = \rho$, so that $\mathbb{P}(X_\infty = 1) = \rho$.

26. Let $x, \theta > 0$. By Jensen's inequality, $X(t) = e^{\theta W(t)}$ is a submartingale. By the maximal inequality,

$$\mathbb{P}\left(\sup_{0\le t\le T} W(t) \ge x\right) = \mathbb{P}\left(\sup_{0\le t\le T} e^{\theta W(t)} \ge e^{\theta x}\right)$$
$$\le e^{-\theta x}\mathbb{E}(e^{\theta W(T)}) = e^{-\theta x}e^{\frac{1}{2}\theta^2 T} = e^{-\frac{1}{2}x^2/T},$$

on setting $\theta = x/T$.

27. Let Z_n denote the elapsed time between the $(n-1)$ and nth claim. The accumulated loss at the moment of the nth claim is $S_n = \sum_{r=1}^{n}(X_r - \rho Z_r)$, and the ruin probability is

$$r = \mathbb{P}(S_n > y \text{ for some } n).$$

Now, S_n is the sum of random variables with mean $\mathbb{E}(X - \rho Z) = -\theta/(\lambda\mu\rho)$ where $\theta = \mu - (\lambda/\rho)$. In particular, if $\theta < 0$, then $S_n \xrightarrow{\text{P}} \infty$ by the law of large numbers, whence $r = 1$. The same conclusion holds by the central limit theorem when $\theta = 0$. We shall therefore concentrate on the case $\theta > 0$, which we assume henceforth.

For $K > 0$, let

$$T = \inf\{n \ge 1 : S_n > y\}, \qquad T(K) = \inf\{n \ge 1 : \text{either } S_n \ge y \text{ or } S_n \le -K\},$$

noting that $T(K) < \infty$ a.s., and $T(K) \uparrow T \le \infty$ as $K \to \infty$.

Observe that, with $\theta = \mu - (\lambda/\rho)$,

$$\mathbb{E}(e^{\theta S_1}) = \mathbb{E}\big(e^{\theta X}e^{-\rho\theta Z}\big) = \frac{\mu}{\mu - \theta}\cdot\frac{\lambda}{\lambda + \rho\theta} = 1.$$

By Wald's identity (12.5.19),

$$\begin{aligned}1 &= \mathbb{E}(e^{\theta S_{T(K)}}) \\ &= \mathbb{E}(e^{\theta S_{T(K)}} \mid S_{T(K)} \ge y)\mathbb{P}(S_{T(K)} \ge y) + \mathbb{E}(e^{\theta S_{T(K)}} \mid S_{T(K)} \le -K)\mathbb{P}(S_{T(K)} \le -K).\end{aligned}$$

Since $\theta > 0$,

$$\mathbb{E}(e^{\theta S_{T(K)}} \mid S_{T(K)} \le -K)\mathbb{P}(S_{T(K)} \le -K) \le e^{-\theta K} \to 0 \qquad \text{as } K \to \infty.$$

In addition, by conditioning on $T(K)$, $S_{T(K)-1}$, and the event $S_{T(K)} > y$, we obtain by the lack-of-memory property of the exponential distribution that the excess, $S_{T(K)} - y$, has the exponential distribution with parameter μ. Therefore, $r = \lim_{K\to\infty}\mathbb{P}(S_{T(K)} \ge y)$ satisfies

$$1 = e^{\theta y}\frac{\mu}{\mu - \theta}\cdot r,$$

as required.

13

Diffusion processes

13.2 Solution. Brownian motion

1. Let $s_0 = a, s_1, \dots, s_n = b$ be a partition of $[a, b]$. The associated quadratic variation $Q_{a,b}$ satisfies

$$Q_{a,b} = \sum_{r=0}^{n-1} [W(s_{r+1}) - W(s_r)]^2 \le \left\{ \max_{0 \le r < n} |W(s_{r+1}) - W(s_r)| \right\} \sum_{r=0}^{n-1} |W(s_{r+1}) - W(s_r)|.$$

Since W has continuous sample paths which are uniformly continuous on bounded intervals, the above maximum converges to 0 as the partition is progressively refined. If the total variation exists, then the quadratic variation of W is 0 on bounded intervals, which contradicts the fact that $Q_{a,b} = b - a$ (recall Exercise (8.5.4)).

In a more direct approach, one may use the fact that the increments $W(s_{r+1}) - W(s_r)$ are independent and normally distributed as $N(0, s_{r+1} - s_r)$. The ensuing calculation is easiest when $s_{r+1} - s_r = (b - a)/n$.

13.3 Solutions. Diffusion processes

1. It is easily seen that

$$\mathbb{E}\{X(t+h) - X(t) \mid X(t)\} = (\lambda - \mu)X(t)h + o(h),$$
$$\mathbb{E}(\{X(t+h) - X(t)\}^2 \mid X(t)) = (\lambda + \mu)X(t)h + o(h),$$

which suggest a diffusion approximation with instantaneous mean $a(t, x) = (\lambda - \mu)x$ and instantaneous variance $b(t, x) = (\lambda + \mu)x$.

2. The following method is not entirely rigorous (it is an argument of the following well-known type: it is valid when it works, and not otherwise). We have that

$$\frac{\partial M}{\partial t} = \int_{-\infty}^{\infty} e^{\theta y} \frac{\partial f}{\partial t}\, dy = \int_{-\infty}^{\infty} \{\theta a(t, y) + \tfrac{1}{2}\theta^2 b(t, y)\} e^{\theta y} f \, dy,$$

by using the forward equation and integrating by parts. Assume that $a(t, y) = \sum_n \alpha_n(t) y^n$, $b(t, y) = \sum_n \beta_n(t) y^n$. The required expression follows from the 'fact' that

$$\int_{-\infty}^{\infty} e^{\theta y} y^n f \, dy = \frac{\partial^n}{\partial \theta^n} \int_{-\infty}^{\infty} e^{\theta y} f \, dy = \frac{\partial^n M}{\partial \theta^n}.$$

3. Using Exercise (13.3.2) or otherwise, we obtain the equation

$$\frac{\partial M}{\partial t} = \theta m M + \tfrac{1}{2}\theta^2 M$$

with boundary condition $M(0, \theta) = 1$. The solution is $M(t) = \exp\{\frac{1}{2}\theta(2m + \theta)t\}$.

4. Using Exercise (13.3.2) or otherwise, we obtain the equation

$$\frac{\partial M}{\partial t} = -\theta \frac{\partial M}{\partial \theta} + \tfrac{1}{2}\theta^2 M$$

with boundary condition $M(0, \theta) = 1$. The characteristics of the equation are given by

$$\frac{dt}{1} = \frac{d\theta}{\theta} = \frac{2\,dM}{\theta^2 M},$$

with solution $M(t, \theta) = e^{\frac{1}{4}\theta^2} g(\theta e^{-t})$ where g is a function satisfying $1 = e^{\frac{1}{4}\theta^2} g(\theta)$. Therefore $M = \exp\{\frac{1}{4}\theta^2(1 - e^{-2t})\}$.

5. Fix $t > 0$. Suppose we are given $W_1(s)$, $W_2(s)$, $W_3(s)$, for $0 \le s \le t$. By Pythagoras's theorem, $R(t+u)^2 = X_1^2 + X_2^2 + X_3^2$ where the X_i are independent $N(W_i(t), u)$ variables. Using the result of Exercise (5.7.7), the conditional distribution of $R(t+u)^2$ (and hence of $R(t+u)$ also) depends only on the value of the non-centrality parameter $\theta = R(t)^2$ of the relevant non-central χ^2 distribution. It follows that R satisfies the Markov property. This argument is valid for the n-dimensional Bessel process.

6. By the spherical symmetry of the process, the conditional distribution of $R(s+a)$ given $R(s) = x$ is the same as that given $W(s) = (x, 0, 0)$. Therefore, recalling the solution to Exercise (13.3.5),

$$\begin{aligned}
&\mathbb{P}\big(R(s+a) \le y \,\big|\, R(s) = x\big) \\
&\quad = \int_{\substack{(u,v,w):\\ u^2+v^2+w^2 \le y^2}} \frac{1}{(2\pi a)^{3/2}} \exp\left\{-\frac{(u-x)^2 + v^2 + w^2}{2a}\right\} du\,dv\,dw \\
&\quad = \int_{\rho=0}^{y} \int_{\phi=0}^{2\pi} \int_{\theta=0}^{\pi} \frac{1}{(2\pi a)^{3/2}} \exp\left\{-\frac{\rho^2 - 2\rho x \cos\theta + x^2}{2a}\right\} \rho^2 \sin\theta \, d\theta\, d\phi\, d\rho \\
&\quad = \int_0^y \frac{\rho/x}{\sqrt{2\pi a}} \left\{\exp\left(-\frac{(\rho - x)^2}{2a}\right) - \exp\left(-\frac{(\rho + x)^2}{2a}\right)\right\} d\rho,
\end{aligned}$$

and the result follows by differentiating with respect to y.

7. Continuous functions of continuous functions are continuous. The Markov property is preserved because $g(\cdot)$ is single-valued with a unique inverse.

8. (a) Since $\mathbb{E}(e^{\sigma W(t)}) = e^{\frac{1}{2}\sigma^2 t}$, this is not a martingale.

(b) This is a Wiener process (see Problem (13.12.1)), and is certainly a martingale.

(c) With $\mathcal{F}_t = \sigma(W(s) : 0 \le s \le t)$ and $t, u > 0$,

$$\begin{aligned}
\mathbb{E}\left\{(t+u)W(t+u) - \int_0^{t+u} W(s)\,ds \,\middle|\, \mathcal{F}_t\right\} &= (t+u)W(t) - \int_0^t W(s)\,ds - \int_t^{t+u} W(t)\,ds \\
&= tW(t) - \int_0^t W(s)\,ds,
\end{aligned}$$

whence this is a martingale. [The integrability condition is easily verified.]

9. (a) With $s < t$, $S(t) = S(s)\exp\{a(t-s) + b(W(t) - W(s))\}$. Now $W(t) - W(s)$ is independent of $\{W(u) : 0 \le u \le s\}$, and the claim follows.
(b) $S(t)$ is clearly integrable and adapted to the filtration $\mathcal{F} = (\mathcal{F}_t)$ so that, for $s < t$,

$$\mathbb{E}\big(S(t) \mid \mathcal{F}_s\big) = S(s)\mathbb{E}\big(\exp\{a(t-s) + b(W(t) - W(s))\} \mid \mathcal{F}_s\big) = S(s)\exp\{a(t-s) + \tfrac{1}{2}b^2(t-s)\},$$

which equals $S(s)$ if and only if $a + \frac{1}{2}b^2 = 0$. In this case, $\mathbb{E}(S(t)) = \mathbb{E}(S(0)) = 1$.

10. Either find the instantaneous mean and variance, and solve the forward equation, or argue directly as follows. With $s < t$,

$$\mathbb{P}\big(S(t) \le y \mid S(s) = x\big) = \mathbb{P}\big(bW(t) \le -at + \log y \mid bW(s) = -as + \log x\big).$$

Now $b(W(t) - W(s))$ is independent of $W(s)$ and is distributed as $N(0, b^2(t-s))$, and we obtain on differentiating with respect to y that

$$f(t, y \mid s, x) = \frac{1}{y\sqrt{2\pi b^2(t-s)}} \exp\left(-\frac{(\log(y/x) - a(t-s))^2}{2b^2(t-s)}\right), \qquad x, y > 0.$$

11. One needs to check that the functions $f_{xy}(t) = f(t, y \mid 0, x)$ satisfy

$$f_{xy}(s+t) = \int_{\mathbb{R}} f_{xz}(s) f_{zy}(t)\, dz, \qquad x, y \in \mathbb{R},\ s, t \ge 0.$$

13.4 Solutions. First passage times

1. Certainly X has continuous sample paths, and in addition $\mathbb{E}|X(t)| < \infty$. Also, if $s < t$,

$$\mathbb{E}\big(X(t) \mid \mathcal{F}_s\big) = X(s)e^{\frac{1}{2}\theta^2(t-s)}\mathbb{E}\big(e^{i\theta\{W(t)-W(s)\}} \mid \mathcal{F}_s\big) = X(s)e^{\frac{1}{2}\theta^2(t-s)}e^{-\frac{1}{2}\theta^2(t-s)} = X(s)$$

as required, where we have used the fact that $W(t) - W(s)$ is $N(0, t-s)$ and is independent of $\mathcal{F}_s$.

2. Apply the optional stopping theorem to the martingale X of Exercise (13.4.1), with the stopping time T, to obtain $\mathbb{E}(X(T)) = 1$. Now $W(T) = aT + b$, and therefore $\mathbb{E}(e^{-\psi T + i\theta b}) = 1$ where $-\psi = ia\theta + \frac{1}{2}\theta^2$. Solve to find that

$$\mathbb{E}(e^{-\psi T}) = e^{-i\theta b} = \exp\left\{|a|b - b\sqrt{a^2 + 2\psi}\right\}$$

is the solution which gives the Laplace transform of a density function.

3. We have that $T \le u$ if and only if there is no zero in $(u, t]$, an event with probability $1 - (2/\pi)\cos^{-1}\{\sqrt{u/t}\}$, and the claim follows on drawing a triangle.

4. By Theorem (13.4.5) and using conditional probability, the required density function is

$$\int_0^\infty \frac{1}{\sqrt{2\pi t}} \exp\left(-\frac{y^2}{2t}\right) \cdot \frac{x}{\sqrt{2\pi t^3}} \exp\left(-\frac{x^2}{2t}\right) dt = \frac{x}{\pi(x^2 + y^2)}.$$

5. This agrees with Theorem (13.4.6) when $a = 0$. When $a < 0$, it suffices to show, by the result of Problem (5.12.18c) or otherwise, that the given density function f has the Laplace transform obtained in Exercise (13.4.2). One makes the change of variables $x = v^{-2}$ to obtain

$$\begin{aligned}\int_0^\infty f(x)e^{-\psi x}\,dx &= \frac{b}{\sqrt{2\pi}}\int_0^\infty 2\exp\left\{-\frac{\psi}{v^2} - \frac{a^2}{2v^2} - ab - \frac{1}{2}b^2v^2\right\}dv \\ &= e^{|a|b}\exp\left\{-b\sqrt{a^2+2\psi}\right\}.\end{aligned}$$

The logarithm of the Laplace transform is $L(\psi) := |a|b - b\sqrt{a^2+2\psi}$, with inverse $\psi(L) = -|a|L/b + \frac{1}{2}(L/b)^2$, which is the cumulant generating function of a normal variable, whence the name.

6. As in the proof of Theorem (13.4.6), for $0 \le w \le m$,

$$\begin{aligned}\mathbb{P}\big(M(t) \ge m,\ W(t) \le w\big) &= \mathbb{P}\big(M(t) \ge m,\ W(t) - W(T(m)) \le w - m\big) \\ &= \mathbb{P}\big(M(t) \ge m,\ W(t) - W(T(m)) \ge m - w\big) \\ &= \mathbb{P}(W(t) \ge 2m - w),\end{aligned}$$

where $T(m)$ is the first passage time to m. The claim follows since $W(t)$ has the $N(0, t)$ distribution.

7. The first passage time $T(x)$ has density function as in Theorem (13.4.5). Make the change of variables $y = \sqrt{t}$, with Jacobian $2y$, to deduce that the density function of $Y = \sqrt{T(x)}$ is

$$f_Y(y) = \frac{2|x|}{y^2\sqrt{2\pi}}\exp\left(-\frac{x^2}{2y^2}\right), \qquad y > 0.$$

Let Z be an $N(0, \sigma^2)$ variable, and $U = 1/|Z|$. The density function of U is found by making the change of variable $u = 1/|z|$ in the $N(0, \sigma^2)$ density function. The inverse map $z \mapsto u$ is two-to-one, whence

$$f_U(u) = \frac{1}{\sigma\sqrt{2\pi}}\cdot\frac{2}{u^2}\exp\left(-\frac{1}{2\sigma^2u^2}\right), \qquad u > 0.$$

Thus $f_Y = f_U$ when $\sigma = 1/|x|$.

8. By Theorem (13.4.5) and a change of variables, $T(x) \stackrel{\text{D}}{=} x^2T(1)$. Self-similarity holds since, for $a > 0$,

$$T(ax) \stackrel{\text{D}}{=} (ax)^2T(1) \stackrel{\text{D}}{=} a^2T(x).$$

By Exercise (13.4.2) with $a = 0$ and $b = x \ge 0$,

$$\mathbb{E}(e^{-\psi T(x)}) = e^{-x\sqrt{2\psi}}, \qquad \psi \ge 0.$$

This equation determines the distribution of $T(x)$ as the Lévy first passage density (recall Exercise (5.10.7)), and furthermore this distribution is $\frac{1}{2}$-stable, by Theorem (8.7.16).

9. Let $\mathcal{F}$ denote the natural filtration. With $\Delta = X(h) - X(0)$, we have $\mathbb{E}(\Delta) = axh + \mathrm{o}(h)$ and $\mathbb{E}(\Delta^2) = bxh + \mathrm{o}(h)$. We derive a backward equation for m by conditioning on events in the interval $[0, h]$ thus:

$$\begin{aligned}m(x) &= \mathbb{E}_x\left[\mathbb{E}_x\left(\exp\left\{-\psi\int_h^T X(u)\,du\right)\exp\left\{-\psi\int_0^h X(u)\,du\right\}\,\Big|\,\mathcal{F}_h\right)\right] \\ &= \mathbb{E}_x\left[m(x+\Delta)\big(1 - \psi xh + \mathrm{o}(h)\big)\right] \\ &= \mathbb{E}_x[m(x+\Delta)]\big(1 - \psi xh + \mathrm{o}(h)\big) \\ &= \big\{m(x) + \mathbb{E}(\Delta)m'(x) + \tfrac{1}{2}\mathbb{E}(\Delta^2)m''(x) + \mathrm{o}(h)\big\}\big(1 - \psi xh + \mathrm{o}(h)\big) \\ &= m(x) + hx\left[\tfrac{1}{2}bm''(x) + am'(x) - \psi m(x)\right] + \mathrm{o}(h),\end{aligned}$$

where we have taken some liberties over rigour, and, for conciseness, we have subsumed the argument ψ of the function m. Cancel where possible, divide by h and take the limit as $h \to 0$.

Subject to the boundary conditions $m(0, \psi) = 1$ and $\lim_{x\to\infty} m(x, \psi) = 0$ for $\psi > 0$, the solution to this second-order equation is

$$m(x, \psi) = \exp\left\{-\frac{ax}{b} - \frac{x}{b}\sqrt{a^2 + 2b\psi}\right\}.$$

This may be seen to be the Laplace transform of the first-passage density of Exercise (13.4.2), where the target line has equation $u\sqrt{b} = av + x$.

13.5 Solutions. Barriers

1. Solving the forward equation subject to the appropriate boundary conditions, we obtain as usual that

$$f^{\mathrm{r}}(t, y) = g(t, y \mid d) + e^{-2md} g(t, y \mid -d) - \int_{-\infty}^{-d} 2me^{2mx} g(t, y \mid x)\, dx$$

where $g(t, y \mid x) = (2\pi t)^{-\frac{1}{2}} \exp\{-(y - x - mt)^2/(2t)\}$. The first two terms tend to 0 as $t \to \infty$, regardless of the sign of m. As for the integral, make the substitution $u = (x - y - mt)/\sqrt{t}$ to obtain, as $t \to \infty$,

$$-\int_{-\infty}^{-(d+y+mt)/\sqrt{t}} 2me^{2my} \frac{e^{-\frac{1}{2}u^2}}{\sqrt{2\pi}}\, du \to \begin{cases} 2|m|e^{-2|m|y} & \text{if } m < 0, \\ 0 & \text{if } m \ge 0. \end{cases}$$

2. Let A be the event that W is absorbed at a. By the result of Problem (12.9.24), we have $\mathbb{P}_x(A) = x/a$ where $\mathbb{P}_x$ denotes the probability measure conditional on $W(0) = x$. The transition probability densities $f(y, t \mid x) := f(y, t \mid x, 0)$ of W and C satisfy

$$\begin{aligned} f_C(y, t \mid x)\, dy &= \mathbb{P}_x\big(W(t) \in (y, y + dy) \,\big|\, A\big) + \mathrm{o}(dy) \\ &= \frac{1}{\mathbb{P}_x(A)} \mathbb{P}_x\big(W(t) \in (y, y + dy)\big) \mathbb{P}_x(A \mid W(t) = y) + \mathrm{o}(dy) \\ &= \frac{\mathbb{P}_y(A)}{\mathbb{P}_x(A)} f_W(y, t \mid x)\, dy + \mathrm{o}(dy), \end{aligned}$$

so that $f_C(y, t \mid x) = (y/x) f_W(y, t \mid x)$. The 'increment' $\Delta = C(t + h) - C(t)$, given $C(t) = c \in (0, a)$, satisfies

$$\mathbb{E}_c(\Delta) = \int_0^a \frac{y}{c} \cdot (y - c) f_W(y, h \mid c)\, dy + \mathrm{o}(h),$$

whence, by equation (13.3.4), $\mathbb{E}_c(\Delta)/h \to 1/c$ as $h \downarrow 0$. We have used the facts that W has instantaneous mean 0 and variance 1. Likewise,

$$\mathbb{E}_c(\Delta^2) = \int_0^a \frac{y}{c} \cdot (y - c)^2 f_W(y, h \mid c)\, dy + \mathrm{o}(h) = h + \mathrm{o}(h),$$

since W has instantaneous third moment 0.

3. We shall bravely use Taylor's theorem without worrying about the analytical details. By conditioning on the increment $\Delta = C(h) - C(0)$, the function $m(x) = \mathbb{E}_x(T)$ satisfies

$$\begin{aligned} m(x) &= h + \mathbb{E}_{x+\Delta}(T) = h + \mathbb{E}m(x + \Delta) \\ &= h + \mathbb{E}\left\{m(x) + \Delta m'(x) + \frac{1}{2}\Delta^2 m''(x) + \mathrm{O}(\Delta^3)\right\} \\ &= h + m(x) + \frac{h}{x} m'(x) + \frac{1}{2} h m''(x) + \mathrm{o}(h). \end{aligned}$$

Divide by h and let $h \downarrow 0$ to obtain

$$\frac{1}{2}m''(x) + \frac{1}{x}m'(x) + 1 = 0,$$

which we integrate, subject to $m(a) = 0$ and $m(x) < \infty$ for $x \in (0, 1)$, to obtain the result. The inequality $m(x) < \infty$ holds since C has a rightward drift, so its mean passage time to 1 is dominated by that of a standard Wiener process (see Theorem (12.4.5)).
[A discrete version of this question appeared at Exercise (3.9.2).]

13.6 Solutions. Excursions and the Brownian bridge

1. Let $f(t, x) = (2\pi t)^{-\frac{1}{2}} e^{-x^2/(2t)}$. It may be seen that

$$\mathbb{P}\big(W(t) > x \,\big|\, Z, W(0) = 0\big) = \lim_{w \downarrow 0} \mathbb{P}\big(W(t) > x \,\big|\, Z, W(0) = w\big)$$

where $Z = \{\text{no zeros in } (0, t]\}$; the small missing step here may be filled by conditioning instead on the event $\{W(\epsilon) = w, \text{ no zeros in } (\epsilon, t]\}$, and taking the limit as $\epsilon \downarrow 0$. Now, if $w > 0$,

$$\mathbb{P}\big(W(t) > x, Z \,\big|\, W(0) = w\big) = \int_x^\infty \{f(t, y - w) - f(t, y + w)\}\, dy$$

by the reflection principle, and

$$\mathbb{P}\big(Z \,\big|\, W(0) = w\big) = 1 - 2\int_w^\infty f(t, y)\, dy = \int_{-w}^{w} f(t, y)\, dy$$

by a consideration of the minimum value of W on $(0, t]$. It follows that the density function of $W(t)$, conditional on $Z \cap \{W(0) = w\}$, where $w > 0$, is

$$h_w(x) = \frac{f(t, x - w) - f(t, x + w)}{\int_{-w}^{w} f(t, y)\, dy}, \qquad x > 0.$$

Divide top and bottom by $2w$, and take the limit as $w \downarrow 0$:

$$\lim_{w \downarrow 0} h_w(x) = -\frac{1}{f(t, 0)} \frac{\partial f}{\partial x} = \frac{x}{t} e^{-x^2/(2t)}, \qquad x > 0.$$

2. It is a standard exercise that, for a Wiener process W,

$$\mathbb{E}\big\{W(t) \,\big|\, W(s) = a, W(1) = 0\big\} = a\left(\frac{1 - t}{1 - s}\right),$$

$$\mathbb{E}\big\{W(s)^2 \,\big|\, W(0) = W(1) = 0\big\} = s(1 - s),$$

if $0 \le s \le t \le 1$. Therefore the Brownian bridge B satisfies, for $0 \le s \le t \le 1$,

$$\mathbb{E}\big(B(s)B(t)\big) = \mathbb{E}\big\{B(s)\mathbb{E}\big(B(t) \,\big|\, B(s)\big)\big\} = \frac{1 - t}{1 - s}\mathbb{E}(B(s)^2) = s(1 - t)$$

as required. Certainly $\mathbb{E}(B(s)) = 0$ for all s, by symmetry.

3. $\widehat{W}$ is a zero-mean Gaussian process on $[0, 1]$ with continuous sample paths, and also $\widehat{W}(0) = \widehat{W}(1) = 0$. Therefore $\widehat{W}$ is a Brownian bridge if it has the same autocovariance function as the Brownian bridge, that is, $c(s, t) = \min\{s, t\} - st$. For $s < t$,

$$\operatorname{cov}\big(\widehat{W}(s), \widehat{W}(t)\big) = \operatorname{cov}\big(W(s) - sW(1), W(t) - tW(1)\big) = s - ts - st + st = s - st$$

since $\operatorname{cov}(W(u), W(v)) = \min\{u, v\}$. The claim follows.

4. Either calculate the instantaneous mean and variance of $\widetilde{W}$, or repeat the argument in the solution to Exercise (13.6.3). The only complication in this case is the necessity to show that $\widetilde{W}(t)$ is a.s. continuous at $t = 1$, i.e. that $u^{-1}W(u-1) \to 0$ a.s. as $u \to \infty$. There are various ways to show this. Certainly it is true in the limit as $u \to \infty$ through the integers, since, for integral u, $W(u-1)$ may be expressed as the sum of $u-1$ independent $N(0, 1)$ variables (use the strong law). It remains to fill in the gaps. Let n be a positive integer, let $x > 0$, and write $M_n = \max\{|W(u) - W(n)| : n \le u \le n+1\}$. We have by the stationarity of the increments that

$$\sum_{n=0}^{\infty} \mathbb{P}(M_n \ge nx) = \sum_{n=0}^{\infty} \mathbb{P}(M_1 \ge nx) \le 1 + \frac{\mathbb{E}(M_1)}{x} < \infty,$$

implying by the Borel–Cantelli lemma that $n^{-1}M_n \le x$ for all but finitely many values of n, a.s. Therefore $n^{-1}M_n \to 0$ a.s. as $n \to \infty$, implying that

$$\lim_{u\to\infty} \frac{1}{u+1}|W(u)| \le \lim_{n\to\infty} \frac{1}{n}\{|W(n)| + M_n\} \to 0 \qquad \text{a.s.}$$

We have that

$$\begin{aligned}\Big\{\sup_{0\le t\le 1} \widetilde{W}(t) > m\Big\} &= \Big\{\sup_{0\le t\le 1} (1-t)W\Big(\frac{t}{1-t}\Big) > m\Big\} \\ &= \Big\{\sup_{0\le s<\infty} W(s) > m(1+s)\Big\} \qquad \text{on setting } s = \frac{t}{1-t} \\ &= \Big\{\sup_{0\le s<\infty} D_{-m}(s) > m\Big\},\end{aligned}$$

where D_{-m} is the Wiener process with drift $-m$. The result follows by Corollary (13.4.14).

5. In the notation of Exercise (13.6.4), we are asked to calculate the probability that W has no zeros in the time interval between $s/(1-s)$ and $t/(1-t)$. By Theorem (13.4.8), this equals

$$1 - \frac{2}{\pi}\cos^{-1}\sqrt{\frac{s(1-t)}{t(1-s)}} = \frac{2}{\pi}\cos^{-1}\sqrt{\frac{t-s}{t(1-s)}}.$$

13.7 Solutions. Stochastic calculus

1. Let $\mathcal{F}_s = \sigma(W_u : 0 \le u \le s)$. Fix $n \ge 1$ and define $X_n(k) = |W_{kt/2^n}|$ for $0 \le k \le 2^n$. By Jensen's inequality, the sequence $\{X_n(k) : 0 \le k \le 2^n\}$ is a non-negative submartingale with respect to the filtration $\mathcal{F}_{kt/2^n}$, with finite variance. Hence, by Exercise (4.3.3) and equation (12.6.2), $X_n^* = \max\{X_n(k) : 0 \le k \le 2^n\}$ satisfies

$$\mathbb{E}(X_n^{*2}) = 2\int_0^\infty x\mathbb{P}(X_n^* > x)\,dx \le 2\int_0^\infty \mathbb{E}(W_t^+ I_{\{X_n^* \ge x\}})\,dx = 2\mathbb{E}\left\{W_t^+ \int_0^{X_n^*} dx\right\}$$
$$= 2\mathbb{E}(W_t^+ X_n^*) \le 2\sqrt{\mathbb{E}(W_t^2)\mathbb{E}(X_n^{*2})} \qquad \text{by the Cauchy–Schwarz inequality.}$$

Hence $\mathbb{E}(X_n^{*2}) \le 4\mathbb{E}(W_t^2)$. Now X_n^{*2} is monotone increasing in n, and W has continuous sample paths. By monotone convergence,

$$\mathbb{E}\Big(\max_{s\le t}|W_s|^2\Big) = \lim_{n\to\infty}\mathbb{E}(X_n^{*2}) \le 4\mathbb{E}(W_t^2).$$

2. See the solution to Exercise (8.5.4).

3. (a) We have that

$$I_1(n) = \tfrac{1}{2}\left\{\sum_{j=0}^{n-1}(V_{j+1}^2 - V_j^2) - \sum_{j=0}^{n-1}(V_{j+1} - V_j)^2\right\}.$$

The first summation equals W_t^2, by successive concellation, and the mean-square limit of the second summation is t, by Exercise (8.5.4). Hence $\lim_{n\to\infty} I_1(n) = \frac{1}{2}W_t^2 - \frac{1}{2}t$ in mean square.

Likewise, we obtain the mean-square limits:

$$\lim_{n\to\infty} I_2(n) = \tfrac{1}{2}W_t^2 + \tfrac{1}{2}t, \qquad \lim_{n\to\infty} I_3(n) = \lim_{n\to\infty} I_4(n) = \tfrac{1}{2}W_t^2.$$

4. Clearly $\mathbb{E}(U(t)) = 0$. The process U is Gaussian with autocovariance function

$$\mathbb{E}\big(U(s)U(s+t)\big) = \mathbb{E}\Big(\mathbb{E}\big(U(s)U(s+t)\,\big|\,\mathcal{F}_s\big)\Big) = e^{-\beta s}e^{-\beta(s+t)}\mathbb{E}\big[W(e^{2\beta s})^2\big] = e^{-\beta t}.$$

Thus U is a stationary Gaussian Markov process, namely the Ornstein–Uhlenbeck process. [See Example (9.6.10).]

5. Clearly $\mathbb{E}(U_t) = 0$. For $s < t$,

$$\mathbb{E}(U_s U_{s+t}) = \mathbb{E}(W_s W_t) + \beta^2\mathbb{E}\left(\int_{u=0}^{s}\int_{v=0}^{t} e^{-\beta(s-u)}W_u e^{-\beta(t-v)}W_v\,du\,dv\right)$$
$$- \mathbb{E}\left(W_t\beta\int_0^s e^{-\beta(s-u)}W_u\,du\right) - \mathbb{E}\left(W_s\int_0^t e^{-\beta(t-v)}W_v\,dv\right)$$
$$= s + \beta^2 e^{-\beta(s+t)}\int_{u=0}^{s}\int_{v=0}^{t} e^{\beta(u+v)}\min\{u,v\}\,du\,dv$$
$$- \beta\int_0^s e^{-\beta(s-u)}\min\{u,t\}\,du - \int_0^t e^{-\beta(t-v)}\min\{s,v\}\,dv$$
$$= \frac{e^{2\beta s}-1}{2\beta}e^{-\beta(s+t)}$$

after prolonged integration. By the linearity of the definition of U, it is a Gaussian process. From the calculation above, it has autocovariance function $c(s,s+t) = (e^{-\beta(t-s)} - e^{-\beta(t+s)})/(2\beta)$. From this we may calculate the instantaneous mean and variance, and thus we recognize an Ornstein–Uhlenbeck process. See also Exercise (13.3.4) and Problem (13.12.4).

13.8 Solutions. The Itô integral

1. (a) Fix $t > 0$ and let $n \geq 1$ and $\delta = t/n$. We write $t_j = jt/n$ and $V_j = W_{t_j}$. By the absence of correlation of Wiener increments, and the Cauchy–Schwarz inequality,

$$\begin{aligned}
\mathbb{E}\left(\left|\int_0^t W_s\,ds - \sum_{j=0}^{n-1} V_{j+1}(t_{j+1}-t_j)\right|^2\right) &= \mathbb{E}\left(\left|\sum_{j=0}^{n-1}\int_{t_j}^{t_{j+1}} (V_{j+1}-W_s)\,ds\right|^2\right) \\
&= \sum_{j=0}^{n-1} \mathbb{E}\left(\left|\int_{t_j}^{t_{j+1}} (V_{j+1}-W_s)\,ds\right|^2\right) \\
&\leq \sum_{j=0}^{n-1}\left\{(t_{j+1}-t_j)\int_{t_j}^{t_{j+1}} \mathbb{E}(|V_{j+1}-W_s|^2)\,ds\right\} \\
&= \sum_{j=0}^{n-1} \frac{1}{2}(t_{j+1}-t_j)^3 = \sum_{j=0}^{n-1}\frac{1}{2}\left(\frac{t}{n}\right)^3 \to 0 \quad \text{as } n\to\infty.
\end{aligned}$$

Therefore,

$$\begin{aligned}
\int_0^t s\,dW_s &= \lim_{n\to\infty}\sum_{j=0}^{n-1} t_j(V_{j+1}-V_j) = \lim_{n\to\infty}\sum_{j=0}^{n-1}\big(t_{j+1}V_{j+1} - t_jV_j - (t_{j+1}-t_j)V_{j+1}\big) \\
&= \lim_{n\to\infty}\left(tW_t - \sum_{j=0}^{n-1} V_{j+1}(t_{j+1}-t_j)\right) = tW_t - \int_0^t W_s\,ds.
\end{aligned}$$

(b) As $n \to \infty$,

$$\begin{aligned}
\sum_{j=0}^{n-1} V_j^2(V_{j+1}-V_j) &= \tfrac{1}{3}\sum_{j=0}^{n-1}\{V_{j+1}^3 - V_j^3 - 3V_j(V_{j+1}-V_j)^2 - (V_{j+1}-V_j)^3\} \\
&= \tfrac{1}{3}W_t^3 - \sum_{j=0}^{n-1}\Big[V_j(t_{j+1}-t_j) + V_j\{(V_{j+1}-V_j)^2 - (t_{j+1}-t_j)\}\Big] - \tfrac{1}{3}\sum_{j=0}^{n-1}(V_{j+1}-V_j)^3 \\
&\to \tfrac{1}{3}W_t^3 - \int_0^t W(s)\,ds + 0 + 0.
\end{aligned}$$

The fact that the last two terms tend to 0 in mean square may be verified in the usual way. For example,

$$\begin{aligned}
\mathbb{E}\left(\left|\sum_{j=0}^{n-1}(V_{j+1}-V_j)^3\right|^2\right) &= \sum_{j=0}^{n-1}\mathbb{E}[(V_{j+1}-V_j)^6] \\
&= 6\sum_{j=0}^{n-1}(t_{j+1}-t_j)^3 = 6\sum_{j=0}^{n-1}\left(\frac{t}{n}\right)^3 \to 0 \qquad \text{as } n\to\infty.
\end{aligned}$$

(c) It was shown in Exercise (13.7.3a) that $\int_0^t W_s\,dW_s = \frac{1}{2}W_t^2 - \frac{1}{2}t$. Hence,

$$\mathbb{E}\left(\left[\int_0^t W_s\,dW_s\right]^2\right) = \tfrac{1}{4}\{\mathbb{E}(W_t^4) - 2t\mathbb{E}(W_t^2) + t^2\} = \tfrac{1}{2}t^2,$$

and the result follows because $\mathbb{E}(W_t^2) = t$.

2. Fix $t > 0$ and $n \geq 1$, and let $\delta = t/n$. We set $V_j = W_{jt/n}$. It is the case that $X_t = \lim_{n\to\infty} \sum_j V_j(t_{j+1} - t_j)$. Each term in the sum is normally distributed, and all partial sums are multivariate normal for all $\delta > 0$, and hence also in the limit as $\delta \to 0$. Obviously $\mathbb{E}(X_t) = 0$. For $s \leq t$,

$$\begin{aligned}\mathbb{E}(X_s X_t) &= \int_0^t \int_0^s \mathbb{E}(W_u W_v)\,du\,dv = \int_0^t \int_0^s \min\{u, v\}\,du\,dv \\ &= \int_0^s \tfrac{1}{2}u^2\,du + \int_0^s u(t-u)\,du = s^2\left(\frac{t}{2} - \frac{s}{6}\right).\end{aligned}$$

Hence $\operatorname{var}(X_t) = \frac{1}{3}t^3$, and the autocovariance function is

$$\rho(X_s, X_t) = 3\sqrt{\frac{s}{t}}\left(\frac{1}{2} - \frac{s}{6t}\right).$$

3. By the Cauchy–Schwarz inequality, as $n \to \infty$,

$$\mathbb{E}\big[\{\mathbb{E}(X_n \mid \mathcal{G}) - \mathbb{E}(X \mid \mathcal{G})\}^2\big] \leq \mathbb{E}\big[\mathbb{E}\{(X_n - X)^2 \mid \mathcal{G}\}\big] = \mathbb{E}[(X_n - X)^2] \to 0.$$

4. We square the equation $\|I(\psi_1 + \psi_2)\|_2 = \|\psi_1 + \psi_2\|$ and use the fact that $\|I(\psi_i)\|_2 = \|\psi_i\|$ for $i = 1, 2$, to deduce the result.

5. The question permits us to use the integrating factor $e^{\beta t}$ to give, formally,

$$e^{\beta t} X_t = \int_0^t e^{\beta s}\frac{dW_s}{ds}\,ds = e^{\beta t} W_t - \beta \int_0^t e^{\beta s} W_s\,ds$$

on integrating by parts. This is the required result, and substitution verifies that it satisfies the given equation.

6. Find a sequence $\boldsymbol{\phi} = (\phi^{(n)})$ of predictable step functions such that $\|\phi^{(n)} - \psi\| \to 0$ as $n \to \infty$. By the argument before equation (13.8.9), $I(\phi^{(n)}) \xrightarrow{\text{m.s.}} I(\psi)$ as $n \to \infty$. By Lemma (13.8.4), $\|I(\phi^{(n)})\|_2 = \|\phi^{(n)}\|$, and the claim follows.

13.9 Solutions. Itô's formula

1. The process Z is continuous and adapted with $Z_0 = 0$. We have by Theorem (13.8.11) that $\mathbb{E}(Z_t - Z_s \mid \mathcal{F}_s) = 0$, and by Exercise (13.8.6) that

$$\mathbb{E}([Z_t - Z_s]^2 \mid \mathcal{F}_s) = \mathbb{E}\left(\int_s^t \frac{X_u^2 + Y_u^2}{R_u^2}\,du\right) = t - s.$$

The first claim follows by the Lévy characterization of a Wiener process (12.7.10).

We have in n dimensions that $R^2 = X_1^2 + X_2^2 + \cdots + X_n^2$, and the same argument yields that $Z_t = \sum_i \int_0^t (X_i/R)\,dX_i$ is a Wiener process. By Example (13.9.7) and the above,

$$d(R^2) = 2\sum_{i=1}^n X_i\,dX_i + n\,dt = 2R\sum_{i=1}^n \frac{X_i}{R}\,dX_i + n\,dt = 2R\,dW + n\,dt.$$

2. Applying Itô's formula (13.9.4) to $Y_t = W_t^4$ we obtain $dY_t = 4W_t^3\,dW_t + 6W_t^2\,dt$. Hence,

$$\mathbb{E}(Y_t) = \mathbb{E}\Big(\int_0^t 4W_s^3\,dW_s\Big) + \mathbb{E}\Big(\int_0^t 6W_s^2\,ds\Big) = 6\int_0^t s\,ds = 3t^2.$$

3. Apply Itô's formula (13.9.4) to obtain $dY_t = (Y_t/t)\,dt + t\,dW_t$. Cf. Exercise (13.8.1).

4. Note that $X_1 = \cos W$ and $X_2 = \sin W$. By Itô's formula (13.9.4),

$$\begin{aligned} dY = d(X_1 + iX_2) &= dX_1 + i\,dX_2 = d(\cos W) + i\,d(\sin W) \\ &= -\sin W\,dW - \tfrac{1}{2}\cos W\,dt + i\cos W\,dW - \tfrac{1}{2}\sin W\,dt. \end{aligned}$$

5. We apply Itô's formula to obtain:
(a) $(1+t)\,dX = -X\,dt + dW$,
(b) $dX = -\frac{1}{2}X\,dt + \sqrt{1-X^2}\,dW$,
(c) $d\begin{pmatrix} X \\ Y \end{pmatrix} = -\frac{1}{2}\begin{pmatrix} X \\ Y \end{pmatrix} dt + \begin{pmatrix} 0 & -a/b \\ b/a & 0 \end{pmatrix}\begin{pmatrix} X \\ Y \end{pmatrix} dW.$

6. Use Itô's simple formula (13.9.4) with $f(t, w) = \sinh(t + w)$.

13.10 Solutions. Option pricing

1. (a) We have that

$$\begin{aligned} \mathbb{E}\big((ae^Z - K)^+\big) &= \int_{\log(K/a)}^{\infty} (ae^z - K)\frac{1}{\sqrt{2\pi\tau^2}}\exp\left(-\frac{(z-\gamma)^2}{2\tau^2}\right) dz \\ &= \int_{\alpha}^{\infty} (ae^{\gamma+\tau y} - K)\frac{e^{-\frac{1}{2}y^2}}{\sqrt{2\pi}}\,dy \quad \text{where } y = \frac{z-\gamma}{\tau},\ \alpha = \frac{\log(K/a)-\gamma}{\tau} \\ &= ae^{\gamma+\frac{1}{2}\tau^2}\int_{\alpha}^{\infty}\frac{e^{-\frac{1}{2}(y-\tau)^2}}{\sqrt{2\pi}}\,dy - K\Phi(-\alpha) \\ &= ae^{\gamma+\frac{1}{2}\tau^2}\Phi(\tau-\alpha) - K\Phi(-\alpha). \end{aligned}$$

(b) We have that $S_T = ae^Z$ where $a = S_t$ and, under the relevant conditional $\mathbb{Q}$-distribution, Z is normal with mean $\gamma = (r - \frac{1}{2}\sigma^2)(T-t)$ and variance $\tau^2 = \sigma^2(T-t)$. The claim now follows by the result of part (a).

2. (a) Set $\xi(t, S) = \xi(t, S_t)$ and $\psi(t, S) = \psi(t, S_t)$, in the natural notation. By Theorem (13.10.15), we have $\psi_x = \psi_t = 0$, whence $\psi(t, x) = c$ for all t, x, and some constant c.
(b) Recall that $dS = \mu S\,dt + \sigma S\,dW$. Now,

$$d(\xi S + \psi e^{rt}) = d(S^2 + \psi e^{rt}) = (\sigma S)^2\,dt + 2S\,dS + e^{rt}\,d\psi + \psi r e^{rt}\,dt,$$

by Example (13.9.7). By equation (13.10.4), the portfolio is self-financing if this equals $S\,dS + \psi r e^{rt}\,dt$, and thus we arrive at the SDE $e^{rt}\,d\psi = -S\,dS - \sigma^2 S^2\,dt$, whence

$$\psi(t, S) = -\int_0^t e^{-ru} S_u\,dS_u - \sigma^2\int_0^t e^{-ru} S_u^2\,du.$$

(c) Note first that $Z_t = \int_0^t S_u\,du$ satisfies $dZ_t = S_t\,dt$. By Example (13.9.8), $d(S_t Z_t) = Z_t\,dS_t + S_t^2\,dt$, whence

$$d(\xi S + \psi e^{rt}) = Z_t\,dS_t + S_t^2\,dt + e^{rt}\,d\psi + re^{rt}\,dt.$$

Using equation (13.10.4), the portfolio is self-financing if this equals $Z_t\,dS_t + \psi re^{rt}\,dt$, and thus we require that $e^{rt}\,d\psi = -S_t^2\,dt$, which is to say that

$$\psi(t, S) = -\int_0^t e^{-ru} S_u^2\,du.$$

3. We need to check equation (13.10.4) remembering that $dM_t = 0$. Each of these portfolios is self-financing.

(a) This case is obvious.

(b) $d(\xi S + \psi) = d(2S^2 - S^2 - t) = 2S\,dS + dt - dt = \xi\,dS$.

(c) $d(\xi S + \psi) = -S - t\,dS + S = \xi\,dS$.

(d) Recalling Example (13.9.8), we have that

$$d(\xi S + \psi) = d\left(S_t \int_0^t S_s\,ds - \int_0^t S_s^2\,ds\right) = S_t^2\,dt + dS_t \int_0^t S_s\,ds - S_t^2\,dt = \xi\,dS_t.$$

4. The time of exercise of an American call option must be a stopping time for the filtration $(\mathcal{F}_t)$. The value of the option, if exercised at the stopping time τ, is $V_\tau = (S_\tau - K)^+$, and it follows by the usual argument that the value at time 0 of the option exercised at τ is $\mathbb{E}_{\mathbb{Q}}(e^{-r\tau}V_\tau)$. Thus the value at time 0 of the American option is $\sup_\tau\{\mathbb{E}_{\mathbb{Q}}(e^{-r\tau}V_\tau)\}$, where the supremum is taken over all stopping times τ satisfying $\mathbb{P}(\tau \le T) = 1$. Under the probability measure $\mathbb{Q}$, the process $e^{-rt}V_t$ is a martingale, whence, by the optional stopping theorem, $\mathbb{E}_{\mathbb{Q}}(e^{-r\tau}V_\tau) = V_0$ for all stopping times τ. The claim follows.

5. We rewrite the value at time 0 of the European call option, possibly with the aid of Exercise (13.10.1), as

$$e^{-rT}\mathbb{E}\Big(\big(S_0\exp\{rT - \tfrac{1}{2}\sigma^2 T + \sigma\sqrt{T}N\} - K\big)^+\Big) = \mathbb{E}\Big(\big(S_0\exp\{-\tfrac{1}{2}\sigma^2 T + \sigma\sqrt{T}N\} - Ke^{-rT}\big)^+\Big),$$

where N is an $N(0,1)$ random variable. It is immediate that this is increasing in S_0 and r and is decreasing in K. To show monotonicity in T, we argue as follows. Let $T_1 < T_2$ and consider the European option with exercise date T_2. In the corresponding American option we are allowed to exercise the option at the earlier time T_1. By Exercise (13.10.4), it is never better to stop earlier than T_2, and the claim follows.

Monotonicity in σ may be shown by differentiation.

13.11 Solutions. Passage probabilities and potentials

1. Let H be a closed sphere with radius $R\ (> |w|)$, and define $p_R(r) = \mathbb{P}\big(G \text{ before } H \,\big|\, |W(0)| = r\big)$. Then p_R satisfies Laplace's equation in $\mathbb{R}^d$, and hence

$$\frac{d}{dr}\left(r^{d-1}\frac{dp_R}{dr}\right) = 0$$

since p_R is spherically symmetric. Solve subject to the boundary equations $p_R(\epsilon) = 1$, $p_R(R) = 0$, to obtain

$$p_R(r) = \frac{r^{2-d} - R^{2-d}}{\epsilon^{2-d} - R^{2-d}} \to (\epsilon/r)^{d-2} \qquad \text{as } R \to \infty.$$

2. The electrical resistance R_n between 0 and the set Δ_n is no smaller than the resistance obtained by, for every $i = 1, 2, \dots$, 'shorting out' all vertices in the set Δ_i. This new network amounts to a linear chain of resistances in series, points labelled Δ_i and Δ_{i+1} being joined by a resistance if size N_i^{-1}. It follows that

$$R(G) = \lim_{n\to\infty} R_n \geq \sum_{i=0}^{\infty} \frac{1}{N_i}.$$

By Theorem (13.11.18), the walk is recurrent if $\sum_i N_i^{-1} = \infty$.

3. Thinking of G as an electrical network, one may obtain the network H by replacing the resistance of every edge e lying in G but not in H by ∞. Let 0 be a vertex of H. By a well known fact in the theory of electrical networks, $R(H) \geq R(G)$, and the result follows by Theorem (13.11.18).

13.12 Solutions to problems

1. (a) $T(t) = \alpha W(t/\alpha^2)$ has continuous sample paths with stationary independent increments, since W has these properties. Also $T(t)/\alpha$ is $N(0, t/\alpha^2)$, whence $T(t)$ is $N(0, t)$.

(b) As for part (a).

(c) Certainly V has continuous sample paths on $(0, \infty)$. For continuity at 0 it suffices to prove that $tW(t^{-1}) \to 0$ a.s. as $t \downarrow 0$; this was done in the solution to Exercise (13.6.4).

If (u, v), (s, t) are disjoint time-intervals, then so are (v^{-1}, u^{-1}), (t^{-1}, s^{-1}); since W has independent increments, so has V. Finally,

$$V(s+t) - V(s) = tW((s+t)^{-1}) - s\{W(s^{-1}) - W((s+t)^{-1})\}$$

is $N(0, \beta)$ if $s, t > 0$, where

$$\beta = \frac{t^2}{s+t} + s^2\left(\frac{1}{s} - \frac{1}{s+t}\right) = t.$$

2. Certainly W is Gaussian with continuous sample paths and zero means, and it is therefore sufficient to prove that $\text{cov}(W(s), W(t)) = \min\{s, t\}$. Now, if $s \leq t$,

$$\text{cov}\big(W(s), W(t)\big) = \frac{\text{cov}(X(r^{-1}(s)), X(r^{-1}(t)))}{v(r^{-1}(s))v(r^{-1}(t))} = \frac{u(r^{-1}(s))v(r^{-1}(t))}{v(r^{-1}(s))v(r^{-1}(t))} = r(r^{-1}(s)) = s$$

as required.

If $u(s) = s$, $v(t) = 1 - t$, then $r(t) = t/(1-t)$, and $r^{-1}(w) = w/(1+w)$ for $0 \leq w < \infty$. In this case $X(t) = (1-t)W(t/(1-t))$.

3. Certainly U is Gaussian with zero means, and $U(0) = 0$. Now, with $s_t = e^{2\beta t} - 1$,

$$\begin{aligned}\mathbb{E}\{U(t+h) \mid U(t) = u\} &= e^{-\beta(t+h)}\mathbb{E}\{W(s_{t+h}) \mid W(s_t) = ue^{\beta t}\} \\ &= ue^{-\beta(t+h)}e^{\beta t} = u - \beta uh + \text{o}(h),\end{aligned}$$

whence the instantaneous mean of U is $a(t, u) = -\beta u$. Secondly, $s_{t+h} = s_t + 2\beta e^{2\beta t}h + \text{o}(h)$, and therefore

$$\begin{aligned}\mathbb{E}\{U(t+h)^2 \mid U(t) = u\} &= e^{-2\beta(t+h)}\mathbb{E}\{W(s_{t+h})^2 \mid W(s_t) = ue^{\beta t}\} \\ &= e^{-2\beta(t+h)}\big(u^2e^{2\beta t} + 2\beta e^{2\beta t}h + \text{o}(h)\big) \\ &= u^2 - 2\beta h(u^2 - 1) + \text{o}(h).\end{aligned}$$

It follows that

$$\begin{aligned}\mathbb{E}\{|U(t+h)-U(t)|^2 \mid U(t)=u\} &= u^2 - 2\beta h(u^2-1) - 2u(u-\beta uh) + u^2 + o(h)\\ &= 2\beta h + o(h),\end{aligned}$$

and the instantaneous variance is $b(t,u) = 2\beta$.

4. Bartlett's equation (see Exercise (13.3.4)) for $M(t,\theta) = \mathbb{E}(e^{\theta V(t)})$ is

$$\frac{\partial M}{\partial t} = -\beta\theta\frac{\partial M}{\partial \theta} + \tfrac{1}{2}\sigma^2\theta^2 M$$

with boundary condition $M(\theta,0) = e^{\theta u}$. Solve this equation (as in the exercise given) to obtain

$$M(t,\theta) = \exp\left\{\theta u e^{-\beta t} + \frac{1}{2}\theta^2\cdot\frac{\sigma^2}{2\beta}(1-e^{-2\beta t})\right\},$$

the moment generating function of the given normal distribution. Now $M(t,\theta) \to \exp\{\frac{1}{2}\theta^2\sigma^2/(2\beta)\}$ as $t \to \infty$, whence by the continuity theorem $V(t)$ converges in distribution to the $N(0,\frac{1}{2}\sigma^2/\beta)$ distribution.

If $V(0)$ has this limit distribution, then so does $V(t)$ for all t. Therefore the sequence $(V(t_1),\dots,V(t_n))$ has the same joint distribution as $(V(t_1+h),\dots,V(t_n+h))$ for all $h, t_1,\dots,t_n$, whenever $V(0)$ has this normal distribution.

In the stationary case, $\mathbb{E}(V(t)) = 0$ and, for $s \le t$,

$$\text{cov}\big(V(s),V(t)\big) = \mathbb{E}\{V(s)\mathbb{E}\big(V(t)\mid V(s)\big)\} = \mathbb{E}\{V(s)^2 e^{-\beta(t-s)}\} = c(0)e^{-\beta|t-s|}$$

where $c(0) = \text{var}(V(s))$; we have used the first part here. This is the autocovariance function of a stationary Gaussian Markov process (see Example (9.6.10)). Since all such processes have autocovariance functions of this form (i.e. for some choice of β), all such processes are stationary Ornstein–Uhlenbeck processes.

The autocorrelation function is $\rho(s) = e^{-\beta|s|}$, which is the characteristic function of the Cauchy density function

$$f(x) = \frac{1}{\beta\pi\{1+(x/\beta)^2\}}, \qquad x \in \mathbb{R}.$$

This observation is due to J. L. Doob.

5. Bartlett's equation (see Exercise (13.3.2)) for M is

$$\frac{\partial M}{\partial t} = \alpha\theta\frac{\partial M}{\partial \theta} + \tfrac{1}{2}\beta\theta^2\frac{\partial M}{\partial \theta},$$

subject to $M(0,\theta) = e^{\theta d}$. The characteristics satisfy

$$\frac{dM}{0} = \frac{dt}{1} = -\frac{2\,d\theta}{2\alpha\theta+\beta\theta^2}.$$

The solution is $M = g(\theta e^{\alpha t}/(\alpha+\frac{1}{2}\beta\theta))$ where g is a function satisfying $g(\theta/(\alpha+\frac{1}{2}\beta\theta)) = e^{\theta d}$. The solution follows as given.

By elementary calculations,

$$\begin{aligned}\mathbb{E}(D(t)) &= \left.\frac{\partial M}{\partial \theta}\right|_{\theta=0} = de^{\alpha t},\\ \mathbb{E}(D(t)^2) &= \left.\frac{\partial^2 M}{\partial \theta^2}\right|_{\theta=0} = \frac{\beta d}{\alpha}e^{\alpha t}(e^{\alpha t}-1) + d^2e^{2\alpha t},\end{aligned}$$

whence $\text{var}(D(t)) = (\beta d/\alpha)e^{\alpha t}(e^{\alpha t} - 1)$. Finally

$$\mathbb{P}(D(t) = 0) = \lim_{\theta \to -\infty} M(t, \theta) = \exp\left\{\frac{2\alpha d e^{\alpha t}}{\beta(1 - e^{\alpha t})}\right\}$$

which converges to $e^{-2\alpha d/\beta}$ as $t \to \infty$.

6. The equilibrium density function $g(y)$ satisfies the (reduced) forward equation

$$(*) \qquad -\frac{d}{dy}(ag) + \frac{1}{2}\frac{d^2}{dy^2}(bg) = 0$$

where $a(y) = -\beta y$ and $b(y) = \sigma^2$ are the instantaneous mean and variance. The boundary conditions are

$$\beta y g + \tfrac{1}{2}\sigma^2 \frac{dg}{dy} = 0, \qquad y = -c, d.$$

Integrate $(*)$ from $-c$ to y, using the boundary conditions, to obtain

$$\beta y g + \tfrac{1}{2}\sigma^2 \frac{dg}{dy} = 0, \qquad -c \le y \le d.$$

Integrate again to obtain $g(y) = Ae^{-\beta y^2/\sigma^2}$. The constant A is given by the fact that $\int_{-c}^{d} g(y)\,dy = 1$.

7. First we show that the series converges uniformly (along a subsequence), implying that the limit exists and is a continuous function of t. Set

$$Z_{mn}(t) = \sum_{k=m}^{n-1} \frac{\sin(kt)}{k} X_k, \qquad M_{mn} = \sup\{|Z_{mn}(t)| : 0 \le t \le \pi\}.$$

We have that

$$(*) \qquad M_{mn}^2 \le \sup_{0 \le t \le \pi} \left|\sum_{k=m}^{n-1} \frac{e^{ikt}}{k} X_k\right|^2 \le \sum_{k=m}^{n-1} \frac{X_k^2}{k^2} + 2\sum_{l=1}^{n-m-1}\left|\sum_{j=m}^{n-l-1} \frac{X_j X_{j+l}}{j(j+l)}\right|.$$

The mean value of the final term is, by the Cauchy–Schwarz inequality, no larger than

$$2\sum_{l=1}^{n-m-1}\sqrt{\mathbb{E}\left(\left|\sum_{j=m}^{n-l-1}\frac{X_j X_{j+l}}{j(j+l)}\right|^2\right)} = 2\sum_{l=1}^{n-m-1}\sqrt{\sum_{j=m}^{n-l-1}\frac{1}{j^2(j+l)^2}} \le 2(n-m)\sqrt{\frac{n-m}{m^4}}.$$

Combine this with $(*)$ to obtain

$$\mathbb{E}(M_{m,2m})^2 \le \mathbb{E}(M_{m,2m}^2) \le \frac{3}{\sqrt{m}}.$$

It follows that

$$\mathbb{E}\left(\sum_{n=1}^{\infty} M_{2^{n-1},2^n}\right) \le \sum_{n=1}^{\infty} \frac{6}{2^{n/2}} < \infty,$$

implying that $\sum_{n=1}^{\infty} M_{2^{n-1},2^n} < \infty$ a.s. Therefore the series which defines W converges uniformly with probability 1 (along a subsequence), and hence W has (a.s.) continuous sample paths.

Certainly W is a Gaussian process since $W(t)$ is the sum of normal variables (see Problem (7.11.19)). Furthermore $\mathbb{E}(W(t)) = 0$, and

$$\operatorname{cov}\big(W(s), W(t)\big) = \frac{st}{\pi} + \frac{2}{\pi}\sum_{k=1}^{\infty} \frac{\sin(ks)\sin(kt)}{k^2}$$

since the X_i are independent with zero means and unit variances. It is an exercise in Fourier analysis to deduce that $\operatorname{cov}(W(s), W(t)) = \min\{s, t\}$.

8. We wish to find a solution $g(t, y)$ to the equation

$$(*) \qquad \frac{\partial g}{\partial t} = \frac{1}{2}\frac{\partial^2 g}{\partial y^2}, \qquad |y| < b,$$

satisfying the boundary conditions

$$g(0, y) = \delta_{y0} \quad \text{if } |y| \le b, \qquad g(t, y) = 0 \quad \text{if } |y| = b.$$

Let $g(t, y \mid d)$ be the $N(d, t)$ density function, and note that $g(\cdot, \cdot \mid d)$ satisfies $(*)$ for any 'source' d. Let

$$g(t, y) = \sum_{k=-\infty}^{\infty} (-1)^k g(t, y \mid 2kb),$$

a series which converges absolutely and is differentiable term by term. Since each summand satisfies $(*)$, so does the sum. Now $g(0, y)$ is a combination of Dirac delta functions, one at each multiple of $2b$. Only one such multiple lies in $[-b, b]$, and hence $g(y, 0) = \delta_{d0}$. Also, setting $y = b$, the contributions from the sources at $-2(k-1)b$ and $2kb$ cancel, so that $g(t, b) = 0$. Similarly $g(t, -b) = 0$, and therefore g is the required solution.

Here is an alternative method. Look for the solution to $(*)$ of the form $e^{-\lambda_n t}\sin\{\frac{1}{2}n\pi(y+b)/b\}$; such a sine function vanishes when $|y| = b$. Substitute into $(*)$ to obtain $\lambda_n = n^2\pi^2/(8b^2)$. A linear combination of such functions has the form

$$g(t, y) = \sum_{n=1}^{\infty} a_n e^{-\lambda_n t} \sin\left(\frac{n\pi(y+b)}{2b}\right).$$

We choose the constants a_n such that $g(0, y) = \delta_{y0}$ for $|y| < b$. With the aid of a little Fourier analysis, one finds that $a_n = b^{-1}\sin(\frac{1}{2}n\pi)$.

Finally, the required probability equals the probability that W^{a} has been absorbed by time t, a probability expressible as $1 - \int_{-b}^{b} f^{\mathrm{a}}(t, y)\,dy$. Using the second expression for f^{a}, this yields

$$\frac{4}{\pi}\sum_{n=1}^{\infty} \frac{1}{n} e^{-\lambda_n t} \sin^3(\tfrac{1}{2}n\pi).$$

9. Recall that $U(t) = e^{-2mD(t)}$ is a martingale. Let T be the time of absorption, and assume that the conditions of the optional stopping theorem are satisfied. Then $\mathbb{E}(U(0)) = \mathbb{E}(U(T))$, which is to say that $1 = e^{2ma}p_a + e^{-2mb}(1 - p_a)$.

10. (a) We may assume that $a, b > 0$. With

$$p_t(b) = \mathbb{P}\big(W(t) > b,\ F(0, t) \,\big|\, W(0) = a\big),$$

we have by the reflection principle that

$$\begin{aligned} p_t(b) &= \mathbb{P}\big(W(t) > b \,\big|\, W(0) = a\big) - \mathbb{P}\big(W(t) < -b \,\big|\, W(0) = a\big) \\ &= \mathbb{P}\big(b - a < W(t) < b + a \,\big|\, W(0) = 0\big), \end{aligned}$$

giving that

$$\frac{\partial p_t(b)}{\partial b} = f(t, b + a) - f(t, b - a)$$

where $f(t, x)$ is the $N(0, t)$ density function. Now, using conditional probabilities,

$$\mathbb{P}\big(F(0, t) \,\big|\, W(0) = a, W(t) = b\big) = -\frac{1}{f(t, b-a)} \frac{\partial p_t(b)}{\partial b} = 1 - e^{-2ab/t}.$$

(b) We know that

$$\mathbb{P}(F(s, t)) = 1 - \frac{2}{\pi} \cos^{-1}\{\sqrt{s/t}\} = \frac{2}{\pi} \sin^{-1}\{\sqrt{s/t}\}$$

if $0 < s < t$. The claim follows since $F(t_0, t_2) \subseteq F(t_0, t_1)$.

(c) Remember that $\sin x = x + \mathrm{o}(x)$ as $x \downarrow 0$. Take the limit in part (b) as $t_0 \downarrow 0$ to obtain $\sqrt{t_1/t_2}$.

11. Let $M(t) = \sup\{W(s) : 0 \le s \le t\}$ and recall that $M(t)$ has the same distribution as $|W(t)|$. By symmetry,

$$\mathbb{P}\Big(\sup_{0 \le s \le t} |W(s)| \ge w\Big) \le 2\mathbb{P}\big(M(t) \ge w\big) = 2\mathbb{P}\big(|W(t)| \ge w\big).$$

By Chebyshov's inequality,

$$\mathbb{P}\big(|W(t)| \ge w\big) \le \frac{\mathbb{E}(W(t)^2)}{w^2} = \frac{t}{w^2}.$$

Fix $\epsilon > 0$, and let

$$A_n(\epsilon) = \big\{|W(s)|/s > \epsilon \text{ for some } s \text{ satisfying } 2^{n-1} < s \le 2^n\big\}.$$

Note that

$$A_n(\epsilon) \subseteq \left\{ \sup_{2^{n-1} < s \le 2^n} |W(s)| \ge 2^{2n/3} \right\} \subseteq \left\{ \sup_{0 \le s \le 2^n} |W(s)| \ge 2^{2n/3} \right\}$$

for all large n, and also

$$\sum_{n=1}^{\infty} \mathbb{P}\left(\sup_{0 \le s \le 2^n} |W(s)| \ge 2^{2n/3} \right) \le \sum_{n=1}^{\infty} \frac{2^{n+1}}{2^{4n/3}} < \infty.$$

Therefore $\sum_n \mathbb{P}(A_n(\epsilon)) < \infty$, implying by the Borel–Cantelli lemma that (a.s.) only finitely many of the $A_n(\epsilon)$ occur. Therefore $t^{-1}W(t) \to 0$ a.s. as $t \to \infty$. Compare with the solution to the relevant part of Exercise (13.6.4).

12. We require the solution to Laplace's equation $\nabla^2 p = 0$, subject to the boundary condition

$$p(\mathbf{w}) = \begin{cases} 0 & \text{if } \mathbf{w} \in H, \\ 1 & \text{if } \mathbf{w} \in G. \end{cases}$$

Look for a solution in polar coordinates of the form

$$p(r,\theta) = \sum_{n=0}^{\infty} r^n \{a_n \sin(n\theta) + b_n \cos(n\theta)\}.$$

Certainly combinations having this form satisfy Laplace's equation, and the boundary condition gives that

$$(*) \qquad H(\theta) = b_0 + \sum_{n=1}^{\infty} \{a_n \sin(n\theta) + b_n \cos(n\theta)\}, \qquad |\theta| < \pi,$$

where

$$H(\theta) = \begin{cases} 0 & \text{if } -\pi < \theta < 0, \\ 1 & \text{if } 0 < \theta < \pi. \end{cases}$$

The collection $\{\sin(m\theta), \cos(m\theta) : m \geq 0\}$ are orthogonal over $(-\pi, \pi)$. Multiply through $(*)$ by $\sin(m\theta)$ and integrate over $(-\pi, \pi)$ to obtain $\pi a_m = \{1 - \cos(\pi m)\}/m$, and similarly $b_0 = \frac{1}{2}$ and $b_m = 0$ for $m \geq 1$.

13. The joint density function of two independent $N(0, t)$ random variables is $(2\pi t)^{-1} \exp\{-(x^2 + y^2)/(2t)\}$. Since this function is unchanged by rotations of the plane, it follows that the two coordinates of the particle's position are independent Wiener processes, regardless of the orientation of the coordinate system. We may thus assume that l is the line $x = d$ for some fixed positive d.

The particle is bound to visit the line l sooner or later, since $\mathbb{P}(W_1(t) < d \text{ for all } t) = 0$. The first-passage time T has density function

$$f_T(t) = \frac{d}{\sqrt{2\pi t^3}} e^{-d^2/(2t)}, \qquad t > 0.$$

Conditional on $\{T = t\}$, $D = W_2(T)$ is $N(0, t)$. Therefore the density function of D is

$$f_D(u) = \int_0^\infty f_{D|T}(u \mid t) f_T(t)\, dt = \int_0^\infty \frac{d}{2\pi t^2} e^{-(u^2+d^2)/(2t)}\, dt = \frac{d}{\pi(u^2 + d^2)}, \qquad u \in \mathbb{R},$$

giving that D/d has the Cauchy distribution.

The angle $\Theta = \widehat{\text{POR}}$ satisfies $\theta = \tan^{-1}(D/d)$, whence

$$\mathbb{P}(\Theta \leq \theta) = \mathbb{P}(D \leq d \tan\theta) = \frac{1}{2} + \frac{\theta}{\pi}, \qquad |\theta| < \tfrac{1}{2}\pi.$$

14. By an extension of Itô's formula to functions of two Wiener processes, $U = u(W_1, W_2)$ and $V = v(W_1, W_2)$ satisfy

$$dU = u_x\, dW_1 + u_y\, dW_2 + \tfrac{1}{2}(u_{xx} + u_{yy})\, dt,$$
$$dV = v_x\, dW_1 + v_y\, dW_2 + \tfrac{1}{2}(v_{xx} + v_{yy})\, dt,$$

where u_x, v_{yy}, etc, denote partial derivatives of u and v. Since ϕ is analytic, u and v satisfy the Cauchy–Riemann equations $u_x = v_y$, $u_y = -v_x$, whence u and v are harmonic in that $u_{xx} + u_{yy} = v_{xx} + v_{yy} = 0$. Therefore,

$$dU = u_x\, dW_1 + u_y\, dW_2, \quad dV = -u_y\, dW_1 + u_x\, dW_2.$$

The matrix $\begin{pmatrix} u_x & u_y \\ -u_y & u_x \end{pmatrix}$ is an orthogonal rotation of $\mathbb{R}^2$ when $u_x^2+u_y^2 = 1$. Since the joint distribution of the pair (W_1, W_2) is invariant under such rotations, the claim follows.

15. One method of solution uses the fact that the reversed Wiener process $\{W(t-s)-W(t) : 0 \le s \le t\}$ has the same distribution as $\{W(s) : 0 \le s \le t\}$. Thus $M(t)-W(t) = \max_{0\le s\le t}\{W(s)-W(t)\}$ has the same distribution as $\max_{0\le u\le t}\{W(u) - W(0)\} = M(t)$. Alternatively, by the reflection principle,

$$\mathbb{P}\big(M(t) \ge x,\ W(t) \le y\big) = \mathbb{P}(W(t) \ge 2x - y) \quad \text{for } x \ge \max\{0, y\}.$$

By differentiation, the pair $M(t)$, $W(t)$ has joint density function $-2\phi'(2x - y)$ for $y \le x,\ x \ge 0$, where ϕ is the density function of the $N(0, t)$ distribution. Hence $M(t)$ and $M(t) - W(t)$ have the joint density function $-2\phi'(x + y)$. Since this function is symmetric in its arguments, $M(t)$ and $M(t) - W(t)$ have the same marginal distribution.

16. The Lebesgue measure $\Lambda(Z)$ is given by

$$\Lambda(Z) = \int_0^\infty I_{\{W(t)=u\}}\,du,$$

whence by Fubini's theorem (cf. equation (5.6.13)),

$$\mathbb{E}(\Lambda(Z)) = \int_0^\infty \mathbb{P}\big(W(t) = u\big)\,dt = 0.$$

17. Let $0 < a < b < c < d$, and let $M(x, y) = \max_{x\le s\le y} W(s)$. Then

$$M(c, d) - M(a, b) = \max_{c\le s\le d}\big\{W(s) - W(c)\big\} + \big\{W(c) - W(b)\big\} - \max_{a\le s\le b}\big\{W(s) - W(b)\big\}.$$

Since the three terms on the right are independent and continuous random variables, it follows that $\mathbb{P}\big((M(c, d) = M(a, b)\big) = 0$. Since there are only countably many rationals, we deduce that $\mathbb{P}\big((M(c, d) = M(a, b)$ for all rationals $a < b < c < d\big) = 1$, and the result follows.

18. The result is easily seen by exhaustion to be true when $n = 1$. Suppose it is true for all $m \le n-1$ where $n \ge 2$.

(i) If $s_n \le 0$, then (whatever the final term of the permutation) the number of positive partial sums and the position of the first maximum depend only on the remaining $n - 1$ terms. Equality follows by the induction hypothesis.

(ii) If $s_n > 0$, then

$$A_r = \sum_{k=1}^{n} A_{r-1}(k),$$

where $A_{r-1}(k)$ is the number of permutations with x_k in the final place, for which exactly $r - 1$ of the first $n - 1$ terms are strictly positive. Consider a permutation $\pi = (x_{i_1}, x_{i_2}, \dots, x_{i_{n-1}}, x_k)$ with x_k in the final place, and move the position of x_k to obtain the new permutation $\pi' = (x_k, x_{i_1}, x_{i_2}, \dots, x_{i_{n-1}})$. The first appearance of the maximum in π' is at its rth place if and only if the first maximum of the reduced permutation $(x_{i_1}, x_{i_2}, \dots, x_{i_{n-1}})$ is at its $(r - 1)$th place. [Note that $r = 0$ is impossible since $s_n > 0$.] It follows that

$$B_r = \sum_{k=1}^{n} B_{r-1}(k),$$

where $B_{r-1}(k)$ is the number of permutations with x_k in the final place, for which the first appearance of the maximum is at the $(r - 1)$th place.

By the induction hypothesis, $A_{r-1}(k) = B_{r-1}(k)$, since these quantities depend on the $n-1$ terms excluding x_k. The result follows.

19. Suppose that $S_m = \sum_{j=1}^{m} X_j$, $0 \le m \le n$, are the partial sums of n independent identically distributed random variables X_j. Let A_n be the number of strictly positive partial sums, and R_n the index of the first appearance of the value of the maximal partial sum. Each of the $n!$ permutations of $(X_1, X_2, \dots, X_n)$ has the same joint distribution. Consider the kth permutation, and let I_k be the indicator function of the event that exactly r partial sums are positive, and let J_k be the indicator function that the first appearance of the maximum is at the rth place. Then, using Problem (13.12.18),

$$\mathbb{P}(A_n = r) = \frac{1}{n!}\sum_{k=1}^{n!} \mathbb{E}(I_k) = \frac{1}{n!}\sum_{k=1}^{n!} \mathbb{E}(J_k) = \mathbb{P}(R_n = r).$$

We apply this with $X_j = W(jt/n) - W((j-1)t/n)$, so that $S_m = W(mt/n)$. Thus $A_n = \sum_j I_{\{W(jt/n)>0\}}$ has the same distribution as

$$R_n = \min\{k \ge 0 : W(kt/n) = \max_{0\le j\le n} W(jt/n)\}.$$

By Problem (13.12.17), $R_n \xrightarrow{\text{a.s.}} R$ as $n \to \infty$. By Problem (13.12.16), the time spent by W at zero is a.s. a null set, whence $A_n \xrightarrow{\text{a.s.}} A$. Hence A and R have the same distribution. We argue as follows to obtain that that L and R have the same distribution. Making repeated use of Theorem (13.4.6) and the symmetry of W,

$$\begin{aligned}\mathbb{P}(L < x) &= \mathbb{P}\Big(\sup_{x\le s\le t} W(s) < 0\Big) + \mathbb{P}\Big(\inf_{x\le s\le t} W(s) > 0\Big)\\ &= 2\mathbb{P}\Big(\sup_{x\le s\le t} \{W(s) - W(x)\} < -W(x)\Big) = 2\mathbb{P}\big(|W(t) - W(x)| < W(x)\big)\\ &= \mathbb{P}\big(|W(t) - W(x)| < |W(x)|\big)\\ &= \mathbb{P}\Big(\sup_{x\le s\le t} \{W(s) - W(x)\} < \sup_{0\le s\le x} \{W(s) - W(x)\}\Big) = \mathbb{P}(R \le x).\end{aligned}$$

Finally, by Problem (13.12.15) and the circular symmetry of the joint density distribution of two independent $N(0, 1)$ variables U, V,

$$\mathbb{P}\big(|W(t) - W(x)| < |W(x)|\big) = \mathbb{P}\big((t-x)V^2 \le xU^2\big) = \mathbb{P}\left(\frac{V^2}{U^2+V^2} \le \frac{x}{t}\right) = \frac{2}{\pi}\sin^{-1}\sqrt{\frac{x}{t}}.$$

20. Let

$$T_x = \begin{cases} \inf\{t \le 1 : W(t) = x\} & \text{if this set is non-empty,}\\ 1 & \text{otherwise,}\end{cases}$$

and similarly $V_x = \sup\{t \le 1 : W(t) = x\}$, with $V_x = 1$ if $W(t) \ne x$ for all $t \in [0, 1]$. Recall that U_0 and V_0 have an arc sine distribution as in Problem (13.12.19). On the event $\{U_x < 1\}$, we may write (using the re-scaling property of W)

$$U_x = T_x + (1 - T_x)\widetilde{U}_0, \quad V_x = T_x + (1 - T_x)\widetilde{V}_0,$$

where $\widetilde{U}_0$ and $\widetilde{V}_0$ are independent of U_x and V_x, and have the above arc sine distribution. Hence U_x and V_x have the same distribution. Now T_x has the first passage distribution of Theorem (13.4.5), whence

$$f_{T_x,\widetilde{U}_0}(\tau, \phi) = \left\{\frac{x}{\sqrt{2\pi\tau^3}}\exp\left(-\frac{x^2}{2\tau}\right)\right\}\left\{\frac{1}{\pi\sqrt{\phi(1-\phi)}}\right\}.$$

Therefore,

$$f_{T_x,U_x}(t,u) = f_{T_x,\widetilde{U}_0}\left(t, \frac{u-t}{1-t}\right) \cdot \frac{1}{1-t},$$

and

$$f_{U_x}(u) = \int_0^u f_{T_x,U_x}(t,u)\,du = \frac{1}{\pi\sqrt{u(1-u)}} \exp\left(-\frac{x^2}{2u}\right), \qquad 0 < x < 1.$$

21. (a) Note that V is a martingale, by Theorem (13.8.11). Fix t and let $\psi_s = \text{sign}(W_s)$, $0 \le s \le t$. We have that $\|\psi\| = \sqrt{t}$, implying by Exercise (13.8.6) that $\mathbb{E}(V_t^2) = \|I(\psi)\|_2^2 = t$. By a similar calculation, $\mathbb{E}(V_t^2 \mid \mathcal{F}_s) = V_s^2 + t - s$ for $0 \le s \le t$. That is to say, $V_t^2 - t$ defines a martingale, and the result follows by the Lévy characterization theorem of Example (12.7.10).
(b) By part (a), the process $B(t) = \int_0^t \text{sign}\, V\, dV = \int_0^t (\text{sign}\, V)^{-1} dV$ is a Wiener process, and writing the outer terms in differential form gives $dV = \text{sign}\, V\, dB$, which is to say that V is a solution of the given SDE.
(c) By Problem (13.12.16), $-\text{sign}\, V\, dB = \text{sign}(-V)\, dB$ except on a null set, so $-V$ also solves the given SDE.
(d) By a slightly extended version of Theorem (13.8.11) (with left-continuous predictable integrand) any solution X of the SDE is a continuous martingale, and since $(\text{sign}\, X)^2 = 1$, its quadratic variation is t. Hence, by Lévy's characterization theorem (12.7.10), it is a Wiener process.

22. The mean cost per unit time is

$$\mu(T) = \frac{1}{T}\left\{R + C\int_0^T \mathbb{P}(|W(t)| \ge a)\,dt\right\} = \frac{1}{T}\left\{R + 2C\int_0^T \left(1 - \Phi(a/\sqrt{t})\right) dt\right\}.$$

Differentiate to obtain that $\mu'(T) = 0$ if

$$R = 2C\left\{\int_0^T \Phi(a/\sqrt{t})\,dt - T\Phi(a/\sqrt{T})\right\} = aC\int_0^T t^{-1}\phi(a/\sqrt{t})\,dt,$$

where we have integrated by parts.

23. Consider the portfolio with $\xi(t, S_t)$ units of stock and $\psi(t, S_t)$ units of bond, having total value $w(t, S_t) = x\xi(t,x) + e^{rt}\psi(t, S_t)$. By assumption,

$$(*) \qquad (1-\gamma)x\xi(t,x) = \gamma e^{rt}\psi(t,x).$$

Differentiate this equation with respect to x and substitute from equation (13.10.16) to obtain the differential equation $(1-\gamma)\xi + x\xi_x = 0$, with solution $\xi(t,x) = h(t)x^{\gamma-1}$, for some function $h(t)$. We substitute this, together with $(*)$, into equation (13.10.17) to obtain that

$$h' - h(1-\gamma)(\tfrac{1}{2}\gamma\sigma^2 + r) = 0.$$

It follows that $h(t) = A\exp\{(1-\gamma)(\frac{1}{2}\gamma\sigma^2 + r)t\}$, where A is an absolute constant to be determined according to the size of the initial investment. Finally, $w(t,x) = \gamma^{-1}x\xi(t,x) = \gamma^{-1}h(t)x^{\gamma}$.

24. Using Itô's formula (13.9.4), the drift term in the SDE for U_t is

$$\left(-u_1(T-t, W) + \tfrac{1}{2}u_{22}(T-t, W)\right) dt,$$

where u_1 and u_{22} denote partial derivatives of u. The drift function is identically zero if and only if $u_1 = \frac{1}{2}u_{22}$.

25. (a) Setting the proof down in its detail is unlikely to be edifying, and so we resort to an informal explanation. The distribution of the process W is invariant under rotations of space around the origin. Therefore, the conditional distribution of the hitting position on the sphere is independent of the value of the hitting time.

(b) (i) By induction on n and part (a), $(R_0, R_1, \dots, R_n)$ and $(W(T_0), W(T_1), \dots, W(T_n))$ have the same distribution for $n \ge 0$.

(ii) The first-passage time T_B to the boundary B is a.s. finite. Therefore, the sequence (T_n) is increasing and a.s. bounded, and hence converges to some a.s. finite limit T_∞. By the continuity of the sample paths of the Wiener process, the sequence (R_n) converges a.s. to some random point $R_\infty \in B \cup C$. It must be the case that $R_\infty \in B$ since otherwise the iterative process would not have terminated. Therefore, $T_\infty = T_B$ a.s., and the claim follows.

(iii) Since $R_n \to R_\infty \in B$ a.s., we have $r(R_n) \to 0$ a.s., so that $S(a)$ is a.s. finite for $a > 0$. For the last part, let the smallest sphere that circumscribes C have diameter d, and note that $d < \infty$.

Let $B(a)$ be the set of points in $B \cup C$ that are within distance a of B. Let $x \in C$, and let $b \in M(x) \cap B(a)$. The intersection $M(x) \cap B(a)$ contains a cap T of $M(x)$ containing b. The area of this cap is a minimum when the part of B containing b is locally a tangent plane to $M(x)$. (We omit the proof of this, but drawing a picture will help). Given that $R_m = x$, the probability $p(x) = \mathbb{P}(R_{m+1} \in B(a))$ satisfies

$$p(x) \ge \frac{|T|}{4\pi r^2} \ge \frac{2\pi r a}{4\pi r^2} = \frac{a}{2r} > \frac{a}{d},$$

where $|T|$ is the area of T and r is the radius of $M(x)$. We have used the circumscribing cylinder theorem to bound $|T|$.

It follows that the number of steps until hitting $B(a)$ is stochastically smaller than a geometrically distributed random variable with mean d/a, and hence the result.

Bibliography

A man will turn over half a library to make one book. Samuel Johnson

Abramowitz, M. and Stegun, I. A. (1965). *Handbook of mathematical functions with formulas, graphs and mathematical tables*. Dover, New York.

Billingsley, P. (1995). *Probability and measure* (3rd edn). Wiley, New York.

Breiman, L. (1968). *Probability*. Addison-Wesley, Reading, MA, reprinted by SIAM, 1992.

Cayley, A. (1875). *Mathematical questions with their solutions*. The Educational Times 23, 18–19. See *The collected mathematical papers of Arthur Cayley* 10, 587–588 (1896), Cambridge University Press, Cambridge.

Chung, K. L. (1974). *A course in probability theory* (2nd edn). Academic Press, New York.

Clifford, P. and Stirzaker, D. R. (2019). *Ulam's random adding process*. `https://arxiv.org/abs/1911.07529`.

Cox, D. R. and Miller, H. D. (1965). *The theory of stochastic processes*. Chapman and Hall, London.

Doob, J. L. (1953). *Stochastic processes*. Wiley, New York.

Feller, W. (1968). *An introduction to probability theory and its applications,* Vol. 1 (3rd edn). Wiley, New York.

Feller, W. (1971). *An introduction to probability theory and its applications,* Vol. 2 (2nd edn). Wiley, New York.

Grimmett, G. R. and Stirzaker, D. R. (2020). *Probability and random processes* (4th edn). Oxford University Press, Oxford.

Grimmett, G. R. and Welsh, D. J. A. (2014). *Probability, an introduction* (2nd edn). Oxford University Press, Oxford.

Hall, M. (1983). *Combinatorial theory* (2nd edn). Wiley, New York.

Han, Y. and Wang, G. (2019). *Expectation of the largest bet size in the Labouchere system*. Electronic Journal of Probability 24, paper 11.

Harris, T. E. (1963). *The theory of branching processes*. Springer, Berlin.

Hofstad, R. van der and Keane, M. S. (2008). *An elementary proof of the hitting time theorem*. American Mathematical Monthly 115, 753–756.

Karlin, S. and Taylor, H. M. (1975). *A first course in stochastic processes* (2nd edn). Academic Press, New York.

Karlin, S. and Taylor, H. M. (1981). *A second course in stochastic processes*. Academic Press, New York.

Laha, R. G. and Rohatgi, V. K. (1979). *Probability theory*. Wiley, New York.

Lindley, D. V. (1961). *Dynamic programming and decision theory*. Journal of the Royal Statistical Society. Series C. Applied Statistics 10, 39–51.

Loève, M. (1977). *Probability theory,* Vol. 1 (4th edn). Springer, Berlin.

Loève, M. (1978). *Probability theory,* Vol. 2 (4th edn). Springer, Berlin.

Moran, P. A. P. (1968). *An introduction to probability theory*. Clarendon Press, Oxford.

Mörters, P. and Peres, Y. (2010). *Brownian motion*. Cambridge University Press, Cambridge.

Roberts, H. E. (1998). *Encyclopedia of comparative iconography*. Fitzroy Dearborn, Chicago & London.

Stirzaker, D. R. (2003). *Elementary probability* (2nd edn). Cambridge University Press, Cambridge.

Stirzaker, D. R. (1999). *Probability and random variables*. Cambridge University Press, Cambridge.

Williams, D. (1991). *Probability with martingales*. Cambridge University Press, Cambridge.

Index

Abbreviations used in this index: c.f. characteristic function; distn distribution; eqn equation; fn function; ineq. inequality; m.g.f. moment generating function; p.g.f. probability generating function; pr. process; r.v. random variable; r.w. random walk; s.r.w. simple random walk; thm theorem.

D

E

F

N

S

T